Contemporary Economic Issues in Asian Countries: Proceeding of CEIAC 2022, Volume 1

An Thinh Nguyen · Thu Thuy Pham · Joon Song ·
Yen-Ling Lin · Manh Cuong Dong
Editors

Contemporary Economic Issues in Asian Countries: Proceeding of CEIAC 2022, Volume 1

Springer

Editors
An Thinh Nguyen
VNU University of Economics and Business
Vietnam National University
Hanoi, Vietnam

Joon Song
Department of Economics
Sungkyunkwan University
Seoul, Korea (Republic of)

Manh Cuong Dong
VNU University of Economics and Business
Vietnam National University
Hanoi, Vietnam

Thu Thuy Pham
World Agroforestry Center
Center for International Forestry Research
Hanoi, Vietnam

Yen-Ling Lin
Department of Economics
Tamkang University
New Taipei City, Taiwan

ISBN 978-981-19-9668-9 ISBN 978-981-19-9669-6 (eBook)
https://doi.org/10.1007/978-981-19-9669-6

This Springer imprint is published by the registered company Springer Nature Singapore Pte Ltd.
The registered company address is: 152 Beach Road, #21-01/04 Gateway East, Singapore 189721, Singapore

Contents

Livelihood Restoration for Thai Ethnic Minority After Resettlement—Impacts in Son La Hydropower Plant

Hoa Dang Thi, Linh Pham Quang, Loi Pham Van, Tram Bui Thi Huong, and Mai Nguyen Thi Tuyet

Abstract Resettlement in hydropower projects always may harm the development of affected people in the migration areas. Son La hydropower plant is the largest hydroelectric project in Vietnam with a great influence on the livelihoods of millions of people, mainly ethnic minorities. The theory of inclusive social development states that economic development and infrastructure investment bring opportunities and equality in prosperity to all population groups, especially vulnerable people such as the poor, ethnic minorities, women and children. But there remains an issue of gender inequality in this economic growth model. This chapter analyzes the obstacles and challenges that households resettled by hydropower projects in stabilizing livelihoods and development. Thai ethnic minority has many difficulties during the relocation and adaption to new living conditions in resettlement sites. There were concerns about livelihood restoration in resettlement sites of Son La hydropower plant, particularly the limited interest and focus on job creation and livelihood support for affected Thai ethnic people. They have faced various obstacles to adapt to a modern lifestyle, which is different from their traditional ethnic culture. Moreover, illiterate people could not find a new job with a stable income.

Keywords Resettlement · Livelihood restoration · Ethnic · Hydropower plant · Vietnam

H. D. Thi (✉)
Institute of Psychology, Vietnam Academy Social Sciences, Hanoi, Vietnam
e-mail: danghoa.ifgs69@gmail.com

L. P. Quang
Institute of Anthropology, Vietnam Academy Social Sciences, Hanoi, Vietnam

L. P. Van
VNU Institute of Vietnamese Studies and Development Sciences, Vietnam National University, Hanoi, Vietnam

T. B. T. Huong
Institute for Family and Gender Studies, Vietnam Academy Social Sciences, Hanoi, Vietnam

M. N. T. Tuyet
Graduate Academy of Social Sciences, Hanoi, Vietnam

A. T. Nguyen et al. (eds.), *Contemporary Economic Issues in Asian Countries: Proceeding of CEIAC 2022, Volume 1*,
https://doi.org/10.1007/978-981-19-9669-6_1

1 Introduction

In order to achieve economic development goals and secure the energy supply for development, a number of developing countries have focused on building hydroelectric plants. Apart from the significant benefits from hydropower projects, such as electricity supply for socio-economic development and budget revenue generation, there are multiple negative impacts that hinder development, especially issues related to involuntary resettlement of displaced communities. There are concerns about issues surrounding forced land acquisition and adverse effects on people whose land is acquired by hydropower projects. Involuntary resettlement and farmland acquisition lead to loss of livelihood assets, less access to public services and disruption of traditional social networks of indigenous peoples, resulting in stress and burdens on displaced people (Gutman, 2003; IFC, 2012; WB, 2004).

An assessment of damages caused by relocation and resettlement in hydropower projects is a mandatory requirement by national regulations and donors. The resettlement plan and social assessment are strongly linked with displaced people; therefore, their interests and satisfaction should be integrated into these documents. The resettlement planning and social impact assessment processes both interact with people's sense of belonging and their wellbeing (Vanclay et al., 2015). Although donors such as the World Bank or International Finance Corporation (IFC) require compliance with social safeguard policies in case of involuntary relocation in projects, governments do not always fully comply by negotiating with the affected people. As a result, resettlement may lead to exposure to poverty and significant mental losses by displaced persons (Frank, 2017).

Various in-depth studies on social impacts of hydropower projects in the world indicated that many aspects are omitted in the planning and delivery of compensation and resettlement support for affected people such as stakeholder's engagement in planning and decision-making process; compensation and resettlement support mechanism is unclear and disadvantaged people are not given adequate attention in the compensation and resettlement policies (Gutman, 2003; Rowan, 2017; Smyth et al., 2015). Among the vulnerable people, ethnic minorities, particularly ethnic women, are exposed to double vulnerability because they have poor development indicators, including high poverty rate, living in disadvantaged areas, limited livelihoods, low resilience to shocks, among others. On the other hand, governments have few comprehensive researches on the impacts of relocation and resettlement, especially for ethnic minorities. Existing studies on social impact assessment of hydropower projects have not covered the adaptation of disadvantaged people to new living conditions, the ability to restore their livelihoods in the long term and particularly the plans to restore the traditional spiritual and cultural values of affected ethnic communities (Colchester, 2000; Gutman, 2003; Nguyen et al., 2017; Ronggang, 2008).

In the research on free-migrant women in the Three Gorges Dam Hydro Electric Power Plant (China), Yan Tan stated that women have less opportunity to participate in the relocation decision-making process. After being relocated, women had difficulties in finding jobs in the non-agricultural sector, although they tried hard to participate in agricultural activities to ensure livelihoods in the new living environment. It is shown that gender perspectives are either incomplete or absent in resettlement planning and livelihood restoration (Tan, 2008: 1–38).

The development view argues that gender participation in the decision-making process and development planning, including resettlement and livelihood restoration plans for involuntarily displaced people by hydropower and infrastructure projects, brings obvious economic benefits (Gutman, 2003). However, the impacts of resettlement in hydropower projects on Vietnamese ethnic minority women have been discussed in few studies. Gender mainstreaming in resettlement policy of hydropower plants and infrastructure projects in general has not been adequately considered. Therefore, it is essential and urgent to analyze the gender mainstreaming in resettlement policy formulation for developing countries.

Vietnam is a developing country in needs of projects to build and upgrade infrastructure such as hydroelectric projects to meet development objectives. Hundreds of large and small hydropower projects have been built, mainly in mountainous areas, where ethnic minorities reside. Although the Government of Vietnam has made great efforts in resettlement planning, socio-environmental impact assessment and livelihood restoration, the immense impacts caused by hydropower projects have significantly affected the lives, livelihoods, cultural and spiritual properties of the affected people, especially the ethnic minorities and women. The social issues related to resettlement and livelihood recovery for displaced people are posing significant challenges to the government as well as the communities affected by hydropower projects.

2 Methodology

The approach in this chapter benefits from an anthropological perspective. The theory of inclusive social development with comprehensive development goals, without low-lying areas for disadvantaged people in society, demonstrates the social capacity to organize productive resources to meet changes in the social transformation, including organizational, infrastructure, physical, social, mental and psychological changes.

Nonetheless, there have been imbalances resulting from alterations, redundancies, deficiencies or disturbances that slow growth or threaten life during the development process. Therefore, inclusive development is viewed as the upward movements of the society with positive changes from lower to higher levels, including infrastructure conditions as well as results in enjoyment and creation, and must bring equal opportunities and prosperity to all. From an anthropological perspective, the issue of policy formulation needs to be seen by perceptions of both government and stakeholders with the uniform participation in the research of policy formulation and implementation (Michael, 1993).

A proper governance in the development of resettlement policies creates substantial power because the stakeholders' engagement is fundamental to success in governance. It requires the voice of stakeholders and the interests of all parties are taken into account in the formulation of resettlement policies. Engagement strategy should have a clear segregation in each stage of design and implementation. In addition, the voices of social organizations and individuals affected by the resettlement policies should be seriously considered (Erdiaw-Kwasie et al., 2014).

Gender mainstreaming in resettlement policies is essential. Gender roles are set by culture, social norms and values. However, gender division and gender-related influencing factors are insignificant in the current development of resettlement policies. There is a significant difference between men and women in need of livelihood restoration (UNDP, 2017; WB, 2004). Meanwhile, the resettlement policy in infrastructure projects often overlooks traditional cultural norms and ignores livelihood restoration factors that tend to be influenced by culture. Vulnerable groups, especially women from ethnic minorities, are always double-vulnerable due to their higher poverty rate, lower education levels, poorer access to social services, and being neglected in policy making. The resettlement policies of hydropower projects need to be improved with gender actions to help women and men restore their livelihoods successfully. Nevertheless, it is evident in Vietnam's hydropower projects that the resettlement plans are designed without taking into account gender mainstreaming. The resettlement plans failed to recognize the different roles of men and women in households as well as equality in decision-making to bring benefits from supportive policies and recovery opportunities to women, especially ethnic ones.

The livelihood of Thai people resettled in Son La hydropower is analyzed in this case study to learn their difficulties, especially women, in restoring their livelihoods. Displacement and involuntary resettlement have resulted in Thai women losing their livelihood assets, reducing their access to public services and disrupting their traditional social networks. Resettlement led to stresses and burdens on displaced people in general and Thai women in particular. The study also recommends how Vietnamese government should observe social safeguards policies in the face of involuntary relocation and resettlement of hydropower projects from the case of Son La hydropower plant.

In this chapter, the authors use field survey data, which was collected for the social impact assessment of resettlement in Son La hydropower project to inform its pre-feasibility study between 2001 and 2014. The authors conducted numerous field trips during the relocation of Thai ethnic communities in the districts of Da river reservoir to their new residence in Son La and Dien Bien provinces. The authors directly engaged in different surveys of major relocations in 2004, 2008 and 2014 with women and men from the displaced Thai communities previously living in Quynh Nhai and Thuan Chau districts (Son La province) and Tua Chua district (Dien Bien province) and currently moving to new houses in Moc Chau, Thuan Chau and Yen Chau districts (Son La province) and Tuan Giao district (Dien Bien province). The survey data was collected in ten years, allowing the research team to return to the study sites several times and replicate the qualitative surveys to assess the ex-ante and ex-post resettlement of selected resettled households.

In addition, the chapter benefited from a desk review of documents and policies of the Government of Vietnam related to resettlement and livelihood restoration for Son La hydropower project. The collected documents are arranged in chronological order and implementation issues of the resettlement plan by phases. They are policy changes and improvements, instead of building resettlement models in Moc Chau district (2003–2005), which were not suitable with Thai culture and expectations, the voluntary resettlement model, which was consulted with the displaced people, and the plan to restore livelihoods and preserve local cultural values in Thuan Chau district (2007–2008); and the consultations with affected Thai women with gender dimensions and participatory decision-making in post-resettlement livelihood restoration plan and indigenous cultural factors of Thai people. In 2004–2005, Dang made two field trips in Son La province to conduct in-depth interviews with 10 Thai women in It Ong commune of Muong La district (where they had to leave) and Tan Lap commune of Moc Chau district (where they had to resettle). In 2007, the research team (Dang and Pham Linh) had two meetings with district officials and 15 in-depth interviews with Thai women at the new resettlement site in the districts of Thuan Chau, Quynh Nhai and Muong La (Son La province). In 2014, Pham Van Loi and Pham Quang Linh received funds from Son La Department of Science and Technology to deliver qualitative surveys, including 10 in-depth interviews, two group discussions and field observations, and a 761-household survey on social impact assessment of resettlement in Son La hydropower project. Household survey data were analyzed by Dang and Pham Linh using SPSS.20 software (see Table 1).

Table 1 Demographic characteristics of the survey sample

		%	N (760)
Sex	Male	71.4	543
	Female	28.5	217
Education	Never been to school	9.2	70
	Primary	38.6	293
	Lower secondary school	37.2	283
	Higher secondary school	10.1	77
	College, university	4.9	27
Occupation	Agriculture	88.7	674
	Trade and service	2.1	16
	Paid employment	5.4	41
	Others	3.9	29
Marriage	Married	90.7	689
	Not married	5.1	39
	Widow, divorced separated	4.2	32
Type of resettlement	In-situ resettlement	40.9	264
	Displacement in resettlement site	59.1	381

Among the surveyed 645 resettled households, 264 households had in-situ resettlement and were relocated in their original commune, and 381 households moved to a new place of residence outside their district. Specifically, most of households in the districts, which were located in the reservoir center such as Muong La and Quynh Nhai, had in-situ resettlement, which means that these households resettled in the village or commune of their homeland. People from the districts far from the reservoir center such as Moc Chau and Mai Son had to move to new resettlement sites, in 50 km to 200 km away from their hometown.

3 Son La Hydropower Plant Resettlement Project Information

Son La hydropower plant is the biggest key project in Vietnam ever. In this important project, 92,301 people from 20,477 households in three provinces of Lai Chau, Dien Bien and Son La were relocated. The relocation and resettlement started in 2003 and lasted until 2013 with many social problems arising during the implementation. Among those affected by the Son La hydropower project, the Thai ethnic group, which is the largest ethnic population living in the floodplain area of the hydropower reservoir, accounts for 80%.

Currently, the population of Thai ethnic group in Vietnam is 1,550,423 people, ranking third in 54 ethnic groups. Thai people have a long history of residence and a unique culture in Son La province. This province is regarded as a cultural cradle of Thai people, especially the "black" Thai group. The affected Thai people accounted for more than 88% of total displaced households in Son La hydropower project.

The resettlement of people living in the floodplain area of the hydroelectric plant in Son La province took place in three stages:

Stage 1: Resettling people to clear land for the plant construction and testing the sample resettlement area in Tan Lap commune (Moc Chau district). This stage took place in 2003 and 2004 to resettle the people in It Ong and Muong Trai communes (Muong La district) and Liep Te commune (Thuan Chau district) and clear land for starting the plant construction by November 2005.

Stage 2: Resettling people living below the 140 m elevation and in the floodplain area for building a phase-1 cofferdam in December 2005.

Stage 3: Resettling people living people in the floodplain area in stage 3 in 2006–2010. This is the largest resettlement, which ended in April 2010. On May 15, 2010, the contractors started filling the reservoir. On November 5, 2010, the water level was 189.3 m to activate the power of Son La hydropower plant (see more Table 2).

Table 2 Resettlement progress in Son La hydropower project by 2010

Province	Total to be displaced population	Number of displaced people of which							Total displaced people by 2010
		Total	2003–2004	2005	2006	2007	2008	2009	
Son La	12,500	11,488	470	1,136	1,630	3,182	2,444	2,626	1,012
Dien Bien	4,436	3,345	200	0	177	161	301	2,506	1,091
Lai Chau	3,324	3,324	200	58	310	1,382	1,168	206	0
Total	**20,260**	**18,157**	**870**	**1,194**	**2,117**	**4,725**	**3,913**	**5,338**	**2,103**

Unit Household
Source Vietnamese Prime Minister's Decision, 2004, 2005, 2006

Thai rice fields of Na Pha, Villages, Muong Trai Commune, Muong La District before relocation and resettlement.

Photo by Dang, June 2007

Thai women catch crabs and snails in It Ong streams Muong La District before relocation and resettlement.

Photo by Dang, June 2007

Summary of Son La Hydropower resettlement in Son La province in each year is presented as follows:

The people in Son La province were resettled by four forms: Rural Resettlement, Urban Resettlement, Voluntary Resettlement and Resettlement under Decree 197. Rural Resettlement includes concentrated resettlement and mixed resettlement. Urban Resettlement consists only concentrated resettlement in Phieng Lanh town (Quynh Nhai district) and Chieng Sinh town (Son La city) (Vietnamese Prime Minister, 2004, 2013, 2014).

Under the Prime Minister's final decision in 2013, the resettlement plan for the people in Son La province was arranged as follows:

People were relocated in the form of concentrated resettlement in 54 sites, 237 locations, of which: rural concentrated resettlement 52 sites, 224 locations; urban concentrated resettlement 2 sites, 13 locations; mixed resettlement in 37 villages of 16 communes, and voluntary resettlement. A total of 12,584 households were relocated, of which: rural concentrated resettlement 9,862 households; urban concentrated resettlement 1,497 households; mixed resettlement 488 households; and voluntary resettlement 737 households. The total area of residential land and productive land is 37,207 ha, of which: residential land 580 ha (resettlement land 389 ha, reserve land 191 ha); farmland 17,900 ha; and forest land 18,726 ha. In the Son La hydropower project, all people subject to mixed resettlement are living in the rural area (Son La Province People's Committee, 2005, 2016; Vietnamese Primer Minister, 2013), (see Tables 3 and 4).

Thai displaced people account for 88% of the total displaced population. Therefore, they are most affected among the ethnic groups to be relocated for Son La hydropower project (see Table 5).

The survey 2014 shows that the life and income of the resettled Thai people have changed significantly after moving to a new place and especially when the project's

Table 3 Resettlement results in Son La province for Son La hydropower project

No	Resettlement form	Number of resettlement sites	Number of resettlement locations	Total number of households	of which	
					Rural	Urban
1	Rural concentrated resettlement	52	224	9,664	9,664	0
2	Rural mixed resettlement	16	37	488	488	0
3	Urban concentrated resettlement	2	13	1,497	0	1,497
4	Voluntary resettlement	–	–	737	737	0
5	Resettlement under Decree 197	–	–	198	198	0
Total		**70**	**274**	**12,584**	**11,087**	**1,497**

Source Vietnamese Prime Minister's Decision, 2013

Table 4 Summary of resettlement forms in Son La province

Resettlement forms	Number of households	
	Quantity	Percentage
Rural concentrated resettlement	9,664	76.80
Urban concentrated resettlement	1,317	10.47
Mixed resettlement (rural)	488	3.88
Voluntary resettlement	737	5.86
Resettlement under Decree 197	198	1.57
Total	**12,584**	**100**

Source Vietnamese Prime Minister's Decision, 2013

Table 5 Ethnic structure of people in floodplain area of Son La hydropower plant

Ethnic	Number of households	Number of people	Percentage (%)
Thai	6,699	42,225	87.95
Khang	307	1,751	3.85
La Ha	430	2,700	5.62
Kinh	280	1,230	2.56
Total	**7,716**	**48,006**	**100**

Source Institute of Agricultural Planning and Design, 1998:8

Table 6 Arable land of Thai people in the surveyed areas

Indicators	Prior to resettlement	After resettlement
Number of family members (person)	4.6	4.7
Number of main laborer (person)	2.4	2.6
Ethnicity (%)		
Thai	98.0	97.6
Others	2.0	2.4
Average land area (m^2/household)		
Land for growing wet rice	2,000	293
Upland land growing rice	11,460	2,000
Land for growing vegetables and fruit trees	700	75,9
Forest land	18,470	2,566
Residential land	350	400
Garden	100	64

Source Field survey results, Pham Van Loi and Pham Quang Linh, 2014

food supports finished in two years. The area of arable land of wet rice cultivation decreased by 2/3 compared with the old place, and the area of upland cultivation also declined by half. Not to mention that the benefits from community forests are no longer the same. The life of the relocated Thai is actually more difficult (see Table 6).

For example, each household has an average of 2,000 m^2 of wet-rice field before relocation, and only 293 m^2 after relocation. Swidden land for upland paddy has also shrunk by 5–7 times. Land for crops and fruit trees before resettlement was 700 m^2 per household, and 72% of households no longer have this type of land after resettlement. Nonetheless, their residential land is larger, because Thai people have the cultural custom to build houses next to each other to visit and care in family and kinship relationships. In new resettlement sites, the housing plots are divided as planned, giving them a little larger residential land, but the family and community relationships are less tight for different reasons. First, households with family and kin relationships are not resettled in one place. Second, households with a close relationship do not build houses near each other because they have to draw lots to pick their residential land plots (see Table 7).

Table 7 Actual production land area of households before and after resettlement

Land category	Area	Prior to resettlement			After resettlement		
		In-situ resettlement (% hhs)	Displacement in resettlement site (% hhs)	N	In-situ resettlement (% hhs)	Displacement in resettlement site (% hhs)	N
Wet-rice land	***				*		
	< 1,000 m^2	**36.3**	**20.4**	164	**90.2**	**83.0**	523
	1,000 m^2–4,000 m^2	48.2	41.3	270	7.8	9.9	55
	4,000–8,000 m^2	4.9	17.9	78	0.8	2.5	11
	> 8,000 m^2	**10.6**	**20.4**	101	**1.2**	**4.7**	20
	Average	*100*	*100*	*609*	*100*	*100*	*609*
Swidden land	< 1,000 m^2	18.1	19.4	115	41.3	45.6	267
	1,000 m^2–4,000 m^2	11.5	9.3	63	14.0	9.6	69
	4,000–8,000 m^2	8.6	13.1	69	8.7	11.5	63
	> 8,000 m^2	**61.7**	**58.2**	363	**36.0**	**33.3**	209
	Average	*100*	*100*	*609*	*100*	*100*	*609*
Crop and fruit tree land	***				*		
	< 1,000 m^2	**68.3**	**46.2**	332	**89.3**	**89.7**	539
	1,000 m^2–4,000 m^2	7.8	13.4	67	4.1	2.2	18
	4,000–8,000 m^2	7.4	12.5	63	1.2	3.6	16
	> 8,000 m^2	**16.5**	**27.9**	140	**5.3**	**4.5**	29
	Average	*100*	*100*	*602*	*100*	*100*	*602*
Forest land	*				***		
	< 1,000 m^2	**58.8**	**64.6**	375	**81.1**	**92.2**	528
	1,000 m^2–4,000 m^2	8.6	5.6	41	4.1	4.2	25
	4,000–8,000 m^2	3.3	7.0	33	1.2	1.9	10

(continued)

Table 7 (continued)

Land category	Area	Prior to resettlement			After resettlement		
		In-situ resettlement (% hhs)	Displacement in resettlement site (% hhs)	N	In-situ resettlement (% hhs)	Displacement in resettlement site (% hhs)	N
	> 8,000 m^2	**29.2**	**22.8**	153	**13.6**	**1.7**	39
	Average	*100*	*100*	*602*	*100*	*100*	*602*
Residential land	*				***		
	< 300 m^2	**51.7**	**46.5**	290	**97.9**	**94.4**	572
	300 m^2–1,000 m^2	46.3	47.9	282	2.1	5.6	25
	> 1,000 m^2	2.1	5.6	25	0	0	0
		100	*100*	*609*	*100*	*100*	*609*
Gardening land	*						
	< 300 m^2	**72.4**	**60.0**	389	**92.2**	**95.5**	564
	300 m^2–1,000 m^2	18.1	23.9	129	5.8	3.1	25
	> 1,000 m^2	9.4	16.0	80	2.0	1.4	10
		100	*100*	*599*	*100*	*100*	*599*

Note Significance level: *** $p < 0.001$; ** $p < 0.005$; * $p < 0.1$
Source Field survey results, Pham Van Loi and Pham Quang Linh, 2014

Burn cultivation along Da River in Nam Gion Cummune, Muong La District. Photo by Dang, 2007

Building a new house in Muong Trai Commune. Photo by Dang, 2007

Households subject to involuntary resettlement have been significantly affected because they lost the stable livelihoods that they used to have in the original place of residence. Besides, resettled households were disconnected from their social relationships (e.g., family and community) and culture (e.g., festivals and traditional customs and practices associated with the forests and land which are the homeland of them and their ancestors). On the ground of the government's policy, Son La province decided to compensate each household by 400 m^2 of residential land, 19,000 m^2 of agricultural land and 13,000 m^2 of forest land. In fact, the households' agricultural land, especially wet rice fields, which is the main livelihood of Thai households, has been largely downsized. In particular, the percentage of households with more than 1,000 m^2 of farmland dropped sharply after resettlement (from 63.8% prior to resettlement to only 8.4%) (Pham Quang Linh, 2016; Pham Van Loi, 2015).

The evaluation of compensation and resettlement policies in Son La hydropower project indicates that resettled people lost their important livelihoods from agriculture, especially land. The government made efforts to compensate affected households to enable them to attain an equivalent or better living standard than they had in the former residence. In reality, the land resource is limited because most agricultural land was allocated to legal owners. The local people do not have much wet rice land to give to their new neighbors. The land for growing vegetables and other crops as well as forest land are also limited because the available land is not productive. As a result, these mounting challenges have put resettled households in difficult situations.

4 Policy Effectiveness and Barriers on Livelihoods of Resettled Households

4.1 Policy Benefits

92.4% of displaced families benefited from the compensation, support and resettlement policies. The majority of them received relocation support (83.2%) and life stabilization support (42.3%). In addition, some households in the survey sample (7.6%) were not supported because they were local residents or teachers' families living in resettlement sites.

Nearly half of the households (49.5%) received one out of nine supports.[1] 27.9% of families got two supports, while a small percentage of resettled people were entitled to three or more supports (under 6.4%). There is no difference in the number of supports between the in-situ resettled households and the households relocated in resettlement sites (Table 8).

The analysis shows that 31.2% of households were better off and 22.2% of households had the same living standards, while nearly half of households were worse off (46.7%). In-situ resettled households had a better standard of living than those in resettlement sites, but the difference is insignificant ($p < 0.05$) (Fig. 1).

Sources of Livelihood

Nearly 90% of resettled households earn their income from agricultural production, so their livelihoods still rely on agriculture after resettlement. 59.1% of families shifted to other crops, mostly maize, while 56% of households changed their livestock structure, focusing more on cattle. Only 9.7% of families opened a local business.

[1] Nine forms of support include: relocation support, life stabilization support, business support, vocational training support, housing rental support, support for ethnic minorities, support for female-headed households, support for illiteracy eradication, and social support.

Table 8 Policy enjoyment by location and type of resettlement (%)

	No support	One support	Two supports	Three or more supports	N
*Location****					
Mộc Châu	5.6	58.2	24.5	11.7	196
Mai Sơn	48.4	17.2	33.6	0.8	122
Mường La	5.0	57.0	29.0	9.0	200
Quỳnh Nhai	6.8	55.6	25.1	12.6	207
Type of resettlement					
In-situ resettlement	4.9	56.8	27.7	10.6	264
Displacement in resettlement sites	5.5	51.4	32.9	11.0	381

Note Significance level: *** $p < 0.001$
Source Field survey results, Pham Van Loi and Pham Quang Linh, 2014

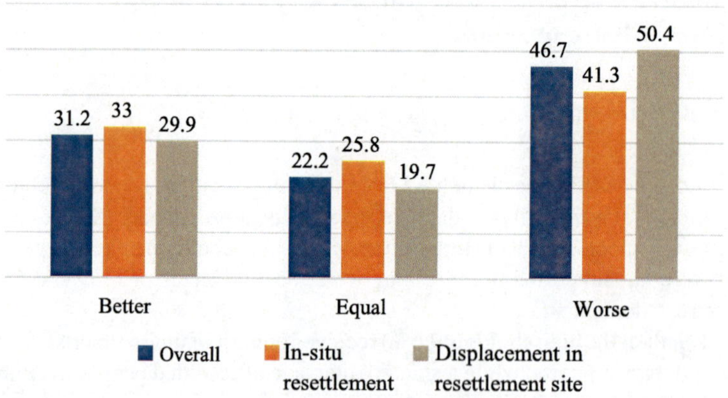

Fig. 1 Post-resettlement living standard by type of resettlement (%) (N = 645). (*Source* Field survey results, Pham Van Loi and Pham Quang Linh, 2014)

Few families earn living from local handicraft, processing or tourism services (less than 3.7%).

Households mainly use the cash support to finance crops on their new land, and partially spend on family consumption, house construction/repair and livestock production. Families in resettlement sites prioritized crops (52.8%) and house construction/repair (42.1%) more than in-situ resettled households (35.9% and 32.4%, respectively) (see Fig. 2).

The survey indicates that most of the resettled people use cash compensation and support to build houses, spend on family consumption and invest in crops and animal husbandry. After resettlement, they have almost no money left to fund businesses.

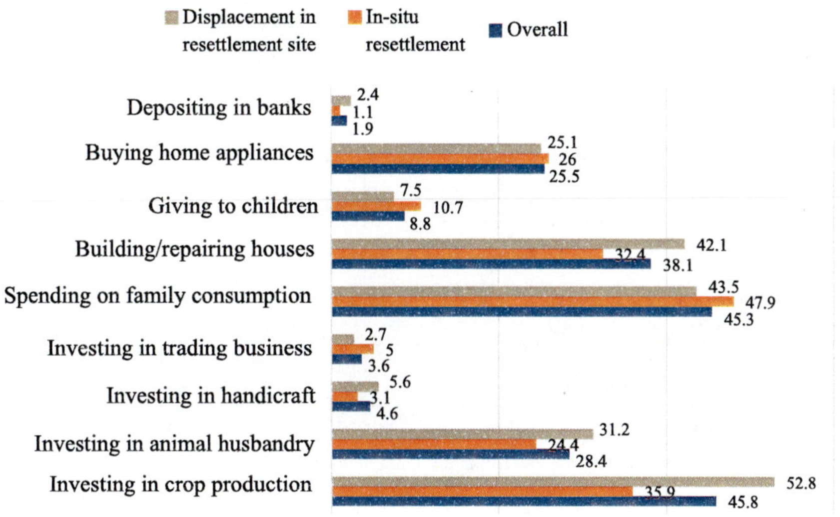

Fig. 2 Use of cash support by type of resettlement (%) (N = 645). (*Source* Field survey results, Pham Van Loi and Pham Quang Linh, 2014)

4.2 Barriers to Livelihoods After Resettlement

To answer questions on policy evaluation opinions, people expressed more interest in occupation-related policies. Most respondents believe that cash support is not sufficient to start a new job or restore the existing livelihood after resettlement (58.7%). 32.6% interviewees believe that the irrigation system fails to meet the needs of agricultural development, while others think that they don't receive any vocational training for the current job (23.9%) (Table 9).

The field survey reveals that the government's policies on supports for resettled households have not been implemented as expected. The support packages on training to change production practices were effective when the infrastructure such as irrigation, transport and other necessary conditions (the inadequate training courses) in new resettlement sites is not sufficient for people to get new jobs or change their working customs.

Currently, the resettled households are challenged by lack of productive land, shortage of food and especially new knowledge to cultivate new crops in new climatic and soil conditions, which are different from their old hometown (see Table 10).

The policy to support resettled households has been prioritized by the government. However, in reality, households do not know how to use these production and credit supports effectively to finance their production and trading activities. The respondents suggested that the production and credit supports should be accompanied by appropriate occupational training and production models.

For example, there are rooms for improvement in selected policies such as credit and employment supports for resettled households, especially poor ethnic women.

Table 9 Opinions on barriers to livelihoods by gender of respondents (%)

	Male (N = 543)	Female (N = 217)	Overall (N = 760)
Don't receive any vocational training for the current job	21.2	30.8	23.9
Previous training is outdated to do the current job	10.6	11.5	10.9
Production practices impede rapid adaptation to the new occupational demands	7.6	7.7	7.6
Irrigation system does not meet the needs of agricultural development	27.3	46.2	32.6
Insufficient funds to start a new job or restore the existing livelihood	57.6	61.5	58.7
Difficult transport conditions	18.2	3.8	14.1

Source Field survey results, Pham Van Loi and Pham Quang Linh, 2014

For example, the government only applied low interest rate credit to poor households. The field survey indicates that the credit policy is not suitable for both women who are able to restore production in a new place and for poor women who are eligible to preferential interest rates.

"There should be changes in loan maturity and value for household borrowers in the credit policy. The current policy allows households, who have to invest in completely new production, to borrow maximum VND 30 million, which is not enough to recover the production as before. Non-poor households like my family are not offered preferences, but the higher interest rates, so we do not dare to borrow. Meanwhile, poor Thai women who are eligible to preferential interest rates do not know how to apply for loans because they are illiterate, so they have to ask husbands or brothers (who are men) to process the borrowing procedures for them. Some do not know what to invest in production and do not dare to borrow either" (In-depth interview with Mrs. L.T.H., 45 years old, Thai ethnic group, Tan Lap commune, Moc Chau district, 2007).

"The land and forest allocation policy also has many shortcomings. There are many complex issues related to the boundaries between villages and communes. The allocation of land and forest caused different disputes and conflicts between resettled people and local residents. We directed the local residents to give land to the resettled people and the latter will plant what the former cultivate. But the new comers are unfamiliar with local climate and soil characteristics, so they usually failed and had no income" (FGD Leader group, Moc Chau District, 2007).

"The production support mechanism is inappropriate. For those who have to migrate to the new area, the allocated land is only a quarter of what they had in the old place. Therefore, there must be technical guidance for them to adapt to new production conditions. It takes a long time for the resettled people to restore their production. In addition, there is no investment or production support for local residents who handed over their land to the migrants, so they have faced a lot of

Table 10 Opinions on challenges of resettled households

	Gender		N = 760	Type of resettlement		N = 645
	Male (N = 543)	Female (N = 217)		In-situ resettlement (N = 264)	Displacement in resettlement sites (N = 381)	
Unsufficient arable land	87.7	86.1	558	88.3	85.8	558
New varieties of crop	***					
	23.7	27.8	173	11.7	34.8	163
New varieties of animal	9.6	9.6	67	6.4	12.4	64
Excess labor, no available jobs	**			***		
	54.9	49.7	368	51.6	58.4	346
Unsufficient wet rice land	***			*		
	75.4	74.3	525	38.4	61.6	560
Shortage of funds	*			**		
	62.5	64.7	446	59.1	66.5	408
Poor irrigation	**					
	51.4	46.5	350	45.8	51.2	315
Unfamiliarity with to new weather, climate, and cropping seasons	*			***		
	7.6	10.2	58	17.4	19.8	121
Inefficient uses of cash support	**			***		
	78.6	21.4	192	41.2	37.9	174
Lost traditional family and kin relationships	**			***		
	47.3	58.8	351	37.4	64.4	340

Note Significance level: *** $p < 0.001$; ** $p < 0.005$; * $p < 0.1$
Source Pham Van Loi, 2015 and Pham Quang Linh, 2016

difficulties. If the inadequate resettlement models are applied, people's livelihood cannot be developed because their production is not stable" (FGD Leader group, Quynh Nhai district, 2007).

As a large project with a massive number of affected people, of which the majority are ethnic minorities, the Son La hydropower project made some adjustments during the course of implementation. The resettlement policies partly satisfy the immediate living needs of relocated people. Basically, Son La hydropower project supported the affected people to have a temporarily stable life. In addition to accommodation, it is necessary to have policies and financial mechanisms to create jobs (for households losing farmland or resettled people). It takes a long time to restore the lives and livelihoods of affected households (Pham Quang Hoan editor, 2012). However, the support policies to restore people's incomes have not been secured by long-term financial resources. The project is limited to compensation for land uses and directly damaged assets, mainly infrastructure, food relief during resettlement. Other indirect and intangible damages such as livelihoods, income, business location, fisheries, forest benefits, traditional cultural value, etc. have not been thoroughly considered in the resettlement plans.

For example, the project's livelihood restoration policy only compensates for losses from the crops in the old residence and pays a lump sum to support vocational training and job change. But in reality, there is no detailed guidance for the people, especially women. As a result, many Thai women have had a hard time after getting paid for resettlement and trying new production activities in new places.

"My family received compensation from the project and invested in livestock in the new place. But myself, as a woman, did not receive any technical guidance in building barns, taking care of livestock, particularly raising cattle and poultry. So I tinkered to find out how to make a garden to raise pigs and ducks. Then, the animals got sick and died. We lost all the money and did not know what to do to earn income" (In-depth interview with Mrs. Vi.T.C, 45 years old, Thai ethnic group, Muong Giang commune, Quynh Nhai district, 2007).

"I heard that there are households raising pigs according to the Biogas model, which is very successful, but it has not been replicated by the project to other households. In this Nghe Tong village, there are many Thai households had to tinker with pigeons and porcupines without advice from technical staff. So they all failed" (In-depth interview with Lo T. N., 38 years old, Nghe Tong village, Muong Giang commune, Quynh Nhai district, 2007).

While implementing the resettlement policy, there have been lots of arising problems from livelihood support for households. One single support policy was applied to the resettlement sites with different socio-economic conditions and different resettlement processes. Therefore, some resettled Thai households did not accept the livelihood restoration supports.

"Currently, there is a slowdown in the development of new resettlement sites. In many places with good conditions like Mai Son district, the migrants are poorer than the local people. Previously, the Thai village had many trees. In the new resettlement village, there are no trees at all. In Tan Lap commune of Moc Chau district, migrants want to return to their old villages because of the change in climate. Previously, they stayed in a place all the time and was topless. Now they have to wear sweaters all day long. Some households lived by rivers and earned their living from fishing. Now they lost the traditional livelihood and suddenly had to switch to tea cultivation. So they want to come back to where they lived before, because the new place is developing into an urban area. The most important thing is that the resettlement and livelihood support policy is not appropriate, and the government's operating mechanism is not suitable" (Staff group discussion, Muong La district, 2007).

Thai ethnic women are familiar with farming activities, working hard and taking care of the family. Before the resettlement, they had a peaceful life, doing farming activities as wet rice cultivation, upland cultivation, animal husbandry and collection of forest products. Although living under self-sufficiency, their livelihood and income are quite stable. The natural conditions in the living environment favor them in a peaceful life, having food and raiment with close and cozy relationships in the community (Dang Thi Hoa, 2012). In addition to upland cultivation of rice, maize, cassava and legume, Thai women have a lot of experiences in growing intercropped vegetables; collecting forest products as food for their daily meals such as forest vegetables, bamboo shoots and mushrooms; catching mollusks and small animals as food for family meals such as snails, crabs, fish, shrimp, etc.

Because of the resettlement, the life of Thai women, which was self-contained with little impact, has changed dramatically. They do not earn any income, apart from the food subsidy provided by the state, and no longer access to natural benefits. They have to manage hard to feed the family. This is a considerable obstacle for Thai women resettled in the Son La hydroelectricity project. According to the results of in-depth interviews with Thai women in their new places, various respondents shared the view that the forest land is much smaller than the old place. Therefore, the gathering of forest products is challenging due to the rapid deterioration of forest. Previously, Thai women in Bia và Co Tran villages (Quynh Nhai district) went to the forest to pick various kinds of vegetables, bamboo shoots and mushrooms in the natural environment for daily meals. The stock of vegetables and mushrooms are considered by them to be plentiful and easy to find. At present, they still go to the forest in their new place, but get very little and not enough for daily meals. Ms. Lo Thi L. in Bia village (Quynh Nhai district) said: "*Before I and other Thai women went to the forest occasionally to pick vegetables, bamboo shoots and mushrooms, caught fishes and took moss from the stream for our meals. Now the streams are deeply flooded, the forest land was divided into small pieces and allocated to households for management, where they planted trees for economic purposes. Even the forests owned by some families are not rich in terms of vegetables and fruits as before. Currently, we rarely go to the forest, and shop in stores or markets instead because many things are available there and shopping is much more convenient. Everything is available but it costs a lot of money, not free from the natural environment like before resettlement*".

While adapting to new living conditions, many Thai households had to alter their livelihoods. Previously, most of them earn their living from two-crop wet rice cultivation and upland cultivation. After resettlement, up to one-third of households shifted to cattle, poultry and aquaculture farming. Some households invested in small businesses. Particularly, 49.2% of households invested in farming, only 8% of households spent money on additional arable land; 30.38% of households invested in cattle, poultry and aquaculture. Most of the compensation were spent on household consumption, purchasing equipment, amenities, building houses and sharing with children.

Thai girls go to school in resettlement sites of Nam Gion commune. Photo by Dang, 2009.

Thai Women go to work as workers in the resettlement site of Phieng Lanh, Quynh Nhai. Photo by Pham Linh, 2014.

At the resettlement site in Phieng Lanh town (Quynh Nhai district), the choice of resettlement in the urban area means that they have only residential land and no more productive land. Then, many people have been looking for new forms of livelihood such as trading, motorbike taxi driving, photocopying, catering, etc. Most of women chose trading, meaning buying vegetables, fruits or food from farmers to sell in markets. Ms. Lo Thi V. informed that she earned VND 200,000–300,000 on a daily basis by doing this business. However, unlike the elderly, many young people are not interested in these jobs because they are ashamed. They prefer occupations which allow them less likely to appear on the streets, such as restaurant waiters or workers in faraway factories. Younger women, who are in good health conditions, usually chose working as workers (Field survey in a town of Quynh Nhai district, 2007).

Data from the survey (Table 11) shows that the main income sources of households from rice cultivation and collection of forest benefits before resettlement shifted to animal husbandry and working for hire after resettlement. These changes also reflect clearly the current situation of resettled Thai people in Son La hydropower plant.

The survey clearly presents the livelihoods of displaced Thai households in 2014 five districts/cities of Son La province after four years of resettlement. The ex-post resettlement issues with regard to livelihood are important challenges to central and provincial governments. The Prime Minister also distinctly indicated the concerns in his speech at the Son La Hydropower Plant Resettlement Review Conference in December 2016. According to the Prime Minister, the ethnic minorities made a tremendous contribution when they agreed to leave the hometown to clear land for the project. As evaluated by the Project Management Unit, the resettled people have earned more than VND 1.2 million/person/month in 2015, 3.92 times higher than their income before the resettlement. The poverty rate decreased by 2.56 times. Therefore, the Prime Minister questioned whether the results match the actual livelihoods of affected people. Obviously, there are rooms for improvements in the compensation policies because income-earning jobs have not been sufficiently created and the displaced people have not been attached to the new land. This key challenge poses a fundamental concern for leaders (Vietnamese Government, Son La province People's Committee, 2016; Vietnamese Prime Minister, 2013).

Table 11 Assessment of current household compared to ex-ante resettlement of Thai ethnic

Indicators	Male (N = 511)	Female (N = 189)	Total (N = 700)
*Household income***			
Much better	6.3	4.2	5.7
Better	30.1	31.2	30.4
Same	25.2	24.3	25.0
Worse	18.0	21.7	19.0
Much worse	18.5	20.4	19.9
*Product conditions***			
Better	5.7	4.8	5.4
Same	16.0	19.3	16.9
Worse	78.3	75.9	77.7
*Current obstacles**			
Lack of arable land	75.4	74.3	75.1
Lack of funds	63.5	64.7	63.8
Lack of water for farming	51.4	46.5	50.1
No training for new jobs	21.2	30.8	23.9
Failure to find a new job	54.0	49.7	52.9
Loss of traditional cultural activities of the community	21.6	33.2	24.7

Unit %
Source Field survey results, Pham Van Loi and Pham Quang Linh, 2014
Note Significance level: **$p < 0.001$, *$p < 0.01$

As a cultural characteristic of Thai families, women usually take care of housework and farming activities. Men represent the households to communicate with the outside society and make decisions about all family affairs in relation to the community and the broader society. The issue of earnings and job seeking for the resettled Thai women is quite complicated. The ex-post resettlement jobs, such as animal husbandry and hired labor, are only suitable for men. Thai women have practically no job or income, and their lives depend substantially on the men in the family.

"For a long time, Thai women in the floodplain area of the hydropower plant as well as elsewhere are not only famous for weaving unique brocade fabrics, but also for their exquisite handicrafts made from fabric, such as dresses, shirts, blankets, pillows, cushions, curtains, towels, etc. Currently, these textile products are also used for daily life and partly for tourists. However, Thai women are less likely to weave brocade now. Many of them discarded the looms in the warehouse or in the garden. Villagers said that weaving is no longer important to them because they can easily buy these products from Thai people from other places in markets" (In-depth interviews with Mrs. Đ.T.N, Thai ethnic, Leader Cultural Department, Quynh Nhai District. 2014).

5 Some Remarks

Son La hydropower plant is a large project of Vietnam, which acquired land and resettled tens of thousands of households in Son La, Dien Bien, and Lai Chau provinces–hometown of Thai ethnic people. The Thai people had to sacrifice their homeland to develop the country.

The government introduced various policies to support, compensate and resettle the affected households. However, households have faced numerous difficulties and challenges after resettlement while restoring livelihoods and stabilizing their lives. The post-resettlement life was not equal to that in the hometown because they have lost important livelihoods from land, especially wet rice cultivation in river valleys.

The field survey in selected communes of Quynh Nhai and Moc Chau districts reveals the fact that a large number of resettled Thai do not have a job. They are not familiar with doing business, services and hiring labor after being resettled. As a result, they become even more vulnerable. New jobs such as working as hired labor or trading business are completely inappropriate with the inherent personality and skills of Thai traditional. Special, resettlement policies fail to accommodate the need of support in skill training for Thai women in particular and ethnic women in general in job conversion to adapt to new living conditions.

Findings from the study demonstrate the limitations of resettlement policies in hydroelectric projects, including Son La hydropower plant, livelihood recovery is challenging, especially for ethnic minority. Thai ethnic have become more disadvantaged because they lived in a close ethnic community and had limited social knowledge, causing their difficulties in adapting to the new community and new social relationships. On the other hand, they have to abandon existing knowledge in farming and collecting natural benefits from rivers, streams and forests to feed the daily meals of their families, or turn their back on experiences of using medicinal plants in health care. Moving to a new place, Thai ethnic must adapt to the new cultural relationships of the community and the new landscape environment. They are exposed to a completely new life, even though they are not fully equipped and supported with new social communication skills. The existing resettlement policies only meet a very small part of demand from the livelihoods and daily life of Thai people. Therefore, it is essential to have more adequate researches and assessments from the perspective of gender to measure the impacts of resettlement on livelihood stability and development opportunities for ethnic minority in hydroelectric projects in general and Son La hydropower plant in particular.

Lessons learned from resettlement policies in Son La hydropower plant have been taken by the government to apply to Trung Son hydropower plant, which is another large project in the Central Region of Vietnam implemented in ten years later. In addition to land and livelihood supports for resettled households, resettlement policies in Trung Son hydropower project have been mainstreamed with gender under the support of the World Bank in Vietnam to achieve the sustainable development.

There is a stronger focus on training in crop production, animal husbandry and forestry development for households, especially women. The household economic development model based on natural capital, human capital, social capital and gender mainstreaming has been adopted.

References

Colchester, M. (2000). Dams, indigenous people and ethnic minorities. *World Commission on Dams Thematic Review Social Issues*, I2.

Erdiaw-Kwasie, M. O., Ackuayi, E. D., Abunyewah, M., Appiah, S.O. (2014). Theoretical and practical motives for participation obstacles in resettlement programs. review from the mining perspective. *Journal ofEconomics and Sustainable Development*; *5*(14), 119–128. Retrieved from https://www.researchgate.net/publication/272481482/. www.iiste.org. ISSN 2222–1700 (Paper) ISSN 2222–2855 (Online)

Frank, V. (2017). Project-induced displacement and resettlement: From impoverishment risks to an opportunity for development? *Impact Assessment and Project Appraisal, 35*(1), 3–21. https://doi.org/10.1080/14615517.2017.1278671

Gutman, P. (2003). Involuntary resettlement in hydropower projects. *Annual Review of Energy and the Environment, 19*(1), 189–210. https://doi.org/10.1146/annurev.eg.19.110194.001201

Hoa, D. T. (2012). Impact of hydropower projects on traditional culture of resettlement Thai people in North West and North Central Region. In *The 6th National Convention on the Thai* (pp. 467–476). Thanh Hoa: World Publishing House.

Hoan, P. Q. (Ed.). (2012). *Ethnic groups' culture in the reservoirs and son la hydropower resettlement area.* Hanoi: Social Science Publishing House.

International Finance Corporation (IFC). (2012). *Policy on environmental and social sustainability.* Retrieved from https://www.ifc.org/wps/wcm/connect/7141585d-c6fa-490b-a812-2ba87245115b/SP_English_2012.pdf?MOD=AJPERES&CVID=kiIrw0g.

Linh, P. Q. (2016). The current livelihood of the Thai resettlers in Quynh Nhai district, son la province. *Anthropology Review, 5*, 29–37.

Michael, C. M. (1993). Social science research and the crafting of policy on population resettlement. *Knowledge and Policy 6*(176). https://doi.org/10.1007/BF02696288

Nguyen, H. T., Pham, T. H., & de Bruyn, L. L. (2017). Impact of hydroelectric dam development and resettlement on the natural and social capital of rural livelihoods in Bo hon village in central Vietnam. *Article of Journal Sustainability, 9*(8), 1422. https://doi.org/10.3390/su9081422

Ronggang, S. (2008). *The three gorge hydropower – complex project: Flood reduction and power supply.* China Intercontinental Press.

Rowan, M. (2017). Aligning resettlement planning and livelihood restoration with social impact assessment: A practitioner perspective. *Impact Assessment and Project Appraisal, 35*(1), 81–93. https://doi.org/10.1080/14615517.2016.1271541

Smyth, E., Steyn, M., Esteves, A. M., Franks, D. M., & Vaz, K. (2015). Five 'big' issues for land access, resettlement and livelihood restoration practice: Findings of an international symposium. *Impact Assessment and Project Appraisal, 33*(3), 220–225. https://doi.org/10.1080/14615517.2015.1037665

Tan, Y. (2008). Displacement and resettlement in the three gorges project: issues confronting women migrants. *Asian and Pacific Migration Journal, 17*(1), 1–31. https://doi.org/10.1177/011719680801700101

UNDP. (2017). *Guidance note for the application of the sustainable livelihoods framework in development projects.* Retrieved from www.undp.org.rblac_livelihoods%20guidance%20note_en-210july2017.pdf

Vanclay, F., Esteves, A. M., Aucamp, I., Franks, D. (2015). Social impact assessment: guidance for assessing and managing the social impacts of projects. *International Association for Impact Assessment*. Retrieved from http://www.iaia.org/uploads/pdf/SIA_Guidance_Document_IAIA.pdf [Google Scholar].

Van Loi, P. (2015). Son la hydropower resettlement in son la province: decline in cultivated land area, some problems and solutions. In *The 7th National Convention on the Thai*, 478–490. Lai Chau: World Publishing House.

Vietnamese Government, Son La province People's Committee. (2005). *Decision of son la province people's committee on the promulgation of regulations specifying the policy of the government on compensation, immigration, resettlement in the son la hydropower project, son la.*

Vietnamese Government, Son La province People's Committee. (2016). *Report on migration project, son la hydropower resettlement, son la province, son la.*

Vietnamese Prime Minister. (2004). *Prime minister's decision on the issuance of compensation, immigrant and resettlement rules in son la hydropower project, Hanoi.*

Vietnamese Prime Minister. (2013). *Decision approving the master plan for resettlement of son la hydropower project, Hanoi.*

Vietnamese Prime Minister. (2014). *Decision approving the list of annexes to the master plan for migration in son la hydropower project resettlement by decision no 2009/QĐ-TTg on 04th November in 2013 of the prime minister, Hanoi.*

World Bank. (2004). *Involuntary resettlement sources book: planning and implementation in development projects.* The International Bank for Reconstruction and Development: Washington, DC, USA. Retrieved from https://openknowledge.worldbank.org/handle/10986/14915.

Factors Influencing the Intention of Using Solar Energy Household Appliances with an Extended TPB Model Approach: Evidence from Vietnam

Thuong Huy Do, Hong Phuong Thi Nguyen, and Thuy Hong Dang

Abstract This study examines factors influencing the intent of purchasing solar energy appliances in Vietnam. In this research, the proposed model is constructed on the TPB paradigm with the addition of context factors (perceived benefit, perceived cost, word of mouth) to analyze factors affecting the intent. A survey of 375 customers was conducted in Hanoi, Bac Ninh, Hai Duong, and Nam Dinh in Vietnam to test the research model and hypotheses. The relation among constructs was evaluated through structural equation modeling. Only attitude and subjective norms were found to be positive contributors to the intention. Besides, the positive relationship between attitude and perceived benefit was also highlighted, but the relationship between attitude and perceived cost was not confirmed. The positive impact of interpersonal influence and e-word of mouth on subjective norms was supported, too. Finally, the study stimulates further research and serves as a reference for managers and scholars interested in solar energy.

Keywords TPB · Purchase intention · Solar energy · Appliances

1 Introduction

Together with economic development, the demand for electricity usage in Vietnam like other developing countries has been escalating in recent years. According to the estimation of Vietnam Electricity Corporation (EVN), the demand for annual electricity usage in Vietnam over the period of 2021–2025 will rise by 8.9%, equivalent to 23.6–30.5 billion kW/year. Meanwhile, the new supplementary output of electricity only reaches 6.1–17.6 billion kW/year. In addition, the energy power source

T. H. Do (✉)
VNU School of Interdisciplinary Studies, Vietnam National University, Hanoi, Vietnam
e-mail: thuonghuydo@yahoo.com

H. P. T. Nguyen
Hanoi University of Natural Resources and Environment, Hanoi, Vietnam

T. H. Dang
Hanoi Industrial Textile Garment University, Hanoi, Vietnam

A. T. Nguyen et al. (eds.), *Contemporary Economic Issues in Asian Countries: Proceeding of CEIAC 2022, Volume 1*,
https://doi.org/10.1007/978-981-19-9669-6_2

mainly comes from coal-fired and hydroelectric power. Coal-fired power represents 50% of the electricity production in 2020 (123 billion kW). Next to it is hydroelectric power with 29.5% (73 billion kW) and gas-fired power with 14% (35 billion kW). Whereas, solar energy power and other renewable energy powers only account for 4% and 1.4% respectively (Manh Duc, 2021; Ali et al., 2020a, 2020b). The utilization of traditional energy sources has seriously impacted the environment and the eco-system (Duong et al., 2017; Nguyen et al., 2017). Being aware of that problem, many people in Vietnam have turned to the use of solar energy power for household appliances.

In recent years, solar energy power has received much attention from practitioners and scholars alike. Thus, there have been various researches on solar energy power from such various aspects as economic feasibility, policy barriers, and the role of propaganda and media (Khalid et al., 2013; Wang et al., 2017). From consumers' perspective, technology (Bandara & Amarasena, 2019), cost (Bandara & Amarasena, 2019; Ayoub et al., 2019; Akroush et al., 2019), ease of use (Stragier et al., 2010; Ali et al., 2020a, b; Bandara & Amarasena, 2019; Ayoub et al., 2019; Hua & Wang, 2019), usefulness or benefits (Stragier et al., 2010; Ali et al., 2020a, 2020b; Ayoub et al., 2019; Hua & Wang, 2019; Akroush et al., 2019; Wang et at., 2017), risk (Pham & Duong, 2020), environmental concern (Bhutto et al., 2022; Tan et al., 2017; Wang et al., 2017), and moral norm (Bhutto et al., 2021; Bhutto et al., 2022) were confirmed to have effects on consumers' behavioral intention in the prior studies.

In spite of the prior researches' theoretical and practical implications for solar energy power and energy-efficient products in general, few studies have centered on solar energy appliances in particular. Moreover, many other factors have not been thoroughly examined, but they potentially impact people's intent of using such appliances. Thus, to fill these gaps, this research is to explore how the factors of perceived benefit, perceived cost, and word of mouth (both traditional and electronic) affect people's intent of using solar energy appliances through behavioral attitude and subjective norms. In addition, this research is also to investigate how such factors influence people's purchase action through their purchase intention.

2 Literature Revision and Hypothesis

2.1 Literature Background

This study employed the Theory of Planned Behavior (TPB) as the underpinning one with the additional factors (perceived benefit, perceived cost, inter-person influence, and word of mouth) to predict and analyze the factors influencing the intent of purchasing solar energy appliances through behavioral attitude and subjective norm on the basis of a massive literature review.

TPB as the underpinning model with three antecedents (attitude, subjective norm, and perceived behavioral control) is employed to foretell and examine the behavior of consumers in various domains (Ajzen, 2010). It has been considered as the backbone for the theoretical framework of various researches when it comes to electricity saving (Wang et al., 2011, 2014), green products (Chen & Tung, 2014; Sun & Wang, 2019), energy-efficient appliances (Akroush et al., 2019; Bhutto et al., 2021; Wang et al., 2017; Hua & Wang, 2019; Tan et al., 2017; Ali et al., 2020a, b; Bhutto et al., 2022), solar energy (Ali et al., 2020a, b; Tsuar et al., 2018), and smart household appliances (Stragier et al., 2010). Despite its widespread acceptance in predicting behavioral intent, the model has been criticized. The major complaint stems from the need for more variables to improve its predictive and explanatory power (Bhutto et al., 2021; Wang et al., 2014). Some researchers claimed that the TPB model failed to handle a sufficient variety of intentions (Ajzen, 2002; Rhode et al., 2003). As a consequence, additional factors can be included in the TPB paradigm if they provide a significant insight into the behavior (Ajzen, 1991, 2020). Therefore, researchers have attempted to improve its explanatory power with additional constructs such as environmental benefit (Bhutto et al., 2021; Bhutto et al., 2022) or environmental concern (Chen & Tung, 2014; Li et al., 2019; Bhutto et al., 2021), warm glow benefits (Bhutto et al., 2021), moral obligation (Bhutto et al., 2022; Chan & Bishop, 2013; Chen & Tung, 2014; Kaiser, 2006; López-Mosquera et al., 2014), knowledge of energy (Wang et al., 2014), publicity of information (Wang et al., 2014), daily habit (Wang et al., 2014), and residue effect (Wang et al., 2017). However, in the modern context with internet development, TPB may be inadequate to predict behavior due to benefit—cost awareness as well as internet impacts. To fill these gaps, some improvements were made in the TPB paradigm with the additional factors of perceived benefit and perceived cost to comprehend consumers' attitude toward the intent of utilizing solar energy appliances in this research. Moreover, word of mouth (both traditional and electronic) was added to the model in order to comprehend how these variables affect consumers' subjective norm.

2.2 *Theoretical Model*

Perceived Benefit

There has been a debate that people are more apt to perform a certain behavior when perceiving more benefits than costs. The positive effect of renewable energy utilization on consumers' attitude was confirmed, but its impact on the intent of adopting renewable energy was not supported (O'Driscoll et al., 2013). This significant finding suggested further examining the relation between perceived benefit and intent of buying energy-effective products in various domains. Leelakulthanit (2014) posited that perceived benefit was the only determinative factor of adopting LED lighting, yet the nature of the benefit sought was characterized in various terms. Quality was also found to be the most prevalent benefit that affected the behavioral intent among low-income families while the energy-saving benefit was found to be the most influential among higher-income ones. Akroush et al. (2019) confirmed that perceived benefit positively affected both consumer attitude and intent of using energy-efficient products. Hence, this study hypothesizes that:

H1a: Perceived benefit positively affects behavioral attitude toward solar energy appliances.

Perceived Cost

The cost for solar energy appliances is inclusive of the initial amount of money to buy appliances and periodic maintenance. The costlier the appliances are, the lower values consumers receive and the lower appliances' utility rate gets (Premkumar et al., 1997). Cost is considered as one of the determinants affecting consumers' intent of adopting solar energy (Bandara & Amarasena, 2019; Ayoub et al., 2019; Akroush et al., 2019; Alam & Rashid, 2012). If people are to employ new technologies, these new technologies must be properly priced in relation with alternatives. If not, consumers' adopting new technologies may not be feasible.

Previous researchers agreed that there was a direct and significant correlation between cost and intent of using products (Bandara & Amarasena, 2019; Ayoub et al., 2019; Alam & Rashid, 2012). Meanwhile, Akroush et al. (2019) assumed that perceived cost was mediated through its effect on attitude toward behavioral intent and price was not confirmed in the research by Wang et al. (2017), who empirically analyzed the factors impacting Chinese urban residents' intent of purchasing energy-efficient appliances on the structural equation model, as well as in the research by Bketi et al. (2022), who examined factors influencing people's intention of using rooftop solar photovoltaic in Indonesia. In this current research, we are to verify the effect of perceived cost on purchase intention through the mediation of attitude toward behavioral intent. Therefore, we develop the hypothesis:

H1b: Perceived cost has a negative correlation with behavioral attitude toward solar energy appliances.

Interpersonal Influence

Word of mouth is known as a non-commercial communication among users about a certain product, service or business (Litvin et al., 2008). Such interpersonal exchanges offer references to the product or service consumption over and above the advertising released by companies. The interpersonal influence is seen as one of the aspects significantly affecting subjective norm (Bhattacherjee, 2000). It is attributed to the behavioral alteration in individuals due to others' perception and comments, which might impact others' adoption. Additionally, the strong influence of friends and family members (e.g., Hsu et al., 2006; Wang et al., 2017) was posited and stressed. However, the relationship between interpersonal influence and subjective norm was rarely investigated in previous studies on appliances. Hence, the hypothesis is suggested:

H2a: Interpersonal influence is positively related with subjective norm regarding solar energy appliances.

Electronic Word of Mouth (EWOM)

In the modern life, a novel form of EWOM can be consumers' positive or negative remarks on a certain product or service online (Hennig-Thurau & Walsh, 2003). EWOM is also regarded as a significant platform that permits customers to convey their thinkings to others (Godes & Mayzlin, 2004). Also, EWOM is argued to be more productive than traditional one on account of its ease of use and its range width (Fong & Burton, 2008). Furthermore, EWOM is also suggested to have effect on subjective norm (Doma et al., 2015; Jalilvand & Samiei, 2012). During the digital age, Vietnamese youngsters spend most of their spare time online (Nielsen, 2018). Thus, the study recommends the hypothesis:

H2b: EWOM has a positive correlation with subjective norm when it comes to solar energy appliances.

Behavioral Attitude

In the prior researches based on the TPB paradigm, one of the determinants shaping the intent is behavioral attitude (Ajzen, 1991), which is inclusive of agreeable or disagreeable evaluation, emotional sensations, and behavioral proneness. Many researches were implemented to scrutinize and confirm the positive attitudinal influence on behavioral intent in various domains (Ajzen, 2010; Chew & Abdul Adis, 2018; Jalilvand & Samiei, 2012; Juniwati, 2014). As for appliances, attitude is also one of the main determinants positively impacting the purchasing intent (Bhutto et al., 2021; Tan et al., 2017; Hua & Wang, 2019; Tsuar & Lin, 2018; Alam & Rashid, 2012; Akroush et al., 2019). The hypothesis is proposed:

H3: Behavioral attitude has a positive correlation with the intent of buying solar energy appliances.

Subjective Norm

In the TPB paradigm, subjective norm is defined as the second determinant of behavioral intent by Ajzen (1991). It is the awareness of pressures from the people around, who are important to an individual in some ways. It captures an individual's perception of social pressure regarding a certain behavior. The influence of subjective norm on behavioral intent has been proved in various domains of food (Shin & Hancer, 2016), health (Caso et al., 2019), advertising (Sanne & Wiese, 2018), green hospitality (Chen & Tung, 2014), shopping online (Hasbullah et al., 2016), and energy-efficient appliances (Bhutto et al., 2021; Bhutto et al., 2022). Hence, the following hypothesis is developed:

H4: Subjective norm positively correlates with the intent of purchasing solar energy appliances.

Perceived Behavioral Control

Perceived behavioral control was introduced into TPB by Ajzen (1991) as a determinative factor of behavioral intent. Perceived behavioral control is referred to an individual's awareness of the ease or difficulty in undertaking a certain behavior (Wang et al., 2011). Concerning solar energy appliances, some external factors such as time, financial cost, and effort may be out of an individual's control (Lindenberg & Steg, 2007). When individuals believe that they have more opportunities and resources as well as expect fewer obstacles, their perceived behavioral control will be stronger and their intent of purchasing solar energy appliances will be greater. In line with prior studies, consumers are more likely to engage in the intent of purchasing solar energy appliances when they perceive that they can control the factors (Hua & Wang, 2019; Tan et al., 2017; Wang et al., 2017). This research is also to verify the effect of perceived behavioral control on the intent. Thus, the hypothesis is developed:

H5: Perceived behavioral control positively correlates with the intent of buying solar energy appliances.

Purchase Intention

Purchase intent is reflective of consumers' social trend. Accordingly, individuals often take specific actions in compliance with such tendency. It can be measured by individuals' willingness to have an attempt to take actions. On the basis of the TPB paradigm, behavioral intent is pointed out to be the best way to foretell actual actions. In another word, the stronger intent of taking certain actions individuals have, the higher likelihood of taking the action will be. Therefore, it can be inferred that consumers' intent of using solar energy appliances can determine the purchase behavior of solar energy appliances. This study proposes the hypothesis (Fig. 1):

H6: Purchase intention has significant and positive correlation with purchase action.

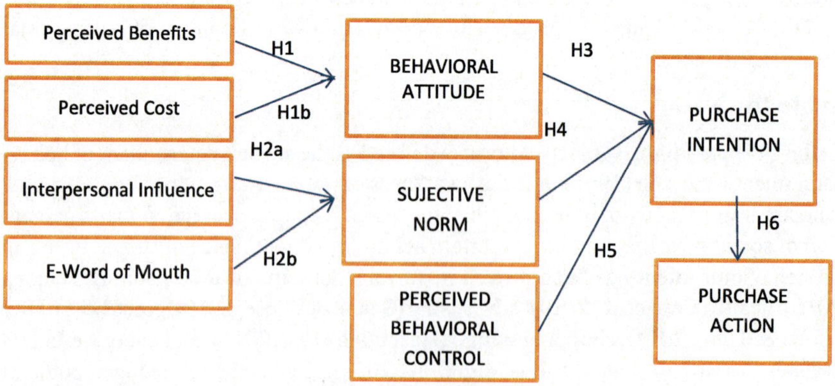

Fig. 1 Proposed paradigm and hypotheses. *Source* Authors

3 Research Method

3.1 Development of Measure and Questionnaire

With the convenient sampling, this research targeted those who have bought solar energy appliances such as lights, watches, heating water systems, fans, pumps, and air conditioners in Hanoi, Bac Ninh, Hai Duong, and Nam Dinh in Vietnam. The questionnaire has two sections: (1) demographics and (2) measurement items on Likert scale. Nine constructs were measured with 33 items. Each scale is composed of 3–5 items determined and adapted from related literature (Table 1). Perceived Benefit, Perceived Cost, and Purchase Intention were measured with the item scales of Akroush et al. (2019). The scales measuring Interpersonal Influence and E-Worth of Mouth were adapted from Kawakami et al. (2013). Subjective norm was measured with the item scale of Hua and Wang (2019) while the item scale of Tan et al. (2017) was taken to measure Perceived Behavioral Control. The item scale for Purchase Action was adapted from Bhutto et al. (2022). All scale items were translated into Vietnamese from English. The 5-point Likert scale from strong disagreement (1) to strong agreement (5) was employed to measure all the items.

3.2 Sample and Data Collection

Forty respondents were used in the pre-test to check the apprehension of survey questions and the time to complete the questions as well as the face validity of measurement scales, which was considered sufficient as recommended by Hair et al. (2010). There were tiny changes in the question-wording. Afterward, the survey questions were delivered to the people who have bought solar energy appliances in Hanoi, Bac Ninh, Hai Duong, and Nam Dinh in Vietnam from April 2021 to March 2022, through the mailing system and in person. The total of collected responses was 402, of which 120 were from the mailing system and 282 from customers directly in person. After eliminating inadequate replies, we found that 375 responses were valid for the analysis in the next steps. This is in alignment with a reasonable sample size to employ SEM (Kline, 2011). SPSS20 and AMOS24 software were employed to process the data in order to check hypotheses. Respondents' demographic information is composed of age, sex, household income, and education background (Table 2).

Table 1 Definition of variables and scale references

Research variables	Conceptual definition	Reference scholars
Purchase action	Action of purchasing solar energy appliances	Bhutto et al. (2022)
Purchase intention	Intention of using solar energy appliances	Akroush et al. (2019)
Behavioral attitude	Positive and negative comments on solar energy appliances	Akroush et al. (2019)
Subjective norm	Social pressure caused by using solar energy appliances	Hua and Wang (2019)
Perceived behavioral control	Assessment of the ability to use solar energy appliances after evaluating personal resources	Tan et al. (2017)
Perceived benefit	Individuals' perception of benefits on using solar energy appliances	Akroush et al. (2019)
Perceived cost	Individuals' perception of the costs related to initial investment in buying/installing solar energy appliances and periodic maintenance	Akroush et al. (2019)
Interpersonal influence	Effects of the opinions of friends, family, co-workers on using solar energy appliances	Kawakami et al. (2013)
E-word of mouth	Effects of comments (positive or negative) made by customers regarding solar energy appliances on the Internet	Kawakami et al. (2013)

Source Authors

4 Results

The scales' reliability and validity were assessed with Cronbach's Alpha and confirmatory factor analysis (CFA). It is because the proposed paradigm in this research was adopted from the related literature. Then, the relation among constructs was evaluated with structural equation modeling (SEM).

4.1 *Reliability and Validity of Scales*

The scales' internal consistence reliability was evaluated with Cronbach's Alpha. The Cronbach's Alpha for each scale was greater than 0.70 and within the range from 0.773 to 0.871, indicating a good level of reliability (Hair et al., 2010). Furthermore, to form the data's convergent validity, we checked the estimated loadings, the assessed average variance (AVE), and the extracted composite reliability (CR) for each indicator (Fornell & Larcker, 1981), too. With the value from 0.689 to 0.910 which was

Table 2 Demographic information

Sample	Category	Frequency	Percentage
Age	21–30	50	13.3
	31–40	66	17.6
	41–50	78	20.8
	51–60	122	32.5
	Above 60	59	15.7
Gender	Male	296	78.9
	Female	79	21.1
Monthly household income	Under 5 million VND	41	10.9
	5–under 10 million VND	112	29.8
	10–under 15 million VND	58	15.4
	15–under 20 million VND	51	13.6
	20–under 25 million VND	46	12.3
	25–30 million VND	37	9.8
	Above 30 million VND	30	8.2
Educational level	High school or less	69	18.4
	2-year or less vocational training	80	21.3
	3-year college degree	84	22.5
	Bachelor degree	85	22.6
	Graduate degree	57	15.2

Source Authors

greater than the cut-off value of 0.60, all the factor loadings for the construct items were of statistical significance ($p < 0.01$) (Table 3). Additionally, all the AVEs were higher than 0.5, and all the CRs were greater than 0.7, indicating that nine constructs achieved the high-level validity of convergence and internal consistence (Hair et al., 2010). Measurement model fit, convergent validity, and discriminant validity of the data were evaluated with Confirmatory factor analysis (CFA) (Hair et al., 2010). The findings of CFA proved the acceptable level of fit: X^2 (Chi-square) = 800.1; CMIN/df = 1.759, $p < 0.001$; GFI (goodness of fit index) = 0.889; CFI (comparative fit index) = 0.934; TLI (Tucker Lewis index) = 0.923; and RMSEA (root mean square error of approximation = 0.045. All the t-tests of observed variables were statically significant at the level of 0.001. All model fit indicators are confirmed to be proper, and the conformity between the paradigm and collected data is proved to be significant.

Table 3 Model fit level

Variables	Items	Factor loading	AVE (> 0.50)	CR (> 0.70)
Purchase action (PA)	PA1	0.880	0.665	0.855
	PA2	0.805		
	PA3	0.910		
Purchase intention (PI)	PI1	0.840	0.642	0.877
	PI2	0.882		
	PI3	0.870		
	PI4	0.831		
Behavioral attitude (BA)	BA1	0.689	0.538	0.777
	BA2	0.803		
	BA3	0.824		
Subjective norm (SN)	SN1	0.727	0.510	0.837
	SN2	0.779		
	SN3	0.845		
	SN4	0.738		
	SN5	0.776		
Perceived behavioral control (PBC)	PBC1	0.730	0.510	0.838
	PBC2	0.784		
	PBC3	0.803		
	PBC4	0.848		
	PBC5	0.693		
Perceived benefit (PB)	PB1	0.834	0.549	0.828
	PB2	0.848		
	PB3	0.813		
	PB4	0.738		
Perceived cost (PC)	PC1	0.855	0.569	0.798
	PC2	0.785		
	PC3	0.793		
Interpersonal influence (II)	II1	0.781	0.543	0.780
	II2	0.850		
	II3	0.847		
E-word of mouth (EWOM)	EWOM1	0.878	0.582	0.804
	EWOM2	0.819		
	EWOM3	0.801		

Source Authors' synthesis from SPSS20 software

4.2 Paradigm and Hypothesis Testing

*Interrelation among constructs

Prior to hypothesis testing, the correlations among the constructs are checked. Table 4 indicates that no serious problem of multi-collinearity was found. Also, the square root of AVE was higher than its highest relation with any other constructs and AVE was greater than MSV, too. Thus, all the variables' validity of discriminant in this paradigm was confirmed. To summarize, appropriate reliability, convergent validity, and discriminant validity were demonstrated by the constructs in this research.

*Analysis of Structural Paths

The paradigm of this research was tested with structural equation modeling (SEM) to make an examination of the overall paradigm fit and the related power of the individual casual paths (Hair et al., 2010). The indicators (GFI = 0.913, CFI = 0.969, RMSEA = 0.045, TLI = 0.965, PCLOSE = 1.000, and Chi-square/df = 1.948) show that the paradigm reached a good level of fit. Therefore, the paradigm is proved to have given significant comprehension of direct and indirect predictors of the intent of purchasing solar energy appliances. On the basis of the standardized beta coefficients, significant at the level of 0.05, the findings prove that the hypotheses (H1a, H2a, H2b, H3, H4, and H6) are supported, except for H1b and H5 (Table 5). The positive and significant correlation between Perceived Benefit and Behavioral Attitude ($\beta = 0.17, t = 2.71$) is confirmed, supporting H1a, but the positive relationship between Perceived Cost and Behavioral Attitude doesn't support H1b. The positive and significant relation between Interpersonal Influence and Subjective Norm ($\beta = 0.19, t = 3.76$) as well as between E-word of Mouth and Subjective Norm ($\beta = 0.30$, $t = 3.57$) is confirmed, giving support to H2a and H2b.

Regarding the TPB paradigm, the positive correlation is confirmed between Behavioral Attitude and Purchase Intent ($\beta = 0.07$, $t = 5.16$) as well as Subjective Norm and Purchase Intention ($\beta = 0.14, t = 7.95$), providing support for H3 and H4. However, the relationship between Perceived Behavioral Control and Purchase Intention is not supported. Furthermore, the positive and significant relationship between Purchase Intention and Purchase Action ($\beta = 0.30, t = 4.35$) is confirmed, supporting H6. Also, Subjective Norm is also indicated to have exerted the strongest effect ($\beta = 0.14$, $t = 7.95$) on the Purchase Intention of solar energy appliances.

Finally, R^2 result of 0.52 indicates that 52 percent of variation in purchase action of solar energy appliances was caused by the paths of Behavioral Attitude-Purchase Intention and Subjective Norm-Purchase Intention.

Table 4 Discriminant validity for the measure paradigm

	AVE	MSV	SN	PBC	PI	PB	PA	PC	EWOM	II	BA
SN	0.510	0.201	**0.714**								
PBC	0.516	0.044	−0.170**	**0.718**							
PI	0.643	0.201	0.448***	−0.004	**0.802**						
PB	0.549	0.093	0.056	−0.150*	−0.030	**0.741**					
PA	0.665	0.057	−0.063	0.013	0.021**	0.156**	**0.815**				
PC	0.569	0.214	0.133	−0.011	0.017	0.305***	−0.052	**0.754**			
EWOM	0.582	0.166	0.221**	0.179**	0.105	−0.011	−0.037	0.164*	**0.763**		
II	0.543	0.167	0.069**	0.133*	0.022	0.016	0.024	0.007	0.408	**0.737**	
BA	0.538	0.214	0.196	0.209	0.163**	0.205**	0.239	0.463	0.244**	0.110**	**0.733**

Note *, *p*-value < 0.1; **, *p*-value < 0.05; ***, *p*-value < 0.001. Significant at the 0.05 level
Source Authors' synthesis from SPSS20 software

Table 5 Estimate of structural equation coefficients

Hypothesis	Structural path	Standardized estimate	*P*-value	Decision
H1a	PB → BA	0.170	0.023	Accepted
H1b	PC → BA	0.202	0.000	**Unaccepted**
H2a	II → SN	0.191	0.008	Accepted
H2b	EWOM → SN	0.302	0.000	Accepted
H3	BA → PI	0.072	0.047	Accepted
H4	SN → PI	0.144	0.008	Accepted
H5	PBC → PI	0.019	0.776	**Unaccepted**
H6	PI → PA	0.315	0.000	Accepted

Source Authors' synthesis from AMOS24 software

5 Discussions and Implications

5.1 Discussions

This research is to provide a thorough insight into antecedents on the intent of purchasing solar energy appliances among Vietnamese people. The TPB paradigm was employed to give a better understanding of consumers' purchasing intent. Six out of the eight tested hypotheses received the confirmation from the data.

In line with TPB, two out of three predictors in our proposed paradigm (behavioral attitude and subjective norm) were positive contributors to the purchase intention and purchase action of solar energy appliances. Of the two direct antecedents, the latter (SN) positively had the strongest influence on customers' purchase intention (PI). This finding is similar to the empirical researches by Bhutto et al. (2021) and Bhutto et al. (2022). The former (BA) also had a positive impact on consumers' purchasing intent, which supports the research results of Bhutto et al. (2021), Tan et al. (2017), Hua and Wang (2019), Tsuar and Lin (2018), Alam and Rashid (2012), Ayoub et al. (2019), and Akroush et al. (2019).

In addition, only perceived benefit (PB) positively and significantly influenced the behavioral attitude toward customers' purchase intent. This is similar to the research findings of Akroush et al. (2019). Meanwhile, perceived cost was not confirmed to affect consumers' behavioral attitude toward the purchase intention of solar energy appliances. This result is different from the findings of Akroush et al. (2019), but supports the research results of Wang et al. (2017) and Bketi et al. (2022). It may be due to the fact that with the average income in the developing country like Vietnam, customers feel that the price of such appliances is quite high and not as their expectation.

Another finding is that subjective norm (SN) was significantly affected by interpersonal influence (II) and e-word of mouth (EWOM). This finding may be partly attributed to the following things. First, as the internet access is convenient and the e-commerce in Vietnam is booming rapidly, people are filled up with a huge amount of information on the internet. Therefore, customers are often easily led by other internet users' remarks or judgements. Second, customers with certain education background often consult others before coming to a decision on purchasing solar energy appliances.

5.2 Implications

The outcomes of this research give useful insights for policymakers and producers to make up appropriate strategies to boost solar energy products for households.

Attitude is proved to have positively influenced consumers' purchase intent and purchase action. For that reason, attempts must be done to entice consumers for a favorable attitude toward such appliances through influential campaigns so that consumers will be well aware of the benefits of such appliances in relation with costs and alternatives of the same types.

Moreover, the significant impacts of subjective norm (interpersonal influence and e-word of mouth) on the purchasing intent signify the requirement for various online advertising and sharing channels to touch customers' purchasing intent in the digital era.

More importantly, costs and behavior control (costs and availability of appliances) seem not to be perceived to impact customers' purchasing intent. Maybe, the price of such appliances seems rather expensive while 40.7% of households have the monthly income of less than 10 million VND. Hence, it can be implied that a structured policy of incentives and replacement (including flexible and attractive schemes for consumers and manufacturers) will boost the use and production of such appliances. Cost incentives for the initial purchase and the exchange of conventional appliances for solar energy ones may enhance consumers' intent of buying such appliances for experience. Also, promotion schemes for retailers, such as tax exemption for selling solar energy appliances, can be productive.

Finally, the proliferation of the appliance manufacturing factories in localities will give rise to the availability of such products. Hence, a policy can be made up in order to boost local manufacturing and startups with the help of technical issues, fiscal incentives, and tax exemptions on importing solar energy technologies as well as related equipment. In this sense, fair transfer of benefits to all related stakeholders is key to ensure the feasibility and success of the policy.

6 Conclusions and Limitations

On the basis of evaluating the relationships among the constructs through structural equation modeling, only behavioral attitude and subjective norm were found to be positive contributors to the purchase action through purchase intention. In addition, the positive relationship between behavioral attitude and perceived benefit was also highlighted, but the relationship between behavioral attitude and perceived cost was not supported. Furthermore, the positive impact of interpersonal influence and e-word of mouth on subjective norm was supported, too.

Despite some contributions to apprehending the likely driver of consumers' purchasing solar energy appliances, a number of limitations still exist in this research. First, our research placed an emphasis on the antecedents of consumers' intent and action of buying solar energy appliances only. There may still be some variations between respondents' stated preference and revealed preference. For that reason, using the combined methods (survey and interview) may give much more robust findings. Second, the data were gathered in a short time span. Thus, further researches should use longitudinal data to provide a deeper insight into the targeted customers' behavior. Finally, with time and budget restrictions, only customers in Hanoi, Bac Ninh, Hai Duong, and Nam Dinh in Vietnam were targeted for the data in this research. Hence, further researches should consider a wider geographic scope in both rural and urban areas, which will help us know how individuals in the rural and urban areas perceive the consumption of such appliances. With more constructs, larger sample, and qualitative analysis, further researches will provide more thorough comprehension of customers' purchase behavior of solar energy appliances in developing economies like Vietnam.

7 Appendix

Construct	Items		References
Perceived benefits	PB1	Solar energy appliances give me extra value such as economic, environmental and social value	Akroush et al. (2019)
	PB2	Solar energy appliances have high utility	
	PB3	Solar energy appliances can meet my requirements	
	PB4	Solar energy appliances give me more benefits than costs	
Perceived cost	PC1	The price of solar energy appliances is not expensive	Akroush et al. (2019)
	PC2	The price of solar energy appliances is not higher than that of ordinary ones	
	PC3	The price of solar energy appliances is not higher than my expectation	

(continued)

(continued)

Construct	Items		References
Behavioral attitude	BA1	It is important to me when making purchases of solar energy appliances	Akroush et al. (2019)
	BA2	If I can choose between solar energy and traditional appliances, I prefer solar energy ones	
	BA3	I have favorable attitude toward purchasing solar energy appliances	
Interpersonal influence	II1	I always refer to the people around me, including my friends, family and colleagues when purchasing solar energy appliances	Kawakami et al. (2013)
	II2	I look at others' suggestions on purchasing solar energy appliances	
	II3	I consider people around me who have recommended solar energy appliances before	
E-word of mouth	EWOM1	I always learn about positive aspects of solar energy appliances from user blogs and user websites	Kawakami et al. (2013)
	EWOM2	I always consult the websites of people who use solar energy appliances	
	EWOM3	I visit community web sites and review online postings when purchasing solar energy appliances	
Subjective norm	SB1	It is pleasing to have solar energy appliances	Hua and Wang (2019)
	SB2	If respectable or important people use solar energy appliances, I would like to use them more	
	SB3	If my family and friends use solar energy appliances, I would like to use them more	
	SB4	If the people around me use solar energy appliances, I would like to use them more	

(continued)

(continued)

Construct	Items		References
	SB5	Using solar energy appliances is a social trend	
Perceived behavioral control	PBC1	I am confident that I would use solar energy appliances even if it is slightly more expensive	Tan et al. (2017)
	PBC2	I am confident that I would use solar energy appliances even if another person advises me to use the conventional one	
	PBC3	I am sure that I would be able to make a difference by using solar energy appliances	
	PBC4	Using solar energy appliances is entirely within my control	
	PBC5	I have resources, knowledge and ability to use solar energy appliances	
Purchase intention	PI1	I like to purchase solar energy appliances	Akroush et al. (2019)
	PI2	I will pay more money on solar energy appliances	
	PI3	I will take solar energy appliances as a first consideration	
	PI4	I will recommend others to purchase solar energy appliances	
Purchase action	PA1	I have switched to solar energy appliances for economic, environmental and social reasons	Bhutto et al. (2022)
	PA2	I switch to the solar energy appliances for the social trend	
	PA3	When I have a choice between the traditional appliances and solar energy ones, I purchase the latter	

References

Alam, S. S., & Rashid, M. (2012). Intention to use renewable energy: Mediating role of attitude. *Energy Research Journal, 3*(2), 37–44. https://doi.org/10.3844/erjsp.2012.37.44

Ali, M. R., Shafiq, M., & Andejany, M. (2020a). Determinants of consumers' intentions towards the purchase of energy efficient appliances in Pakistan: An extended model of the theory of planned behavior. *Sustainability, 13*(565). https://doi.org/10.3390/su13020565.

Ali, S., Poulova, P., Akaba, A., Javed, H. M. U., & Danish, M. (2020b). Determining the influencing factors in the adoption of solar photovoltaic technology in Pakistan: A decomposed technology acceptance model approach. *Journal of Economies*. https://doi.org/10.3390/economies8040108

Akroush, M. N., Zuriekat, M. I., Al Jabali, H. I., & Asfour, N. A. (2019). Determinants of purchasing intentions of energy-efficient products: The roles of energy awareness and perceived benefits. *International Journal of Energy Sector Management, 13*(1), 128–148. https://doi.org/10.1108/IJESM-05-2018-0009

Ajzen, I. (1991). The theory of planned behavior. *Organizational Behavior and Human Decision Processes, 50*(2), 179–211. https://doi.org/10.1016/0749-5978(91)90020-T

Ajzen, I. (2002). *Constructing a TPB questionnaire: Conceptual and methodological considerations*. CiteSeer.

Ajzen, I. (2010). *TPB Questionnaire construction*. University of Massachusetts-Amherst.

Ajzen, I. (2020). The theory of planned behavior: Frequently asked questions. *Human Behavior and Emerging Technologies, 2*, 314–324.

Ayoub, S., Dastgir, G., & Waqas, M. (2019). Factors affecting consumer purchase intentions for solar energy application at domestic level. *International Journal of Economics, Commerce and Management, 7*(10), 349–363.

Bandara, U. C., & Amarasena, T. S. M. (2019). Impact of perceived ease of use, awareness and perceived cost on intention to use solar energy technology in Sri Lanka. *Journal of International Business and Management, 3*(4), 01–13

Bhattacherjee, A. (2000). Acceptance of e-commerce services: The case of electronic brokerages. *IEEE Transactions on Systems, Man, and Cybernetics Part A: Systems and Humans, 30*(4), 411–420. https://doi.org/10.1109/3468.852435

Bekti, D. B. M., Prasetyo, Y. T., Redi, A. A. N. P., Budiman, A. S., Mandala, I. M. P. L., Putra, A. R., Persada, S. F., Nadlifatin, R., & Young, M. N. (2022). Determining factors affecting customer intention to use rooftop solar photovoltaics in Indonesia. *Sustainability, 14*(1), 280. https://doi.org/10.3390/su14010280

Bhutto, M. Y., Liu, X. H., Soomro, Y. A., Ertz, M., & Baeshen, Y. (2021). Adoption of energy-efficient home appliances: Extending the theory of planned behavior. *Sustainability, 13*(250). https://doi.org/10.3390/su13010250.

Bhutto, M. Y., Soomro, Y. A., & Yang, H. (2022). Extending the theory of planned behavior: Predicting young consumer purchase behavior of energy efficient appliances (evidence from developing economy). *SAGE Journals*, 1–14. https://doi.org/10.1177/21582440221078289

Caso, D., Carfora, V., Starace, C., & Conner, M. (2019). Key factors influencing italian mothers' intention to vaccinate sons against HPV: The influence of trust in health authorities, anticipater regret and past behavior. *Sustainability, 11*(23), 6879. https://doi.org/10.3390/su11236879

Chan, L., & Bishop, B. (2013). A moral basis for recycling: Extending the theory of planned behavior. *Journal of Environmental Psychology, 36*, 96–102. https://doi.org/10.1016/j.jenvp.2013.07.010

Chen, M. F., & Tung, P. J. (2014). Developing an extended theory of planned behavior model to predict consumers' intention to visit green hotels. *International Journal of Hospitality Management, 36*, 221–230.

Chew, B. E., & Abdul Adis, A. A. (2018). A study on Malaysian consumers' attitude and behavioral intention towards print advertising. *International Journal of Marketing Studies, 10*(2), 132. https://doi.org/10.5539/ijms.v10n2p132

Doma, S. S., Elaref, N. A., & Abo Elnaga, M. A. (2015). Factors affecting electronic word-of-mouth on social networking websites in Egypt—An application of the technology acceptance model. *Journal of Internet Social Networking and Virtual Communities, 2015*, 1873–1902. https://doi.org/10.5171/2015.280025

Duong, M. H., Truong, A. H., Nguyen, H. N., & Trinh Nguyen, H. A (2017). Synthesis report on socio-environmental impacts of coal, and coal-fired power plants in Vietnam. *Technical Report.* Vietnam Sustainable Energy Alliance, 16–17, retrieved from https://hal-enpc.archives-ouvertes.fr/hal-01441680 (Accessed on February 18, 2022).

Fong, J., & Burton, S. (2008). A cross-cultural comparison of electronic word-of-mouth and country-of-origin effects. *Journal of Business Research, 61*(3), 233–242.

Fornell, C., & Larcker, D. F. (1981). *Structural equation models with unobservable variables and measurement error: Algebra and statistics*. Sage Publications.

Godes, D., & Mayzlin, D. (2004). Using online conversations to study word-of-mouth communication. *Marketing Science, 23*(4). https://doi.org/10.1287/mksc.1040.0071

Hair, J., Black, W., Babin, B., & Anderson, R. (2010). *Multivariate data analysis* (7th ed.). Prentice-Hall.

Hasbullah, N. A., Abdullah, O., Abdullah, S., Salahuddin, S. N., Ramlee, N. F., & Soha, H. M. (2016). The relationship of attitude, subjective norm and website usability on consumer intention to purchase online: An evidence of Malaysian youth. *Procedia Economics and Finance, 35*, 493–502. https://doi.org/10.1016/S2212-5671(16)00061-7

Hennig-Thurau, T., & Walsh, G. (2003). Electronic word-of-mouth: Motives for and consequences of reading customer articulations on the internet. *International Journal of Electronic Commerce, 8*(2), 51–74. https://doi.org/10.1080/10864415.2003.11044293

Hua, L., & Wang, S. (2019). Antecedents of consumers' intention to purchase energy-efficient appliances: An empirical study based on the technology acceptance model and theory of planned behavior. *Sustainability, 11*, 2994. https://doi.org/10.3390/su11102994

Hsu, M., Yen, C., Chiu, C., & Chang, C. (2006). A longitudinal investigation of continued online shopping behavior: An extension of the theory of planned behavior. *International Journal of Human-Computer Studies, 64*(9), 889–904.

Jalilvand, M. R., & Samiei, N. (2012). The impact of electronic word of mouth on a tourism destination choice: Testing the theory of planned behavior (TPB). *Internet Research, 22*(5), 591–612. https://doi.org/10.1108/10662241211271563

Juniwati. (2014). Influence of perceived usefulness, ease of use, risk on attitude and intention to shop online. *European Journal of Business and Management, 6*(27), 1905–2222.

Kawakami, T., & Parry, M. E. (2013). The impact of word-of-mouth sources on the perceived usefulness of an innovation. *Journal of Product Innovation Management, 30*(6), 1112–1127. https://doi.org/10.1111/jpim.12049

Kaiser, F. G. (2006). A moral extension of the theory of planned behavior: Norms and anticipated feelings of regret in conservationism. *Personality and Individual Differences, 41*, 71–81.

Khalid, A., & Junaidi, H. (2013). Study of economic viability of photovoltaic electric power for Quetta—Pakistan. *Renewable Energy, 50*, 253–258. https://doi.org/10.1016/j.renene.2012.06.040

Kline, R. B. (2011). *Principles and practice of structural equation modeling* (3rd ed.). Guilford.

Leelakulthanit, O. (2014). The factors affecting the adoption of LED lamps. *International Business and Economics Research Journal (IBER), 13*(4), 757–768.

Li, G., Li, W., Jin, Z., & Wang, Z. (2019). Influence of environmental concern and knowledge on households' willingness to purchase energy-efficient appliances: A case study in Shanxi China. *Sustainability, 11*, 1073. https://doi.org/10.3390/su11041073

Lindenberg, S., & Steg, L. (2007). Normative, gain, and hedonic goal frames guiding environmental behavior. *Journal of Social Issues, 63*, 117–137.

Litvin, S. W., Goldsmith, R. E., & Pan, B. (2008). Electronic word-of-mouth in hospitality and tourism management. *Tourism Management, 29*, 458–468. https://doi.org/10.1016/j.tourman.2007.05.011

López-Mosquera, N., Garcia, T., & Barrena, R. (2014). An extension of the theory of planned behavior to predict willingness to pay for the conservation of an urban park. *Journal of Environment Management, 135*, 91–99. https://doi.org/10.1016/j.jenvman.2014.01.019

Manh Duc (2021). Why is the power source abundant, but electricity still lacking? *VnEconomy*, 22–23, retrieved from https://vneconomy.vn/vi-sao-nguon-thi-thua-dien-van-thieu.htm. (in Vietnamese) (Accessed September 15, 2021).

Nielsen (2018). Connect with Vietnamese young generations. Nielsen Retrieved from https://www.nielsen.com/vn/vi/insights/report/2018/insight-article-generation-z-invietnam/ (Accessed on April 29, 2020).

Nguyen, H. T., Pham, H. T., & de Bruyn, L. L. (2017). Impact of hydroelectric dam development and resettlement on the natural and social capital of rural Livelihood in Bo Hon Village in Central Vietnam. *Sustainability, 9*(8), 1422. https://doi.org/10.3390/su9081422

O'Driscoll, A., Claudy, M., & Peterson, M. (2013). Understanding the attitude-behavior gap for renewable energy systems using behavioral reasoning theory. *Journal of Macro Marketing, 33*(4), 273–287.

Pham, H. M., & Duong, V. S. (2020). Analyzing the factors affecting the intention to use solar energy technology of households in Phan Rang—Thap Cham city. *Industry and Trade Magazine, 4*, 166–175. (in Vietnamese).

Premkumar, G., Ramamurthy, K., & Crum, M. R. (1997). Determinants of EDI adoption in the transportation industry. *European Journal of Information Systems, 6*(2), 107–121.

Rhodes, R. E., & Courneya, K. S. (2003). Investigating multiple components of attitude, subjective norm, and perceived control: An examination of the theory of planned behavior in the exercise domain. *British Journal of Social Psychology, 42*, 129–146.

Shin, Y. H., & Hancer, M. (2016). The role of attitude, subjective norm, perceived behavioral control, and moral norm in the intention to purchase local food products. *Journal of Food-Service Business Research, 19*(4), 338–351. https://doi.org/10.1080/15378020.2016.1181506

Stragier, J., Hauttekeete, L., & De Marez, L. (2010). Introducing smart grids in residential contexts: consumers' perception of smart household appliances. *Conference Paper, Ghent University*. https://doi.org/10.1109/CITRES.2010.5619864

Sanne, P. N. C., & Wiese, M. (2018). The theory of planned behavior and user engagement applied to Facebook advertising. *SA Journal of Information Management, 20*(1), 1–10. https://doi.org/10.4102/sajim.v20i1.915

Sun, Y., & Wang, S. (2019). Understanding consumers' intentions to purchase green products in social media marketing context. *Asian Pacific Journal of Marketing and Logistics*, November, 1355–5855. https://doi.org/10.1108/APJML-03-2019-0178

Tan, C. S., Ooi, H. Y., & Goh, Y. N. (2017). A moral extension of the theory of planned behavior to predict consumers' purchase intention for energy-efficient household appliances in Malaysia. *Energy Policy, 107*, 459–471.

Tsaur, R. C., & Lin, Y. H. (2018). Exploring the consumer attitude of building-attached photovoltaic equipment using revised technology acceptance model. *Sustainability, 10*(4177). https://doi.org/10.3390/su10114177

Wang, Z. H., Zhang, B., Yin, Z. H., & Zhang, Y. X. (2011). Determinants and policy implications for household electricity-saving behavior: Evidence from Beijing China. *Energy Policy, 39*(6), 3550–3557. https://doi.org/10.1016/j.enpol.2011.03.055

Wang, Z., Zhang, B., & Li, G. (2014). Determinants of energy-saving behavioral intention among residents in Beijing: Extending the theory of planned behavior. *Journal of Renewable and Sustainable Energy, 6*(5), 45–54. https://doi.org/10.1063/1.4898363

Wang, Z. H., Wang, X. M., & Guo, D. X. (2017). Policy implications of the purchasing intentions towards energy-efficient appliances among China's Urban residents: Do subsidies work? *Energy Policy, 102*, 430–439. https://doi.org/10.1016/j.enpol.2016.12.049

Digitalizing the Container Terminal to Meet the Demand of the Stakeholders in the Transportation Supply Chain: Technology Acceptance Model Extended Approach Case Study in the Southeast Region of Vietnam

Hung Thanh Nguyen

Abstract Most of the world's port operators are aiming to adopt advanced technologies to survive and thrive in the context of the Fourth Industrial Revolution and under the severe impact of the COVID-19 pandemic. Thus, container terminals in the Vietnam Southeast region are undergoing digital transformation. The study sheds light on whether the digitization of the service supply processes of these terminals meets the needs of the stakeholders in the transport supply chain. The study examines and analyzes the relationship between technology complexity and compatibility to the continuance to use of container terminals through an extended technology acceptance model (TAM). The study mainly uses quantitative methods with the analysis technique of structural equation modeling (SEM). Survey subjects from 222 respondents are stakeholders in the transport supply chain, including: shipping lines, forwarding companies, logistics service providers and trucking companies in the Vietnam Southeast region. This study contributes to enriching and expanding the TAM on container terminal digitization. Research results show that the compatibility and usefulness of the technology have a significant influence on the continuance to use of container terminals. Container terminal managers should consider developing technology and operating systems with the appropriate level of utility and compatibility and according to the needs of the stakeholders in the transport supply chain.

Keywords Terminal digitization · Technology acceptance model · Transportation supply chain

H. T. Nguyen (✉)
University of Finance—Marketing, Ho Chi Minh City, Vietnam
e-mail: nguyenhung@ufm.edu.vn

A. T. Nguyen et al. (eds.), *Contemporary Economic Issues in Asian Countries: Proceeding of CEIAC 2022, Volume 1*,
https://doi.org/10.1007/978-981-19-9669-6_3

1 Introduction

The construction and development of container terminals in ports group 4 in the Vietnam Southeast region (according to Decision No. 1579/QD-TTg of the Prime Minister: Approving the master plan on development of Vietnam's seaport system. In the period of 2021–2030, with a vision to 2050, issued on September 22, 2021, Vietnam's seaport system consists of 5 groups, and Group 4 includes 5 seaports: Ho Chi Minh City seaport, Dong Nai seaport, Ba Ria-Vung Tau seaport, Binh Duong seaport and Long An seaport) have increased the competitiveness of the port industry in recent years. Competition is increasingly evident for ports operated by state-owned companies and private port operators. The key to winning this competition is the loyalty of the stakeholders in the transportation supply chain. The loyalty of the stakeholders shows that, first of all, one of the stakeholders will refer the other partners. Second, stakeholders will repeatedly use the services they trust (Yang & Peterson, 2004). The more loyal the parties in the transport supply chain, the greater the chance of winning the market. Container ports in ports group 4 in the Southeast region have had innovations in port development, and this study discusses one of those innovation aspects, also the needs of the stakeholders in the 4.0 era: port digitization. The research is to assess whether the digital content of service provision of container terminals in ports group 4 can meet the needs of container terminal users. This assessment is based on a review and analysis of the relationship between technology complexity and compatibility to continued use of container terminals through an extended technology adoption model. This study contributes to fill the gap in TAM for container terminal digitization. Thus, the study added variables for technology complexity and compatibility in addition to perceived usefulness and perceived ease of use and examined their impact on continue or intend to use a digitized container terminal, thereby studying to improve the operational efficiency of container terminals in Number 4 port group, evaluating the content of policies on terminal digitization in the Southeast region and filling in the research gap on digital transformation in the field of port management.

2 Research Overview, Hypotheses and Methodology

2.1 Technology Acceptance Model (TAM)

In 1989, inheriting the Theory of Reasonable Action, Davis, Bogozzi and Warshaw established the TAM. The purpose of the model is to explain the factors affecting the acceptance of computers, thereby explaining the behavior of computer technology users (Yang & Peterson, 2004). In 2000, Venkatesh and Davis developed TAM, which is widely known as TAM2. TAM2 considers that the user's mental assessment that combines the performance of work goals and the results of the performance of work tasks using the system is the basis for forming perceptions about usefulness of the system (Lai, 2017). Therefore, Venkatesh and Davis (2000) develop TAM2 by adding factors outside the model not shown in previous studies including social processes (subjective standards, voluntariness, image) and cognitive tool processes (relevance to work, quality of output, performance results, perceived ease of use). Venkatesh and Bala (2008) combined TAM2 and perceived ease-of-use determinants to develop a technology acceptance integration model called TAM3 (Venkatesh & Davis, 2000). The authors developed TAM3 by adding factors that influence "perceived ease of use" (computer effectiveness, perceived external control, computer anxiety, computer excitement, comfort, usability). TAM3 poses three newly tested relationships: (i) "perceived usefulness" and "perceived ease of use"; (ii) concerns about computers and "perceived ease of use"; and (iii) "perceived ease of use" and intention to use. This study uses the TAM3 as the basic analytical framework, thereby adjusting and supplementing it to suit the context of digital transformation of the port operation industry and the transport supply chain in the Vietnam Southeast region.

2.2 Research Model and Hypotheses

Container terminals in Vietnam Southeast region are a modern system based on automated information technology equipment and systems. The TAM does not fully answer the influence of perceived utility variables and perceived ease of use on the attitude to continue using the digitized port because it is only concerned with the issue of application or not. The complexity of the technology used in implementing digital transformation is considered a factor that needs to be considered when transacting at this regional container port as port when users are generally apprehensive about the complexity of technology. Likewise, with synchronous and user-friendly technologies, the study assumes that policies to digitize operational processes can influence service users to consider continuing transactions at this container terminals or not. This refers to the compatibility of a system or technology with its transport supply chain stakeholders, viewed from the extent to which the technology or system meets the needs of the transportation supply chain stakeholders.

Complexity is the degree to which innovation is considered relatively confusing and difficult to use. In terms of technological complexity, it can be understood as the degree of innovation of a new technology or system, which introduces new elements that lead to simplicity, unusual and unusual in the application of technology or system. In implementing container port digital transformations, the complexity of technology used includes automation and technological equipment. In addition, transactions are carried out online 24 h 7 days using information technology-based applications. Several previous studies, albeit with different subjects, have found that technology complexity affects perceived usefulness and perceived ease of use (Chin & Lin, 2015; Somang et al., 2019). From there, the study proposes the following hypothesis:

H1: Technology complexity affects perceived usefulness
H2: Technology complexity affects perceived ease of use.

Compatibility can be understood as conformity. In the case of integration with a technology or system, the compatibility of a system or technology is the gap between existing facilities or resources and the new technology or system. If the variance of a system is greater than the existing facilities and resources, the system will certainly be difficult to accept by the stakeholders in the transport supply chain. Rogers in Mazhar (2014) defines compatibility as the degree of consistency between technology and the new needs of the stakeholders in the transport supply chain. Some studies by Chin and Lin (2015), Somang et al. (2019), Di Pietro et al. (2015), and Shakrokh (2019) indicate that compatibility significantly affects usability and perceived ease of use. However, Di Pietro et al. (2015) argue that the technological complexity variable does not significantly affect the perceived usefulness. Meanwhile, Shakrokh (2019) concluded that compatibility has a significant influence on perceived usefulness and perceived ease of use:

H3: Compatibility affects perceived usefulness
H4: Compatibility affects perceived ease of use.

According to Venkatesh and Davis in Devi and Suartana (2014), perceived usefulness is an individual's degree of belief that using a system or technology can improve performance. Likewise, Lisa et al. (2017) suggested that perceived ease of use has a significant influence on perceived usefulness. The TAM concludes that people will rate the usefulness and ease of using a new technology or system when a new technology or system is adopted. The studies mentioned above also support this conclusion.

According to Ramayah and Ignatius (2005), perceived usefulness includes: (1) Efficiency is the perception of time savings resulting from the use of a technology or system; (2) faster completion describes the extent to which a job can be done faster using a system or technology; (3) utility describes the extent to which a technology or system can be useful to someone, especially in corporate operations; (4) advantages describe the benefits of using the system or technology.

As for the perceived ease of use, Dewi (2013) states that it is "the belief that using technology will not take much effort". It can be understood that perceived ease of use is the degree to which a person's confidence in using a system or technology affects ease of work.

Referring to the TAM of Davis (1993), TAM2 of Venkatesh and Davis (2000), research by Lee et al. (2011) on the acceptance of e-commerce use and perceived ease of use is that the transport supply chain stakeholders think that using a digitized container terminals will not require much effort.

According to Ramayah and Ignatius (2005), perceived ease of use includes: (1) Ease is the ease of use of a system or technology; (2) clarity and intelligibility are the degree of clarity and intelligibility of the system or technology used; (3) ease of learning is how easy a system or technology is for someone to learn or apply; (4) overall ease is the overall ease of use of a system or technology. From several observational studies, perceived usefulness and perceived ease-of-use variables are important factors influencing intention and continued use (Lisa et al., 2017; Weng &

et al., 2017; Taufik & Hanafiah, 2019; Wu & Chen, 2016; Ho et al., 2020; Cheng, 2015; Ashfag et al., 2020; Hamid et al., 2016). Accordingly, the study proposes the following hypotheses:

H5: Perceived usefulness affects the intention to continue using the container terminal
H6: Perceived ease of use affects the intention to continue using the container terminal
H7: Perceived ease of use affects perceived usefulness.

According to Davis in Teng and Chen (2008), intention to use continuity is a person's preference or desire to continue using a system or technology. Meanwhile, according to Bhattacherjee in Islam and Mantymaki (2014), intention to use continuity can be understood as an individual's interest to continue to participate or participate in a particular system.

According to research done by Anderson and Sullivan in Hung and Hsu (2011), aspects of intention to continue using are: (1) Repurchase likelihood is the probability that a person will reuse the system or technology; (2) number of acquisitions refers to the preference to use the online service system and the ability to buy or transact in the online system or technology.

Studies by Chin and Lin (2015), Shakrokh (2019), Mndzebele (2013), and Akinnuwesi et al. (2016) suggest that technology complexity and compatibility have a significant influence on continuous usability. Another study suggested that the complexity of technology has an indirect effect on continued usability (Somang et al., 2019). Technological complexity affects perceived usefulness (HI), and HI has an effect on continuous usability. Accordingly, the study proposes the following hypothesis:

H8: Technological complexity affects intention or ability to continue using the container terminal
H9: Compatibility affects intention or ability to continue using the container terminal.

The research model and hypotheses are shown in Fig. 1 (research model).

2.3 Research Methodology

In this study, the authors applied the method of structural equation model (SEM), which used SPSS 22.0 software and AMOS version 20 software, via four steps: Cronbach's alpha reliability analysis, exploratory factor analysis (EFA), confirmatory factor analysis (CFA), structural equation modeling (SEM) analysis.

The study applies the structural equation modeling (SEM) to analyze the multidimensional relationship between many variables in the proposed research model. Thereby, the research can visually examine the relationships that exist between the variables to prioritize resources to better serve customers. According to Tho (2009),

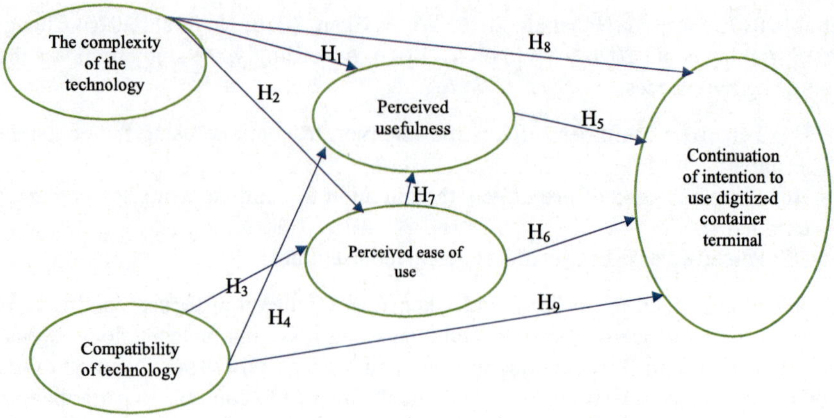

Fig. 1 Research model. *Source* Author

determining the required sample size for the study depends on many factors such as the processing method (regression, EFA, CFA, SEM), the necessary reliability. Besides, according to Bolen (1989), there must be a minimum of 5 observations per estimator (ratio 5:1). The number of observed variables used in the formal study is 22, so the minimum sample size should be 110. This study uses the CB-SEM tool for the model test step, and the hypothesis with the sample size of 222 (over 200) is meeting the sample size requirements of Hsu et al. (2006) and Henseler et al. (2009).

The sampling technique in this study is convenience and target-oriented sampling. All samples in this study are stakeholders in the transport supply chain of container terminals in ports group 4 with the criteria: (1) transactions with container stakeholders at least twice a day; (2) minimum transaction is 100 containers per month; (3) still being the stakeholders in the transport supply chain of the container terminal system in the Vietnam Southeast region (shown in the feature that the latest shipment of the enterprise through the container terminal of port group 4 is after January 2022); and (4) Vietnamese enterprises.

3 Research Result

3.1 Sample

About 250 questionnaires were sent to outbound shipping lines (12 companies), inbound shipping lines (8 companies), forwarding agents, logistics service providers (200 companies) and 30 trucking companies. A total of 246 questionnaires were

Table 1 Results of the scale reliability evaluation

Factors	Number of variables	Cronbach's alpha	Correlation coefficient	Conclusion
The complexity of the technology	4	0.801	(0.640; 0.662)	Yes (eliminate PT4)
Compatibility of technology	4	0.815	(0.605; 0.652)	Yes
Perceived usefulness	4	0.823	(0.637; 0.661)	Yes
Perceived ease of use	6	0.849	(0.504; 0.686)	Yes
Continuation of intention to use digitized container terminal	4	0.830	(0.626; 0.686)	Yes

Source Author

answered. Twenty-four questionnaires did not meet the requirements of the respondent's criteria, so the respondent data was 222, in which the classification of respondents based on the respondent's position in the company was divided into 5 groups, including: CEO: 173 respondents (77.9%), General Director: 7 respondents (3.2%), Branch Director: 15 respondents (6.8%), Representative of the Board of Directors: 3 respondents (1.4%) and Director of Information Technology: 24 respondents (10.8%).

3.2 *Model Analysis*

3.2.1 Evaluating the Scale Reliability by Cronbach's Alpha

The results of Cronbach's alpha analysis (Table 1: results of the scale reliability evaluation) show that most of the scales have Cronbach's alpha coefficient greater than 0.6 and the total correlation coefficients greater than 0.3 (Hair et al., 2009). Therefore, with standard scales, reliability assurance will be included in the exploratory factor analysis.

3.2.2 Exploratory Factor Analysis (EFA)

From the results of KMO and Bartlett's test, we have a KMO coefficient of 0.912, greater than 0.5, so factor analysis is appropriate and Sig. (Bartlett's test) is 0.000 less than 0.005, showing that the observed variables are correlated with each other in the population. The total variance explained results show that the total variance explained is 55.578%, greater than 50%, and the pattern matrix results show that all factor loading factors are greater than 0.5 (after removing the DD6). Therefore, these

Table 2 Results of exploratory factor analysis

Factors	KMO	Sig	Average variance extracted	Loading
The complexity of the technology (PT) Compatibility of technology (TT) Perceived usefulness (HI) Perceived ease of use (DD) Continuation of intention to use digitized container terminal (SD)	0.912	0.000	55.578	(0.720; 0.813) (0.684; 0.783) (0.626; 0.819) (0.602; 0.757) (0.571; 0.794)

Source Author

observed variables contribute significantly to the model, and the specific content is shown in Table 2 (results of exploratory factor analysis).

Therefore, all the criteria in EFA are met to implement CFA and SEM.

3.2.3 Confirmatory Factor Analysis (CFA)

Considering model fit, according to Hu and Bentler (1999), we find: CMIN/df = 1,188 (less than 3) is good. TLI = 0.981 (greater than 0.9) is good. CFI = 0.984 (greater than 0.9) is fine. GFI = 0.923 and RMSEA = 0.029 (less than 0.08) are good, and PCLOSE is good because 0.992 (>0.05). Thus, the model fits the data. The results are illustrated in Fig. 2 (CFA result).

All observed variables are significant in the model because the p-value is less than 0.05 (Hu & Bentler, 1999). CR values are both greater than 0.7, and AVE is greater than 0.5, so the scales are all convergent (Hair et al., 2009). The square root of AVE is larger than the correlations between the latent variables, and the MSV value is smaller than AVE, so the discriminant is guaranteed (Fornell & Larcker, 1981). The research model is consistent with the research data.

3.2.4 Structural Equation Modeling (SEM)

3.2.4.1. Test Models and Hypotheses

The research model estimation results show that the values of CMIN/df = 1,188 (less than 3) are good. CFI = 0.984 (greater than 0.9) is fine. GFI = 0.923, satisfying GFI above 0.9. RMSEA = 0.029 (less than 0.08) is fine. At the same time, CMIN = 190.054 and df = 160 with p-value less than 0.05 are satisfactory. The content of model testing and hypotheses is illustrated in Fig. 3 (SEM model results).

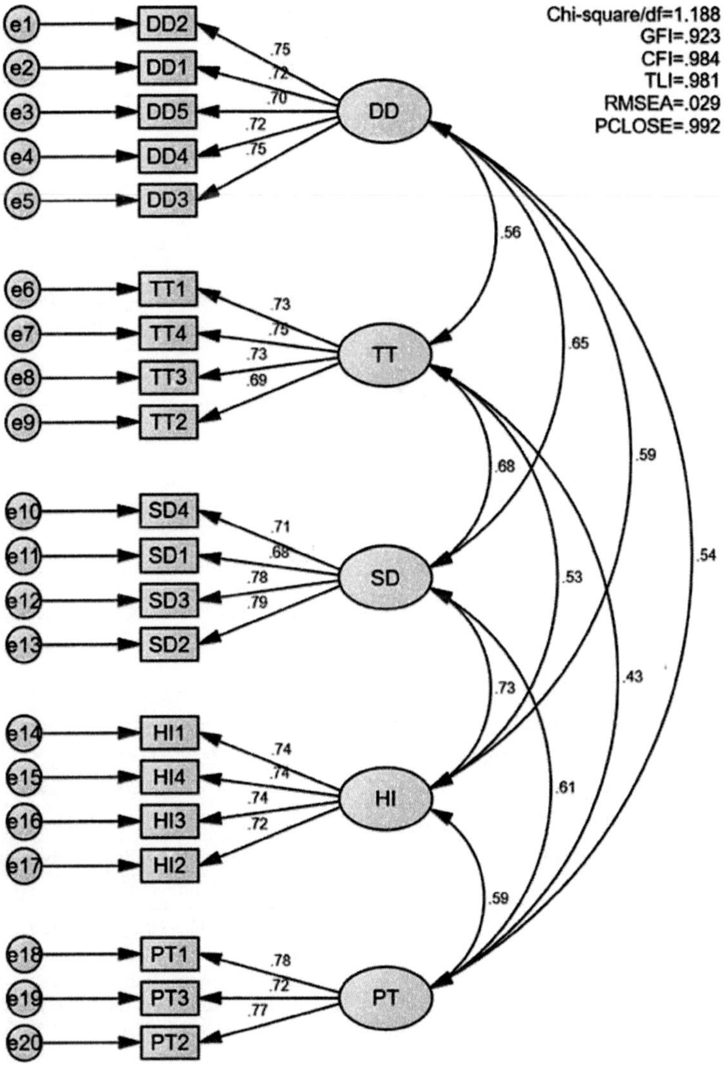

Fig. 2 CFA result. *Source* Author

3.2.4.2. Test Hypotheses Using SEM

Estimation results of the research model according to the SEM show that the hypothesized relationships in the formal research model have statistical value because the highest p-value less than 0.05, reaching the required significance at the 95% confidence level, according to Table 3 (regression coefficient of the formal research model).

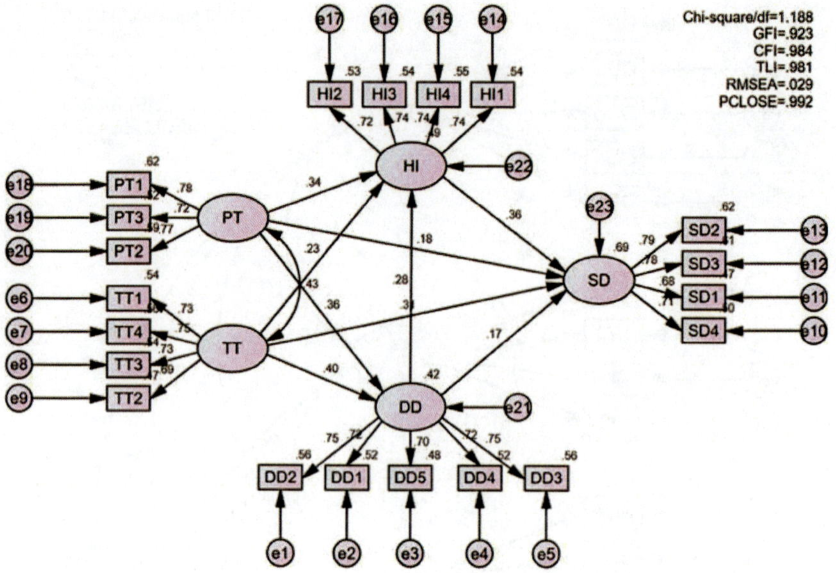

Fig. 3 SEM model results. *Source* Author

Table 3 Regression coefficient of the formal research model

Relationship			Unstandardized coefficients	Standardized coefficients	SE	CR	*p*-value	Conclusion
DD	←	PT	0.259	0.364	0.060	4.328	0.000	Discriminant
DD	←	TT	0.342	0.403	0.073	4.702	0.000	Discriminant
HI	←	PT	0.261	0.335	0.070	3.726	0.000	Discriminant
HI	←	TT	0.212	0.229	0.082	2.580	0.010	Discriminant
HI	←	DD	0.309	0.283	0.105	2.940	0.003	Discriminant
SD	←	PT	0.124	0.177	0.057	2.172	0.030	Discriminant
SD	←	TT	0.263	0.314	0.069	3.826	0.000	Discriminant
SD	←	HI	0.325	0.360	0.084	3.856	0.000	Discriminant
SD	←	DD	0.166	0.168	0.083	1.985	0.047	Discriminant

Source Author

Technology complexity and compatibility have a significant effect on perceived usefulness, ease of use and continued usability. Technological complexity is the degree to which innovation of a technology or system is considered to be relatively difficult to understand and use, where the more complex the technological innovation, the lower the level of acceptance. Research results demonstrate that the complexity of technology has a positive and significant effect on perceived usefulness. This means that H1 is accepted. This finding is related and supports previous studies by Chin and Lin (2015), and Somang et al. (2019), which stated that technology complex variable affects perceived usefulness. It can be concluded in this study that the complexity of container terminal technology (automation equipment and digital-based operating systems in container terminals) affects perceptions of transport supply chain stakeholders on the benefits of port digitization. Technological complexity was also found to have a positive effect on ease of use. This means that H2 is accepted. It can be said that stakeholders in the transport supply chain of container terminals in ports group 4 find the port's innovative digital transformation very easy to use and deploy. This finding is also relevant and supports previous studies by Chin and Lin (2015), Somang et al. (2019), and Akinnuwesi et al. (2016), which stated that the complexity of technology has a significant effect on ease of use.

This study also found that technological complexity directly and significantly affects continued use. This means H8 is accepted. This finding is relevant and supports research conducted by Chin and Lin (2015), and Mndzebele (2013), which suggested that the complexity of technology directly affects continuity of use. However, this finding is not consistent with the previous finding of Akinnuwesi et al. (2016), which suggested that complexity does not directly affect continuity of use.

Compatibility is the degree of consistency between new technology and the needs of transport supply chain stakeholders, daily habits, experiences and values (Rogers in Mazhar (2014)). Research results show that compatibility has a positive and significant effect on perceived usefulness. This means that H3 is accepted. This finding is related and supports previous studies by Chin and Lin (2015), Somang et al. (2019), Di Pietro et al. (2015), and Shakrokh (2019), which stated that compatibility has a positive and significant effect on perceived usefulness. It can be concluded in this study that the compatibility level of digital transformation (automation equipment and digital-based operation system) in the container terminals in the Southeast region to the needs of the stakeholders in the transport supply chain is deemed appropriate and useful.

Furthermore, this study also found that compatibility has a positive and significant impact on ease of use. This means that H4 is accepted. This result is related and similar to the studies by Chin and Lin (2015), and Shakrokh (2019), which stated that compatibility has a positive and significant impact on ease of use. It can be said that digital transformation implemented by container terminals in the Vietnam Southeast region is considered to be suitable for the needs of stakeholders in the transport supply chain and easy to use. At the same time, this study found that compatibility also directly affects continued use. This means that H9 is accepted. It can be concluded that the level of digital transformation (automation equipment and digital-based operating systems) implemented by the container terminal system in the Southeast region is in line with the needs and values of the stakeholders in transport supply chain.

Perceived usefulness and perceived ease of use are intrinsic factors in the TAM. The results of this study indicate that perceived ease of use has a significant influence on perceived usefulness. Meanwhile, perceived usefulness and perceived ease of use have a positive and significant influence on continued and frequent use of the digitized port. This means H5, H6 and H7 are accepted. It can be concluded that the perception of the stakeholders in the transport supply chain about the benefits of digital transformation implemented by container terminals in ports group 4 is affected by ease of use. Next, the ability to continue and regularly use the digitized terminal (automation equipment and digital-based operating systems) is positively and significantly impacted by perceived usefulness and ease of use. Continued use and intention to use container terminals in ports group 4 to be digitized in a sustainable way depend on the usefulness and ease of use of the technology for stakeholders in the transport supply chain. This finding is related and similar to some previous studies by Somang et al. (2019), Di Pietro et al. (2015), Lisa et al. (2017), Weng et al. (2017), Taufik and Hanafiah (2019), Wu and Chen (2016), Cheng (2015), Ashfaq et al. (2020), and Choi and Park (2015).

4 Conclusion and Discussion

Digital transformation implemented by container terminals in ports group 4 is researched in this study which used an expanded TAM with variables of perceived usefulness and ease of use affecting intention to continue using container ports with added technology complexity and compatibility. The results of this study indicate that the impact of perceived usefulness factor is 36%, technology compatibility factor is 31.4%, technological complexity factor that affects the continued use of port is 17.7%, and perceived ease is 16.8% to the continued use of the port. In theory, this study enriches and expands the TAM, in which the intention to continue using the port with digital transformation in ports group 4 is affected by the usefulness and interoperability. Predominantly liked to be perceived, technological complexity and ease of perception have also been shown to have a significant influence. Those are important factors that supply chain stakeholders consider to continue using ports that have undergone digital transformation. The results of this study are also meaningful for the management of container terminals to determine the appropriate development strategy for the stakeholders in the transport supply chain, related to the issues of digital transformation. Based on the research findings, the container terminal managers should consider the usefulness and compatibility of the technology or system developed. Digitization or digital transformation is now only at the port that have applied software to their operational processes. The port needs to develop to a new level of connecting, collecting and processing customer—market data, operational data, personnel data to support managers in decision-making, develop and improve new products/services, look for business opportunities or, more importantly, change the port operation business model. The limitation of the study is that it only expands the TAM and the research sample in the Vietnam Southeast region, whereby each model has its advantages and disadvantages, and each research area has its own particularities. Future studies can be conducted in the direction of combining technology application models and expanding the scope of research.

References

Akinnuwesi, B. A., Uzoka, F. E., Okwundu, O. S., & Fashoto, G. (2016). Exploring biometric technology adoption in a developing country context using the modified UTAUT. *International Journal of Business Information Systems, 23*(4), 482–521; Available from: https://doi.org/10.1504/IJBIS.2016.080219

Ashfaq, M., Yun, J., Yu, S., & Loureiro, S. M. (2020). Modeling the determinants of users' satisfaction and continuance intention of ai-powered service agents. *Journal Telematics and Informatic, 54*, 66–69.

Bollen, K. A. (1989). *Structural equations with latent variables.* John Wiley and Sons Inc.

Cheng, Y. M. (2015). Towards an understanding of the factors affecting M-learning acceptance: roles of technological characteristics and compatibility. *Journal Asia Pacific Management Review, 20*, 109–119.

Chin, J., & Lin, S. (2015). Investigating users' perspectives in building energy management system with an extension of technology acceptance model: A case study in Indonesian manufacturing companies. *Journal Procedia Computer Science, 72*, 31–39.

Choi, J. H., & Park, J. W. (2015). A study on factors influencing 'cyber airport' usage intention: An incheon international airport case study. *Journal of Air Transport Management, 42*, 21–26.

Davis, F. D. (1993). User acceptance of information technology: System characteristics, user perceptions and behavioral impacts. *International Journal of Man-Machine, 38*, 475–487.

Devi, N. L. N. S., & Suartana, I. W. (2014). Analysis of technology acceptance model (TAM) on the use of information systems at Nusa Dua Beach Hotel and Soa. *Accounting e-Journal of Udayana University, 6*(1), 167–184.

Dewi, W. (2013). Pengaruh struktur modal, profitabilitas, dan ukuran perusahaan pada nilai perusahaan. *E-Jurnal Akuntansi Universitas Udayana, 4*(2), 358–372. (in Indonesian).

Di Pietro, L., Mugion, R. G., Mattia, G., Renzi, M. F., & Toni, M. (2015). The integrated model on mobile payment acceptance (IMMPA): An empirical application to public transport. *Journal Transportation Research Part, 5*(6), 463–479.

Fornell, C., & Larcker, D. F. (1981). Evaluating structural equation models with unobservable variables and measurement error. *Journal of Marketing Research, 18*(1), 39–50; Available from: https://doi.org/10.2307/3151312

Hamid, A. A., Razak, F. Z. A., Bakar, A. A., & Abdullah, W. S. W. (2016). The effects of perceived usefulness and perceived ease of use on continuance intention to use e-government. *Procedia Economics and Finance, 35*, 644–649.

Hair, J. F., Black, W. C., Babin, B. J., & Anderson, R. E. (2009). *Multivariate data analysis*, 7th ed. Upper Saddle River, NJ: Pearson Prentice Hall, 401.

Henseler, J., Christian, M.R., & Rudolf, R.S. (2009*)*. The Use of Partial Least Squares Path Modeling in International Marketing. Advances in International Marketing. Emerald JAI Press; Available from:https://doi.org/10.1108/S1474-7979(2009)0000020014

Ho, J. C., Wu, C. G., & Lee, C. S. (2020). Factors affecting the behavioral intention to adopt mobile banking: An international comparison. *Journal Technology in Society, 7*, 45–59.

Hung, C. L., & Hsu, H. (2011). Effects of guided writing strategies on students' writing attitudes based on media richness theory. *Turkish Online Journal of Educational Technology, 10*, 148–164.

Hsu, M. H., Yen, C. H., Chiu, C. M., & Chang, C. M. (2006). A longitudinal investigation of continued online shopping behavior: An extension of the theory of planned behavior. *International Journal of Human-Computer Studies, 64*, 889–904.

Hu, L., & Bentler, P. M. (1999). Cutoff criteria for fit indexes in covariance structure analysis: Conventional criteria versus new alternatives. *Structural Equation Modeling: A Multidisciplinary Journal, 6*, 1–55; Available from: https://doi.org/10.1080/10705519909540118

Islam, N. A. K. M., & Mäntymäki, M. (2014). Social virtual world continuance among teens: uncovering the moderating role of perceived aggregate network exposure. *Behaviour and Information Technology, 1*, 23–32; Available from: https://doi.org/10.1080/0144929X.2013.872190

Lai, P. C. (2017). The literature review of technology adoption models and theories for the novelty technology. *Journal of Information Systems and Technology Management, 1*(1), 21–38.
Lee, D., Park, J., & An, J. H. (2011). On the explanation of factors affecting E-Commerce adoption. In *Twenty-second international conference on information systems*, Oct 15–19; Seoul, Korea.
Lisa, R. A., Lehmuskallio, A., Thielmann, T., & Abend, P. (2017). Introduction: Mobile Digital Practices. Situating People, Things, and Data. *Mobile Digital Practices, 3*(2), 5–18; Available from: https://doi.org/10.25969/mediarep/13507
Mazhar, F., Muhammad, R., Umar, F., & Sobia, I. (2014), An investigation of factors affecting usage and adoption of internet and mobile banking in Pakistan. *International Journal of Accounting and Financial Reporting, 4*(2), 34–45; Available from: https://doi.org/10.5296/ijafr.v4i2.6586
Mndzebele, N. (2013). The effects of relative advantage, compatibility and complexity in the adoption of EC in the hotel industry. *International Journal of Computer and Communication Engineering, 2*(4), 35–46.
Ramayah, T., & Ignatius, J. (2005). Impact of perceived usefulness, perceived ease of use and perceive enjoyment on intention to shop online. *ICFAI Journal of Systems Management, 3*(3), 1–16; Available from: http://www.ramayah.com/journalarticlespdf/impactpeu.pdf
Shakrokh, N. (2019). Factors driving the adoption of smart home technology: An empirical assessment. *Journal Telematics and Informatics, 45*, 21–34.
Somang, M., Kevin, K. F. S., & Miyoung, J. (2019). Consumer adoption of the uber mobile application: Insights from diffusion of innovation theory and technology acceptance model. *Journal of Travel and Tourism Marketing, 7*, 46–57.
Taufik, N., & Hanafiah, M. (2019). Airport passengers' adoption behaviour towards self-check-in Kiosk services: The roles of perceived ease of use, perceived usefulness and need for human interaction. *Journal Heliyon, 5*, 78–85.
Teng, Y. Y., & Chen, S. Y. (2008). Social tagging in digital archives. Buchanan, G., Masoodian, M., & Cunningham, S. J. (Eds.), *ICADL*, Springer, Heidelberg. LNCS, 3562, pp. 414–415.
Tho, N.Đ. (2009). *Scientific research methods in business.* Ho Chi Minh: Financial Publishing House, p. 594 (in Vietnamese)
Venkatesh, V., & Davis, F. D. (2000). A theoretical extension of the technology acceptance model: Four longitudinal field studies. *Management Science, 2*(2), 186–204.
Venkatesh, V., & Bala H. (2008). Technology acceptance model 3 and a research agenda on interventions. *Decision Sciences, 2*, 39; Available from: https://doi.org/10.1111/j.1540-5915.2008.00192.x
Weng, G. S., Zailani, S., Iranmanesh, M., & Hyun, S. S. (2017). Mobile taxi booking application service's continuance usage intention by user. *Journal Transportation Research, 57*, 207–216.
Wu, B., & Chen, X. (2016). Continuance intention to use MOOCs: Integrating the technology acceptance model (TAM) and task technology fit (TTF) model. *Journal Computers in Human Behavior, 67*, 221–232.
Yang, Z., Jun M., & Peterson R.T. (2004). Measuring customer perceived online service quality: Scale development and managerial implications. *International Journal of Operations and Production Management, 24*, 1149–1174; Available from: https://doi.org/10.1108/01443570410563278

The Role of Foreign Direct Investment in Boosting Tourism Development: A Study of Vietnam for 1990–2020 Interval

Thao Thi Phuong Hoang

Abstract The vital purpose of this research is to use the ARDL model in order to determine the relationship between foreign direct investment (FDI) and tourism development in Vietnam from 1990 to 2020. The outcomes reveal that tourism development and FDI have a positive short-term and long-term interaction. In other words, FDI strongly causes tourism development in this study interval. This paper supports policymakers to perfect the development strategy for Vietnam's tourism industry. Apart from creating a favorable business environment to attract more FDI to the tourism industry in the future, it is also necessary to improve the efficiency of using this FDI capital to increase its contribution to tourism growth.

Keywords Finance direct investment · Tourism industry · ARDL model · Granger test · Vietnam

1 Introduction

Attracting foreign direct investment (FDI) associated with sustainable development has become a crucial goal in Vietnam's socio-economic development strategy in recent years. The FDI section has been a significant contributor to Vietnam's economic development and international economic integration (Anwar & Nguyen, 2010).

With an attractive business environment, FDI inflows into Vietnam have increased sharply over the years. More recently, the new facilitation regulation for FDI and Trade has obviously promoted Vietnam's integration into the global economy. As a consequence, Vietnam has become an important member of regional production networks. Strongly, innovations in policy have resulted in creating more jobs and improving the revenue and the national balance of payments. Parallel with the direct

T. T. P. Hoang (✉)
University of Pécs, Pécs, Hungary
e-mail: hoangphuongthao16@gmail.com

Yersin University, Da Lat, Vietnam

A. T. Nguyen et al. (eds.), *Contemporary Economic Issues in Asian Countries: Proceeding of CEIAC 2022, Volume 1*,
https://doi.org/10.1007/978-981-19-9669-6_4

benefits that can be seen through the growth, FDI capital has generated significant indirect benefits, namely, technology transfer, business experience sharing, and workforce skills development (Ministry of Planning and Investment, World Bank, 2018).

According to empirical studies, the greater the FDI attraction, the better the ability for economic growth, which results in attracting tourists and promoting tourism development in the future. Endo (2006) investigates the pattern and scale of global FDI in the tourism industry using data aggregation. The scholar confirmed that investment in the hospitality sector has increased significantly over time, but the scale remains small in comparison to global FDI. Due to favorable policies and infrastructure, developed countries tend to attract more FDI capital for tourism. Meanwhile, for some developing countries, these investments play a much larger role than in other traditional economic activities. Fauzel (2020a, 2020b) confirms that FDI has been playing a crucial factor in the tourism development of economies on 17 islands from 1995 to 2018, and the two-way causal relationship between FDI and tourism development has been confirmed as well. Cró and Martins (2020) investigated the determinants of the attraction and efficient use of FDI in the French tourism sector in the period between 2000 and 2017. This study contributed many important recommendations for enhancing the efficiency of FDI in this country. In general, there have been numerous studies analyzing the role of FDI in economic development in the overview. It is obvious that not much research has been investigated on the relationship between FDI and tourism, while the tourism industry is now expanding markedly and is also a potential section for attracting FDI.

The global foreign investment direction has shifted as a result of the health crisis hitting. By 2020, Vietnam had emerged as a bright spot in terms of attracting FDI inflows. Foreign investors are putting great prospects to Vietnam's link in the global supply chain based on considering many advantages.

Vietnamese national data express that the total revenue in the tourism sector recorded a boom when it increased from 1.23 billion USD in 2000 to 34.3 billion USD in 2019. With this outstanding growth, Vietnam had attracted approximately 15 billion USD of FDI invested in tourism by the end of 2019, making it one of the most appealing fields for domestic and foreign investors. However, quantitative studies have not examined the positive effect that FDI has on tourism growth. In other words, studies on the role of FDI in tourism development in Vietnam with econometric models and time-series data are limited.

Developing on research results from Sadi and Henderson (2008), which investigated the FDI context of Vietnam in the 1990s, this study will expand the study period from 1990 to 2020. The main aim is to find out the driving forces between FDI and tourism in the short and long term. The outcomes will support policymakers to perfect the development strategy for Vietnam's tourism industry. More specifically, policies focus on consolidating and enhancing the image of Vietnam's destination by attracting FDI and improving the efficiency of FDI capital benefit.

The tourism industry is currently a key economic sector, and as announced in many early papers, FDI is one of the sources of capital to support upgrading the quality of infrastructure, quality of accommodation, diversification of amusement parks, and development opportunities for potential new lands. In the preliminary analysis, research on tourism FDI in Vietnam has not attracted many scholars, therefore the number of studies is limited. This study will be a significant supplement research data on the topic of FDI and create an impetus for future research. To demonstrate the relationship between FDI and tourism in Vietnam, the observed variables, as well as appropriate research methods, are inherited and developed based on many previous research models. These contents are presented in Sects. 2 and 3. Section 4 covers the reality results found in Vietnam. The paper closes with a conclusion and policy recommendations.

2 Theoretical Framework

2.1 The Link Between FDI and Tourism Development

Tourism, with its outstanding potential and benefits, is one of the most impressive-growing industries nowadays (Garza-Rodriguez, 2019). Numerous nations that are rich in tourism resources have taken advantage of all opportunities for economic development with the tourism-led growth strategy (Ozturk, & Acaravci, 2009; Spinthiropoulos et al., 2020). Ivanovic (2011) addressed not only the contribution of FDI to economic growth but also the positive influence of FDI on Croatia's tourism industry.

Norlida Hanim et al. (2011) used the ARDL model to investigate the relationship between the development of the tourism industry and FDI capital in some Asian countries for the period 1980–2008. A bidirectional interaction between tourism and FDI in Hong Kong is found, a one-way positive effect from tourism to FDI in Malaysia and Thailand, while there is no relationship between these two variables in Singapore and China.

Using the VECM model, Samimi et al. (2013) demonstrated a two-way causal relationship in the long run between FDI in tourism and tourism development from 1995 to 2008. Fauzel (2020b) affirmed that FDI is a key factor for tourism development in 17 island countries. The aim of Perić and Radić (2015) is to measure the contribution of FDI capital to the progress of Croatia's tourism industry. Using data from 2000 to 2012, and the Cobb–Douglas production function model, the findings confirm that FDI is a significant component of Croatia's tourism development.

The relationship between foreign investment and tourism development was investigated by using panel data from seven countries from 1980 to 2012. Işik (2015) proved that the tourism industry has become an attractive sector for foreign investment in D7. The recommendation of this study has important implications for tourism sustainable development policy, and it calls for further attention in future studies. With the aim of stimulating domestic and international investment in tourism, the government of India has allowed the most foreign direct investment projects into construction development including hotel construction, resorts, and recreational facilities. Therefore, Kaur (2019) revealed a strong and positive influence of foreign direct investment in the Indian tourism industry as well as its impact on the Indian economy. And, Amin et al. (2020) found a long-run mutual positive effect of FDI and tourism in Bangladesh using time series data from 1972 to 2017.

Rakhmatullaeva (2021) asserted the importance of investments in tourism. Investment in tourism is the process of creating new tourist facilities with the help of large capital from huge and professional corporations. In addition, tourism has emerged as an important economic sector in Uzbekistan, contributing significantly to the country's regional economic development. Modern techniques are also exchanged through capital investment, thus the quality of tourism products is improved (Nguyen, 2021). Sou and Vinnicombe (2021) investigate the link between governance, FDI, and tourism demand. According to the regression results, the effectiveness of governance increases the ability to attract FDI, and thereby FDI has a positive influence to promote tourism development.

Sheng Yin and Hussain (2021) investigate the connection between FDI inflows, environmental footprint, and economic growth in Southeast Asian countries from 1990 to 2017. The findings support the existence of a one-way causal from FDI inflows to tourist arrivals. As a result, ASEAN countries are encouraged to improve the performance of the tourism sector, which is currently driving FDI attraction.

Since 1986, the Vietnamese government has improved the Law on Foreign Investment, encouraging FDI in Vietnam, particularly in the tourism industry. The initial study on FDI and Vietnam's tourism industry was carried out in 1999. At this time, due to constraints and challenges in economic development and tourism, section was not yet focused, so there was not much attraction to FDI (Sadi and Henderson, 2008). Suntikul et al. (2010) investigate the impact of FDI on tourist accommodation patterns in Vietnam. The beneficial effects of FDI have improved the quality of Vietnam's tourist accommodations, increased the attractiveness of tourism services, and thus increased the attraction of international visitors to Vietnam. Thence, this study is conducted for the period 1990–2020 as an examination of the role of FDI in the growth of tourism. In other words, measuring the FDI's positive influence and testing whether it is one of the vital factors that influence the growth of tourism in Vietnam today is the main aim of this empirical research.

2.2 FDI Perspective and Economic Development in Vietnam

Due to the movement restrictions and border closures from the COVID-19 pandemic hitting, global FDI inflows in 2020 decreased by 35%. In which, FDI capital into emerging economies and developing countries fell by 8%. However, developing economies still accounted for two-thirds of total global FDI by 2020, up from nearly half in 2019. According to geographic region analysis, FDI has decreased globally except for Asia. This region accounted for half of the total global FDI in 2020, with an FDI growth rate of 4%.

Table 1 Results of attracting FDI from 1990 to 2020

Period	Number of projects	Total registered capital (million USD)	Total realized capital (million USD)
1990–1999	3.164	42.7293	18.2695
2000–2010	10.473	171.6433	60.8767
2011–2020	20.495	239.6464	180.3267

Source Statistical Office of Vietnam, 2021

Thanks to the transition from a centrally planned economy to a market economy model, political stability, young and low-cost labor force, Vietnam has become a potential destination for foreign investors. In 1990, the amount of FDI registered in Vietnam was only 2.07 billion USD. The accession to the World Trade Organization in 2007 greatly increased the registered FDI in Vietnam up to now. However, the amount of capital increased significantly after 2000, and the efficiency of capital use was highly valued.

From 1988 to 2019, realized FDI accounted for approximately 47% of registered capital (Table 1). FDI contributed 25.7% to Vietnam's economic growth from 2011 to 2019.

Over the last ten years, the FDI sector has grown at a faster rate than the country's GDP (Table 2). GDP growth rate was 5.9% per year from 2011 to 2015 and 6.75% per year from 2016 to 2019. Meanwhile, the FDI sector grew at a rate of about 8.4% in 2010 and 10.6% in 2019.

By 2019, 6.1 million people were working in FDI enterprises. Labor productivity in the FDI sector increased at a rate of 8.7% per year, far exceeding that of domestic enterprises, which increased at a rate of 4.6% per year. Therefore, the average income per employee of the FDI enterprise sector is 1.2 times higher than the average of the economy.

So far, Vietnam has received FDI from 141 countries and territories worldwide. East Asian countries are the largest and longest-term partners. For many years, Korea, Japan, and Singapore have been the primary sources of capital for Vietnam. These three countries' investment capital accounts for half of total FDI capital (as of 2020). Figure 1 lists the top 15 countries with the most FDI inflows into Vietnam today.

Vietnam has been being an attractive destination for real estate investors around the world. In particular, real estate investment in the tourism industry, namely, high-class hotels, resorts, and restaurants, which brings high profits in the long run. From the

Table 2 Vietnam's economic growth rate from 2011 to 2020

Economic sectors	2011–2015 (%)	2016–2020 (%)
Government economy	4.9	4.1
Domestic private economy	6.1	6.9
Foreign-invested economy	8.4	10.6

Source Statistical Office of Vietnam, 2021

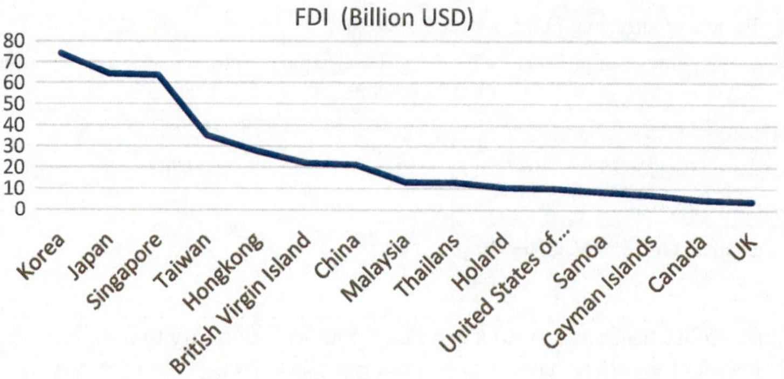

Fig. 1 Vital FDI partners in Vietnam. *Source* Statistical Office of Vietnam, 2021

tourism development data, it is undeniable that the role and contribution of FDI in the field of HORECA (Hotel, Restaurant, and Coffee shop) and amusement parks. This is the earliest source of capital present in Vietnam right after Vietnam promulgated the Law on Foreign Investment in Vietnam in 1987. Up to now, the amount of FDI in this area is up to 12%, by the end of the year 2019. The structure of FDI in Vietnam's economic sectors is shown in Fig. 2.

FDI in the tourism industry started to increase sharply in 1998 and peaked in 2003, 2013, and 2019 (Fig. 3).

An empirical study in Vietnam was conducted using listed data from the World Bank and the National Investment Agency for the period 1990–2020 with the ARDL model to investigate the impact and role of FDI on the country's tourism development. This study will fill the current research gap on FDI and tourism in Vietnam. It lays the groundwork for future research also.

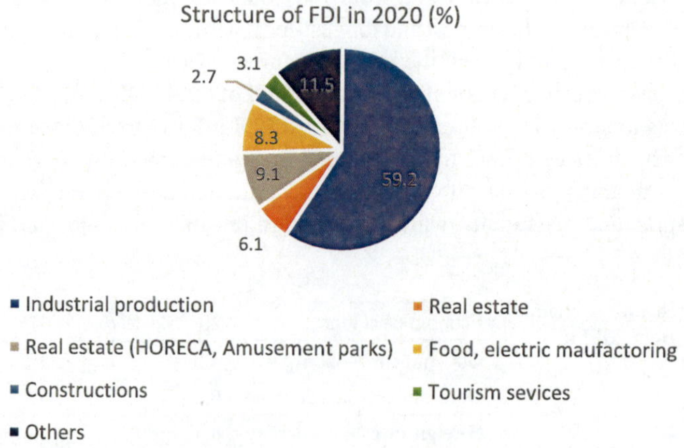

Fig. 2 Structure of FDI in 2019. *Source* Statistical Office of Vietnam, 2021

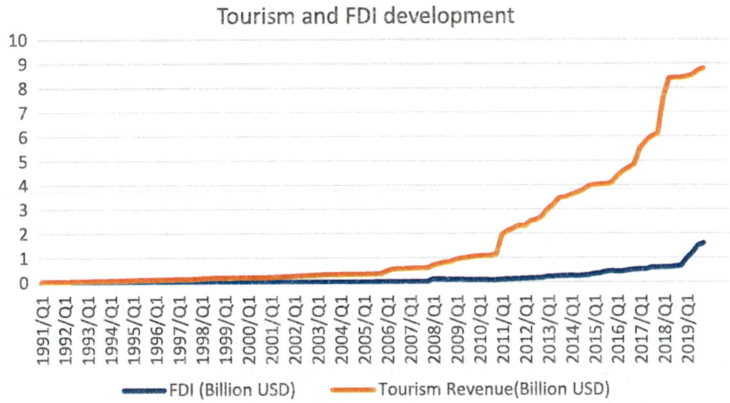

Fig. 3 Tourism and FDI development from 1991 to 2019. *Source* Statistical Office of Vietnam, 2021

3 Methodology and Data Collection

This empirical study investigates the role of FDI in Vietnam's tourism development for the 1990–2020 interval. Eviews 12 is used to process the data. Based on the outcomes, strategies for tourism development in the future will be discussed. The model of this study is inherited from the results of previous studies, namely, Central and Capital, Ivanovic (2011), and Işik (2015). In this sense, the empirical model is formed:

Function: $\text{TR} = f(\text{FDI})$
where

- TR = Tourism revenue (in billion USD).
- FDI = Foreign direct investment in the tourism industry (in billion USD).

The dependent variable is tourist revenue, and data is obtained from the Vietnamese statistics office. In terms of an independent variable, this is FDI in tourism, and it is collected and calculated based on data from World Bank and Vietnam Investment Department. Quarterly data is collected between 1990 and 2020.

The result includes testing the both long and short-term correlation between these observed variables. The linear logarithm form is posed:

$$\ln \text{TR}_{it} = \alpha_i + \beta_i \ln \text{TR}_{it} + \delta_i \ln \text{FDI}_{it} + u_{it}$$

3.1 Unit Root Test

Unit root test is to check if the time series is stationary or not. Stationary series is an essential category for the estimation process in economics. If a series is nonstationary, the results are not generalizable. Also, for forecasting purposes, the non-stationary series has no practical application value. This study uses a test developed by Dickey and Fuller, called the Dickey-Fuller statistic.

Test hypothesis:

H0: $\beta = 0$ (Yt is a non-stationary data series).

H1: $\beta < 0$ (Yt is a stationary data series).

"In particular, if the absolute value of the calculated value is greater than the absolute value of the critical value, hypothesis H0 will be rejected, that is, the data series is stationary, and vice versa, the hypothesis H0 is accepted, the data series is non-stationary" (Wong, 2018).

3.2 Bound Test

AutoRegressive Distributed Lag (ARDL) is a combination of the vector autoregression VAR) model and the least-squares regression (OLS) model. ARDL is considered an efficient, flexible, and easy-to-apply model for the analysis of multivariable time series (Nkoro & Uko, 2016; Wong, 2018). The ARDL model allows for determining the impact of established variables on the dependent variable (Kreishan, 2015). To ensure the reliability of the ARDL model, tests need to be performed to ensure that the time series variables are stationary, the lag is optimal, the model is fitness, and there is no autocorrelation and Heteroscedasticity (Kumar et al., 2016; Wong, 2018).

3.3 Causality Test

Causal analysis is applied to understand the impact of macroeconomic variables. For time-series macro data, the most commonly used causal analysis is Granger Causality (White and Pettenuzzo, 2010; Foresti, 2007).

4 Analysis of Findings

4.1 Correlation Statistics Between Variables

Table 3 describes the correlation between these observed variables. The value of r has statistical significance if the sig value is less than 0.05. The closer to 1 of r is, the

Table 3 Correlations matrix

Correlation	LTR	LFDI
LTR	1	0.947**(0.000)
LFDI	0.947**(0.000)	1

Note *, **, *** significance at 1%, 5%, and 10% level, respectively
Source Analysis results from Eviews 12

stronger the linear correlation is. The correlation results show that two variables are highly positively correlated with each other, when one variable changes, the other variables also change in the same direction.

4.2 Unit Root Test

All the observed variables are stationary at (I), which is shown in detail in Table 4. Therefore, we can continue with the test for co-integration between observed variables.

As the result from Table 5, following the Akaike information criterion (AIC), the optimal lag of the observed variables is at (I).

Table 4 Results of unit root test

Variable with Intercept	ADF test				
	Level		1st difference		Decision
	t-statistic	*p*-value	*t*-statistic	*p*-value	
TR	−2.430404	0.1357	−10.33485	0.0000	(I)
FDI	−0.524018	0.8814	−11.10144	0.0000	(I)

Source Analysis results from Eviews 12

Table 5 Results of optimal lags

Lag	LogL	LR	FPE	AIC	SC	HQ
0	−13.59046	NA	0.076258	0.264245	0.311206	0.283313
1	*136.8429*	*293.2176**	*0.06058**	*−2.268524**	*−2.198083**	*−2.239923*
2	137.2992	0.881630	0.006114	−2.259309	−2.165387	−2.221174

Source Analysis results from Eviews 12

Table 6 Results of ARDL—cointegration testing

Equation model	F-statistic	SC lag-length criteria	Bound critical values (at 1% significance level)		Decision
			I(0)	I(1)	
TR = f(FDI)	8.097876	(1, 0)	6.84	7.84	Co-intergrated

Source Analysis results from Eviews 12

Table 7 Long-run estimate from the ARDL model

Variable	Coefficient	Standard error	t-statistic	Prob.
LFDI	0.655092	0.038773	16.89555	0.0000

Source Analysis results from Eviews 12

4.3 ARDL—Cointegration Testing

The tests show the bounds test for co-integration (Table 6) and the result for the series model (Table 4). In the model, TR = f(FDI), at a 1% significance level, the F-statistic is 8.09 and is higher than the upper bound critical value. These results show that the variables share a relationship in the long term. From this base, we can proceed with the next estimation.

In long-term results from (Table 7), FDI increase is positively and strongly associated with tourism development: a 1% improvement in FDI can be linked to a 0.65% increase in tourism development.

4.4 ECM—Short-Term Relationship

Then, to acquire the assessments of the short-run correlation between tourism development and finance direct investment in Vietnam, the error correction approach was utilized. The ECM model measures the dynamics of the short-run model, and the speed for the model is adjusted to an equilibrium whenever a shock occurs. The results demonstrate that a 1% improvement in FDI can be linked to a 0.11% increase in tourism development, and the speed of the variables return to the equilibrium in the long term is 9.3% at a 1% significance level. The details are shown in Table 8 below:

Table 8 Results of ECM

Variable	Coefficient	Std. error	*t*-Statistic	Prob.
LFDI	0.117059	0.036760	3.184439	0.0019
*CointEq(−1)**	*−0.093657*	*0.023172*	*−4.041854*	*0.0001*
R-squared	*15.5%*			

Source Analysis results from Eviews 12

Table 9 Granger causality outcomes

Null hypothesis	F-statistic	Prob.	Result
Tourism development does not granger cause FDI	3.37558	0.0687	Reject H0 at 10%
FDI does not granger cause tourism development	6.40925	0.0127	Reject H0 at 5%

Source Analysis results from Eviews 12

4.5 Granger Causality Test

Hypothesis H0 is rejected when the probability of variables values is less than 0.05 (5%); the details are presented in Table 9. The results indicate that due to Prob statistics being significant, FDI does lead to a change in tourism development in Vietnam for the study period. Furthermore, at the 10% probability level, tourism also promotes the growth of FDI.

4.6 Diagnostic Tests for the Model

The regression model undergoes rigorous and complete tests to ensure the correct results:

- Serial correlation LM test with the null hypothesis is a regression function that does not occur autocorrelation;
- Heteroskedasticity with the null hypothesis is a model without variance change.
- The CUSUM test will insist on the stability of model.

The outcomes are illustrated in Table 10.

Table 10 Diagnostic test for the model

Test	Statistic	Value	Prob.	Figure 4
Serial correlation LM	Chi-Square	1.51763	0.2180	
Heteroskedasticity (Breush-Pagan Godfrey)	Chi-Square	5.65247	0.1298	

Source Analysis results from Eviews 12

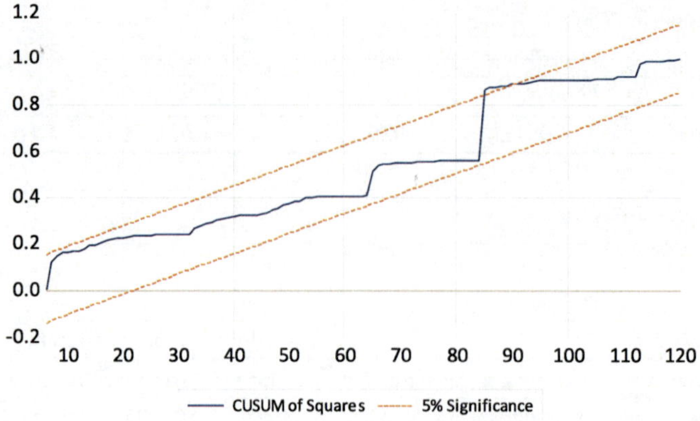

Fig. 4 The plot of CUSUM of square test for TR model

Table 10 and Fig. 4 confirm the stability of the TR model through the stability of all the parameters, and the values of the coefficients are within the critical bound values. However, the graph of CUSUM of square is slightly deviated from the upper bound at the 5% significance level.

5 Conclusions

The present study examines the relationship between tourism development and FDI attraction using 31-year data from 1990 to 2020 and the ARDL and the ECM model. The model has fully implemented the process, which includes: the unit root test, determining optimal latency, and measuring the correlation between variables. The following rigorous testing steps are carried out: checking the phenomenon of variance, autocorrelation, and the correct model to confirm that this is a suitable model and that the obtained results are reliable.

According to empirical findings, tourism development and FDI have a positive interaction in both the short-run and long-run. Also, there is a causal relationship between FDI to Tourism development, which is the same findings in the study of Kaur (2019), Amin et al. (2020), Rakhmatullaeva (2021), Sou and Vinnicombe (2021), and Sheng Yin and Hussain (2021). In other words, foreign investment is a vital part of Vietnam's tourism economic development over a period of 30 years of research.

Thus, the top goal in the future is to increase the efficiency of using FDI in order to increase the tourism development rate. The latter is, that despite impressive FDI inflow, Vietnam is still not the most attractive nation for foreign investors in the ASEAN region. Obviously, huge multinational corporations now tend to prioritize Thailand, Malaysia, and Indonesia... for investment due to the most competitive investment environment. In Vietnam, the majority of FDI capital comes from countries with medium technology and do not hold high technology accounting for a large proportion. As a consequence, Vietnam needs a solution to attract FDI investors from developed countries, holding leading technology, having financial potential, and having a large market. To do this, the investment policy should be improved to increase Vietnam's competitiveness in the ASEAN region, not only for achieving socio-economic development goals but also for the tourism industry in long run. Some recommendations focus on the following policies:

Firstly, priority should be given to completing the legal framework on foreign investment to ensure a more open environment and conditions for investors, while remaining consistent with Vietnamese law.

Second, to simplify and neatly improve administrative procedures for foreign investment projects. Administrative procedures must be made transparent, and widely publicized in the media to ensure that information reaches foreign investors most conveniently.

Third, upgrade the information system and connect with numerous foreign investment sources. Prioritize FDI sources in areas with abundant tourism resources that have yet to be properly invested in. Effective allocation ensures equitable development across Vietnam's territories while also providing investors with a competitive advantage. Furthermore, the investment attraction mechanism also needs to be flexible between localities, which results in increasing investment opportunities.

Fourth, having high-quality human resources to meet the needs of FDI enterprises is essential. Evoking and encouraging innovative ideas, improving multi-skilled and multi-lingual ability in the workforce is the priority.

Fifth, policies should be strengthened to encourage tourism development, increase the contribution rate to GDP growth, and maintain tourism as one of the key economic sectors. Many other economic sectors are being boosted as a result of the development of tourism, and the economy as a whole is growing comprehensively. In this age of globalization, the priority is to upgrade the digital economic platform, particularly in the tourism sector. In recent years, Vietnam has established a highly effective network, which becomes the ground for the promotion of the strengths of tourism entrepreneurship. Briefly, the government needs to quickly have plans and solutions to improve the quality of the entrepreneur ecosystem and digital ecosystem. This is now the most important factor in attracting investment.

Future research could expand on this paper's limitation by including COVID-19-affected period data to see how external financial shocks affect tourism activities. In addition, experimenting with the VAR model for further study of shocks will also provide more diverse results in terms of tourism and foreign investment connection.

References

Amin, S., et al. (2020). An empirical Investigation between Foreign Direct Investment (FDI) and tourism in Bangladesh. *Journal of Tourism Management Research, 7*(1), 110–121. https://doi.org/10.18488/journal.31.2020.71.110.121

Anwar, S., & Nguyen, L. P. (2010). Foreign direct investment and economic growth in Vietnam. *Asia Pacific Business Review, 16*(1–2), 183–202. https://doi.org/10.1080/10438590802511031

Central, I. and Capital, G. F. (no date) 'Does Foreign Direct Investment generate regional economic growth ? The case of Emerging Europe', pp. 1–23.

Cró, S. and Martins, A. M. (2020) 'Foreign Direct Investment in the tourism sector: The case of France', *Tourism Management Perspectives, 33*. https://doi.org/10.1016/j.tmp.2019.100614

Endo, K. (2006). Foreign direct investment in tourism—Flows and volumes. *Tourism Management, 27*(4), 600–614. https://doi.org/10.1016/J.TOURMAN.2005.02.004

Fauzel, S. (2020a). FDI and tourism futures: A dynamic investigation for a panel of small island economies. *Journal of Tourism Futures, 7*(1), 98–110. https://doi.org/10.1108/JTF-05-2018-0026/FULL/PDF

Foresti, P. (2007) 'Testing for Granger causality between stock prices and economic growth'.

Garza-Rodriguez, J. (2019) 'Tourism and poverty reduction in Mexico: An ARDL cointegration approach'. *Sustainability (Switzerland), 11*(3). https://doi.org/10.3390/su11030845

Influence of Foreign Direct Investment on Tourism|JournalNX—A Multidisciplinary Peer Reviewed Journal (no date). Available at: https://repo.journalnx.com/index.php/nx/article/view/696 (Accessed 11 April 2022).

Işik, C. (2015). Foreign direct investment in tourism: Panel data analysis of D7 countries. *Athens Journal of Tourism, 2*(2), 93–103. https://doi.org/10.30958/ajt.2-2-2

Ivanovic, Z. (2011). *Influence of foreign direct investment on tourism in Croatia*, vol. 2, issue 1, pp. 21–28.

Kaur, B. (2019). Foreign direct investment: A growth engine for tourism. *SSRN Electronic Journal*. https://doi.org/10.2139/ssrn.3385163

Kieu Anh, B., et al. (1995) Foreign Direct Investment (FDI) in Vietnam Economy. https://doi.org/10.4236/tel.2019.94064

Kreishan, F. M. (2015). Empirical study on tourism and economic growth of Bahrain: An ARDL bounds testing approach. *International Journal of Economics and Finance, 7*(11), 1. https://doi.org/10.5539/ijef.v7n11p1

Kumar, R. R., et al. (2016). Exploring the nexus between tourism and output in cook islands: An ARDL bounds approach. *Social Indicators Research, 128*(3), 1085–1101. https://doi.org/10.1007/s11205-015-1070-y

Nguyen, H. T., Nguyen, H. V., & Meyer, K. E. (no date). 9 Foreign Direct Investment in Vietnam.

Nguyen, Q. H. (2021). Impact of investment in tourism infrastructure development on attracting international visitors: A nonlinear panel ARDL approach using Vietnam's data. *Economies, 9*(3), 131. https://doi.org/10.3390/ECONOMIES9030131

Nkoro, E., & Uko, A. K. (2016). Autoregressive distributed lag (ARDL) cointegration technique: Application and interpretation. *Journal of Statistical and Econometric Methods, 5*(4), 63–91.

Norlida Hanim, M. S., Redzuan, O., & Tamat, S. (2011). An analysis of the relationships between tourism development and foreign direct investment: An empirical study in selected major Asian countries. *International Journal of Business and Social Science, 2*(17), 250–258.

Ozturk, I., & Acaravci, A. (2009) On the causality between tourism growth and economic growth: Empirical evidence from Turkey. *Transylvanian Review of Administrative Sciences*. Available at: https://www.rtsa.ro/tras/index.php/tras/article/view/193 (Accessed: 31 March 2021).

Perić, J., & Radić, M. N. (2015). Tourism productivity and tourism FDI in Croatia. *Mediterranean Journal of Social Sciences, 6*(5S2), pp. 425–433. https://doi.org/10.5901/mjss.2015.v6n5s2p425

Rakhmatullaeva. (2021). Influence of foreign direct investment on tourism. *JournalNX - A Multidisciplinary Peer Reviewed Journal*. Available at: https://repo.journalnx.com/index.php/nx/article/view/696 (Accessed: 11 April 2022).

Sadi, M. A., & Henderson, J. C. (2008). *Tourism and foreign direct investment in Vietnam*, vol. 2, issue 1, pp. 67–90. https://doi.org/10.1300/J149V02N01_04

Samimi, A. J., Sadeghi, S., & Sadeghi, S. (2013). The relationship between foreign direct investment and tourism development: Evidence from developing countries. *Institutions and Economies, 5*(2), 59–68.

Sheng Yin, X., & Hussain, J. (2021). *The implication of technological innovation and tourism development on FDI-growth-environment nexus in Association of Southeast Asian countries: A simultaneity modeling analysis*, vol. 16, issue 9, pp. 878–902. https://doi.org/10.1080/15567249.2021.1971801

Sou, J. P. U., & Vinnicombe, T. (2021) Does governance quality matter for FDI-led tourism development? A supply-side perspective. https://doi.org/10.1177/13548166211052814

Spinthiropoulos, K., Nikas, C., & Zafeiriou, E. (2020). Tourism and economic growth in Greece: An ARDL bound testing approach in a Kaldorian framework. *Acta Oeconomica, 70*(2), 215–227. https://doi.org/10.1556/032.2020.00011

Suntikul, W., Butler, R., & Airey, D. (2010). The influence of foreign direct investment on accommodation patterns in Vietnam as a result of the open-door policy. *Current Issues in Tourism, 13*(3), 261–277. https://doi.org/10.1080/13683500902937628

White, H., & Pettenuzzo, D. (no date). *Granger causality, exogeneity, cointegration, and economic policy analysis.*

Wong, K. (2018) Guideline Peasaran et al. (2001) Bound Test and ARDL cointegration Test. *Journal of Applied Econometrics, 16*

Website: Statistical Office of Vietnam: Ministry of Planning and Investment—World Bank: https://www.gso.gov.vn/en; https://www.worldbank.org/en

Vietnam's Seafoood Imports Under the Impact of the United Kingdom–Vietnam Free Trade Agreement

Chu Tien Minh and Nguyen Ngoc Diep

Abstract This paper assesses the potential impacts of the UK-Vietnam Free Trade Agreement (UKVFTA) on Vietnam's seafood imports from the UK by adopting the SMART model based on two scenarios. The simulation results reveal that the UKVFTA will result in a significant increase in Vietnam's seafood imports from the UK, implying that the UK will be still one of the most important seafood for Vietnam in the upcoming time. When Vietnam also extends its coverage of tariff elimination to ASEAN+5 including China Japan, South Korea, India, Australia—New Zealand having signed an FTA with ASEAN to which Vietnam is a country member, the reduction in Vietnam's seafood imports from the UK will occur but it will be very inconsiderable. Besides, in both scenarios, trade creation effects will be higher than trade diversion effects meaning that the UKVFTA can raise the welfare of Vietnam. Based on these results, the paper ends by drawing out some implications for the Vietnamese government to be better implementing the ambitious UKVFTA.

Keywords Imports seafood · SMART model · FTA · UKVFTA · Vietnam

1 Introduction

Vietnam is one of the largest seafood producers in the world, with a high growth rate. According to the Vietnam Association of Seafood Exporters and Producers (VASEP), the country produced over 8.15 million tons (t) of finfish and shellfish in 2019 (VASEP, 2020), of which capture fisheries accounted for 46% and aquaculture accounted for 54% of total volume. Aquaculture accounted for about 75% of the total value and its farming volumes are growing faster than that of fisheries (Fig. 1).

Namely, over the last decade (2009–2019), production increased sharply; with output increasing from 4.9 million tons in 2009 to 8.15 million tons in 2019, an average annual growth rate of 5%. During the same period, aquaculture production

C. T. Minh · N. N. Diep (✉)
Thuongmai University, Hanoi, Vietnam
e-mail: diep.nn@tmu.edu.vn

A. T. Nguyen et al. (eds.), *Contemporary Economic Issues in Asian Countries: Proceeding of CEIAC 2022, Volume 1*,
https://doi.org/10.1007/978-981-19-9669-6_5

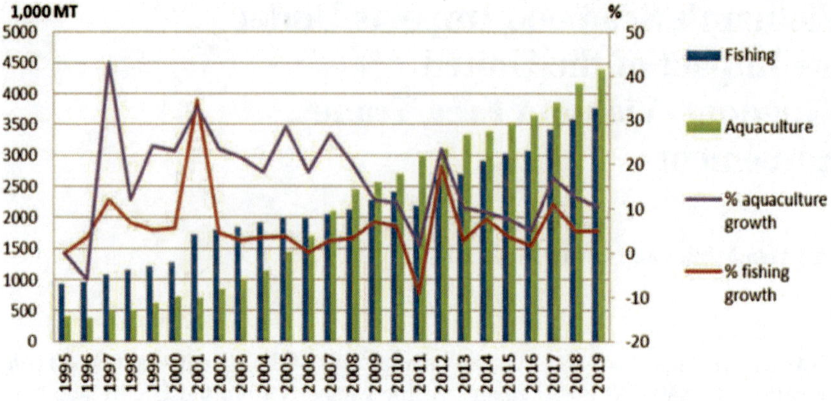

Fig. 1 Vietnam's capture fisheries and aquaculture production, 1995–2019. *Source* VASEP (2020)

increased significantly, from 2.6 million tons in 2009 to 4.4 million tons in 2019, resulting in an average annual growth rate of 6 percent (Fig. 1).

Driven by growing global demand for food and for seafood in particular, Vietnamese seafood exports have increased significantly. The value of exported seafood increased from USD 1.8 billion in 2000 to nearly USD 8.6 billion in 2019 (VASEP, 2020). This has made Vietnam the world's third largest exporter of seafood, after China and Norway, according to ITC Trade map (Fig. 2).

Despite being among top seafood exporters in the world, Vietnam has recently imported more and more seafood products for domestic consumption and for processing to serve re-export purpose. Namely, in the period of 2010–2018, Vietnam's seafood imports witnessed a more than 21.10% growth rate per annum, from USD 328,854,000 to USD 1,522,080,000. Moreover, during the outbreak of the Covid-19, 2019–2020, Vietnam's seafood imports continued to enjoy a steady rise

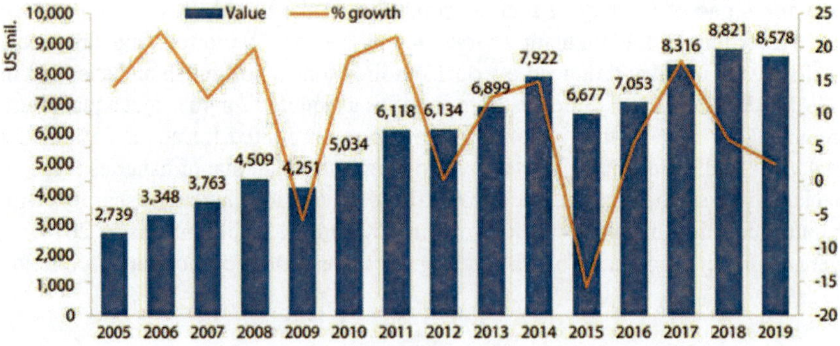

Fig. 2 Vietnam's export value of seafood, 2005–2019. *Source* VASEP (2020)

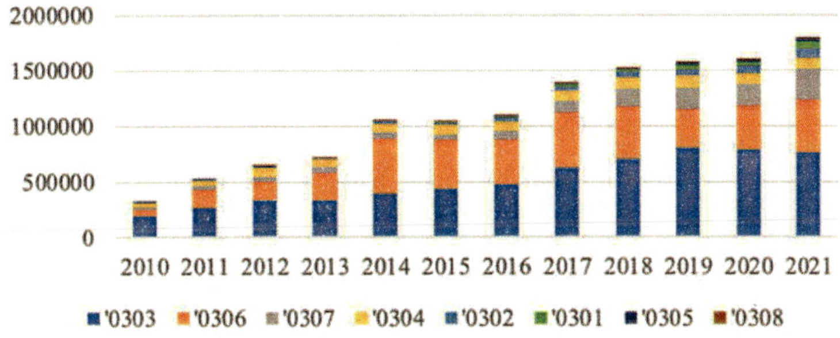

Fig. 3 Vietnam's seafood import value according to HS 4-digit. *Source* ITC Trade Map

at a rate of 3.3% and 4.7% compared to 2018, respectively. Then the import turnover even soared to a new peak at USD 1,800,002,000 in 2021 (ITC Trade map) (Fig. 3).

Imports of seafood growing gradually on a year-over year basis could be explained by the following reasons:

Firstly, the average seafood consumption per capital in Vietnam grew steadily from 1975 and even surpassed the consumption of meat in the early 2000s and continued to go up rapidly. According to the Agro Processing and Market Development Authority at the Ministry of Agriculture and Rural Development (MARD), average seafood consumption in Vietnam was 31 kg (kg) per capita in 2017. Seafood consumption in the domestic market was predicted to increase sharply, with the annual average expected to hit 33–35 kg per person by 2020 (Fig. 4).

Generally, the rise in overall demand for fish and fishery products is caused by two main factors, population growth and income growth. In Viet Nam, population growth has been quite stable with the rate of about 1% recently. Income growth which is measured by growth in GDP per capita was high from 2015 to 2019 with annual growth levels between 5 and 6.1% but then collapsed to 2% and lower in 2020

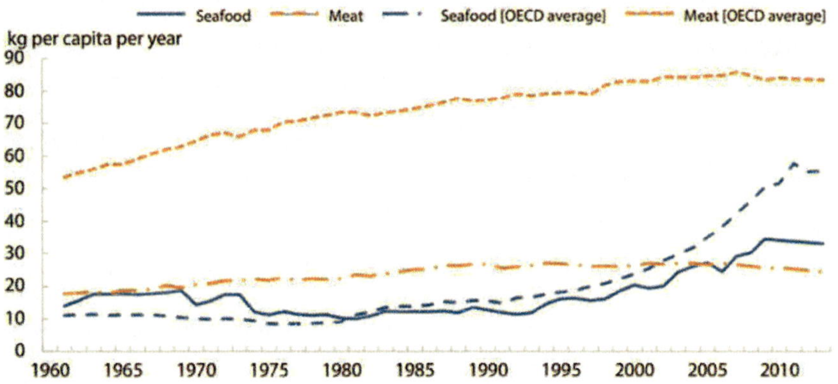

Fig. 4 Seafood and meat consumption in Vietnam and OECD countries, 1961–2013. *Source* FAO (2016)

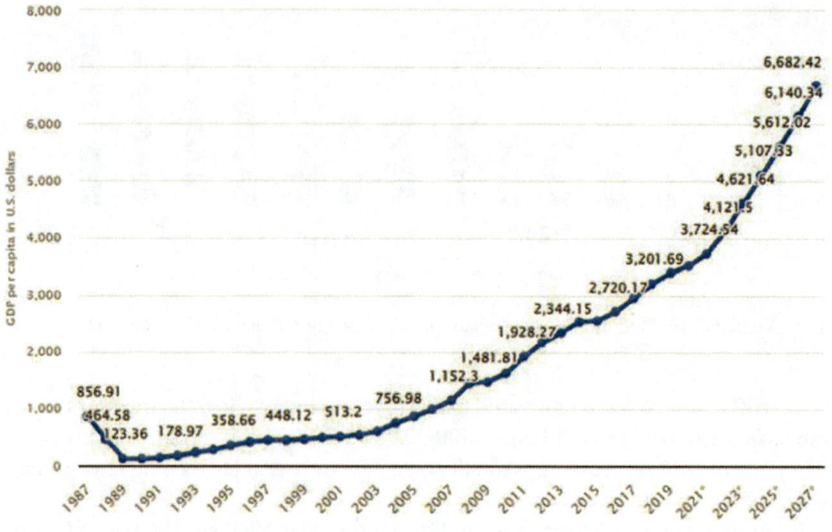

Fig. 5 GDP per capita in current prices from 1987 to 2027*. *Source* Statista

and 2021 because of the Covid-19 (World Bank, 2021). However, income growth is expected to remain quite strong for the medium term making it more likely to expect a strong and sustained demand for fish and fishery products from the Vietnamese population (Fig. 5).

Secondly, there was an upward trend in the number of foreign visitor arrivals to Vietnam, from nearly 8 million in 2015 to more than 18,000,000 in 2019 (World Bank, 2019). Although this figure suffered a sharp decrease to just more than 3.8 million in 2020 due to the outspread Covid-19, it could lead to a growing need of imported fresh/chilled whole fish, especially Atlantic and other salmon and some frozen fish, particularly frozen salmon for use by the higher-end food service industry and high-end retail outlets.

Thirdly, the country's current capacity still cannot meet the demand for fish fries in aquaculture production so fish fries are imported overseas. The UK has been one of the largest and most important trade partners of Vietnam in the European market. In the fishing industry, in 2019, the UK used to be the 2nd biggest seafood exporter to Vietnam among other EU countries (EU was listed in the top 10 fish and seafood exporting countries to Vietnam and Denmark and the UK were dominant—ITC Trade Map). In 2020, after the UK officially left EU on January 31, 2020, the UK has become an independent trade partner of Vietnam and was ranked as one of twenty countries from which Vietnam imported fish and fishery products the most.

Although there was an upward trend in the UK's exports to Vietnam in general since 2010 when Vietnam and the UK signed the Joint Declaration to establish the UK-Vietnam Strategic Partnership, which was a remarkable achievement in the bilateral trade relation with the two countries, the UK's export value experienced an inconsiderable share which was less than 1% in the total import value of Vietnam from 2010 to 2020. Especially, regarding Vietnam's seafood imports from the UK, despite a 7% growth rate, they only met from 1–2% of the Vietnam's demand. That is the reason why the UK-Vietnam free trade agreement (UKVFTA), which has taken into effect since May 1, 2021, is expected to promote bilateral trade between the two countries, especially Vietnam and its imports of fish and fishery products from the UK as they still have an enormous potential (Fig. 5).

With a view to maintaining bilateral trade relation between Vietnam and the UK after the Brexit, Vietnam and the UK actively negotiated and signed the UKVFTA on December 29, 2020. The agreement took effect temporarily on January 1, 2021, right after the EU-Vietnam free trade agreement (EVFTA, 2021) no longer applied to the UK. The UKVFTA, which has been officially and fully effective since May 1, 2021, is an important milestone in the bilateral trade relation between Vietnam and the UK after Brexit.

The UKVFTA is a "new generation" bilateral agreement aimed at a high degree of liberalization and a wide range of commitments from traditional ones such as trade in goods, trade in services, investment and nontraditional ones such as public procurement, state-owned enterprises, competition policy, ... The UKVFTA also includes commitments on non-commercial aspects which are closely related to trade and play an important role in the prosperous and sustainable development of the two economies such as labor, environment, social responsibility...

The UKVFTA is negotiated on the basis of the EVFTA, therefore, almost all of the EVFTA's commitments on trade in goods, trade in services, investment were basically inherited by the UKVFTA and then open to adjustments appropriate to bilateral trade relation between Vietnam and the UK. The UKVFTA consists of nine articles, one appendix amending a number of sentences in articles of the EVFTA, one bilateral letter, one protocol on rules of origin and two explanatory notes.

Regarding the UKVFTA's commitments on trade in goods, Vietnam's import tariffs on fish and fishery products from the UK will be eliminated in 2 years after the Agreement's effective day. As the UK is among Vietnam's largest seafood import markets, tariff on seafood imported from the EU is now very high (at 11.65% on average—SMART model), this tariff elimination is likely to affect considerably Vietnam's seafood imports and industry.

On these grounds, this paper, by adopting the SMART model and Partial Equilibrium theory to analyze and evaluate the impacts of tariff elimination under UKVFTA on seafood products imported from Vietnam imports from the UK in both scenarios. From there, the article gives some implications for the Vietnamese government in better implementation of UKVFTA, especially when this agreement has been signed.

2 Literature Review

It is not an exaggeration to say that policy making in connection with FTAs should start and end with impact assessment. At the initial stages of creating an FTA, an assessment of the potential costs and benefits of the prospective FTA is a prerequisite for shaping the FTA's objectives, informing consultations with public and private stakeholders and formulating effective negotiating strategies. In contrast, after the FTA is implemented, an assessment of the FTA's actual versus projected impact is necessary for determining whether the FTA's objectives have been met and what adjustments are needed. Correspondingly, two methods used to assess the impact of an FTA are ex-ante evaluation and ex-post evaluation which are usually complementary to each other.

Ex-post evaluation of an FTA's impact often applies gravity model and panel data to quantify the effect of tariff and non-tariff abolishment on nations' trade flow within a certain period.

Evaluating the performance FTAs by analyzing the determinants of trade flows of Asian economies for a panel of 31 countries during 2007–2014 using a gravity model. Along with the OLS2, fixed and random estimators, PPML was used to arrive at the robust empirical findings. The results revealed that there were certain distortions which existed in the operation of FTAs within Asia. Variables such as GDP and population were significant in explaining total trade flows in the region. Thus, the creation of economies of scale could result in lowering of the multilateral resistance enabling lesser trade costs and more efficient trade flows. This could then lead to a higher level of intra-regional trade within the Asian region. Another feature that could be inferred from the results was that high trade costs occurred due to 'distance' between two trading partners. In case of better connectivity routes developed within these trading partners in Asia, there could be a better realization of the trade potentials within the region (Sunder et al. 2020).

A gravity model analysis for trade between the GCC and developed countries. Gravity model was used to explain the bilateral flow of trade determined by GDP per capita, population and distance. It was assumed that trade flow between the two countries was positively related to their economic size and population. The gravity model was analyzed in six developed countries concerning trade with GCC countries from 2001 to 2012. The study concluded that GDP per capita and population for GCC and destination countries was significant and suggested that the trade barriers among the countries must be eradicated for better trade flow as well (Sahar, 2019).

With the application of gravity model, evaluated the trade patterns of China with the member countries of OPEC from the period 1990–2016 and confirmed that China's bilateral trade with OPEC members positively impacted on GDP, income (GDP per capita) and trade openness in China and the WTO member countries in OPEC while negatively influenced on distance (trade cost) (Irshad et al., 2017).

Using gravity model and panel data of India trade aimed at determining the trade impact of India-ASEAN FTA. The results showed the possibility of greater trade between India and ASEAN countries through FTA. Since the initial tariff levels were higher in India compared to ASEAN, ASEAN was likely to gain more in the short term. For India to exploit the trade potential with ASEAN, the FTA should be operationalized in the services and investment domain (Chandran, 2018).

Investigating the effect of the China-Pakistan FTA (CPFTA) on Pakistan and China agricultural trade using PPML and trade data between the 2 countries and 110 partners. The results suggested an increasingly agricultural exports from Pakistan to China but not vice versa under the CPFTA (Lateef & Riaz, 2018).

Main strengths of the gravity model in evaluating an FTA are that it allows analysts to control for other trade-related variables and quantify any changes in a country's trade due to the FTA. However, the model may yield misleading results if the data is inaccurate or important variables are omitted from the estimation. Moreover, the usage of ex-post analyses is limited by the availability of data as the gravity model demands data of trade flow under the FTA over time which is generally unavailable with new FTA. This problem gives rise to a range of ex-ante researches which are based on simulation rather than estimation to evaluate the potential impact of new FTA.

The ex-ante evaluation was mainly based on the GE and PE theory with GTAP and SMART model, respectively.

The GTAP model, originally formulated by Hertel (1997), is the most widely used CGE model for analyzing trade policy. The model is multi-market, with markets for final goods, intermediate goods, traded goods and factors of production. It is also multiregional, with a region representing a country or a group of countries. The quantity of endowments such as land, skilled labor, unskilled labor, natural resources and initial capital in each region is fixed exogenously within the GTAP model. The main agents in this model are producers, consumers and the government.

Using CGE and GTAP models and simulations including partial and total liberalization to analyze both qualitative and quantitative the impact of free trade cooperation on the formation of ASEAN+5 FTAs (AJFTA, AIFTA, ACFTA, AKFTA and AANZFTA) (Ana et al., 2016). The study showed many positive influences of these FTAs. Namely, free trade cooperation among ASEAN members with Japan, India, China, Korea, Australia and New Zealand in the schema ASEAN+5 FTAs provided bigger benefits in full liberalization scenario for all regions, except Cambodia and Australia-New Zealand. The formation of ASEAN+5 FTA formed a trade creation in the form of less efficient domestic production transfer, which was replaced by more effective import among FTA member countries. Welfare, real GDP, international trade and the investment of all countries joined in ASEAN+5 FTAs witnessed an upward trend. India was a country which experienced the highest increase in welfare, while Vietnam experienced the highest increase in real GDP, international trade and direct investment. In the sectoral analysis, partner countries' balance of trade (Japan, India, China, Korea, Australia and New Zealand) compared to ASEAN member countries balance of trade, the condition was better than the vice versa. The effect of resources usage allocation (land, labor and capital) for ASEAN countries was more focused on the sector of agriculture product, food, textile and some extractive industries and technologies, while for the partner countries was more focused on heavy industry, technology, equipment, construction and services.

Investigating the economic impacts of possible FTAs among China, Japan, and Korea and then compared the GDP and welfare effects of those FTAs by employing a multi-region, multi-sector static CGE model with 6 regions, 12 sectors and 5 endowments, (Yoon et al., 2009) authors assume that land, unskilled labor and natural resource were not mobile among the regions and that skilled labor and capital were mobile among them. Considering four possible FTA scenarios, authors examined the economic effects of the FTAs and then ranked them to show which form of FTA was most beneficial for each country. Their results pointed out that Korea would establish an FTA with China, China would form a trilateral FTA with Korea and Japan and Japan would choose to form a trilateral FTA with China and Korea.

In another study analyzing the economic effects of the ongoing FTA, China–Japan–Korea Trilateral FTA and several other prospective FTAs including Northeast Asia Preferential FTA, Northeast Asia plus the Eurasian Economic Union (EAEU) Preferential FTA and Northeast Asia plus the Regional Comprehensive Economic Partnership (RCEP) plus the EAEU Preferential FTA on the basis of CGE model and GTAP Data Base 9.0a. The results revealed that all parties of the agreements would benefit from the formation of these FTAs, having welfare gains and real GDP expansions regardless of international capital mobility status. Moreover, the results indicated that for the NEA region as a whole, the NEA FTA was preferable to the China–Japan–Korea Trilateral FTA alone and it would be even better off with the formation of wider free trade areas, such as with the other RCEP and EAEU members (Shagdar & Nakajima, 2019).

With a view to estimating the effects of trade agreements on countries linked to global value chains (GVCs) more accurately, (Itakura & Lee, 2019) authors utilized the GTAP database and inter-country input–output tables to construct a global CGE model that disaggregates imports of intermediate products by country of origin. Basing on the CPTPP and the RCEP, which are two mega-regional trade agreements (MRTAs) involving Asian countries, they suggested that while incorporating the GVC structure did not significantly affect the overall welfare results, the magnitudes of changes in sectoral output became considerably greater in several industries.

Similarly, in the case of estimating the impact of different mega-RTAs including EVFTA, TPP, CPTPP (TPP without the US), RCEP and FTAAP as well, but on Vietnam's economy (Kikuchi et al. 2018). Their results showed that EVFTA increased Vietnam's real GDP by 8.1% while TPP increased it by 13.2%. Vietnam was found to be more adversely affected by the US withdrawal from TPP than other economies in the region. CPTPP increased real GDP by 6.5% while RCEP increases it by 9.2%. The economic benefits from RCEP appeared to be relatively small given a lesser degree of liberalization for tariffs and NTMs. The increase in real GDP was the largest at 27.1% for FTAAP1 based on the TPP template while it was 19.4% for FTAAP2 based on the RCEP template. Regarding impacts by sector, EVFTA expanded Vietnam's exports to the EU mainly for light-manufactures, especially leather products. TPP led to remarkable growth in textiles, apparel and leather products and surprisingly, without the US market, CPTPP resulted in much lower export growth for textiles, apparel and leather goods. Meanwhile, RCEP promoted substantial growth in capital-intensive sectors such as chemicals, metals, motor vehicles and transport equipment, electronics, and machinery. And finally, for the FTAAP scenarios, FTAAP1 leads to much higher growth than FTAAP2.

GTAP model uses GE theory helping it to account for economic changes in all sectors. However, on the other hand, the GTAP model faces the same limitations, as follows: (i) it is constrained by the availability of data and a lack of data may severely compromise the scope and relevance of a study and the researcher's ability to model certain trade policies; (ii) it involves many parameters, which may be difficult to estimate and validate and (iii) it contains assumptions or characteristics that may not reflect real-world features.

Contrary to the GTAP model, SMART model based on PE theory only requires data for the trade flows, the trade policy (tariff) and a couple of behavioral parameters (elasticities) which are always available. Another advantage (which follows directly from the minimal data requirement) is that it permits an analysis at a fairly disaggregated (or detailed) level. In general, by virtue of their simplicity, SMART model tends to be more transparent, easy to implement and results can be easily explained as well.

In evaluating the role of the East African community customs unions on trade, total revenue and welfare of Uganda (Othieno & Shinyekwa, 2011). In addition, in another study, the authors incorporated both SMART and gravity model to find out the impact of ASEAN-India FTA on export of plantation products including coffee, tea and pepper in India (Veeramani & Saini, 2011). In other research, using SMART to analyze the potential economic impact of RCEP on Vietnam's automobile sector (Anh & Ngoc, 2011). Another study also utilized SMART model as the framework to evaluate the impact of South Asia FTA (Kumar & Ahmed, 2014). Meanwhile, Llano et al. (2019) developed research and evaluated the impact of US tariffts on the iron and steel industry on different economic sectors; using SMART to evaluate the impact of 10–25% of aluminum, iron and steel tariff import of US on Spanish regions and indicated that US import tariff might induce a total job loss of 185,000 employees worldwide, while in Spain, it may cost 3500 jobs.

Regarding the UKVFTA, there are a limited number of papers assessing its contribution to the Vietnam's economic development. Two reasons explain this situation might be that the UKVFTA's commitments are similar to the ones of EVFTA and the UKVFTA has only taken in effect for nearly two years. In research of Hoi (2021) analyzed the potential opportunities given by the UKVFTA to some of Vietnam's main exports including textiles and garments, seafood, rice, wood, vegetables and fruits and footwear. Meanwhile, Pham et al. (2022) evaluates the influence of the UKVFTA on the sustainable development of Vietnam's wood processing sector and suggested Vietnam's government to provide investment incentives on technology innovation and promotion strategy to enterprises.

Due to no papers measuring the impact of the UKVFTA on Vietnam's imports of seafood, this study focuses on evaluating the UKVFTA's potential impact on Vietnam's seafood industry and some recommendations for Vietnam's domestic seafood to compete with the UK's seafood which are entitled to enjoy tariff elimination.

3 Methodology and Data

3.1 PE

PE is a model that equates supply and demand in one or more markets in order to analyze and assess a particular market under the changes of the policy (increase or decrease of the tariff) or others affecting the demand or supply in a good. PE model often eliminates the impacts of changes in relevant or replaceable sectors and suggests that the sector that analyzed was a small niche in economy, so the changes of the sector could not affect to others. The nature of the model could make its usage and assessment easier. However, the weakness of the model is to miss important interactions and feedbacks between various markets. In particular, the PE approach tends to neglect the important inter-sectoral input/output (or upstream/downstream) linkages that are the basis of GE analyses. It also misses the existing constraints that apply to the various factors of production (labor, capital, land, …) and their movement across sectors.

An imported country J imports a commodity i of an exported nation K. The demand curve would be as below:

$$M = \alpha \text{MPM}\varepsilon$$

αM is a constant (αM > 0) and $\varepsilon < 0$ is elasticity of demand for imports. Similarly, the supply curve would be as below:

$$X = \alpha \text{XPX}\eta$$

αM > 0 is a constant and $\eta > 0$ is elasticity of supply for exports. When the exported nation has a small economy or the import turnover of i is smaller than the total international import turnover, η is the same as positive infinity.

The equal condition of the model requests: $M = X$.

Difference between import and export prices is caused by tariff as a below curve:

$$\text{PM} = \text{PX}\,(1 + \text{T}/100)$$

T is tariff on import (%).

The changes in the gains from exchange to the importing and exporting countries are:

$$\Delta\text{WM} = \int_{\text{PM}_0}^{\text{PM}_1} \alpha_{\text{M}}\text{PM}^{\varepsilon}\,\text{dPM} = \frac{\alpha_{\text{M}}}{\varepsilon + 1}\left(\text{PM}_0^{\varepsilon+1} - \text{PM}_1^{\varepsilon+1}\right)$$

$$\Delta\text{WX} = \int_{\text{PX}_0}^{\text{PX}_1} \alpha_{\text{X}}\text{PX}^{\eta}\,\text{dPX} = \frac{\alpha_{\text{X}}}{\eta + 1}\left(\text{PM}_0^{\eta+1} - \text{PM}_1^{\eta+1}\right)$$

These expressions complete a basic partial equilibrium model that we can use to simulate the economic impact of tariff changes in a single sector.

3.2 SMART Model

SMART is known as a PE model that can be used in assessing the trade, tariff revenue and welfare effects of an FTA. This model and the simulation tools are part of the World Integrated Trade Solution (WITS) database and software suite provided jointly by the World Bank and the United Nations Conference on Trade and Development. The strengths of the model are that it is easily implemented together with the WITS database, it yields important quantitative results on the trade and tariff revenue effects of an FTA and the analysis can be performed at the most disaggregated level of trade data. However, the main limitation of the model is that it is a partial equilibrium model, which means the results of the model are limited to the direct effects of a trade policy change only in one market.

The demand side of the market in SMART is based on the Armington assumption that commodities are differentiated by their country of origin. This assumption implies that, for a particular commodity, imports from one country are an imperfect substitute for imports from another country. Thus, even though an FTA entails preferential trade liberalization, import demand does not completely shift to a source from within the FTA. The SMART model also assumes that consumers' demand is decided in a two-stage optimization process that involves allocating their spending by commodity and by national variety. At the first stage, consumers decide how much to spend on the commodity given changes in the price index of this commodity. The relationship between changes in the price index and the impact on import demand for this commodity is determined by a given import demand elasticity. At the second stage, the chosen level of spending for this commodity is allocated among the different national varieties, depending on the relative price of each variety. The extent of the between-variety response to a change in the relative price is determined by the substitution elasticity.

3.3 Data

SMART model calculates on 3 elasticity such as demand elasticity, supply elasticity and Armington elasticity. The exactness of the model depends on selecting the elasticity for the model. Import elasticity is 1.5 in SMART. The export elasticity is defined as the changes of the export supply when the price changes. As the seafood imports of Vietnam are slight and the seafood market in Vietnam is also subtle compared to other exporters, the assumption of perfectly elastic demand (99) is used in SMART.

The Armington elasticity or the elasticity of substitution is depended on the changes of import demand on commodity that made from many export nations in the changes of prices (Kapuscinski & Warr, 1999). The selection of the Armington elasticity plays a very important role in deciding the exactness of measure in SMART model. This study would use the Armington elasticity of Hertel et al. (2007). Hertel and his partner used the regression analysis to measure the Armington elasticity for some imports and compared between the elasticity and the Armington elasticity recommended by GTAP. According to Hertel and partners, the Armington elasticity for fishing would be 2.5 that is used in the study.

This paper adopts the HS (Harmonized System) classification and assesses the impact of the UKVFTA on Vietnam's imports of fish and fishery products namely HS03 (fish and crustaceans, molluscs and other aquatic invertebrates). In 2020, the assumption was established so the data of bilateral trade turnover between the Vietnam and partners was in that year. The data of import turnover of seafood HS03 of Vietnam from the UK could be extracted from data of UNCTADTRAINS by SMART model.

3.4 Scenarios

Two scenarios are constructed based on Vietnam's seafood-related commitments under the UKVFTA as well as the current pace of Vietnam's integration in this sector with ASEAN+5 including China, Japan, South Korea, India, Australia—New Zealand.

- Scenario 1: Vietnam eliminates tariff on seafood imported from the UK without taking into consideration Vietnam's other FTAs.
- Scenario 2: This scenario includes FTAs of ASEAN+5 in simulation, in which Vietnam eliminates tariffs for fish and fishery products imported from both the UK, China, Japan, South Korea, India, Australia—New Zealand.

With ASEAN+5, Vietnam signed many FTAs as a member state of ASEAN including ASEAN—China FTA (ACFTA), ASEAN—Japan Comprehensive Economic Partnership (AJCEP), ASEAN—Korea FTA (AKFTA), ASEAN—India FTA (AIFTA) and ASEAN—Australia—New Zealand FTA (AANZFTA). Vietnam also has two FTAs with Japan (Vietnam—Japan Economic Partnership Agreement—VJEPA) and South Korea (Vietnam—Korea FTA—VKFTA) individually. In these FTAs, Vietnam commits to eliminate seafood import tariffs but some types of seafood which are imported from the India are still subject to a certain level of tariff rates. Therefore, this scenario aims at evaluating how ASEAN+5 will keep up with the pace of liberalization in the UKVFTA.

4 Results

On the basis of SMART model and the assumed scenarios, results on Vietnam's imports of fish and fishery products from the UK are given as follows:

Under the enactment of the UKVFTA, there will be a surge of USD 13,628.045 in seafood imports of Vietnam from the UK, corresponding to a more than 100% increase in total. Overall, Vietnam's import value of fish and fishery products from the UK will reach approximately USD27.1 million. The changes in import duties on seafood between Vietnam and the UK are captured, namely, fish and fishery products under HS03 are open 11.65% of tariff on average (SMART model) and then entitled to tariff elimination to 0%, leading to the total trade effect of more than USD13.5 million. As Vietnam is the importer, the total trade effect is the sum of trade creation and trade diversion effect which are nearly USD10.3 million and USD3.4 million, respectively (Table 1).

In comparison with Scenario 1, there will be a decline of USD 1,067,287 in the import value of Vietnam in Scenario 2, which directly corresponds to the decrease in the trade diversion effect. In scenario 2, the imports will grow at a lower rate of 93.3% because when Vietnam also removes tariffs for ASEAN+5, the seafood prices of ASEAN+5 relative to that of the UK would be lower, making Vietnam transfer a part of its imports from the UK to the ASEAN+5.

Overall, there will be some reductions in the investigated indexes in Scenario 2 compared to Scenario 1. However, these reductions are insignificant, which are only nearly 4% meaning that Vietnam's participants in FTA with ASEAN+5 will not much affect Vietnam's seafood import growth from the UK.

Table 2 shows that Vietnam does not import as many fish and fishery products from the UK as many as it does from ASEAN+5. However, the UK is the only country getting huge benefits in both Scenario 1 and Scenario 2 which are a more than 100% and 93% soar in seafood exports, respectively. Meanwhile, ASEAN+5 will face a fall in seafood export value, from −0.04 to −0.6% (Scenario 1). Even when Vietnam's FTAs with both the UK and ASEAN+5 are taken into account (Scenario 2), China, Japan, South Korea, Australia—New Zealand will still experience a drop in exports

Table 1 Change in trade indicators of seafood imports of Vietnam from the UK in two scenarios

Indicators	Scenario 1	Scenario 2
Initial seafood import value (in 1000 USD)	13,466.926	13,466.926
Final seafood import value (in 1000 USD)	27,094.971	26,027.684
Total import value change (in 1000 USD)	13,628.045	12,560.758
Total export value change (%)	101.2	93.3
Trade creation effect (in 1000 USD)	10,245.223	10,245.223
Trade diversion effect (in 1000 USD)	3,382.822	2,315.535
Total trade effect (in 1000 USD)	13,628.045	12,560.758

Source The author's calculation in SMART model

Table 2 Change in import value of Vietnam from 7 specified trading partners

Country	Scenario 1			Scenario 2		
	Initial import value (in 1000 USD)	Total import change (in 1000 USD)	Total import change (%)	Initial import value (in 1000 USD)	Total import change (in 1000 USD)	Total import change (%)
The UK	13,466.926	13,628.045	101.2	13,466.926	12,560.758	93.3
China	110,666.369	−94.838	−0.1	110,666.369	−211.706	−0.2
India	195,917.384	−1,146.487	−0.6	195,917.384	73,643.048	37.6
Japan	154,726.418	−383.912	−0.2	154,726.418	−700.059	−0.5
South Korea	50,036.401	−19.699	−0.04	50,036.401	−177.877	−0.4
Australia	41,017.984	−79.365	−0.2	41,017.984	−1,899.465	−4.6
New Zealand	19,046.796	−20.568	−0.1	19,046.796	−67.366	−0.4

Source The author's calculation in SMART model

to Vietnam and only India's fish and fishery exports will ascend by more than 37% (Table 2).

Actually, seafood exports of China, Japan, South Korea, Australia—New Zealand are already entitled to enjoy tariff elimination thanks to the FTAs with Vietnam meaning that they are open to 0% of import duty in both two scenarios (SMART model). India, by contrast, are suffering from 8.86% of tariff for seafood products under HS03 on average (SMART model). That is the reason why when it comes to FTAs of Vietnam with all mentioned-above partners, Vietnam will shift seafood imports from the UK and ASEAN+4 to India. Overall, in Scenario 2, regarding absolute change in export value of fish and fish products to Vietnam, India will be superior to the UK. However, in terms of relative change, UK's exports will far exceed the one of India (Table 2).

The above results show that Vietnam's imports of different types of seafood will positively be influenced by the UKVFTA in both two scenarios as they will experience a considerable increase in value. It is because Vietnam's seafood imports from the UK suffer from quite high rate of tariff. Namely, seafood under HS030614 faces the lowest tariff which is 1.5% but the remaining types are open to higher one ranging from 10 to 15%. Then reduction of tariff to 0% will result in a climb in their import turnover (Table 3).

Besides, Vietnam should take much more attention to the changes in imports of products under HS030354 as they will enjoy the highest rise in absolute import value (more than USD 8 million). They will also make up for more than 60% in the total change of Vietnam's imports from the UK. Moreover, when Vietnam's FTAs with ASEAN+5 are taken into consideration, Vietnam's imports of seafood under HS030354 will not change, making the UK still the second largest exporter to Vietnam (after Japan) (Table 3).

Table 3 The impact of the UKVFTA on import value of seafood under 6-digit HS codes from the UK to Vietnam

Product (HS code)	Initial import value (in 1000 USD)	Scenario 1		Scenario 2	
		Import value change (in 1000 USD)	Proportion in total change (%)	Import value change (in 1000 USD)	Proportion in total change (%)
030313	618.638	173.626	1.3	173.626	1.4
030349	30.712	163.148	1.2	163.148	1.3
030354	2,772.812	8,300.002	60.9	8,300.002	66.1
030355	388.496	111.771	0.8	111.771	0.9
030369	177.738	51.485	0.4	51.263	0.4
030481	1,430.873	764.638	5.6	764.638	6.1
030499	645.008	247.878	1.8	241.538	1.9
030541	72.423	34.246	0.3	34.246	0.3
030572	52.272	68.82	0.5	68.82	0.5
030579	111.029	209.743	1.5	209.743	1.7
030614	506.101	23.802	0.2	23.778	0.2
030617	6,660.823	3,478.886	25.5	2,418.185	19.1

Source The author's calculation in SMART model

The second biggest product in terms of absolute import value increase will be HS030617 accounting for 25.5% of total additional seafood imports into Vietnam from the UK in Scenario 1 and over 19% in Scenario 2 (Table 3). In Scenario 2, the integration of Vietnam with ASEAN+5 will lead to a significant fall in Vietnam's imports of this type of product from the UK which will be replaced by imports from India (Table 2).

With regard to relative change, HS030349 will see a sharp rise by over five times compared to the import turnover before the UKVFTA has come into effect, followed by HS030354 (nearly three times) and HS030579 (nearly twofold) (Table 3).

Overall, HS030354 will benefit the most in terms of both absolute and relative change in import turnover in both scenarios.

As above-mentioned, the total trade effect of the UKVFTA on Vietnam's imports of fish and fishery products from the UK consists of trade creation and trade diversion effect. Trade creation means that Vietnam's domestic production will be replaced by more efficient imports from the UK, leading to an increase in economic benefits of both Vietnam and the UK. Namely, the UK will enjoy specialization in production based on its comparative advantage, meanwhile Vietnam's domestic customers will be able to buy cheaper seafood and Vietnam's domestic producers will compete against foreign rivals resulting in their production efficiency improvement. Trade diversion, by contrast, means that Vietnam will import seafood from the UK due to elimination of import duty leading to a lower relative price of seafood from the UK compared to one from rest of the world. This will cause lower welfare as more efficient production of rest of the world will be replaced by the UK's one in Vietnam's market.

According to the SMART model, the trade creation effect will be bigger than the trade diversion effect in both two scenarios meaning that the UKVFTA will increase the welfare of Vietnam. In scenario 1, the trade creation effect will account for 75.2% of the total trade effect. In Scenario 2, with the combination of both Vietnam's FTAs with the UK and ASEAN+5, the trade creation effect will make up for a higher proportion which will be 81.6% due to a decreasing of the trade diversion effect. This also implies that the UKVFTA's impact on Vietnam's seafood imports will be affected by ASEAN+5 in a moderate manner (Table 4).

When Vietnam only eliminates import duty on fish and fishery products from the UK in Scenario 1, Vietnam will shift seafood imports from ASEAN+5 to the UK leading to big losses of India and Japan. However, when tariff reduction to 0% is applied on both seafood imported from the UK and ASEAN+5 in Scenario 2, India will get the biggest advantage, followed by the UK. Meanwhile, China, Japan, South Korea, Australia and New Zealand will even suffer from bigger losses compared to Scenario 1.

5 Conclusion and Recommendations

By adopting the SMART model, this paper analyzes the impact of tariff elimination under the UKVFTA on Vietnam's imports of seafood from the UK in two scenarios which are constructed based on Vietnam's commitments on tariff elimination under the UKVFTA and the integration of Vietnam with ASEAN+5 in seafood sector. In Scenario 1, Vietnam eliminates tariffs only for fish and fishery products imported from the UK while Scenario 2 enlarges the scope of tariff reduction to also include ASEAN+5. The results reveal that Vietnam's seafood imports from the UK will significantly increase in both scenarios, implying that the UK will still be among important seafood sources for the Vietnam's market in the future. However, Vietnam's FTAs with ASEAN+5 will result in a reduction in the Vietnam's import turnover of fish and fishery products from the UK. In Scenario 1, Vietnam's import value of seafood from the UK will shoot up by more than 100%, equivalent to USD 13,628,045. Meanwhile Scenario 2 will experience a fall to 93.3% and USD 12,560,758, respectively (Table 2).

Moreover, an uneven distribution in Vietnam's seafood imports from the UK will occur when the UKVFTA has come into effect. Most of absolute import value increases will focus on HS030354 and HS030617. Meanwhile, the most remarkable relative change will be registered in HS030349 and HS030354. Overall, HS030354 will benefit the most in terms of both absolute and relative change in import turnover in both scenarios (Table 2).

Besides, the UKVFTA will result in a considerable trade creation effect which will be higher than the trade diversion effect in both two scenarios. Namely, trade creation effect will account for 75.2% of the total trade effect in Scenario 1. This

Table 4 Trade creation and trade diversion effect of the UKVFTA

Country	Scenario 1				Scenario 2			
	Trade creation (in 1000 USD)	Proportion of trade creation (%) in total trade effect	Trade diversion (in 1000 USD)	Proportion of trade diversion (%) in total trade effect	Trade creation (in 1000 USD)	Proportion of trade creation (%) in total trade effect	Trade diversion (in 1000 USD)	Proportion of trade diversion (%) in total trade effect
The UK	10,245.223	75.2	3,382.822	24.8	10,245.223	81.6	2,315.535	18.4
China	0	0	−94.838	100	0	0	−211.706	100
India	0	0	−1,146.487	100	60,946.853	82.8	12,696.196	17.2
Japan	0	0	−383.912	100	0	0	−700.059	100
South Korea	0	0	−19.699	100	0	0	−177.877	100
Australia	0	0	−79.365	100	0	0	−1,899.47	100
New Zealand	0	0	−20.568	100	0	0	−67.366	100

Source The author's calculation in SMART model

figure will go up to 81.6% in Scenario 2, meaning that the UKVFTA will bring about an increasing welfare for Vietnam and Vietnam's integration with both the UK and ASEAN+5 will even lead to a higher social welfare.

In addition to some foregoing main points drawn from the SMART model, the paper would like to introduce some recommendations as follows:

Firstly, when it comes to the combination of UKVFTA and Vietnam's FTAs with ASEAN+5, India will cause Vietnam's seafood imports from the UK to reduce (in Scenario 2). Vietnam will shift imports of fish and fishery products from the UK and ASEAN+4 to India. Therefore, with a view to promoting the UKVFTA and trade relation with the UK, Vietnam should keep its commitments in existing FTAs with ASEAN+5, especially India in seafood sector (Table 2).

Secondly, since the effective day of the UKVFTA, Vietnam's domestic seafood enterprises has faced more fierce competition not only from ASEAN+5 but also the UK's enterprises. Especially, Vietnam's domestic seafood enterprises focusing on fish and fishery products under HS030354, HS030617 and HS030349 will be competed the most as these types of products will see a surge in both absolute and relative import change. Therefore, Vietnam's businesses should increase their quantity and quality of seafood to serve the domestic market better. The Vietnam's Government should support domestic enterprises to invest in technology innovation, distribution system building, brand building and development to catch up with the UK's enterprises.

Thirdly, with a population of nearly 100 million people, Vietnam's domestic market indeed has huge potential to promote the consumption of aquatic products but has not been fully and effectively exploited by Vietnam's domestic enterprises. It is because seafood export has some advantages over domestic consumption, namely: (i) selling products to domestic supermarkets, enterprises do not get paid immediately but in about a month, meanwhile the payment transactions in seafood export can be completed very quickly and conveniently; (ii) the preservation of aquatic products which are almost frozen in domestic supermarkets matters as it can negatively affect quality of products and reputation of enterprises and (iii) because of domestic supermarkets' promotions, seafood enterprises find it difficult to determine their selling price which can greatly affect their revenue, meanwhile seafood export activities are transparent on the basis of quality assurance, food safety and hygiene and disease safety. Due to the above-mentioned reasons, seafood enterprises often take advantages of their all resources to promote export activities and penetrate foreign markets rather than focus on domestic one.

However, it is clear that if enterprises are capable of well exploiting the domestic market, their pressure on export activity will reduce, especially in case of great volatility in the world market due to the Covid-19 epidemic.

With a view to encouraging seafood enterprises' participation in development of the domestic market, the Vietnam's Government should promulgate appropriate policies to solve the aforesaid barriers including counterfeit or poor-quality products, payment transaction time, aquatic product preservation or sales and promotions of supermarkets, …

Regarding seafood manufacturers, in addition to building their core competitiveness on the basis of investment in technological innovation to produce quality and highly differentiated products and meet domestic customers' needs as mentioned above, they should invest in market building and expanding and developing a strong distribution network as well.

It is also important to strengthen the connection between seafood manufacturers, suppliers and retailers in information sharing, strategic planning, brand building and development and investment cooperation.

References

Ana, S., & et al. (2016). ASEAN free trade agreements: Global trade analysis project (GTAP) model. IJER *Serials Publications, 13*(3), 1141–1168.

Anh, T. T., & Ngoc, L. M. (2011). An assessment of the potential economic impacts of RCEP on Vietnam automobile sector1. The World, 2013.

Chandran. (2018). Trade impact of the India-ASEAN free trade agreement (FTA): An augmented gravity model analysis. Retrieved June, 2022, from https://doi.org/10.18488/journal.aefr.2018.81.103.116

Hertel, T. W. (1997). Global Trade Analysis: Modeling and Applications, Cambridge University Press.

Hertel, T., Hummels, D., Ivanic, M., & Keeney, R. (2007). How confident can we be of CGE-based assessments of free trade agreements? *Economic Modelling, 24*(4), 611–635.

Hoi H. V. (2021). Opportunities and challenges for Vietnam's merchandise exporting following Vietnam-UK free trade agreement. *VNU Journal of Science: Economics And Business, 37*(2).

Irshad, M. S., Xin, Q., Shahriar, S., & Arshad, H. (2017). A panel data analysis of Chinas trade pattern with OPEC members: Gravity model approach. *Asian Economic and Financial Review, 8*(1), 103–116.

Itakura, K., & Lee, H. (2019). Estimating the effects of the CPTPP and RCEP in a general equilibrium framework with global value chains.

ITC Trade Map. (2020). Retrieved from June 2022, https://www.trademap.org/

Kapuscinski, C. A., & Warr, P. G. (1999). Estimation of Armington elasticities: An application to the Philippines. *Economic Modelling, 16*(2), 257–278.

Kikuchi, T., Yanagida, K., & Vo, H. (2018). The effects of mega-regional trade agreements on Vietnam. *Journal of Asian Economics, 55*, 4–19.

Kumar, S., & Ahmed, S. (2014). Impact of sensitive lists under SAFTA: Quantitative assessment using a partial equilibrium modeling. *European Journal of Globalization and Development Research, 10*(1).

Lateef, T., & Riaz. (2018). Exploring the gravity of agricultural trade in China–Pakistan Free Trade Agreement. *The Chinese Economy, 51*(6), 522–533.

Llano, C., et al. (2019). Global and regional effects of the US tariffs on iron, steel and aluminium: A SMART combination of models with a focus on Spain. *Regional Science Policy and Practice, 11*(3), 525–547.

Nguyen. A. T., & Pham. T., T. (2022). Chuỗi giá trị và thương mại ngành hàng nông lâm sản Việt Nam: Trong bối cảnh hội nhập quốc tế và khủng hoảng sinh thái. 291–304.

Othieno, L., & Shinyekwa, I. (2011). Trade, revenue and welfare effects of the East African Community Customs Union Principle of Asymmetry on Uganda: An application of Wits-Smart simulation model.

Pham, T. T., et al. (2022). Tác động của UKVFTA đối v´ơi sự phát triển bền v˜ưng của ngành công nghiệp chế biến gỗ Việt Nam.

Sahar, H. K. (2019). A gravity model analysis for trade between the GCC and developed countries. *Cogent Economics and Finance, 7*(1).

Shagdar, E., & Nakajima, T. (2019). Economic Effects of Free Trade Agreements in Northeast Asia: CGE Analysis with the GTAP 9.0 a Data Base.

Sunder, R., et al. (2020). Evaluating Asian free trade agreements: What does gravity model tell us? *SAGE Journals, 56*(1).

VASEP. (2020). https://vasep.com.vn/gioi-thieu/tong-quan-nganh

Veeramani, C., & Saini, G. K. (2011). Impact of ASEAN-India FTA in the plantation sector: A quantitative assessment. *Economic and Political Weekly, XLVI, 10*, 83–92.

World Bank. (2019). https://data.worldbank.org/indicator/ST.INT.ARVL?locations=VN, Vietnam's foreign arrivals.

World Bank. (2021). https://data.worldbank.org/indicator/NY.GDP.PCAP.CD?locations=VN, Vietnam's GDP per capita 2015–2021.

Yoon, Y. M., et al. (2009). A CGE analysis of free trade agreements among China, Japan, and Korea. *Journal of Korea Trade, 13*, 45–64.

Post-Covid-19 and the Change in Income of Employees

Thanh Thuy Cu and Thi Yen Le

Abstract The purpose of this study was to analyze the change in the income of labor in the post-Covid-19 situation. The difference in income is approaching when comparing the two periods: the outbreak period and the period after the epidemic is gradually controlled (based on the field of work and job position undertaken). Data for the study was gathered through a survey of 477 employees working in the aviation and tourism/hospitality sectors in Vietnam. In the study, the multivariate regression method and tests for differences were applied. The employee's income is determined by several factors, such as education level, gender, job position, the field of employment, and personal investment. However, when the context changes, the degree of education and employment position have a significant influence on employees' incomes (The coefficient of management position variable is 1.484136; the coefficient of the graduate education variable is 4.181241). At the same time, the research findings suggest that the income gap between the two groups of workers with management positions and those without manager positions between the two periods of the pandemic is approximately 9.367 million VND per month. The research findings serve as the foundation for developing solutions to boost the adaptability of workers in the new situation.

Keywords Employee · Jobs · Income · Aviation · Tourism/hospitality · Covid-19

1 Introduction

In the process of conducting a survey with research subjects who are workers in the service sector about their experiences during the disease outbreak and the period following Vietnam's gradual adaptation to the "New Context," the economy gradually recovered and safely adapted to the new context, and workers reflected: (1) Workers have never experienced such a difficult period in their lives; when the epidemic broke out, the income of workers in the tourism and aviation sectors dropped by

T. T. Cu · T. Y. Le (✉)
Hanoi Architectural University, Hanoi, Vietnam
e-mail: Lethiyenktdt@gmail.com

A. T. Nguyen et al. (eds.), *Contemporary Economic Issues in Asian Countries: Proceeding of CEIAC 2022, Volume 1*,
https://doi.org/10.1007/978-981-19-9669-6_6

approximately 70–90%, and some workers are temporarily unemployed due to the employer's inability to pay wages during the disease outbreak. (2) After March 2022, when the Vietnamese government decided to change its policy in response to the Covid-19 pandemic, all production and business activities recovered, workers' incomes gradually stabilized, and life is less difficult than it was during the Covid-19 pandemic. Workers' incomes had recovered to about 70% of normal levels at the time of the data survey. Workers' opinions have revealed the significant impact of the Covid-19 epidemic on them, and once the epidemic is gradually controlled, workers will face new challenges and will need to safely adapt to ensure the best possible life.

For Supply Chain Operations, if just one link in the supply chain becomes inefficient or disrupted, the entire supply chain suffers. Indeed, when employers face difficulties during an epidemic outbreak, it directly affects workers, who lose their jobs and income, affecting their quality of life. Studies have shown that the negative effects of the Covid-19 outbreak have caused difficulties for actors in the economy (ILO, 2020; World Bank, 2020). In the study of Shen et al. (2020); Pattiruhu & Paais (2020); Obrenovic et al (2020) pointed out the negative impacts of the epidemic on enterprises and Vietnam is no exception, according to the statistical results of GSO (2021a, b), when surveying 126,565 firms on the effects of Covid-19, about 85.7% of enterprises surveyed answered that they were affected by the Covid-19 epidemic; those effects are expressed through the shutdown of activities due to the effects of social distancing, the impact of halting the exchange of goods, limiting international trade activities, etc. Among those surveyed enterprises, the enterprises operating in the field of aviation, tourism, and hotel are the most impacted of the surveyed enterprises, with influence rates of 100% and 97.1%, respectively.

When businesses are affected by the epidemic, workers are also directly affected; workers will experience temporary unemployment, and reduced income, and life will become much more difficult as a result. According to statistical results from GSO (2022a, b), Vietnam had more than 16.9 million people of working age affected by the Covid-19 epidemic in the first quarter of 2022; however, this number has decreased by approximately 7.8 million people compared to the fourth quarter of 2021. There 0.9 million people lost their jobs out of the 16.9 million workers who were negatively impacted by the pandemic; approximately 5.1 million workers lost their jobs temporarily; about 5.7 million workers had their working hours reduced, and roughly 13.7 million workers lost their income. That can be seen as the negative effects that Covid-19 has on workers, and it is also a shock that causes damage to workers.

People's lives in Vietnam have gradually returned to normal since March 2022, when the Vietnamese government decided to adjust policy in response to the Covid-19 epidemic; however, the pressure on workers remains extremely high. According to the actual situation of the employers, some workers were unable to find new jobs, and others were forced to cut unwanted working hours. This forces workers to adapt safely to maintain life stability.

This study will be carried out to examine the change in workers' income in the new environment when the Covid-19 epidemic is under control, as well as safe opening policies, which will be adopted in Vietnam in the current context in comparison to the previous one. The research will focus on the sectors most affected by the Covid-19 epidemic, which include the tourism, hotel, and aviation sectors to study this change. The change in income will show the ability to respond to the new context of workers. These research findings will serve as the foundation for developing new livelihood strategies for employees in the new situation.

2 Research Overview

According to the DFID research framework (1999), the income of workers is affected by five groups of resources (human resources, physical resources, social resources, financial resources, and natural resources), as well as changing external factors (epidemic, natural disasters, policies). Previous research has also used this sustainable livelihoods analysis framework as a foundation for conducting studies to quantify the impact of factors on people's/employees' incomes when there are external influences.

The group of studies on the effects of changing urbanization and industrialization contexts has been studied by Tran (2013); Tran and Vu (2014); Jansen et al. (2006), these studies have estimated the change in people's income and employment when the context changes. The data used in these studies is gathered from people who are directly affected when the context changes. A multivariate regression model is used to help studies quantify the relationship between factors. According to the findings of these studies, people have income vulnerabilities when the context changes; these damages are caused by job loss and loss of means of production. At the same time, studies have shown that the inability to adapt to the new context has caused vulnerabilities to the income and employment of the research subjects (Cu & Nguyen, 2021; Nguyen & Bui, 2011).

Meanwhile, using the framework of sustainable livelihood analysis and data collected from research subjects, studies focus on calculating adaptive capacity and livelihood vulnerability through indicators VI and ACI to assess people's adaptive capacity when the new context changes (Adger et al., 2001; Nelson et al, 2010; Rohan et al., 2005). These studies use survey data to calculate people's adaptive capacity in relation to each resource group, which includes physical resources, natural resources, human resources, social resources, and financial resources. When calculating the adaptive capacity of each resource, the studies will take into account the general adaptive capacity of the survey participants as well as the vulnerability that they face when the context changes (Cu et al., 2020; Tu et al., 2015). The findings of the research are consistent with the view that people who can respond well and make good use of the new changing context will adapt better, and their lives will be better. However, if they do not adjust to the new environment, life will become difficult, and people will face income and employment issues, lowering their quality of life (Tu et al, 2015; Vo & Nguyen, 2012).

When the Covid-19 pandemic broke out, studies seemed to support the view that the outbreak of the disease as an external shock affects people in general and workers in particular when considering aspects of the epidemic's impact on income and employment (Qian & Fan, 2020; Bezerra et al., 2020). The information was gathered through a survey of people affected by the Covid-19 pandemic in China. According to research findings, the negative impact of the epidemic has damaged people through estimated variables that have a negative relationship with people's income, such as unemployment and epidemics. Furthermore, studies support the view that, during an outbreak, education levels, and gender are also important factors determining employee income when directly affecting employees' ability to maintain jobs (Buheji et al, 2020; Paul et al, 2021; Suratman et al., 2021). The research findings are completely consistent with the studies of Dao et al. (2022); Nguyen and Tran (2021), which argue that workers face significant challenges in terms of employment when the epidemic breaks out, thereby changing workers' income. At the same time, studies have shown that the working sector has a significant impact on workers' employment and income. People in the public sector appear to be more stable and less affected by the Covid-19 pandemic than those in the private sector (Le et al., 2021; Qian & Fan, 2020).

In short, shocks have caused difficulties for workers in particular and people in general; however, when the new context changes, changing methods and behaviors to adapt to new conditions is also a requirement for workers. Thus, it is necessary to execute research on the new context (the period of adaptation to the epidemic), and changes in workers' income are required; from there, it can be observed that workers' adaptability after the Covid-19 pandemic is controlled and nations are progressively adjusting and "living together" with the epidemic is needed.

3 Research Methods

The survey on the impact of Covid-19 on employees of GSO (2021a, b) contains questions directly related to the research team's problem, so, inheriting the survey form of GSO (2021a, b), the research team designed a survey form to assess the effects of Covid on income and employment of workers. Based on those assessments, it will examine the change in workers' income between the two periods when the epidemic broke out and the post-Covid-19 periods.

The design survey form is divided into two sections: (1) The first piece of content is the demographic questions asked of the survey subjects: (2) The second content is the research questions about the change in worker income and employment. This content will be divided into two phases: the period reflecting worker income and employment when Covid-19 breaks out, selected before March 2022, and the period after March 2022, when the Vietnamese government adjusts its policies to adapt to the Covid-19 epidemic in the context of the "New Normal".

Selected survey subjects: Workers working in the tourism and aviation sectors were chosen as survey subjects to collect data for the study both during the outbreak and after the epidemic was under control. The goal of such a survey is to ensure that changes in income and employment of the same subject can be compared across two different periods of the Covid-19 epidemic.

For employees working in the aviation sector: Research and survey for employees at Vietjet Air, a major airline in Vietnam (This is a group of workers working outside the State sector) and employees of Vietnam Airlines (These are employees working in the State sector).

To ensure a balanced distribution across the country's regions for those working in the tourism and hotel sectors, surveys were administered to workers at popular tourist destinations in Vietnam. Survey locations include: Ha Noi capital; Quy Nhon city—Binh Dinh province; and Ho Chi Minh city. These are all appealing destinations for both domestic and international tourists visiting Vietnam. Employees working at hotels and tourism service providers in the three survey locations were surveyed.

In terms of the survey sample size: The study was carried out and delivered 600 survey questionnaires to two groups of employees working in the aviation and tourism and hotel sectors. Following the survey, the study team made the following observations: For the group of workers working in the aviation sector: 257 survey responses were collected. For the group of workers in the field of tourism and hotel, 220 survey questionnaires were collected. As a result, the total number of survey questionnaires obtained from the study is 477, which corresponds to 477 observations for statistical operations. According to Nguyen (2014), the minimal observation size for statistical procedures is 100. As a result, with 477 observations acquired by the study team, they are qualified to execute statistical operations.

Survey time: The research team carried out the survey from April to May 2022.

To analyze the change in employees' income with the new context, the study selects the scales for analysis. The scales are described as follows as Table 1.

The study used a multivariate regression model to examine the impact of several variables on employee earnings. The estimation is based on data collected post-Covid-19 because this is the time when workers are adjusting to their new surroundings. Additionally, the research makes use of the information it has gathered to examine the differences in employment and income between the two groups of workers to assess the vulnerabilities and adaptability of workers in the context of change.

4 Results

Following an analysis of the income disparity between the two sectors of aviation and tourism/hotels, the findings of a study of the income disparities between these two sectors are as follows as Table 2.

The study's findings demonstrate that the average wage of workers in the two fields of tourism/hotels and aviation differs significantly. Comparison by research phase: When the Covid-19 epidemic came out, worker pay dropped dramatically, especially for those in the aviation sector. This is reasonable when, due to the spread

Table 1 Research variables

Items	Explain	Basis of selection
Dependent variable		
Employee's income (Y)	This is the variable that represents the employee's earnings. Income is compared between the two times when Covid-19 broke out and post-Covid-19. Income is calculated according to the amount that the employee actually receives each month, the unit of calculation is: million dong	Tran and Vu (2014), Cu and Nguyen (2021), Dao et al (2022), and Qian and Fan (2020)
Independent variables		
Employee's gender (Gend)	This is a variable that represents the gender of the surveyed person. The gender is encoded. If the person is male, the code will be 0, and if the person is female, the code will be 1	Qian and Fan (2020), Bezerra et al. (2020), Siegel (2005), and Huynh and Mai (2011)
Education level (Edu)	The level of education indicates the employee's trained capability. Employees who complete high school will be coded as 1, college completion will be coded as 2, have a university degree coded as 3, and a graduate level will be coded as 4	Tran and Vu (2014), Qian and Fan (2020), and Le et al. (2021)
Employee's work experience (Exper)	Calculated by years of work of employees in the field of work	Dao et al. (2022), Nguyen and Bui (2011), Le et al (2021), Burki and Abbas (1991), Bhatti (2013), and Vernon (2002)
Working position (Posit)	This is the variable used to consider the effect of the employee's job position, if the employee is the manager, it will be coded as 1, and if it is an employee, it will be encoded as 0	Brunello et al. (2000), and Campos and Jolliffe (2002)

(continued)

Table 1 (continued)

Items	Explain	Basis of selection
Working area (Work)	The research chose two fields, aviation, and tourism/hotel, to conduct surveys and check the adaptation of workers in these two areas. Aviation workers are coded as 1, and the field of tourism and hotel coded is 0	Qian and Fan (2020), Bezerra et al. (2020), and Dao et al. (2022)
Personal investment (Invest)	This is the scale used to determine the worth of individual employee investments in order to produce additional revenue for employees. Unit is million dong	Qian and Fan (2020), Dao et al. (2022), and Bezerra et al. (2020)
Cost of living (Cost)	This is a scale to measure the monthly living expenses of employees, calculated in millions of VND/month	Qian and Fan (2020) and Bezerra et al. (2020)

Source Author's summary

Table 2 Descriptive statistics of employees' incomes by field of work

Work		N	Mean	Std. deviation	Std. error mean
Y_Post	Avi	257	19.71	8.16	0.51
	Tou	220	15.42	7.00	0.47
Y_Covid	Avi	257	3.78	2.11	0.13
	Tou	220	6.26	4.14	0.28
Y_Post–Y_Covid	Avi	257	15.92	6.21	0.39
	Tou	220	9.17	3.66	0.25

Source Author's data processing results

of the disease, all flights are halted and workers will fall into forced unemployment at the employer's request. During the epidemic, the monthly average pay for workers in the aviation sector was around 3.78 million VND. Meanwhile, the employee group working in the tourist and hotel industries earns more than the aviation sector, earning an average of 6.26 million VND per month. In reality, during an outbreak of the disease, there are times when the pandemic is under control, and tourism activities and domestic hotel services may still take place; therefore, the income for this group of workers is still higher than that of the aviation industry.

When the epidemic is gradually controlled, policies to deal with the Covid-19 pandemic have changed, employees' income has increased, and the salary of the group of workers working in the aviation sector has increased by about 70% compared to the period prior to the outbreak of the Covid-19 epidemic, some workers even have a stable income and receive additional non-wage income when the aviation sector reopens international routes and passenger demand increases. The average monthly income is expected to be 19.71 million VND per month. Meanwhile, the group of tourist and hotel employees earned around 15.42 million VND per month (Table 3).

The results of examining the income difference between the two groups of workers show that during the outbreak of the Covid-19 epidemic, the difference in the average income of survey respondents for two groups of workers working in the aviation and tourism sectors is estimated to be around 2.47 million VND. However, after the epidemic is under control, the operation of aviation is gradually restored, and the rise in the income of the labor group in the aviation sector is larger, with the average income of the labor group in the aviation sector being roughly 4.28 million dong/month more than the labor group in the tourist and hotel sectors. Meanwhile, the income difference of the two groups during the outbreak of the Covid-19 epidemic and when the epidemic is under control is estimated at about VND 6.76 million.

When workers working in different occupations have differences in income, there are also various differences for workers in different employment positions. As follows as Table 4.

The survey results show that the average value of workers' income after the epidemic is under control is much higher than at the time of the outbreak. As follows:

In terms of the income of employees in management positions, following stabilization, the average manager's salary is about 24.722 million VND per month; however, while during the outbreak of the epidemic, the income of this group of workers is only around 7.826 million VND per month. Moreover, when the epidemic broke out, the income of the group of workers holding staff positions was only about 3.938 million VND per month; however, after the epidemic stabilized, the economy recovered, and production and business activities returned to normal, the income of this group increased significantly, averaging about VND 15.356 million per month.

Examine the difference in income between the two groups of workers in management positions and staff positions (Table 5).

The difference test findings suggest that while the pandemic is controlled, the salary disparity between workers in various occupations is significantly larger. During the epidemic's peak, the income disparity between the two groups was around 3.888 million VND per month. Meanwhile, after the outbreak is under control, the difference between these two groups is expected to be over 9.367 million VND per month.

Reduced working hours, wage cutbacks, and job cuts from businesses owing to financial difficulties in the context of the pandemic all contribute to the shift in workers' income. According to the survey results, the number of employees affected by employment in the context of the pandemic outbreak is as follows as Table 6.

According to research findings, workers have difficulty lowering working hours (approximately 46.75% of the total number of employees); workers have problems with temporary leave and receiving a basic salary (accounting for 31.87%), and workers take unpaid leave. These difficulties have a direct impact on the stability of employees' lives.

Table 3 Independent samples test

		Levene's test for equality of variances		t-test for equality of means						
		F	Sig	T	Df	Sig. (2-tailed)	Mean difference	Std. error difference	95% Confidence interval of the difference	
									Lower	Upper
Y_Post	Equal variances assumed	8.00	0.00	6.10	475.00	0.00	4.28	0.70	2.90	5.66
	Equal variances not assumed			6.17	475.00	0.00	4.28	0.69	2.92	5.65
Y_Covid	Equal variances assumed	86.98	0.00	−8.39	475.00	0.00	−2.47	0.29	−3.05	−1.90
	Equal variances not assumed			−8.01	314.08	0.00	−2.47	0.31	−3.08	−1.87
Y_Post–Y_Covid	Equal variances assumed	107.07	0.00	14.17	475.00	0.00	6.76	0.48	5.82	7.69
	Equal variances not assumed			14.71	424.59	0.00	6.76	0.46	5.85	7.66

Source Author's data processing results

Table 4 Income of employees by work position

Position		N	Mean	Std. deviation	Std. error mean
Y_Post	Management	121	24.722	4.488	0.408
	Employee	356	15.356	7.430	0.394
Y_Covid	Management	121	7.826	3.741	0.340
	Employee	356	3.938	2.696	0.143
Y_Post-Y_Covid	Management	121	16.897	4.888	0.444
	Employee	356	11.418	5.966	0.316

Source Author's data processing results

Workers had new livelihood alternatives during this period to meet their income during the challenging period when the epidemic broke out; yet, these livelihood alternatives are maintained until employment is restored, and demonstrating employees' good adaptability in the changing setting.

Estimated level of influence of factors on the income of employees in post-Covid-19 (Table 7).

The results show that the selected model is suitable; the R-Square coefficient is about 0.8487. The regression parameters of the model are as follows as Table 8.

According to the findings of the research, the gender of the employee affects the employee's salary, particularly in the difficult circumstances of the pandemic. Employers are more likely to use male workers. For female workers, the coefficient = −1.036534 also reflects that. Meanwhile, when employees are well-trained and have professional qualifications, stability in jobs will be higher, as does worker flexibility, which has a positive impact on worker's income. The coefficient for employees with a graduate degree is around 4.181241; the coefficient for workers with a university degree is approximately 1.070547. The research findings fully support the views of Qian and Fan (2020) and Le et al. (2021) that the greater the education, the better the job responsiveness, and the higher the employee's earnings. Businesses reviewed and pushed a lot of workers to resign, especially when the pandemic broke out, so having professional qualifications, and skills will help employees secure their positions in difficult times.

Job position also has an influence on an employee's income; the higher the position, the greater the pressure, and the more tasks, hence the income will be larger than in other positions. The coefficient of the variable is 1.484136 reflects this as well. The findings of the study support the viewpoints of Brunello et al. (2000), Campos and Jolliffe (2002), Qian and Fan (2020) and Dao et al. (2022). Along with that, the field of work has a considerable effect on employees' income; workers in the aviation sector, which has a quick recovery rate, have a higher influence on income fluctuations, and the adaptability of workers in this group will be better. The coefficient of this variable is about 1.118245. The research results completely support the views of Qian and Fan (2020); Bezerra et al. (2020); Dao et al. (2022).

Table 5 Independent samples test

		Levene's test for equality of variances		t-test for equality of means						
		F	Sig	T	Df	Sig. (2-tailed)	Mean difference	Std. error difference	95% confidence interval of the difference	
									Lower	Upper
Y_Post	Equal variances assumed	39.436	0.000	13.075	475.000	0.000	9.367	0.716	7.959	10.774
	Equal variances not assumed			16.519	346.160	0.000	9.367	0.567	8.251	10.482
Y_Covid	Equal variances assumed	13.667	0.000	12.337	475.000	0.000	3.888	0.315	3.269	4.507
	Equal variances not assumed			10.540	164.389	0.000	3.888	0.369	3.160	4.616
Y_Post-Y_Covid	Equal variances assumed	2.469	0.117	9.113	475.000	0.000	5.479	0.601	4.297	6.660
	Equal variances not assumed			10.046	250.609	0.000	5.479	0.545	4.405	6.553

Source Author's data processing results

Table 6 Vulnerability to the employment of workers during an epidemic outbreak

Criteria	Quantity (person)	Rate (%)
1. Workers have reduced working hours	223	46.75
2. Workers on leave with minimum wage	152	31.87
3. Workers leave without a salary	102	21.38
Total	477	100

Source Author's data processing results

Table 7 Results of the model test

Linear regression	Number of obs	=	477
	F(10, 466)	=	521.19
	Prob > F	=	0
	R-squared	=	0.8487
	Root MSE	=	3.1172

Source Author's data processing results

Table 8 Regression parameters of the model

Robust					
Y_Post	Coef	Std. Err	*t*	*P* > |*t*|	Beta
Gend					
1. Female	−1.036534	0.4432729	−2.34	0.020	−0.0654198
Edu_Post					
2. College	1.596011	0.5307503	3.01	0.003	0.0643445
3. University	1.070547	0.3022383	3.54	0.000	0.0617889
4. Post Graduate	4.181241	0.7269813	5.75	0.000	0.1394163
Posit_Post					
1. Management	1.484136	0.4918888	3.02	0.003	0.0815134
Work_Post					
1. Avi	1.118245	0.3043209	3.67	0.000	0.0703643
Exper_Post	0.2199327	0.1040973	2.11	0.035	0.1078855
Invest_Post	0.0342722	0.0048338	7.09	0.000	0.2280339
Cost_Post	1.011918	0.1069266	9.46	0.000	0.4170914
Work_invest_Post	0.0192732	0.0053836	3.58	0.000	0.1104369
_cons	4.460743	0.5591726	7.98	0.000	

Source Author's data processing results

The study findings also suggest that the transition to a new livelihood strategy has had a positive influence on workers in the changing context, as have private investments made by employees in difficult times, which have generated income to serve the life of workers. This demonstrates that workers' flexibility is rather excellent. The research findings also suggest that the coefficient of the private investment variable is around 0.0342722, representing increases in worker adaptability. Employees utilize this investment in financial investments and self-employment investments through Internet platforms for a stable income in that challenging environment. The findings support the views of Qian and Fan (2020), Dao et al (2022) and Bezerra et al. (2020).

5 Discussion

According to research findings, workers' adaptability capabilities in the setting of the epidemic are reasonably excellent. Once the epidemic is under control, workers' income and employment will have improved to satisfy living necessities. Some suggestions are given to respond to changes brought about by the current changing circumstances:

First, research demonstrates that, during the outbreak and the time after the epidemic is under control, competent and qualified workers appear to more successfully secure their positions, earning more consistent income than other workers. This is especially critical when force majeure conditions, such as epidemics or production/business unit challenges, compel the reduction of human resources. As a result, personnel must have the experience, certifications, and abilities to be able to adapt to any circumstance.

Second, diversifying revenue creation sources will assist employees in ensuring a steady living. At the same time, income diversity will assist workers in being able to adjust flexibly to varied circumstances, particularly during the adverse phase caused by the external environment.

Third, employers must have special orientations to help workers meet the needs imposed by the real work of the units. From there, employers will then be proactive in increasing their qualifications, allowing them to adapt securely to any situation.

6 Conclusions

When comparing the change in workers' pay during the outbreak phase and after the epidemic is under control, the research results met the specified target, utilizing data from a survey of 477 people working in the aviation and tourism/hospitality sectors. According to research, workers in the aviation sector are more resilient than those in the tourist and hotel sector. The income gap between these two groups before and after the pandemic is approximately 6.76 million VND/month. When the epidemic first broke out, the difference between employment positions was not very substantial, but after the epidemic is under control, the difference between job positions becomes relatively large, and the income of the group of workers who are managers is about 9.367 million VND/month higher than average compared to the group of employees who are employees. The study results also demonstrate that once the pandemic is under control, employees' salaries would be consistent and greater provided they have good qualifications, have good adaptability (through the diversification of income-generating options; investment is one of those income-generating channels).

References

Adger, W. N., Kelly, P. M., & Ninh, N. H. (2001). *Environment, society and precipitous change, in living with environmental change: Social vulnerability, adaptation and resilience in Viet Nam.* Routlege.

Bezerra, A. C. V., Silva, C. E. M. D., Soares, F. R. G., & Silva, J. A. M. D. (2020). Factors associated with people's behavior in social isolation during the Covid-19 pandemic. *Ciencia and saude coletiva, 25*, 2411–2421. https://www.scielosp.org/article/csc/2020.v25suppl1/2411-2421/en/#

Bhatti, S. H. (2013). Estimation of the mincerian wage model addressing its specification and different econometric issues (Doctoral dissertation). Retrieved from https://halshs.archivesouvertes.fr/tel-00780563/

Brunello, G., Comi, S., & Lucifora, C. (2000). The return to education in Italy: New look at evidence Discussion. IZA Discussion Paper No. 130. Retrieved from http://papers.ssrn.com/sol3/papers.cfm?abstract_id=224239

Buheji, M., da Costa Cunha, K., & Beka, G., Bartola Mavrić, Yuri Leandro do Carmo de Souza, Simone Souza da Costa Silva, Mohammed Hanafi, & Tulika Chetia Yein. (2020). The extent of Covid-19 pandemic socio-economic impact on global poverty. A global integrative multidisciplinary review. *American Journal of Economics, 10*(4), 213–224. https://doi.org/10.5923/j.economics.20201004.02

Burki, A. A., & Abbas, Q. (1991). Earnings functions in Pakistan's Urban informal sector: A case study. *The Pakistan Development Review, 30*, 695–706.

Campos, N. F., & Jolliffe, D. (2002). After and before during: Returns to education in the Hungarian transition. William Davidson Working Paper Number 475. Retrieved from http://deepblue.lib.umich.edu/handle/2027.42/39860

Cu, T. T., & Nguyen, T. A. (2021). Development of industrial parks and a new livelihood strategy for the people in VietNam. *Problems and Perspectives in Management, 19*(1), 221–230. https://doi.org/10.21511/ppm.19(1).2021.19

Cu, T. T., Hoang, T. H. H., & Le, T. Y. (2020). Impacts of industrial park development on the surroundings' Livelihood. *The Journal of Asian Finance, Economics, and Business, 7*(9), 737–746. https://doi.org/10.13106/jafeb.2020.vol7.no9.737

Dao, Q., Le, T., & Pham, V. (2022). Changes in income distribution during the COVID-19 pandemic: Empirical evidence from Vietnam. *The Journal of Asian Finance, Economics and Business, 9*(1), 241–248. https://doi.org/10.13106/jafeb.2022.vol9.no1.0241

DFID. (1999). *Sustainable livelihood guidance sheets.* London: DFID. Retrieved from https://www.livelihoodscentre.org/-/sustainable-livelihoods-guidance-sheets

GSO. (2020). Report on the impact of the covid-19 pandemic on Vietnamese enterprises. www.gso.gov.vn

GSO. (2021a). *Report the influence of Covid-19 on economic*. Retrieved from https://www.gso.gov.vn/du-lieu-va-so-lieu-thong-ke/2021a/12/thong-cao-bao-chi-ve-tinh-hinh-kinh-te-xa-hoi-quy-iv-va-nam-2021a

GSO. (2021b). *Survey on the impact of Covid-19 on enterprises and labors*. General Statistics Office. Retrieved from https://www.gso.gov.vn/du-lieu-va-so-lieu-thong-ke/2021b/09/bao-cao-lao-dong-viec-lam-quy-1-nam-2021b/

GSO. (2022a). *Report the influence of Covid-19 on laborers' income and employment*. Retrieved from https://www.gso.gov.vn/du-lieu-va-so-lieu-thong-ke/2022a/01/thong-cao-bao-chi-tinh-hinh-lao-dong-viec-lam-quy-iv-nam-2021-va-chi-so-phat-trien-con-nguoi-viet-nam-2016-2020/

GSO. (2022b). *Report on the impact of the Covid-19 epidemic on labor and employment situation in the first quarter of 2022b*. Retrieved from https://www.gso.gov.vn/du-lieu-va-so-lieu-thong-ke/2022b/04/bao-cao-tac-dong-cua-dich-covid-19-den-tinh-hinh-lao-dong-viec-lam-quy-i-nam-2022b/

Huynh, T. D. X & Mai, V. N. (2011). Phan tich cac yeu to anh huong den thu nhap cua ho chan nuoi gia cam o dong bang song cuu long (Analysis of factors affecting the income of poultry farmers in the Mekong Delta), *Scientific jouRnals, 17*(b), 87–96.

ILO (2020). *Danh gia nhanh tac dong cua dai dich COVID-19 toi doanh nghiep va nguoi lao dong trong mot so nganh kinh te chinh: Ung pho, dieu chinh va kha nang phuc hoi [Quick impact assessment of COVID-19 pandemic on the key economic sectors: Responses, adjustment and resilience of businesses and workers]*. (In Vietnamese). Retrieved from https://www.ilo.org/wcmsp5/groups/public/---asia/---ro-bangkok/---ilo-hanoi/documents/publication/wcms_757928.pdf

Jansen, H. G. P., John, P., Amy, D., Willem, W., & Rob, S. (2006). Policies for sustainable development in the hillside areas of Honduras: A quantitative livelihoods approach. *Agricultural Economics, 34*, 141–153. https://doi.org/10.1111/j.1574-0864.2006.00114.x

Le, T., Dao, Q., & Doan, V. (2021). Effects of the Covid-19 pandemic on employee income: A case study of the Vietnamese aviation industry. *Problems and Perspectives in Management, 19*(4), 186–197. https://doi.org/10.21511/ppm.19(4).2021.16

Nelson, R., Kokic, P., Crimp, S., Martin, P., Meinke, H., Howden, S. M., Voil, P. D., & Nidumolu, U. (2010). The vulnerability of Australian rural communities to climate variability and change: Part II—integrating impacts with adaptive capacity. *Environmental Science and Policy., 13*(1), 18–27. https://doi.org/10.1016/j.envsci.2009.09.007

Nguyen, H. N., & Tran, M. D. (2021). The effect of perceived organizational support on employee engagement during the COVID-19 pandemic: An empirical study in Vietnam. *The Journal of Asian Finance, Economics and Business, 8*(6), 415–426. https://doi.org/10.13106/jafeb.2021.vol8.no6.0415

Nguyen, V.T. (2014). *Giao trinh thuc hanh nghien cu trong kinh te va quan tri kinh doanh* (Curriculum of research practice in economics and business administration), National Economics University publishing house.

Nguyen, Q. N., & Bui, V. T. (2011). Cac yeu to anh huong den thu nhap cua nguoi dan toc thieu so o vung dong bang song cuu long (Factors affecting the income of ethnic minorities in the Mekong Delta). *Scientific Journals, 18*(a), 240–245.

Obrenovic, B., Du, J., Godinic, D., Tsoy, D., Khan, M. A. S., & Jakhongirov, I. (2020). Sustaining enterprise operations and productivity during the Covid-19 pandemic: "Enterprise Effectiveness and Sustainability Model." *Sustainability, 12*(15), 5981.

Pattiruhu, J. R., & Paais, M. (2020). Effect of liquidity, profitability, leverage, and firm size on dividend policy. *The Journal of Asian Finance, Economics, and Business, 7*(10), 35–42. https://doi.org/10.13106/JAFEB.2020.VOL7.NO10.035

Paul, A., Nath, T., & Mahanta, J., et al. (2021). Psychological and livelihood impacts of COVID-19 on Bangladeshi lower income people. *Asia Pacific Journal of Public Health, 33*(1), 100–108. https://doi.org/10.1177/1010539520977304

Qian, Y., & Fan, W. (2020). Who loses income during the Covid-19 outbreak? Evidence from China. *Research in Social Stratification and Mobility, 68*, 100522, https://www.sciencedirect.com/science/article/pii/S0276562420300585?via%3Dihub#bib0055

Rohan, N., Philip, K., Elliston, L., & King, J. A. (2005). Structural adjustment: A vulnerability index for Australian broadacre agriculture. *Australian Commodities, 12*(1), 171–179.

Shen, H., Fu, M., Pan, H., Yu, Z., & Chen, Y. (2020). The impact of the Covid-19 pandemic on firm performance. *Emerging Markets Finance and Trade, 56*(10), 2213–2230.

Siegel. (2005). Using an asset-based approach to identify drivers of sustainable rural growth and poverty reduction in Central America: A conceptual framework. *World Bank Policy Research Working Paper, 3475*. https://doi.org/10.1596/1813-9450-3475

Suratman, A., Suhartini, S., Palupi, M., Dihan, F. N., & Muhlison, M. B. (2021). The impact of psychological climate and self-resilience on employee performance during the COVID-19 pandemic: An empirical study in Indonesia. *The Journal of Asian Finance, Economics and Business, 8*(5), 1019–1029. https://doi.org/10.13106/jafeb.2021.vol8.no5.1019

Tran, Q. T. (2013). Livelihood strategies for coping land loss among households in VietNam's sub-urban areas. *Asian Social Science., 9*(15), 33–46. https://doi.org/10.5539/ass.v9n15p33

Tran, Q. T., & Vu, V. H. (2014). The impact of land loss on household income: The case of HaNoi's sub-urban areas, Vietnam. *International Journal of Business and Society, 15*(2), 339–358.

Tu, V. H., Mitsuyasu, Y., Nguyen, T. T., & Huynh, V. K. (2015). Adaptive capacity assessment of rural out migrants: A case study of An Giang province, Vietnam. *Journal Odd the Faculty of Agriculture Kuyshu University., 60*, 265–271.

Vernon, V. K. (2002). Return to Human Capital in transitional Russia. *SSRN Electronic Journal*. https://doi.org/10.2139/ssrn.308619

Vo, H. T., & Nguyen, T. T. (2012). Tinh ton thuong sinh ke cua nong ho bi anh huong lu tai tinh An Giang va cac giai phap ung pho (Vulnerability of farmers affected by floods in An Giang province and response solutions). *Scientific Journal, 22*(b), 294–303.

World Bank. (2020). Impact of Covid-19 on firms in Vietnam. The quick survey report on enterprises and Covid-19, Report No. 1.

Analysis of Factors Affecting the Financial Capacity of Vietnam Commercial Banks After M&A by CAMEL Standards

Nguyen Thi Bich Vuong and Nguyen Phi Dat

Abstract After implementing the project on restructuring the credit institution system in period 1 (2011–2015) and period 2 (2016–2020), banking M&A activities in Vietnam as of 2020 have not been professional, small number, sometimes spontaneous, sometimes due to the pressure of mechanisms and regulations in legal documents, not yet derived from the economic interests of the bank and the economy, and thus lack of experience and information. Moreover, after restructuring, new commercial banks were formed, which is the result of M&A deals. But after a while, how these commercial banks develop and how effective they are, it is a difficult problem that bank administrators must continue to solve. Therefore, the question for commercial banks after M&A is how to improve financial capacity to maintain stability after M&A and the bank to still operate effectively. Therefore, the article used Probit binary regression model to analyze the factors affecting the financial capacity of commercial banks after M&A in Vietnam in the period 2012–2021 to see whether they meet CAMEL standards or not. This will be the basis for bank managers to come up with solutions to improve the financial capacity of commercial banks after M&A in Vietnam in the near future.

Keywords Financial capacity · Probit binary regression model · CAMEL standard · Commercial bank after M&A

1 Introduction

Probit binary regression model is a model in which the dependent variable is the predictor variable that receives only 2 values, 1 and 0. In this study on assessing the financial capacity of commercial banks after M&A in Vietnam, it is necessary to predict whether banks will meet CAMEL standards or not. Therefore, this study will

N. T. B. Vuong (✉)
Hanoi Technology Economics College, Hanoi, Vietnam
e-mail: Violet1072007@gmail.com

N. P. Dat
Banking Academy, Hanoi, Vietnam

A. T. Nguyen et al. (eds.), *Contemporary Economic Issues in Asian Countries: Proceeding of CEIAC 2022, Volume 1*,
https://doi.org/10.1007/978-981-19-9669-6_7

use the Probit binary regression model to predict the probability P in the interval (0, 1), where the value of $y = 1$ (meet CAMEL standard) or $y = 0$ (failure to meet CAMEL standard). Based on the collected data of 8 commercial banks after M&A in Vietnam, the calculation of criteria according to CAMEL standard includes 15 criteria, assessed according to CAMEL standard according to each criterion of passing (=1) or failing (=0), thereby assessing the financial capacity of commercial banks in each year after M&A by averaging and then performing Probit binary regression analysis with the dependent variable being Financial capacity (NLTC) with 2 values 0 and 1 and 15 independent variables in the model affecting the dependent variable.

2 Research Overview

The topic of financial capacity of commercial banks is a topic that has been studied a lot in the world, in which there have been many research works using Dupont, DEA, Capm, Probit, Logistic methods…to assess the financial capacity of a commercial bank, whether it meets CAMEL or Basel standards? Alton Gilbert et al. (2002) studied the financial capacity of US commercial banks according to CAMEL standard compared with SEER standard of commercial banks in the US in the period 1990–1998. The author's study used the Probit binary model to assess the financial capacity of US commercial banks according to the set of 5 criteria C, A, M, E, and L based on financial criteria such as capital size, financial leverage, profitability, asset quality, management quality, and liquidity of assets. Tatom (2011) studied the causes of the failure of commercial banks. The main reason is the financial capacity of these banks. The author's research also shows that the assessment of financial capacity of banks can be affected by C, A, M, E, and L, from which the author conducts regression according to Proxy to determine the influencing factors. The results show that the financial capacity of commercial banks is influenced by factors such as capital size, profitability, asset quality, management quality, and liquidity of assets. Then, the author uses downgrade method to predict the probability of future failure of commercial banks in the period 2003–2007. Nguyen Viet Hung (2008) has studied the performance of Vietnamese commercial banks in the period 2000–2005. In the study, the author has determined the factors affecting the financial capacity of banks according to CAMEL standards, then regression with Tobit. Research results show that factors such as bank's assets, loan/deposit ratio, ROA, and bad debt ratio have an influence on the financial capacity of the commercial banks during that period. Nga (2013) has deeply researched and assessed the financial capacity of 28 Vietnamese commercial banks according to CAMEL standards in the period 2003–2012. The author used Probit model to test 13 factors affecting the financial capacity of Vietnamese commercial banks, including size of equity; financial leverage; minimum capital adequacy ratio; outstanding debt/total assets; bad debt/ total outstanding debt (noxau_duno); ROA; ROE; NIM; operating cost index; asset liquidity ratio; deposit guarantee factor; short-term liquidity ratio; and loans/deposits. All of the above factors have a certain impact on the financial capacity of Vietnamese commercial banks. Research "Applying CAMEL standards in financial analysis at Bank for Investment and Development of Vietnam" by Hoang Van Thang has focused on studying the financial capacity of Bank for Investment and Development of Vietnam through using the criteria system of CAMEL standards to analyze in the form of descriptive statistics the secondary data that the author collected in the period 2003–2008, including: Tier 1 Capital/Total Risky Assets, Equity/Total Assets, Equity/Risk Assets—CAR, BIS Ratio (Own Equity/Risk Assets), Financial leverage ratio (Total liabilities/equity), Internal capital generation ratio (Undivided profit/Tier 1 capital), NPLs/Total outstanding loans, Provision for credit risks/Debts bad, Provision expense ratio (Provision for debt loss/Average outstanding loans), Ability to cover bad debts (Provision for loss of debts/NPLs), ROA, ROE, Non-interest income/Total assets, Net Margin, Net Interest Income/Total Operating Income, Operating Expenses/Total Assets…. (Hoang Van Thang, 2009).

In summary, research overview shows that there have been many studies on the financial capacity of commercial banks, including the use of different evaluation analysis methods (Capm, DEA, Dupon, Probit, Proxy, Tobit, Logistic) and based on different standards (CAMEL, Basel1, Basel 2) but they are all commercial banks in general, not commercial banks after M&A. Therefore, the author directs his research to analyze the factors affecting the financial capacity of commercial banks after M&A in Vietnam according to CAMEL standards.

3 Research Model

The author builds the research model as follows:

$$
\begin{aligned}
Y_i = {} & \beta_0 + \beta_1 X_1 + \beta_2 X_2 + \beta_3 X_3 + \beta_4 X_4 + \beta_5 X_5 + \beta_6 X_6 + \beta_7 X_7 + \beta_8 X_8 \\
& + \beta_9 X_9 + \beta_{10} X_{10} + \beta_{11} X_{11} + \beta_{12} X_{12} + \beta_{13} X_{13} + \beta_{14} X_{14} + \beta_{15} X_{15}
\end{aligned}
$$

Dependent variable (Y): To forecast the probability that commercial banks after M&A meet CAMEL standards or fail to meet CAMEL standards, call the dependent variable y = NLTC (Financial capacity):

- NLTC = 1 if the bank meets CAMEL standard
- NLTC = 0 if the bank does not meet CAMEL standard.

Based on data collected from the annual report of commercial banks after M&A in Vietnam, analyzing the financial capacity of commercial banks after M&A according to CAMEL's standards based on 15 criteria. For each bank in each research year, evaluate each criterion according to CAMEL standards if rating 1 is achieved and if not rating 0 is achieved and then averaging 15 indicators.

- If the average of 15 indicators of each bank in each year is 0.5, it is CAMEL standard (NLTC = 1)

- If the average of 15 indicators of each bank in each year is <0.5, it is not CAMEL standard (NLTC = 0).

To forecast the probability $P(y = 1)$ while the variable NLTC only takes 2 values of 0 and 1, so the probability of not defaulting $P(y = 0) = 1 - P(y = 1)$.

Independent variable (X) consists of 15 variables, which are 15 criteria to assess the financial capacity of banks according to CAMEL criteria as follows:

- X1: Size of equity
- X2: Financial leverage coefficient
- X3: Equity/Total Assets ratio
- X4: Minimum capital adequacy ratio (CAR)
- X5: Loan balance/Total assets
- X6: Bad debt ratio
- X7: Ratio of provision expenses
- X8: Profit growth rate
- X9: Credit growth rate
- X10: Return on Assets (ROA)
- X11: Return on Equity (ROE)
- X12: Net Interest Margin (NIM)
- X13: Non-net Interest Margin (NNIM)
- X14: Deposit/Total assets ratio
- X15: Loan/Deposit ratio.

From the above research model, the thesis also proposes research hypotheses including (Table 1).

Based on this model, the study was conducted to test the suitability of the model by the following tasks:

The first: Autocorrelation test is the phenomenon when the errors in the model have a relationship with each other, the cause of using time data, the delay of the data, and the inertia of the data. As a result, the estimate is biased. To test the phenomenon of autocorrelation, the thesis uses the Estat Dwatson function to calculate the coefficient d (Durbin-Watson), if $1 < d < 3d$ and d is close to 2, the model does not have autocorrelation.

The second: Testing for multicollinearity is a phenomenon where the independent variables in the model are correlated with each other, leading to no regression or incorrect model results. To test to determine the correlation coefficient (ri) between variables through the Corr function, if $ri < \pm 0.5$, it is considered that there is no multicollinearity phenomenon.

The third: Testing the phenomenon of variance is because the hypothesis of constant variance of the model is violated, due to the nature of economic quantities, or because of advances in measurement and data processing. It will affect the unbiased result of the estimate. To test the variable variance, the study uses the Hettest function from which to determine the P value, if the P value <5%, the phenomenon of variance does not occur and vice versa.

Table 1 Summary of research hypotheses

Hypotheses	Content
H1	The larger the scale of equity, the stronger the financial capacity of Vietnamese commercial banks after M&A
H2	Using financial leverage as possible will make the financial capacity of Vietnamese commercial banks better after M&A
H3	The higher the equity/total assets ratio, the better the financial capacity of Vietnamese commercial banks after M&A
H4	The higher the minimum capital adequacy ratio (CAR), the better the financial capacity of Vietnamese commercial banks after M&A
H5	The higher the loan balance/total assets, the lower the financial capacity of Vietnamese commercial banks after M&A
H6	The higher the bad debt ratio, the lower the financial capacity of Vietnamese commercial banks after M&A
H7	The higher the ratio of provision expenses, the lower the financial capacity of Vietnamese commercial banks after M&A
H8	The higher the profit growth rate, the stronger the financial capacity of Vietnamese commercial banks after M&A
H9	The higher the credit growth rate, the stronger the financial capacity of Vietnamese commercial banks after M&A
H10	The higher the return on assets (ROA), the better the financial capacity of Vietnamese commercial banks after M&A
H11	The higher the return on equity (ROE), the better the financial capacity of Vietnamese commercial banks after M&A
H12	The higher the Net interest margin (NIM), the better the financial capacity of Vietnamese commercial banks after M&A
H13	The higher the Non-net interest margin (NIM), the better the financial capacity of Vietnamese commercial banks after M&A
H14	The larger the ratio of deposits/total assets, the higher the financial capacity of Vietnamese commercial banks after M&A will be improved
H15	The higher the loan/deposit ratio, the lower the financial capacity of Vietnamese commercial banks after M&A

Source Suggested by the author

The fourth: LR(chi2) test means testing the hypothesis that at least one regression coefficient of an independent variable in the model is different from 0.

4 Characteristics of the Analyzed Data

After collecting data, calculating the criteria to evaluate the financial capacity of commercial banks after M&A, assessing whether they pass or fail according to CAMEL standards in each bank, using the Stata16.0 software to perform the

Table 2 Characteristics of the analyzed data

Unweighted cases		N	Percent
Selected cases	Included in analysis	47	100.0
	Missing cases	0	0.00
	Total	47	100.0
Unselected cases		0	0.00
Total		47	100.0

Source Author's data analysis results

Table 3 Coding of dependent variable

Original value	Internal value
KhongdattieuchuanCAMEL	0
DattieuchuanCAMEL	1

Source Author's data analysis results

Probit binary analysis. Research data includes 8 commercial banks after M&A with data collected: LienVietPostBank, SCB (2012–2019); SHB (2013–2019), HDBank, PVcomBank (2014–2019); Sacombank, BIDV, and Maritimebank (2016–2019) include a total of 47 observations.

Table 2 shows the data characteristics included in the Probit binary analysis of this study, including 47 observations, none of which were missing data, and none of which were unselected.

Table 3 shows that the dependent variable has 2 values, "not meeting CAMEL standard" is coded as 0, and "meeting CAMEL standard" is encoded as 1.

5 Data Analysis Results

After checking the analysis data of the Probit binary regression model with no missing observations, no unselected observations, and coding the dependent variable, the next step is to test the phenomenon of autocorrelation, test the phenomenon of multicollinearity, test the phenomenon of variance, and test LR(chi2).

5.1 Test the Phenomenon of Autocorrelation

To test the autocorrelation phenomenon of residual variables in the model, the study uses the function Estat dwatson to determine the value of the coefficient d (Durbin-Watson). If $1 < d < 3$ and d is close to 2, then the variables are considered to have no autocorrelation. The results are as follows: Durbin-Watson coefficient d-statistic (8.72) = 1.384798; this value ranges from $1 < d < 3$. Therefore, it can be

Table 4 Calculation results of variance exaggeration factor VIF

Model variables	Coefficient VIF	1/VIF
Chovay/Tongtaisan	2.80	0.357751
Duno/Tiengui	2.20	0.453587
VonCSH/Tongtaisan	1.84	0.544303
VonCSH	1.57	0.635788
NNIM	1.47	0.681754
Noxau	1.35	0.738009
CAR	1.29	0.776560
Mean VIF	1.79	

Source Test results from Corr function on Stata 16.0

confirmed that the model does not have autocorrelation of residuals. This means that the regression model does not violate the assumption of error independence.

5.2 Test the Phenomenon of Multicollinearity

The phenomenon of multicollinearity occurs when the independent variables are strongly correlated and expressed as a function, and this affects the results of the regression function. Therefore, testing whether or not multicollinearity contributes significantly to the conclusion that the regression results are appropriate or not. In this study, the author uses the Corr function on Stata 16.0 software to analyze the correlation between the independent variables of the model. The results show that the variables with correlation coefficients all reach statistical significance ($-0.5 < r < 0.5$). However, to ensure that the variables do not have multicollinearity, the author also uses the variance exaggeration factor VIF to overcome. The result is as follows:

The results of Table 4 show that the variance exaggeration coefficient VIF of all variables is lower than 10 and the average value is 1.79, so it can be concluded that the model does not have multicollinearity; this means that the variables in the model are closely correlated with each other.

5.3 Test the Phenomenon of Variance

To test the phenomenon of variance, the author uses the Hettest function on Stata 16.0 software with results as follows:

Breusch-Pagan / Cook-Weisberg test for heteroskedasticity Ho: Constant variance Variables: fitted values of NLTC chi2(1) = 1.75 Prob > chi2 = 0.1853

Source Test results from Hettest function on Stata 16.0

The results show: Prob > Chi2 = 0.1853 > 0.05, so the hypothesis that the variance is fixed, in other words, can reject the hypothesis that the variance varies.

5.4 Test LR (Chi2)

Another test that needs to be done is to test the hypothesis that at least one regression coefficient of an independent variable in the model is different from 0, where the test is done through the LR (chi2) statistic with degrees of freedom is 7. The test results are as follows:

LR chi2 (7) = 42.60

Prob > chi2 = 0.0000.

The results show that the coefficient Prob > chi2 = 0.0000 < 0.05, so it can be confirmed that the model does not occur at the same time, and the regression coefficients are zero.

In short, after testing the fit of the model with the results obtained, the research can answer: The built regression model is suitable, there is no autocorrelation and no multicollinearity, variance does not change, and there is no case where the regression coefficients are equal to 0.

5.5 The Results of the Regression Analysis

The results of the regression analysis show that the variables with statistical significance show the impact on the bank's financial capacity, including the following variables: size of equity; equity/total assets ratio; minimum capital adequacy ratio (CAR); loan balance/total assets; bad debt ratio; loan/deposit ratio; and non-net interest margin (NNIM). Probit regression has no equivalent to R squared found in OLS regression, but Pseudo R2 values close to 1 are still considered good for the model. The result Pseudo R2 = 0.5253 > 0.5, thus also showing the model, has high reliability.

The results of Table 5 show that all variables in the model show a statistically significant impact, here the author chooses an error of less than 10% to be accepted, because the research sample size is limited in number. Checking the model's marginal effects, it can be seen that the difference is not large between the coefficients of the variables in the model when keeping the remaining variables unchanged. In addition, the correctly classified level of the model also has a high rate of 86.11%, and they show that the model can correctly explain 86.11% of the cases of the dependent variable, which is a high percentage. Then, the regression model is written as follows:

$$Y = -9.9307340 + 0.0005101 * \text{VonCSH} + 0.1761097 * \text{VonCSH/Tongtaisan} + 0.4686083 * \text{CAR} - 0.1062785 * \text{Chovay/Tongtaisan} - 0.5944837 * \text{Noxau} + 0.5051688 * \text{NNIM} - 0.0495970 * \text{Duno/Tiengui}$$

According to the regression results, it can be seen that because of the Probit function, the impact results of the factors are measured by probability and the level of impact is determined in Table 5; therefore, the factors that really affect the financial capacity of Vietnamese commercial banks after M&A include:

Size of equity: If the size of a bank's equity is larger, the financial capacity of the bank will be stronger with the impact that when equity increases by 1 unit, the probability that the bank's financial capacity will reach CAMEL standard increased to 3.86E−06%. This result is consistent with the studies of previous authors such as Johnston (2009), Alton Gilbert et al. (2002), Tatom (2011), Wirnkar and Tanko (2007), and Godfrey Cadogan (2011), and this is also consistent with the reality in Vietnam when a bank with large equity will help them be financially self-sufficient; at the same time, there are many opportunities to implement big projects and also easily withstand when there are risks. Indeed, among the 8 Vietnamese commercial banks after M&A, BIDV is the bank before and after the M&A implementation that always has the largest scale of equity and is ranked as one of the 4 commercial banks with strong financial capacity in banking system in Vietnam.

Equity/Total Assets: The larger the equity/total assets ratio, the better the financial capacity of Vietnamese commercial banks after M&A, with the increase in the equity/total assets ratio by 1 unit, and they can increase the probability that the bank meets the CAMEL standard to 0.1331%. This result is consistent with the studies of previous authors such as Frank Heid (2007) and Hoang Van Thang (2009), and this is also consistent with the reality in Vietnam when a bank with only owning capital and large assets will help the bank to have good financial autonomy to implement large projects and easily support when the risk occurs.

Bad debt ratio: If the bad debt ratio of any bank is higher, the financial capacity of that bank will decrease. When the bad debt ratio increases by 1 unit, the probability that the financial capacity meets the CAMEL standard will decrease by 0.449%. This result is consistent with the studies of previous authors such as Berger et al. (1997), Alton Gilbert et al. (2002), and Nguyen Viet Hung (2008), and this is also consistent with the reality in Vietnam when a bank has a high bad debt ratio; the possibility of capital loss will be greater, especially in the current period when many businesses have difficulty doing business. With a lot of difficulties and losses, the increasing number of bankrupt enterprises will make it more difficult to collect debts of the

Table 5 Results of regression analysis

Variable	Coef	Std. Err	z	P > |z|	[95% conf. interval]
VonCSH	0.0005101	0.0001407	3.62	0.000	0.0002343 0.0007860
VonCSH/TongTS	0.1761097	0.1036263	1.70	0.089	−0.0269941 0.3792135
CAR	0.4686083	0.2314109	2.03	0.043	0.0150513 0.9221653
Chovay/TongTS	−0.1062785	0.0413386	−2.57	0.010	−0.1873006 – 0.0252563
Noxau	−0.5944837	0.1927318	−3.08	0.002	−0.9722310 – 0.2167363
NNIM	0.5051688	0.2178985	2.32	0.020	0.0780956 0.9322420
Duno/Tiengui	−0.0495970	0.0287323	−1.73	0.084	−0.0067168 0.1059119
_cons	−9.9307340	4.0047090	−2.48	0.013	−17.779820 – 2.0816470

Variable	dy/dx	Std. err	z	P > |z|	[95% C.I.]	X
VonCSH	3.86E-06	0.00001	0.46	0.644	−1.2E-05 0.00002	16,924.70
VonCSH/TongTS	0.001331	0.00297	0.45	0.654	−0.00449 0.007150	7.565570
CAR	0.003542	0.00746	0.47	0.635	−0.01107 0.018156	10.77420
Chovay/TongTS	−0.00080	0.00173	−0.47	0.642	−0.00419 0.002578	55.16680
Noxau	−0.00449	0.00984	−0.46	0.648	−0.02378 0.014795	2.708750
NNIM	0.003818	0.00814	0.47	0.639	−0.01214 0.019780	2.619220
Duno/Tiengui	−0.00037	0.00081	0.46	0.645	−0.00122 0.001970	78.28090

Sensitivity	Pr(+|D)	88.64%
Specificity	Pr(−|~D)	82.14%
Positive predictive value	Pr(D|+)	88.64%
Negative predictive value	Pr(~D|−)	82.14%
False + rate for true ~D	Pr(+|~D)	17.86%
False − rate for true D	Pr(−|D)	11.36%
False + rate for classified	Pr(~D|+)	11.36%
False − rate for classified	Pr(D|−)	17.86%
Correctly classified		86.11%

Source Probit regression results for independent and dependent variables from Stata 16.0

bank, leading to adverse effects on the financial situation of the bank. Therefore, banks are increasingly showing their efforts in thoroughly solving outstanding bad debts and minimizing the arising of new bad debts with stricter regulations in lending activities.

Non-net interest margin (NNIM): If Non-net interest margin of bank is larger, the financial capacity of that bank will be more likely to meet the CAMEL standard, with the ratio when NNIM increases by 1 unit, the probability of achieving CAMEL increases to 0.3818%. This result is consistent with the studies of Barnes and Lopez (2005) and Nga (2013), and this is also consistent with reality in Vietnam when a bank has a larger difference between non-interest income and non-interest expenses, the more it contributes to increasing profits and increasing business efficiency and increase the financial potential of the bank.

Loan balance/total assets: The higher the loan balance/assets of any bank, the lower the financial capacity of that bank. When loan balance/assets increase by 1 unit, the probability of achieving CAMEL decreases by 0.08%. This result is consistent with the study of Wagner (2007) and Tatom (2011), and this is also consistent with the reality of business at commercial banks because when the loan balance is too high compared to the total assets of the bank, it is easy for banks to use borrowed funds at high cost to cover the difference, or it will cause the possibility of an imbalance in payments, causing strong risks to the bank's operations.

Loan/deposit: The larger the loan/deposit balance of any bank, the lower the financial capacity of that bank. When the loan/deposit balance increases by 1 unit, the probability of achieving CAMEL decreases by 0.037%. This result is consistent with the study of Wagner (2007), Tatom (2011), and Nga (2013). This is also consistent with the reality in Vietnam when a bank with a higher loan/deposit balance means that the bank has made a larger loan than the amount of customer deposits that the bank has mobilized. In fact, the loan is a loan that the bank will periodically collect according to the agreed credit contract. Banks do not arbitrarily collect early unless the customer voluntarily repays the loan before the due date and the deposit can be withdrawn at any time. Therefore, if the bank lends more than the mobilized amount, it will lead to liquidity risk, especially in the current period, which can make the bank insolvent if this ratio is too high.

Minimum capital adequacy ratio (CAR): The higher the minimum capital adequacy ratio (CAR), the better the financial capacity of Vietnamese commercial banks after M&A. When CAR increases by 1 unit, the probability of achieving CAMEL increases to 0.3542%. This result is consistent with the studies of previous authors such as Heid (2007), Brown and Kevin Davis (2009), Barnesa and Lopez (2005), Wirnkar and Tanko (2007), Cadogan (2011), and Nga (2013). This result is also confirmed in practice not only in Vietnam but also around the world, that is, the higher the capital adequacy ratio of banks, the higher their financial capacity is.

Table 6 Individual regression coefficients of each factor

Factors	Code	z	Sig
X1: size of equity	VonCSH	3.62	0.000
X2: financial leverage coefficient	DonbayTC	No impact	0.314
X3: equity/total assets ratio	VonCSH/Taisan	1.70	0.089
X4: minimum capital adequacy ratio	CAR	2.03	0.043
X5: loan balance/total assets	DunoCV/Taisan	−2.57	0.010
X6: bad debt ratio	Noxau	−3.08	0.002
X7: ratio of provision expenses	ChiphiDP	No impact	0.322
X8: profit growth rate	TangtruongLN	No impact	0.964
X9: credit growth rate	TangtruongTD	No impact	0.829
X10: return on assets	ROA	No impact	0.878
X11: return on equity	ROE	No impact	0.890
X12: net interest margin	NIM	No impact	0.280
X13: non-net interest margin	NNIM	2.32	0.020
X14: deposit/total assets ratio	Tiengui/Taisan	No impact	0.298
X15: loan/deposit ratio	DunoCV/Tiengui	−1.73	0.084

Source Summary of separate regression results on Stata 16.0 with each factor

5.6 The Result of Hypothesis Testing

To test the research hypothesis, the study conducted a separate regression of each independent variable affecting the financial capacity of Vietnamese commercial banks after M&A, and the results are summarized as follows (Table 6).

Hypothesis 1 (H1):

Does the size of equity have a positive effect on the financial capacity of Vietnamese commercial banks after M&A?

The test results in $z = 3.62$ and the P value (Sig) of VonCSH (X1) $= 0.000 <$ 10%. From that, it can be concluded that the hypothesis H1 is accepted with high confidence (X1 is statistically significant) and equity has a positive influence on the financial capacity of the bank. In other words, the larger the scale of equity, the stronger the financial capacity of Vietnamese commercial banks after M&A.

Hypothesis 2 (H2):

Does Financial leverage coefficient have a positive effect on the financial capacity of Vietnamese commercial banks after M&A?

The test gives the result P value (Sig) of DonbayTC (X2) $= 0.314 >$ 10%. From that, it can be concluded that hypothesis H2 is not accepted. Therefore, it is not possible to confirm the impact relationship between the financial leverage coefficient and the financial capacity of Vietnamese commercial banks after M&A.

Hypothesis 3 (H3):

Does the equity/total assets ratio positively affect the financial capacity of Vietnamese commercial banks after M&A?

The test results in $z = 1.7$ and P value (Sig) of VonCSH/Taisan $= 0.089 < 10\%$. From that, it can be concluded that the hypothesis H3 is accepted with high confidence (X3 is statistically significant) and the equity/total assets ratio has a positive influence on the bank's financial capacity. In other words, the larger the equity/total assets ratio, the better the financial capacity of Vietnamese commercial banks after M&A.

Hypothesis 4 (H4):

Does the minimum capital adequacy ratio (CAR) positively affect the financial capacity of Vietnamese commercial banks after M&A?

The test results in $z = 2.03$ and P value (Sig) of CAR $= 0.043 < 10\%$. From that, it can be concluded that the hypothesis H4 is accepted with high confidence (X4 is statistically significant) and the minimum capital adequacy ratio (CAR) has a positive effect on the financial capacity of the bank. In other words, the higher the minimum capital adequacy ratio (CAR), the better the financial capacity of Vietnamese commercial banks after M&A.

Hypothesis 5 (H5):

Does loan balance/total assets have a negative effect on the financial capacity of Vietnamese commercial banks after M&A?

The test results in $z = -2.57$ and P value (Sig) of DunoCV/Taisan $= 0.010 < 10\%$. From that, it can be concluded that the hypothesis H5 is accepted (X5 is statistically significant) and the loan balance/total assets have a negative effect on the financial capacity of the bank. In other words, the higher the loan balance/total assets, the lower the financial capacity of Vietnamese commercial banks after M&A.

Hypothesis 6 (H6):

Does the bad debt ratio have a negative effect on the financial capacity of Vietnamese commercial banks after M&A?

The test results in $z = -3.08$ and Noxau's P value (Sig) $= 0.002 < 10\%$. From that, it can be concluded that the hypothesis H6 is accepted with high confidence (X6 is statistically significant) and the bad debt ratio has a negative effect on the financial capacity of the bank. In other words, the higher the bad debt ratio, the lower the financial capacity of Vietnamese commercial banks after M&A.

Hypothesis 7 (H7):

Does the ratio of provision expenses have a negative effect on the financial capacity of Vietnamese commercial banks after M&A?

The test gives the results P value (Sig) of ChiphiDP $= 0.322 > 10\%$. From that, it can be concluded that hypothesis H7 is not accepted. Therefore, it is not possible to confirm the impact relationship between ratio of provision expenses and financial capacity of Vietnamese commercial banks after M&A.

Hypothesis 8 (H8):

Profit growth rate has a positive influence on the financial capacity of Vietnamese commercial banks after M&A.

Test results in P value (Sig) of TangtruongLN = 0.964 > 10%. From that, it can be concluded that hypothesis H8 is not accepted. Therefore, it is not possible to confirm the impact relationship between the profit growth rate and the financial capacity of Vietnamese commercial banks after M&A.

Hypothesis 9 (H9):

Credit growth rate has a positive influence on the financial capacity of Vietnamese commercial banks after M&A.

Test results in P value (Sig) of TangtruongTD = 0.829 > 10%. From that, it can be concluded that hypothesis H9 is not accepted. Therefore, it is not possible to confirm the impact relationship between credit growth rate and financial capacity of Vietnamese commercial banks after M&A.

Hypothesis 10 (H10):

Return on Assets (ROA) has a positive influence on the financial capacity of Vietnamese commercial banks after M&A.

The test results in P value (Sig) of ROA = 0.878 > 10%. From that, it can be concluded that hypothesis H10 is not accepted. Therefore, it is not possible to confirm the impact relationship between ROA and financial capacity of Vietnamese commercial banks after M&A.

Hypothesis 11 (H11):

Return on Equity (ROE) has a positive influence on the financial capacity of Vietnamese commercial banks after M&A.

The test results in P value(Sig) of ROE = 0.890 > 10%. From that, it can be concluded that hypothesis H11 is not accepted. Therefore, it is not possible to confirm the impact relationship between ROE and financial capacity of Vietnamese commercial banks after M&A.

Hypothesis 12 (H12):

Net Interest Margin (NIM) has a positive effect on the financial capacity of Vietnamese commercial banks after M&A.

The test results in P value (Sig) of NIM = 0.280 > 10%. From that, it can be concluded that hypothesis H12 is not accepted. Therefore, it is not possible to confirm the impact relationship between NIM and the financial capacity of Vietnamese commercial banks after M&A.

Hypothesis 13 (H13):

Non-net Interest Margin (NIM) has a positive effect on the financial capacity of Vietnamese commercial banks after M&A.

The test results in $z = 2.32$ and P value (Sig) of NNIM = 0.020 < 10%. From that, it can be concluded that the hypothesis H13 is accepted with high confidence (X13 is statistically significant) and the Non-net Interest Margin (NNIM) has a positive influence on the financial capacity of the bank. In other words, the larger the NNIM, the better the financial capacity of Vietnamese commercial banks after M&A.

Hypothesis 14 (H14):

Deposit/Total assets ratio has a positive effect on the financial capacity of Vietnamese commercial banks after M&A.

The test results in P value (Sig) of Tiengui/Taisan = 0.298 > 10%. From that, it can be concluded that hypothesis H14 is not accepted. Therefore, it is not possible to confirm the impact relationship between the deposit/Total assets ratio and the financial capacity of Vietnamese commercial banks after M&A.

Hypothesis 15 (H15):

The Loan/Deposit ratio has a negative effect on the financial capacity of Vietnamese commercial banks after M&A.

The test results in $z = -1.73$ and P value (Sig) of DunoCV/Tiengui = 0.084 < 10%. From that, it can be concluded that the hypothesis H15 is accepted with high confidence (X15 is statistically significant) and the loan/deposit ratio has a negative effect on the financial capacity of the bank. In other words, the higher the loan/deposit ratio, the lower the financial capacity of Vietnamese commercial banks after M&A.

6 Conclusion

Probit regression analysis results for each factor show that there are 11 factors that positively affect the financial capacity of Vietnamese commercial banks after M&A and 4 factors are bad debt ratio, Ratio of provision expenses, loan balance/total assets, Loan/deposit ratio which have the opposite effect. These results are consistent with previous studies of authors such as Wagner (2007), Tatom (2011), Barnes and Lopez (2005), Nguyen Viet Hung (2008), Van Thang (2009), and Nga (2013) as well as in line with the actual operation and financial capacity of Vietnamese commercial banks in general and Vietnamese commercial banks after M&A in particular.

However, according to the standardized estimation results of the theoretical model, out of 15 factors, only 7 factors are statistically significant, showing the influence on the bank's financial capacity, including size of equity, equity/total assets ratio, minimum capital adequacy ratio (CAR), loan balance/total assets, bad debt ratio, Non-net Interest Margin (NNIM), and loan/deposit ratio. The remaining 8 factors that are not statistically significant are excluded from the model. The results of testing the causal relationship between concepts in the standardized model show that the importance of factors affecting the financial capacity of Vietnamese commercial banks after M&A is very different. However, 15 factors also reflect 86.11% of the problem of the bank's financial capacity whether or not it meets the CAMEL standards. Therefore, there will be other factors that may affect the financial capacity of Vietnamese commercial banks after M&A but have not been covered by this study in the current research model. The regression results show that the role of factors in contributing to improving the financial capacity of the bank is clearly hierarchically defined through the regression equation.

References

Alton Gilbert, R., Meyer, A. P., & Vaughan, M. D. (2002, January/February). Could a CAMELS downgrade model improve off-site surveillance? *Federal Reserve Bank of St. Louis Review, 84*(1).

Annual reports of banks LPB, SCB, SHB, HDBank, Pvcombank, Sacombank, BIDV, Maritimebank in the years 2012–2021.

Berger, A. N., Saunders, A., Scalise, J. M., & Udell, G. F. (1997). The effects of bank mergers and acquisitions on small business lending. *Finance and Economics Discussion Series 1997–1928*. Board of Governors of the Federal Reserve System (US).

Barnes, M. L., Lopez, J. A. (2005, September 23). Alternative measures of the Federal Reserve Banks cost of equity capital. *FRB of Boston Public Policy Discussion*, Paper No. 05-2.

Brown, C., & Davis, K. (2009). Capital management in mutual financial institutions. *Journal of Banking & Finance, 33*(3), 443–455.

Cadogan, G. (2011). *A theory of asset pricing and performance evaluation for minority banks with implications for bank failure prediction, compensating risk, and CAMELS rating*. Working Paper.

Heid, F. (2007). The cyclical effects of the Basel II capital requirements. *Journal of Banking & Finance, 31*(12), 3885–3900.

Hung, N. V. (2008). *Operational efficiency of Vietnam commercial bank in the period 2000–2005*. Doctoral Thesis in Economics, Banking Academy.

Johnston, M. (2009). Extending the Basel II approach to estimate capital requirements for equity investments. *Journal of Banking & Finance, 33*(6), 1177–1185.

Nga, P. T. H. (2013). *Financial capacity of Vietnamese commercial banks*. Doctoral Thesis in Economics. Banking University City. Ho Chi Minh.

Tatom, J. (2011, December 6). Predicting failure in the commercial banking industry. Networks Financial Institute at Indiana State University. Working Paper, No. 2011-WP-27.

Van Thang, H. (2009). *Applying CAMEL model in financial analysis at bank for investment and development of Vietnam*. Foreign Trade University, Hanoi.

Wagner, W. (2007). The liquidity of bank assets and banking stability. *Journal of Banking & Finance, 31*(1), 121–139.

Wirnkar, A., & Tanko, M. (2007). Camel(s) and banks performance, evaluation: the way forward. *SSRN Electronic Journal*.

Behavioral Intention and Behavior of Using E-Commerce Platforms for Online Purchases and Payments by Vietnamese Consumers

Van Duong Ha

Abstract The study was conducted to determine the factors affecting the behavioral intention and behavior of using e-commerce platforms for online purchases and payments by Vietnamese consumers, through the application of the unified theory of acceptance and use of technology (UTAUT2) model. The methods of Cronbach's alpha test, exploratory factor analysis (EFA), confirmatory factor analysis (CFA), and structural equation modeling analysis (SEM) were used with the SPSS and AMOS software. Besides inheriting the UTAUT2 model, this study has supplemented the UTAUT2 model by adding other related consumer constructs. The findings of the study indicate that there are four factors that have a positive effect on behavioral intention, such as effort expectancy, social influence, hedonic motivation, and autonomy. Besides, performance expectancy has a negative impact on behavioral intention. The factors that have a positive impact on use behavioral that are behavioral intention and facilitating conditions. The factors that have a negative impact on use behavioral that are habits and e-payment. This study has some important managerial implications, and it was proposed to primary stakeholders. This study may prove diagnostically useful to the behavioral intention and behavior of using e-commerce platforms for online purchases and payments of Vietnamese consumers.

Keywords Behavioral intention · Behavior of using · E-commerce · Price value · Use behavior

1 Introduction

E-commerce refers to economic activity that takes place online that includes all business models, such as retail shopping, banking, investments, and rentals. E-commerce applies technologies to marketing over the Internet, performing mobile commerce transactions, electronic payments, online transaction management and

V. D. Ha (✉)
Saigon Institute of Economics and Technology, Ho Chi Minh City, Vietnam
e-mail: dhv05@yahoo.com

A. T. Nguyen et al. (eds.), *Contemporary Economic Issues in Asian Countries: Proceeding of CEIAC 2022, Volume 1*,
https://doi.org/10.1007/978-981-19-9669-6_8

processing, electronic data interchange, electronic data management systems, automated inventory management, and data collection systems (Niranjanamurthy et al., 2013). Commerce is an exchange transaction between two parties, as the number of transactions conducted by electronic means has increased significantly since the spread of the Internet (Kaur & Joshi, 2012). E-commerce is usually associated with commercial transactions over the Internet, or the performance of any transaction involving the transfer of ownership or right to use goods or services through online transactions. E-commerce is also the use of electronic communication and digital technology in commercial business transactions to create, transform, and redefine relationships to create value between organizations, and between organizations and individuals (Gupta, 2014). Therefore, e-commerce is a business model that includes electronic payments and other activities related to online commercial transactions, it allows organizations and individuals to do their business over the Internet. This refers to economic activity that occurs online, which is not only buying and selling goods and products over the Internet, but also various business models within organizations and individuals.

In Vietnam, changes in consumers' behavioral intention and behavior of using e-commerce platforms for online purchases and payments promise to promote the Vietnamese e-commerce market to a new height. Major trends in Vietnam's e-commerce market include the trend of personalizing the buying experience, grocery shopping on e-commerce platforms, multi-channel retailing as well as joining e-commerce platforms, using available online marketing tools, etc. Many Vietnamese consumers spend more time on the e-commerce platforms, and they are willing to place orders of greater quantity and value. At the same time, diversifying payment methods brings convenience and safety for consumers, e-money payments are becoming more popular and are likely to dominate over payment on delivery.

The contribution of this study is to add to the UTAUT2 model other relevant consumer constructs. Besides inheriting the UTAUT2 model, the results of this study have added autonomy and electronic payment to the new research model. This study has several important management implications, and it has been suggested to key stakeholders. At the same time, this study proves the usefulness in terms of diagnostics for the behavioral intention and behavior of using e-commerce platforms to purchase and pay online of Vietnamese consumers.

2 Literature and Hypotheses

2.1 The Unified Theory of Acceptance and Use of Technology (UTAUT2) Model

The theoretical model of UTAUT was developed to represent the consumers' behavioral intentions and usage behavior for information technology applications. The UTAUT model conducted a test on four organizations for a period of six months, after reviewing and comparing eight competing models with thirty-two factors, and the UTAUT model was established. This is a combined model of eight previous models based on the most common view of studying user acceptance of a new information system. Eight models include Theory of Reasoned Action (TRA), Technology Acceptance Model (TAM), motivation model (MM), theory of planned behavior (TPB), a model combining TAM and TPB—model of combining TAM and TPB (C-TAM-TPB), model of PC utilization (MPCU), innovation diffusion theory (IDT), and social cognitive theory (SCT). The UTAUT model is built with four main factors determining acceptance and use the technology. According to this theory, four factors play a direct role in consumer acceptance and usage behavior for information technology, including: performance expectancy; effort expectancy; social influence; and facilitation conditions. There are also extrinsic factors (age, gender, experience, and voluntariness of use) that influence the consumers' behavioral intentions. This model is considered as integrating essential elements of other models, considering the influence of factors on the consumers' behavioral intentions and usage behavior for information technology applications, and it has been tested and proven to be superior to other models (Venkatesh et al., 2003). However, the literature lacked the evidence of user behavior model that could explain consumer use of technology. Given these limitations, proposed an extension of UTAUT, named UTAUT2. The goal of this model was to predict the technology acceptance and use behavior of an organization or individual, and integrated with hedonic motivation, price value, and habits into the original UTAUT model. In addition, The UTAUT2 model removed voluntariness of use from the demographic variables in the original UTAUT model as can be seen in Fig. 1 (Venkatesh et al., 2012).

2.2 Research Model

From the practical conditions of using e-commerce platforms by Vietnamese consumers, based on the theoretical basis of the UTAUT2 model, synthesizing the theoretical basis related to the behavioral intention and behavior of using e-commerce platforms for online purchases and payments by Vietnamese consumers to propose a research model. At the same time, the authors also combined the discussion with experts on the observed variables, and this study extended the UTAUT2 model, adding a new explanatory variable which is variable electronic payment. Because e-commerce refers to economic activity that includes electronic payments. Therefore, the research model of the behavioral intention and behavior of using e-commerce platforms for online purchases and payments by Vietnamese consumers is proposed as shown in Fig. 2.

Performance expectancy (PE), which is described as the level of belief in the ability of the consumers to use the system to achieve benefits in performing certain activities (Venkatesh et al., 2012). The theoretical background of this factor comes from usefulness perceptions, extrinsic motivation, and job fit. Performance expectancy has positive significant influence on behavioral intention to use e-commerce (Mansur

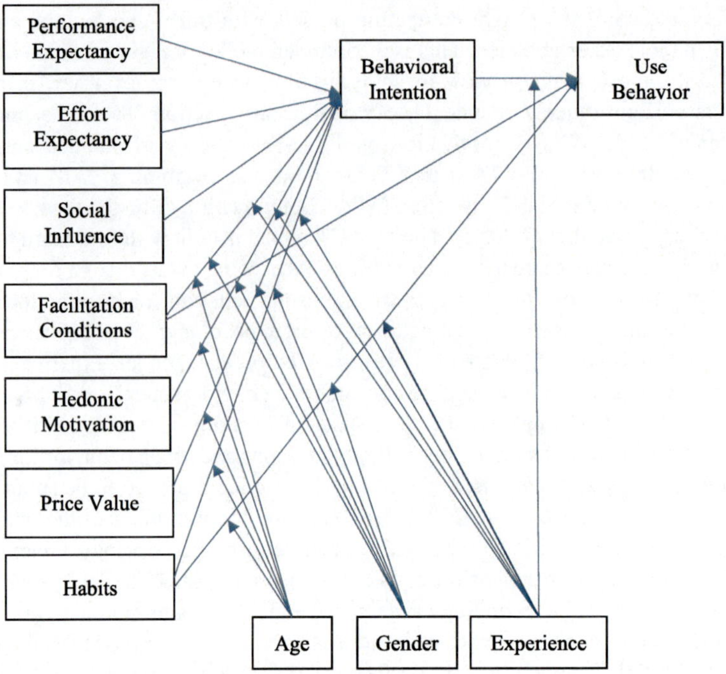

Fig. 1 UTAUT2 model. *Source* Venkatesh et al. (2012)

et al., 2019). Performance expectancy positively influences the behavioral intention of online transaction systems in e-commerce (Cabrera-Sánchez et al., 2020). At the same time, performance expectancy refers to consumers' belief that their purchases from an e-commerce platform will improve shopping efficiency and reduce costs (Chen et al., 2021) as well as there is a result study to show that the behavioral intention of e-commerce adoption is significantly influenced by performance expectancy (Ezennia & Marimuthu, 2022). Therefore, we formed a hypothesis:

Hypothesis 1 (H1). Performance expectancy has a positive effect on the behavioral intention of using e-commerce platforms for online purchases and payments by Vietnamese consumers.

Effort expectancy (EE), it is defined as the degree of facility associated with consumers' use of technology (Venkatesh et al., 2012). Effort expectancy refers to the ease and convenience for the consumers to purchase on the e-commerce platform. Effort expectancy has positive significant influence on behavioral intention to use e-commerce (Mansur et al., 2019). Consumers easily find the products they need on e-commerce platforms, and they have quick solutions to specific problems. There is research showing that effort expectancy significantly influences consumers' behavioral intention to purchase online via localization of e-commerce mobile applications or websites (Hungilo & Setyohadi, 2020). On the other hand, a significant relationship exists between efforts expectancy and e-shopping intention (Rehman

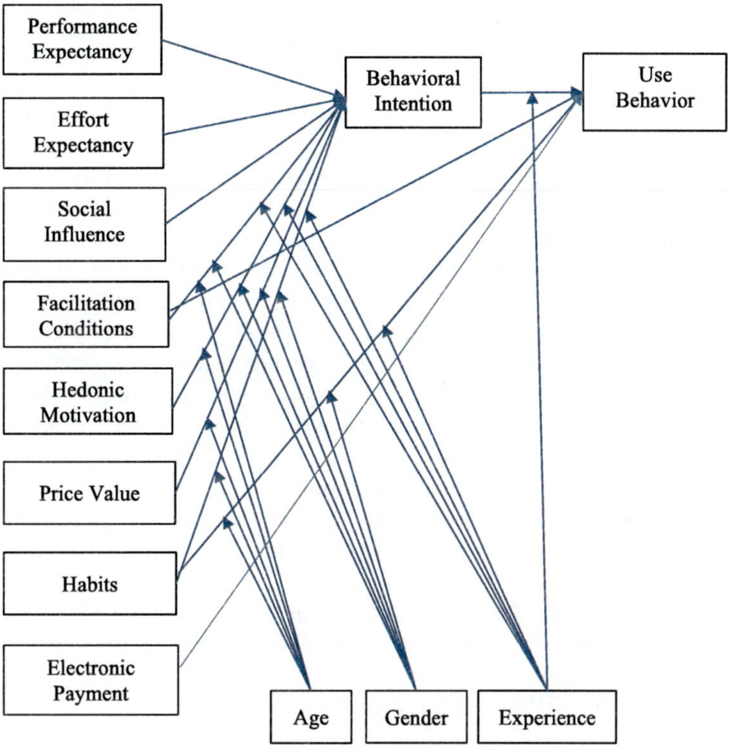

Fig. 2 Proposed model. *Source* Venkatesh et al. (2012) and author's supplement

et al., 2022) as well as there is a result study to show that the behavioral intention of e-commerce adoption is significantly influenced by effort expectancy (Ezennia & Marimuthu, 2022). Thus, we propose the following hypothesis:

Hypothesis 2 (H2): Effort expectancy has a direct positive and significant impact on the behavioral intention of using e-commerce platforms for online purchases and payments by Vietnamese consumers.

Social influence (SI), it measures the degree to which consumers are influenced by what others (e.g., family and friends), and an individual thinks the others should use a particular technology (Venkatesh et al., 2012). Social influence has positive significant influence on behavioral intention to use e-commerce (Mansur et al., 2019). Social influence has a significantly positive effect on consumers' behavioral intention of using an e-commerce platform (Chen et al., 2021). On the other hand, there is a result study to show that the behavioral intention of e-commerce adoption is significantly influenced by social influence (Ezennia & Marimuthu, 2022).Thus, the social influence for consumers' behavioral intention of using an e-commerce platform is the extent to which consumers think that those they consider important, believe that the person should use an e-commerce platform. Hypothesis H3 is stated as follows:

Hypothesis 3 (H3): Social influence has a positive effect on the behavioral intention of using e-commerce platforms for online purchases and payments by Vietnamese consumers.

Facilitating conditions (FC), these are defined as the extent to which consumers believe the organizations' infrastructure exists to support technological usage, the consumers are aware of the resources and supports available to perform their behavior (Venkatesh et al., 2012). Facilitating conditions showed a significant association with behavioral intention and facilitating conditions significantly influence the behavioral intention of consumers to adopt e-commerce (Pobee, 2021). A significant relationship exists between facilitating conditions and e-shopping intention (Rehman et al., 2022), and there is a result study to show that the behavioral intention of e-commerce adoption is significantly influenced by facilitating conditions (Ezennia & Marimuthu, 2022). Therefore, facilitating conditions for consumers' behavioral intention of using e-commerce platforms for online purchases and payments are the organization's degree of technological readiness or technical support for the use of e-commerce platforms. Therefore, two hypotheses were formed:

Hypothesis 4a (H4a): Facilitating conditions have a positive effect on the behavioral intention of using e-commerce platforms for online purchases and payments by Vietnamese consumers.

Hypothesis 4b (H4b): Facilitating conditions have a positive effect on the behavioral of using e-commerce platforms for online purchases and payments by Vietnamese consumers.

Hedonic motivation (HM), which is defined as the pleasure obtained from the use of technology. Hedonic motivation plays an important role in the acceptance and use of technology, and it has been shown to directly influence the acceptance and use of technology (Venkatesh et al., 2012). The consumers obtain a feeling of pleasure in utilizing e-commerce, and this revealed that hedonic motivation would positively affect behavioral intention to use e-commerce (Naruetharadhol et al., 2022). There are the study results show that behavioral intention of e-commerce adoption is significantly influenced by hedonic motivation (Ezennia & Marimuthu, 2022) as well as there is the study result which revealed customers' impulsive buying behavior is significantly affected by hedonic motivation on e-commerce platforms (Kamalia et al., 2022). We make the hypothesis H5 as following:

Hypothesis 5 (H5): Hedonic motivation has a positive effect on the behavioral intention of using e-commerce platforms for online purchases and payments by Vietnamese consumers.

Price value (PV), it refers to the perceived benefits of using technology and the monetary cost of using it. Besides, there is also a cost and pricing structure, and it may have a significant impact on consumers' technology use (Venkatesh et al., 2012). Price value is a key concern for consumers, and they would tend to purchase online through e-commerce websites when they believe that they are ending up paying less and reasonably priced in addition to all other benefits (Singh et al., 2017). There is research showing that price value significantly influences consumers' behavioral intention to purchase online via localization of e-commerce mobile applications or websites (Hungilo & Setyohadi, 2020), as well as there are the study results show that behavioral intention of e-commerce adoption is significantly influenced by price value (Ezennia & Marimuthu, 2022). Hypothesis H6 is stated as follows:

Hypothesis 6 (H6): Price value has a negative effect on the behavioral intention of using e-commerce platforms for online purchases and payments by Vietnamese consumers.

Habits (HA), they play an important role in the use of technology, which have been described various fundamental processes and habits that influence the use of technology. Because, habits are defined as the degree to which people tend to perform behaviors automatically as a result of learning (Venkatesh et al., 2012). There are the researches to show that habit influences consumers' use behavior for e-commerce (Mansur et al., 2019; Yoga & Triami, 2021). In other words, habits for e-commerce have a positive impact on consumers' behavioral intention as well as habits have a positive impact on consumers' use behavior (Wulandari et al., 2022). Therefore, two hypotheses were formed:

Hypothesis 7a (H7a): Habits have a positive effect on the behavioral intention of using e-commerce platforms for online purchases and payments by Vietnamese consumers.

Hypothesis 7b (H7b): Habits have a positive effect on the behavioral of using e-commerce platforms for online purchases and payments by Vietnamese consumers.

Electronic payment or e-payment (EP), which is a type of payment that allows consumers to pay for the products and services via electronic methods, such as Internet banking, mobile banking, card payments, e-wallet, and mobile wallet. (Josephine, 2021). The use of e-payment as a means of a payment system on e-commerce platforms causes consumers to buy more online and increasingly frequent transactions (Halim et al., 2020). E-payment offers a payment method which allows consumers to buy products or services via the e-commerce platforms with the delaying of payment (Pratika, 2021). E-payment is an important aspect of e-commerce, and it is a new driver of online shopping via the e-commerce platforms (Gupta et al., 2022). E-payment is a system that provides tools for payment of products and services in the e-commerce transactions. Using the e-payment has many benefits for consumers and contributes to increase behavior of using the e-commerce platforms. Therefore, we make the hypothesis as following:

Hypothesis 8 (H8): E-payment has a positive effect on the behavioral of using e-commerce platforms for online purchases and payments by Vietnamese consumers.

Behavioral intention (BI), this is defined as an individual subjective probability that person would like to use the technology continuously in the future. Therefore, the consumers' behavioral intention determines their technology use behavior (Venkatesh et al., 2012). There is a research to show that the acceptance and use of e-commerce was influenced by behavior intention. Researchers observed relationship between consumers' behavioral intention and use behavior for e-commerce and revealed this relationship between behavioral intention and the use behavior was proved significant (Yoga & Triami, 2021). Researchers also examined the behavioral intention correlates with use behavior and found a significant relationship between behavioral intention and actual adoption of the e-commerce platforms (Pobee, 2021). Therefore, hypothesis was formed:

Hypothesis 9 (H8): Behavioral intention has a positive effect on the behavioral of using e-commerce platforms for online purchases and payments by Vietnamese consumers.

Use behavior (UB), which refers to activity frequency to use the technology and the real conditions for the technology use. This is an important factor in determining the use of technology by consumers (Venkatesh et al., 2012). This case is using the e-commerce platforms, and the consumers' e-commerce use behavior in Vietnam is affected by the use behavioral intention of e-commerce.

The demographic factors, there are gender, age, and experience. These factors moderate the impact of facilitating conditions on behavioral intentions in the early stages of experience with the e-commerce platforms. They moderate the impact of hedonic motivation on behavioral intentions, so the effect is stronger in young consumers in the early stages of their experience with the e-commerce platforms. The demographic factors moderate the impact of price value on behavioral intentions, so the impact will be stronger for older consumers. These factors also moderate the impact of habit on behavioral intentions and habit on e-commerce use, so the impact will be stronger for older consumers with whom they have a high level of experience with the e-commerce platforms. The experience moderates the impact of behavioral intention on e-commerce use, so this impact will be stronger for consumers with less experience.

3 Research Methodology

3.1 Research Design

Qualitative research is used to explore the components of behavioral intention and behavior of using e-commerce platforms for online purchases and payments by Vietnamese consumers. This research is carried out through one-on-one discussions with experts on observed variables. From there, the study adjusts and supplements observed variables used to measure in the research model.

For quantitative research, this is research based on the positivism philosophy, used quantitative data which were collected from primary sources to examine certain populations or samples. Through the application of the UTAUT2 model, this study was conducted to determine the factors affecting the behavioral intention and behavior of using e-commerce platforms for online purchases and payments by Vietnamese consumers. The independent variables and dependent variable are presented in Table 1.

Table 1 Independent, moderating, and dependent variables in the research

No.	Code	Item
	PE	Performance expectancy
1	PE1	I use e-commerce platforms because it is easy for me to shop and pay online anywhere
2	PE2	I use e-commerce platforms because I can understand the interaction of online shopping and e-payment easily and clearly
3	PE3	I believe that the e-commerce platforms brings about the expected effect in online shopping and e-payment
4	PE4	By using e-commerce platforms, it will allow me to do online shopping and e-payment more rapidly
5	PE5	The use of e-commerce platforms is in accordance with all aspects of my work, and it would be beneficial in my everyday life
6	PE6	Using e-commerce platforms will increase my comfort in online shopping and e-payment transactions
7	PE7	Using e-commerce platforms will bring me more convenience, because I can online shopping and e-payment in 24 h
	EE	Effort expectancy
8	EE1	The use of e-commerce platforms can help me speed up online shopping and e-payment transactions
9	EE2	The use of e-commerce platforms can significantly improve the quality of the products as well as e-payment transaction results, because it does not take much time
10	EE3	The use of e-commerce platforms can help me make online shopping and e-payment more secure
11	EE4	The use of e-commerce platforms makes it easy for me to find out online shopping and e-payment information
12	EE5	The use of e-commerce platforms helps me always have full online shopping and e-payment information
13	EE6	The e-commerce platforms are practical, and they help me increase efficiency in online shopping and e-payment transactions
	SI	Social influence
14	SI1	Many people that have influence over my conduct advised me that I should use e-commerce platforms in online shopping and e-payment transactions
15	SI2	People who influence me think that I should use e-commerce platforms in online shopping and e-payment transactions
16	SI3	People who are familiar with me think that I should use e-commerce platforms in online shopping and e-payment transactions
17	SI4	I use e-commerce platforms for online shopping and e-payment transactions, because my coworkers and friends use it
18	SI5	My family supports me by using e-commerce platforms for online shopping and e-payment transactions

(continued)

Table 1 (continued)

No.	Code	Item
19	SI6	My neighborhood helped me with using e-commerce platforms for online shopping and e-payment transactions
	FC	Facilitating conditions
20	FC1	I have control over the use of e-commerce platforms in online shopping and e-payment transactions
21	FC2	I have the necessary knowledge to adopt e-commerce platforms in online shopping and e-payment transactions
22	FC3	I am certain that my online shopping and e-payment transactions are guaranteed the transaction conditions when I use e-commerce platforms
23	FC4	I possess the requisite resources to engage in online shopping and e-payment transactions
24	FC5	I am certain that online shopping and e-payment transactions are safe when I use e-commerce platforms
25	FC6	I have a smartphone and have received support from the e-commerce platform suppliers to use the e-commerce platforms in online shopping and e-payment transactions
	HM	Hedonic motivation
26	HM1	I feel comfortable when I use e-commerce platforms for online shopping and e-payment transactions
27	HM2	I feel blessed when I use e-commerce platforms for online shopping and e-payment transactions
28	HM3	I sincerely revel when I use e-commerce platforms for online shopping and e-payment transactions
29	HM4	I feel happy when I use e-commerce platforms for online shopping and e-payment transactions
30	HM5	I feel entertained when I use e-commerce platforms for online shopping and e-payment transactions
	PV	Price value
31	PV1	I can save time in doing online shopping and e-payment transactions when I use e-commerce platforms
32	PV2	I can save a lot of online shopping and e-payment transaction costs when I use e-commerce platforms
33	PV3	By using e-commerce platforms, I agree to pay the costs incurred for Internet registration
34	PV4	I do not have to pay fees to check my transactions when I use e-commerce platforms for online shopping and e-payment transactions
35	PV5	By using e-commerce platforms, I do not have to pay any extra costs in my transactions
		Habits
36	HA1	I regularly use the e-commerce platforms for online shopping and e-payment transactions

(continued)

Table 1 (continued)

No.	Code	Item
37	HA2	I can autonomously use the e-commerce platforms for online shopping and e-payment transactions
38	HA3	The usage of the e-commerce platforms for online shopping and e-payment transactions has turned out to be an addiction for me
39	HA4	I have received direct guidance from the e-commerce platform suppliers related to online shopping and e-payment transactions
40	HA5	I can use the e-commerce platforms for online shopping and e-payment transactions when no one is simulating
	EP	Electronic payment
41	EP1	I believe that e-payment always ensures security and privacy for all consumers when they use the e-commerce platforms for online shopping
42	EP2	I realize that e-payment always ensures the correct verification of information among the involved when I use the e-commerce platforms for online shopping
43	EP3	I believe that the e-payment system always has a plan to deal with attacks on security when I pay for my online shopping via the e-commerce platforms
44	EP4	I believe that e-payment has developed in parallel with the application of e-commerce platforms for online shopping
45	EP5	I believe my private data and e-payment transactions to be kept confidential when I use the e-commerce platforms for online shopping
46	EP6	I realize that the privacy of e-payment always ensures me to keep information undisclosed, free from unauthorized access when I use the e-commerce platforms for online shopping
		Behavioral intention
47	BI1	I intend to continue to use the e-commerce platforms for online shopping and e-payment transactions
48	BI2	I will use e-commerce platforms if I need to do online shopping and e-payment transactions
49	BI3	I would recommend e-commerce platforms for online shopping and e-payment to other people
		Use behavior
50	UB1	There are certain people and communities to help me when there are difficulties in using the e-commerce platforms for online shopping and e-payment transactions
51	UB2	I am able to carry out online shopping and e-payment via the e-commerce platforms without help from other people
52	UB3	I am able to use the e-commerce platforms for online shopping and e-payment transactions even though I have never used it beforehand

Source Venkatesh et al. (2012) and the authors' suggestions

3.2 *Sample and Data*

The population in this study was all consumers who were doing online shopping and e-payment transactions via the e-commerce platforms in the southern regions of Vietnam. The southern region of Vietnam is the most important center of economic activity in the country. This region is a leading zone for advanced manufacturing and is encouraging investments in knowledge-based and high-tech industries and services as well as the most important financial and trading hub of the whole country. Regarding the sample size, which is determined if the number of the population is unknown, the sampling is done using the size 5–10 multiplied by the question item, since the number of question items is 52, so the sample size in this study ($10 \times 52 = 520$) respondents. Therefore, the number of 650 samples collected from customers ensures a sufficient sample size for this study.

Explanatory research design will be used to analyze the data which is collected from the Vietnamese consumers. A necessary and feasible condition to participate in the survey was that the consumers who use the e-commerce platforms for online purchases and payments by Vietnamese consumers, and the method of purposive sampling was used. The questionnaire was derived from the constructs as shown in Table 1, and the data were collected through a structured questionnaire measured on a 5-point Likert scale (1 = strongly disagree; 5 = strongly agree).

SPSS 25.0 and AMOS 24.0 are two statistical software which were used and this study did a descriptive analysis, reliability test, and validity test. Descriptive statistics and demographic presentation of 650 sample data were analyzed using SPSS 25.0 software. The analysis method through Cronbach's alpha coefficient, EFA, and CFA analysis was also analyzed using SPSS 25.0 software. The SEM was established, CFA diagram and hypothesis verification analysis were conducted using AMOS 24.0 software.

4 Research Results

4.1 *Demographic Statistics*

A total of 650 respondents are consumers who completed the whole survey. The results of the hypothesized analysis of each path obtained from the path analysis results using SPSS 25.0 software are as follows as Table 2.

In terms of age composition, the respondents' age ranges from 20 years old and up, and a large number of them are a young generation with the age range of 20–40 years. In terms of gender, it was found that 60.62% of the respondents were female. This shows that females' participation is high when it comes to online shopping and e-payment transactions. Because women show more interest in online shopping and e-payment transactions on e-commerce platforms. The characteristics of respondents

Table 2 Demographic profile

Item	Optional	Frequency	Percentage
Age	20–30 year old	238	26.62
	31–40 year old	162	24.92
	41–50 year old	136	20.92
	51–60 year old	58	8.92
	61 years old and up	56	8.62
Gender	Male	256	39.38
	Female	394	60.62
Experience	Start–6 months	72	11.08
	Over 6 months–1 year	118	18.15
	Over 1–2 years	322	49.54
	Over 2 year	138	21.23
Education level	Vocational college graduation	159	24.46
	College graduation	485	74.62
	Postgraduate level	6	0.92
Income level	600 USD–800 USD/month	345	53.08
	Over 800 USD–1000 USD/month	176	27.08
	Over 1000 USD–1500 USD/month	62	9.54
	Over 1500 USD–2000 USD/month	55	8.46
	Over 2000 USD/month	12	1.84

Source Primary data, processed in 2022

when viewed from their experience of using e-commerce platforms for online shopping and e-payment transactions turned out to be the biggest answer at over 1–2 years, or around 49.54%. In terms of education, with 74.62% of respondents are at a college graduation, this percentage represents the majority of respondents. The respondents also found that 24.46% are at a vocational college graduation, and the rest at post-graduate level. In terms of the monthly disposable income, the respondents' income range is from 600 USD and up per month and 53.08% of them are over 600 USD to 800 USD per month.

4.2 *Cronbach's Alpha Reliability Analysis*

To conduct factor analysis, it is necessary to first conduct reliability analysis through Cronbach's alpha coefficient and total variable correlation coefficient. A scale with Cronbach's alpha coefficient >= 0.60 is acceptable in terms of reliability. Variables with a total correlation coefficient less than 0.3 will be excluded (Hulin et al., 2001). The results of the reliability analysis of this study show that the variables have an alpha coefficient greater than 0.6 and a total correlation coefficient greater than 0.3, as shown in Table 3. All scales of this study are the same eligible to perform EFA.

4.3 Exploratory Factor Analysis

The study implements the EFA method to test the influencing factors and identify the factors that are considered suitable for inclusion in the CFA. For independent variables, the results of Kaiser–Meyer–Olkin (KMO) and Bartlett's test show that KMO coefficient = 0.817 (greater than 0.5), so EFA is suitable for the reality data. At the same time, Sig = 0.000 < 0.05 shows that the observed variables are correlated with each other in terms of the population.

This study performed factor analysis for the independent variables with the principal components and varimax rotation (absolute value below: 0.3). The results show that 46 observed variables were initially grouped into ten groups. The total value of variance extracted is 59.371%, which is greater than 50% and EFA is considered satisfactory. It can be said that these ten factors explain 59.731% of the variability of the data. The eigenvalues of the factors are all high (>1). The tenth factor has the lowest eigenvalues of 1.114 > 1. Therefore, there are ten factors that are defined as shown in Table 4.

Ten factors were found after performing the rotation (absolute value below: 0.3) for the independent variables together with the factor loading factors all that are greater than 0.5. This result also showed there were two new factors (with the observed variables: HA1 and HA2; EP5 and EP6) to be explored, as can see in Table 5. With the characteristics of two observed variables in the factor HA associated with customers' autonomy (HA1 and HA2 as shown in Table 1) in using the e-commerce platforms for online shopping and e-payment transactions. Therefore, this new factor is named autonomy (AU). The hypothesis for AU is similar to the hypothesis for HA. AU has a positive effect on consumers' behavioral intention of using the e-commerce platforms for online shopping and e-payment transactions, and AU has a positive effect on consumers' behavior of using the e-commerce platforms for online shopping and e-payment transactions. At the same time, with the characteristics of two observed variables in the factor EP associated with security and privacy (EP5 and EP6 as shown in Table 1). Therefore, this new factor is named security and privacy (SP). The hypothesis for SP is similar to the hypothesis for EP. SP has a positive effect on consumers' behavior of using e-commerce platforms for online shopping and e-payment transactions.

For dependent variables, the results of Kaiser–Meyer–Olkin (KMO) and Bartlett's test show that KMO coefficient = 0.678 (greater than 0.5), so EFA is suitable for the reality data. At the same time, Sig = 0.000 < 0.05 shows that the observed variables are correlated with each other in terms of the population. This study also performed factor analysis for the dependent variables with the principal components and varimax rotation. The results show that six observed variables were initially grouped into two groups. The total value of variance extracted is 65.162%, which

Table 3 Item reliability test

Item	Scale mean if item deleted	Scale variance if item deleted	Corrected item-total correlation	Cronbach's alpha if item deleted
Performance expectancy: Cronbach's alpha = 0.873				
PE1	20.71	16.251	0.718	0.846
PE2	20.70	16.353	0.681	0.851
PE3	20.40	19.070	0.599	0.864
PE4	20.50	18.204	0.579	0.864
PE5	20.56	17.230	0.623	0.859
PE6	20.65	15.945	0.775	0.837
PE7	20.61	17.548	0.608	0.860
Effort expectancy: Cronbach's alpha = 0.828				
EE1	18.07	10.629	0.689	0.781
EE2	17.85	11.870	0.538	0.815
EE3	17.94	11.413	0.634	0.793
EE4	18.02	11.735	0.581	0.804
EE5	18.08	11.486	0.593	0.802
EE6	17.96	11.584	0.556	0.810
Social influence: Cronbach's alpha = 0.838				
SI1	18.19	15.490	0.669	0.800
SI2	18.13	15.591	0.606	0.815
SI3	17.78	17.448	0.599	0.815
SI4	18.02	17.391	0.584	0.818
SI5	17.98	17.163	0.576	0.819
SI6	18.28	15.778	0.663	0.801
Facilitating conditions: Cronbach's alpha = 0.826				
FC1	12.91	7.096	0.676	0.786
FC2	12.89	7.183	0.580	0.807
FC3	12.76	7.331	0.605	0.801
FC4	12.83	7.547	0.586	0.805
FC5	12.82	7.431	0.603	0.801
FC6	12.85	7.585	0.550	0.812
Hedonic motivation: Cronbach's alpha = 0.816				
HM1	14.55	3.613	0.543	0.798
HM2	14.54	3.565	0.523	0.805
HM3	14.41	3.352	0.668	0.761
HM4	14.46	3.491	0.575	0.789
HM5	14.46	3.266	0.729	0.742

(continued)

Table 3 (continued)

Item	Scale mean if item deleted	Scale variance if item deleted	Corrected item-total correlation	Cronbach's alpha if item deleted
Price value: Cronbach's alpha = 0.778				
PV1	14.17	3.249	0.557	0.736
PV2	13.91	2.871	0.632	0.710
PV3	13.93	3.291	0.669	0.702
PV4	14.92	3.751	0.413	0.779
PV5	13.91	3.583	0.515	0.750
Habit: Cronbach's alpha = 0.678				
HA1	14.47	3.529	0.285	0.695
HA2	14.71	3.508	0.378	0.649
HA3	14.70	3.259	0.503	0.597
HA4	14.45	3.034	0.507	0.591
HA5	14.50	3.224	0.508	0.594
Security and privacy: Cronbach's alpha = 0.663				
EP1	17.70	7.486	0.410	0.623
EP2	17.72	7.582	0.428	0.623
EP3	17.75	7.739	0.367	0.636
EP4	17.69	7.435	0.462	0.613
EP5	17.78	5.355	0.463	0.604
EP6	17.69	5.169	0.448	0.621
Behavioral intention: Cronbach's alpha = 0.759				
BI1	7.31	1.326	0.592	0.676
BI2	7.54	1.531	0.578	0.692
BI3	7.32	1.369	0.602	0.662
Use behavior: Cronbach's alpha = 0.700				
UB1	7.36	1.267	0.484	0.648
UB2	7.27	1.176	0.529	0.591
UB3	7.22	1.216	0.536	0.584

Source The authors' calculation from SPSS 25.0

is greater than 50% and EFA is considered satisfactory. It can be said that these two factors explain 65.162% of the variability of the data. The eigenvalues of the factors are all high (>1). The second factor has the lowest eigenvalues of 1.572 > 1. Therefore, there are two factors that are defined as shown in Table 6.

Two factors were found after performing the rotation (absolute value below: 0.3) for the dependent variables together with the factor loading factors all that are greater than 0.5, as can be seen in Table 7.

Table 4 Exploratory factor analysis for independent variables

Component	Initial eigenvalues			Extraction sums of squared loadings			Rotation sums of squared loadings	
	Total	% of variance	Cumulative %	Total	% of variance	Cumulative %	Total	Cumulative %
1	6.417	13.950	13.950	6.417	13.950	13.950	4.066	8.839
2	3.537	7.689	21.639	3.537	7.689	21.639	3.464	16.369
3	3.277	7.123	28.762	3.277	7.123	28.762	3.322	23.590
4	2.851	6.197	34.960	2.851	6.197	34.960	3.299	30.762
5	2.540	5.521	40.480	2.540	5.521	40.480	2.986	37.253
6	2.465	5.359	45.839	2.465	5.359	45.839	2.744	43.219
7	2.199	4.780	50.619	2.199	4.780	50.619	2.318	48.258
8	1.651	3.590	54.209	1.651	3.590	54.209	2.137	52.903
9	1.260	2.740	56.950	1.260	2.740	56.950	1.498	56.161
10	1.114	2.421	59.371	1.114	2.421	59.371	1.477	59.371
11	0.922	2.005	61.376					
Extraction method: principal component analysis								

Source The authors' calculation from SPSS 25.0

4.4 *Confirmatory Factor Analysis*

In the CFA, the results of Kaiser–Meyer–Olkin (KMO) and Bartlett's test show that KMO coefficient = 0.808 (greater than 0.5), so CFA is suitable for real data. At the same time, Sig = 0.000 < 0.05 shows that the observed variables are correlated with each other in terms of the population. This study performed factor analysis for the variables with the principal axis factoring and promax rotation (absolute value below: 0.3). The results of CFA showed that the number of observations is 650 as shown in Fig. 3. After linking e4 and e5, e4 and e6, e10 and e12, e21 and e24, e28 and e30, and e38 and e39 to correct for covariance show that this model has a chi-square = 2183. 213, with 1204 degrees of freedom (df); chi-square/df = 1813 < 3 with p-value = 0.000 and other indicators such as CFI = 0.915; TLI = 0.906; GFI = 0.888; RMSEA = 0.035 < 0.06; and PCLOSE = 1000 > 0.05. Thus, the model is completely consistent with the market data, and it can be concluded that the factors are convergent values because the standardized and unstandardized coefficients are greater than 0.5, and the total variance values are greater than 0.5 in this study.

4.5 *Structural Equation Modeling*

The results of performing SEM model analysis through AMOS software show the suitability of the research model. The estimated results of the proposed model are

Table 5 Rotated component matrix for independent variables

Variables	Component									
	1	2	3	4	5	6	7	8	9	10
PE6	0.829									
PE6	0.833									
PE1	0.780									
PE7	0.726									
PE2	0.720									
PE3	0.719									
PE5	0.711									
PE4	0.632									
SI6		0.773								
SI1		0.758								
SI3		0.717								
SI4		0.685								
SI2		0.685								
SI5		0.652								
EE1			0.800							
EE3			0.769							
EE5			0.711							
EE4			0.708							
EE6			0.702							
EE2			0.637							
FC1				0.788						
FC3				0.741						
FC5				0.732						
FC4				0.708						
FC2				0.707						
FC6				0.684						
HM5					0.853					
HM3					0.807					
HM4					0.728					
HM1					0.702					
HM2					0.681					
PV3						0.818				
PV2						0.780				
PV1						0.732				
PV5						0.686				

(continued)

Table 5 (continued)

Variables	Component									
	1	2	3	4	5	6	7	8	9	10
PV4						0.590				
EP2							0.781			
EP4							0.739			
EP3							0.731			
EP1							0.642			
HA5								0.839		
HA4								0.805		
HA3								0.760		
HA1									0.822	
HA2									0.776	
EP6										0.780
EP5										0.736

Extraction method: principal component analysis. Rotation method: varimax with Kaiser normalization

Source The authors' calculation from SPSS 25.0

Table 6 Exploratory factor analysis for dependent variables

Component	Initial eigenvalues			Extraction sums of squared loadings			Rotation sums of squared loadings	
	Total	% of variance	Cumulative %	Total	% of variance	Cumulative %	Total	Cumulative %
1	2.338	38.968	38.968	2.338	38.968	38.968	2.032	33.863
2	1.572	26.195	65.162	1.572	26.195	65.162	1.878	65.162
3	0.656	10.937	76.100					

Extraction method: principal component analysis

Source The authors' calculation from SPSS 25.0

shown in Fig. 4. The indicators include chi-square = 2216,852; df = 1211; p = 0.000; chi-square/df = 1,831; CFI = 0.913; TLI = 0.905; GFI = 0.886; RMSEA = 0.036; and PCLOSE = 1000. Thus, the proposed research model achieves compatibility with the market data.

Using 95% confidence standard, the sig of FC on BI is 0.103 > 0.05, variable FC has no effect on BI; sig of PV on BI is 0.388 > 0.05, variable PV has no effect on BI; sig of HA on BI is 0.091 > 0.05, variable HA has no effect on BI; sig of AU on UB is 0.103 > 0.05, variable AU has no effect on UB; sig of SP on UB is 0.198 > 0.05, variable SP has no effect on UB. The remaining variables all have sig < 0.05, so these relationships are significant as can be seen in Table 8.

Table 7 Rotated component matrix for dependent variables

Variable	Component	
	1	2
BI3	0.830	
BI1	0.815	
BI2	0.808	
UB3		0.809
UB2		0.788
UB1		0.763
Extraction method: principal component analysis		
Rotation method: varimax with Kaiser normalization		

Source The authors' calculation from SPSS 25.0

The variables that have a positive effect on BI are EE, SI, HM, and AU, in which the variable SI is the variable that has the strongest impact on BI. Besides, there is a variable that has a negative impact on BI that is the PE variable. The variables have a positive impact on UB that are BI and FC, where BI is the variable that has the strongest impact on UB. At the same time, there are two variables that have a negative impact on UB, which decrease in the following order: HA and EP as shown in Table 9.

The bootstrap method is used to estimate summary statistics such as the mean or standard deviation to evaluate the robustness of the theoretical model. This study used the number of replicate samples $N = 1500$. Compare the C.R value with 1.96 (since 1.96 is the value of the normal distribution at 0.9750). The results show that C.R < 1.96, p-value $> 5\%$, so the estimated model can be reliable as can see in Table 10.

4.6 Testing for Differences in Demographic Characteristics

The results of testing the differences according to demographic variables for behavioral intention and behavior of using e-commerce platforms are presented in Table 11. Accordingly, gender has sig value in independent sample T-test is greater than 0.05, which shows that there is no statistically significant difference in behavioral intention for the respondents of different genders.

Age and experience have sig value in the Levene test higher than 0.05, the variance between the choices of these variables is not different for the behavioral intention of using e-commerce platforms. The Anova test has a sig value greater than 0.05, this shows that there is no statistically significant difference in behavioral intention of using e-commerce platforms for the respondents of different age groups.

At the same time, experience has sig value in the Levene test higher than 0.05, the variance between the choices of these variables is not different for the behavior of

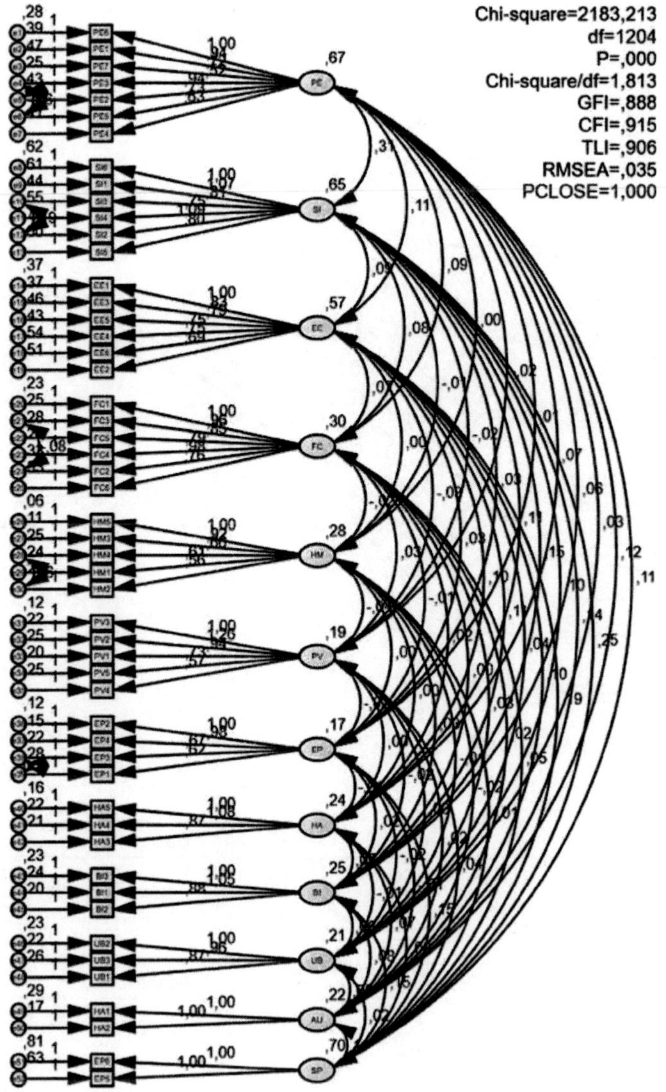

Fig. 3 Confirmatory factor analysis. *Source* The authors' calculation from AMOS 24.0

using e-commerce platforms. The Anova test has a sig value greater than 0.05, and this shows that there is no statistically significant difference in the behavior of using e-commerce platforms for the respondents of different age groups.

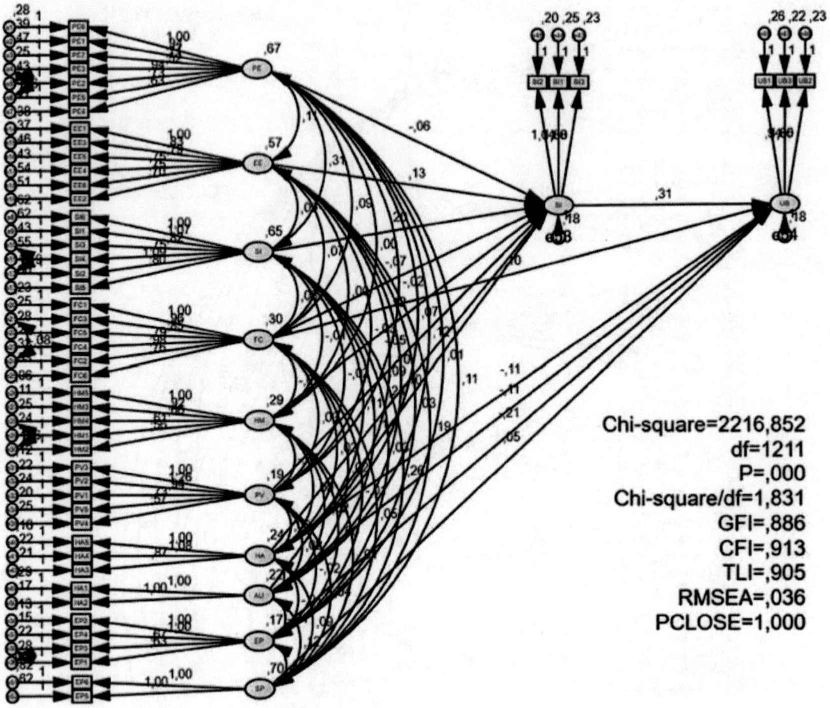

Fig. 4 Structural equation modeling. *Source* The authors' calculation from AMOS 24.0

Table 8 Regression weights

			Estimate	S.E	C.R	*P*
BI	←	PE	−0.065	0.032	−2.022	0.043
BI	←	EE	0.132	0.034	3.915	***
BI	←-	SI	0.203	0.036	5.597	***
BI	←	FC	−0.070	0.043	−1,631	0.103
BI	←	HM	0.116	0.042	2.791	0.005
BI	←	PV	−0.047	0.054	−0.863	0.388
BI	←	HA	0.090	0.053	1.689	0.091
BI	←	AU	0.235	0.067	3.513	***
UB	←	BI	0.312	0.059	5.260	***
UB	←	FC	0.104	0.042	2.444	0.015
UB	←	HA	−0.113	0.055	−2.051	0.040
UB	←	AU	−0.111	0.068	−1.632	0.103
UB	←	EP	−0.205	0.076	−2.719	0.007
UB	←	SP	0.054	0.042	1.287	0.198

Source The authors' calculation from AMOS 24.0

Table 9 Standardized regression weights

			Estimate
BI	←	PE	−0.106
BI	←	EE	0.199
BI	←	SI	0.326
BI	←	FC	−0.077
BI	←	HM	0.124
BI	←	PV	−0.041
BI	←	HA	0.088
BI	←	AU	0.219
UB	←	BI	0.338
UB	←	FC	0.124
UB	←	HA	−0.120
UB	←	AU	−0.112
UB	←	EP	−0.184
UB	←	SP	0.098

Source The authors' calculation from AMOS 24.0

Table 10 Bootstrap method on SEM

Parameter			SE	SE-SE	Mean	Bias	SE-Bias	C.R = Bias/SE-Bias
BI	←	PE	0.061	0.001	−0.105	0.001	0.002	0.5
BI	←	EE	0.055	0.001	0.197	−0.001	0.001	−1.0
BI	←	SI	0.063	0.001	0.328	0.002	0.002	1.0
BI	←	FC	0.053	0.001	−0.078	−0.001	0.001	−1.0
BI	←	HM	0.046	0.001	0.124	0.000	0.001	0
BI	←	PV	0.050	0.001	−0.040	0.001	0.001	1.0
BI	←	HA	0.052	0.001	0.089	0.001	0.001	1.0
BI	←	AU	0.074	0.001	0.220	0.001	0.002	0.5
UB	←	BI	0.068	0.001	0.339	0.001	0.002	0.5
UB	←	FC	0.051	0.001	0.124	0.000	0.001	0
UB	←	HA	0.063	0.001	−0.122	−0.002	0.002	−1.0
UB	←	AU	0.078	0.001	−0.111	0.001	0.002	0.5
UB	←	EP	0.072	0.001	−0.185	−0.001	0.002	−0.5
UB	←	SP	0.087	0.002	0.099	0.002	0.002	1.0

Source The authors' calculation from AMOS 24.0

Table 11 Testing for differences in demographic characteristics

Demographic characteristics	Sig value		
	Levene	T-test	Anova
Difference in behavior intention			
Gender		0.886	
Age	0.073		0.643
Experience	0.278		0.109
Difference in use behavior			
Experience	0.681		0.058

Source The authors' calculation from SPSS

5 Discussions Results

5.1 *Factors Affecting Behavioral Intention*

Performance expectancy is a factor that has a negative impact on the behavioral intention of using e-commerce platforms for online purchases and payments by Vietnamese consumers. This result disagreed with the expected sign and hypotheses, and it was not consistent with the findings of previous researchers (Venkatesh et al., 2012; Mansur et al., 2019; Cabrera-Sánchez et al., 2020; Chen et al., 2021; Ezennia & Marimuthu, 2022). Performance expectancy is a factor that has a negative impact on the behavioral intention of using e-commerce platforms for online purchases and payments by Vietnamese consumers. This result disagreed with the expected signs and hypotheses, and it was not consistent with the findings of previous researchers. In Vietnam, e-commerce has been regulated in a number of legal documents in the past years, but these regulations still need to be revised, supplemented, and perfected in recent years. On the other hand, logistics services are still a challenge for e-commerce businesses in Vietnam due to the lack of qualified businesses and the small and fragmented logistics industry. Although it has been and is being improved, Vietnam's infrastructure is still considered inadequate, and moving between provinces is still expensive. The road network as well as the main retail and distribution network is one of the major challenges for e-commerce businesses, which has affected the use of e-commerce platforms. Therefore, performance expectancy has a negative impact on the behavioral intention of using e-commerce platforms for online purchases and payments by Vietnamese consumers.

Effort expectancy is a factor that has a positive impact on the behavioral intention of using e-commerce platforms for online purchases and payments by Vietnamese consumers. These findings are relevant and support previous studies (Venkatesh et al., 2012; Mansur et al., 2019; Ezennia & Marimuthu, 2022; Hungilo & Setyohadi, 2020; Rehman et al., 2022), and they agree with the expected signs and hypotheses in this study. The effort expectancy of good e-commerce platforms can affect the behavioral intention of using e-commerce platforms for online purchases and payments by Vietnamese consumers. This shows that consumers' clear understanding of the interaction between e-commerce with online shopping and e-payment, as well as the use of e-commerce platforms can significantly improve the quality of the products and e-payment transactions, because it does not take much time.

The relationship of social influence toward the behavioral intention of using e-commerce platforms for online purchases and payments by Vietnamese consumers is proved to be significant. These findings are the following research conducted by previous researchers (Venkatesh et al., 2012; Mansur et al., 2019; Chen et al., 2021; Ezennia & Marimuthu, 2022) on using e-commerce platforms, which shows that social influence can explain behavioral intention. This happens because consumers are very interested in the recommendations and attitudes of family, friends, neighbors, and colleagues in forming intentions to use e-commerce platforms.

Hedonic motivation is a factor that has a positive impact on the behavioral intention of using e-commerce platforms for online purchases and payments by Vietnamese consumers. This result agreed with the expected signs and hypotheses, and it was consistent with the findings of previous researchers (Venkatesh et al., 2012; Ezennia & Marimuthu, 2022; Naruetharadhol et al., 2022; Kamalia et al., 2022). This indicates that the comfortable, excited, blessed, happy, and entertained of using e-commerce platforms for online purchases and payments can encourage Vietnamese consumers to intend to use the platform.

Autonomy has a positive effect on consumers' behavioral intention of using the e-commerce platforms for online shopping and e-payment transactions. This result agreed with the expected signs and hypotheses in this study. This shows that the e-commerce platforms, absorption, and autonomy of online shopping and e-payment transactions positively enhance the sense of behavioral intention, and that it helps consumers meet their psychological needs, which ultimately affects consumers' behavioral intention of using the e-commerce platforms for online shopping and e-payment transactions.

5.2 Factors Affecting Use Behavioral

There was a complementing relationship that exists between the behavioral intention and behavioral of using e-commerce platforms for online purchases and payments by Vietnamese consumers. This result is in line with research conducted by previous researchers, and it agreed with the expected signs and hypotheses in this study (Venkatesh et al., 2012; Pobee, 2021; Yoga & Triami, 2021). This means that good behavioral intentions can influence the behavior of using e-commerce platforms for online shopping and e-payment transactions by Vietnamese consumers. The more the consumers' behavioral intention of using the e-commerce platforms, the greater the desire to use e-commerce platforms of Vietnamese consumers.

Facilitating conditions are a factor that has a positive impact on the behavior of using e-commerce platforms for online purchases and payments by Vietnamese consumers. These findings are consistent with previous research (Venkatesh et al., 2012; Ezennia & Marimuthu, 2022; Rehman et al., 2022). This shows that good facilitating conditions of e-commerce platforms can influence the use behavior of e-commerce platforms. The consumers have enough facilities for using e-commerce platforms, facilitating settings and resources connected to the usage of Internet data can have a favorable influence on the behavioral of using e-commerce platforms for online purchases and payments by Vietnamese consumers.

Habit has a negative effect on consumers' behavior of using e-commerce platforms for online purchases and payments by Vietnamese consumers. This result is not in line with research conducted by previous researchers (Venkatesh et al., 2012; Mansur et al., 2019; Yoga & Triami, 2021; Wulandari et al., 2022), and it disagreed with the expected signs and hypotheses in this study. The e-commerce platform is one of the options for modern and convenient purchases and payment methods. However, in Vietnam, e-commerce has been regulated in a number of legal documents in the past years, but these regulations still need to be revised, supplemented, and perfected in recent years. Besides, the habit of using cash of Vietnamese consumers is also a big barrier to the development of e-commerce. Although the government has made many efforts to promote non-cash payment, the rate of cash use in Vietnam is still high compared to other countries in the region. Therefore, habits have a negative impact on the behavior of using e-commerce platforms for online purchases and payments by Vietnamese consumers.

Electronic payment is a factor that has a negative impact on the behavior of using e-commerce platforms for online purchases and payments by Vietnamese consumers. This result disagreed with the expected signs and hypotheses, and it was not consistent with the findings of previous researchers (Josephine, 2021; Halim et al., 2020; Pratika, 2021; Gupta et al., 2022). As analyzed, e-payment is a type of payment that allows consumers to pay for products and services via electronic methods. It offers a payment method which allows consumers to buy products or services via e-commerce platforms. However, the payment transactions in Vietnam are mainly in cash, which has affected the use of e-payment in payment transactions. This is also a big barrier to the development of e-commerce. Therefore, e-payment has a negative impact on the behavior of using e-commerce platforms for online purchases and payments by Vietnamese consumers.

6 Conclusions and Recommendations

To find out the factors affecting the behavioral intention and behavior of using e-commerce platforms for online purchases and payments of Vietnamese consumers, this study used the UTAUT2 model with its purpose being to improve understanding of UTAUT2 and how to apply this model in different fields. Besides inheriting the UTAUT2 model, this study has supplemented the UTAUT2 model by adding other related consumer constructs such as e-payment. The theoretical contribution of this study is the addition and discovery of new factors such as e-payment and autonomy to the UTAUT2 model. At the same time, the present study critically examined the behavioral intention and behavior of using e-commerce platforms for online purchases and payments of Vietnamese consumers. The findings of the study indicate that there are four factors that have a positive effect on behavioral intention, such as effort expectancy, social influence, hedonic motivation, and autonomy. Besides, performance expectancy has a negative impact on behavioral intention. The factors that have a positive impact on use behavioral that are behavioral intention and facilitating conditions. On the other hand, the factors that have a negative impact on use behavioral that are habits and e-payment. This study has some important managerial implications, and it was proposed to primary stakeholders. This research helps managers to develop their e-commerce platforms, offers valuable insights for e-commerce founders, and the approach taken in this study may prove diagnostically useful to the behavioral intention and behavior of using e-commerce platforms for online purchases and payments of Vietnamese consumers. At the same time, based on the study results, this study provides a relevant recommendation to increase the behavioral intention and behavior of using e-commerce platforms for online purchases and payments by Vietnamese consumers as follows.

Firstly, the social influence exerts a significantly positive effect on the behavioral intention of using e-commerce platforms for online purchases and payments of Vietnamese consumers. So, e-commerce founders and managers should make e-commerce platforms prevalent and fashionable, such as advertising and promotion. They need to implement marketing campaigns that employ public influencers and should timely and quickly update attractive consumer care programs on online channels. This will help the e-commerce platforms be spread to many consumers, and it is very important to maintain a consumer base as well as to attract new consumers.

Secondly, effort expectancy in the use of e-commerce platforms can help consumers speed up online purchases and payment transactions, can significantly improve the quality of online purchases and payment transaction results because it does not take much time, can make the online purchases and payment transactions more secure, helps customers always have full information and increase efficiency in the online purchases and payment transactions. The e-commerce providers should research and develop e-commerce platforms with high compatibility with many devices, diversify the utility of online purchases and payment transactions to suit many consumers, as well as the ability to expand and develop e-commerce platforms in the future. Therefore, the improvement in the use of e-commerce platforms by the e-commerce providers will help consumers to achieve higher efficiency without much effort in online purchases and payment transactions and increase the behavioral intention of using e-commerce platforms for online purchases and payments by Vietnamese consumers.

Thirdly, the e-commerce founders and managers should understand the extent of the hedonic motivation toward the behavioral intention and behavior of using e-commerce platforms for online purchases and payments of Vietnamese. They should regularly upgrade the e-commerce platforms to ensure that the transaction is done quickly, accurately in the online purchases and payment transactions, ready to support consumers when needed as well as immediately handle problems arising if any, and meet all the needs of consumers. These help many consumers feel entertained, excited, blessed, comfortable, and happy when they use the e-commerce platforms for online purchases and payment transactions. On the other hand, the e-commerce providers should research to understand the needs of consumers to take care of the consumers in a timely manner, help them beyond merely providing online purchases and payment transactions, and make them feel as comfortable and efficient as possible. It was these things that have made the consumers increase their behavioral intention of using e-commerce platforms for online purchases and payments in Vietnam.

Fourth, the e-commerce platforms are the virtual markets, where secure electronic transactions take place. Therefore, consumers and e-commerce providers meet in this virtual space to do online shopping and e-payment transactions. Consumers use e-commerce platforms to speed up the exchange of information, gain improved service levels, and reduce transaction costs. At the same time, autonomy plays an important role in the consumers' behavioral intention of using the e-commerce platforms for online shopping and e-payment transactions. It implies that the e-commerce platforms need to not only improve its online shopping and e-payment transactions but also present the improvement and its effort to increase the autonomy of consumers. On the one hand, the e-commerce founders and managers should contribute to consumer well-being by making consumer choices easier, more practical, and more efficient. On the other hand, they should also contribute to increasing consumers' sense of autonomy and enhancing the sense of behavioral intention, and that it helps consumers meet their psychological needs, which ultimately affects consumers' behavioral intention of using the e-commerce platforms for online shopping and e-payment transactions in Vietnam.

Fifthly, the good facilitating conditions of e-commerce platforms can influence the use behavior of e-commerce platforms. Hence, the e-commerce founders and managers should provide good facilitating conditions for the consumers. These include management of the e-commerce platforms and resources connected to the usage of e-commerce platforms. Therefore, consumers already have facilities in the form of a smartphone and adequate settings will boost their behavior of using e-commerce platforms for online purchases and payment transactions.

Sixthly, behavioral intention was considered as one of the important factors influencing consumers' behavior of using e-commerce platforms for online purchases and payment transactions by Vietnamese consumers. Therefore, the e-commerce founders and managers should improve the usefulness of online purchases and payment transactions, focus on the information about buying and selling, products, and services via the e-commerce platforms, ensure online purchases and payment transactions at cost, and increasingly expand their product portfolio and capabilities. These are the best ways for the e-commerce providers to respond to the consumers' needs and bring more convenience. Thereby, it also cultivates relationships with consumers and designs products that meet consumer needs well. These results are the top concern and contribute to promoting the behavior of using e-commerce platforms for online purchases and payment transactions by Vietnamese consumers.

Seventhly, consumers are faced with more choices and more information about e-commerce platform options than ever before. The growth of e-commerce platforms will help consumers search and select options that best suit their needs, allowing them to reduce search costs and enhance the utility they get from their options. Hence, e-commerce providers should enhance customer experience, design e-commerce application platforms that meet optimal standards of service quality so that consumers can easily and conveniently log in and make online purchases and payment transactions. It is necessary to use an easy to understand and simple language, and Internet-connected devices must be secure and safe in online purchases and payment transactions. At the same time, they should provide specific and detailed instructions for consumers on how to use the e-commerce platforms in writing, video, online, or by phone when requested by consumers. Thereby, helping consumers become more and more familiar with using e-commerce platforms in Vietnam.

Eighthly, marketers, researchers, policy-makers research to promote the development of e-commerce platforms with synchronous mechanisms and policies. Regulations are needed to ensure that for online retailers specializing in standardized products, and the new e-commerce platform is expected to improve product quality, improve delivery and refunds, enhance payment security, and more. At the same time, it is necessary to have an incentive policy to empower consumers and increase consumer welfare as well as increase efficiency and lower costs brought by the e-commerce platforms as a decisive factor to attract consumers in using e-commerce platforms in Vietnam.

References

Cabrera-Sánchez, J.-P., Ramos-de-Luna, I., Carvajal-Trujillo, E., & Villarejo-Ramos, Á. F. (2020). Online recommendation systems: Factors influencing use in e-commerce. *Sustainability, 12*(21), 8888.

Chen, L., Rashidin, M. S., Song, F., Wang, Y., Javed, S., & Wang, J. (2021). Determinants of consumer's purchase intention on fresh e-commerce platform: perspective of UTAUT model. *SAGE Open, 11*(2), 21582440211027876.

Ezennia, C. S., & Marimuthu, M. (2022). Factors that positively influence e-commerce adoption among professionals in Surulere, Lagos, Nigeria. *African Journal of Science, Technology, Innovation and Development, 14*(2), 405–417.

Gupta, A. (2014). E-commerce: Role of e-commerce in today's business. *International Journal of Computing and Corporate Research, 4*(1), 1–8.

Gupta, S., Kiran, R., & Sharma, R. K. (2022). Validating the role of digital payment mode as a new driver of online shopping: A modified UTAUT2 model. *Journal of Public Affairs, 22*(2), e2434.

Halim, E., Januardin, R., & Hebrard, M. (2020). The impacts of e-payment system and impulsive buying to purchase intention in e-commerce. In *2020 International Conference on Information Management and Technology (ICIMTech)*, Bandung, Indonesia.

Hulin, C., Netemeyer, R., & Cudeck, R. (2001). Can a reliability coefficient be too high? *Journal of Consumer Psychology, 10*(1/2), 55–58.

Hungilo, G. G., & Setyohadi, D. B. (2020). Factors influencing acceptance of online shopping in Tanzania using UTAUT2. *Journal of Internet Banking and Commerce, 25*(1), 1–23.

Josephine, C. O. (2021). The effects of e-payment system on the efficiency of banks in Nigeria. *International Research Journal of Management, IT and Social Sciences, 8*(6), 548–558.

Kamalia, D., Djajadinata, M., Gunawan, F. H., & Gunadi, W. (2022, March 7–10). The role of hedonic motivation and FOMO on the Impulsivity of e-commerce users during COVID-19 Pandemics in Indonesia. In *Proceedings of the International Conference on Industrial Engineering and Operations Management Istanbul*, Turkey.

Kaur, P., & Joshi, M. M. (2012). E-commerce in india: A review. *IJCST*, *3*(1), 802–804.

Mansur, D., Sule, E., Kartini, D., & Oesman, Y. (2019). Understanding Factors Influence E-Commerce Use Behavior in Indonesia. *Opción: Revista de Ciencias Humanas y Sociales, 35*(20), 742–754.

Naruetharadhol, P., Wongsaichia, S., Zhang, S., Phonthanukitithaworn, C., & Ketkaew, C. (2022). Understanding consumer buying intention of e-commerce airfares based on multivariate demographic segmentation: A multigroup structural equation modeling approach. *Sustainability, 14*(15), 89–97.

Niranjanamurthy, M., Kavyashree, N., Jagannath, S., & Chahar, D. (2013). Analysis of e-commerce and m-commerce: Advantages, limitations and security issues. *International Journal of Advanced Research in Computer and Communication Engineering, 2*(6), 2360–2370.

Pobee, F. (2021). Modeling e-commerce adoption factors among Gen-Z in a developing country: The case of Ghana. *Marketing and Menedzsment, 55*(1), 81–94.

Pratika, Y. (2021). UTAUT model: Identifying the driving factors of the intention to use paylater. *Jurnal Bisnis Dan Manajemen, 8*(2), 345–352.

Rehman, A. U., Bashir, S., Mahmood, A., Karim, H., & Nawaz, Z. (2022). Does e-shopping service quality enhance customers'e-shopping adoption? An extended perspective of unified theory of acceptance and use of technology. *PLoS ONE, 17*(2), e0263652.

Singh, A., Alryalat, M. A. A., Alzubi, J. A., & Sarma, H. (2017). Understanding jordanian consumers' online purchase intentions: Integrating trust to the UTAUT2 framework. *International Journal of Applied Engineering Research, 12*(20), 10258–10268.

Venkatesh, V., Morris, M. G., Davis, G. B., & Davis, F. D. (2003). User acceptance of information technology: Toward a unified view. *MIS Quarterly, 27*(3), 425–478.

Venkatesh, V., Thong, J. Y., & Xu, X. (2012). Consumer acceptance and use of information technology: Extending the unified theory of acceptance and use of technology. *MIS Quarterly, 36*(1), 157–178.

Wulandari, A., Junipriansa, D., Suryawardani, B., & Marcelino, D. (2022). Integration the UTAUT2 model: Adoption of e-commerce as solution for fashion industry in Bandung facing the COVID-19 pandemic. *PalArch's Journal of Archaeology of Egypt/Egyptology, 19*(1), 1609–1630.

Yoga, I. M. S., & Triami, N. P. S. (2021). The online shopping behavior of Indonesian generation X toward e-commerce. *Journal of Economics, Business, and Accountancy Ventura, 23*(3), 441–451.

Mechanism of State Regulation of Regional Economy Competitiveness

Huy Quang Phan and Marina Eduardovna Buyanova

Abstract Globalization has fostered many development opportunities to developing countries, including Vietnam. One of those possibilities is that countries will be able to cultivate their competitive edge in international economic relations if they show initiative, flexibility, and creativity in integration. This implies that regional policy has also significantly evolved, with competitive capacity emerging as a more significant objective, prompted by concerns that regional policy is facing challenges with the gap in implementation. Therefore, the objective of this chapter is to develop a state regulation of the regional economic competitiveness in Vietnam, taking into account the effectiveness of policies. By synthesizing the scientific literature on competitiveness, regional policy, and statistics, the results indicated three main problems that hinder economic development and an increase in regional competitiveness capacity. Firstly, the apparent spatial disparity in Vietnam's regional competitiveness could be the result of long-term socio-economic processes (such as economic growth and investment attraction), and the quality of the business environment depends on the operating capacity of the authorities (via PCI Rank Up). Secondly, the mechanism of innovation and scientific development is incomplete. Finally, there are some potential "weaknesses" of industrial clustering.

Keywords Regional competitiveness · State support · Policies

1 Introduction

Globalization is the process of facilitating the more dynamic movement of production factors between countries in order to create a borderless open world economy. Information, knowledge, and technology are the source of efficiency and productivity. As a result, groups of countries with corresponding comparative advantages

H. Q. Phan (✉) · M. E. Buyanova
Volgograd State University, Volgograd, Russian Federation
e-mail: huyquang.phan@gmail.com

H. Q. Phan
The University of Danang, Danang, Vietnam

A. T. Nguyen et al. (eds.), *Contemporary Economic Issues in Asian Countries: Proceeding of CEIAC 2022, Volume 1*,
https://doi.org/10.1007/978-981-19-9669-6_9

will be divided to complement each other and promote cooperation and development. This emphasizes robust opportunities for development in developing countries like Vietnam. Therefore, improving the country's competitiveness as a factor of economic development in modern conditions is of particular interest to policy makers.

This concern has been and is being driven by the distribution of policy making powers from national, regional to local levels in search of budget balance (Diaz-Serrano & Rodriguez-Pose, 2015), investment attraction (Cui et al., 2020), economic recovery during the COVID pandemic (Phan & Buyanova, 2021a), development of clusters and innovation systems (Kogut-Jaworska & Ociepa-Kicińska, 2020), as well as develop a smart and sustainable specialization strategy (Palinchak et al., 2021). However, on the one hand successful policy implementation depends on creating strategies that are broad enough to have a significant impact but are tailored to specific contextual issues (Khan, 2016), on the other hand this depends on good linkages between levels of government and stakeholders (Teng & Gu, 2007), hence any shortfall would lead to an "implementation gap". Accordingly, without effective policies, there is a risk of regional unrest. Rich regions are more likely to incorporate production factors into the economy and concentrate productive activities, thereby creating economic growth points, while poor regions are not immune to reduced investment in production, population decline, desertification, and ultimately economic and social crises.

Seeking to address this gap, the purpose of this chapter is to develop a mechanism of the state regulation on Vietnam's regional economic competitiveness, taking into account the effectiveness of the implementation policies. This includes the disclosure of the concept of competitiveness, the relationship between regional economic development and policies (Sect. 2). For the development of the state regulatory mechanism of regional economic competitiveness in Vietnam, Sect. 3 has applied a combination of methods including methods of descriptive statistics, analysis and synthesis, and quantitative comparative. It deals with the availability of relevant data, including scientific articles, monographs and periodicals of domestic and foreign economists, documents and analysis reports of the World Economic Forum (WEF), World Intellectual Property Organization (WIPO), World Bank (WB), General Statistics Office of Vietnam, Ministry of Planning and Investment, Vietnam Chamber of Commerce and Industry (VCCI) provincial competitiveness data. To begin with, define the goals, tasks, and priorities for the development of the region. Subsequently is an assessment of the state's policies in the field of enhancing regional economic competitiveness. Finally, the conclusions that can be drawn from the chapter outlined in Sect. 4. The results of the analysis can be used in the development and implementation of policies to improve national and regional competitiveness in modern circumstances.

2 Competitiveness and Policies of the Regional Economy

Competitiveness is a frequently used but widely controversial term in economic policy debates. The concept of competitiveness first appeared in the 1980s as a solution to the economic crisis in the USA (Bristow, 2010a) and became famous in the early 1990s, when Japan emerged to challenge the economic dominance of the USA (Porter, 1990). Agreeing with Porter's point of view, Krugman (1994) affirms productivity as an ultimate measure of competitiveness. This is also consistent with economic theory, which considers productivity growth as the basis for economic growth and therefore, productivity is seen as a determinant of a high and increasing standard of living for the citizens of a country.

The essence of competitiveness is not only a macroeconomic issue. From this spatial perspective, the main contribution of scientists is understanding at the micro level when it comes to the presence of conditions that allow businesses to compete in their chosen markets and based on the value these enterprises create, with a focus on regional competitiveness (Porter, 2000; Huggins et al., 2014; Sadki et al., 2020). Regional competitiveness differs from concepts at the national level in two main respects: firstly, a local "specific advantage" depends crucially on its ability to provide favorable business, institutional, social, technological, and infrastructural facilities with which local businesses could maintain a long-term competitive advantage as they are difficult to imitate by other regions (Storpper, 1997; Camagni, 2002); secondly is the important role of interrelationships between businesses, entrepreneurship, local organizations, and spatial linkages with other regions (Fujita & Thisse, 2002; Martin, 2000). Therefore, competitiveness can vary depending on geographic spatiality, depending on the growth drivers of regions developing at different rates (Audretsch & Keilbach, 2004).

In general, the region determines the economic situation of the country, the state of the market for services, resources, goods and currencies, the structure of expenditures, the price levels and welfare, the costs, and the technological structure. Each region will adapt to major development trends, structural changes in the national economic system, roughly reflecting its positive and negative fluctuations. This indicates that there are significant regional variations in the rates of strategic and opportunity change, the degree of adaptive change, the mechanisms for finding ways out of economic distress, and also the extent to which degree risk could be controlled.

The logical implication of applying the concept of regional competitiveness is a policy system that focuses on regional economic development. As Martin (2005) has demonstrated, concerns about competitive capacity have been filtered down to the regional, city and provincial levels, particularly the role of regional policy interventions in enhancing competitive capacity. Regional policy, by definition, is territorially focused on interventions targeted at regions or sub regions, with varying degrees of selectivity (Bachtler et al., 2014, p. 42). The subject of regional research is very diverse that not only the factors of economic and regional growth, economic stability, convergence, and divergence at the regional level, but also the determinants regarding industrial location and the impact of regional situation on the number of local companies (Chrobocińska, 2020). Regional competitiveness is one of the pillars of European Union (EU) regional policy. It is systematically monitored and regularly surveyed for this using the Regional Competitiveness Index (RCI). The design of the RCI was inspired by the World Economic Forum's Global Competitiveness Index (see Annoni & Dijkstra, 2019). The RCI is an excellent tool for the EU regions to develop a heterogeneous set of policies with regard to potential impacts on regional competitiveness, including science, innovation, education, infrastructure, competition, regulatory, and environmental policy. Meanwhile in Vietnam, policies related to competitiveness are mainly explored in the provincial space (Phan & Buyanova, 2021b; Tinh & Ha, 2021) focused on addressing issues in improving a favorable business environment as an endogenous factor of long-term socio-economic growth. The issue of enhancing regional competitiveness has not been adequately developed and needs to be improved. Therefore, the development of a state regulatory mechanism to improve regional competitiveness in Vietnam is presented in the next section.

3 The State's Regulatory Mechanism on the Competitiveness Capacity of the Regional Economy

3.1 Identify Goals, Tasks, and Targets for Regional Development

The current socio-economic development orientations of the region and Vietnam are dominated by economic integration trends and 4.0 industrial revolution in the era of globalization. The result of these processes is the strengthening division of territories in terms of socio-economic development levels due to the unequal distribution of production and income factors. Under such circumstances, the ability of a territory to attract economic resources and factors of production plays a decisive role in determining the competitiveness of that territory. This depends mainly on the position of that region in the international and inter-regional division of the labor system, as well as on the attractiveness of the conditions created on its territory. These are complex and long-term tasks that must be systematically addressed through the formulation and implementation of regional economic policy (Fig. 1).

Vietnam is an economy that successfully transitioned to a market-oriented economy as a result of the gradual economic reform starting in the mid-1980s. Market institutions have prepared for the transition in a careful and orderly manner. As a result, Vietnam has entered the group of the leading economic growth and high-performing countries in recent decades (Nguyen, 2021). However, with the globalization of the world economic process, Vietnam is facing many challenges in maintaining sustainable economic growth. These include the economic downturn, credit boom, the rise of protectionism around the world, and the risks from market opening (Nguyen et al., 2018). They have exacerbated the internal problems of the national economy, forcing the search for new formats of relations between the state, region, province/city, and business.

Under these circumstances, in an idealized form, Vietnam needs to develop strategies for the development and economic integration of its regions as active and meaningful participants in the system of production, commodity, financial, scientific, technological, and cultural processes. This will allow the country and regions

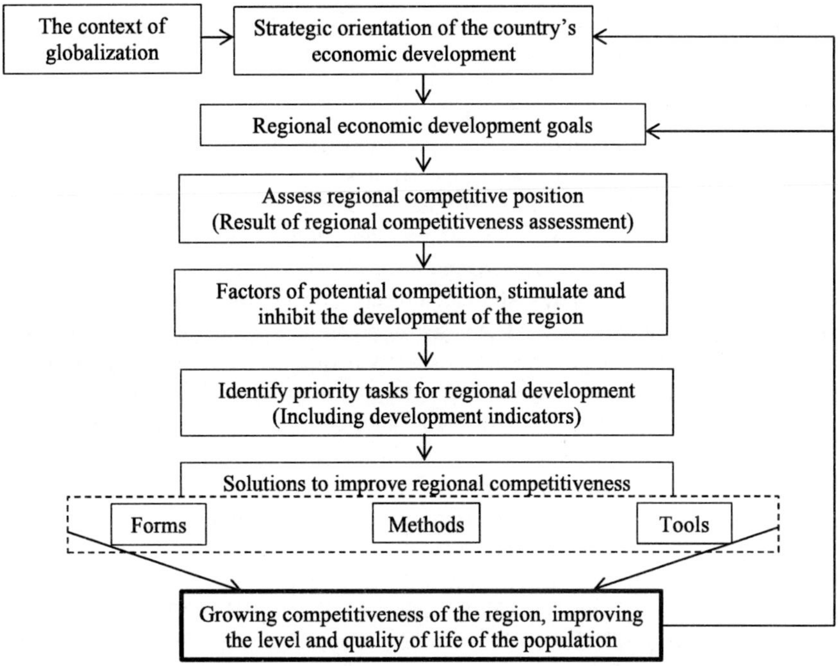

Fig. 1 Regional competitiveness enhancement mechanism. *Source* Developed by the authors

to successfully integrate into the international division of labor, while securing an important position within the domestic and foreign economic system.

Creating favorable conditions for the market to develop dynamically, forming competitive capacity is a key element of national and regional priorities in any country, and is the most important function of the state regulating the economy (Shibarshina, 2017). Therefore, Resolution 05-NQ/TW dated November 1, 2016, of the 12th Central Executive Committee of Vietnam is considered the first and most comprehensive resolution on economic competitiveness, clearly expressing the point of view, *"Renovating the growth model, improving the quality of growth, labor productivity and competitiveness of the economy are fundamental, important and decisive tasks for the realization of rapid and sustainable development goal of the country in the current period"*. The resolution has set forth the specific objective of narrowing the gap in national competitiveness with the leading ASEAN-4 countries. Based on how this goal is determined, the country's competitiveness can be examined according to the criteria of the World Economic Forum's Global Competitiveness Index report and the business environment report of the World Bank.

Subsequently, the state's stance on competitiveness, especially in the two most recent congresses, are the foundational basis for the creation of a series of state policies and mechanisms in improving national competitiveness, such as the National Assembly promulgated the 2013 Constitution, Enterprise Law, 2014 Investment Law,

Land Law 2013, Bidding Law 2013, and Law on SME Support 2017. The government promulgates decrees, resolutions, and circulars to realize the goal of improving the competitiveness of the economy, especially Resolutions No. 19/NQ-CP from 2014 to 2018 and recently converted into Resolution No. 02/NQ-CP from 2019 to 2022 of the Vietnamese Government on the main tasks and solutions to improve the business environment and enhance national competitiveness.

Success in competition is primarily associated with the development of those countries whose regions as socio-economic and cultural entities are built on the basis of common external social and economic life outside administrative boundaries. Thus, each region forms its own strategic objectives under the positive influence of external and internal challenges. The basis for planning six socio-economic regions is based on similarities in geographical location, natural, and social conditions in 63 provinces, including the Red River Delta, the Northern Midlands and Mountains, the North Central and Central Coast, the Central Highlands, the Southeast and the Mekong River Delta. This is clearly stated in Decree No. 92/2006/NĐ-CP of the government dated September 7, 2006.

On the basis of a common assessment on the mutual influence of these many factors, regions independently regulate the processes of economic and social development taking place within, defining tasks and formulating criteria for assessing the extent to which strategic goals have been achieved (Piven, 2014). Vietnam's regions have had socio-economic development strategies approved since 2001. Most strategies cover a period of 8–10 years, although some of them are designed for as long as 20 years. In the process of implementing the regional development strategy, the general development direction for six socio-economic regions was issued by the Politburo's Resolutions and then the Conclusions were drawn for each region. On that basis, the socio-economic development master plans of six regions in the next period are established to orient the development of branches, fields, territories, and organize the implementation.

In 2022, implementing the state's policy on enhancing competitiveness in the new context, regional governments have rebuilt their socio-economic development strategies on the basis of organizing conferences to review the implementation of the socio-economic development strategy, ensuring national security, and defense of the regions for the period 2010–2020.

As a result, the government issued Resolution No. 78-NQ/CP dated June 18, 2022, on the Action Program to on the implementation of Politburo Resolution No. 13-NQ/TW of April 2, 2022, on socio-economic development direction and securing the defense of the Mekong Delta by 2030 and security, and a vision to 2045. The Red River Delta Regions; Northern Midlands and Mountains; the North Central and Central Coast; the Central Highlands and Southeast are in the process of being built.

Taking into account the approaches described above, the following priorities and objectives are currently identified within the framework of Viet Nam's regional development strategies: (i) improving the quality of life of the population in the region, creating conditions for human capital development; (ii) implementing an institutional policy in the region, to ensure the formation of competitive advantages in the region; (iii) achieving sustainable high growth rates of the region's economic indicators; (iv) creating conditions to produce potential for formation and development of clusters; (v) developing inter-regional and international cooperation; (vi) completing the regional socio-economic development strategic planning; (vii) improving the efficiency of regional development management, optimizing the power structure, and governing body. Clearly, not all of the prioritized fields will ensure the region's competitiveness capacity as described above; nevertheless the main fields are highlighted.

Therefore, in order to comprehensively assess the current status and development of regions, a comprehensive view of competitiveness and a system of indicators are determined.

- The framework for assessing the implementation of socio-economic development goals in regions and provinces was promulgated in Decision No. 288/QD-TTg of the prime minister dated February 28, 2022. In which, the indicators on socio-economic development have an impact on the competitiveness of the region and the province: (1) population; (2) gross regional domestic product (GRDP); (3) economic growth; (4) average GRDP per capita; (5) average income per capita; (6) export turnover; (7) budget revenue; (8) investment; and (9) science and technology.
- Provinces/cities have issued a master plan to implement Resolution No. 02/NQ-CP dated January 10, 2022, of the Vietnamese Government on key tasks and solutions to improve the business environment, as well as enhance regional and national competitiveness in 2022. In order to achieve this, the provinces have developed solutions to improve the scores and rankings of the provincial competitiveness index, related to ten components: (1) market entry; (2) access to land; (3) transparency; (4) time costs; (5) informal expenses; (6) fair competition; (7) the dynamism of the provincial government; (8) business support services; (9) labor training; and (10) legal institutions and security.

3.2 Evaluation of the State's Policies in Enhancing Competitiveness of the Regional Economy

In the context of globalization, enhancing competitive capacity is a top leading and clear policy objective at all levels (enterprise, local, regional, and national) of countries around the world, including Vietnam. State policies aimed at increasing the competitive capacity of the regional economy often involve creating the conditions under which economic activities can prosper (Bristow, 2010b). The success of a local or regional economy is determined by the socio-economic system and entrepreneurial capacity; local and institutional factors of production; mutual relations between local actors, creating favorable conditions for the absorption of accumulated knowledge; decision-making capabilities enable local social and economic actors to lead development and innovation processes (Huggins, 2016; Pietrzak et al., 2017).

Outlining and implementing state policies to enhance competitiveness in this context is a complex, dynamic, and imperative innovative process, with the main reason being that there is no sole and optimal policy system (Magro & Nauwelaers, 2015). An effective competitive strategy involves the ability of governments in the region to inquire from the impacts of the economic policy, especially through methods based on regional and international comparisons (Huggins, 2010). Therefore, benchmarking and comparative measures of socio-economic performance in the region have become a particularly common tool for regional economic policymaking in recent years.

The context of regional development is placed in the general context of the whole country in the period 2011–2020 with optimism for an early economic recovery. However, in the early years of implementing the 2011–2020 strategy, our country faced many difficulties and challenges during the global economic fluctuations, the COVID pandemic, along with the internal weaknesses of the economy.

Throughout this strategic period was the mission to restructure the economy and renew the growth model. In addition to the three central focuses identified in the first phase, which includes the restructuring of investment, financial-banking system, enterprise, tasks, as well as building and developing an appropriate regional economic structure is also mentioned as one of the main orientations of the economic restructuring process. Regarding the results of restructuring the economic region during this period, Resolution No. 24/2016/QH14 dated November 8, 2016, of the National Assembly of Vietnam stated, *"The restructuring of the economic region has not been paid enough attention, the development space is still divided by administrative boundaries, there is a lack of institutions to create links between localities in the region and between regions"*. In order to overcome this situation, the resolution has set out a number of orientations to modernize the regional economic structure, focusing on planning, building outstanding institutions for a number of regions to play a leading role in development, strengthening regional linkages and perfecting the urban economic development model. Furthermore, policy makers in Vietnam have sought to improve the general business environment that affects all businesses. This occurs through regional development policies such as macroeconomic stability, tax policies that encourage savings, investment in basic research and development (R&D), institutional development, public support for colleges and universities in human resource training, infrastructure investment, and antitrust regulations.

The results of socio-economic analysis of the regions as well as regional competitiveness based on the provincial competitiveness index indicate the positive impact of the above policies in the field of improving the competitiveness of the regional economy. This task is very significant in analyzing potential growth areas, determining the location and role of the region in the whole country, this is clearly demonstrated by some examples below:

Firstly, regions across the country have had positive developments. The average GRDP growth rate of the six regions is approximately the same as the gross domestic product (GDP) growth rate of the entire economy. The Red River Delta region is considered as a region with high competitive attractiveness in terms of economic factors due to the level of economic development. The three out of eight indicators ranked as first are population, GRDP growth, and investment capital. The remaining indexes are ranked two out of six regions. In which, the average GRDP per capita is VND 81.30 million, just behind the Southeast region (see Tables 1 and 2).

On the contrary, according to five out of eight economic growth indicators ranked six out of six regions, the Central Highlands region has lower competitiveness than other regions. These are accessed in terms of population, GRDP, export turnover, budget revenue, and investment capital. This could be due to several reasons: firstly, the Central Highlands has a total population of 5.87 million people (accounting for 6.1% of the country's population), of which ethnic minorities have 2.2 million (accounting for 37.7%)—have different living and production habits such as primarily self-sufficient; heavy on natural exploitation, weak economic potential, low level of education and labor (Vu et al., 2020); secondly, the region has many advantages for agricultural development, however, due to climate change, epidemics, pests, diseases, and the decline in prices of some agricultural products during the period 2015–2018 have strongly impacted the production and export system of agricultural products in the region.

Table 1 Statistics of some socio-economic indicators of regions in 2018

Regions	Population (whole country: million persons. region: %)	GRDP (whole country: trillion dongs. region: %)	GRDP growth (%)	GRDP per capita (million dongs)	Income per capita (million dongs)	Export Turnover (whole country: billion USD. Region: %)	Budget revenue (whole country: trillion dongs. region: %)	Investment capital (Billion dongs)
The whole country	94.67	5542.332	7.02	68.18	46.51	283.383	1424.914	1,857,061.0
Red river delta	21.57	31.63	10.23	81.30	58.01	33.41	37.84	834,891.2
Northern Midlands and Mountains	12.29	9.69	7.53	43.72	29.46	16.11	3.60	241,020.2
North Central and Central Coast	20.06	17.37	9.47	48.01	36.18	4.82	11.96	481,974.4
Central Highlands	5.87	4.91	7.87	46.43	34.75	1.27	1.42	93,201.1
The South East	17.07	39.55	8.23	128.4	68.51	37.23	39.01	713,755.8
Mekong River Delta	17.81	14.85	8.69	46.23	43.06	7.20	6.14	279,856.6

Source Calculated by the authors based on the socio-economic data of 63 provinces in 2020 (GSO, 2020)

Table 2 Ranking of regions by some basic indicators in 2018

Regions	Population	GRDP	GRDP growth	GRDP per capita	Income per capita	Export turnover	Budget revenue	Investment capital
Red river delta	1	2	1	2	2	2	2	1
Northern Midlands and Mountains	5	5	6	6	6	3	5	5
North Central and Central Coast	2	3	2	3	4	5	3	3
Central Highlands	6	6	5	4	5	6	6	6
The South East	4	1	4	1	1	1	1	2
Mekong River Delta	3	4	3	5	3	4	4	4

Source Calculated by the authors

Based on the ranking of eight basic indicators, the North Central and Central Coast regions are considered to be quite attractive, with a growth rate of GRDP of 9.23% (ranked two out of six regions). This area continues to be successful thanks to the operation of businesses and investment in the industry–construction sector. This is the region with the third largest contribution to the overall growth of the country after the Southeast region and the Red River Delta. There are projects with great momentum in the region, promoting economic growth and restructuring, as well as increasing revenue for the local budget: Dung Quat oil refinery, Chu Lai complex–Truong Hai manufactures cars, Formosa steel factory, etc.

Secondly, the economic structure of the region has shifted positively toward exploiting the potentials, strengths, comparative advantages, enhancing competitiveness, and improving socio-economic efficiency. The Red River Delta, the Northern Midlands and Mountains, the North Central Coast and the Central Coast focus on developing construction industries; the Southeast region focuses on developing service sectors; The Mekong Delta and Central Highlands regions focus on developing agricultural and processing industries. Each region has its own characteristics, advantages and in recent years, these regions have made efforts to utilize the potential of the region for economic development.

Third, industrial and construction development was the main driver for growth of most regions in 2018 and except for the Southeast region (Table 3). The economic

Table 3 Economic structure by sector industry–construction of regions compared to the whole country

Regions	2010	2015	2018
Red river delta	24.76	27.99	28.79
Northern Midlands and Mountains	4.84	7.50	8.54
North Central and Central Coast	11.87	13.66	13.41
Highlands	1.67	2.01	2.01
South East	48.69	41.26	38.96
Mekong Delta	8.17	7.57	8.29
The whole country	100.00	100.00	100.00

Source Calculated by the authors based on the socio-economic data of 63 provinces in 2020 (GSO, 2020)

structure of the region tends to shift strongly between two industrial sectors—construction and services. The proportion of the service industry increased rapidly from 36.6% in 2010 to 47% in 2018; the proportion of industry–construction decreased to 48% in 2018; the share of agriculture remained stable in the range of 5–5.5% in the period 2010–2018.

In addition, policies to develop economic space in Vietnam, especially industrial clusters, have been concretized into major guidelines, policies and orientations of the state. Resolution 23-NQ/TW dated March 22, 2018, of the Politburo on the directions of the national industrial development policies to 2030 and the vision to 2045 emphasized the formation of industrial zones, clusters, parks, production networks, and industrial value chains, in which industrial clusters are the central focus. The government issued Decision No. 32/QD-TTg on January 13, 2015, approving a synchronization program to develop and upgrade clusters, as well as value chains of competitively manufactured products including: electronics and information technology; textiles; food processing, agricultural machinery; tourism and related services. Currently, some regions and localities have formed a number of clusters in priority areas to promote competition (Nguyen, 2015), such as

- Electronic industry cluster: In the Red River Delta region including Hanoi, Bac Ninh, Thai Nguyen, Hai Phong, Vinh Phuc, Hai Duong, Hung Yen, Bac Giang, Binh Duong with core multinational companies such as Canon, Samsung, Nokia, LG, Panasonic, and many satellite businesses. Multinational companies play the role of centers to create spillover effects, forming clusters of the electronics industry.
- High-tech industry clusters, especially microchips and electronic technology in Ho Chi Minh and neighboring provinces with Intel play a leading role.
- Clusters of automobile and motorcycle assembly and manufacturing industries in Ha Noi, Hai Duong, Bac Ninh, Hung Yen, Vinh Phuc, Hai Phong, Quang Ninh). This cluster contains many multinational companies assembling automobiles and

motorcycles such as Toyota, Honda, Ford, Yamaha, Vinaxuki, 3–2 Automobiles, and supporting industrial enterprises for the automobile and motorcycle industry.
- The textile and garment industry cluster in the Red River Delta, especially in Hung Yen, Nam Dinh, the suburbs of Ha Noi, Hai Duong, and in the Southern key economic region including Dong Nai, Ho Chi Minh City, Binh Duong, Long An, Tien Giang.

Fourthly, the infrastructure system that connects the regions and inter-regions has developed a new step, creating conditions to promote exchanges and linkages between localities and regions. In which, it is worth mentioning that the Noi Bai–Lao Cai Expressway came into operation in 2014 has great significance for the Northern Midlands and Mountains, helping to connect the disadvantaged and underdeveloped economic regions of five localities along the route with dynamic economic regions. In the Red River Delta, the construction of important highways such as Hanoi–Hai Phong, Hoa Lac–Hoa Binh, and Ha Long–Van Don has been completed. In the Southeast region, Ho Chi Minh City–Long Thanh–Dau Giay Expressway has been put into operation. Ben Luc–Long Thanh Expressway is under construction.

The North Central Coast and Central Coast also have significant regional connection constructions such as Deo Ca road tunnel, on National Highway 1A connecting Khanh Hoa with Phu Yen; opened to traffic on the entire Da Nang–Quang Ngai Expressway (belonging to the North–South Expressway to the east). In the Central Highlands, Ho Chi Minh Road (National Highway 14) was completed and upgraded in 2015. The traffic infrastructure in the Mekong Delta region has also drawn interest for investment: completing the South Song Hau road (National Highway 91C), Co Chien Bridge, Nam Can Bridge, Dam Cung Bridge, My Loi Bridge, and Cao Lanh Bridge.

Fifthly, investment is a method to promote economic growth. Investment policy is one of the decisive factors in the process of economic restructuring in order to achieve rapid growth for regions. The localities in the economic zones have actively developed programs to attract investment capital in accordance with the proposed socio-economic development planning. The total realized investment capital of the whole society in 2018 of the Red River Delta region ranked one out of six regions, reaching 834.9 trillion VND, accounting for 44.95% of the total realized investment capital of the whole country. The Central Highlands region ranked six out of six regions, reaching only 93.2 thousand billion VND, accounting for 5.02% of the total realized investment capital of the whole country.

The policy of foreign direct investment (FDI) creates more favorable conditions for investors, the process of capital disbursement is promoted, and therefore, the realized foreign direct investment capital tends to increase over the years. The Southeast region is leading the country in attracting foreign direct investment with 159.25 billion USD, accounting for 41.47% of the country's FDI capital (Table 4). Ho Chi Minh City, Binh Duong, Ba Ria–Vung Tau, and Dong Nai are the provinces/cities with the largest total FDI in the country, contributing a large proportion in the region's FDI attraction. The main reason is that these localities have built up strategies to

Table 4 Total FDI capital (as of December 20, 2020) and ranking of FDI attraction by regions

Regions	Total registered capital (million USD)	Ranking
The whole country	384,044.21	
Red river delta	111,972.56	2
Northern Midlands and Mountains	20,132.89	5
North Central and Central Coast	60,024.77	3
Central Highlands	1087.83	6
The South East	159,253.39	1
Mekong Delta	28,804.08	4

Source Compiled and calculated by the authors based on data of Vietnam Foreign Investment Agency in 2021 (Department of Foreign & Investment, 2022)

attract investment selectively, targeting high-tech fields and service industries with great economic value and efficiency.

Competitiveness is also one of the advantages for economic regions to attract domestic and foreign investment. If we consider regions as a "locality", the calculation results of the regional PCI show that the Red River Delta ranks first, followed by the Northern Midlands and Mountains. Table 5 also demonstrates that the business environment of some regions has improved markedly, changing the ranking position in 2021 compared to 2020 such as the Red River Delta and the Mekong Delta. Based on the analysis results of the PCI component indexes of the regions, the business environment of the regions has improved markedly thanks to the institutional reform efforts of local authorities in areas such as market entry costs, land access and tenure, informal charges, business support services, and law and order (Phan & Buyanova, 2022).

In the process of implementing state policies in the field of improving the competitive capacity of the regional economy, regions have shown an unstable and unequal dynamism, which is especially noticeable when comparing development indicators

Table 5 PCI rankings of regions in 2020 and 2021

Regions	PCI 2020	Ranking	PCI 2021	Ranking
Red River Delta	63.84	2	67.65	1
Northern Midlands and Mountains	62.71	6	62.89	6
North Central and Central Coast	63.67	3	64.46	4
Central Highlands	63.12	4	64.20	5
The South East	65.02	1	66.63	2
Mekong Delta	63.11	5	64.74	3

Source Calculated by the authors based on PCI data for the period 2020–2021 (VCCI, 2020, 2021)

across regions. Aside from positive results, a number of issues related to the implementation efficiency of these policies may hinder economic development and the region's competitive capacity. Specifically:

- The gap difference in living standards is increasingly wide between localities and regions, causing potential risks toward social instability.
- The speed of migration from poor and rural areas to big cities such as Hanoi and Ho Chi Minh City is increasing, while the infrastructure has not been invested in time, leading to the overloading of the urban infrastructure system and the decline in the quality of the living environment.
- The current cluster models only promote the advantage scale of geographical concentration, while the economic links as well as promoting participation in the value chain production network of the cluster are still weak. New clusters only exploit the advantages of location (connecting to the market) and infrastructure (seaport) to form in place of the factors that enhance business linkages (McCarty et al., 2005). The foundation to attract FDI enterprises and form these clusters is unsustainable when relying mainly on cheap labor. Moreover, FDI enterprises in the textile, garment, footwear, and electronics industries are not interested in training workers and only keep wages and working conditions to a minimum (Hoang, 2011).

The available statistics in Vietnam do not permit factor analysis of the region's competitive capacity in all areas of policy. According to international theory and experience, innovation is an unlimited driver of growth and is the key to helping some East Asian countries overcome the "middle-income trap". However, the economic growth of Vietnam and other regions has not relied much on knowledge, science and technology.

Therefore, besides analyzing the PCI index of regions, the authors use official reports of domestic and foreign organizations, scientific articles–research related to creative innovation, science and technology to identify the main problems hindering its development. On international rankings, Vietnam's technology and innovation are still low-ranking fields. Details are presented below:

- Analysis of WEF's Future Manufacturing Readiness Report (WEF and Kearney, 2018) can identify some more bottlenecks in Vietnam's innovation such as low rank (90/100) in terms of technology and innovation level, foundation technology (92th), innovation capacity (77th) with investment in research and development account for only 0.2% of GDP (84th). Therefore, the value of mid-to-high-end technology products in Vietnam only accounts for 30% of the total export value, while countries in the region account are 80%, as low as the Philippines also account for 50%.
- According to the assessment of WEF (Schwab, 2019), the overall ranking of Vietnam's Competitive Capacity 4.0 in 2019 is 67/141 (increased by 10 ranks compared to 2018), but only ranks 76th in innovation, in which multi-cooperation in innovation ranked 80th; scientific publication 59th. The growth of innovative companies is also ranked 68th.

- According to WIPO (2022), a number of innovation indicators in Vietnam are still low. For example, in 2021, ease of resolving insolvency (106/132); tertiary inbound mobility (102th); environmental performance (110th); investment (111th); knowledge-intensive employment (110th); ICT services imports (129th); intellectual property receipts (106th); and ICT services exports (115th).

4 Conclusion and Recommendations

Over the past several decades, the discourse on competitiveness in the analysis of territorial development has resonated particularly at the regional level, and today policymakers often claim to enhance competitiveness as a factor driving economic development. The range of policies related to regional competitiveness is very wide, on one hand marked by the determinants of the productivity of firms located within a territory, and on the other hand depends on linkages between levels of government and stakeholders.

Based on the information reviewed, a mechanism is presented as a way to influence the process of formation and development of regional competitiveness, which has a complex and effective structure, contributing to the achievement of goals and activating existing reserves. The essence of the state-regulated mechanism for regional competitiveness is the integration of objectives, purposes, principles, agents, objects, forms, methods, tools, and aggregative effects are reflected in the service quality development criteria, and the scope is determined by the management entities. The arrangement between levels of government is inevitable in the context of regional development characterized by complex interactions and incentives between national and local actors (Keating, 2003). Therefore, effective management of policy instruments implemented within administrative boundaries plays a significant role, especially on the basis of policy effectiveness. Moreover, to further broaden the regional competitiveness policy agenda to combine sectorial goals with multiple issues (administrative and legal, economic and financial, programmatic orientation, information news–communication) and this not only demonstrates the adequacy of regional development spending from national financial allocations but also provides additional impetus to ensure a consistent and effective system of policy implementation.

The most important directions for increasing regional and subsequent national competitiveness are to promote state support measures in addressing three fundamental issues: Firstly, the apparent spatial disparity in regional competitiveness in Vietnam could be the result of long-term socio-economic processes (such as economic growth and investment attraction), and the quality of the business environment depends on the operating capacity of the authorities (via PCI rank up). Secondly, the mechanism of innovation and scientific development is not perfect due to limited investment and awareness. Finally, there are some potential "weaknesses" of industrial clustering arising from the inherent structural constraints of Vietnamese firms. In that context, in-depth studies play a crucial role in verifying and confirming the authors' empirical conclusions.

Acknowledgements This research is part of a Ph.D. project, which was funded by the Governments of the Russian Federation and Vietnam (scholarship under the agreements). We sincerely thank the anonymous reviewers for their useful comments and suggestions that improved the chapter.

References

Annoni, P., & Dijkstra, L. (2019). *EU regional competitiveness index 2019* (p. 42p). Publications Office of the European Union.

Audretsch, D., & Keilbach, M. (2004). Entrepreneurship and regional growth: An evolutionary interpretation. *Journal of Evolutionary Economics, 14*(5), 605–616.

Bachtler, J., Mendez, C., & Vironen, H. (2014). *Regional development and policy in Europe contributions for the debate in Latin America.* Collection Studies No. 2. EURO Social Programme, p. 88.

Bristow, G. (2010a). *Critical reflections on regional competitiveness: Theory, policy and practice* (p. 200). Routledge.

Bristow, G. (2010b). Resilient regions: Re-"place"ing regional competitiveness. *Cambridge Journal of Regions, Economy and Society, 3*(1), 153–167.

Camagni, R. (2002). On the concept of territorial competitiveness: Sound or misleading? *Urban Studies, 39*(13), 2395–2411.

Chrobocińska, K. (2020). Selected factors in the location of service activities conducted by small enterprises. *Zesz. Nauk. Uniw. Przyr. Humanist. Siedlcach. Seria Adm. Zarządzanie, 52(125)*, 27–36.

Cui, L., Fan, D., Li, Y., & Choi, Y. (2020). Regional competitiveness for attracting and retaining foreign direct investment: A configurational analysis of Chinese provinces. *Regional Studies, 54*(5), 692–703.

Department of Foreign and Investment, 2022. Situation of attracting foreign investment in 2021. Available: https://fia.mpi.gov.vn/Detail/CatID/f3cb5873-74b1-4a47-a57c-a491e0be4051/NewsID/5d476094-8272-4d9d-b810-1609ce7b67b3/MenuID (accessed 10 August 2022).

Diaz-Serrano, L., & Rodriguez-Pose, A. (2015). Decentralization and the welfare state: What do citizens perceive. *Social Indicators Research, 120*, 411–435.

Fujita, M., & Thisse, J. F. (2002). *Economics of agglomeration: Cities, industrial location and regional growth* (p. 466p). Cambridge University Press.

General Statistic Office. (2020). *Socio-economic statistical data of 63 provinces and cities.* Statistical Publishing House, 1065p. Available: https://www.gso.gov.vn/en/data-and-statistics/2020/05/socio-economic-statistical-data-of-63-provinces-and-cities/ (accessed 22 August 2022).

Hoang, V. H. (2011). Developing industrial clusters in Vietnam in the context of globalization and international economic integration. *Economy and Development (in Vietnamese), Special issue*, pp. 54–61.

Huggins, R. (2010). Regional competitive intelligence: Benchmarking and policy-making. *Regional Studies, 44*(5), 639–658.

Huggins, R., Izushi, H., Prokop, D., & Thompson, P. (2014). Regional competitiveness, economic growth and stages of development. *Zb Rad Ekon Fak Rij, 32*(2), 255–283.

Huggins, R. (2016). Capital, institutions and urban growth systems. *Cambridge Journal of Regions, Economy and Society, 9*(2), 443–463.

Keating, M. (2003). Social inclusion, devolution and policy divergence. *Political Quarterly, 74*, 429–438.

Khan, A. R. (2016). Policy implementation: Some aspects and issues. *Journal of Community Positive Practices, 3*, 3–12.

Kogut-Jaworska, M., & Ociepa-Kicińska, E. (2020). Smart specialisation as a strategy for implementing the regional innovation development policy—Poland case study. *Sustainability, 12*(19), 7986.

Krugman, P. (1994). Competitiveness: A dangerous obsession. *Foreign Affairs, 73*, 28–44.

Magro, E., & Nauwelaers, C. (2015). Reconciling territorial strategy goals and means: Towards smart competitiveness policies. In J. M. Valdaliso & J. R. Wilson (Eds.), *Strategies for shaping territorial competitiveness* (pp. 73–93). Routledge.

Martin, R. L. (2000). Institutional approaches in economic geography. In E. Sheppard & T. J. Barnes (Eds.), *A companion to economic geography* (pp. 77–94). Blackwell.

Martin, R. (2005). *Thinking about regional competitiveness: Critical issues.* Report Commissioned by the East Midlands Regional Development Agency. East Midlands Regional Development Agency, p. 47.

McCarty, A., Record, R., & Riedel, J. (2005). Competition and cooperation: Vietnam. In A. Kuchiki & M. Tsuji (Eds.), *Industrial clusters in Asia.* IDE-JETRO Series. Palgrave Macmillan, pp. 27–108.

Nguyen, D. P., Ho, V. T., & Vo, X. V. (2018). Challenges for Vietnam in the globalization era. *Asian Journal of Law and Economics, 9*(1), 1–3.

Nguyen, N. S. (2015). *Industrial cluster development in the context of globalization and international economic integration* (p. 300p). National Political Publishing House.

Nguyen, N. S. (2021). Improving the quality of economic growth in Vietnam. *Vietnam Studies (in Russian), 2*(1), 52–78.

Palinchak, M., Tsalan, M., Brenzovych, K., Kucher, A., Kajánek, T., & Grešš, M. (2021). Competitiveness as the basis of EU regional policy: Smart specialization and sustainability. *European Journal of Sustainable Development, 10*(4), 227.

Phan, H. Q., & Buyanova, M. E. (2021a). Ways to improve Vietnam's competitiveness and economic recovery in the context of a pandemic. *Journal of Volgograd State University Economics (in Russian), 23*(3), 112–122.

Phan, H. Q., & Buyanova, M. E. (2021b). Analysis on regional competitiveness concepts: Case of the Da Nang province. *Regional Economy and Management: Electronic Scientific Journal (in Russian), 4*(68), 6807.

Phan, H. Q., & Buyanova, M. E. (2022). Competitive position of Vietnamese regions: Analysis of results of provincial competitiveness index. *Regional Economy, South of Russia, 10*(3), 78–87.

Pietrzak, M. B., Balcerzak, A. P., Gajdos, A., & Arendt, Ł. (2017). Entrepreneurial environment at regional level: The case of Polish path towards sustainable socio-economic development. *Entrepreneurship and Sustainability Issues, 5*(2), 190–203.

Piven, I. G. (2014). Public-private partnership in the strategic management of regions socio-economic development. *π-Economy (in Russian), 6(209)*, 50–58.

Porter, M. E. (1990). *The competitive advantage of nations* (p. 855). Palgrave Macmillan.

Porter, M. E. (2000). Location, clusters, and company strategy. In G. L. Clark, M. P. Feldman, & M. S. Gertler (Eds.), *The Oxford handbook of economic geography* (pp. 253–274). Oxford University Press.

Sadki, T., Nmili, M., & Jalal, C. (2020). Regional competitiveness: A critical theoretical review and managerial and politician implications. *Revue Alternatives Manageriables Et Economiques, 2*(4), 194–216.

Schwab, K. (2019). *The global competitiveness report 2019.* World Economic Forum, p. 666. Available: https://www3.weforum.org/docs/WEF_TheGlobalCompetitivenessReport2019.pdf (accessed 20 August 2022).

Shibarshina, O. Yu. (2017). *State regulation of economic growth.* Collection of scientific articles by young teachers and graduate students "Actual problems of the development of society, economics and law". Witte Moscow University (In Russian), pp. 93–99.

Storper, M. (1997). *The regional world: Territorial development in a global economy*. Guilford, p. 338.

Teng, F., & Gu, A. (2007). Climate change: National and local policy opportunities in China. *Environmental Sciences, 4*(3), 183–194.

Tinh, D. T., & Ha, N. T. T. (2021). A comparative study of the competitiveness of Nam Dinh province in the Red River Delta region of Vietnam. *European Journal of Business and Management Research, 6*(4), 327–331.

VCCI. (2017, 2018, 2019, 2020, 2021). *Provincial competitiveness index-PCI 2017, 2018, 2019, 2020, 2021: Evaluating quality of economic governance to promote business development*. Hanoi: VCCI.

Vu, T. T., Vu, D., & Nguyen, T. M. L. (2020). Production organization capacity of ethnic minorities in northwestern region of Vietnam. *E3S Web of Conferences 210*, 16004, 6–9.

WEF & Kearney, A. T. (2018). *Readiness for the future of production report 2018*. World Economic Forum, p. 266. Available: https://www3.weforum.org/docs/FOP_Readiness_Report_2018.pdf (accessed 20 August 2022).

WIPO. (2022). *Global innovation index 2021 profile Vietnam*. World Intellectual Property Organization, p. 10. Available: https://www.wipo.int/edocs/pubdocs/en/wipo_pub_gii_2021/vn.pdf (accessed 20 August 2022).

The Adaptability of Vietnamese Worker to the Industry 4.0

Vu Thi Bich Ngoc, Nguyen Manh Thang, and Nguyen Chu Du

Abstract The Fourth Industrial Revolution, often known as Industry 4.0, is being hailed as a technological miracle. It helps individuals, cuts down on the amount of work that needs to be done, and initiates a chain of beneficial developments in engineering and technology. As a consequence of this, the job market, and more specifically the labor force, is faced with both possibilities and challenges. In 6 provinces and cities, including Hanoi, Bac Ninh, Ha Tinh, Nghe An, Tien Giang, and Can Tho, the polling method was utilized to collect 750 votes (640 employees and 110 corporate and trade union officials). The respondents worked in firms in the garment, footwear, and electronics industries. The results of a study indicate that workers have partially adapted to Industry 4.0.

Keywords Workers · Employees · Adaptability · Industry 4.0

1 Introduction

The Fourth Industrial Revolution, often known as Industry 4.0, is being hailed as a technological miracle. It helps individuals, cuts down on the amount of work that needs to be done, and initiates a chain of beneficial developments in engineering and technology. Accordingly, there are both opportunities and challenges for the workforce in general, and specifically for the human resources of education. Workers in Vietnam are also being impacted by the Industry 4.0. In particular, in addition to the appearance of new occupations and the creation of an unprecedented number of new jobs, current jobs typically decrease gradually, or the jobs that are currently being carried out by humans will be taken over by machines in the near future. In

V. T. B. Ngoc (✉) · N. M. Thang · N. C. Du
Trade Union University, Hanoi, Vietnam
e-mail: ngocvtb@dhcd.edu.vn

N. M. Thang
e-mail: thangnm@dhcd.edu.vn

N. C. Du
e-mail: dunc@dhcd.edu.vn

A. T. Nguyen et al. (eds.), *Contemporary Economic Issues in Asian Countries: Proceeding of CEIAC 2022, Volume 1*,
https://doi.org/10.1007/978-981-19-9669-6_10

particular, industries such as textiles and garments, leather and footwear, and others like these primarily use low-skilled employees and have a tendency to demand labor to change, gradually adapting to new trends, the trend that artificial intelligence has recently come to dominate.

2 Literature Reviews

The concept of Industry 4.0 comes from Germany—a country with the most competitive manufacturing industry in the world and even a global leader in the field of equipment manufacturing (Rojko, 2017). According to the findings of the study (Khan & Turowski, 2016), one of the opportunities that the Industry 4.0 brings with it is the incorporation of technology into production and the application of the Internet in various industrial procedures. Consequently, this will result in the creation of new industrial prospects in the future. Technology and services provided through the cloud will make it possible for systems and partners to communicate from any location. In addition, Industry 4.0 offers integration across an entire company, as well as optimization and decision making in real time, even in the situation of businesses that have several manufacturing plants, either centralized in one country or dispersed over multiple countries and continents.

Some other studies such as *"The Future of Jobs Employment, Skills and Workforce Strategy for the Fourth Industrial Revolution"* (World Economic Forum, 2016) or *"Industry 4.0: New Challenges and Opportunities for the Labor Market"* (Kergroach, 2017) or *"A Skills Revolution: from consumer of work to builders of talent"* (Manpower Group, 2017) discussed future jobs, strategies, skills, and workforce before Industry 4.0. Specifically, the authors argue how the employment impact will be the appearance of many new opportunities with increased labor productivity and worker emancipation; yet, the government will face rising unemployment and increased inequality as a result of this. Aside from the potential benefits of growing economic power and women's power, the most substantial threat is the development of occupations that can be performed from a remote location on a global scale. Lower-skilled workers will likely lose their jobs. Future generations of workers will need to acquire digital fluency as well as the ability to continue their education throughout their working lives in order to prevent the threat of being laid off. In addition, prospective employees have a responsibility to improve skills such as problem-solving abilities, the ability to persuade others, self-organization, management, teamwork, and communication abilities. The efforts made to retrain and train personnel will not be successful since the employer will not profit from them. The lack of skills and the absence of human resource strategy in businesses are the obstacles that workers will face during Industry 4.0.

In another study on the challenges of the Industry 4.0 to workers, especially industrial engineers (Sackey & Bester, 2013), it is said that the impact of Industry 4.0 on manufacturing systems will be higher efficiency and better robots, which will mean fewer workers are needed and jobs are threatened in some areas. Similar to what was stated above, Dimitrieska et al. (2018) contend that the Industry 4.0 will bring about more inequality and will wreak havoc on the labor market; workers may be replaced by robots in the future, but machines will never have a conscience.

There has been discussion over both the positive and negative effects that the Fourth Industrial Revolution will have on the labor market. According to the findings of a study, one strategy for workers who wish to remain employable in the Industry 4.0 economy is to broaden their professional horizons and acquire other skill sets. Strengthen tenacity, adaptability, and the capacity for learning (Manpower Group, 2017). Additionally, it is necessary for workers to receive training and education right at the enterprises (Dimitrieska et al., 2018). In the future of the Fourth Industrial Revolution, employees will need to be proficient in both interpersonal communication and programming in order to be employed in any capacity (Alhloul & Kiss, 2022). According to the International Labor Organization (ILO), 2018, cooperation between the state, businesses, and educational institutions is required in order to train a labor force that can match the demand of the market. This cooperation is needed between those who make policies and those who train people to meet business needs. Cooperation between businesses and schools also makes it easier for universities to send students to work at businesses, where they can learn about technology and practice skills they'll need for work in the Industry 4.0.

3 Results

3.1 Worker Awareness About the Industry 4.0

The research team initially investigated the workers' comprehension of the Industry 4.0 before moving on to learn about the workers' ability to adapt to the changes brought about by the Industry 4.0. With the question, *"What media do you study and know about the Industry 4.0 through?,"* the findings indicate that the majority of workers have heard of the Industry 4.0 through the medium of the Internet, with a rate of 63.8%. This rate is followed by the mediums of television, radio, books, and newspapers, which have a rate of 51.9%. Only 23.8% of businesses are communicating with their staff members about the Industry 4.0, which is a figure that is considered to be relatively low. This demonstrates that firms are not very interested in spreading propaganda about the Industry 4.0 among their staff members. When it comes to disseminating knowledge regarding the Industry 4.0, the Internet has shown to be an excellent medium. Employees can gain some insight into what they need to perform in order to meet future job obligations with the help of such information.

The research team discovered that "automatic machines" received 49.6% of workers' understanding of the Fourth Industrial Revolution, "Internet of things" received 48.4%, "robots with connections" received 38.7%, "Human–machine integration" received 39.7%, "3D printing technology" received 17.1%, and "cloud computing" received 8.8%. Workers who only have experience with automatic machines may appear more frequently in the workplace; yet, it is possible that they do not fully comprehend the advantages that the Industry 4.0 delivers to either workers or enterprises. In the same vein, they are not yet aware of the difficulties that are associated with their own work.

If businesses start to make use of artificial intelligence and robots, only then will support for these technologies become truly feasible. When that time comes, the robot will take over the tasks previously performed by humans in a number of the various stages. The findings of the research suggest that in response to the question, *"What steps has your company made in preparation for the Industry 4.0?,"* the vast majority of employees believe that business owners are still in the learning phase (46.0%). The following item on the list is "applied automatic machinery to production 39.9%." The fact that businesses are beginning to use automated machinery in their production processes is evidence that they are also beginning to think about gradually replacing human labor in some stages of production. However, businesses have not paid a lot of attention to taking measures that could be ways to restructure labor, modernize production equipment, and invest methodically to increase the qualifications of employees. These kinds of actions include: The fact that the skill level of workers needs to be enhanced even higher in order to keep up with new technology is something that business owners have not realized.

The adoption of artificial intelligence in place of manual labor will lead to an increase in the number of people without jobs. On the other hand, this is not the only reason. The workers who participated in this research also highlighted some of the following causes as ones that will put them out of work before the trend of the Industry 4.0 arrives (Fig. 1).

If one's state of health is a vital aspect, then being healthy is the pinnacle of achievement. However, the employees here are concerned about their qualifications and ability levels. They are concerned that if their abilities are not enhanced, they will be replaced by machines and lose their employment. As a result, "Insufficient

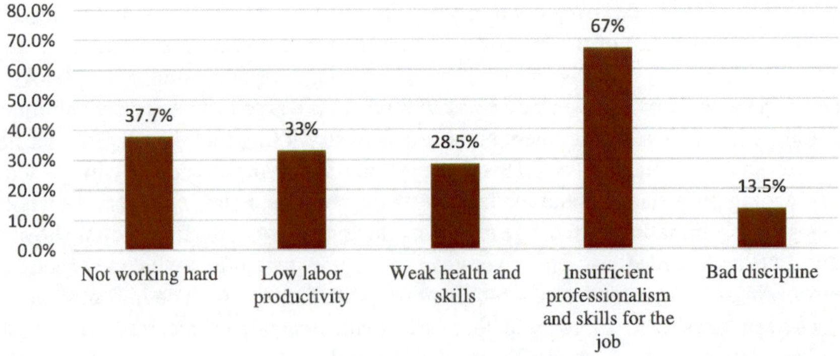

Fig. 1 Comments of workers on the causes of job loss and unemployment on the context of the Industry 4.0

professionalism and skills for the job" is cited as the leading cause, accounting for 67%. The next reason for this percentage, which is 37.7%, is "Not working hard."

Skills and knowledge are key considerations for employees, as they play a significant role in determining their income and wage. When businesses invest in newer and more advanced machinery technology, the number of employees they need also grows. It is necessary to have a more in-depth knowledge of how modern machinery operates. However, the work force in our country is rather low at the moment. According to the report on the labor and employment situation in the second quarter and the first six months of 2021, the number of trained workers from the "beginner" level or higher in the first six months of 2021 was estimated to be 13.3 million people, accounting for 26.1%, an increase of 0.9% points compared to the same period last year (General Statistics Office of Vietnam, 2021). This demonstrates that the competence level of workers in general is extremely low. As a result, it is easy to comprehend why they are concerned about the potential factors that led to their unemployment prior to the advent of the Industry 4.0.

When it comes to the company, workers have the opinion that implementing the advancements made during the Industry 4.0 presents difficulties for organizations. That comes out to "large technology investment costs," which accounts for 73.8% of the total, followed by "pressure on improving the qualifications of workers," which accounts for 54.2% of the total, "pressure on increasing labor productivity," which accounts for 33.8% of the total, "declining advantages cheap labor," which accounts for 11.3%, and "job conflicts," which accounts for 10.8%. This demonstrates that the Industry 4.0 not only affects workers, but it also places pressure on businesses to increase workers' credentials so that they can keep up with technological advancements. On the commercial side of things, it is essential to cultivate people and actively train human resources. It is possible for businesses to form partnerships with educational institutions in order to propose, associate, and train skilled people resources to suit the production needs of businesses. This is the gap that many businesses in Vietnam face in the present day.

3.2 *Adaptability to the Industry 4.0*

The team posed the question, *"Are employees capable of adapting in any way to the Industry 4.0?"* in order to obtain information regarding the manner in which workers are able to adjust to the upcoming changes. The findings of the research indicate that employees rank "Proactive, creative and technical improvement" as the ability that will allow them to adapt the best to the Industry 4.0. This ability was followed by "Flexible perception of new technology," which had a rate of 49.1%, and "Skills to work independently," which had a rate of 17%.

Workers also determined that in order to prepare themselves for the Industry 4.0, they need to equip themselves with "creative skills or learning strategies," with the rate of 30.9% being the highest. This was followed by "difficult problem-solving skills," with the rate of 25.6%, and the rate of "communication and behavior," with the rate of 23.1%. Creativity, problem solving, the ability to work together as a team, communication skills, the ability to self-organize, managerial abilities, and so on are some of the skills and abilities that workers need to be prepared to adapt to the Industry 4.0. The workers who participated in this study placed a high importance on a variety of talents, including the ability to be proactive and creative in the workplace. So, based on what they think, they are afraid of losing their jobs because they don't have enough skills and can't use modern production lines. So, workers think that if they want to avoid being laid off, they need to create and adapt to the Industry 4.0. This shows that, compared to other countries in the area, Vietnamese workers don't have as many skills, but they take the initiative and expect a lot from their work.

The question *"How does the Industry 4.0 effect your employment?"* was posed by the research team in order to get insight into the implications that Industry 4.0 has for people in terms of their employment. According to the findings of the research, the most common response given by respondents was "Both positive and negative effects," which received a rate of 41.1%. Following it, with a rate of 37.8%, we have "positive impact." 8.4% said "no influence." The good aspects include the introduction of new, cutting-edge technology, and the reduction of the need for human labor; nevertheless, in certain respects, it results in the displacement of people with lower levels of education and training.

The following inquiry is *"What do you intend to do in the event that you are made redundant as a result of the incorporation of robots into the production process at your place of employment?"* The following is an illustration of the obtained results: The response "I am thinking about it" was given by 80.8% of respondents, making up the bulk of people's responses. Even 13.8% of respondents stated they would "go back to my hometown," while just 3.2% said they would "go on strike." Based on these findings, it is clear that employees, despite their belief that they should be innovative and adaptable in order to succeed in their jobs under new circumstances, are actually suffering and are unsure of where to begin if there has been no prior preparation in terms of strategy, either on the part of the state, the enterprises, or themselves. This demonstrates that it is vital for the state to provide workers with information and professional skills on a macro-level in order to equip them.

It is required to have experience and high skills with a rate of 51.6% in order to keep jobs in the face of technological advancements when artificial intelligence is used in production, according to employees. This is the case because artificial intelligence is utilized in production. In addition, they believe that it is essential to innovate and be creative with a percentage of 50.9% of the population. Particularly, employees think that they need more ability to comprehend the characteristics and functioning mechanism of the device (47.9%), the capacity to change the working approach (36%), and the ability to work anywhere and at any time (26.7%). Therefore, in order for workers to maintain their positions in light of the intense rivalry that exists between workers for job security, workers now have to compete with machines and robots. Employees have demonstrated the ability to adjust to working with new procedures, even moving in many various job roles over the course of their careers. Therefore, the spirit of self-change in order to adapt to the job is clearly indicated in the awareness of the employees that participated in this study.

The workers who participated in this research anticipate, furthermore, the possibility to get professional training that will be made available to them by the Industry 4.0. According to 59.1% of respondents, the Industry 4.0 would be characterized by "Trained and skilled people." Next up, with a rate of 57%, is the statement that "Employees master science and technology." Learning how to make the most of science and technology is one of the most important things that workers can do. If you are skilled in science and technology, your work productivity will go up, which will lead to an increase in your revenue. However, in order for workers to achieve mastery in science and technology, they require training that is both methodical and conducted in a professional manner. Yet, in order to accomplish this goal, efforts are required from both the individual and the business. On the other hand, not all companies are prepared to shell out cash to train or retrain their workforce in order to keep up with their technical goals. Because the maximization of profits remains the primary objective of commercial enterprises, the strategy entails the reduction of expenses to the greatest extent possible.

Employees need to put in a lot of work in order to adapt to the Industry 4.0, which is currently underway. With the inquiry, *"In your opinion, in order for workers to enhance their adaptability before the Industry 4.0, what factors should they meet?,"* the findings indicate that a rate of 60.5% of respondents believe that it is vital to increase the "capacity to generate and apply new technologies." The following most popular response, with a rate of 51.9%, is "having education and professional qualifications." The next most popular option, with a selection rate of 49.3%, is "be in good health" and last, the "soft skills competency" category, which has a rate of 33%. Despite the fact that employees in any industry require a significant amount of focus on their soft skills. On the other hand, employees in this survey put a strong emphasis on creative problem solving and the implementation of emerging technologies. Because they are also concerned that the introduction of new technologies will cause them to lose their occupations and therefore their source of income.

In addition to professional qualifications, the factors that workers need to adapt to the Fourth Industrial Revolution are those shown in Fig. 2.

An overwhelming majority of workers (68.7%) believe that having a "high sense of responsibility" is the single most significant aspect in successfully adapting to Industry 4.0. Following that, with a rate of 50.1%, comes "social understanding,"

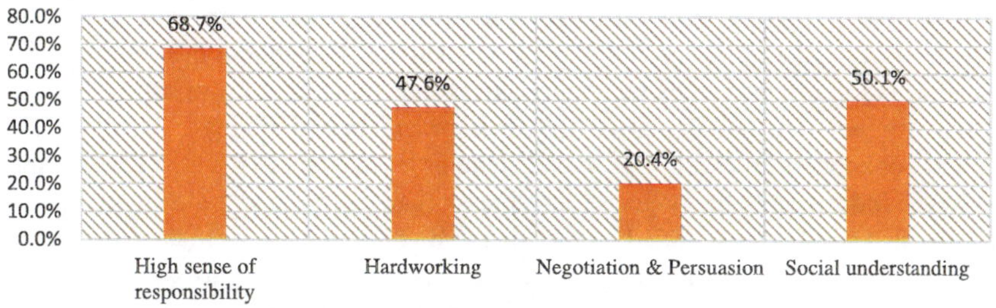

Fig. 2 Employees' perceptions of factors that help them adapt to the Industry 4.0

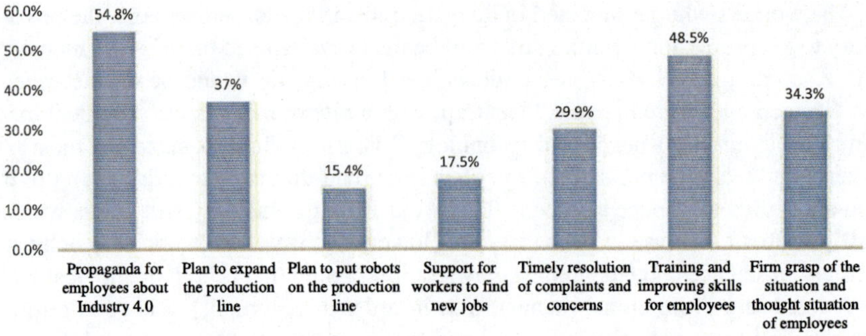

Fig. 3 Cooperation between trade union and enterprises for employees before the impact of the Industry 4.0

which is followed by "hardworking," "negotiation and persuasion," and so on. It's possible that Vietnamese workers recognize a sense of responsibility as an essential quality for them to have in order to thrive in the increasingly professional environment in which they now find themselves. This demonstrates that we should anticipate an increasingly highly qualified workforce in addition to a sense of responsibility while they are on the job.

The information presented above not only indicates how workers see the Industry 4.0, but it also demonstrates how they evaluate the factors that are necessary to adapt to it. In addition, the authors of this study gained insight into the role that grassroots trade unions play in assisting workers in adjusting to the new jobs that are being created in this era. The following is a presentation of the findings from the study (Fig. 3).

According to the findings of the research, there is support for workers coming from the union side. Trade unions can help workers in a variety of ways; however, the most significant contribution they can make is in the area of "Propaganda for employees about Industry 4.0," which has a rate of 54.8%. Following closely behind with a rate of 48.5% is the labor union working in conjunction with the business to provide "Training and improving skills for employees." Even though, when compared with propaganda, the percentage of trade unions is lower than that of working with businesses to help workers improve their abilities. This is because trade unions tend to focus on promoting their own interests. But this demonstrates that the union has also provided employees with appropriate support throughout the course of the transformation brought on by technological advancement, thereby assisting them in improving their abilities and preventing unemployment and job loss.

It is not enough for employees to change in order to adapt to the changes brought on by the Industry 4.0; the policies of business organizations and trade union organizations also need to be modified, as the adaptation of workers is a crucial component of these changes. Trade unions seek to assist workers in adjusting to work in its new context; types of propaganda can make full use of social networks to reach workers;

workers' comprehension of the advantages and downsides that the Fourth Industrial Revolution will bring to them can be improved.

4 Conclusions

Industry 4.0 is also presenting numerous issues to policy managers in Vietnam, such as the need to create an environment that is conducive to business, ensure that interests are in harmony with respect to existing service business models, govern the transparency of information, and so on. But the most difficult obstacle is to increase the quality of workers, which will need for a shift in mindset as well as innovative solutions. Despite the fact that the professional and technical credentials of the people who work in our country are still rather low, labor productivity is not particularly high, and both hard and soft skills are still lacking. The findings of this research indicate that employees have a good comprehension of the Fourth Industrial Revolution, albeit not a very thorough one. However, they are also aware of the necessity to improve their skills in addition to developing a sense of work and a sense of discipline in order to prevent themselves from becoming unemployed as technology continues to advance day by day, and artificial intelligence increasingly tends to replace humans.

References

Alhloul, A., & Kiss, E. (2022). Industry 4.0 as a challenge for the skills and competencies of the labor force: A bibliometric review and a survey. *Science, 4*(3), 34. https://doi.org/10.3390/sci4030034

Dimitrieska, S., Stankovska, A., & Efremova, T. (2018). The fourth industrial revolution-advantages and challenges. *Economics and Management, XV*(2), 182–187.

General Statistics Office of Vietnam. (2021). *Press release on labor and employment situation in the second quarter and the first 6 months of 2021*. https://www.gso.gov.vn/du-lieu-va-so-lieu-thong-ke/2021/07/thong-cao-bao-chi-tinh-hinh-lao-dong-viec-lam-quy-ii-va-6-thang-dau-nam-2021/#:~:text=Lao. động đã qua đào,v´ơi cùng kỳ năm trư´ơc.&text=có việc làm-,Lao động t`ư 15 tuổi trở lên có việc làm,v´ơi cùng kỳ năm trư´ơc

International Labour Organisation (ILO). (2018). *Industrial revolution (IR) 4.0 in Viet nam: what does it mean for the labour market? May*. http://www.ilo.org/global/topics/future-of-work/WCMS_448448/lang

Kergroach, S. (2017). (Sackey & A. Bester2, 2013). *Foresight and STI Governance, 11*(4), 6–8. https://doi.org/10.17323/2500-2597.2017.4.6.8

Khan, A., & Turowski, K. (2016). A survey of current challenges in manufacturing industry and preparation for Industry 4.0. *Proceedings of the First International Scientific Conference "Intelligent Information Technologies for Industry" (IITI'16), 450*, 15–27. https://doi.org/10.1007/978-3-319-33609-1

Manpower Group. (2017). *A skills revolution: From consumers of work to builders of talenT*. http://www.manpowergroup.com/wps/wcm/connect/d05472a8-cf68-42fb-90b0-bed9d82bb08b/MG_2017_Future_Forces_Skills_Revolution_lo+%281%29.pdf?MOD=AJPERES&CONVERT_TO=url&CACHEID=d05472a8-cf68-42fb-90b0-bed9d82bb08b

Rojko, A. (2017). Industry 4.0 concept: Background and overview. *International Journal of Interactive Mobile Technologies (IJIM)*, *11*(5), 77. https://doi.org/10.3991/ijim.v11i5.7072

Sackey, S. M., & Bester, A. (2013). Industrial engineering curriculum in Industry 4.0 in a South African context. *The South African Journal of Industrial Engineering*, *24*, 175–189. https://doi.org/10.7166/27-4-1579

World Economic Forum. (2016). The future of jobs employment, skills and workforce strategy for the fourth industrial revolution. *Growth Strategies*, *January*, 2–3. https://doi.org/10.1177/1946756712473437

The Role of Using Processes and Production Methods in International Trade Law to Achieve the Sustainable Consumption and Production Goals: Challenges and Opportunities for Asian Countries

Thu Minh Do

Abstract The Sustainable Consumption and Production (SCP) Goals focus on minimizing environmental impacts and maximizing socio-economic benefits. The shift toward SCP, therefore, has been promoted after the pandemic, particularly in low- and middle-income countries (LMICs). Nevertheless, LMICs are far behind in the race. Since the majority of Asian countries are LMICs, of which 13 countries are low-income countries, the shift toward SCP would be tremendously significant. One of the most effective methods to achieve SCP would be the Processes and Production Methods (PPMs) which is also a crucial issue in the relationship between trade and the environment. The paper will analyze the importance of PPMs in the shift toward SCP. Recently, numerous PPMs-related issues have been witnessed in Asia, such as the issue of genetically modified organisms (GMOs) products approval as the greater demand for food sources in the most crowded countries–China and India) and the most recent hotly-debated law case between European Union and Indonesia about biofuels. The paper will scrutinize the legality of trade-related PPMs measures in the 1994 General Agreement on Tariffs and Trade, Oct. 30, 1947, 61 Stat. A-11, 55 U.N.T.S. 194 [hereinafter GATT 1994]. (the GATT 1994), the Sanitary and Phytosanitary Agreement (The SPS Agreement), and the Technical Barriers to Trade Agreement (the TBT Agreement) to shed light on obstacles Asian developing countries are facing in practicing PPMs in the pathway toward SCP patterns. Finally, the paper will propose some recommendations on how to produce sound PPMs measures for Asian developing countries, such as promulgating consultation with other stakeholders before implementing PPMs, enhancing technical aid and fund from developed countries, and setting regional PPMs-related standards.

Keywords Sustainable consumption and production (SCP) · Processes and production methods (PPMs) · Like products · Asian countries

---

T. M. Do (✉)
Macquarie University, Sydney, Australia
e-mail: minh.thu-do@students.mq.edu.au

A. T. Nguyen et al. (eds.), *Contemporary Economic Issues in Asian Countries: Proceeding of CEIAC 2022, Volume 1*,
https://doi.org/10.1007/978-981-19-9669-6_11

1 Introduction

In 1972, a study found that unsustainable development can lead to the collapse of both the global economy and the environment (Meadows, 1982). Therefore, sustainable development is identified as "without alternatives" for the survival of humankind (Seiffert & Loch, 2005). To the best practice of sustainable development (SD), integrating consumption and production systems with SD has been considered and has become one of the core sustainable development goals (SDGs) (Akenji et al., 2017). In response to this, the Johannesburg Plan of Implementation in 2002 calls for all states to promote SCP patterns, with developed countries taking the lead and all states benefiting equally from the shift toward SCP patterns (Von Frantzius, 2004).

After years of moving toward SCP patterns, it can be seen that SCP strategies of developing/under-developing countries appear to be different from that of developed countries (Wang et al., 2019). While developed countries, with technology and economic strength, undoubtedly seem better at altering over-consumption patterns and enhancing national participation in reducing material and energy intensity, moving faster to the SCP goals, developing/under-developing countries are taking slower with the priority of poverty alleviation and with the attitude "grow first, clean up later". (Staniškis, 2012). In Asian countries, rapidly increasing population and increasing economic growth results in the over-consumption of natural resources and thus unsustainable consumption and production patterns (Rock & Angel, 2007). The unsustainable pattern makes a negative impact on the way to completing SCP goals in the Asia area (based on Asia–Pacific Sustainable Development Goals Progress Report in 2022). Even the Asia–Pacific Sustainable Development Goals Progress Report in 2022 states that "*Widening disparities amid COVID which reveals that the progress to SCP in Asian countries has not been occurring at a satisfactory rate, even worse alarming regression on responsible consumption and production has occurred*". With firm commitments to SDGs, Asian countries are being urged to decouple environmental considerations in economic recovery strategies after the pandemic, achieving SCP targets.

To complete the SCP goals, harmony between trade and the environment is vital; however, international legal frameworks and policies show opposite tendencies, hindering countries from achieving their SCP goals (Jackson, 1992). From some environmentalists' ideas, international trade has become an obstacle to the contemporary environmental protection effort (Gaines, 2002). Specifically, the Agenda 21 of the United Nations Conference on Environment and Development states that the continued deterioration of the global environment is caused by unsustainable consumption and production patterns. States, hence, are responsible for setting set standards to regulate PPMs to produce goods and services, so environmental deterioration is eliminated (Jere, 2017). In this context, PPMs are considered the most intense topic in the trade-environment debate (Potts, 2008). It has been argued that in most PPMs-related cases, trade liberalization has been considerably prioritized over the environment.

Part I of this paper will examine the role of practicing PPMs in achieving SCP goals to prove that practicing PPMs could be a massive opportunity for all countries, particularly Asian developing countries, to achieve SCP goals progressively. Part II will scrutinize the legality status of PPMs under WTO law to shed light on challenges Asian developing countries are facing. Part III covers some proposals about how to build PPMs-related trade-restrictive measures to push SCP races in Asia faster.

2 Sustainable Consumption and Production Goals and Environmental PPMs

2.1 Sustainable Consumption and Production Goals

Chapter IV of the Agenda 21, United Nations Conference on Environment & Development identified two objectives for promoting SCP patterns: (1) "*reducing the negative impact and pressure on our environment*" and (2) "*understanding better the role of consumption in sustainable development and seek feasible approaches to move successfully toward sustainable production*". One of the recommendations of Agenda 21 points out that using economic instruments such as environmental charges and taxes as approaches to influence customers' and producers' behaviors should be promoted. It calls for all states to replace unsustainable consumption and production patterns and use efficient natural resources to minimize depletion and reduce pollution.

The importance of the SCP goals continued to be reaffirmed after the Rio Summit, which focuses on enhancing the role of international trade in promoting sustainable development, and ten years later in the Johannesburg Plan of Implementation (the JPOI), which focuses on changing unsustainable patterns of consumption and production (set out in Chapter III). The JPOI provides that changing the consumption and production pattern is indispensable for achieving SDGs and encourages all states to change to SCP patterns (Johannesburg declaration on sustainable development and plan of implementation of the world summit on sustainable development, 2002). The JPOI also states that moving toward SCP patterns by decoupling economic growth with environmental factors through improving efficiency and sustainability in using natural resources and production processes in constructing a domestic 10-Year Framework of Programs.

Numerous points are being made about the entailment of SCP goals, such as the level of focusing on consumers and lifestyle, emphasizing the difference between sustainable consumption and sustainable production, and changing the aggregate level of consumption (Katila et al., 2019). Nevertheless, SCP is closely related to the efficient and sustainable use of resources and energy to produce consumer goods and services (Farber, 2012). In "Sustainable Consumption & Production Indicators for The Future SDGs"–UNEP Discussion Paper, SCP is defined as "*the use of services and related products, which respond to basic needs and bring a better quality of life while minimizing the use of natural resources...as well as the emissions of waste and pollutants over the life cycle of the service or product so as not to jeopardize the needs of further generations"*. (Farber, 2012). The Declaration of the United Nations Conference on Sustainable Development–The Future We Want, held in Rio in 2012, states that achieving SCP patterns can mitigate environmental and natural resource depletion caused by economic activities. As reaffirmed in numerous international documents, the SCP goals will play a crucial role in achieving SDGs formulated in the United Nations Summit in New York in 2019 (so-called The 2030 Agenda for Sustainable Development).

To sum up, SCP goals aim to maintain human beings' use of natural resources and pollution intensity within the environment's capacity (Dernbach, 1998). SCP patterns mitigate environmental damage during the life cycle of goods and services (Dernbach, 1998). SCP goalkeepers thus should balance the population and the population's demand for goods by using technologies and efficiently utilizing renewable and non-renewable resources (Shibin et al., 2016).

With all the benefits of SCP stated above, achieving SCP is paramount because we are witnessing the population boom and demand for materials in the ten years (based on Asia–Pacific Sustainable Development Goals Progress Report in, 2022: Widening disparities amid COVID). In terms of the progress toward the SCP goals in Asia, although in Asia–Pacific Sustainable Development Goals Progress Report, 2022, data shows a reduction in hazardous waste generated, an improvement in renewable energy capacity, and compliance with hazardous waste conventions, these trends, nevertheless, show insufficient progress. In general, the region has regressed in responsible consumption and production goals.

2.2 Processes and Production Methods (PPMs) and SCP Goals

PPMs link closely to consumption and production processes. Additionally, PPMs are regulated by international trade law and environmental law tools because countries tend to impose PPMs measures both inside territories and outside their sovereignty practices (on their trade partners) to protect the environment.

Environmental policies, laws, and regulations dealing with environmental degradation and internalizing all costs during the life cycle of products for trade purposes can be categorized into three broad types: (1) product characteristics, (2) resources assessment regulation, and (3) PPMs regulation (Gaines, 2002). PPMs are the most common and diverse class among these types (Gaines, 2002). Based on the paper prepared by the Organization for Economic Co-operation and Development about PPMs analyzing possible methods to enforce PPMs regulations, the PPMs can be defined as methods in which products are produced or processed or natural resources are harvested or extracted.

PPMs, if implemented properly for environmental purposes, can benefit both developed and developing countries making progress toward SCP. Under international trade law, two types of PPMs identified: product-related-PPMs (pr-PPMs) and non-product-related-PPMs (npr-PPMs) (Silveira et al., 2013, and Nielsen, 2007). The former is closely relevant to product characteristics or physical and chemical properties of products (e.g., any mandatory type of packaging and containers, waste disposal, retrieval, and recycling of products can be categorized as PR-PPMs). The latter does not show any trace in the quality of products (they can be regulations about animal welfare, labor treatment, the methods to extract natural resources, the methods of hunting or fishing) (Gaines, 2002).

PPMs enable countries to keep consumers informed about green choices and remind consumers to use efficient renewable and non-renewable resources, hence formulating an SCP lifestyle among consumers (so-called green consumerism stated in the United Nations 10-Year Framework). On the consumption side, environmental PPMs can be used as market-oriented tools influencing consumers' choices. With sufficient product information, consumers can easily recognize which is environmentally friendly and make wise choices (Gaines, 2002). This information helps to enhance consumers' awareness about environmental protection and change consumers' preferences and behaviors toward sustainability. A great demand for green goods and services would also become a strong driving force to encourage or enforce producers/suppliers to apply environmental PPMs to produce and process more green products. The number of green products thus would be increasing, driving down the relative prices, making green goods and services more affordable and accessible, and creating a closed cycle. Consumers will gain their pursuing power in the market during this cycle. The appearance of green goods and services in the market will reinforce the SCP lifestyle.

Implementing environmental PPMs requires the participation of governments (through "top-down" sustainable efforts by policymakers) and other stakeholders in society ("bottom-up activities by companies). Therefore, companies play an essential role in achieving SCP goals (Wang et al., 2019). On the production side, since environmental PPMs include waste disposal management, implementing environmental PPMs enables corporations to internalize the life-cycle approach to product management fully, wherein a company's responsibility would be extended to waste disposal and recycling of end-use products (Gaines, 2002). That means the polluting output of one set of industrial processes can be utilized to feed others (Gaines, 2002). That also helps corporations manage waste disposal, reducing hazardous waste and cost-saving by not dealing with high amounts of toxic waste. In addition, PPMs measures (such as carbon taxes and other carbon pricing schemes) have been proposed and applied for domestic producers and foreign exporters. Applying environmental PPMs also helps producers precisely calculate the amount of carbon dioxide embedded in their products during their life cycle (Kuik & Gerlagh, 2003). Proper carbon calculation indirectly eliminates the unsustainable subsidies for fossil fuels, alleviates carbon leakage, and encourages the shift to renewable resources in the long term (Kuik & Gerlagh, 2003).

Every corporation wishes to attract more investors and financing and shine its reputation in the eyes of the public. The more they can contribute to society and the environment, the more trustworthy they are, leading to increased investment attraction and public trust. Therefore, applying environmental PPMs also helps corporations to easily integrate sustainability information into their sustainability reports (ESG report as an example–Environment, Society and Governance Reports) which tools to assess corporations' trustworthiness, social responsibility, and environmental responsibility (Plastun et al., 2020).

In developing countries, the lack of resources constitutes the primary obstacle to the shift to SCP. Governmental policies in developing countries tend to address poverty alleviation and accelerate economic growth, paying less attention to environmental issues, even though they have been observed to have severe environmental failures. The excuse for this might be the so-called environmental Kuznets curve which is conventional thinking in developing countries (Cole et al., 1997). Based on this conventional school of thought, Kuznets curve shows a hypothesized inverted U-shape relationship between economic growth and environmental quality (Cole et al., 1997). The developing countries are persuaded to think that they must go through a high environmental degradation and pollution phase to concentrate on their economic development before they become more affluent and can overcome environmental challenges (Cole et al., 1997). This attitude formulates the so-called grow first and clean up after. However, with SDGs agreed upon by the international community, this conventional thinking becomes no longer valid and is even a mistake. With the prevalence of PPMs in the future, developing countries, especially active Asian developing countries, can be able to utilize funds and technology aids from developed countries, making a leapfrog to more sustainable modes of production and consumption by decoupling economic and environmental objectives at the same time. In other words, by practicing PPMs, they can be able to complete two objectives (economic and environmental objectives) at the same time.

3 WTO Law-Compatible PPM Measures in SCP Focus Areas

3.1 PPMs in Food Safety, Security, and Nutrition

*** The Context of GM (Genetic Modified) Products in Asia**

Gene-editing or genome modification covers a wide range of genetic engineering techniques used to modify the genetics of organisms (Cole et al., 1997). The modification involves inserting, deleting, and editing DNA sequences of genomes in any living organism (Marris, 2001). The method to produce and consume a GM product falls under the spectrum of PPM. GMOs are mainly for livestock and other animal feed. In Asia, the amount of GM crop materials imported into Asia for processing food and animal feed has grown substantially as almost Asian countries import and develop GM crops (Anderson et al., 2004).

GM products so far are treated as a bullet point to address poverty, particularly in low and middle-income countries like Asia. Poverty has been considered an urgent task in many developing countries. In 2008, a report pointed out that the surging cost of rice and wheat threatened many Asia citizens, pushing them into poverty (Laxman & Ansari, 2011). From 2008 to 2019, the poverty rate was alleviated; however, as the pandemic happened, it set back the fight of Asian countries against poverty (according to figures from 'Pandemic Sets Back Fight Against Poverty in Asia by At Least 2 Years, Has Likely Hurt Social Mobility' published by Asia Development Bank) (Asian Development Bank, 2022). For now, while most economic indicators show strong recovery, many people seem harder to get out of poverty (Tsatsakis et al., 2017). The pandemic also influences the perspective of some countries about sustainable food resources. For example, after the dreadful COVID-19 pandemic and trade war with the USA, China–the most crowded country in Asia–has come to acknowledge the importance of GM products might be an alternative and available food source (Herring, 2009) (Tsatsakis et al., 2017). In India–the second most crowded country, the government has paid much attention to GM products in response to the booming population (Tsatsakis et al., 2017). Additionally, many other Asian countries are dealing with the pressure of food and economic growth, which is multiplied by shrinking farmland, rising labor costs, and a shortage of farm workers (Anderson et al., 2004). Not only does poverty, but also some argue that biotechnology enhances the capacity to resist hostile weather and enhance the taste and nutrition of products. That has been inextricably tied to SCP goals (food nutrition and environment fall under the SCP spectrum). In such a case, using GM products as an alternative source seems to be the ideal method to ensure food security in Asia's most challenging time (Lang, 2002).

The question of staying competitive in global trade is one primary driver for many Asian developing countries to invest in biotechnology (Thomas, 2003). Indeed, most Asian developing countries look to biotech as a future driver of economic growth (Thomas, 2003). Strong domestic regulation will be a prerequisite for the biotechnology industry to thrive. More crucially, a proper biosafety regulation will act as a tool to avoid trade disputes in the future and step steadily on the successful SCP path (Sengupta, 2016).

Confronting consumers' concerns about potential allergenic and long-term toxic effect rooting from GM products and other concerns about the negative effect that GM products causing damage to the environment and inmate animals and plants, (Marris, 2001) many countries, including Asian developing countries, deem to be cautious in building up their legal framework regulating GM products, such as conducting risk assessment models, mandatory labeling schemes and quarantine (Marris, 2001). The stringency level of regulations varies. Many countries also have regulatory schemes for commercializing and approving food safety of GM crops, such as Japan and South Korea, which get tough on GM crops; (Thamali & Jayawardana, 2022) other countries, such as Malaysia, Philippines, Thailand, and Vietnam, are more "welcome" to GM crops (Thamali & Jayawardana, 2022). However, one of Asia's challenges is building domestic biotechnology regulations in sync with complex international trade and environment regulations.

*** Considering Some Legal Issues in Building Domestic Biosafety Regulation**
As stated above, producing GM products raises several concerns about the environment, and human, animal, and plant health. Since GM products are listed as goods traded globally, GMOs are under cover of both WTO law and environmental legal framework, more specifically the GATT 1994, the SPS Agreement, the TBT Agreement, and some international environmental agreements such as the Cartagena Protocol on Biosafety. Both frameworks encourage country members to establish their domestic biotechnology regulation as their sovereignty practice. In practice, a country may be able to draft and implement broad regulations that contain some requirements covered by the TBT Agreement (i.e., grading and mandatory labeling scheme) and others covered by the SPS Agreement (treatment of GM products to prevent diseases).

The SPS Agreement has been in correlation with the Cartagena Protocol on Biosafety regarding GMOs. The Protocol reaffirms the rights of ratifying countries to set up their domestic regulations but also respects the conditions that countries accept importing GM products according to their socio-economic situation (Bail et al., 2014). The Protocol also reflects the acceptance of the precautionary principle, allegedly reflected in Article 5.7 of the SPS Agreement. Besides, the SPS Agreement makes import restrictions subject to many requirements. In particular, any restrictive measures on GMOs for sanitary and phytosanitary purposes (namely, to protect human, animal, and plant health) must be based on scientific evidence of a risk assessment that conforms to specific standards in the Agreement (WTO, 1995). In case of uncertain risks, article 5.7 of the SPS Agreement allows state members to apply trade measures if relevant scientific evidence is insufficient, based on available pertinent information, and in a reasonable period. In the SPS Agreement, there is no "like product analysis" because the focus is the "scientific assessment" (Ansari & Mahmod, 2008). According to these articles, an import ban on a GM product would have to meet the risk assessment criteria of the SPS Agreement, and scientific justification would have to be made if risks are in excess of what sets out in recommended international standards (WTO, 1995). WTO emphasizes that members must consider related processes and production methods when assessing risks (WTO, 1995).

The TBT Agreement establishes specific criteria that technical regulations such as mandatory labeling schemes have to fulfill and might apply to restrictions on GMOs import. Similarly to the SPS Agreement, the TBT Agreement is linked to the Cartagena Protocol, requiring the modified organisms for direct use as food, feed, or processing. The Protocol requires a label stating that "the product may contain such GMOs" (Eggers & Mackenzie, 2000). The difference between the SPS Agreement and the TBT Agreement, in the TBT Agreement, the "likeness" concept is significant and applied. Whether the difference in PPMs to create GM products can differentiate it from other conventional counterparts remains unclear as mixed arguments exist.

The SPS and TBT Agreement encourage the use of recommended international standards, guidelines, and recommendations from expertise organizations, such as the realm of Codex Alimentarius (the FAO's international standards for GM products). Take Vietnam as an example of using Codex Alimentarius standards and drafting biosafety regulations based on this realm (FAO GM Foods Platform). However, states also have the right to set domestic standards higher than recommended international ones. International harmonization of regulations is currently under discussion in several forums.

There are some issues identified:

- The "protecting human, animal, and plant health" objective is of paramount importance under the SPS Agreement. When applying trade-restrictive measures to protect human, animal, and plant health, a risk assessment based on scientific evidence must be conducted to assess products. Different countries have different approaches when treating GM products (Sheldon, 2002). Among them, the precautionary principle approach would be the most controversial. Countries following this approach conduct risk assessment and manage GMOs based on this precautionary principle as they revise their regulations and adopt rules and guidelines for the mandatory labeling of GMOs and food containing GM ingredients (Sheldon, 2002). The European Union (the EU), Japan, South Korea, and Sri Lanka are examples of the second approach (Sheldon, 2002). In the EU, the prudence principle (similar to the precautionary principle) is prevalent and applied in any legal framework, including food safety regulation (Andorno, 2004). Since the EU has been a critical trade partner with most of Asian developing countries, the EU's stringent regulations on GM products concerning compulsory label schemes and risk assessments have become hurdles to Asian developing countries. That is because risk assessments are conducted based on the prudence principle (the precautionary principle), which is allegedly broader than risk assessment under article 5.7 of the SPS Agreement (based on the statement of Panel and Appellate Body in the EU biotech case) (Winham, 2009).
- Similarly, with the labeling schemes under the TBT agreement. The label scheme is varied between countries. For example, Vietnam applied the label requirement to any product containing above or equal to 5 percent of GMOs in its ingredient list ; China is one of the first countries to mandate labeling both GMOs and non-GMOs. On the contrary, owning to the prudence principle applied, the labeling scheme in the EU legal framework is stringent: A food or feed containing or produced from GMOs must present "this product contains GMO", the name of the GMO, the name and address of the representative of manufacturers producing this GMO, and the method to obtain public information of producers' registration (Zheng & Wang, 2021). GM product exporters from Asian developing countries are not likely to afford the onerous cost generated by the labeling scheme.
- In the context of lacking compulsory international standards on GM products' safety, countries with technology strength impose relatively stringent standards, putting their poorer counterparts in a weak position. In the case of not yet international harmonization of food safety regulations imposed, this would increase trade disputes, particularly between wealthy and poorer countries. The trade disputes typically take a long time to process and are money-consuming, specifically for

developing countries, including in Asia. The first formal complaint to the WTO over GMO import regulation concerned a prohibition on imports by Egypt from Thailand of tuna containing GM soybean oil (WT/DS205/1) (Josling & Sheldon, 2002). While the dispute has now been settled, it indicates the likelihood of conflict over GMO regulations as they impact international trade.

- The core of the SPS Agreement is still "risk assessment" and "scientific evidence". Food manufacturers conduct a risk assessment, considering any impact of PPMs on consumers' health and environment. Assessing the possible impact of PPMs requires the strength of research and development capacity, experts, and funds. The lack of thereof in Asian developing countries could lead to insufficient or incomplete scientific evidence. As stated above, Asian countries are majorly low and middle income. They rarely have the full ability as developed countries to conduct and gather more scientific evidence, let alone GMOs themselves are an uncertain aspect and hard to know scientifically precisely, even in developed countries.

3.2 PPMs in Environmental Protection Objective

*** Carbon Tax and Other PPMS Measure Environmental Objectives**
The concept of PPMs in international trade is also relevant to the objective of environmental protection. One of the targets of SCP is reducing hazardous emissions released into the air, water, and soil (Akenji & Bengtsson, 2014). Consumption and production patterns should be developed in the way that chemicals used in consumption and production are in the way to minimize the significant adverse effect on human health and the environment. In international trade, the important is how to regulate PPMs to eradicate hazardous emissions during the consumption and production stage and ensure sustainable resources for future generations.

Asia–Pacific is one of the most crucial areas to combat over-consumed energy, and carbon emissions play an essential role in achieving net zero and change to SCP patterns (Fernandes & Rezaei, 2021). Recently with the commitments to net zero under the Paris Agreement and COP 26 (held in Glasgow, United Kingdom). While most Asian countries develop carbon pricing schemes (or emission trade schemes), such as Japan, Singapore, China, Thailand, and Vietnam, the unilateral effort seems insufficient, highly leading to carbon leakage. In response to carbon leakage, a stringent method has been proposed and applied in several countries–carbon border tax. The efficiency of imposing a carbon tax worldwide to stop carbon leakage has not been proved in practice.

Nevertheless, scientific data shows that if a carbon border tax is imposed globally, it will reverse the carbon leakage cycle. The EU is one of the pioneers in introducing a carbon scheme (so-called the Green Deal), including imposing a carbon tax on heavy-polluted output and highly traded industries such as steel and cement (Siddi, 2020). This Green Deal will be enacted officially in 2023. Despite its emphasis on environmental protection, many countries that are exposed to newly introduced carbon schemes argue that this new legal framework can be considered "protectionism", enabling the EU to recover its economy, which was damaged by the pandemic (Siddi, 2020). The top traders exposed heavily to the new EU's scheme consist of low, middle-income, and high-income countries. In Asia, China, India, Brazil, South Korea, and especially Mozambique, a top exporter of aluminum to the EU, are called out loud in the list of countries heavily affected by the new EU policy (Teevan et al., 2021). The imposition of a carbon border tax on the products exported from those countries put these countries in a significantly vulnerable position, given that the exportation of industries is an engine of their economies (Teevan et al., 2021). The compliance cost, such as hazardous waste monitoring, could burden them.

Along with environmental objectives, Indonesia raised a PPMs-related case recently against the EU with its new policy (Mayr et al., 2021). According to the EU's recast Renewable Energy Directive (RED II), followed by the RED I, intending to eradicate biofuel produced by palm oil which allegedly releases carbon dioxide emissions three times greater than biofuels produced by other plants, the EU declares that biofuel produced by palm oil will be phased out. In contrast, biofuels produced by other ingredients, such as corn and sugar, will only reduce to 7 percent of use (Mayr et al., 2021). In this case, Indonesia, the biggest palm exporter to the EU, is likely to get hurt by the new RED II imposed by the EU (Mayr et al., 2021). The EU also claims that the increase in the demand for biofuels produced by palm oil led to land-use change in Indonesia, resulting in significant deforestation (lands replaced to grow palm forests), phasing out palm oil to avoid deforestation (Mayr et al., 2021). The palm oil case between Southeast Asia and the EU raises some questions about PPMs in the "likeness" concept and exceptions under Article XX of GATT.

*** PPMs and the "Like Products" Concept and Exemption Under the GATT**

The carbon border tax measure and the phasing-out policy imposed by the EU are nr-PPMs measures. Any trade measures must adhere to the WTO law, with the "likeness" being the main focus. GATT 1994 allows its members to treat different products. It comes down to whether the carbon embedded in products or ingredients used to produce products make them different. The "like product" test remains unfixed and varies on a case-by-case basis. In the Japan Alcoholic case, the "like product" test contains four criteria: (1) physical characteristics of products; (2) different tariff categories; (3) the end-use product; and (4) consumer's habit and preference (Horn & Mavroidis, 2004). Unlike pr-PPMs, nr-PPMs are still controversial among WTO members (Charnovitz, 2002). With nr-PMs, the physical characteristics of products produced by either environmentally friendly or non-environmentally friendly methods rarely make them different (Horn et al., 2004). For example, many argue that steels produced by unclean technology are similar to those produced by clean technology. Likewise, biofuels produced mainly by palm oil are majorly similar to biofuels produced by other plants, even though the efficiency of biofuels by palm oil is more remarkable than those produced by other plants. The second and third criteria do not seemingly play an important role in comparing products. The glimpse of hope lies in the last criteria. Despite rising environmental concerns among consumers, the product difference is still hard to tell consumers. The nr-PPMs cannot make products different.

However, GATT 1994 allows states to justify their trade measures under Article XX. In other words, these trade measures must satisfy at least one of the objectives in Article XX (a) to (g); and the Chapeau of Article XX. Under Article XX (b) and (g), states justifying their trade measures under these sub-articles have to prove their trade measures imposed to protect human health, animal and plant health. Next, the Chapeau of Article XX states that a trade measure must not be applied in a manner which would constitute "a means of arbitrary or unjustifiable discrimination between countries where the same conditions prevail", and is not "a disguised restriction on international trade". In many cases, the word "arbitrary" and "unjustifiable discrimination" and "same condition prevails" have been defined differently. In the case of Indonesia and EU regarding biofuels produced by palm oil and other plants, although EU's objective can be justified under Article XX (g), there is nothing to be sure EU can pass the Chapeau test. Indonesia states that under the "same conditions" prevails, biofuels produced by palm oil and other vegetables should not be treated differently. In other words, other plants' biofuels should also be phased out under RED II (Mayr et al., 2021).

As stated above, with the loopholes of GATT explanations, PPMs measures can be easily used under cover of protecting the environment to protect domestic industry. The question of "environment protection" or "protectionism" will be raised, and eventually, developing countries will still get hurt.

4 Recommendations

The harmonized international regulation on safety biotechnology is still lacking, leading to the increased likelihood of trade disputes. Asian developing countries and other countries inside or outside the region should establish more mutual or regional agreements regulating standards and categories of products differentiated by their PPMs. The standards can be based on recommended international standards or members' objectives and priorities in their territory. These agreements are expected to tighten relationships between members, enabling them to utilize funds and technology from richer member countries to develop PPMs and achieve SCP goals. The standards on PPMs can also be included in free trade agreements. The likelihood of trade disputes will be reduced, and free trade is less likely to be interrupted. However, openness can hardly be seen in several regional organizations. ASEAN is a typical example. Being known as the "ASEAN Way", (Yukawa, 2018) a harmonized regulatory over PPMs seems hard to achieve. The ASEAN Way emphasizes the sovereignty rights of a country. Many researchers thus argue that the ASEAN Way established long ago seems obsolete now, and it is time to change (Yukawa, 2018). In the case of Asian countries, many countries being wary of PPMs in global trade seem to find it harder to accept a standard on PPMs. That would slow down the progress of establishing the harmonization and even further the progress of SCP goals, among other sustainability goals.

The WTO/GATT objects to achieving and maintaining negotiated levels of market access but, at the same time, to ensure its members practice their sovereignty entirely right over their regulations. A possible solution here might be that countries have the right to set up their domestic standards and even a bound tariff (carbon tax) on exporters. That could be stricter than international standards recommended by international experts. At the same time, they have to grant greater market access to other exporters who meet domestic standards. That can be called "balancing market access" (Bredahl et al., 1990). Balancing the market also allows countries to increase sovereignty over their domestic regulatory choices and still oblige trading obligations. However, "rebalancing market access" is not the most optimal method because countries still face high costs (e.g., labeling non-GM products will require identity preservation of non-GM ingredients). This cost will be passed on to consumers (Bredahl et al., 1990).

In terms of policy recommendations, products from Asian developing countries brought on trade disputes are mostly the targeted industries of countries, generating high levels of profitability. For example, Indonesia and Malaysia count on palm oil industries; most Asian countries are exposed to new carbon schemes (Lim, 2018). Developing countries must reconsider the level of profitability generated from targeted industries. Seeking alternative industries that ensure Asian developing countries' development is not to be hindered and provide as many quality employment opportunities as altered might take time (Lim, 2018). Asian developing countries do not always have enough resources for research and development, especially in the case of new environmentally friendly PPMs. Asian countries should collaborate with developed countries via free trade agreements. The EU, which has been in trade relationships with numerous Asian developing countries, should work together and see if there are other industries in other countries worth investing time in. Aiding developing countries is an obligation of developed countries set out in many documents, including the GATT and multilateral environmental agreements. In addition, before applying any trade-PPMs-related measure, it is mandatory to make advance notice to those who would be possibly exposed to it.

The negotiation must be conducted in good faith, avoiding any trade disputes. In many cases, advance consultation promptly with stakeholders can be helpful. In the case of Indonesia and the EU, given that the free trade agreement between the EU and Indonesia has been a rough ride to implement (Team, 2021), there is still room for EU and Indonesia with regards to palm oil only if palm oil practices are kept "green" and "clean" (Lim, 2018).

5 Conclusion

As PPMs is still not a panacea for addressing environmental and trade law conflicts, they are just a temporary method. PPMs-related SCP goals still cover many issues, such as illegal/legal wildlife trade and wildlife consumption. In the extent of this paper, PPMs presenting the conflict between trade and the environment have been described. The prevalence of carbon tax, indirect/indirect land use, GMO labeling schemes, and scientific evidence (PPMs measures) has been seen clearly. As stated above, despite the benefits PPMs could bring in the path toward SCP and SD in general, Asian developing countries face several difficulties and challenges from implementing environmental PPMs in trade and even being exposed to stringent PPMs measures from developed countries. Asian developing countries cannot be outside of the game. In the future, we must develop other better methods to keep trade and the environment in harmony. That is also the meaning of fulfilling SDGs.

References

Agenda 21, U.N. Conference on Environment and Development, U.N. Doc. AIConf. 151/26/Rev.1 paragraph 4.3.

Akenji, L., & Bengtsson, M. (2014). Making sustainable consumption and production the core of sustainable development goals. *Sustainability, 6*(2), 513–529.

Akenji, L., Bengtsson, M., & Schroeder, P. (2017). Sustainable consumption and production in Asia—aligning human development and environmental protection in international development cooperation. In Sustainable Asia: Supporting the transition to sustainable consumption and production in Asian Developing Countries, pp. 17–43.

Anderson, K., Nielsen, C. P., & Robinson, S. (2004). Estimating the economic effects of gmos: The importance of policy choices and preferences. *Economic and Social Issues in Agricultural Biotechnology*, 359–391. https://doi.org/10.1079/9780851996189.0359

Andorno, R. (2004). The precautionary principle: A new legal standard for a technological age. *Journal of International Biotechnology Law, 1*(1). https://doi.org/10.1515/jibl.2004.1.1.11

Ansari, A. H., & Mahmod, N. A. K. N. (2008). Biosafety protocol, SPS agreement and export and import control of LMOs/GMOs. *Journal of International Trade Law and Policy.*

Asian Development Bank (Ed.). (2022). Pandemic sets back fight against poverty in Asia by at least 2 years, has likely hurt Social Mobility. In *Asian Development Bank*. Retrieved from https://www.adb.org/news/pandemic-sets-back-fight-against-poverty-asia-least-2-years.

Asia-Pacific sustainable development goals progress report in 2022: Widening disparities amid COVID. (2022). Retrieved from https://www.unescap.org/kp/2022/asia-and-pacific-sdg-progress-report-2022.

Bail, C., Falkner, R., & Marquard, H. (2014). The Cartagena protocol on biosafety: Reconciling trade in biotechnology with environment and development.

Bredahl, M. E., Josling, T. E., Miner, W. M., Rossmiller, G. E., Tangermann, S., & Warley, T. K. (1990). Bringing agriculture into the GATT: The comprehensive proposals for negotiations in agriculture (No. 938–2016–74512).

Charnovitz, S. (2002). The law of environmental PPMs in the WTO: Debunking the myth of illegality. *Yale J. Int'l L., 27*, 59.

Cole, M. A., Rayner, A. J., & Bates, J. M. (1997). The environmental Kuznets curve: An empirical analysis. *Environment and Development Economics, 2*(4), 401–416.

Dernbach, J. C. (1998). Sustainable development as a framework for national governance. *Case W. Res. L. Rev., 49*, 1.

Eggers, B., & Mackenzie, R. (2000). The Cartagena protocol on biosafety. *Journal of International Economic Law, 3*(3), 525–543.

FAO GM foods platform. Retrieved from https://www.fao.org/food/food-safety-quality/gm-foods-platform/browse-information-by/country/country-page/en/?cty=VNM.

Farber, D. A. (2012). Sustainable consumption, energy policy, and individual well-being. *Vand. L. Rev., 65*, 1479.

Fernandes, E., & Rezaei, F. (2021). Climate actions, COP 26 and Implications on public health for Asia Pacific region. *Epidemiology International, 6*(4), 1–2. E-ISSN: 2455–7048

Gaines, S. E. (2002). Processes and production methods: How to produce sound policy for environmental PPM-based trade measures. *Colum. J. Envtl. L., 27*, 383.

General Agreement on Tariffs and Trade, Oct. 30, 1947, 61 Stat. A-11, 55 U.N.T.S. 194 [hereinafter GATT 1994].

Herring, R. (2009). China, rice, and GMOs: Navigating the global rift on genetic engineering. *The Asia-Pacific Journal: Japan Focus, 7*, 1–12.

Horn, H., & Mavroidis, P. C. (2004). Still hazy after all these years: The interpretation of national treatment in the GATT/WTO case-law on tax discrimination. *European Journal of International Law, 15*(1), 39–69.

Jackson, J. H. (1992). World trade rules and environmental policies: Congruence or conflict. *Wash. & Lee L. Rev., 49*, 1227.

Jere, M. (2017). *Trade and sustainable development: Regulating PPMs in the WTO.* Master's thesis, University of Cape Town.

Johannesburg declaration on sustainable development and plan of implementation of the world summit on sustainable development A/CONF.199/L.7, 2002.

Josling, T., & Sheldon, I. (2002). *Biotechnology Regulations and the WTO.* Working Paper 02–2. The Ohio State University.

Katila, P., Colfer, C. J. P., De Jong, W., Galloway, G., Pacheco, P., & Winkel, G. (Eds.). (2019). *Sustainable development goals.* Cambridge University Press.

Kuik, O., & Gerlagh, R. (2003). Trade liberalization and carbon leakage. *The Energy Journal, 24*(3), 97–100.

Lang, T. (2002). Can the challenges of poverty, sustainable consumption and good health governance be addressed in an era of globalization? *The nutrition transition*, 51–70.

Laxman, L., & Ansari, A. H. (2011). GMOs, safety concerns and international trade: Developing countries' perspective. *Journal of International Trade Law and Policy, 10*(3), 281–307.

Lim, S. (2018). EU-Indonesia FTA relations: Palm oil–in for a rough ride?. EU-Asia at a Glance.

Marris, C. (2001). Public views on GMOs: Deconstructing the myths. *EMBO Reports, 2*(7), 545–548.

Mayr, S., Hollaus, B., & Madner, V. (2021). Palm oil, the RED II and WTO law: EU sustainable biofuel policy tangled up in green? *Review of European, Comparative & International Environmental Law, 30*(2), 233–248.

Meadows, D. H. (1982). *The limits to growth: A report for the club of Rome's project on the predicament of mankind.* Universe Books.

Nielsen, L. (2007). *The WTO, animals and PPMS.* https://doi.org/10.1163/ej.9781571051868.i-355

Plastun, A., Makarenko, I., Khomutenko, L., Osetrova, O., & Shcherbakov, P. (2020). SDGs and ESG disclosure regulation: Is there an impact? Evidence from top-50 world economies. *Problems and Perspectives in Management, 18*(2), 231–245. https://doi.org/10.21511/ppm.18(2).2020.20

Potts, J. (2008). The legality of PPMs under the GATT: Challenges and opportunities for Sustainable Trade policy. *International Institute for Sustainable Development.*

Rock, M. T., & Angel, D. P. (2007). Grow first, clean up later? Industrial transformation in East Asia. *Environment: Science and Policy for Sustainable Development, 49*(4), 8–19.

Seiffert, M. E. B., & Loch, C. (2005). Systemic thinking in environmental management: Support for sustainable development. *Journal of Cleaner Production, 13*(12), 1197–1202.

Sengupta, R. (2016). International trade and the 2030 agenda for sustainable development. *Spotlight on Sustainable Development*, pp. 130–139.

Sheldon, I. M. (2002). Regulation of biotechnology: Will we ever 'freely'trade GMOs? *European Review of Agricultural Economics, 29*(1), 155–176.

Shibin, K. T., Gunasekaran, A., Papadopoulos, T., Dubey, R., & Mishra, D. (2016). Sustainable consumption and production: need, challenges and further research directions. *International Journal of Process Management and Benchmarking, 6*(4), 447–468.
Siddi, M. (2020). The European green deal: Asseasing its current state and future implementation. *FIIA Working Paper 114.*
Silveira, L. D. D. O., & Obersteiner, T. (2013). The scope of the TBT Agreement in light of recent WTO case law. *Global Trade and Customs Journal, 8*(4), 112–120.
Staniškis, J. K. (2012). Sustainable consumption and production: How to make it possible. *Clean Technologies and Environmental Policy, 14*(6), 1015–1022.
Sustainable consumption and production indicators for the future SDGs' UNEP discussion paper.
Team, E. I. A. S. (2021). EU-indonesia FTA relations: Palm oil – in for a rough ride? In *EIAS*. Retrieved from https://eias.org/policy-briefs/eu-indonesia-fta-relations-palm-oil-in-for-a-rough-ride/.
Teevan, C., Medinilla, A., & Sergejeff, K. (2021). The green deal in EU foreign and development policy. *ECDPM Briefing Note No. 131.*
Thamali, K. I. S., & Jayawardana, N. U. (2022). The current status of national biosafety regulatory systems in South Asia. *Environment Sustenance and Food Safety: Need for More Vibrant Policy Initiatives for Sri Lanka, 198*, 198.
The future we want – declaration of the un conference on sustainable development. United Nations General Assembly Resolution A/RES/66/288.
Thomas, J. (2003). Recent developments and future needs in developing countries of Southeast Asia. In *International Symposium on Biopesticides for Developing Countries*, Vol. 187.
Tsatsakis, A. M., Nawaz, M. A., Tutelyan, V. A., Golokhvast, K. S., Kalantzi, O. I., Chung, D. H., Chung, G., et al. (2017). Impact on environment, ecosystem, diversity and health from culturing and using GMOs as feed and food. *Food and Chemical Toxicology, 107*, 108–121.
Von Frantzius, I. (2004). World summit on sustainable development Johannesburg 2002: A critical analysis and assessment of the outcomes. *Environmental Politics, 13*(2), 467–473.
Wang, C., Ghadimi, P., Lim, M. K., & Tseng, M. L. (2019). A literature review of sustainable consumption and production: A comparative analysis in developed and developing economies. *Cleaner Production, 206*, 741–754.
Winham, G. R. (2009). The GMO panel: Applications of WTO law to trade in agricultural biotech products. *European Integration, 31*(3), 409–429.
WTO. (1995). *Agreement on sanitary and phytosanitary measures. (hereafter the SPS Agreement)*
Yukawa, T. (2018). The ASEAN way as a symbol: An analysis of discourses on the ASEAN norms. *The Pacific Review, 31*(3), 298–314.
Zheng, Q., & Wang, H. H. (2021). "Do consumers view the genetically modified food labeling systems differently "contains GMO" versus "Non-GMO" labels. *The Chinese Economy, 54*(6), 376–388.

The Impact of Job Satisfaction on Organizational Commitment: An Empirical Study in Vietnam

Sinh Hoang Nguyen

Abstract This research aims at identifying the impact of job satisfaction on organizational commitment in companies in Vietnam due to the fact that the subject of organizational commitment is one of the most important issues of many companies' development. Despite the essential importance of job satisfaction on organizational commitment, many enterprises still do not pay sufficient attention to this issue. A qualitative research was carried out by designing specific research questions then developing a conceptual model. A quantitative research was employed to test the conceptual model using online survey via Google Forms which was sent to over various occupations at companies in Vietnam. Overall, there were 173 respondents of which substantial proportion is female (64.16%). Using SPSS software to conduct statistical tests, the results of the study have shown that job satisfaction, salary, leadership, relationship, and working environment are key factors to organizational commitment of employees. All these factors have shown a very significant impact on the level of organizational commitment. Thus, the management of any organization should consider these factors and give serious attention to them in order to improve the organizational commitment.

Keywords Organizational commitment · Job satisfaction · Salary · Leadership · Relationship · Working environment

1 Introduction

Job satisfaction is one of the most important factors for all organizations no matter whether in public or private organizations or working in advanced or underdeveloped countries. If employees are satisfied with the organization's policies including salary, working conditions, relationships at the workplace, etc., then they will stay for a long time with the organization. So, the organizational commitment and job satisfaction

S. H. Nguyen (✉)
Ho Chi Minh City Open University, HCMC, Vietnam
e-mail: sinh.nh@ou.edu.vn

A. T. Nguyen et al. (eds.), *Contemporary Economic Issues in Asian Countries: Proceeding of CEIAC 2022, Volume 1*,
https://doi.org/10.1007/978-981-19-9669-6_12

are key contributors in any organization and the extent of success of the organization depends on these two factors.

Job satisfaction is a positive feeling about a job resulting from an evaluation of its characteristics (Robbins, 1998). There are a variety of factors that can influence a person's level of job satisfaction. Some of these factors include the level of pay or salary, the quality of the working conditions, leadership, and relationships. The happier people are within their job, the more satisfied they are said to be and they will stay along with the organization. With the development of advanced technology, people easily access a lot of information sources and can search the work vacancies on the Internet, which means employees can find new jobs anytime and anywhere if they are no longer satisfied with their current job. That leads to a high rate of employee turnover. To avoid the bad effects caused by employee turnover, organizations need to maintain their current employee's commitment. In other words, the employees should be satisfied their needs and requirements so that the managers should pay attention to their satisfaction. Hence, the more incentive on factors related to employee satisfaction at work, the more company can reduce the turnover rate and retain employees. The main purpose of this study is to examine the impacts of selected factors: independent variables including salary, leadership, relationship, working environment, and the mediator variable job satisfaction on the dependent variable organizational commitment. This study could be helpful for organizations that are still concerned with the strategy to keep their talented employees. With the research findings, we can know which factors play important roles in a successful organizational commitment strategy.

Following the research background, the research questions of this study are: (1) How do salary, leadership, relationship, working environment impact job satisfaction? (2) What is the degree of job satisfaction contributed to the level of organizational commitment?

The key research objectives of this study are to: (1) Find out factors that impact job satisfaction that leads to organizational commitment, and quantify the impact of these factors: salary, leadership, relationship, working environment. (2) Investigate the relationship between job satisfaction and organizational commitment.

2 Literature Review

2.1 Job Satisfaction

In the current modern era, keeping the workers in the organization is very hard, and different researchers have demonstrated that job dissatisfaction induces people to give up on leaving their job; the dissatisfied employees want to quit their job. Because of this reason, many previous studies set the important point to express the special interest in the aspects of job satisfaction. The other reason for the popularity of job satisfaction over the recent decades is its impact on employee's productivity.

In the literature, studies have demonstrated the positive relationship between job satisfaction and productivity. Job satisfaction has an important role in enhancing productivity. Additionally, satisfied employees contribute to enhancing long-term success and organizational effectiveness.

Job satisfaction is a crucial issue for all organizations regardless of whether public or private organization (Locke, 1976). One of the reasons is that satisfied staff are accounted for as conferred laborers and responsibility is a sign for organizational output and useful operations have been characterized in a wide range of ways. Job satisfaction is related to the degree to which people like or dislike aspects of their work, such as work for themselves, pay, promotion of opportunities, control, and cooperation (Spector, 1997).

Skaalvik and Skaalvik (2011) defined job satisfaction as the feeling an employee holds toward the job. They argued that when the expectation from the job matches with the real outcome, job satisfaction occurs. Togia et al. (2004) defined job satisfaction as feelings and thoughts of employees about their work and place of work. According to Armstrong (2006), job satisfaction is the attitude and feeling people have about their work. Similarly, Kim et al. (2005) defined job satisfaction as the feeling or a general attitude of the employees in relation with their jobs and the job components such as the working environment, working conditions, equitable rewards, and communication with the colleagues. Other scholars measured job satisfaction in terms of need fulfillment, discrepancies, value attainment, equity, and dispositional or genetic components models (Kinicki & Kreitner, 2007).

2.2 *Organizational Commitment*

The organization must have the full commitment from the employee for long-term development (Mouhamadou, 2015). So increasing the organizational commitment will improve the productivity of the employees, and it is the main reason for organizational success. Nowadays, there is no organization that can perform at the top level without the employee who focuses on the organization's goal. And employee productivity is usually improved whenever employees are more satisfied with their work. In today's competitive world, each organization is confronting new difficulties in regards to maintaining efficiency and creating a committed workforce (Andavar et al., 2020).

Muthuveloo and Rose (2005) defined employee commitment as the ability of employees to be loyal and identify with the organization in relation to the duties and responsibilities. According to Aghdasi et al. (2011), individuals with strong attachment to the organization will feel cohesive with it and get pleasure from being a member of organization.

2.3 Salary

Pay is vital, but the awareness of individual about salary is more vital. People with positive attention seem to be much satisfied with pay relative to people with negative affectivity. Organ (1994) claimed that both positive and negative affectivity stands as key factors of job satisfaction. But the individuals who have extraordinary negative affectivity are nervous and worried (the emotional state that may designate an anxious personality), but that does not suggest across-the-board displeasure. High-negative affectivity persons may be discontented with characteristics of their jobs, but that does not essentially mean that they would be more disappointed with their salary. Organizational possession is also an imperative reason in determining pay satisfaction and contentment of employees. Solomon (1986) suggested that public sector executives experience lower intensities of job and pay satisfaction. Low performance of employees may be a consequence of low levels of satisfaction with their salary. A pay motivation scheme is a different measurement of pay satisfaction. In modern ages, there remained a cumulative trend for public and private administrations to implement additional inspired and creative procedures of wage enticement such as group incentives and profit distribution schemes. Carrell and Dittrich (1978) also stated that motivation plans that used many distribution rules would move toward different magnitudes of pay satisfaction. So, it is anticipated that distributive justice will affect satisfaction with motivational incentive plans.

For making pay satisfaction and job satisfaction, organizations have to encourage a strategy of perception of pay-for-performance. Insight of pay-for-performance is a positive stimulus on pay satisfaction. Omar and Ogenyi (2006) observed that perceived associations between pay and performance account for additional changes in pay increase satisfaction than entire demographic variables put together. Consequently, launching a pay-for-performance salary system may be the greatest effective technique to encourage salary level satisfaction.

Employees enjoying high salaries are more likely to observe enticement recompenses as supportive. According to Maslow's (1943) hierarchy of needs, additional income may assistance in the contentment of esteem needs for the reason that high salary suggests high capability and overall individual value. So, even after satisfaction of basic physiological and security needs are not an issue, a number of people drive value high salary as an indicator of proficiency and individual value (Malka & Chatman, 2003). Performance salary is assessed because it is problematic to measure quantitatively, that it can encourage individuals to emphasis too barely, that it can weaken intrinsic interest, which monetary rewards only work for selected people, that it is harmful to teamwork and cooperation and that general pay costs can upsurge earlier if not strongly measured (Armstrong & Murlis, 2005).

2.4 Leadership

The literature reveals that leadership is the process involving the influences that occurs within group which motivates the employees toward goal attainment. When a leader transfers his/her emotions and thoughts to employees in a strong way, it affects employees' emotions, performance (Goleman, 2002), and job satisfaction of the employees. This is because leaders have an effect not only on employees' emotions and thoughts but on also their motivations-values, needs demands, and desires-expectations, which are instrumental in their actions toward identified objectives (Burns, 2007).

Sparrowe (1994) has explored the antecedents and outcomes of empowering leadership style in the hospitality industry, and his findings reveal that psychological empowerment among hospitality employees is worthwhile which will increase job satisfaction among employees and job turnover will decrease. The researcher has explored employee empowerment in the hospitality industry with a study on antecedents and outcomes. The results reveal that employees feel more satisfied with their empowerment and also their intention of leaving their job will also decrease (Sparrowe, 1994).

Stogdill and Shartle (1948) have researched methods for determining patterns for leadership behavior in relation to organization structure and objectives. The results show that the persons working in group-oriented work in an organization tend to follow a pattern of the consultative or participative style of leadership.

Leadership is one of factors of job satisfaction so this research focuses on what level of leadership affects the job satisfaction that leads to organizational commitment.

2.5 Relationship

In general, the success of an organization is the way of connecting people in a team for a special goal, implementing the work, then getting the result. In that way, the relationship among team members is very important, especially between managers and their subordinates. The relationship is built up in trust and pleasure will enhance the internal strengthening of the organization. In a trusted environment or trust spirit among organizations, the people will feel the working station is like home then it increases their commitment with the organization.

Awais et al. (2015) classified interpersonal relationships in the workplace into relationships with superiors and colleagues. The level of support obtained from the supervisor and the way of work control by the supervisor creates relations with superiors. In contrast, the relations between colleagues arise through the interaction of cooperation and conflict.

A different approach to the relationship can be found in the leader-member exchange (LMX) theory, which assumes that managers build bilateral case by case basis relations with subordinates (Graham & Van, 2010; Ilies et al., 2007), and they do not treat the subordinates with the same deference (Turek et al., 2014). The quality of interpersonal relationships with colleagues and managers with subordinates has a positive impact on the commitment to work (Kulikowski, 2015).

An important element in building positive relationships with employees is trust, which is an essential element of management. According to Królik (2015), the trust that employees have toward managerial staff can provide the company the prevalence over rivals. Undoubtedly, this situation has a positive effect on the development of the company.

According to Moczydłowska (2013), making mistakes in building relationships with employees can be a source of internal crises in the enterprise, which indicates the importance of this management area. The researcher discusses the concept of employee relationship management (ERM) and indicates the source of the crisis around customer relationships management with employees. The manifestation of this may be a loss of valuable employees. One of the factors that affect the retention of employees in an organization is a good supervisor–subordinate relation. The author concludes by pointing out that good employee relations can become the foundation of a rapid overcoming of the difficulties, and even using the crisis to the further development of the company.

2.6 Working Environment

Comfortable people are more productive for a better working environment. However, comfort is one of those words that are easy to use and difficult to define. People feel comfortable when they feel comfortable. It is a state of mind that depends on physical feelings and emotional states. Creating an effective personal environment should take into account these two elements as well as cost and technology constraints.

In the coming years, companies will successfully or unsuccessfully depend on their ability to hire and retain highly skilled workers, said Hoskins and his employees. Companies have realized the importance of comfort in the workplace by improving ergonomic functional parts to maintain quality personnel, increase productivity and maintain a competitive edge. The quality of the employees' working environment has the greatest impact on the level of motivation and subsequent productivity. The way they relate to the organization, especially in the immediate environment, has a major impact on their error rate, level of innovation and collaboration with other employees, and is not involved and, ultimately, how long they work.

Tio (2014) has examined and analyzed the influence of the working environment condition toward employee job satisfaction and to provide insight on how to increase employee job satisfaction through the work environment. The result is that there are significant influences of work environment toward the job satisfaction of the employees. Therefore, it can be concluded that in order to improve the company employee job satisfaction, the company needs to pay attention to the work environment they provide for the employees.

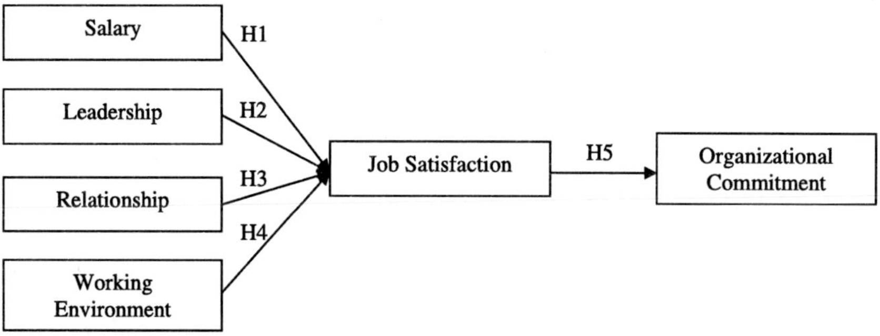

Fig. 1 Conceptual model

3 Methodology

3.1 Conceptual Model

Based on the literature review, the conceptual model is formulated for the independent variables and the dependent variables as shown in Fig. 1.

Based on the conceptual model, the following hypotheses are proposed.

H1: Salary has positively related to job satisfaction.

H2: Leadership has positively related to job satisfaction.

H3: Relationship has positively related to job satisfaction.

H4: Working environment has positively related to job satisfaction.

H5: Job satisfaction has positively related to organizational commitment.

3.2 Data Collection

The survey is chosen as the main method with a questionnaire to collect primary data. The questionnaire is divided into two sections: Perceived relationship among the factors and demographic details. The questionnaires were sent to participants through the Internet using Google Forms via email and social media such as Facebook. There are 173 respondents who participated in the survey, spreading from 18 to over 50 years old.

The measurement is employed based on the previous study (Tatar, 2020), using the 5-point Likert scales (Table 1).

Table 1 Scales of independent and dependent variables

Salary	
1	Commensurate
2	Paid on time (no delay)
3	Extra payment for overtime work
4	Don't do any other work to increase salary
5	Equity
6	Sufficient to accommodate employees' needs
Leadership	
7	Respect and interest
8	Helps me to apply the modern management methods
9	Support
10	Not exposed to criticism
11	Treat all employees in the same manner
12	Thoughts and opinions are taken into consideration
13	Clearly defines the key performance criteria
Relationship	
14	Mutual respecting and understanding with colleagues
15	Problems with colleagues
16	Support and mutual cooperation
17	Pressures affect my personal life
18	Human and social relations between managers and employees
19	Teamwork
Work environment	
20	Appropriate facilities to achieve outstanding performance and functionality
21	Security and safety elements
22	Maintenance of equipment and tools for the work
23	Comfortable and safe
Job satisfaction	
24	Job condition
25	Working happy hours
26	No boring working hours
27	Appropriate job in comparison
28	Happy work in comparison
Organizational commitment	
29	Working policy
30	Conviction to policy
31	Communication policy

(continued)

Table 1 (continued)

32	System that addresses employee problems and complaints
33	Institution's objectives
34	Homogeneity between personal values and the organization values
35	Feeling of importance
36	Effort to achieve the organizational goals
37	Organization's fate interest
38	Responsibilities
39	Secure feeling
40	Belonging feeling
41	Moral obligation to continue to work
42	Freedom of expressing opinion

Source Tatar (2020)

4 Results

4.1 Descriptive Statistics

Figure 2 below describes the number of respondents ($N = 173$) and the percentage of gender and income.

In terms of gender, it could represent a slight bias in responses, as the greater part of employees who took the surveys is female (64.16%). It is nearly double in comparison to the number of males who took part in the surveys (34.10%). Nevertheless, it has been found women are more likely to participate in online surveys than men (Curtin et al., 2000), which may therefore explain this imbalance.

In terms of income, most respondents earned more than 4 million VND. Specifically, employees earned 4–9 million VND (32.37%), more than 9 to 15 million VND

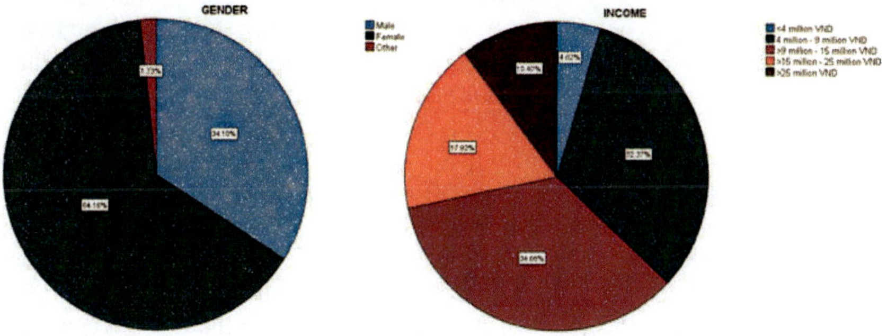

Fig. 2 Gender and income

Table 2 Mean of factors

Factors	Mean	Standard Deviation
Salary	3.85	0.876
Leadership	3.88	0.914
Relationship	3.96	0.689
Working environment	4.13	0.695
Job satisfaction	3.79	0.834
Organizational commitment	3.84	0.807

(34.68%), more than 15 to 25 million VND (17.92%), and more than 25 million VND (10.40%).

Salary
The descriptive statistics in Table 2 show that at least one participant struggles with the salary of finding organizational commitment as the minimum is one. The mean is 3.85 which suggests that most participants struggle with salary. The standard deviation tells us how clustered or spreads the distribution is around the mean value (Malhotra, 2015). The low standard deviation of 0.876 indicates that the data is clustered closely around the mean, thus meaning we can trust that the data is reliable.

Leadership
The descriptive statistics show that at least more than one participant struggles with the leadership of finding organizational commitment as the minimum is 1.14. The mean is 3.88 which suggests that most participants struggle with leadership. The standard deviation of 0.914 is relatively low, thus meaning we can trust the data is reliable.

Relationship
The descriptive statistics show that at least one and a half participants struggle with the relationship of finding organizational commitment as the minimum is 1.5. The mean is 3.96 which suggests that most participants struggle with relationship. The standard deviation tells us how clustered or spreads the distribution is around the mean value (Malhotra, 2015). The low standard deviation of 0.689 suggests that the majority of responses sit closely to the mean, thus meaning we can trust that the data is reliable.

Working environment
The descriptive statistics show that at least two participants struggle with the working environment of finding organizational commitment as the minimum is two. The mean for working environment was the highest at 4.13, followed by relationship at 3.96. Thus, meaning participants pay their attention to this factor the most, a favorable sign for companies to attract employees' commitment. The low standard deviation of 0.695 indicates that the data is clustered closely around the mean, thus meaning we can trust that the data is reliable.

Job satisfaction

The descriptive statistics show that at least more than one participant struggles with the job satisfaction of finding organizational commitment as the minimum is 1.2. The mean is 3.79 which indicates that the majority of participants are neither satisfied nor dissatisfied with the current works. However, as this value is over three, it suggests that participants are leaned marginally toward satisfied. The low standard deviation of 0.834 indicates that the data is clustered closely around the mean, thus meaning we can trust that the data is reliable.

Organizational commitment

The descriptive statistics show that at least one and a half participants struggle with organizational commitment as the minimum is 1.5. The mean is 3.84 which suggests that most participants struggle with organizational commitment. The low standard deviation of 0.807 indicates that the data is clustered closely around the mean, thus meaning we can trust that the data is reliable.

4.2 Hypothesis Testing

According to results (Table 3), the correlations between the organizational commitment and the five other variables, which are the influential factors, are positive. There are existing relationships between the dependent variable and each independent variable (ps < 0.001), and the relationships are really strong due to the fact of Pearson's correlation from 0.66 to 0.84.

There is a correlation among 6 factors (all ps < 0.001). Especially, there is a very strong correlation between job satisfaction and organizational commitment ($r = 0.843$).

The test was found to be statistically significant as $p < 0.001$ is less than 0.05 at the 95% level of significance, indicating a significant relationship between these two variables including job satisfaction and organizational commitment. As shown on the result with $r = 0.843$, this relationship was found to be positive. Thus, H5 is accepted.

Table 3 Correlations

Factors	1	2	3	4	5	6
1. Salary	–					
2. Leadership	0.610**	–				
3. Relationship	0.554**	0.663**	–			
4. Working environment	0.588**	0.659**	0.668**	–		
5. Job satisfaction	**0.687****	**0.725****	**0.677****	**0.731****	–	
6. Organizational commitment	0.669**	0.810**	0.690**	0.712**	**0.843****	–

** $p < 0.001$

The test found the relationship between working environment and job satisfaction to be significant. This is confirmed by $p < 0.001$. With $r = 0.731$, this relationship was found to be positive. So, H4 is accepted.

The test found the relationship between relationship and job satisfaction to be significant. This is confirmed by $p < 0.001$. With $r = 0.677$, this relationship was found to be positive. Hence, H3 is accepted.

The test found the relationship between leadership and job satisfaction to be significant. This is confirmed by $p < 0.001$. With $r = 0.725$, this relationship was found to be positive. Therefore, H2 is accepted.

The test found the relationship between salary and job satisfaction to be significant. This is confirmed by $p < 0.001$. With $r = 0.687$, this relationship was found to be positive. Thus, H1 is accepted.

Quantitative research indicated a range of factors may have influenced job satisfaction. To analyze the effect of salary, leadership, relationship, and working environment on job satisfaction, a multiple linear regression was conducted.

According to the ANOVA statistics, it can be seen that $p = 0.01 < 0.05$, then we can make a conclusion that the built linear regression model fits the population. In other words, the independent variables are linearly related to the dependent variable, and the confidence is 95%. There are salary (beta = 0.187), leadership (beta = 0.474), relationship (beta = 0.141), and working environment (beta = 0.195) impacts on job satisfaction (ps < 0.01). Job satisfaction is explained by salary, leadership, relationship, and working environment about 74% ($R^2 = 0.74$). The R^2 value shows the variation of the job satisfaction variable which is explained 74% because of the affection of the independent variables in the conceptual framework. There is 26% which is explained be the other factors outside of the research model.

4.3 *Further Analysis*

A one-way ANOVA test requires an independent categorical variable and a dependent variable. This test was performed to see if income, gender have an impact on organizational commitment.

The ANOVA test found the relationship between income and organizational commitment to be significant ($p < 0.05$). The difference among income ranges regarding organizational commitment. Specifically, there is a significant difference in organizational commitment between income groups of < 4 million VND and 4–9 million VND ($p < 0.001$): M = 3.52 < M = 4.20, respectively.

The difference between males and females is no significant in organizational commitment ($p > 0.05$).

5 Conclusion and Policy Implications

5.1 Conclusion

The main purpose of this research was to investigate the impact of job satisfaction on organizational commitment through salary, leadership, relationship, and working environment in the companies in Vietnam. The results from the quantitative statistics identified that there was sufficient evidence to make affirmative answers for the questions of the research.

The quantitative method was used to survey 170 employees in various occupations, focusing on the impact of job satisfaction, salary, leadership, relationship, working environment. As a result, the statistical testing found all the variables were the predictors of organizational commitment (all ps < 0.001). Especially, there was a very strong correlation between job satisfaction and organizational commitment ($r = 0.843$) which was followed by the strong correlation between leadership and organizational commitment ($r = 0.810$).

Thus, the data collected from the analysis provides some managerial implications to increase organizational commitment.

5.2 Policy Implications

After a thorough analysis of the quantitative data that have been collected, it is recommended that the companies in Vietnam should create a campaign surrounding job satisfaction to recruit and maintain talented employees by keeping their organizational commitment. There are as follows.

Salary
Besides the maintenance of the current based base salary, companies should make a balance between internal and external competitiveness. Likewise, companies should pay attention to additional welfare measures for employees. The companies may allow employees to buy preferred shares of the companies, or housing purchase assistance, along with improving the efficiency of labor unions in companies to protect more interests of employees. In addition, the companies can diversity their benefits by offering preferential rates services to the companies' employees.

Leadership
According to statistics, leadership is the strongest factor in improving employee job satisfaction. Hence, companies need to pay close attention to this one. This factor needs to be improved in order of priority: receiving suggestions/opinions of employees; politeness and gentleness in communication as well as in work; always instruct and support staff professionally; fair treatment with all of the staff.

Relationship
Companies need to always create conditions to promote employees' coworker relationships to tighten the spirit of solidarity internally. To achieve that, companies need to regularly organize field activities, foreign affairs, sightseeing, travel, extra-curricular activities, and organization of communication contests ownership between departments. These activities are not only extremely meaningful important in bringing together members of the organization, but also in energy for employees to relieve pressure as well as an exercise both physically and mentally for the staff.

Working Environment
Companies need to improve a cleaner, cooler, and safer working environment. Besides, enterprises also need to create conditions for employees to study and improve their working knowledge and skills. Additionally, companies should structure a system to create promotion opportunities for qualified people as well as employees who are lower than overall job satisfaction.

Further, if the managers want to increase the job satisfaction level, research findings highlight certain features deemed important by respondents that there are differences in individuals' interests. People with different incomes do not have the same desired job satisfaction. Likewise, no strategy fits all the organizations to improve their employees' commitment. In order words, the companies and their managers should conduct further research on another factor, and then align it to their organizational structures and strategies.

Therefore, this research is a piece of good evidence for the companies and their executives to use in order to understand the importance of job satisfaction and its' influencers including salary, leadership, relationship, and working environment. Moreover, the research also provides a framework for the managers to build up a suitable human resource plan for their companies to increase their organizational commitment.

References

Aghdasi, S., Kiamanesh, A. R., & Ebrahim, A. N. (2011). Emotional intelligence and organizational commitment: Testing the mediatory role of occupational stress and job satisfaction. *Procedia-Social and Behavioral Sciences, 29*, 1965–1976.

Andavar, V., Ali, B. J., & Ali, S. A. (2020). Rainwater for water scarcity management: An experience of woldia university (Ethiopia). *Journal of Business, Economics and Environmental Studies, 10*(4), 29–34.

Armstrong, M. (2006). *A handbook of human resource management practice* (10th ed., p. 264). Kogan Page Publishing.

Armstrong, M., & Murlis, H. (2005). *Reward management: A handbook of remuneration strategy and practice* (p. 100). Kogan Page.

Awais, M., Malik, M. S., & Qaisar, A. (2015). A review: The job satisfaction act as mediator between spiritual intelligence and organizational commitment. *International Review of Management and Marketing, 5*(4), 203–210.

Burns, J. D. (2007). *Analyses of transactional and transformational leadership on job satisfaction of college faculty*. Northcentral University, pp. 50–55.

Carrell, M. R., & Dittrich, J. E. (1978). Equity theory: The recent literature, methodological considerations, and new directions. *Academy of Management Review, 3*(2), 202–210.

Curtin, R., Presser, S., & Singer, E. (2000). The effects of response rate changes on the index of consumer sentiment. *Public Opinion Quarterly, 64*(4), 413–428.

Graham, L. N., & Van, W. A. (2010). *Leader-member exchange communication frequency and burnout*. Discussion Paper, Utrecht School of Economics, Tjalling C. Koopmans Research Institute, pp. 1–40.

Ilies, R., Nahrgang, J. D., & Morgeson, F. P. (2007). Leader-member exchange and citizenship behaviors: A meta-analysis. *Journal of Applied Psychology, 92*(1), 269–277.

Kim, W. G., Leong, J. K., & Lee, Y. K. (2005). Effect of service orientation on job satisfaction, organizational commitment, and intention of leaving in a casual dining chain restaurant. *International Journal of Hospitality Management, 24*(2), 171–193.

Kinicki, A., & Kreitner, R. (2007). *Employee behavior* (pp. 40–55). McGraw-Hill.

Królik, G. (2015). *Role of trust in interpersonal relationships as efficiency determinant of the team.* Economic Studies. Scientific Papers, the University of Economics in Katowice, pp. 71–82.

Kulikowski, K. (2015). Interpersonal relationship with associates: A commitment to work – analysis within the framework of requirements, theory and labour resources, around culture. *Scientific Quarterly, 4*(24), 128–137.

Locke, E. A. (1976). The nature and causes of job satisfaction. In M. D. Dunnette (Ed.), *Handbook of industrial and organizational psychology* (pp. 1297–1349). Rand McNally.

Malhotra, N. K. (2015). *Essentials of marketing research: A hands-on orientation.* Essex: Pearson, p. 101.

Malka, A., & Chatman, J. A. (2003). Intrinsic and extrinsic work orientations as moderators of the effect of annual income on subjective well-being: A longitudinal study. *Personality and Social Psychology Bulletin, 29*(6), 737–745.

Maslow, A. H. (1943). A theory of human motivation. *Psychological Review, 50*(4), 370–382.

Moczydłowska, J. M. (2013). Errors in employee relationship management as an internal crisis source in the organization. *Strategies of action in crisis conditions* (pp. 346–355). Publishing House Kul Lublin.

Mouhamadou, T. S. (2015). *Relationship between organizational commitment and turnover intentions among healthcare internal auditors.* Published Dissertation, Walden University, p. 65.

Muthuveloo, R., & Rose, R. C. (2005). Typology of organisational commitment. *American Journal of Applied Science, 2*(6), 1078–1081.

Omar, O. E., & Ogenyi, V. O. (2006). Determinants of pay satisfaction of senior managers in the Nigerian civil service. *International Journal of Public Sector Management, 19*(7), 687–701.

Organ, D. W. (1994). Personality and organizational citizenship behavior. *Journal of Management, 20*(2), 465–478.

Robbins, S. P. (1998). *Organizational behavior* (p. 54). Prentice Hall.

Skaalvik, E. M., & Skaalvik, S. (2011). Teacher job satisfaction and motivation to leave the teaching profession: Relations with school context, feeling of belonging, and emotional exhaustion. *Teaching and Teacher Education, 27*(6), 1029–1038.

Solomon, E. E. (1986). Private and public sector managers: An empirical investigation of job characteristics and organizational climate. *Journal of Applied Psychology, 71*(2), 247–261.

Sparrowe, R. T. (1994). Empowerment in the hospitality industry: An exploration of antecedents and outcomes. *Journal of Hospitality & Tourism Research, 17*(3), 51–73.

Spector, P. E. (1997). *Job satisfaction: Application, assessment, causes, and consequences* (pp. 33–40). Sage Publications.

Stogdill, R. M., & Shartle, C. L. (1948). Methods for determining patterns of leadership behavior in relation to organization structure and objectives. *Journal of Applied Psychology, 32*(3), 286–291.

Tatar, A. (2020). *Impact of job satisfaction on organizational commitment* (pp. 43–47). Akdeniz University.

Tio, E. (2014). The impact of working environment towards employee job satisfaction: A case study in PT. X. *Ibuss Management, 2*(1), 1–5.

Togia, A., Koustelios, A., & Tsigilis, N. (2004). Job satisfaction among Greek academic librarians. *Library & Information Science Research, 26*(3), 373–383.

Turek, D., Wojtczuk, T. A., & Horodecka, A. (2014). Selected determinants impact determinants of the effects of justice on the attitudes and behaviour of employees in the organization. Comparative Surveys - Poland and New Zealand. *Organization and Management, 3*(163), 113–140.

Prioritizing the Development Strategy of Ethnic Minority Farms in Central Highlands of Vietnam Using SWOT-AHP-TOWS Analysis

Thi Ngoc Loi Nguyen and Inna Vasilievna Mitrofanova

Abstract The increasing division of farms, caused by internal and external circumstances, requires the formulation of a farm economic development strategy that ensures access to a trajectory of sustainability and enhancing competitive capacity, especially in ethnic minority farms. In order to achieve this, a methodology to estimate the impact of factors and prioritize alternative strategies on development of ethnic minority farms is presented and applied to the Central Highlands of Vietnam. The study combines a SWOT-AHP-TOWS analysis, in which the SWOT technique examines internal and external factors of these farms and then prioritizes the use of analytical hierarchy process (AHP). Finally, the TOWS matrix is built with nine strategies to comprehensively prioritize each alternative in relation to the SWOT factors. Hence, factors that pertain to weakness are becoming prevalent such as low educational attainment, lack of access to credit, as well as cultural, tradition, and language barriers. Along with the orientation of farm economic development and long-term goals, it is important to minimize weaknesses to avoid threats to the development of ethnic minority farms. Therefore, in such context the following policy considerations are significant: (1) Developing and issuing specific policies for ethnic minority farms; (2) Encouraging alliances between farms; (3) Promoting the "four-house" linkage model; (4) Optimizing of production resources, and (5) Developing human resources.

Keywords Ethnic minority farm · SWOT · AHP analysis · TOWS matrix · Vietnam

T. N. L. Nguyen (✉) · I. V. Mitrofanova
Volgograd State University, Volgograd, Russian Federation
e-mail: ngocloi99@gmail.com

T. N. L. Nguyen
The University of Danang–Campus in Kontum, Kontum, Vietnam

I. V. Mitrofanova
Federal Research Centre Southern Scientific Centre of the Russian Academy of Sciences (SSC RAS), Rostov-on-Don, Russian Federation

A. T. Nguyen et al. (eds.), *Contemporary Economic Issues in Asian Countries: Proceeding of CEIAC 2022, Volume 1*,
https://doi.org/10.1007/978-981-19-9669-6_13

1 Introduction

Agriculture is the basis and the most integral component of the Vietnamese economy (Song et al., 2020). During more than 30 years of comprehensive economic reform, significant changes in formal institutions, the introduction, and improvement of the institution of private ownership of land, the incorporation of new models of transactions into management practices led to the formation of many small agricultural organizations, a significant part of which began their activities in the form of farms. It plays a leading role not only in advancing agricultural development but also in addressing important social problems, particularly of the villages: creating jobs for rural people; contributing in preserving traditional lifestyles; acting as an additional source for local budgets (Bui, 2015). However, the period of intensive growth of farm's number, which took place during the years of active state support for farming, was replaced by a phase of a remarkable reduction in the number of farms the background of a significant increase in their size. As a result, increasing economic efficiency in the operation of farms and their adaptation to a constantly changing environment becomes an objective necessity (Nguyen & Mitrofanova, 2022). Concerns about the development of minority farms, in particular, should be addressed through the production capacity constraints of minority households, such as nomadic farming, rudimentary production tools, low labor productivity, small-scale, concentrated production (Vu et al., 2020), their resilience to climate change is relatively low (Sen et al., 2020) and there is always a development gap with the main group (Fujii, 2017).

Furthermore, the collection of relevant literature in Vietnam is lacking in-depth analysis on issues regarding the development of ethnic minority farms. Meanwhile, current studies based on various qualitative and quantitative methods such as regression equations, mathematical programming, descriptive analysis, synthesis, and scientific reports focus on proposing general farm development strategies. This includes optimal allocation of inputs for farms (Nguyen et al., 2020), enhancing the application of 4.0 agriculture (Tran et al., 2020), stimulating digital transformation, improving the quality of labor resources, providing credit support, and promoting farms to join the global value chain (Nguyen & Mitrofanova, 2021a). However, these methods do not take into account all factors affecting the development of farms, such as the history of the establishment, the development of local farms, the support of the state, pests and diseases, or damaged crops. Furthermore, in such an uncertain environment, the interdependencies among these factors, as well as the judgments of decision makers, have also not been addressed in those studies. In addition, depending on the material, socio-cultural, economic, and ecological issues and contexts, specific strategies need to be developed for each geographical region.

Therefore, a systematic approach is necessary in the strategy of developing ethnic minority farms. SWOT is an acronym for strengths, weaknesses, opportunities, and threats, which is a powerful tool that assists decision-making and helps to identify internal and external influences that may have a positive or negative impact on the achievement or development of goals (Kolotov, 2018). In the agricultural sector, this matrix is also well known for assessing bottlenecks and identifying potentials of programs and projects in developing countries. For example, in Ukraine, the SWOT analysis is employed to develop strategies for sustainable development in rural areas (Shcherbak et al., 2020). Meanwhile in Nepal, the analysis was used to formulate agricultural policy (Arun & Ghimire, 2018). Whilst in Iran it was implemented to determine the agricultural development strategy, especially the management of the farm system (Ommani, 2011). However, the main limitation of this traditional tool is that it does not quantify the significance of each factor in the decision-making process, and it is also difficult to assess which factor has more influence on the strategic decision (Pesonen et al., 2001).

The decision-making process is the act of choosing the most suitable tactics to accomplish desired tasks and goals. Thereby, the establishment of the multi-criteria decision making (MCDM) aims to improve the SWOT analysis method. In particular, the Analytical Hierarchy Process (AHP) is a common technique within the MCDM method used to prioritize factors affecting decision-making in order to choose an alternative based on the relative importance of such factors (Saaty, 2013). The SWOT-AHP integration is the leading method that most decision-makers refer to in order to make more reliable and effective choices in agricultural and rural development, proven in the scientific work by Ali et al. (2021). A case study demonstrating the feasibility of SWOT and AHP analysis to incorporate stakeholder priorities in strategic decision-making on beef farm development in Indonesia (Budi et al., 2020). Particularly as follows: after identifying strengths, weaknesses, opportunities, and threats in the beef farm business; followed by group weights and SWOT factors calculated by AHP method; the result is that three priority strategies have the highest scores.

On the other hand, the SWOT matrix is a planning tool, whilst the TOWS matrix is an action tool. Because based on the relationship between threats, opportunities, weaknesses and strengths indicated by a precursory SWOT, the TOWS matrix is developed to reveal alternative strategies (Weihrich, 1982). In other words, the TOWS analysis is applied to develop the tactics needed to execute the strategies and to find more specific actions that support these tactics. Despite the advantage of a systematic approach to creating strategies, TOWS analysis has inherent drawbacks like SWOT (Jafari et al., 2013). Therefore, the SWOT-AHP-TOWS analysis is a more in-depth combination of analytical tools to further improve the decision-making process and develop policies based on the results of SWOT-AHP analysis. This model allows for integrated analysis, prioritization of individual factors, and development of appropriate policies. The strength of this model is to provide structured, systematic support and analysis that combines both qualitative and quantitative attributes. The integral SWOT-AHP-TOWS model is a multi-step process in which AHP is based on a previous SWOT analysis and its subsequent further strategic choices are derived using the TOWS matrix. Insights into this process were proposed in many countries during the design of agricultural and rural development strategies. Examples include identifying transition strategies of Turkey's sustainable food system (Seçkin & Özdil, 2022), optimal decision-making to develop adaptation strategies among small-scale farmers in drought conditions in Iran (Savari & Amghani, 2022), and agricultural development measures in the context of drought in

Vietnam (Nguyen & Truong, 2022). Consequently, quickly and objectively demonstrate countermeasures to overcome agricultural crises is considered an advantage of the SWOT-AHP-TOWS analysis.

With the considerations presented, the aim of this chapter is to develop a scientific and methodological approach to the formulation and selection of ethnic minority farm development strategies applicable in the Central Highlands of Vietnam. Therefore, an integrated SWOT-AHP-TOWS analysis serves as a platform to identify comprehensive factors affecting ethnic minority farms in the Central Highlands, determining its development strategy, prioritizing, and identifying the best strategies in the current context. The rest of this chapter is structured as follows: Sect. 2 describes the location and methods of the study. Section 3 presents the results of the SWOT-AHP-TOWS analysis and discusses the policy implications, as well as conclusions summarized in Sect. 4.

2 Methodologies

2.1 Study Area

The Central Highlands of Vietnam locates in the southeast of the Indochina Peninsula, sharing borders with Laos and Cambodia, adjacent to the Southeast region. This location has made the Central Highlands a place with great potential and many advantages for development in renewable energy, agriculture, and tourism. In terms of administrative boundaries, the Central Highlands consist of five provinces: Kon Tum, Gia Lai, Dak Lak, Dak Nong, and Lam Dong. Total natural area is 54,508 sq km (accounting for 16.46% of the whole country); population is 5.93 million people (6.1% of the national population); the average population density is 109 people per sq km (GSO, 2022). The Central Highlands has a large proportion of ethnic minorities, accounting for about 38% of the total population. The Jarai, Bahnar, Ede, Tay, and Nung are the largest ethnic groups. This is also the region with the second highest percentage of poor and near-poor ethnic minority households in the country, comes after the Northern Midlands and Mountains with 35.5% (GSO & CEMA, 2019).

Farm economy in the Central Highlands was formed and established under specific conditions: firstly, due to the policy mechanism of State, creating a new management mechanism, new awareness in agriculture, rural areas, and farmers. Subsequently, because the Central Highlands is a region with many comparative advantages in terms of climate, land, soil, water resources, resources, as well as the people for agricultural and forestry production (Nguyen & Mitrofanova, 2021b). It is a necessary trajectory that contributes to increasing arable land, creating more jobs, eradicating hunger, reducing poverty, increasing commodity production, and transforming the economic structure of agriculture in the Central Highlands (Nguyen, 2019). This highlights how the farm economy has been discovered as a viable poverty reduction intervention strategy. The application and execution of guidelines and policies of the Central Highlands provinces on land, forests, and development of commodity production have been implemented early and effectively. From 2020, implementing the Circular No. 02/2020/TT-BNNPTNT dated February 28, 2020 of the Ministry of Agriculture and Rural Development, the number of farms in the Central Highlands decreased to 1,740 farms, accounting for 7.4% of farms nationwide (only 193 ethnic minority farms) as determined by new criteria (GSO, 2021). Many farms do not meet both criteria in terms of scale and value, while they are still operating well by contributing to promoting agricultural production in the direction of large commodity production (Nguyen & Mitrofanova, 2021b). According to statistics in 2020, the Central Highlands has 1,018,260 households (439,699 ethnic minority households) producing agricultural goods with a scale of 0.5–2 hectares/household; more households than two hectares—this is the core force, some of which will thrive and become farms (GSO, 2021). Currently, a number of restrictions (information, prices, markets, investment, resources, management, etc.) have hindered the development of the farm economy, reducing its stability and competitive capacity (Nguyen, 2019; Nguyen & Mitrofanova, 2021b). Especially for ethnic minority households, agricultural production development is affected by natural conditions, economic infrastructure, social culture, customs, and policies of the government as well as internal factors (Duong & Truong, 2022). These are good bases for determining strategies to develop ethnic minority farms in the Central Highlands, contributing to the successful implementation of national target programs for socio-economic development in ethnic minority and mountainous areas in the period of 2021–2030.

2.2 Data Collection

The issue of ensuring the sustainability of farm operations at the present time and forecasting future trends in Vietnam has been outlined on the basis of an assessment of the main stages of farm institutionalization in the Vietnamese agricultural system (Nguyen & Mitrofanova, 2022). Along with these data, the trends and performance characteristics of ethnic minority farms in the Central Highlands were specifically analyzed based on a survey of 193 ethnic minority farms to better describe the prospects for developing these farms and their respective challenges. The above information will serve as an input for the analysis of strengths, weaknesses, opportunities, and threats to the development of ethnic minority farms in the Central Highlands.

Then, 12 experts were purposefully selected, from the University of the Central Highlands, the University of Danang Campus in Kon Tum, the Institute of Social Sciences in the Central Highlands, the Committee for Ethnic Minority Affairs and the Department of Science and Technology of Lam Dong Province, the Agricultural Extension Center of Dak Lak Province, and the Sub-Department of Rural Development of Kon Tum Province. All participants are over 40 years old and have more than 15 years of research and work experience in agriculture and policy. The data required for this study was collected over two periods through two different questionnaires from March to June 2022. In stage 1, using an open-ended questionnaire, experts were asked to identify the strengths, weaknesses, opportunities, and threats of the Central Highlands selected for ethnic minority farms. Combined with the information gathered previously, a list of the most relevant SWOT with sub-factors is displayed. This is also an important basis for building the TOWS strategy matrix. In stage 2, they were asked to score strengths, weaknesses, opportunities, and threats in pairwise comparisons as well as evaluate the priority of suggested strategies.

2.3 Research Model and Analytical Steps

Based on the above analysis, an integrated SWOT-AHP-TOWS analysis is necessary to identify priority areas in the development of ethnic minority farms in the Central Highlands as presented in the diagram in Fig. 1. This process is carried out in the following four steps: (1) Building strategic information and assessing the situation through internal factors and external factors; (2) Comparing each pair of criteria to calculate the weight in the SWOT factors (strengths, weaknesses, opportunities, and threats); (3) Using AHP analysis to prioritize each criterion in the four analysis groups; and (4) Using the TOWS matrix to rank and choose the optimal strategy.

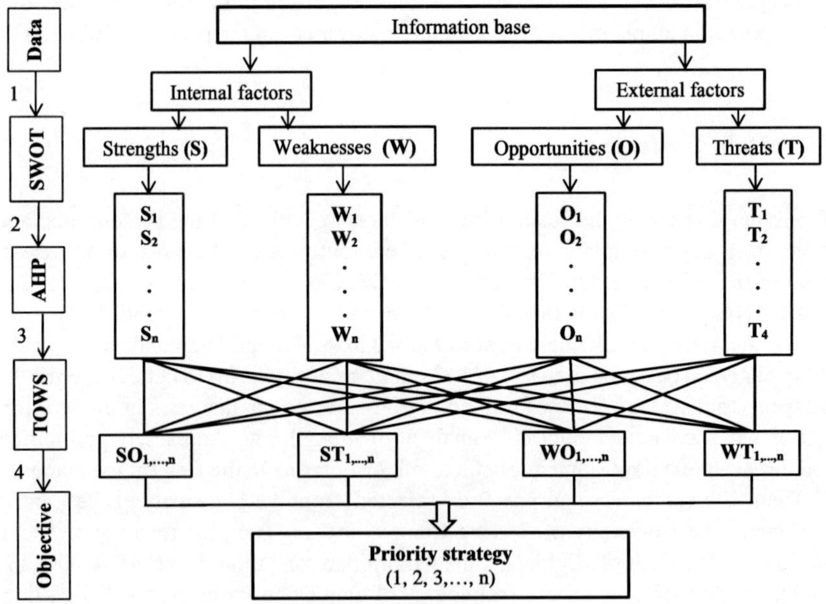

Fig. 1 Research model (*Source* Authors' elaboration)

3 Results and Discussion

Step 1: Situational Assessment (SWOT Analysis)
In order to identify possible development strategies for ethnic minority farms in the Central Highlands, a regional situational SWOT analysis was carried out on the potential for further development of these farms. Participants' responses to the SWOT analysis were listed, weighed, selected, and sorted into a number of relevant and meaningful subgroups. Although there are benefits to considering many factors, the calculation is very complicated because the number of pairwise comparisons in the AHP increases exponentially with the number of factors. Therefore, the current process offers seven strengths, seven weaknesses, four opportunities, and four threats for the future development of ethnic minority farms in the Central Highlands as shown in Table 1.

Step 2: Calculate the Weight of each SWOT Factor Group Using AHP Analysis
Figure 1 shows the hierarchy used in the study. Determining the best strategic plan to the economic development of ethnic minority farms in the Central Highlands, therefore, is the ultimate goal. The lowest level contains the specified criteria for each group of strengths (*S*), weaknesses (*W*), opportunities (*O*), and threats (*T*).

The weights of the factors and the SWOT criteria are determined by considering the degree of influence on the development of ethnic minority farms in the Central Highlands. The weights reflect the relative importance of each factor and must therefore be chosen carefully (Mishra et al., 2015). The weights of all the SWOT factors and the TOWS matrix were determined using the AHP method. In order to assign weights to different criteria (i.e., different factors considered in the SWOT analysis), a pairwise comparison matrix is then derived using Eq. (1), where A is the pairwise matrix, assuming that the element a_{ij} is equal to $1/a_{ij}$, so if i equals j, a_{ij} is also equal to 1. The intensity-Si weighting can also vary from 1 to 9, with 1/1 representing equal importance and 9/1 absolute importance (Satty, 1990).

$$A = \left(a_{ij}\right) = \begin{bmatrix} 1 & S_1/S_2 & \dots S_1/S_n \\ S_2/S_1 & 1 & \dots S_2/S_n \\ \vdots & \vdots & \dots \vdots \\ S_n/S_1 & S_n/S_2 & \dots 1 \end{bmatrix} \tag{1}$$

Some inconsistencies may arise when making sensible pairwise comparisons between decision-makers' determinants. In the case where A contains inconsistencies (provided by an expert), the estimated preference can be obtained using the matrix given in Eq. (1) as input to the eigenvalue technique shown in Eq. (2) (Tafida & Fiagbomeh, 2021).

$$(A - \lambda_{\max} I)q = 0 \tag{2}$$

Table 1 Identified attributes of SWOT factors

Factors			Criteria / sub-factors
Internal factors	Strengths (*S*)	S_1	Farm owners have a lot of experience
		S_2	Large area of agricultural production land
		S_3	Abundant and cheap labor resources
		S_4	The locality has many preferential policies for farm development
		S_5	Natural conditions are suitable for agricultural production
		S_6	Convenient rural transport infrastructure
		S_7	Branded agricultural products with great export potential
	Weaknesses (*W*)	W_1	The level of education and expertise of farmers is still very low
		W_2	Lack of capital to expand production and mechanization in production is low
		W_3	Lack of linkages between farms
		W_4	Insufficient product sales channel due to lack of information
		W_5	Failing to manage the quality of inputs and products
		W_6	Farm owners are not actively participating in technical assistance and related training programs
		W_7	Barriers to language, culture, production practices of ethnic minorities
External factors	Opportunities (*O*)	O_1	The demand for agricultural products is large and growing
		O_2	The State's policy to support the development of farm economy
		O_3	The development of science and technology has created new varieties of plants and new techniques
		O_4	Many guidelines and policies for ethnic minorities
	Threats (*T*)	T_1	Unable to keep up with advanced technology in production
		T_2	Environmental pollution and resource depletion
		T_3	Climate change, natural disasters, epidemics
		T_4	Requires high product quality

Source Authors' elaboration

In Eq. (2), λ_{max} is the largest eigenfactor of a matrix *A* of size *n*; *q* is the coherence vector, and *I* represent the identity matrix of size *n*. Saaty (1980) proved that when comparing SWOT factors, λ_{max} must be equal to n to satisfy the consistency condition.

Consistency in the decisions of experts is controlled by the magnitude of acceptable values, called the consistency ratio (CR). It is calculated by the quotient of the

Table 2 Value of random consistency index (RI)

n	1	2	3	4	5	6	7	8	9	10	11	12
RI	0	0	0.58	0.90	1.12	1.24	1.32	1.41	1.45	1.49	1.51	1.56

Source Saaty (1980, 2008)

consistency index (CI) value and the random inconsistency (RI) (Ali et al., 2021; Budi et al., 2020; Nguyen &Truong., 2022), shown in Eq. (3)

$$CR = CI/RI \tag{3}$$

$$CI = (\lambda_{max} - n)/(n - 1) \tag{4}$$

Equation (4) represents the CI value, where λ_{max} is the largest eigenvalue of the pairwise comparison matrix and n is the number of elements found in each matrix. RI denotes the random index, which depends on the order of the matrix specified by Saaty (1980) in Table 2. If the value of CR is ≤ 0.10 or 10%, it is acceptable and used for further analysis. Conversely, if the CR value is > 0.1, the assessment process needs to repeat and improve.

The identification of priority strategies is done by looking at the results of a joint assessment of experts. Through the AHP method, quantitative values of each expert from the group of factors and SWOT criteria are used to calculate the geometric mean in Eq. (5), resulting in an overall rating.

$$GM = \sqrt[n]{(X_1)(X_2)\ldots(X_n)} \tag{5}$$

where GM is the geometric mean of a series containing (n = 12), experts' assessment is the nth root of the product of the values, and X is the value of each element in the pairwise comparison matrix (1–9 scale). This combined assessment generates new weights for each factor and SWOT attributes (Budi et al., 2020; Saaty, 2008).

The Excel hierarchical analysis procedure was used to support the calculations in this study. Therefore, the results of the wise pairwise comparison for weighting of each group in the SWOT analysis are presented below.

Table 3 shows the results of pairwise comparisons between the four SWOT factors that make up the second level of the hierarchy. The consistency ratio (CR) of this evaluation group was extracted as 0.0565, which guarantees the consistency of the answers. Using the weighting of the relative priorities of the four factors, the calculated weaknesses were 0.3966, which was identified as having the greatest effect on ethnic minority farms in the Central Highlands. Whilst opportunities are calculated as having the second highest impact with a weight of 0.3360, followed by strengths and threats with weights of 0.1453 and 0.1221, respectively. In terms of the magnitude of the influence factors on the development of farms, strengths are nearly two and three times stronger than strengths and weaknesses, respectively, but only 0.061 points stronger than opportunities. This indicates that minimizing the weaknesses is

Table 3 Priorities of SWOT factors (CR = 5.65%)

Factors	*S*	*W*	*O*	*T*	Factor weights	Ranking
Strengths (*S*)	1	1/4	1/3	2	0.1453	3
Weaknesses (*W*)	4	1	1	3	0.3966	1
Opportunities (*O*)	3	1	1	2	0.3360	2
Threats (*T*)	1/2	1/3	1/2	1	0.1221	4

Source Authors'calculation

very important in supporting the development of ethnic minority farms in the Central Highlands.

In each group of strengths, weaknesses, opportunities, and threats, the local weights of factors with criteria directly or indirectly related to the development of ethnic minority farms in the Central Highlands has been calculated. The ranking of strength criteria shows that S_2 (0.2704), S_3 (0.2664), and S_1 (0.2404), with a consistent ratio of CR = 0.0531, as the top three influencing factors in seven criteria are considered (Table 4). Similarly, at CR = 0.0525, the criteria W_1 (0.2886), W_2 (0.1784), and W_7 (0.1541) are the most important weaknesses, with their associated factor weights (Table 5), which are likely to have the maximum effect on achieving the goals of improving the competitive capability of farms.

From the comparison weights, the chance factors are ranked as O_1 (0.3585), O_3 (0.3200), and O_2 (0.1807) with a CR of 0.0435 (Table 6). A pairwise comparison of threat criteria with a consistency ratio of 0.0161 ranked T_1 (0.4228), T_4 (0.2709), and T_3 (0.1623) as the first, second, and third threats pose the greatest challenge to the operation of ethnic minority farms (Table 7). In addition, the ranking of the most integral strengths (S_2), weaknesses (W_1), opportunities (O_1), and threats (T_1) indicates that the failure to keep up with scientific and technical advances in agricultural production (T_1) is ranked as the most important factor affecting the performance of farms. Moreover, high demand for agricultural products (O_1), low education level (W_1), and large agricultural land area (S_2) are ranked second, third, and fourth,

Table 4 Priorities of Strengths factor (CR = 5.31%)

Criterias	S_1	S_2	S_3	S_4	S_5	S_6	S_7	Local weights	Ranking
S_1	1	1/2	1/2	2	4	3	2	0.2404	3
S_2	2	1	1	2	3	2	2	0.2704	1
S_3	2	1	1	2	2	2	3	0.2664	2
S_4	1/2	1/2	1/2	1	1	2	3	0.1547	4
S_5	1/4	1/3	1/2	1	1	0.5	2	0.1042	6
S_6	1/3	1/2	1/2	1/2	2	1	2	0.1261	5
S_7	1/2	0.33	0.33	0.33	0.5	0.5	1	0.0781	7

Source Authors' calculation

Table 5 Priorities of Weaknesses factor (CR = 5.25%)

Criterias	W_1	W_2	W_3	W_4	W_5	W_6	W_7	Local weights	Ranking
W_1	1	3	3	2	4	2	2	0.2866	1
W_2	1/3	1	2	2	2	2	2	0.1784	2
W_3	1/3	1/2	1	2	2	2	1/2	0.1218	4
W_4	1/2	1/2	1/2	1	2	2	1/2	0.1071	5
W_5	1/4	1/2	1/2	1/2	1	2	1/2	0.0789	6
W_6	1/2	1/2	1/2	1/2	1/2	1	1/2	0.0731	7
W_7	1/2	1/2	2	2	2	2	1	0.1541	3

Source Authors'calculation

respectively, are the most notable influencing criteria shaping overall performance results.

Step 3: Prioritize Each Criterion Applied to the Factors After Comparing Each Pair Using the AHP Method

In this step, the weights of the global priorities of the 22 criteria are generated by multiplying each criterion's local priority by the weight of its main SWOT factor. Equation (6) is given below:

$$\text{Global weights} = \text{Factor weights} \times \text{Local weights of each factor} \quad (6)$$

By carefully reviewing global priority rankings, two criteria from opportunities ranked in the top five criteria according to the level of importance that have significant

Table 6 Priorities of Opportunities factor (CR = 4.35%)

Criterias	O_1	O_2	O_3	O_4	Local weights	Ranking
O_1	1	3	1	2	0.3585	1
O_2	1/3	1	1/2	2	0.1807	3
O_3	1	2	1	2	0.3200	2
O_4	1/2	1/2	1/2	1	0.1408	4

Source Authors'calculation

Table 7 Priorities of Threats factor (CR = 1.61%)

Criterias	T_1	T_2	T_3	T_4	Local weights	Ranking
T_1	1	3	2	2	0.4228	1
T_2	1/3	1	1	1/2	0.1440	4
T_3	1/2	1	1	1/2	0.1623	3
T_4	1/2	2	2	1	0.2709	2

Source Authors'calculation

impact on the farm economy of ethnic minorities. That is the large demand for agricultural products ($O_1 = 0.1205$) and the strong development of science and technology ($O_3 = 0.1075$). The three criteria of weakness are low education level (W_1) with the proportion of 0.1137, followed by lack of access to credit (W_2), and cultural and language barriers (W_7) with the proportions of 0.0708 and 0.0611, respectively.

On the other hand, most of the threat criteria have a low rank, indicating a low level of impact on farms. However, failing to keep up with scientific and technical advances in agricultural production (T_1) ranked 7th, with a proportion of 0.0516. Thus, in order to improve the productivity of current ethnic minority farms, it is necessary to accelerate the application of new and efficient production techniques.

Two criteria of strength have the bottom rank of the rankings. In which, local preferential policies ($S_5 = 0.0151$) and export potential ($S_7 = 0.0113$) are ranked at the 21st position and followed by the last position (22nd). This indicates that these two criteria are considered as attributes that need to be promoted last to increase the competitiveness of farms.

In the SWOT criteria, most of the weaknesses and opportunities are more influential than the strengths and threats. Table 8 therefore shows that the results of the SWOT criteria analysis have the same meaning as the results of the SWOT factors.

Step 4: Ranking and Selecting Priority Strategies, Constructed in the TOWS Matrix

The basic objective of strategy formulation is to transform existing conditions or reconstruct the disrupted image in the region into desired scenarios (Prager & Wiebe, 2021; Wickramasinghe & Takano, 2009). In fact, the agricultural sector can propose diverse farm economic development strategies; however, limited financial and other resources or internal and external environmental conditions make it difficult for all the proposed strategies to work, especially for ethnic minorities. Therefore, it is imperative to choose which strategy should be prioritized, and which should be scaled down. In order to form the combined strategies, a TOWS matrix (Table 9) was performed on the basis of the strengths, weaknesses, opportunities, and threats for the study area identified in the above analysis.

Weihrich (1982) developed TOWS matrix with four different types of strategies: (1) SO strategy: take advantage of opportunities to develop strengths, (2) ST strategies: use strengths to avoid external threats, (3) WO strategies: reduce weaknesses to utilize opportunities, and (4) WT Strategy: reduce weaknesses to prevent threats. This does not mean that a particular strategy is influenced only by two respective groups. It can be supported or influenced by other SWOT factors. For example, a strategy of optimizing production resources on top of an SO strategy may receive support from weaknesses and threats to a lesser extent than strengths and opportunities.

The procedure for prioritizing alternative strategies has been established in this step. In order to better represent the impact of strategies on the objective, a verbal/linguistic ranking is provided to display the relationships between SWOT factors and suggested strategies in Table 10.

Table 8 The importance of the factors and criterias of the SWOT—AHP analysis

SWOT factors		Factor weights	SWOT criterias	Local weight	Global weight	Ranking
Internal factors	Strengths (S)	0.1453	S_1	0.2404	0.0349	13
			S_2	0.2704	0.0393	11
			S_3	0.2664	0.0387	12
			S_4	0.1547	0.0225	17
			S_5	0.1042	0.0151	21
			S_6	0.1261	0.0183	19
			S_7	0.0781	0.0113	22
	Weaknesses (W)	0.3966	W_1	0.2866	0.1137	2
			W_2	0.1784	0.0708	4
			W_3	0.1218	0.0483	8
			W_4	0.1071	0.0425	10
			W_5	0.0789	0.0313	15
			W_6	0.0731	0.0290	16
			W_7	0.1541	0.0611	5
External factors	Opportunities (O)	0.3360	O_1	0.3585	0.1205	1
			O_2	0.1807	0.0607	6
			O_3	0.3200	0.1075	3
			O_4	0.1408	0.0473	9
	Threats (T)	0.1221	T_1	0.4228	0.0516	7
			T_2	0.1440	0.0176	20
			T_3	0.1623	0.0198	18
			T_4	0.2709	0.0331	14

Source Authors'calculation

The strategic relationship represents the contribution of factors (strengths and opportunities) contribute to the implementation of the strategy and the expected improvements to the factors (weaknesses and threats) when a particular strategy is implemented. The weights of these relationships were quantified using the pairwise matrices in the usual manner as presented earlier (Table 11).

With the verbal/linguistic relationships as an input, the weight of the proposed strategies is calculated as an explicit value using Eq. (7) below (Nguyen & Truong, 2022; Wickramasinghe & Takano, 2009).

$$S_i = \sum_{j=1}^{n} G_j R_{ij} \tag{7}$$

In which (S_i) is the sum of the weights of the ith strategy, (G_j) is the global weight of the jth SWOT criterion, (R_{ij}) is the degree of relationship between the ith strategy and the jth SWOT criterion, (n) is the number of SWOT criteria.

Table 9 TOWS matrix for ethnic minority farm development strategy in the central highlands

SO strategies (SO_1)–combination of $S_1, S_2, S_3, S_4, O_2, O_3, O_4$: Forming a strategy to optimize production resources, defining a model farm (SO_2)–combination of $S_4, S_5, S_6, S_7, O_1, O_2$: Forming a market development strategy	ST strategies (ST_1)–combination of $S_2, S_3, S_4, S_6, S_7, T_1, T_3$: Forming an in-depth investment strategy (ST_2)–combination of $S_1, S_2, S_4, S_5, T_1, T_2, T_3, T_4$: Forming a strategy to build a technical information system
WO Strategies (WO_1)–combination of $W_1, W_5, W_6, W_7, O_2, O_3, O_4$: Forming human resource development strategy. The focus is on capacity building training for ethnic minority farm owners (WO_2)–combination of $W_2, W_3, W_4, W_5, W_6, W_7, O_1, O_2, O_3, O_4$: Form an alliance strategy between farms	WT Strategies (WT_1)–combination of W_1, W_2, W_5, T_1, T_4: Forming a planning strategy for a concentrated production area (WT_2)–combination of $W_1, W_2, W_3, W_4, W_5, T_2, T_3$: Forming a strategy of "four houses" linking the state–researchers–farmers–enterprises (WT_3)–combination of $W_1, W_2, W_3, W_4, W_5, W_6, W_7, T_1, T_2, T_3, T_4$: formulating strategies for formulating and promulgating specific policies for ethnic minority farms

Source Authors' elaboration

Table 10 Degree of relationship evaluation

Number	0	1	2	3	4	5
Level of relationship	None	Very weak	Weak	Moderate	Stong	Very strong

Source Nguyen and Truong (2022); Wickramasinghe and Takano (2009)

Table 11 Weights of relationship between strategy and SWOT factors

Levels of relationship	Very weak	Weak	Moderate	Strong	Very strong	Weights of relationship
Very weak	1	1/2	1/3	1/4	1/5	0.06
Weak	2	1	1/2	1/3	1/4	0.10
Moderate	3	2	1	1/2	1/3	0.16
Strong	4	3	2	1	1/2	0.26
Very strong	5	4	3	2	1	0.42

Source Authors' calculation

Subsequently, the normalized value of the strategy weights is calculated using Eq. (8), where (N_i) is the normalized weight of strategy i and (m) is the number of strategies.

$$N_i = \frac{S_i}{\sum_{i=1}^{m} S_i} \tag{8}$$

The proposed strategies with the different relationship weights of the SWOT factor are described in Table 12. Along with the orientation of farm economic development and long-term goals, the order of implementation of the economic development strategies for ethnic minority farms in the Central Highlands is divided into two phases. The preferred strategy could be established to select general or individual strategies.

The first phase focuses on ethnic minority farms. The first priority is to implement the WT_3 strategy—develop and issue-specific policies for ethnic minority farms, with

Table 12 Priority strategy evaluation matrix for ethnic minority farms in Central Highlands

SWOT criterias	Strategy with the degree of relationship								
	SO_1	SO_2	ST_1	ST_2	WO_1	WO_2	WT_1	WT_2	WT_3
S_1	4	2	0	5	2	2	0	3	0
S_2	5	0	2	4	0	2	3	2	4
S_3	5	1	5	5	3	2	2	3	4
S_4	3	5	3	4	0	0	0	0	0
S_5	4	0	0	3	0	0	0	0	0
S_6	1	4	5	0	0	0	2	0	0
S_7	0	5	0	0	0	0	1	0	0
W_1	3	0	3	3	5	0	4	5	4
W_2	2	0	2	3	0	3	3	4	5
W_3	1	4	0	0	0	5	0	5	3
W_4	0	3	0	0	0	5	0	4	5
W_5	0	4	0	0	5	4	4	3	4
W_6	0	1	0	0	3	2	0	0	3
W_7	3	0	1	1	4	3	0	0	4
O_1	0	5	0	0	0	4	0	0	0
O_2	5	4	2	4	4	5	3	4	4
O_3	4	2	3	2	4	5	2	3	3
O_4	3	1	2	0	5	5	3	4	5
T_1	3	2	5	5	0	3	4	0	4
T_2	0	0	4	3	0	2	0	4	5
T_3	2	0	0	4	0	0	0	5	5
T_4	2	0	5	4	0	0	5	0	3
Total weights (S_i)	0.1630	0.1391	0.1253	0.1474	0.1486	0.2216	0.1170	0.1764	0.2603
Normalized (N_i)	0.1094	0.0933	0.0840	0.0989	0.0997	0.1432	0.0785	0.1183	0.1746
Ranking	4	7	8	6	5	2	9	3	1

Source Authors' calculation

a weighting factor of 0.1746. Evidently, experts also agree that minimizing weaknesses is the most important factor. Furthermore, low level of education, lack of access to credit, as well as barriers in culture, language, and customs are basic characteristics of ethnic minority farms. Therefore, it is extremely important to add a separate policy group or integrate more details into the general policies—specific solutions directly related to ethnic minority farms. Thereafter, the preferred strategy must focus on alliances between farms ($WO_2 = 0.1432$); development of a "four houses" model ($WT_2 = 0.1183$), optimization of production resources ($SO_1 = 0.1094$), and development of human resources ($WO_1 = 0.0997$).

The next phase focuses on general agricultural policies such as building a technical information system ($ST_2 = 0.989$), market development ($SO_2 = 0.0933$), in-depth investment strategy ($ST_1 = 0.0840$), and concentrated production area planning ($WT_1 = 0.785$).

4 Conclusion

This chapter has successfully introduced a systematic approach and methods of SWOT-AHP-TOWS analysis to select the best strategies for developing ethnic minority farms in the Central Highlands. This is an unexplored area of research in the agricultural literature for vulnerable ethnic minority farmers. However, the decisions based on individual factors without considering the network of criteria are the limitation of the study. Certainly, further research is required to address this issue in order to improve and enhance reliability in the future.

Given the size of the normalized values (Step 4), the final order of strategies is considered as follows: $WT_3 \rightarrow WO_2 \rightarrow WT_2 \rightarrow SO_1 \rightarrow WO_1 \rightarrow ST_2 \rightarrow SO_2 \rightarrow ST_1 \rightarrow WT_1$. The first task is to develop and issue-specific policies for ethnic minority farms. Due to the low starting point and the change in the criteria for determining farms in the Circular of the Ministry of Agriculture and Rural Development No. 02/2020/TT-BNNPTNT dated February 28, 2020, the number of ethnic minority farms that were unable to obtain farm certificates increased sharply. Therefore, it is necessary to have criteria to define separate farms in the direction of reducing output value compared to the prescribed average level to better suit the characteristics of ethnic minority farms. At the same time, it is necessary to amend and concretize the conditions for ethnic minority farms to benefit more from the State's incentive policies, which are stipulated in the Resolution on farm economy No. 03/2000/NQ-CP dated February 2, 2000. Finally, the results of this study will help policymakers and minority farmers to plan appropriate policies to improve the situation in the Central Highlands and address their weaknesses. The methodological achievements of this chapter will also be useful in relation to Vietnam's ethnic minority regions.

Acknowledgements We would like to express our sincere gratitude to the Agreement Scholarship Program between the Governments of Vietnam and the Russian Federation supported the first author to pursue her PhD and this work. We sincerely thank experts from universities, institutes, departments in the Central Highlands and ethnic minority farmers for their support and participation.

References

Ali, E. B., Agyekum, E. B., & Adadi, P. (2021). Agriculture for sustainable development: A SWOT-AHP assessment of Ghana's planting for food and jobs initiative. *Sustainability, 13*, 628.

Arun, G. C., & Ghimire, K. (2018). A SWOT analysis of Nepalese agricultural policy. *International Journal of Agriculture, Environment and Food Sciences, 2*(4), 119–123.

Bui, Q. D. (2015). Concepts of "peasants" and "small-scale agricultural society" in Vietnam: The foundations for the study on rural development. *Vietnam Social Sciences, 4*, 37–47.

Budi, S., Eko, P. B. W. H., & Hartuti, P. (2020). The priority of beef cattle farm development strategy in Semarang Regency using AHP and SWOT (A'WOT) method. *Journal of Sustainability Science and Management, 15*(6), 125–136.

Duong, T. A. N., & Truong, V. (2022). Developing agricultural production of ethnic minority households in the context of climate change (Lak district, Dak Lak province, Central Highland of Vietnam). In A. T. Nguyen & L. Hens (Eds.), *Global changes and sustainable development in Asian emerging market economies* (Vol. 2, pp. 433–449). Springer.

Fujii, T. (2017). *Has the development gap between the ethnic minority and majority groups narrowed in Vietnam?: Evidence from household surveys.* The World Economy, Special issue, pp. 1–35.

General Statistics Office of Vietnam (GSO) & Committee for Ethnic Minorities Affairs (CEMA). (2019). *Report on survey results on information income on socio-economic status of 53 ethnic minorities*. Hanoi: Statistical Publishing House (in Vietnamese), p. 102.

General Statistics Office of Vietnam (GSO). (2021). *Result of mid-term rural and agricultural survey 2020*. Hanoi: Statistical Publishing House, p. 552.

General Statistics Office of Vietnam (GSO). (2022). Socio-economic statistical data in Vietnam. Retrieved from https://www.gso.gov.vn/. Accessed on 22 Aug 2022.

Jafari, H., Jafari, H., & Loyes, C. (2013). Strategic analysis of seaports using multi-criteria decision-making methods and TOWS. *Asian Business Strategy, 3*(7), 182–191.

Kolotov, V. N. (2018). SWOT analysis of the renovation policy in Vietnam and the problems of Russian-Vietnamese relations. *The Russian Journal of Vietnamese Studies, 2*(4), 5–18.

Mishra, A. K., Deep, S., & Choudhary, A. (2015). Identification of suitable sites for organic farming using AHP & GIS. *Egyptian Journal of Remote Sensing and Space Science, 18*(2), 181–193.

Nguyen, T. T. H. (2019). Solutions to develop farm economy in the central highlands. *Journal of Science, Technology and Environment (in Vietnamese), 1*, 30–34.

Nguyen, T. T. T., Le, H. H., Ho, T. M. H., Dogot, T., Burny, P., Bui, T. N., & Lebailly, P. (2020). Efficiency analysis of the progress of orange Farms in Tuyen Quang province, Vietnam towards sustainable development. *Sustainability, 12*(8), 3170.

Nguyen, T. N. L., & Mitrofanova, I. V. (2021a). Farming economy of Vietnam: Trends and ways of development in the Mid-Term. *Economics: Yesterday, Today and Tomorrow (in Russian), 11*(7A), 22–32.

Nguyen, T. N. L., & Mitrofanova, I. V. (2021b). Farm development in Tay Nguyen economic region: Specificity of formation and development. *Regional Economy South of Russia, 9*(4), 88–97.

Nguyen, T. N. L., & Mitrofanova, I. V. (2022). Updating approaches to farm development in Vietnam: An institutional aspect. *Regional Economics South of Russia, 10*(3), 67–77.

Nguyen, H. T., & Truong, T. C. (2022). Integral SWOT-AHP-TOWS model for strategic agricultural development in the context of drought: A case study in Ninh Thuan, Vietnam. *International Journal of the Analytic Hierarchy Process, 14*(1), 1–30.

Ommani, A. R. (2011). Strengths, weaknesses, opportunities and threats (SWOT) analysis for farming system businesses management: Case of wheat farmers of Shadervan district, Shoushtar township, Iran. *African Journal of Business Management, 5*(22), 9448–9454.

Pesonen, M., Kurttila, M., Kangas, J., Kajanus, M., & Heinonen, P. (2001). Assessing the priorities using SWOT among resource management strategies at the Finnish forest and park service. *Forest Science, 4*(7), 534–541.

Prager, S. D., & Wiebe, K. (2021). Strategic foresight for agriculture: Past ghosts, present challenges, and future opportunities. *Global Food Security, 28*, 100489.

Saaty, T. L. (2008). Decision making with the analytic hierarchy process. *International Journal of Services Sciences, 1*(1), 83–98.

Saaty, T. L. (2013). The modern science of multicriteria decision making and its practical applications: The AHP/ANP approach. *Operations Research, 61*(5), 1101–1118.

Saaty, T. L. (1980). *The analytic hierarchy process: planning, priority setting, resource allocation* (p. 287). New York; London: McGraw-Hill International Book Company.

Savari, M., & Amghani, M. S. (2022). SWOT-FAHP-TOWS analysis for adaptation strategies development among small-scale farmers in drought conditions. *International Journal of Disaster Risk Reduction, 67*, 102695.

Seçkin, E., & Özdil, G. G. (2022). Determination of the transition strategies to a sustainable and resilient city-region food system using the AHP-SWOT-TOWS methodology for Istanbul. *Megaron, 17*(2), 209–220.

Sen, L. T. H., Bond, J., Winkels, A., Linh, N. H. K., & Dung, N. T. (2020). Climate change resilience and adaption of ethnic minority communities in the upland area in Thua Thien Hue province, Vietnam. *NJAS - Wageningen Journal of Life Sciences, 92*, 100324.

Shcherbak, V., Ganushchak-Yefimenko, L., Nifatova, O., Fastovets, N., Plysenko, G., Lutay, L., & Ptashchenko, O. (2020). Use of key indicators to monitor sustainable development of rural areas. *Global Journal of Environmental Science and Management, 6*(2), 175–190.

Song, N., Phuong, N., Cuong, H., Diep, N., Diep, D., Huyen, V., Huyen, V., Tiep, N., & Trang, T. (2020). Vietnamese Agriculture before and after opening economy. *Modern Economy, 11*, 894–907.

Tafida, I., & Fiagbomeh, R. F. (2021). Multi-criteria SWOT-AHP analysis for the enhancement of agricultural extension services in Kano State, Nigeria. *Journal of Dryland Agriculture, 7*(5), 77–87.

Tran, N. L. D., Rañola, R. F., Ole Sander, B., Reiner, W., Nguyen, D. T., & Nong, N. K. N. (2020). Determinants of adoption of climate-smart agriculture technologies in rice production in Vietnam. *International Journal of Climate Change Strategies and Management, 12*(2), 238–256.

Vu, T. T., Vu, D., & Nguyen, T. M. L. (2020). Production organization capacity of ethnic minorities in northwestern region of Vietnam. *E3S Web of Conferences, 210*, 16004.

Weihrich, H. (1982). The TOWS matrix-a tool for situational analysis. *Long Range Planning, 15*(2), 54–66.

Wickramasinghe, V., & Takano, S. (2009). Application of combined SWOT and analytic hierarchy process (AHP) for tourism revival strategic marketing planning: A case of Sri Lanka tourism. *Eastern Asia Society for Transportation Studies, 8*, 954–969.

Determinants of the Trade Balance in an Asian Emerging Economy: An ARDL Bounds Testing Approach

Le Thanh Tung

Abstract This study examines the main factors that affect the trade balance in Vietnam, an Asian emerging economy. The ARDL methodology is employed to identify the cointegration and to assess the relationship among variables. The quarterly database is collected in the period of 1995–2020. The bounds test result confirms that there exists a cointegration relationship between the trade balance, national output, money demand, and exchange rate of Vietnam's economy. Furthermore, in the long run, the national output and money demand are found to have negative and significant effects on the trade balance, whereas the exchange rate has a positive and significant impact on the dependent variable. Additionally, in the short run, the national output has a significant negative impact and the money supply has a significant positive impact on the trade balance, while the lag of the trade balance variable has a significant negative effect on itself. Finally, the J-curve phenomenon is confirmed in Vietnam when the coefficient of the exchange rate is negative in the short run and positive in the long run. The empirical evidence suggests that the government needs to use monetary policy, in particular money supply, as an effective policy tool to support the Vietnamese trade balance in the future.

Keywords Trade balance · International trade · Monetary policy · Emerging economy · ARDL · Bounds test

1 Introduction

Trade and trade development are important macroeconomic indicators used to assess the status of an economy. Meanwhile, the role of international trade in supporting an economy is increasing in the context of globalization (Akbas & Sancar, 2021; Amiti & Konings, 2007; Brueckner & Lederman, 2015; Oloyede et al., 2021; Siddiqui, 2015, 2016; Singh, 2010; Thirlwall, 2000; Wong & Yip, 2002). The trade balance also reflects a country's level of participation in the global supply and production chains

L. T. Tung (✉)
Faculty of Economics and Public Management, Ho Chi Minh City Open University, Ho Chi Minh City, Vietnam
e-mail: tung.lt@ou.edu.vn

A. T. Nguyen et al. (eds.), *Contemporary Economic Issues in Asian Countries: Proceeding of CEIAC 2022, Volume 1*,
https://doi.org/10.1007/978-981-19-9669-6_14

(Costinot et al., 2013; Ge & Luo, 2015; Jensen & Shin, 2014). Therefore, empirical studies on the factors affecting the trade balance are important in terms of exploring the economic issues related to this macroeconomic object. Moreover, policymakers can make normative economic decisions based on the obtained evidence. Several studies focusing on the determinants of the trade balance were carried out in both developed countries (Bahmani-Oskooee & Ratha, 2007; Beck, 2020; Gürtler, 2019; Sasaki & Yoshida, 2018) and developing countries (Ahad, 2017; Bahmani-Oskooee & Kanitpong, 2017; Brueckner & Lederman, 2015; Duasa, 2007; Khan & Hossain, 2012; Marin, 2020; Shubaita et al., 2020; Waliullah et al., 2010). However, the results are quite diverse, even contradictory. Hence, further analysis and assessment of the factors affecting the trade balance can help countries to achieve better participation in the international supply chains, making such research necessary.

Since "Doi Moi" economic reforms began in 1986, Vietnam has been known as an emerging economy as well as ranked among the most successful East Asian countries (Abbott & Tarp, 2012; Barker & Üngör, 2019; Bentzen & Tung, 2021). Over the past three decades, Vietnam has followed the path toward developing international trade relations in the direction of diversification. This economy has a proactive expansion strategy of international trade with global partners striving to ensure that all parties share the benefits (Abbott et al., 2009; Abbott & Tarp, 2012; Barker & Üngör, 2019; Deprez, 2018; Thanh, 2005). Hence, international trade has played an increasingly important role in the national economic development plan (Abbott & Tarp, 2012; Thanh, 2005). Since the beginning of the 2000s, Vietnam's foreign trade turnover has increased rapidly with the speeding up of both export and import flows (Abbott & Tarp, 2012). In addition to having a successful attractive strategy for foreign direct investment, the Vietnam government also make many free trade agreements with leading economies worldwide that help to robustly increase the international trade flows of this economy (see Abbott et al., 2009; Phan & Jeong, 2016; Thanh, 2005; Thanh et al., 2019; Tung & Bentzen, 2022). Obviously, the trade openness indicator, which is popularly denoted by the total value of exports plus imports over the gross domestic products (Brueckner & Lederman, 2015), has been significantly increasing in the previous decade (Barker & Üngör, 2019). In detail, the trade openness was 0.64 in 1995, then it significantly increased to 1.65 in 2005, 1.75 in 2015, and 1.98 in 2020, respectively (GSO, 2022). Development of international trade and joining the international production chains are the key targets for this country to maintain a high rate of economic growth in the next decades. However, these targets are not easy when they have faced some challenges in recent years (Baccini et al., 2019; Phan & Jeong, 2016; Tung, 2018; Tung & Thanh, 2015). Therefore, this current study is expected to provide a new evidence of related issues for policymakers in Vietnam as well as other developing countries. On the other hand, the findings from this study are obviously helpful for academics when this empirical evidence is great to explain in detail related the trade balance as a source for supporting the economic development in developing countries such as Vietnam. In this study, the ARDL bounds test approach is applied with a quarterly database collected from the General Statistics Office of Vietnam (GSO, 2022) from 1995 to 2020.

This study includes five sections. Section 2 is a compilation of relevant studies. Section 3 presents the methodology and data source. The result and discussion are shown in Sect. 4. Finally, a conclusion is summarized in Sect. 5.

2 Literature Review

The increase in globalization shows the important role of international commercial activities in the economic development of countries. Hence, there are some approaches to analyze the fluctuation of the trade balance in a country. The exchange rate is one of the elements that have a robust impact on the export–import activities of an economy. The impact of the exchange rate on international trade is explained through the Marshall-Lerner Condition (MLC), in which MLC shows the impact of the currency devaluation on the trade balance (eg., Mahmud et al., 2004). The study of Magee (1973) investigates the increase in exchange rate improving the trade balance described by a J-Curve (eg., Rose & Yellen, 1989). In the gross output analysis, the total number of national products is also a factor affecting the trade balance. Through spending flows, the change in the aggregate demand will be reflected in the total national output as well as the national trade balance (Brueckner & Lederman, 2015). Another approach is the monetary policy, where the change in money supply is also a variable regarding the trade balance (Bahmani-Oskooee & Kanitpong, 2017). There are several experimental results that show the impact of these factors on the balance of international trade in countries.

In an empirical study in Malaysia, Duasa (2007) estimates elasticities between trade balance, national output, money supply, and exchange rate in this country. The findings conclude that the Marshall-Lerner condition does not confirm in the long run for the economy. Waliullah et al. (2010) investigate the relationship between the trade balance, gross domestic product, money supply, and real exchange rate in the economy of Pakistan from 1970 to 2005. The bounds test confirms a significant long-term relationship between the balance of international trade and independent variables. The empirical results show that the exchange rate has a positive relationship to the trade balance in the long and short run. Besides, the results imply that money supply and national output have important roles in determining the fluctuation of the trade balance. The exchange rate policy helps to increase the trade balance much stronger than the national output and money supply. Khan and Hossain (2012) develop a new model to analysis the trade balance in Bangladesh. Based on an empirical test for Bangladesh's trade, the authors find the existence of a cointegration relationship between this economy and its major trading partners. On the other hand, a significant relationship between variables of the empirical research model is confirmed in the long run. Kodongo and Ojah (2013) do a study focusing on the intertemporal causal relationships between the real exchange rate, the balance of international trade, and external capital flows in the African region. The study sample included nine major African countries from 1993 to 2009. The empirical result revealed that the net effect of a depreciation of the domestic currency can have an enhancement in the domestic country's international trade status in the short run.

Ahad (2017) investigates the relationship between financial development, exchange rate, inflation, and trade balance in developing countries. The result shows that three independent variables have significant effects on the trade balance in the long run. However, in the short run, there are only the exchange rate and inflation have statistically significant effects on the dependent variable. Bahmani-Oskooee and Kanitpong (2017) use a nonlinear econometric model to analyze the additional asymmetric effects in an Asian regional sample. The authors conclude that there is empirical evidence of long-term and short-term asymmetric effects of exchange rate fluctuation on the balance of international trade. Sasaki and Yoshida (2018) analyze Japan's trade balance which has reached a deficit after a long period of its surplus. The findings show that the national trade has gone through a structural change in both income and exchange rate pass-through elasticity which made the fluctuation of the trade balance.

From Africa, Ahiakpor et al. (2019) examined the relationship between trade openness and the tools of monetary policy in Ghana using a monthly time-series database from 2002 to 2017. The estimated results confirm that higher trade openness leads to the monetary policy (including the exchange rate) becoming more effective. Hoang et al. (2020) investigate the determinants of international trade between ASEAN countries and Taiwan. The study indicates that the economic scale and per capita income of ASEAN countries have stronger impacts than those of Taiwan in aggregate trade as well as in the degree of manufacturing. Gürtle (2019) investigated the J-curve situation in the Czech economy with quarterly data from 2000 to 2014. The study limits the research data on the economic relationship of a small open economy. The result shows that the real effective exchange rate has a strongly negative impact on the balance of international trade in the short run and a positive one in the long run. The evidence confirms the J-curve phenomenon in the study situation of the Czech economy. On the other hand, the economic growth rate has a significantly higher effect on the trade balance than in other countries.

In a recent study, Marin (2020) compared the affecting factors of trade activities between South Korea and Colombia. The variables are added in the quantitative model such as the gross domestic products of the importer and exporter countries. The results of the empirical econometric model revealed that the free trade agreements (FTAs) have an insignificant impact on international trade in South Korea. Besides, the empirical evidence suggests that the agricultural sector of Colombia has potential opportunities to compete in the Asian region. Shubaita et al. (2020) analyzed the relationship between trade balance, exchange rate, and real gross domestic product in Tunisia from 1980 to 2018. Although the bounds test confirms a cointegration between variables, the estimated result shows that this country does not follow the Marshall-Lerner framework in the long run.

Following the previous results, the quantitative effects of related variables on trade balance are much more varied. The empirical results depend on the characteristics of each economy. To the best of our knowledge, there is no evidence related to the factors affecting the trade balance in Vietnam. Hence, in the next section, the paper presents the quantitative methodology and the database to analyze the determinants of the trade balance in Vietnam in recent years.

3 Methodology

The empirical research model has four variables. The dependent variable is the trade balance, besides, there are some independent variables are included in the function including national output, money demand, and nominal exchange rate. The research function is performed in a simple form as follows.

$$\mathrm{TB} = \mathrm{f}(\mathrm{Y}, \mathrm{M}, \mathrm{EXCH}) \quad (1)$$

where:

TB = Trade balance

Y = National output, trillion VND

M = Money demand, trillion VND

EXCH = Exchange rate, VND/USD

The trade balance indicator is usually calculated by the total value of exports minus the total value of imports. However, in this paper, the trade balance variable is computed by the rate of exports value divides by imports value. This solution is a good idea because the values of the variable are not dependent on the unit of measurement. The national output and money demand values are adjusted from the nominal to the real values by a constant price. This kind of calculation was employed in some previous studies such as Onafowora (2003) and Duasa (2007). Based on the suggestions from related studies (see Baek, 2013; Bahmani-Oskooee & Kanitpong, 2017; Duasa, 2007; Hoang et al., 2020; Waliullah et al., 2010), a linear function will be well presented if all time-series variables transformed to the natural logarithm values, which obtain direct elasticities among variables. Using the natural logarithm of all variables, the function (1) can be rewritten in the below format.

$$\mathrm{LnTB_t} = \beta_0 + \beta_1 \mathrm{LnY}_t + \beta_2 \mathrm{LnM}_t + \beta_3 \mathrm{LnEXCH}_t + \varepsilon_t \quad (2)$$

where β_1, β_2, and β_3 are evaluated values. β_0 is a constant value that presents the trade balance elasticity value that is not explained by the independent variables. Furthermore, ε presents the error term, and t denotes the period.

The long-term cointegration relationship between the independent variables and trade balance is checked by the Autoregressive Distributed Lag (ARDL) method and the bounds values (Pesaran et al., 2001). The methodology is chosen because of some advantaged points including it helps to conclude with a mixture of $I(0)$ and $I(1)$ variables. The test can be employed in a single-equation format, and finally, a smaller observation is required than others (Baek, 2013; Bahmani-Oskooee & Kanitpong, 2017; Duasa, 2007). Hence, the ARDL bounds test is a powerful methodology for testing the cointegration relationship of variables in a linear functional form.

Table 1 The hypothesis of the ARDL bounds test

Null hypotheisis (H_0)	Alternative hypothesis (H_1)	Function
$\psi 1 = \psi 2 = \psi 3 = \psi 4 = 0$	$\psi 1 \neq 0$; or $\psi 2 \neq 0$; or $\psi 3 \neq 0$; or $\psi 4 \neq 0$	f(LnTB/LnY,LnM,LnEXCH)

Source Calculated from the study data

$$\begin{aligned} \Delta LnTB_t = \varphi_0 + \sum_{i=1}^{k} \varphi_{1i} \Delta LnTB_{t-i} + \sum_{i=0}^{k} \varphi_{2i} \Delta LnY_{t-i} + \\ + \sum_{i=0}^{k} \varphi_{3i} \Delta LnM_{t-i} + \sum_{i=0}^{k} \varphi_{4i} \Delta LnEXCH_{t-i} + \\ + \psi_1 LnTB_{t-1} + \psi_2 LnY_{t-1} + \psi_3 LnM_{t-1} + \psi_4 LnEXCH_{t-1} + \varsigma_t \end{aligned} \tag{3}$$

In the first step of the bounds test, the ARDL function will be calculated by the ordinary least squares (OLS) regression, and then chosen the optimal lag length by the Akaike Info Criterion (AIC) indicator (Pesaran et al., 2001). In the second step, the long-term cointegration relationship between the variables is checked by Wald's test (F-statistics). There are two hypotheses including the null hypothesis of no cointegration (H_0) against the alternative hypothesis (H_1) of a long-term cointegration between the variables. The alternative hypothesis means at least one of $\psi 1$–$\psi 4$ is different than zero. The hypothesis of the ARDL bounds test is performed as the Table 1.

There are three potential conclusions of the ARDL bounds test. First, if the F-statistic value is below the lower limit of the bound values, we cannot be rejected the null hypothesis of no long-term cointegration relationship. Second, if the F-statistic value is higher than the upper limit of the bound values, the null hypothesis (H_0) can be rejected, and the testing result confirms the existence of a long-term cointegration relationship among variables. Finally, the cointegration cannot be confirmed in the case the F-statistic value is between the lower limit and upper limit of the bound values.

If the ARDL bounds test confirms there is a long-term cointegration relationship between the variables, we will employ the error correction model (ECM) to explore the short-term dynamic relationship. The ECM model for identifying the short-term relationship among variables is written as the Eq. (4).

$$\begin{aligned} \Delta LnTB_t = \delta_0 + \sum_{i=1}^{k} \delta_{1i} \Delta LnTB_{t-i} + \sum_{i=0}^{k} \delta_{2i} \Delta LnY_{t-i} + \sum_{i=0}^{k} \delta_{3i} \Delta LnM_{t-i} \\ + \sum_{i=0}^{k} \delta_{4i} \Delta LnEXCH_{t-i} + \gamma ECM_{t-1} + \mu_t \end{aligned} \tag{4}$$

This paper uses a quarterly database form that was collected from 1995 to 2020. The total observation is 104. The data on the trade balance and yield are directly downloaded and calculated from the General Statistics Office of Vietnam (GSO, 2022). The data on money demand and nominal exchange rate are sourced from the State Bank of Vietnam. All the time-series variables are transformed to the natural logarithm values before putting in the quantitative process. The statistical descriptive analysis of the variables is shown in the Table 2, (Fig. 1)

Table 2 A statistical summary of the variables

Variable	Max	Min	Mean	Std. Dev	Obs
LnTB	0.146	− 0.595	− 0.106	0.157	104
LnY	6.113	3.782	4.830	0.618	104
LnM	7.855	3.543	6.051	1.267	104
LnEXCH	10.06	9.307	9.751	0.230	104
ΔLnTB	0.335	− 0.373	0.002	0.101	103
ΔLnY	0.317	− 0.563	0.023	0.231	103
ΔLnM	0.306	− 0.145	0.041	0.051	103
ΔLnEXCH	0.053	− 0.017	0.007	0.012	103

Source Calculated from the study data

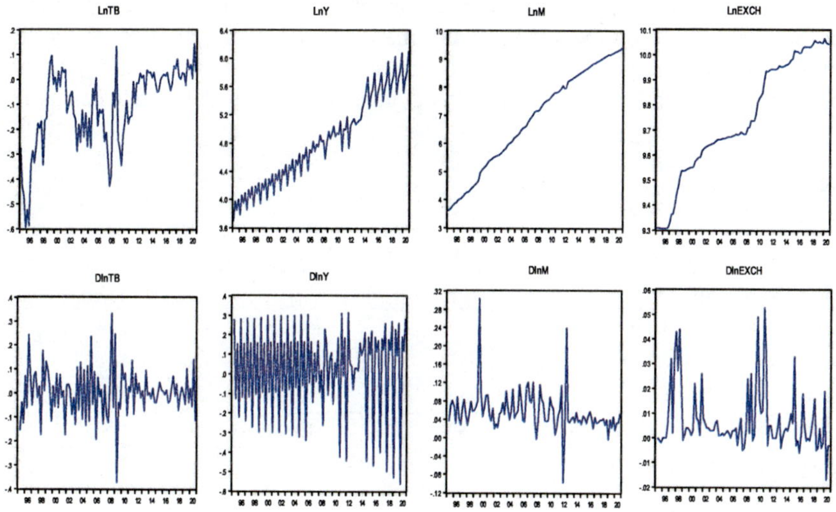

Fig. 1 The graphs of the variables in the ARDL function (*Source* Calculated from the study data)

4 Results

Before estimating the ARDL model to explore the determinants of the balance of foreign trade in Vietnam, the time-series properties of the variables are checked for the unit root by two methodologies including the augmented dickey-fuller (ADF) and the Phillips-Perron (P-P) test. In detail, the null hypothesis (H_0) of the presence of unit root is tested against the alternative hypothesis (H_1) of no unit roots. The testing results of the unit root test confirm that the variables at the level are received both stationary and non-stationary. However, all variables are stationary at their first difference values with both two testing methodologies (Table 3).

Following the bounds testing approach literature, the Akaike Info Criterion (AIC) value is employed to choose the optimal lag length of the general ARDL model (Pesaran et al., 2001). Based on the comparison of the received AIC values, the ARDL (3,2,0,1) is selected for the bounds testing process of the long-term cointegration relationship. The Wald test is done for examining the H_0 hypothesis of the ARDL model. The calculated F statistic = 5.387381 is much higher than the upper bounds critical value = 4.37 at a 1% significance level. The F-statistic value from the Wald test is, in this case, positive and significant; hence, the null hypothesis (H_0): $\psi 1 = \psi 2 = \psi 3 = \psi 4 = 0$ has been rejected at a 1% significance as well. The bounds test concludes that there is a long-term cointegration relationship among variables in the testing function (see Pesaran et al., 2001) (Table 4).

The bounds test has confirmed a long-term cointegration among trade balance and its determinants when trade balance is employed as the dependent variable in the function. Based on the testing result, the quantitative analysis is done by the estimation of both short-term and long-term elasticities using Eqs. (2) and (4). In the long-term result, independent variables significantly affected the trade balance. In

Table 3 The unit root test for the variables

Variable	ADF test			P-P test		
	t-statistic	CV	Conclusion	t-statistic	CV	Conclusion
Time series at the level						
LnTB	– 3.116	– 4.052	NS	– 4.275	– 4.049	*S*
LnY	– 2.340	– 4.053	NS	– 11.80	– 4.049	*S*
LnM	– 1.121	– 4.019	NS	– 1.160	– 4.049	NS
LnEXCH	– 1.606	– 4.050	NS	– 1.605	– 4.049	NS
Time series at the first difference						
ΔLnTB	– 8.792	– 4.052	*S*	– 27.50	– 4.050	*S*
ΔLnY	– 24.19	– 4.052	*S*	– 52.23	– 4.050	*S*
ΔLnM	– 8.189	– 4.051	*S*	– 8.736	– 4.050	*S*
ΔLnEXCH	– 6.845	– 4.050	*S*	– 6.986	– 4.050	*S*

Notes NS = non-stationary; *S* = Stationary; CV = Critical value at 1%
Source Calculated from the study data

Table 4 The Wald test result of the cointegration relationship

F-statistic value	Significance level (%)	Bounds critical values	
		$I(0)$	$I(1)$
5.387381 (P-value 0.0006)	1	3.43	4.37
	2.5	3.13	4.05
	5	2.86	3.78
	10	2.57	3.46

Source Calculated from the study data

detail, the estimated result of the long-run model shows negative impacts of national output (at 10% significance) and money supply (at a 1% significance) on the balance of trade. The negative sign of the coefficient from the national output can be explained when Vietnam had many years with the sum of imports was larger than the sum of exports (a trade deficit phenomenon), meanwhile, the national output has increased during the period (see Abbott & Tarp, 2012; Thanh, 2005). In the future, Vietnamese policymakers need to enhance the domestic demand shifts from foreign products to domestic production of substitution products as a reaction to the higher prices of imports. This solution can help a trade balance improvement. The negative effect of money demand on the trade balance is consistent with previous studies (for example, Duasa, 2007; Waliullah et al., 2010). On the other hand, the estimated coefficients mean a 1% increase in the gross domestic product can reduce the trade balance by 0.091%, while a 1% increase in money supply reduces the balance of foreign trade by 0.123%, respectively.

The long-term analysis demonstrates that the exchange rate variable positively and significantly affects the trade balance when a 1% rise in the exchange rate has caused a 1.336% increase in the trade balance over the study period. The estimated result indicates that in the long run, the exchange rate contributes most considerably to the increase of the trade balance in Vietnam. Furthermore, by finding a positive and significant effect of the exchange rate to trade balance in the long run, the J-curve phenomenon is expected to conclude in the short-term estimation. The short-term result will be identified by the error correction term technique computed from the long-term regression. Besides, it is important to note that the estimated coefficients in Table 5 merged both the short-term effect and the long-term effect of independent variables on the trade balance.

The short-term analysis investigates that the gross output significantly causes trade balance (at a 1% significance); for example, a 1% rise in the gross output causes a 0.159% decrease in the trade balance. Obviously, the short-term and long-term estimations confirm the significant positive effects of the gross domestic product on the trade balance. It is a serious problem when economic growth is merged with the trade deficit. In the future, Vietnam must put a position for itself at the center of the global production process and world value chain. Based on the foreign trade improvement strategy, the economy can approach and join high-value manufactured products in high-income markets related to the free trade agreements (FTAs) where Vietnamese

Table 5 The ARDL long run and short-run analysis

ARDL long-run analysis. Dependent variable = LnTB			
Variable	Coefficient	t-Statistic	P-value
LnY	– 0.091*	– 1.672	0.097
LnM	– 0.123***	– 3.093	0.002
LnEXCH	1.336***	6.494	0.000
Constant	– 11.95***	– 6.776	0.000
R-squared = 0.5342 Adjusted R-squared = 0.5203			
ARDL short-run analysis. Dependent variable = ΔLnTB			
ΔLnTB(−1)	0.034	0.373	0.709
ΔLnTB(−2)	0.148	1.624	0.107
ΔLnTB(−3)	– 0.173*	– 1.929	0.056
ΔLnY	– 0.047	– 1.060	0.291
ΔLnY(−1)	– 0.017	– 0.334	0.738
ΔLnY(−2)	– 0.159***	– 3.549	0.000
ΔLnM	0.248*	1.666	0.099
ΔLnEXCH	0.949	1.525	0.130
ΔLnEXCH(−1)	– 0.335	– 0.530	0.597
ECM(−1)	– 0.383***	– 4.447	0.000
Constant	– 0.004	– 0.375	0.708
R-squared = 0.5598 Adjusted R-squared = 0.5104			
Short-run diagnostic test Normality test (Jarque–Bera) = 4.506 (P-value = 0.105); Breusch-Godfrey Serial Correlation = 1.648 (P-value = 0.184); Heteroskedasticity: Breusch-Pagan-Godfrey = 0.877 (P-value = 0.557); Ramsey Reset = 0.563 (P-value = 0.574)			

Notes ***, **, * display significance at 1%, 5%, and 10% levels, respectively
Source Calculated from the study data

products enjoy preferential market access. With these solutions, Vietnam can receive more of the added value and put the domestic products up in the added value of the international production chain. On the other hand, the Vietnamese enterprises must improve the competitiveness level of domestic products to substitute foreign products as a solution to reduce imports. This strategy also helps to enhance the balance of foreign trade in the future.

The money supply has a significant positive effect on the trade balance at the 10% level. It means that a 1% increase in money supply is significant to a 0.248% increase in the short-term trade balance. The lag variable of the trade balance is negative and significant at a 10% level, which means the international trade shows a harmful effect on itself in the short run. Over three decades after the definition of the Marshall-Lerner condition was presented and the J-Curve evidence came into academic existence. In detail, the J-curve phenomenon shows how a devaluation of a national exchange rate can affect the trade balance. Furthermore, the J-curve is understood as a dynamic view of the Marshall-Lerner condition by the elasticities approach. There is evidence of the J-curve in Vietnam when the coefficient of the exchange rate variable is negative (but insignificant) in the short run, while it shows a significant positive impact in the long run (see Halicioglu, 2007; Nusair, 2016; Onafowora, 2003; Shubaita et al., 2020). The weak negative effect of the exchange rate on the trade balance in the short run can consider a lucky phenomenon in the case of devaluation of the currency in Vietnam. The empirical evidence confirms that the Marshall-Lerner Condition holds in the long run in the Vietnamese economy, as a result, it explains how Vietnam could robustly increase its international trade in previous studies (see Abbott & Tarp, 2012; Barker & Üngör, 2019; Deprez, 2018; Gürtle, 2019; Thanh, 2005). In addition, the estimated result is consistent with the Marshall-Lerner condition in some developing economies (see Bahmani et al., 2013; Chebbi & Olarreaga, 2019; Mahmud et al., 2004; Sastre, 2012).

The coefficient of the ECM_{t-1} variable indicates the speed of adjustment of variables return to equilibrium; based on the literature, it is a statistically significant value with a negative sign. In the short-run estimation, the lagged variable of the error term (ECM_{t-1}) is found to have a negative sign and is statistically significant at a 1% level. The result ensures that long-run equilibrium can be attained. The significance of the error correction term implies the evidence of adjusting level from short run to long run (Pesaran et al., 2001). The coefficient of the ECM_{t-1} is -0.383 showing a high rate of convergence to equilibrium. After estimating the short-term elasticities according to the ARDL method, an important note for the result is the conclusion of free from problems of econometrics, for the estimated coefficients to be consistent. The normality test (Jarque–Bera = 4.506 with P-value = 0.105), hence, the residuals of the regressive estimation follow the normal distribution. The Breusch–Godfrey testing value = 1.648, with P-value = 0.184, which confirms that there is no autocorrelation problem in the estimated result. The Breusch-Pagan-Godfrey test result = 0.877 (with P-value = 0.557) shows that the model is free of the heteroskedasticity phenomenon. The Ramsey Reset test = 0.563 (P-value = 0.574) helps to conclude the correct form used in the research model. Therefore, the diagnostic tests imply that the results of the ARDL estimation are stable and ensure reliability conditions (see Pesaran et al., 2001).

5 Conclusions

In this paper, we employ the ARDL bounds testing approach developed by Pesaran et al. (2001) to analyze the determinants of trade balance in Vietnam using a quarterly database from 1995 to 2020. The estimated results have contributed to several notable findings. First, there is evidence of a long-term cointegration relationship between the trade balance and independent variables. Second, the coefficients of the independent variables have statistically significant impacts on the trade balance in the long-term model. In detail, the coefficients of the national output and money supply are negative; on the other hand, the coefficient of the exchange rate is positive. Third, the short-term estimated result confirms the effect of the exchange rate on the balance of foreign trade followed by the J-curve theory. Besides, in the short-term analysis, national output has a negative and statistically significant impact on the trade balance, while the supply of money confirms a significant positive effect on the dependent variable.

The empirical evidence suggests policymakers need to reconsider the management ways of monetary policy in the coming time. Because the increase in money supply has a short-term positive effect but it changes to a negative side in the long run. Hence, monetary policy should be studied so that the expansion mechanisms can promote the trade balance in both the short run and the long run. It is necessary to continue implementing the exchange rate policy for supporting the increase of the trade balance; besides, currency devaluation only has a negative effect in the short run (as a J-curve phenomenon). The increase in national output leads to a long-term decline in the balance of trade; it is evidence for policymakers to design policies in domestic consumption (decrease imports) as well as boost the production capacity of enterprises to increase exports. Consequently, the increase in national output can be enhanced with the surplus of the Vietnamese trade balance in the future.

References

Abbott, P., Bentzen, J., & Tarp, F. (2009). Trade and development: Lessons from Vietnam's past trade agreements. *World Development, 37*(2), 341–353.

Abbott, P., & Tarp, F. (2012). Globalization crises, trade and development in Vietnam. *Journal of International Commerce, Economics and Policy, 3*(1), 1240006.

Ahad, M. (2017). Impact of financial development on trade balance: An ARDL cointegration and causality approach for Pakistan. *Global Business Review, 18*(5), 1199–1214.

Ahiakpor, F., Cantah, W., Brafu-Insaidoo, W., & Bondzie, E. (2019). Trade openness and monetary policy in Ghana. *International Economic Journal, 33*(2), 332–349.

Akbas, Y. E., & Sancar, C. (2021). The impact of export dynamics on trade balance in emerging and developed countries: An evaluation with middle income trap perspective. *International Review of Economics & Finance, 76*, 357–375.

Amiti, M., & Konings, J. (2007). Trade liberalization, intermediate inputs, and productivity: Evidence from Indonesia. *American Economic Review, 97*(5), 1611–1638.

Baccini, L., Impullitti, G., & Malesky, E. J. (2019). Globalization and state capitalism: Assessing Vietnam's accession to the WTO. *Journal of International Economics, 111*, 75–92.

Baek, J. (2013). Does the exchange rate matter to bilateral trade between Korea and Japan? Evidence from commodity trade data. *Economic Modelling, 30*, 856–862.

Bahmani, M., Harvey, H., & Hegerty, S. W. (2013). Empirical tests of the Marshall-Lerner condition: A literature review. *Journal of Economic Studies, 40*(3), 411–443.

Bahmani-Oskooee, M., & Ratha, A. (2007). Bilateral S curve between Japan and her trading partners. *Japan and the World Economy, 19*, 483–489.

Bahmani-Oskooee, M., & Kanitpong, T. (2017). Do exchange rate changes have symmetric or asymmetric effects on the trade balances of Asian countries? *Applied Economics, 49*(46), 4668–4678.

Barker, T., & Üngör, M. (2019). Vietnam: The next Asian tiger? *North American Journal of Economics and Finance, 47*, 96–118.

Beck, K. (2020). What drives international trade? Robust analysis for the European Union. *Journal of International Studies, 13*(3), 68–84.

Bentzen, J., & Tung, L. T. (2021). Regional income convergence in an emerging Asian economy: Empirical evidence from Vietnam. *Post-Communist Economies, 33*(1), 64–78.

Brueckner, M., & Lederman, D. (2015). Trade openness and economic growth: Panel data evidence from Sub-Saharan Africa. *Economica, 82*, 1302–1323.

Chebbi, H. E., & Olarreaga, M. (2019). Investigating exchange rate shocks on agricultural trade balance: The case of Tunisia. *Journal of International Trade & Economic Development, 28*(5), 628–647.

Costinot, A., Vogel, J., & Wang, S. (2013). An elementary theory of global supply chains. *American Economic Review, 80*(1), 109–144.

Deprez, S. (2018). The strategic vision behind Vietnam's international trade integration. *Journal of Current Southeast Asian Affairs, 37*(2), 3–38.

Duasa, J. (2007). Determinants of Malaysian trade balance: An ARDL bound testing approach. *Global Economic Review, 36*(1), 89–102.

Ge, S., & Luo, W. (2015). Multinational enterprise entry and industrial structural of manufacturing sector in China: Based on global value chain perspective. *Economic Research Journal, 11*, 34–48.

General Statistics Office of Vietnam (GSO). (2022). *Data and statistics*. Retrieved from https://www.gso.gov.vn/en/homepage/.

Gürtler, M. (2019). Dynamic analysis of trade balance behavior in a small open economy: The J-curve phenomenon and the Czech economy. *Empirical Economics, 56*, 469–497.

Halicioglu, F. (2007). The J-curve dynamics of Turkish bilateral trade: A cointegration approach. *Journal of Economic Studies, 34*, 103–119.

Hoang, N. T. T., Truong, H. Q., & Dong, C. V. (2020). Determinants of trade between Taiwan and ASEAN countries: A PPML estimator approach. *SAGE Open, 10*(2), 1–13.

Jensen, N. M., & Shin, M. J. (2014). Globalization and domestic trade policy preferences: Foreign frames and mass support for agriculture subsidies. *International Interactions, 40*(3), 305–324.

Khan, M. Z. S., & Hossain, M. I. (2012). Determinants of trade balance of Bangladesh: A dynamic panel data analysis. *The Bangladesh Development Studies, 35*(2), 45–65.

Kondogo, O., & Ojah, K. (2013). Real exchange rates, trade balance and capital flows in Africa. *Journal of Economics and Business, 66*, 26–46.

Magee, S. P. (1973). Currency Contracts, pass-through, and devaluation. *Brookings Papers on Economic Activity, 1*, 303–325.

Mahmud, S. F., Ullah, A., & Yucel, E. M. (2004). Testing Marshall Lerner condition: A nonparametric approach. *Applied Economics Letters, 11*(4), 231–323.

Marin, S. D. A. (2020). Determinants of trade flows between Colombia and South Korea. *Artículo De Investigación, 39*(70), 75–105.

Nusair, S. A. (2016). The J-curve phenomenon in European transition economies: A nonlinear ARDL approach. *International Review of Applied Economics, 31*(1), 1–27.

Oloyede, B. M., Osabuohien, E. S., & Ejemeyovwi, J. O. (2021). Trade openness and economic growth in Africa's regional economic communities: Empirical evidence from ECOWAS and SADC. *Heliyon, 7*(5), e06996.

Onafowora, O. (2003). Exchange rate and trade balance in East Asia: Is there a J-curve? *Economic Bulletin, 5*(18), 1–13.

Pesaran, M. H., Shin, Y., & Smith, R. J. (2001). Bounds testing approaches to the analysis of level relationships. *Journal of Applied Econometrics, 16*, 289–326.

Phan, T. H., & Jeong, J. Y. (2016). Potential economic impacts of the Vietnam-Korea free trade agreement on Vietnam. *East Asian Economic Review, 20*(1), 67–90.

Rose, A. K., & Yellen, J. L. (1989). In there a J-Curve? *Journal of Monetary Economics, 24*, 53–68.

Sasaki, Y., & Yoshida, Y. (2018). Decomposition of Japan's trade balance. *International Review of Economics & Finance, 56*, 507–537.

Sastre, L. (2012). Simultaneity between export and import flows and the Marshall-Lerner condition. *Economic Modelling, 29*, 879–883.

Shubaita, E., Mar'i, M., & Seraj, M. (2020). Investigating in the J-curve phenomenon in Tunisia-ARDL bound test approach. *Journal of Economics and Behavioral Studies, 12*(5J), 23–32.

Siddiqui, K. (2015). Trade liberalization and economic development: A critical review. *International Journal of Political Economy, 44*(3), 228–247.

Siddiqui, K. (2016). International trade, WTO and economic development. *World Review of Political Economy, 7*(4), 424–450.

Singh, T. (2010). Does international trade cause economic growth? A Survey. *The World Economy, 33*(11), 1517–1564.

Thanh, V. T. (2005). Vietnam's trade liberalization and international economic integration: Evolution, problems, and challenge. *ASEAN Economic Bulletin, 22*(1), 75–91.

Thanh, S. D., Canh, N. P., & Schinckus, C. (2019). Impact of foreign direct investment, trade openness and economic institutions on growth in emerging countries: The case of Vietnam. *Journal of International Studies, 12*(3), 243–264.

Thirlwall, A. P. (2000). Trade agreements, trade liberalization and economic growth: A selective survey. *African Development Review, 12*(2), 129–160.

Tung, L. T., & Thanh, P. T. (2015). Threshold in the relationship between inflation and economic growth: Empirical evidence in Vietnam. *Asian Social Science, 11*(10), 105–112.

Tung, L. T. (2018). The effect of fiscal deficit on economic growth in an emerging economy: Evidence from Vietnam. *Journal of International Studies, 11*(3), 191–203.

Tung, L. T., & Bentzen, J. (2022). The relationship between income growth and inequality: Evidence from an Asian emerging economy. *Economics and Sociology, 15*(2), 95–109.

Waliullah, Kakar, M. K., Kakar, R., & Khan, W. (2010). The determinants of Pakistan's trade balance: An ARDL cointegration approach. *The Lahore Journal of Economics, 15*(1), 1–26.

Wong, K.-Y., & Yip, C. K. (2002). Industrialization, economic growth, and international trade. *Review of International Economics, 7*(3), 522–540.

Derterminants of FDI Inflows to Developing and Least Developed Countries

Dao Thi Bich Thuy

Abstract The purpose of this paper is to explore the determinants of FDI inflows to developing and least-developed countries. The empirical study consists of 102 developing and least developed countries as classified by the United Nations in the period from 2000 to 2019. The fixed effect model with the Driscoll and Kraay standard errors estimation is used and the results reveal that market size, infrastructure development, financial development, economic freedom, and economic globalization play as FDI stimulus factors while social and political globalization have a negative effect on FDI inflows in all countries. Quality of labor force, while having a positive effect on inward FDI in the developing country group, shows to be a drag in FDI inflows in the least developed country group. There is a strong evidence of market-seeking FDI in all countries, natural-seeking FDI in least developed countries, and efficiency-seeking FDI is more evident in developing countries.

Keywords Foreign direct investment · FDI determinants · Developing and least developed countries

1 Introduction

It is a wide belief that foreign direct investment (FDI) plays an important role in economic growth of developing countries. FDI does not only contribute direct capital stock to the host countries but is also considered as an effective source for local labor training, the demonstration of managerial skills and expertise as well as a valuable channel for the transfer of advanced technologies (Abbes et al., 2015). Attraction of FDI to the countries has become the important goal that governments in developing countries wish to pursue. For the governments in developing countries to successfully achieve their goal, it is important that they need to know what are the factors that multinational firms consider when investing overseas.

D. T. B. Thuy (✉)
VNU University of Economics and Business, Vietnam National University, Hanoi, Vietnam
e-mail: thuydaokt@vnu.edu.vn

A. T. Nguyen et al. (eds.), *Contemporary Economic Issues in Asian Countries: Proceeding of CEIAC 2022, Volume 1*,
https://doi.org/10.1007/978-981-19-9669-6_15

The determinants of FDI are complex covering a wide range from traditional to institutional factors. In traditional views, factors affecting FDI include the host country's market size and its potential growth, infrastructure, technology, human factor, geographic location, and natural resources. Foreign investors are either motivated by interest in accessing and exploiting natural resources (natural resource-seeking FDI), serving foreign markets (market-seeking FDI) or gaining higher productivity or lower costs of production (efficiency-seeking FDI) (Wadhwa & Reddy, 2011). The more recent views look at institutional quality, such as the provision of business-friendly environment or the degree of the country's openness (Sabir et al., 2019).

There are an extensive number of studies aimed at finding the determinants of FDI inflows, and the majority focused on developing countries. Those studies are conducted at the individual country level, the group of countries by region, or the pool of developing countries worldwide. However, in the so-called developing countries, the level of development varies among them. Based on human development and socioeconomic development, the United Nations (World Economic Situation and Prospects, 2018) distinguish least-developed countries from developing countries. According to the United Nations, least-developed countries are low-income countries with low levels of human assets and suffer high economic vulnerability. Despite the distinction between developing and least-developed countries, there is lack of studies on determinants of FDI inflows to the developing world taking into consideration of this issue. The purpose of this paper is to fill this gap where we seek to explore the determinants of inward FDI in developing countries in comparison to that in least developed countries.

The rest of the paper is organized as follows. Literature review is provided in Sect. 2 and followed by methodology in Sect. 3. Section 4 discusses the results, and finally, Sect. 5 is the conclusion.

2 Literature Review

Market size is considered as the most crucial factor for attracting foreign direct investment due to high expected demand and prospects of economies of scale. The market size hypothesis claims that efficient utilization of resources and exploitation of economies of scale demand for a large market (Balassa, 1966; Scaperlanda & Mauer, 1969). The implementation of superior technologies is effective for a large scale of production which in turn predetermines the size of the market. Foreign direct investment will be taken in a country only when the country meets the market requirements in terms of size. When the market size increases, the influx of FDI is stimulated to meet the growth in demand. There thus exists a positive relationship between foreign capital inflows and GDP growth of the host country. There are a number of empirical studies that supported for market size being a determinant of inward FDI. Hornberger et al. (2011) by summing up a set of 30 empirical studies on developing and transition economies conducted since 2000 found that the size and growth potential of markets are significantly associated with foreign direct investment inflows. Studies by Kok and Ersoy (2009) for 24 developing countries and Wadhwa and Reddy (2011) for 10 Asian countries in the period 1991–2008 also confirmed a positive relationship between foreign direct investment and GDP. A more recent study by Gabriel et al. (2016) for Nigeria in the period 1970–2011 found that economy size has a statistically positive effect on foreign direct investment. Similarly, Petrović-Ranđelović et al. (2017) concluded that GDP growth fosters FDI inflows in the six countries of the Western Balkans region in the period 2007–2015.

In the traditional view, FDI is attracted to developing countries with the abundance of labor to take advantage of cheap labor cost. However, Pfeffermann and Madarassy (1992) argued that new technological advances have reduced the labor content and increased the knowledge content of production. For this reason, FDI has gradually moved away from low-cost and low-skilled labor-intensive industries toward more capital and knowledge-intensive industries. The availability of well-educated pool of workforce in the host countries has become an increasingly important factor attracting multinational corporations. Empirical studies supported for both views of cheap labor costs and human capital being determinants of FDI inflows. Donaubauer and Dreger (2018) reasoned that the increase in wages in China has caused a shift of FDI away from China to low-wage countries in the Asian region. In examining the determinants of FDI in selected Central and Eastern European Countries (CEECs) for the period of 1995–2003, Bellak et al. (2008) found that higher labor costs affect FDI negatively, whereas the impact of higher labor productivity on FDI is positive. Similarly, Abbas et al. (2021) concluded that both cheap and skilled labor in the host country play as the driven factor to attract FDI. Noorbakhsh et al. (2001) claimed that human capital is one of the most important determinants of FDI inflows to developing countries and its importance has become increasingly greater through time. A positive relationship between human capital and the inflows of FDI to developing countries is found in studies by Karimi et al. (2013) and Kheng et al. (2017).

Financial development plays the role in influencing FDI in two ways: direct and indirect effects. As argued by Desbordesa and Wei (2017), to engage in FDI foreign firms must incur substantial upfront fixed costs such as the cost of establishing or purchasing a production facility in the host country. Besides, due to the nature of projects with a large initial scale, a long gestation but a short harvest period, or the continuing investment requirement, firms may find it unable to finance the upfront fixed costs with internal funds (Rajan & Zingales, 1998). In such the case, firms must rely on external funds and their access to external finance depends on financial development. Higher financial development in the source countries would enable the firms to get external funds, and increase the volume of outward FDI from the source to the host countries. However, when the firms face with constraint of getting external funds from source countries' financial institutions, they may choose to use the source of external funds in the host countries if local financing conditions are favorable (Desai et al., 2004). Therefore, higher financial development in the host country can have a positive direct effect on the volume of inward FDI. Studies by Feinberg and Phillips (2004), Desai et al. (2006), and Bilir et al. (2019) supported the positive direct external finance effect of the host country's financial development when they found that the expansion of the activities of U.S. foreign affiliates is constrained in host countries where external finance is relatively limited and expensive.

Beside the direct effect, higher financial development in the host country can also create an indirect effect on FDI through its promotion of competition and overall economic activity (Rajan & Zingales, 1998; Klapper et al., 2006; Manova, 2013). Higher financial development in the host country induces the entry of both local and foreign firms which raises the competition in the host country's market. Higher competition results in higher prices of local inputs while lower market share and thus a decline in potential sale volume and profits for foreign firms. This obviously creates a bad effect on foreign firms targeting the host country's market and makes a country a less attractive destination than before. The negative indirect competition effect of the host country's financial development is evidence in the finding by Ju and Wei (2010).

The country's openness is considered to be a key determinant of FDI inflows. A considerable amount of empirical studies has been conducted on this relationship and in these studies, trade openness measured by the ratio of trade volume to GDP is used as a proxy for the openness of the country. The majority of studies supported for a positive relationship between trade openness and FDI inflows. Among them are Neumayer and Soysa (2004), Liargovas et al. (2012), Makoni (2018), and Zaman et al. (2018). According to these studies, foreign direct investment is more attracted to those countries with lower restrictions on import and export activities. A negative relationship between trade openness and FDI inflows is found in other studies (Cantah et al., 2018; Khan & Hye, 2014 and Mudiyanselage et al., 2021). And yet studies by Ho et al. (2013) and Wickramarachchi (2019) found that trade openness had no significant impact on FDI inflows.

As defined by Gwartney et al. (2017), economic freedom refers to the extent to which economic activities are "coordinated by personal choice, voluntary exchange, open markets, and clearly defined and enforced property rights". Economic freedom shapes the business climate in which foreign firms operate and thus affects foreign investors' decision in choosing the location to invest. The relation between economic freedom and FDI inflows has captured attention from many researchers. The majority found evidence that economic freedom has a positive impact on FDI inflows. Among these works are Bengoa and Sanchez-Robles (2003) for 18 Latin-American countries, Quazi (2007) for 7 East Asian countries, Nasir and Hassan (2011) for South Asian countries, Moussaa et al. (2016) for 156 countries, Hossain (2016) for 79 developing countries, Imtiaz and Bashir (2017) for 20 South Asian countries, Barua et al. (2017) for 81 countries, and Sooreea et al. (2020) for 40 Sub-Saharan African countries.

3 Methodology

The determinants of FDI inflows to developing and least-developed countries are considered to include market size, labor factor, infrastructure development, financial development, natural resource, economic freedom, and the degree of the country's openness (Fig. 1)

The empirical study is conducted with 102 countries including 67 developing and 35 least-developed countries as classified by the United Nations in the period from 2000 to 2019. The chosen period of study is mainly constrained by the availability of data. Based on the conceptual framework, the regression equation is written as

$$\mathrm{FDI}_{i,t} = c + \beta_1 \mathrm{GDP}_{i,t} + \beta_2 \mathrm{POP}_{i,t} + \beta_3 \mathrm{QLF}_{i,t} + \beta_4 \mathrm{URB}_{i,t} + \beta_5 \mathrm{FIN}_{i,t} + \beta_6 \mathrm{NAR}_{i,t} + \beta_7 \mathrm{EFR}_{i,t} + \beta_8 \mathrm{EGI}_{i,t} + \beta_9 \mathrm{SGI}_{i,t} + \beta_{10} \mathrm{PGI}_{i,t} + \alpha_i + \mu_t + e_{i,t}$$

where subscript i denotes country and t denotes time in year.

Variable description.

FDI inflow (FDI): this variable is measured as net inflows of foreign direct investment in percentage of GDP. Data for net inflows of foreign direct investment is taken from World Development Indicators (World Bank).

Market size (GDP): this variable is measured as the growth rate of GDP. Data for annual GDP growth rate is taken from World Development Indicators (World Bank).

Size of labor force (POP): this variable is measured as the natural logarithm of total population. Data for total population is taken from World Development Indicators (World Bank).

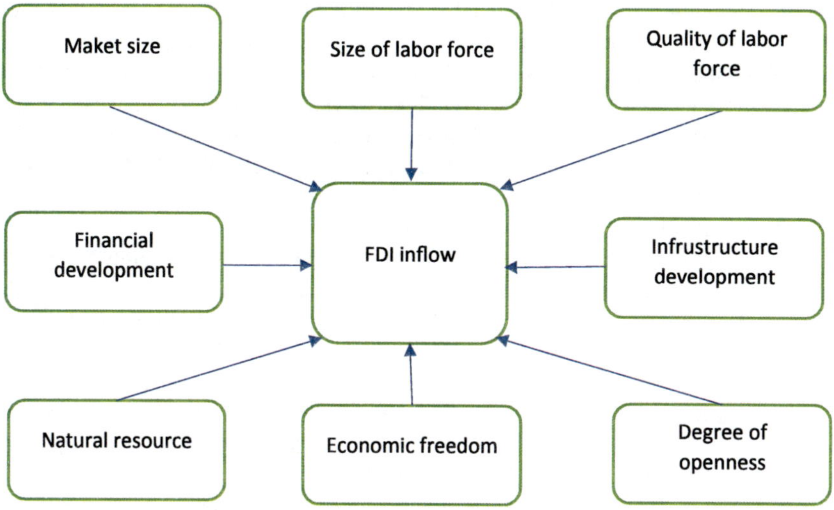

Fig. 1 Conceptual framework on determinants of FDI inflows to developing and least-developed countries (*Source* Author's proposed conceptual framework based on literature reviews)

Quality of labor force (QLF): a good proxy for the measurement of labor quality is education level which is provided by the education index (UNDP Human Development Reports). The education index is calculated as an average of adult's mean years of schooling and children's expected years of schooling.

Infrastructure development (URB): this variable is measured by the rate of urbanization where the rate of urbanization is the percentage of total population who are living in urban area. Data for the rate of urbanization is taken from World Development Indicators (World Bank).

Financial development (FIN): this variable measures the level of a country's financial development. The Financial Development Index which is developed by the International Monetary Fund is used as it captures the complex multidimensional nature of financial development. The index has the maximum score of 1.0, and a higher score means a higher degree of financial development.

Natural resource (NAR): this variable is measured as total natural resources rents in percentage of GDP. Data for total natural resources rents is taken from World Development Indicators (World Bank).

Economic freedom (EFR): this variable measures the extent to which government provides a business-friendly environment conducive for the development of private business sector. Economic freedom is measured by the Index of Economic Freedom published by the Heritage Foundation. The maximum score for the index is 100, and a higher score means a higher level of economic freedom.

Degree of openness: The KOF Globalization Index (KGI) is used to measure the country's degree of openness. Globalization is defined as "a process that erodes national boundaries, integrates national economies, cultures, technologies, and governance, and produces complex relations of mutual interdependence" (Gygli et al., 2018). The index comprises of three dimensions which are economic globalization (EGI), social globalization (SGI), and political globalization (PGI). The score for each sub-index ranges from 0 to 100 with a higher score means the increase in the country's level of openness.

Since data for all countries is collected in the same time period, a strongly balanced panel data is provided. Firstly, the Hausman test is conducted which shows that fixed effect model is more appropriate than random effect model (chi2(19) = 68.51, Prob > chi2 = 0.000). Next, various diagnostics tests are followed. Modified Wald test for groupwise heteroskedasticity in fixed effect regression model shows there exists heteroskedasticity (chi2(103) = 2.2e + 05, Prob > chi2 = 0.000). The Wooldridge test for autocorrelation in panel data confirms serial correlations (F(1, 100) = 17.40, Prob > F = 0.000). With the presence of serial correlations and heteroskedasticity problems in data, the Driscoll and Kraay standard errors estimation is used as suggested by Torres-Reyna (2007). In the model, dummy variables are included to distinguish between developing and least-developed countries.

4 Results

The regression results are reported in Table 1 which are used to summarize the effect of FDI determinants in developing and least developed countries in Table 2.

As can be seen from Table 2, with a positive value ($\beta = 0.15$), market size fosters the inflows of FDI in both groups of developing and least developed countries. This is an evidence of market-seeking FDI where foreign firms are attracted to countries experiencing high economic growth. High economic growth results in larger market size, a higher demand for firms' products, and thus an expansion of business prospect for foreign firms in the host countries.

There is a positive relation between quality of labor force and FDI inflows ($\beta =$ 7.86) in the developing country group. Better quality of workers means higher labor productivity. One of the advantages for foreign firms is their access to superior technologies but the implementation of these superior technologies in the host country is subject to the ability of local workers to handle it. More educated and skilled local workers can be easier to get trained to work with advanced technologies. This is not only help foreign firms to gain competitive advantage over domestic firms but also enable them to increase their labor productivity leading way to earn higher profits. However, unlike developing countries, quality of labor force shows to have a negative effect on inward FDI ($\beta = -15.96$) in the least developed country group. This may suggest that quality of the workforce is not FDI driven factor but instead cheap labor costs attract FDI. Since the level of human capital in least developed countries is rather low, foreign firms are not ready to implement advanced technologies and may prefer to use labor-intensive technologies which enable them to exploit cheap labor costs. The increase in workers' quality may be associated with the requirement for higher wages and reduce the possibility of exploiting cheap labor costs for foreign firms, and thus discourage the inflows of FDI.

Infrastructure development stimulates the inflows of FDI in developing and least developed countries ($\beta = 0.06$). The availability and quality of infrastructure strongly influence business and production operations. For business activities to operate effectively and efficiently, there is the need for good-quality highways and railroads, street lighting, reliable telecommunication, stable power, and water supply. Better infrastructure would reduce the costs of production for firms, such as transportation costs or wasting time costs incurred because of production closedown due to abrupt failure of power or water supply.

Financial development shows to have a positive effect on FDI inflows in all countries ($\beta = 2.67$). Higher financial development in the host country enables foreign firms to gain more access to external funds allowing for the expansion of their activities in the host country. Moreover, when foreign firms operate in the host countries they may establish forward and backward linkages with local firms. Local firms in downstream industries can buy inputs produced by foreign firms and local firms in upstream industries can supply their output to be used as inputs by foreign firms. With higher financial development, local firms are easier to get external funds which boost up their business activities and in turns stimulates foreign firms' business.

Table 1 Determinants of FDI inflows to developing and least developed countries

Explanatory variables	Coefficient	Standard error
GDP: Market size	0.145	0.065**
DUMMY–Least developed countries	0.049	0.054
POP: Size of labor force	0.269	0.325
DUMMY–Least developed countries	– 0.491	0.194**
QLF: Quality of labor force	7.858	2.865**
DUMMY–Least developed countries	– 23.817	3.686***
URB: Infrastructure development	0.056	0.009***
DUMMY–Least developed countries	– 0.005	0.030
FIN: Financial development	2.673	1.070**
DUMMY–Least developed countries	22.649	14.215
NAR: Natural resource	– 0.101	0.017***
DUMMY–Least developed countries	0.125	0.070*
EFR: Economic freedom	0.069	0.025**
DUMMY–Least developed countries	– 0.070	0.055
EGI: Economic globalization	0.125	0.021***
DUMMY–Least developed countries	0.148	0.060**
SGI: social globalization	– 0.174	0.044***
DUMMY–Least developed countries	0.121	0.032***
PGI: Political globalization	– 0.164	0.034***
DUMMY–Least developed countries	0.126	0.038***
CONSTANT	– 1.370	3.771
Year dummies	Yes	
Number of observations	1933	

Dependent variable FDI inflows
Note ***$p < 0.01$, **$p < 0.05$, *$p < 0.1$
Source Author's estimation (see Appendix)

Table 2 Effect of FDI determinants in developing and least developed countries

FDI determinants	Developing countries	Least developed countries
Market size	0.145	0.145
Size of labor force	None	– 0.491
Quality of labor force	7.858	– 15.959
Infrastructure development	0.056	0.056
Financial development	2.673	2.673
Natural resource	– 0.101	0.024
Economic freedom	0.069	0.069
Economic globalization	0.125	0.273
Social globalization	– 0.174	– 0.053
Political globalization	– 0.164	– 0.038

Source Author's calculation from estimation result

The effect on FDI inflows of natural resource is positive in least developed countries ($\beta = 0.02$) and negative in developing countries ($\beta = -0.1$). This is evidence that foreign firms are motivated to undertake investment in least-developed countries for exploitation of natural resources. Countries which are endowed with richer natural resources can attract more foreign investment. However, it would not be the case with developing countries. As the level of development increases, countries are more concerned with the protection of natural resources. In the developing country group, countries with less natural resources can attract more FDI since they can offer other competitive advantages such as good levels of infrastructure and high skilled workforce that efficiency-seeking FDI is looking for.

In all countries, economic freedom plays as FDI stimulus factor ($\beta = 0.07$). Higher economic freedom results in an improvement in the business environment in terms of higher government integrity and lower prevalence of corruption, stronger legal system for safer investment climate, less regulations to free up business productivity and profitability, and more freedom for individuals and businesses in the labor market. Such favorable business environment would induce more foreign investment flowing to the country.

Economic globalization has a positive effect on FDI inflows, though the magnitude of the effect is different with least developed countries having a higher size effect ($\beta = 0.27$) compared to developing countries ($\beta = 0.13$). Economic globalization frees up international trade and investment which allows larger flows of goods and capital among countries. In developing countries, efficiency-seeking FDI is attracted to countries due to their good levels of infrastructure and high skilled workforce.

In their production chain, multinational firms are able to allocate different stages of production in different developing countries in order to exploit the country comparative advantages. Since different stages of production are allocated in different developing countries, output produced in one country can become input of the production process in another country. In each country, foreign firms need to import materials from overseas and export products abroad. This requires them to involve in lots of import and export activities across nations. For this reason, FDI favors countries with a high level of openness regarding to international trade. Free trade through the reduction of trade barriers will facilitate the movement of goods including inputs and output. In least developed countries, foreign investment may favor exploitation of natural resources or primary industries that require low level of technologies and less skilled labor. The exportation of natural resources may depend on the country's degree of trade freedom. Reductions in trade restrictions would make it easier for foreign firms to export their products and contribute significantly to the expansion of the country' exports.

The relationship between social globalization and FDI inflows is found to be negative (β =−0.17 for developing countries and β =−0.05 for least developed countries). Social globalization realizes its effect through the spread of ideas, information flows, and movement of people. Since deeper social globalization increases the rate of technology transfer, knowledge diffusion, and ideas exchange via personal contact and cultural proximity, domestic firms can gain more access to the world's stock of technologies. As a result, foreign firms may lose their advantages in superior technologies over domestic firms. Besides, there are types of products need to be consumed at the place where they are produced such as education or healthcare services. Domestic residents face with the choice between foreign services provided at home by foreign firms or abroad. Social globalization frees up the movement of people across borders. As more domestic residents are able to go abroad to seek for foreign education and healthcare, foreign investors have less incentive to build up their physical presence in the host country. The magnitude of the FDI effect of social globalization may depend on the country's absorption capability. Since developing countries are more capable in absorbing advanced technologies than least-developed countries, social globalization has a larger negative effect on FDI inflows in developing countries than in least-developed countries.

Political globalization plays as an impediment to inward FDI in all countries (β =−0.16 for developing countries and β =−0.04 for least developed countries). One possible explanation for the negative relation between political globalization and FDI inflows is that political globalization is characterized by a diffusion of cooperative government policies. With stronger political cooperation, a country would have more access to international aid and financial support. However, easier access to international financial borrowing causes the accumulation of the country's external debts. The overhang of higher external debts raises the country's risk which discourages foreign investors from undertaking investment in the country.

5 Conclusion

Good understanding of the determinants of FDI is important for it would help governments to provide effective measures for FDI attraction. Toward this end, the empirical study is conducted at the group of country level including developing and least developed countries. In all countries, economic growth plays as an FDI stimulus factor, and this provides a strong evidence of market-seeking FDI. High economic growth results in larger market size and rising demand which stimulate the inflow of FDI. Natural resources-seeking FDI is found in least developed countries where legal exploitation of natural resources lures foreign investors. Efficiency-seeking FDI is more evident in developing countries where good levels of infrastructure and quality of the labor force are FDI-driven factors. In all countries, financial development makes it easier for business firms to acquire external funds. This helps foreign firms to directly increase their business opportunities and also indirectly through the business expansion of local firms with whom foreign firms establish a business relationship. Foreign investors are concerned with the business environment in the host country. More business-friendly environment conducive for the development of the private business sector attracts more foreign investment. With respect to the country's degree of openness, while economic globalization has a positive effect on FDI inflows, social and political globalization show to be a drag in the inflow of FDI. Economic globalization frees up international trade which facilitates the movement of goods including inputs and output across countries. Multinational firms are more able to allocate different stages of production in different developing countries in order to exploit the country comparative advantages. Besides, less trade restrictions would make it easier for foreign firms to export natural resources from least developed countries. In contrast, higher social globalization increases the rate of technology transfer and knowledge diffusion via personal contact and cultural proximity which allows domestic firms to gain more access to the world's stock of technologies and thus weakens the advantage of foreign firms in superior technologies over domestic firms. Finally, stronger political cooperation would enable a country to have more access to the international financial source. However, the likely of higher external debts raises the country's risk and thus reduces the investment incentive for foreign investors.

Appendix: Estimation Result

Regression with Driscoll-Kraay standard errors	Number of obs	=	1933
Method: Pooled OLS	Number of groups	=	101
Group variable (i): Country	F(39, 19)	=	862,995.44

(continued)

(continued)

Maximum lag: 2				Prob > F	=	0.0000
				R-squared	=	0.2227
				Root MSE	=	6.0085
Drisc/Kraay						
FDI	Coef	Std. Err	*t*	*P* > *t*	[95% Conf. Interval]	
GDP	0.1451905	0.0650064	2.23	0.038	0.0091306	0.2812504
NAR	− 0.101055	0.0170187	− 5.94	0.000	− 0.1366755	− 0.0654345
POP	0.2691776	0.3253338	0.83	0.418	− 0.4117538	0.950109
QLF	7.858766	2.865029	2.74	0.013	1.862192	13.85534
URB	0.0568598	0.0091672	6.20	0.000	0.0376727	0.0760469
FIN	2.673074	1.070286	2.50	0.022	0.4329387	4.913209
EFR	0.0693888	0.0256548	2.70	0.014	0.0156926	0.123085
EGI	0.1250753	0.0211133	5.92	0.000	0.0808847	0.1692659
SGI	− 0.1746615	0.0443933	− 3.93	0.001	− 0.2675776	− 0.0817453
PGI	− 0.1644122	0.0342494	− 4.80	0.000	− 0.236097	− 0.0927274
DGDP	0.0497695	0.0547514	0.91	0.375	− 0.0648264	0.1643654
DNAR	0.1249063	0.0706231	1.77	0.093	− 0.0229096	0.2727222
DPOP	− 0.4913292	0.1945079	− 2.53	0.021	− 0.8984389	− 0.0842194
DQLF	− 23.8172	3.686166	− 6.46	0.000	− 31.53243	− 16.10196
DURB	− 0.0056559	0.0306146	− 0.18	0.855	− 0.069733	0.0584211
DFIN	22.64924	14.21523	1.59	0.128	− 7.10358	52.40205
DEFR	− 0.0708646	0.0550942	− 1.29	0.214	− 0.1861781	0.0444489
DEGI	0.1487772	0.0600209	2.48	0.023	0.023152	0.2744023
DSGI	0.1213403	0.0324793	3.74	0.001	0.0533604	0.1893202
DPGI	0.1263573	0.0387995	3.26	0.004	0.0451491	0.2075655
Year						
2000	0	(empty)				
2001	− 0.0554523	0.1363833	− 0.41	0.689	− 0.3409059	0.2300012
2002	0.1035832	0.165828	0.62	0.540	− 0.2434988	0.4506652
2003	0.0341557	0.0990158	0.34	0.734	− 0.1730867	0.2413981
2004	0.6022564	0.1377376	4.37	0.000	0.3139682	0.8905446
2005	1.602516	0.1723858	9.30	0.000	1.241708	1.963324
2006	2.398849	0.1874885	12.79	0.000	2.006431	2.791267
2007	2.85799	0.2184015	13.09	0.000	2.40087	3.31511
2008	3.529832	0.2813555	12.55	0.000	2.940949	4.118716
2009	2.848086	0.4178758	6.82	0.000	1.973462	3.72271
2010	3.661122	0.2709578	13.51	0.000	3.094001	4.228243
2011	4.588828	0.2935543	15.63	0.000	3.974412	5.203245

(continued)

(continued)

2012	4.295941	0.3227629	13.31	0.000	3.620391	4.971492
2013	3.621735	0.3459834	10.47	0.000	2.897583	4.345886
2014	3.165573	0.3603703	8.78	0.000	2.411309	3.919837
2015	3.640611	0.4422091	8.23	0.000	2.715056	4.566165
2016	2.497468	0.3983013	6.27	0.000	1.663813	3.331122
2017	3.22089	0.3865793	8.33	0.000	2.41177	4.03001
2018	2.733226	0.3865554	7.07	0.000	1.924157	3.542296
2019	2.391849	0.4000215	5.98	0.000	1.554594	3.229103
_cons	− 1.370504	3.771525	− 0.36	0.720	− 9.264398	6.523389

References

Abbas, A., Moosa, I., & Ramiah, V. (2021). The contribution of human capital to foreign direct investment inflows in developing countries. *Journal of Intellectual Capital, 23*(1), 9–26.

Abbes, S., Mostéfa, B., Seghir, G., & Zakarya, G. (2015). Causal interactions between FDI, and economic growth: Evidence from dynamic panel co-integration. *Procedia Economics and Finance, 23*, 276–290.

Balassa, B. (1966). American direct investments in the common market. *Banca Nazionale Del Lavoro Quarterly Review, 77*, 121–146.

Barua, S., Naym, J. & Hazera-Tun-Nessa. (2017). Economic climate, infrastructure and FDI: Global evidence with new dimensions. *International Journal of Business and Economics, 16*(1), 31–48.

Bellak, C., Leibrecht, M., & Riedl, A. (2008). Labour costs and FDI flows into Central and Eastern European countries: A survey of the literature and empirical evidence. *Structural Change and Economic Dynamics, 19*(1), 17–37.

Bengoa, M., & Sanchez-Robles, B. (2003). Foreign direct investment, economic freedom and growth: New evidence from Latin America. *European Journal of Political Economy, 19*(3), 529–545.

Bilir, L. K., Chor, D., & Manova, K. (2019). Host-country financial development and multinational activity. *European Economic Review, 115*, 192–220.

Cantah, G., Brafu-Insaidoo, G., Wiafe, E., & Adams, A. (2018). FDI and trade policy openness in Sub-Saharan Africa. *Eastern Economic Journal, 44*, 97–116.

Desai, M., Foley, C., & Hines, J. (2004). The costs of shared ownership: Evidence from international joint ventures. *Journal of Financial Economics, 73*(2), 323–374.

Desai, M., Foley, C., & Hines, J. (2006). Capital controls, liberalizations, and foreign direct investment. *The Review of Financial Studies, 19*(4), 1433–1464.

Desbordesa, R., & Wei, S. (2017). The effects of financial development on foreign direct Investment. *Journal of Development Economics, 127*, 153–168.

Donaubauer, J., & Dreger, C. (2018). The end of cheap labor: Are foreign investors leaving China? *Asian Economic Papers, 17*(2), 94–107.

Feinberg, S. & Phillips, G. (2004). Growth, capital market development and competition for resources within MNCs. *NBER Working Paper No. 9252.*

Gabriel, O., Chigozie, A., & Awara, E. (2016). Foreign direct investment in Nigeria: Reassessing the role of market size. *International Journal of Academic Research in Business and Social Sciences, 6*(2), 95–104.

Gwartney, J., Lawson, R. & Hall, J. (2017). Economic freedom of the world: 2016 annual report. Retrieved from https://www.fraserinstitute.org/studies/economic-freedom.

Gygli, S., Haelg, F. & Sturm, J. (2018). The KOF globalization index – revisited. *KOF Working Paper No. 439*.

Ho, C., Amir, K., Nasaruddin, L. & Abidin, N. (2013). Openness, market size and foreign direct investment. In *14th Malaysian Finance Association Conference Penang*, Malaysia, pp. 1–12.

Hornberger, K., Battat, J. & Kusek, P. (2011). Attracting FDI: How much does investment climate matter? *The World Bank Group, Financial and Private Sector Development Vice Presidency, Note No. 327*.

Hossain, M. (2016). Foreign direct investment, economic freedom and economic growth: Evidence from developing countries. *International Journal of Economics and Finance, 8*(11), 200–214.

Imtiaz, S., & Bashir, M. (2017). Economic freedom and foreign direct investment in South Asian countries. *Theoretical and Applied Economics, 24*(2), 277–290.

Ju, J., & Wei, S. (2010). Domestic institutions and the bypass effect of financial globalization. *American Economic Journal: Economic Policy, 2*(4), 173–204.

Karimi, M., Yusop, Z., Hook, L., & Chin, L. (2013). Effect of human capital on foreign direct investment inflows. *Journal of Economic Research, 18*, 1–23.

Khan, R., & Hye, Q. (2014). Foreign direct investment and liberalization policies in Pakistan: An empirical analysis. *Cogent Economics and Finance, 2*(1), 944667. https://doi.org/10.1080/23322039.2014.944667

Kheng, V., Sun, S., & Anwar, S. (2017). Foreign direct investment and human capital in developing countries: A panel data approach. *Economic Change and Restructuring, 50*, 341–365.

Klapper, L., Laeven, L., & Rajan, R. (2006). Entry regulation as a barrier to entrepreneurship. *Journal of Financial Economics, 82*(3), 591–629.

Kok, R., & Ersoy, A. (2009). Analyses of FDI determinants in developing countries. *International Journal of Social Economics, 36*(1/2), 105–123.

Liargovas, P. G., & Skandalis, K. S. (2012). Foreign direct investment and trade openness: The case of developing countries. *Social Indicators Research, 106*(2), 323–331.

Makoni, P. (2018). FDI and trade openness: The case of emerging African economies. *Journal of Accounting and Management, 8*, 141–152.

Manova, K. (2013). Credit constraints, heterogeneous firms, and international trade. *The Review of Economic Studies, 80*(2), 711–744.

Moussaa, M., Çahab, H., & Karagöz, M. (2016). Review of economic freedom impact on FDI: New evidence from fragile and conflict countries. *Procedia Economics and Finance, 38*, 163–173.

Mudiyanselage, R., Mayoshi, M., Epuran, G., & Tescasiu, B. (2021). Causal links between trade openness and foreign direct investment in Romania. *Journal of Risk and Financial Management, 14*, 90. https://doi.org/10.3390/jrfm14030090

Nasir, Z., & Hassan, A. (2011). Economic freedom, exchange rates stability and FDI in South Asia. *The Pakistan Development Review, 50*(4), 423–432.

Neumayer, E., & Soysa, I. (2004). Trade openness, foreign direct investment and child labor. *World Development, 1*(33), 43–63.

Noorbakhsh, F., Paloni, A., & Youssef, A. (2001). Human capital and FDI inflows to developing countries: New empirical evidence. *World Development, 29*(9), 1593–1610.

Petrović-Ranđelović, M., Janković-Milić, V., & Kostadinović, I. (2017). Market size as a determinant of the foreign direct investment inflows in the Western Balkans. *Economics and Organization, 14*(2), 93–104.

Pfeffermann, G. & Madarassy, A. (1992). Trends in private investment in developing countries. *IFC Discussion Paper No. 14*. Washington, DC.: The World Bank.

Quazi, R. (2007). Economic freedom and foreign direct investment in East Asia. *Journal of the Asia Pacific Economy, 12*(3), 329–344.

Rajan, R., & Zingales, L. (1998). Financial dependence and growth. *American Economic Review, 88*(3), 559–586.

Sabir, S., Rafique, A. & Abbas, K. (2019). Institutions and FDI: Evidence from developed and developing countries. *Financial Innovation, 5*(8). https://doi.org/10.1186/s40854-019-0123-7

Scaperlanda, A., & Mauer, L. (1969). The determinants of U.S. direct investment in the EEC. *American Economic Review, 59*, 558–568.

Sooreea, B., Rasool, U., & Sooreea, R. (2020). Does economic freedom matter to foreign direct investment in Sub-Saharan Africa. *International Journal of Economics and Financial Issues, 10*(3), 195–207.

Torres-Reyna, O. (2007). Panel data analysis fixed and random effects using Stata. Retrieved from https://www.princeton.edu/~otorres/Panel101.pdf.

Wadhwa, K., & Reddy, S. (2011). Foreign direct investment into developing Asian countries: The role of market seeking, resource seeking and efficiency seeking factors. *International Journal of Business and Management, 6*(11), 219–226.

Wickramarachchi, V. (2019). Determinants of foreign direct investment (FDI) in developing countries: The case of Sri Lanka. *International Journal of Business and Social Science, 10*, 76–88.

World Economic Situation and Prospects. (2018). Available at https://www.un.org/development/desa/dpad/publication/world-economic-situation-and-prospects-2018/.

Zaman, Q., Donghui, Z., Yasin, G., & Zaman, S. (2018). Trade openness and FDI inflows: A comparative study of Asian countries. *European Online Journal of Natural and Social Sciences, 7*, 386–396.

The Effect of Foreign Direct Investment on Human Well-Being: A Study of the Relationship Between the Carbon Intensity of Human Well-Being and Foreign Direct Investment in Nine Lower Middle-Income Countries

Hai Bac Dang, Thuan Nguyen, Diem Nga Thi Tran, and Oanh Hoa Thi Su

Abstract Economic growth is the primary purpose for human development of all nations in the world; however, it is needed to put into question whether the growth has brought more benefits to human-being or caused costly consequences for the environment and societies. The paper has studied the carbon intensity of human well-being (CIWB), the level of anthropogenic carbon emissions per unit of human well-being, to measure the relationship between economic growth and human well-being in terms of sustainability. Approach from the theory of ecologically unequal exchange, the study applied Prais–Winsten technique to analyze the function of CIWB and economic variables in nine lower middle-income countries in the period of 2000–2018. We found that economic activities have significantly enhanced human well-being, which hasn't supported the theory. The variables of foreign direct investment (FDI) over time and gross domestic product (GDP) showed negative effects on CIWB, which implies the more value of FDI and GDP, the less CIWB value. However, these countries should be cautioned on strategies for sustainable development, the FDI effect on decreasing CIWB over time is low and unstable. Moreover, the study revealed that energy consumption (EPC) positively impacts the value of CIWB.

Keywords Ecologically unequal exchange · Carbon intensity well-being (CIWB) · Foreign direct investment (FDI) · Lower middle-income countries

1 Introduction

The primary purpose of economics is to enhance human well-being (Dalziel et al., 2018). Though economic growth is seen as the main vehicle for improvements in

---

H. B. Dang · D. N. T. Tran · O. H. T. Su (✉)
University of Natural Resources and Environment, Ho Chi Minh City, Vietnam
e-mail: stohoa@hcmunre.edu.vn

T. Nguyen
Open University, Ho Chi Minh City, Vietnam

A. T. Nguyen et al. (eds.), *Contemporary Economic Issues in Asian Countries: Proceeding of CEIAC 2022, Volume 1*,
https://doi.org/10.1007/978-981-19-9669-6_16

well-being like higher incomes or better living conditions, it's apparent that economic growth for its own sake can result in environmental degradation or widen social inequalities (Bunker, 1984, 1985, 2019; Jorgenson, 2006; Martínez-Alier, 2002). Therefore, the concept of well-being has been transformed from welfare to human well-being at personal and collective levels (McGregor & Pouw, 2016). In the context of sustainable development, numerous measures of well-being have been developed to reflect a broader concept in consideration of the environment (Dietz et al., 2009).

Carbon intensity of well-being (CIWB), which is defined as anthropogenic carbon emissions per life expectancy, has been created in an effort of considering the trade-off between economic growth and the environment in a measure (Jorgenson, 2014). The CIWB is a new indicator of human well-being, which is relative to the harm we do to the environment while developing. Therefore, the success of economic growth is to achieve high levels of human well-being while placing relatively little stress on the environment, which means a low value of CIWB.

Jorgenson (2014) in the cross-national comparative research on economic development and CIWB across regions over time has shown surprising findings that just only the development in Africa has brought a decrease in CIWB over time, while it went into reverse for the remaining regions. Givens (2015) gave the same results that the impact of development on human well-being in terms of the environment from the high-income countries is positive and contradicted in low-income countries. In the case of lower middle-income countries (lower MICs), we found it interesting because they are facing a dilemma of rapid economic growth and have emitted a high amount of CO_2, hence, without appropriate sustainable development strategies, they may fall into a high value of CIWB which could damage the success of a development. However, in the body of literature on CIWB, there are no thorough studies on this group of countries.

2 Literature Review

The term middle-income countries (MICs) are defined by grouping nations according to their income. Lower MIC is a group of countries which just be over the threshold of low income and with a GNI per capita in between US$1,036 and 4,045 range. As to be one of the key sources of economic growth, the lower MICs have attracted FDI to boost their economies and enhance well-being (Forte & Moura, 2013; Sharma & Gani, 2004). However, the inward stocks of foreign investment often lead them toward export-oriented production and in many cases that has caused costly environmental externalities (Giljum & Eisenmenger, 2004; Grimes & Kentor, 2003; Jorgenson, 2005). By adopting a multidisciplinary approach, the theory of ecologically unequal exchange was proposed to describe the global economic structures and relationships that shape the unequal distribution of natural resources and environmental degradation (Bunker, 1984; Givens et al., 2019; Jorgenson, 2006).

The theory of ecologically unequal exchange suggests the existence of structural relationships that facilitate unequal material flows in which wealthier and more powerful developed countries have greater power to access natural resources and emit pollution in the lesser-developed countries (Bunker, 2019; Ciccantell, 2019; Givens et al., 2019; Jorgenson, 2006). In other words, the lesser-developed countries are structured in a weaker position to fulfill the demands for raw materials and a sink for waste from the more-developed countries within the world economic system of extraction, production, consumption, and disposal (Jorgenson, 2006). In the case study of middle-income countries, Rooij and McAllister (2014) explained that their economic interests greatly overwhelm environmental interests, resulting in undermining various forms of legal, political, and social action against environmental degradation. Therefore, a combination of weak enforcement and pervasive noncompliance tends to create a vicious circle that works against pollution mitigation and control.

The origins of ecologically unequal exchange theory can be traced to the studies of Bunker (1984, 1985, 2019). He introduced the idea of the theory by building on earlier structural analyses of unequal economic exchange. He explained the under-development of lesser-developed countries that they are preferring short-term benefits from the extractive economy. Emmanuel (1972) inherited from Marx's labor theory of value, then he added wage differentials to the basis of unequal exchange. In addition, Hornborg (1998) stated that the theory is influenced not only by labor as Emmanuel (1972) but also by natural factors. He also emphasized that while raw materials have greater productive potential, their price is lower than processed goods, a distorted valuation that drives global ecological degradation and inequalities.

Over the last two decades, the theory of ecologically unequal exchange has marked an advance in development when related studies have adopted quantitative analytic techniques to test world system theory and expanded to various topics (Ciccantell, 2019). Most of the studies have confirmed the theory by studying the relationship between trade and environmental indicator as deforestation (Austin, 2010; Jorgenson, 2006). Moreover, Givens (2018) also revealed the supported results of the theory that uneven global trade affects the CIWB of nations through a longitudinal study. Therefore, in line with these studies, the paper has chosen the FDI for the replacement of trade to investigate the relationship with CIWB to broaden topics for the theory.

For our case study of nine lower middle-income countries, we do an effort to test the hypothesis that whether the existence of ecologically unequal exchange in these countries when they promoted and attracted flow stock of FDI for development. The relationship between CIWB and economic variables such as GDP, FDI, and EPC will be the answer for the efficiency of economic performance activities for human well-being in consideration of global climate change. Moreover, the impact of FDI on CIWB through time will give us a whole picture of FDI effect on well-being during the period of studies in the lower middle-income nations. If unequal exits, it will make the increase in the value of CIWB and be gradually erasing the success of development by the high cost in environmental externalities.

3 Methodologies

3.1 Data Collection

We collected the cross-nations data in the period of 2000–2018 for nine lower middle-income countries including: Algeria, Bangladesh, Egypt, India, Morocco, Pakistan, Philippines, Uzbekistan, and Vietnam. Though they are diverse by size and population, they have many common characteristics which are aligned with their level of income as belonging to a rapid increase of GDP, population, life expectancy, and emerging serious environmental issues. In the progress of development, these countries have been enjoying increasing of economic growth by transforming their economies from primary agriculture to industry. As a result, they've got a quick increase in production-based CO_2 emissions, which is the consequence of the development of the consumption of fossil fuels (Hachaichi & Baouni, 2021). Our study has included 171 observations in the panel data set for testing the relationship between CIWB and economic variables as FDI, GDP, and EPC. In which, CIWB is converted by a division of production-based CO_2 per capita (CO_2) and average life expectancy (LE). The data was collected from many sources of information such as World bank, UNCTAD, Country economy, and Our world in data to ensure the complete data set in the study period.

For the calculation of CIWB, the coefficient of variation of CO_2 (the numerator) is much larger than that of LE (denominator); therefore, it is necessary to modify the value CIWB by adding a correction factor (CF) to the CO_2 numerator as the below equation (Dietz et al., 2012; Jorgenson, 2014):

$$\mathbf{CIWB = ((CO_2 + CF)/LE)*100} \quad (1)$$

where CF is inherited by studies of Dietz et al. (2012), using subscripts CO_2 and LE to indicate CO_2 emissions and life expectancy, respectively, the correction factor is

$$\mathbf{CF = ([S_{CO2} * M_{LE}]/S_{LE}) - M_{CO2}}$$

where S is standard deviation and M is mean value.

3.2 Methods

In order to review the relationship between human well-being related to environment and the economics variables as FDI, GDP, and EPC, we developed a function as below:

$$\mathbf{CIWB_{it} = f(FDI_{it}, GDP_{it}, EPC_{it})} \quad (2)$$

According to Jorgenson and Clark (2012), the effect of economic performance activity on the environment may change over time and in a unique way for each nation. Therefore, to analyze the effects of economic development on CIWB over time, we've estimated by Prais–Winsten models with panel-corrected standard errors. This is an appropriate method because it has enabled solving the problems of serially (i.e., temporally) correlated, spatially (i.e., contemporaneously) correlated, and characterized by heteroscedasticity in comparative time series cross-sectional data (Beck & Katz, 1995). Moreover, we included country-specific and year-specific intercepts, making the model equivalent to a two-way fixed effects model to examine country-specific and year-specific effects (Givens, 2015; Jorgenson, 2014). As with a fixed effects model, this technique estimates effects within countries over time and controls for variation between countries. This model construction is especially well suited to hypothesis testing, as it controls out all period-specific and country-specific variations. The independent variables for economic development (FDI, GDP, and ECP) and FDI interaction variables over time have been analyzed for the impact of the variables on the CIWB, and changes in the effect of FDI on the CIWB refer to the impact value of the reference year (2000) over the period of study 2000–2018.

The estimated model of study is

$$\begin{aligned}\mathbf{CIWB_{it}} &= \boldsymbol{\theta_1}\mathbf{FDI_{it}} + \boldsymbol{\theta_2}\mathbf{FDI_{it}} * \mathbf{year2001} + \ldots. \\ &+ \boldsymbol{\theta_{19}}\mathbf{FDI_{it}} * \mathbf{year2018} + \boldsymbol{\theta_{19}}\mathbf{year2001} + \ldots \\ &+ \boldsymbol{\theta_{37}}\mathbf{year2018} + \boldsymbol{\theta_{38}}\mathbf{GDP_{it}} + \boldsymbol{\theta_{39}}\mathbf{ECP_{it}} + \mathbf{u}_i + \mathbf{e_{it}}\end{aligned}$$

where

$\mathrm{CIWB_{it}}$	a metric of human well-being in the consideration of the environment, which is measured as a ratio of production-based CO_2 emission per capita to life expectancy.
θ_1	the estimated parameter of FDI, reflected the impact of FDI on CIWB in the reference year, 2000.
$(\theta_2;\ldots;\theta_{19})$	the estimated parameter of FDI, reflected the impact of FDI on CIWB over time from 2001 to 2018 and refer to the value of the reference year, 2000.
$(\theta_{19}\mathrm{year2001} + \ldots + \theta_{37}\mathrm{year2018})$	the year-specific intercepts.
θ_{38}	the estimated parameter of GDP, reflected the overall impact of GDP on CIWB.
θ_{39}	the estimated parameter of EPD, reflected the overall impact of EPC on CIWB.
u_{i}	the country-specific intercepts.
e_{it}	the error term for each country for each time point.

The estimated parameter of FDI (θ_1) is the unit change in the dependent variable ($\mathrm{CIWB_{it}}$) in the reference year (2000). For the other time points, the effect, if significant, is the sum of the coefficient for FDI (θ_1) and the coefficient for the interaction term; if the interaction term is not significant, the coefficient is the same as the reference year. In the case of a non-significant effect for the reference year, but a later significant interaction with time, the reference year is interpreted as being not significantly different from zero (Allison, 2009; Givens, 2017).

4 Results

4.1 Descriptive Statistics

Descriptive statistics for all variables included can be found in Table 1 and Figs. 1, 2, 3, and 4. According to Kline (2015), variables with skewness > ± 3 and kurtosis > ± 10, the data has problems with the normal distribution. The distribution of skewness and kurtosis for all our variables were within the acceptable ranges, we do not need to convert the data by using the natural logarithm.

The summary statistics of the model's independent variables is presented in Table 1. On average, nine nations in lower middle-income group which have 1.9614 USD gross domestic product per capita receive 38.2254 USD annual foreign direct investment. These nations use an average 7.5651 thousand Kwh energy consumption per capita and reach 34.0878 in value of CIWB.

As for the $CIWB_{it}$ ratio, our data shows that the coefficient of variation of CO_2 is 0.64 and ranges from 0.2 to 5.04. Meanwhile, the coefficient of variation of life expectancy (LE) is 0.049, and the range is 62.5–76.69. The coefficient of variation of CO_2 (the numerator) is much larger than that of the lifetime (denominator), because the coefficient of variation of the numerator affects the variation of this ratio. Therefore, it is necessary to resolve this problem before considering the CIWB ratio as a dependent variable. Following the studies of Dietz et al. (2012) and Jorgenson (2014), we added a constant to the numerator. The constant value calculated as Eq. (1) in

Table 1 Descriptive statistics of all the independent variables in the model

Variables	GDP US dollar at current and constant (2010) price	ECP (thousand Kwh per capita)	FDI US dollar at current and constant (2010) price
Mean	1.9614	7.5651	38.2254
Std. Dev.	1.1132	5.4536	36.1727
Min	0.5249	1.0652	– 14.7132
Max	4.8301	24.0247	162.2256
Skewness	0.9799	1.2178	1.157
Kurtosis	3.2083	3.7254	3.7308

Source Authors' computation

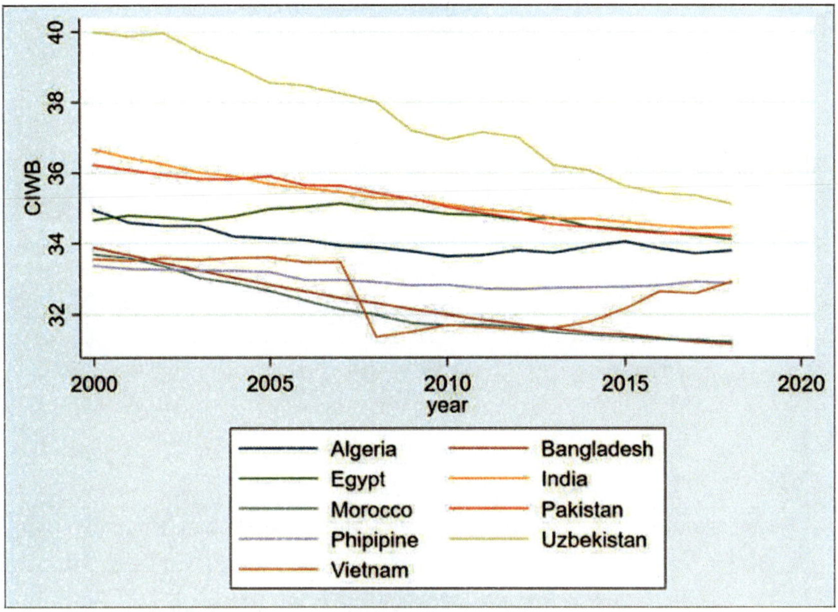

Fig. 1 CIWB for nine nations during 2000–2018 *(Source* Author's own*)*

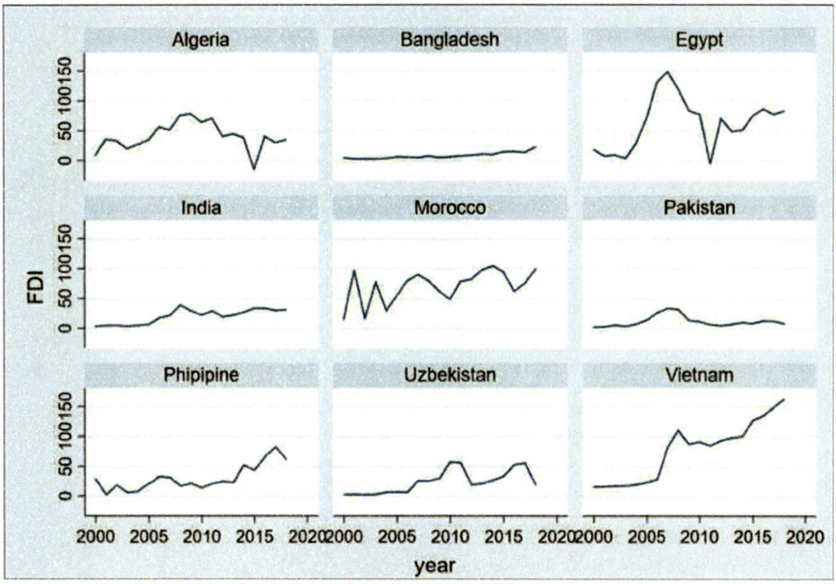

Fig. 2 FDI per capita measures for nine nations 2000–2018 *(Source* Author's own*)*

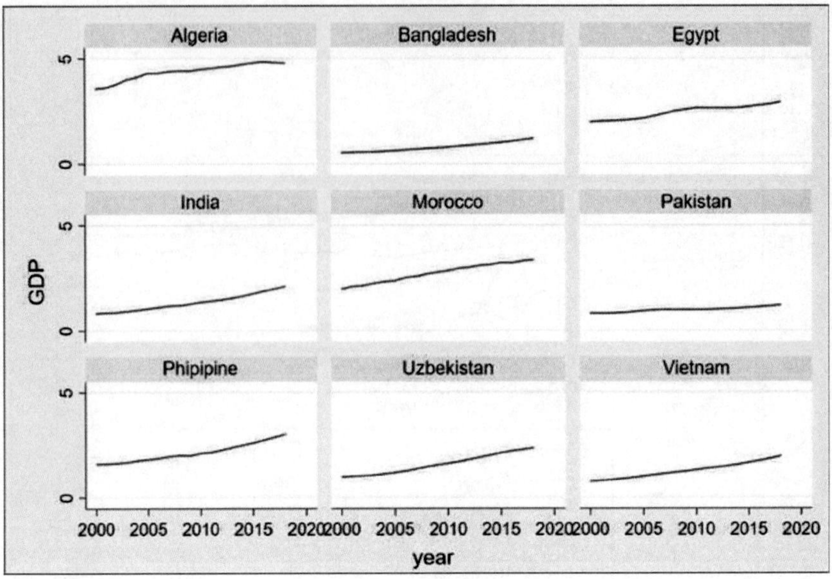

Fig. 3 GDP per capita measures for nine nations 2000–2018 (Source Author's own)

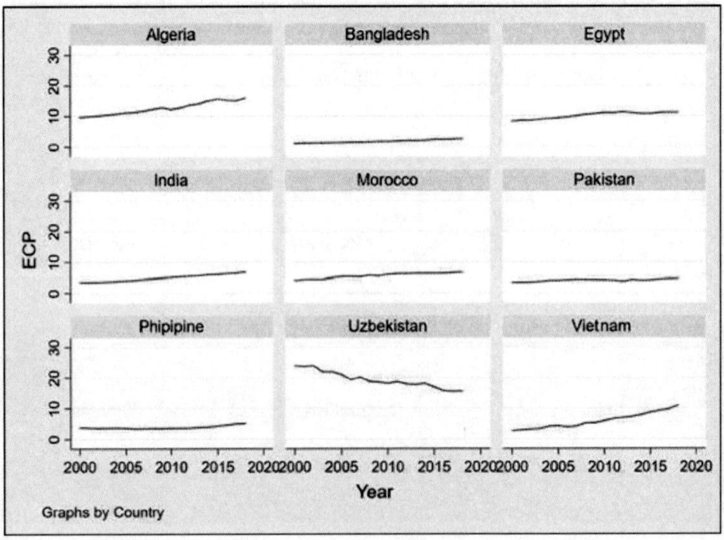

Fig. 4 ECP per capita measures for nine nations 2000–2018 (*Source* Author's own)

Sect. 3.1 from our sample data was 21.98 and added to the value of CO_2. The adjusted value of CIWB is in Fig. 1.

Figure 1 shows the overall downtrend of CIWB for nine lower middle-income countries in the study period with a value from 32 to 40. Vietnam is an exception with a slightly down and gradual uptrend, though its value in a lower level (32–36) but the plunge in 2008, it continuously increases until 2018. Uzbekistan got attention by a successful decrease from a higher to lower value level of CIWB as well as India and Pakistan. The good examples of a better CIWB value are Morocco and Bangladesh with a gradual decrease, while other countries noted a slight fall in the study period.

Besides that, Fig. 2 presented the mixed direction in changing FDI for nine nations from 2000 to 2018. Bangladesh, India, and Pakistan experienced a slightly upward trend in this period. Particularly, Bangladesh is the one attracting FDI which is the smallest in lower middle-income group. In contrast, the FDI value of Algeria, Philippines, and Uzbekistan which grew markedly through time. However, from 2015 to 2018, these nations had a fall in appealing to FDI. The nations which rapidly increased FDI are Egypt, Morocco, and Vietnam. Also, FDI of Egypt dramatically went down in 2011, but it went up in 2012 and continued until 2018, whereas the FDI of Morocco experienced fluctuation in the last 18 years. For Vietnam, FDI value remained stable at the beginning of this stage, in 2007, it surged and then following the upward trend until 2018.

The results from the regression in Table 2 show that the effect of FDI in the reference year 2000 is negative (−0.0223), and the interaction of FDI and the yearly dummy variables (2001) is positive and statistically significant. In the remaining years, the interactions are positive but not statistically significant, which means keeping the effect value in the reference year, 2000 (because there is an insignificant change in the value of interaction variables referring to the reference year). Thus, the relationship between CIWB and FDI in these countries is negative. However, it appeared to move upward in 2001, then went down in 2002 and became flat in the research period, which is demonstrated by Fig. 5.

International scholars argued that the effect of economic performance activities on the environment may change over time (Dietz et al., 2012; Grossman & Krueger, 1995; Jorgenson & Clark, 2012; Knight & Rosa, 2011). The results of our study show that for those lower middle-income countries, the effect of FDI on CIWB is negative in reference year (2000) but all the interaction coefficients got positive, resulting in a lower impact of FDI on a decrease of CIWB. Though just only the interaction coefficient in 2001 got positive and statistic significant, it also shows unstable sustainable strategies of attracting FDI in those countries. These countries should be caution on selecting the flow stock of FID in to enhance human well-being with less stress on environment.

Figure 5 shows the elasticity coefficients for FDI of lower middle-income countries over 18 years from reference year, 2000. The plotted line in Fig. 5 shows the effect of FDI on CIWB, which reached a peak in 2001 by an increase of 0.0218 in changing the effect, afterward it becomes flat as same as the reference year (2000) at 0.0223.

Table 2 Effects of FDI on CIWB over time from 2000 to 2018

Independent variables	Coefficients	Panel corrected standards errors	$P > \|z\|$
FDI	**– 0.0223***[*]	0.0127	0.079
FDI_2001	**0.0218***[*]	0.0131	0.097
FDI_2002	0.0195	0.0137	0.155
FDI_2003	0.0212	0.0137	0.107
FDI_2004	0.0166	0.0147	0.257
FDI_2005	0.0195	0.0135	0.148
FDI_2006	0.0191	0.0130	0.143
FDI_2007	0.0194	0.0129	0.132
FDI_2008	0.0112	0.0129	0.384
FDI_2009	0.0138	0.0130	0.289
FDI_2010	0.0170	0.0130	0.190
FDI_2011	0.0186	0.0130	0.151
FDI_2012	0.0163	0.0132	0.216
FDI_2013	0.0175	0.0133	0.191
FDI_2014	0.0173	0.0133	0.193
FDI_2015	0.0192	0.0130	0.138
FDI_2016	0.0203	0.0132	0.124
FDI_2017	0.0188	0.0133	0.156
FDI_2018	0.0188	0.0133	0.158
GDP	**– 0.6217***[***]	0.0627	0.000
ECP	**0.2657***[***]	0.0146	0.000
Constant	**34.590***[***]	0.1810	0.000
R^2	0.9923		
Rho	0.7937		

Note * significant at 10%; ** significant at 5%; *** significant at 1%
Source Authors' computation

Besides, the coefficient of GDP is also negative and statistically significant, which reflects the positive effect of development on CIWB over time. With a unit increase in GDP per capita, it will lead to a 0.6217 unit decrease in CIWB, the result is in line with the findings of Jorgenson and Clark (2012); Jorgenson and Givens (2015) for non-high income countries. Overall, in the case of these lower middle-income countries, the benefit of economic development for human well-beings is over the cost in terms of environment.

Moreover, the study shows the same positive effect of energy consumption (EPC) making an increase of CIWB (Nguyen & Dang, 2021), which implies the current growth pathway based on fossil consumption will make a cost for these countries and lead them to be less sustainable. In other words, it indicates that energy consumption

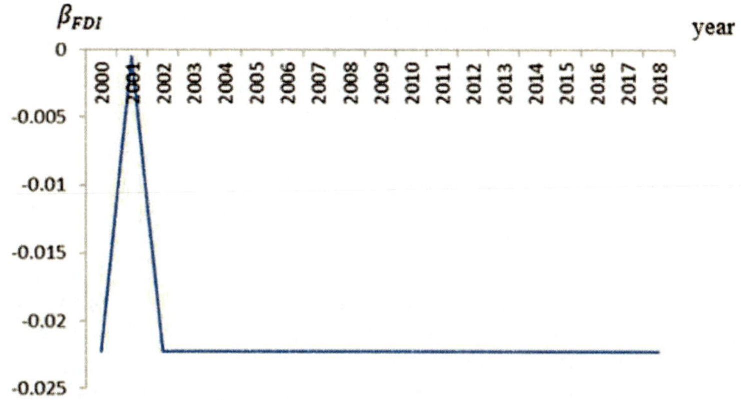

Fig. 5 Coefficients of FDI on CIWB through 2000 to 2018 (*Source* Author's own, *Note* The coefficient of FDI in 2002–2018 is as same as the one in 2000 because rejected hypothesis of the coefficient is different to the reference year in 2000 (sig > 0.05))

has been creating more stress on the environment due to an enormous amount of CO_2 emissions in the air.

In summary, though there is a sign of unstable sustainable strategies to attract FDI flows for nine lower middle-income countries, the inward stocks of foreign investment have still brought a positive effect on both economic development and human well-being. However, if their economic growth has been mainly based on natural resource extraction and the consumption of fossils, the reverse effects may exist. They will be the victims of an ecologically unequal exchange system; therefore, environmental protection and sustainable development strategies should be more cautioned to protect and attain economic achievements.

5 Conclusions

Adopted the thinking that the purpose of economic growth is not only increasing welfare but also human well-being, the study has chosen the case study of nine lower middle-income countries to test whether they have developed sustainably. They are special cases thanks to just getting out from low-income group to lower middle level, but they are facing a dilemma of rapid economic growth and starting a significant pressure on the environment. They are also taking the risk of ecologically unequal exchange while being influenced by more powerful nations on economic development. Without a special caution on environment protection and sustainable development strategies, they are likely to fall into a high value of CIWB, which could lead to environmental degradation and damage the economic achievements.

The results of the study show that these countries still got benefits in term of human well-being on the cooperation with higher-income nations. The negative effects of

FDI on CIWB over time imply that the more inward stocks of foreign investment, the more value of CIWB decreases. In addition, the GDP variable has also given the same results which are in line with studies on CIWB in Africa and in non-high income countries (Givens, 2017; Jorgenson & Givens, 2015). However, the variable on energy consumption is in reverse effect, which has caused stress on the environment by a positive effect on CIWB. This can be explained by the consumption of non-renewable energy, resulting in a huge rate of CO_2 emissions within this group of countries.

In order to decouple the economic growth and environment, maintaining a low value of CIWB is necessary. Givens (2017) presented the findings of associations between world society, world polity, and CIWB value. The more level of engagement in International Government Organizations such as World bank, UN, and WTO, the more decrease in CIWB value the nations can get. Though the study has just shown significant empirical evidence of decreasing CIWB value among less developed nations in the early 1990s and more developed nations in the late of 2000s, we still have an idea that active involvement in such organizations will help the nations develop sustainability.

References

Austin, K. (2010). The "hamburger connection" as ecologically unequal exchange: A cross-national investigation of beef exports and deforestation in less-developed countries. *Rural Sociology, 75*(2), 270–299. https://doi.org/10.1111/j.1549-0831.2010.00017.x

Beck, N., & Katz, J. N. (1995). What to do (and not to do) with time-series cross-section data. *The American Political Science Review, 89*(3), 634–647. https://doi.org/10.2307/2082979

Bunker, S. G. (1984). Modes of extraction, unequal exchange, and the progressive underdevelopment of an extreme periphery: The Brazilian Amazon, 1600–1980. *American Journal of Sociology, 89*(5), 1017–1064. https://doi.org/10.1086/227983

Bunker, S. G. (1985). *Underdeveloping the Amazon: Extraction, unequal exchange, and the failure of the modern state* (pp. 5–21). University of Chicago Press.

Bunker, S. G. (2019). Toward a theory of ecologically unequal exchange. In R. S. Frey, P. K. Gellert, & H. Dahms (Eds.), *Ecologically unequal exchange* (pp. 13–48). Palgrave Macmillan.

Ciccantell, P. S. (2019). Ecologically unequal exchange and raw materialism: The material foundations of the capitalist world - economy. In R. S. Frey, P. K. Gellert, & H. Dahms (Eds.), *Ecologically unequal exchange* (pp. 49–73). Palgrave Macmillan.

Dalziel, P., Saunders, C., & Saunders, J. (2018). From economic growth to wellbeing economics. In P. Dalziel, C. Saunders, & J. Saunders (Eds.), *Wellbeing economics: The capabilities approach to prosperity* (pp. 1–21). Springer International Publishing.

Dietz, T., Rosa, E. A., & York, R. (2009). Environmentally Efficient well-being: Rethinking sustainability as the relationship between human well-being and environmental impacts. *Human Ecology Review, 16*(1), 114–123.

Dietz, T., Rosa, E. A., & York, R. (2012). Environmentally efficient well-being: Is there a Kuznets curve? *Applied Geography, 32*(1), 21–28. https://doi.org/10.1016/j.apgeog.2010.10.011

Emmanuel. (1972). *Unequal exchange: A study of the imperialism of trade* (pp. 54–72). Monthly Review Press.

Forte, R., & Moura, R. U. I. (2013). The effects of foreign direct investment on the host country's economic growth: Theory and empirical evidence. *The Singapore Economic Review, 58*(03), 135–175. https://doi.org/10.1142/S0217590813500173

Giljum, S., & Eisenmenger, N. (2004). North-south trade and the distribution of environmental goods and burdens: A biophysical perspective. *The Journal of Environment & Development, 13*(1), 73–100. https://doi.org/10.1177/1070496503260974

Givens, J. E. (2015). Urbanization, slums, and the carbon intensity of well-being implications for sustainable development. *Human Ecology Review, 22*(1), 107–128.

Givens, J. E. (2017). World society, world polity, and the carbon intensity of well-being, 1990–2011. *Sociology of Development, 3*(4), 403–435. https://doi.org/10.1525/sod.2017.3.4.403

Givens, J. E. (2018). Ecologically unequal exchange and the carbon intensity of well-being, 1990–2011. *Environmental Sociology, 4*(3), 311–324. https://doi.org/10.1080/23251042.2018.1436878

Givens, J. E., Huang, X., & Jorgenson, A. K. (2019). Ecologically unequal exchange: A theory of global environmental injustice. *Sociology Compass, 13*(5), 126–133. https://doi.org/10.1111/soc4.12693

Grimes, P., & Kentor, J. (2003). Exporting the greenhouse: Foreign capital penetration and CO? Emissions 1980?1996. *Journal of World-Systems Research, 9*(2), 261–275.

Grossman, G. M., & Krueger, A. B. (1995). Economic growth and the environment. *The Quarterly Journal of Economics, 110*(2), 353–377. https://doi.org/10.2307/2118443

Hachaichi, M., & Baouni, T. (2021). Virtual carbon emissions in the big cities of middle-income countries. *Urban Climate, 40*, 109–126. https://doi.org/10.1016/j.uclim.2021.100986

Hornborg, A. (1998). Towards an ecological theory of unequal exchange: Articulating world system theory and ecological economics. *Ecological Economics, 25*(1), 127–136. https://doi.org/10.1016/S0921-8009(97)00100-6

Jorgenson, A. K. (2005). Unpacking international power and the ecological footprints of nations: A quantitative cross-national study. *Sociological Perspectives, 48*(3), 383–402. https://doi.org/10.1525/sop.2005.48.3.383

Jorgenson, A. K. (2006). Unequal ecological exchange and environmental degradation: A theoretical proposition and cross-national study of deforestation, 1990–2000. *Rural Sociology, 71*(4), 685–712. https://doi.org/10.1526/003601106781262016

Jorgenson, A. K. (2014). Economic development and the carbon intensity of human well-being. *Nature Climate Change, 4*(3), 186–189. https://doi.org/10.1038/nclimate2110

Jorgenson, A. K., & Clark, B. (2012). Are the economy and the environment decoupling? A comparative international study, 1960–2005. *American Journal of Sociology, 118*(1), 1–44. https://doi.org/10.1086/665990

Jorgenson, A. K., & Givens, J. (2015). The changing effect of economic development on the consumption-based carbon intensity of well-being, 1990–2008. *PLoS One, 10*(5). https://doi.org/10.1371/journal.pone.0123920

Kline, R. B. (2015). *Principles and practice of structural equation modeling* (4th ed., pp.126–173). New York: The Guilford Press.

Knight, K. W., & Rosa, E. A. (2011). The environmental efficiency of well-being: A cross-national analysis. *Social Science Research, 40*(3), 931–949. https://doi.org/10.1016/j.ssresearch.2010.11.002

Martínez-Alier, J. (2002). *The environmentalism of the poor: A study of ecological conflicts and valuation* (pp. 34–69). Edward Elgar.

McGregor, J. A., & Pouw, N. (2016). Towards an economics of well-being. *Cambridge Journal of Economics, 41*(4), 1123–1142. https://doi.org/10.1093/cje/bew044

Nguyen, T., & Dang, B. H. (2021). The impact of energy consumption on Carbon Intensity of Human Well-Being (CIWB). *Ho Chi Minh City Open University Journal of Science, 11*(1), 19–28. https://doi.org/10.46223/HCMCOUJS.econ.en.11.1.1360.2021

Paul D. A. (2009). *Fixed effects regression models* (Vol. 160, pp. 38–72). SAGE Publications, Inc.

Rooij, B., & McAllister, L. K. (2014). Environmental challenges in middle-income countries. In R. Peerenboom & T. Ginsburg (Eds.), *Law and development of middle-income countries: avoiding the middle-income trap* (pp. 288–306). Cambridge University Press.

Sharma, B., & Gani, A. (2004). The effects of foreign direct investment on human development. *Global Economy Journal, 4*(2), 185–215. https://doi.org/10.2202/1524-5861.1049

Differences in Total Factor Productivity Growth Between Industry and Services: A Case Study of Vietnamese Businesses Using Heterogeneous Panel Data

Quang Hai Nguyen

Abstract Total factor productivity (TFP) reflects the efficient use of all factors and constitutes an important source of long-term sustainable growth. Therefore, TFP growth is a key indicator to evaluate the growth and development model. This study identifies TFP growth for enterprises in industrial and service sectors to examine the differences in TFP growth between these sectors in Vietnam. The estimation is used by the Cobb–Douglas production function with two basic inputs, capital and labor, and panel data for the period 2008–2020 of business in the tier-2 industry. Research results show that in the industrial sector, capital plays a much more important role than labor in explaining output value, but in the service sector, labor plays a slightly more important role. Output growth of enterprises is mainly due to capital increase, especially in the industrial sector. During 2008–2020, both the industrial and service sectors experienced TFP growth, but it tended to slow down. Overall, the service sector has a higher average annual TFP growth than industry for most of the study period. In addition, implications for enterprises in two sectors are also proposed from the research results.

Keywords Total factor productivity growth · Industrial and service sectors · Enterprises

1 Introduction

TFP growth has become a key resource for growth and labor productivity, especially in the context of scarce resources because "it reflects efficiency in the use of all factors and constitutes an important source of sustained long-term growth" (Tan & Virabhak, 1998, p: 392). Therefore, studying and measuring TFP growth has received the attention of researchers in different scopes such as at the national level (see Coelli & Rao, 2005; Felipe, 1999; Haider et al., 2021); sector level (see Botrić et al., 2017;

Q. H. Nguyen (✉)
University of Economics and Law, HCMC, Vietnam
e-mail: nhquang@uel.edu.vn

Vietnam National University, HCMC, Vietnam

A. T. Nguyen et al. (eds.), *Contemporary Economic Issues in Asian Countries: Proceeding of CEIAC 2022, Volume 1*,
https://doi.org/10.1007/978-981-19-9669-6_17

Coelli & Rao, 2005; Kong & Tongzon, 2006; Mahadevan, 2002; Nguyen, 2021); industry level (see Avila & Evenson, 2010; Edquist & Henrekson, 2017; Nguyen, 2017, 2019a, 2019b; Tan & Virabhak, 1998;); and enterprises level (see Mai et al., 2019). These studies have identified the factors that influence TFP and used different methods to estimate TFP and/or TFP growth in countries or sectors or industries over different periods. They have enriched the theory of TFP growth and its manifestation in different contexts. However, there is still a lack of research comparing TFP growth between two large sectors in an economy, industry and services, to see the difference in TFP growth between them. Comparing TFP growth between these two fields helps us gain meaningful insights into the disproportionate growth between them and it is the gap for this study to cover.

As a developing country, Vietnam's economic structure is undergoing positive changes. Industry and services are playing a key role in economic development and contributing an increasing proportion to the gross domestic product (GDP). According to the GSO (2022), in 2011 industry and services contributed 78.27% of GDP (industry: 29.12%, services: 49.15%), by 2020 the contribution of these two sectors have increased to 81.35% (industry: 30.75%, service: 50.60%). In which, the business sector plays a key role, in 2020 they contribute more than 60% to GDP (Nguyen, 2021). Therefore, this study selects Vietnam as an empirical study on TFP growth between these two sectors and selected enterprises as research units.

The main objective of this study is to identify the different roles of key inputs (capital and labor) and TFP in explaining output growth of Vietnamese firms in the industrial and service sectors. From there, find out the difference in TFP growth between these two sectors, as a basis for enterprises development policy implications in the research sectors.

The rest of the paper is structured into four sections: Sect. 2 presents the theoretical basis; next, Sect. 3 describes the research methods, samples and data; then, the research findings and discussions are reported in Sect. 4; finally, a summary of the results, the implications and limitations of the study are given in Sect. 5.

2 Theoretical Basis

The role of TFP and TFP growth in growth models is a classic topic and has been vigorously debated by economists for decades. Conceptually, productivity is seen as the difference between output and combined growth in inputs (Wu, 2011). According to Mahadevan (2002) and Kong and Tongzon (2006), the methods of measuring TFP growth can be divided into two basic approaches, frontier and non-frontier. Each approach includes parametric and nonparametric estimation methods. However, the literature review shows that two commonly used methods are parameter estimation in the non-frontier approach and nonparametric estimate in the frontier approach.

The parameter estimation method requires a specific econometric model. It helps to provide an econometric relationship between outputs and inputs and allows the use of statistical techniques to select the model that fits the data. Based on this relationship, TFP growth is specifically estimated. Two popular models used by researchers in the parameter estimation of the non-frontier approach are the production function (see Avila & Evenson, 2010; Edquist & Henrekson, 2017; Felipe, 1999; Mahadevan, 2002; Mai et al., 2019; Nguyen, 2019a, 2019b, 2021) and translog production function (see Tan & Virabhak, 1998). Although both of these models give good mathematical properties, the elasticity of the inputs in the production function allows for easier interpretation than the translog production function (Mai et al., 2018). In addition, estimation techniques in translog models often suffer from a collinearity problem among the regressors (Kinda et al., 2011). Therefore, the production function is commonly applied to calculate TFP or measure the growth rate of TFP(see Felipe, 1999; Le et al., 2020; Mahadevan, 2002; Mai et al., 2018, 2019; Nguyen, 2017, 2019a, 2019b, 2021).

When considering TFP growth across different industries or sectors, a production function with two basic inputs of capital and labor is often chosen because it is the most common inputs for all enterprises in industries or sectors (see Felipe, 1999; Mahadevan, 2002; Nguyen, 2017, 2019a, 2021; Tan & Virabhak, 1998). In parametric estimation approach, TFP growth is usually measured as output growth minus the contribution to output growth due to growth in inputs (see Felipe, 1999; Felipe & McCombie, 2004; Mahadevan, 2002; Nguyen, 2017, 2019a, 2019b, 2021).

In contrast to the parametric method which requires a specification model such as a production function or a translog production function, the nonparametric method does not require explicitly specifying a complete econometric model. It compares feasible input and output combinations based on the available data only. In the nonparametric method of the frontier approach, the technique of data envelope analysis (DEA) by Malmquist index analysis is often applied (see Coelli & Rao, 2005; Haider et al., 2021; Kong & Tongzon, 2006; Vu et al., 2019). In this approach, the decision making units (DMUs) that are most efficient establish a production frontier, and the DMUs are compared against this frontier to determine if they are effective. Since efficient DMUs are on the frontier, their technical efficiency score is equal to 1. Inefficient DMUs are under the frontier so their technical efficiency score is less than 1. The popularity of DEA is due to its ability to normalize multidimensional inputs and outputs as well as its ease of computation. Although a complete specification model is not required, the DEA does require production assumptions, so choosing the production assumption is important in the DEA's research process. Moreover, DEA does not provide an econometric relationship between outputs and inputs from which to calculate the specific value of TFP growth, so it also has certain limitations.

In addition to the above two basic methods, based on the original proposal of Olley and Pakes (1996), Wooldridge (2009) and Petrin and Levinsohn (2012) developed a semi-parametric model to overcome the problem of endogeneity between inputs and unobserved productivity. Van Biesebroeck (2007) emphasizes that the robustness to measurement errors is also an advantage of this method. It is therefore also used to estimate unobserved productivity at the enterprises level (see Nguyen, 2017).

3 Methodologies and Data

3.1 Model Specification

Consistent with the data source collected, the parametric approach by Cobb–Douglas production is chosen in this study. Furthermore, it is also suitable for measuring TFP growth across sectors and allows the estimation of specific coefficients to calculate TFP growth. Next, to measure inputs and outputs across different industries requires the use of homogenous factors, so this study chooses two basic inputs, capital and labor, as in many studies (see Felipe, 1999; Mahadevan, 2002; Nguyen, 2017, 2019a, 2019b, 2021; Tan & Virabhak, 1998). For output value, value added (VA) is chosen because it reflects the most common output value when considering enterprises across different industries and sectors. Using value added as a proxy for output has been widely applied in studies of economic growth (see Edquist & Henrekson, 2017; Felipe, 1999; Haider et al., 2021; Mahadevan, 2002; Mai et al., 2019; Nguyen, 2017, 2021; Tan & Virabhak, 1998). Hence, the Cobb–Douglas production function applied in this study is shown in Eq. (1)

$$\mathrm{VA}_{\mathrm{ij},t} = A_{i,t} K_{\mathrm{ij},t}^{\alpha_i} L_{\mathrm{ij},t}^{\beta_i} \tag{1}$$

where VA denotes the added value; A denotes the TFP; K denotes the amount of capital; L denotes the amount of labor; α, β and are the contribution coefficients of capital and labor, respectively; i denotes the sector ($i = 1$ for industry, $i = 2$ for service); and t denotes year t.

TFP growth is measured as the VA growth minus the contribution to VA growth due to growth in inputs (capital and labor). The formula for calculating TFP growth for two inputs, capital and labor, has been widely applied (see Felipe, 1999; Mahadevan, 2002; Nguyen, 2017, 2019a, 2019b, 2021) and has the following form:

$$G_{\mathrm{TFP}_{i,t}} = G_{\mathrm{VA}_{i,t}} - \alpha_i G_{K_{i,t}} - \beta_i G_{L_{i,t}} \tag{2}$$

where G_{TFP} denotes TFP growth; G_{VA} denotes value-added growth rate; G_K denotes capital growth rate; and G_L denotes labor growth rate.

In Eq. (2), αG_K and βG_L are the contributions to VA growth due to capital increase and increase in labor. Since $\alpha + \beta = 1$ or $\beta = 1 - \alpha$, when we divide both sides of Eq. (1) by L, we get Eq. (3) as follows:

$$\frac{VA_{\mathrm{ij},t}}{L_{\mathrm{ij},t}} = \frac{A_{i,t} K_{ij,t}^{\alpha_i} L_{\mathrm{ij},t}^{\beta_i}}{L_{\mathrm{ij},t}} = \frac{A_{i,t} K_{\mathrm{ij},t}^{\alpha_i} L_{\mathrm{ij},t}^{1-\alpha_i}}{L_{\mathrm{ij},t}} = A_{i,t} \left(\frac{K_{\mathrm{ij},t}}{L_{\mathrm{ij},t}} \right)^{\alpha_i} \rightarrow va_{\mathrm{ij},t} = A_{\mathrm{ij},t} k_{\mathrm{ij},t}^{\alpha_i} \tag{3}$$

where va = VA/Ldenotes value added per labor and $k = K/L$ denotes capital per labor.

Taking the natural logarithm of both sides of Eq. (3) to get Eq. (4) to estimate the contribution coefficient of capital (α) as follows:

$$\mathrm{Log}\big(\mathrm{va}_{\mathrm{ij},t}\big) = \log\big(A_{\mathrm{ij},t}\big) + \alpha_i \log\big(k_{\mathrm{ij},t}\big) + \varepsilon_{\mathrm{ij},t} \quad (4)$$

where $\varepsilon_{\mathrm{ij},t}$ denotes the residual.

After estimating the contribution coefficients of capital (α) and labor (β) to VA for each sector, Eq. (2) is applied to estimate TFP growth.

3.2 Measure Variables

Under the income method, the VA of an enterprise is determined by the sum of net profits, interest, taxes, employee's income, and depreciation. That is, it includes employee's income, depreciation, and profit before tax and interest. Because data on depreciation and interest expense are not available, the total profit before tax and employees' income of enterprises in the industry is determined as a proxy for VA as in some studies applied (see Nguyen, 2021; Tran et al., 2020). The amount of capital and labor is the average of the total annual capital or number of employees of enterprises in the industry. It is calculated as the average of the beginning of the year and the end of the year, or the average of the ending numbers of two consecutive financial years (see Nguyen, 2019a, 2021).

3.3 Sample and Data

Annual data on average capital, number of employees at the end of the year, annual profit before tax and annual employees' income of enterprises in various industries are collected from the Vietnam Statistical Yearbook. According to availability, data is collected for enterprises in tier-2 industries of Decision 27/2018/QD-TTg (Vietnam Prime Minister, 2018) for industry and services sector. Data ranges from 2008 to 2020, except for the number of employees at the end of the year, it is additionally collected for 2007 to calculate the average annual number of employees. Where the sum of pre-tax profit and employee's income is not positive or too small, it is not used to ensure data stability. Samples designed for the two sectors, industry and services, are summarized in Table 1.

The collected data is processed to calculate the average annual number of employees as well as the proxy value for the VA and the capital per employee. The key descriptive statistics of variables and correlation coefficients in the studied sectors are summarized in Table 2.

As given in Table 2, the industrial sector uses more capital per labor than the service sector and also generates more output per labor than the service sector. Furthermore, there is a large variation in the size of the subjects in the enterprises in each sectors,

Table 1 Sample design

Sector	Name of tier-1 industries	Number of tier-2 industries (*N*)	Observation
Industry	Mining and quarrying	5	61
	Manufacturing	24	294
	Electricity, gas, steam, and air conditioning supply	1	13
	Water supply; sewerage, waste management and remediation activities	3	39
	Total	**33**	**407**
Service	Wholesale and retail trade; repair of motor vehicles and motorcycles	3	39
	Transportation and storage	5	61
	Accommodation and food service activities	2	25
	Information and communication	6	75
	Financial, banking, and insurance activities	3	38
	Real estate activities	1	13
	Professional, scientific, and technical activities	6	75
	Administrative and support service activities	6	74
	Education and training	1	12
	Human health and social work activities	1	13
	Arts, entertainment, and recreation	4	49
	Other service activities	2	24
	Total	**40**	**498**

Source Author's design

Table 2 Descriptive statistics of the data

Sector	Variable	Descriptive statistics				Correlation (*k*)
		Mean	Maximum	Minimum	Std. Dev	
Industry	*k* (*K* /*L*)	3.1426	41.2900	0.0680	6.5411	1.0000
	va (VA/*L*)	0.4494	15.3070	0.0210	1.4153	0.7422***
Service	*k* (*K* /*L*)	2.8850	40.7190	0.0240	5.5129	1.0000
	va (VA/*L*)	0.1970	3.7760	0.0100	0.3299	0.4119***

Note *** for significant at the 0.01 level
Source From the results of descriptive statistical analysis

and it shows strong heterogeneity between the objects in the panel data. In addition, Table 2 also shows that the dependent variable (va) and the explanatory variable (k) have a strong correlation with each other in both sectors and it is a solid basis for estimation.

4 Results

4.1 Stationarity Test

To select an estimator that fits the data, stationarity tests for the variables are performed, and the panel unit root test is chosen as it is a popular tool. Since the table data is not balanced and heterogeneous, tests with assumes "individual unit root process" are applied. Test methods include: Im, Pesaran and Shin (IPS), Augmented Dickey–Fuller (ADF), and Philips–Perron (PP). It is implemented for both "Intercept" and "Intercept and trend" patterns. Automatic lag length selection based on Schwarz Information Criteria (SIC) with Newey–West automatic bandwidth selection and Bartlett kernel. The results of the stationarity test are reported in Table 3.

According to Table 3, variables are stationary at 1 or 5% level in all tests, except IPS test with "intercept" patterns for lnk in industry sector, it is significant at 10% level. Based on the results in large numbers, it can be concluded that the variables are stationarity at the level for both the industrial and service sectors. Therefore, it is appropriate to estimate Eq. (4) by the Least Squares (LS) panel method.

Table 3 Stationarity test results by panel unit root test

Sector	Test type	Lnk		Lnva	
		Intercept	Intercept and trend	Intercept	Intercept and trend
Industry	IPS	– 1.59*	-2.51***	– 6.33***	– 4.42***
	ADF	96.93***	96.59***	158.89***	123.56***
	PP	152.56***	122.57***	251.72***	146.54***
	Conclusion	*I(0)*		*I(0)*	
Service	IPS	– 2.25**	-8.40***	– 2.69***	– 3.89***
	ADF	116.84***	199.86***	114.22***	143.18***
	PP	130.97***	241.80***	111.39***	205.23***
	Conclusion	*I*(0)		*I*(0)	

Note *, ** and *** for significant at the 0.10, 0.05, and 0.01 levels, respectively
Source From unit test results

4.2 Estimate the Coefficient of Capital Per Labor

Both the fixed effects model (FEM) and the random effects model (REM) estimators are first performed, and then, the more appropriate model is selected by Hausman test. Table 4 below reports the results estimated using the panel LS method.

As seen in Table 5, the F-statistics of the models are all very large and the p value of F-statistics are all zero, so the estimated models are fit the data. The coefficients of capital per labor and constant are significant at the 1% level in all models. Hausman test results show that the Chi-squared statistic of cross-section random is 0.0302 and 0.0247, respectively, for two samples, industrial and service sectors, with significance is all greater than 10%. So REM is more suitable than FEM for both samples, and they are shown in the final model.

To test robustness, the results estimated by the FEM model are compared with the results from the alternative estimation method. With the research data set, there can be many alternative estimation methods for the LS panel such as panel generalized method of moments (GMM) and the panel autoregressive distributed lag model (ARDL). GMM approach is considered appropriate method for panel data with N > T as in this study. However, this method creates an endogenous variable and can affect the coefficient estimate of the explanatory variable (the coefficient of capital to labor). Hence, the panel ARDL approach is chosen to check the robustness of the panel REM. ARDL was proposed by Pesaran and Shin (1995), and it allows to apply in case the variables exist at I(0), or I(1), or both $I(0)$ and $I(1)$ (Nkoro & Uko, 2016). Accordingly, the Pooled Mean Group (PMG) estimator was chosen because it provides aggregate coefficients in the long run and object-specific coefficients in the short run. Table 5 below reports the results of the panel ARDL approach.

Table 4 Results of the estimated coefficients of capital per labor

Variables		Industry sector		Service sector	
		FEM	REM	FEM	REM
Lnk	Coefficient	0.7414	0.7374	0.4676	0.4657
	t-Statistic	16.88***	19.70***	14.93***	16.18***
Constant	Coefficient	– 1.9740	– 1.9875	– 2.1629	– 2.1757
	t-Statistic	– 93.39***	– 23.57***	– 130.85***	– 24.56***
R-squared		0.8727	0.4900	0.8471	0.3460
Adj R-squared		0.8614	0.4888	0.8337	0.3447
F-statistic		77.4852	389.1528	63.2727	262.45
Prob (F-statistic)		0.0000	0.0000	0.0000	0.0000
Hausman test (Chi-Sq. Statistic)		0.0302		0.0247	

Note Lnva is the dependent variable; *** for statistically significant at the 0.01 level; Coefficients are the estimates of the contribution of capital (α) in Eq. (4)
Source From the results of panel LS regression estimation

Table 5 Estimation results by the panel ARDL approach

Variables	Industry sector		Service sector	
	Coefficient	t-Statistic	Coefficient	t-Statistic
Long-Run Equation				
Lnk	0.7270	10.19***	0.5217	15.90***
Short-Run Equation				
COINTEQ01	– 0.6678	– 11.98***	– 0.6322	– 11.53***
ΔLnk	– 0.0377	– 0.22	– 0.1734	– 1.51
C	– 1.2925	– 7.93***	– 1.3863	– 9.34***
Statistics				
S.E. of regression	0.2834		0.3161	
Sum squared residual	22.004		37.6633	
Log likelihood	219.6679		64.2083	
Akaike info criterion	– 0.4259		0.2281	
Lag length	1		1	

Note Lnva is the dependent variable; *** for statistically significant at the 0.01 level; Coefficients are the estimates of the contribution of capital (α) in Eq. (4)
Source From the results of panel ARDL estimation

According to Table 6, the large Log likelihood and small Akaike info criterion values in the two samples show that the estimates are reliable. The coefficients of the coint equations are significant at the 1% level, and values between 0 and – 1 indicate a long-run equilibrium relationship for both samples. The convergence rate to the equilibrium of the two samples is quite similar, 0.67 for industry sector and 0.63 for service sector. The coefficients of lnk are significant at the 1% level in the long-run equations, but they are not significant in the short-run equations, and the coefficients of the constant are both significant at the 1% level. Therefore, the estimated results by the panel ARDL approach are quite similar to the results from the panel LS method (see Table 6). According to Table 6, the estimated coefficients of the two methods have a rather small deviation, only – 1.41% in the industrial sector and 12.02% in the service sector for lnk. This is evidence to support the robustness of the results estimated by panel LS method.

Table 6 Differences in coefficients estimated by FEM and ARDL methods

Variables	Industry sector			Service sector		
	FEM	ADRL	Bias (%)	FEM	ADRL	Bias (%)
Lnk	0.7374	0.7270	– 1.41	0.4657	0.5217	12.02
C	– 1.9875	– 1.2925	– 34.97	– 2.1757	– 1.3863	– 36.28

Source Estimation results from panel FEM and panel ARDL
Source Calculation results from Tables 4 and 5

The estimated results show that the contribution coefficients of capital and labor in the industrial sector are 0.7374 and 0.2626, respectively; in the service sector are 0.4657 and 0.5343, respectively. It proves that in the industrial sector capital plays a more important role in the output growth of enterprises. By contrast, this role belongs to labor in the service sector, but there is a big difference. To be more specific, the elasticity of capital is much larger than the elasticity of labor in the industrial sector, while the elasticity of labor is insignificantly larger than the elasticity of capital in the service sector.

4.3 Comparing TFP Growth Between Industrial and Service Sectors

From calculating the growth of inputs (capital and labor), output growth (VA) and the contribution to output growth due to capital and labor growth, TFP growth for enterprises in both sectors is determined by Eq. (2). The TFP growth calculation results for the industrial sector are reported in Table 7 and the service sector are reported in Table 8.

Tables 7 and 8 show that the average VA growth of enterprises in the two sectors, industry and services, is fairly balanced over the study period, averaging 15.1% in the industry sector and 15.6% in the service sector. However, VA growth tends to slow down between periods. It is evident in the industrial sector with the CAGR for the periods 2008–2010, 2011–2015, and 2016–2020 of 21.2, 17.11, and 10.81%, respectively. As for the service sector, although VA growth was still maintained at a double-digit rate, it decreased quite deeply in the period 2011–2015 compared to the period 2008–2010 and increased slightly in the period 2016–2020. The CAGR for the periods is 36.26, 11.65, and 12.08%, respectively.

VA growth of enterprises during the study period was mainly due to capital increase, especially in the industrial sector, capital increase contributed to 87.55% of the annual VA growth of enterprises, while the increase in labor only contributed to VA growth of 9.47%. In the service sector, the gap in the roles of these two factors has been significantly reduced, and the proportion contributing to annual VA growth due to capital and labor increase is 57.56% and 27.63%, respectively.

Over the entire study period, enterprises in both the industrial and service sectors experienced growth in TFP. With a CAGR of 2.31%/year, the TFP of enterprises in the service sector has a much higher growth than the industrial sector, and it is only 0.45% per year. They have partly helped enterprises in the service sector have the contribution proportion of TFP to VA growth of 14.81% compared to 2.98% in the industrial sector. However, TFP growth also tends to decrease gradually in the service sector with the CAGR for the periods of 2008–2010, 2011–2015, and 2016–2020 at 9.81, 1.30, and 0.72%, respectively. In the industrial sector, TFP decreased by an average annual rate of 4.03% for the period of 2008–2010, but in the period of 2011–2015, it increased significantly at 2.23%/year and decreased to 0.31%/year

Table 7 Contribution of factors to the growth of enterprise in industry sector

Year/period	Quantity			VA growth (%)	VA growth by (%)		TFP growth (%)
	VA (trillion VND)	*K* (trillion VND)	*L* (person)		*K*	*L*	
2008	257,643	1,673,150	4,208,886				
2009	334,712	2,278,056	4,410,906	29.91	26.66	1.26	1.99
2010	378,458	2,924,275	4,697,573	13.07	20.92	1.71	– 9.55
2011	427,525	3,560,045	5,005,514	12.96	16.03	1.72	– 4.79
2012	523,277	4,190,259	5,315,016	22.40	13.05	1.62	7.72
2013	688,985	4,889,609	5,670,758	31.67	12.31	1.76	17.60
2014	789,235	5,696,593	6,082,194	14.55	12.17	1.91	0.48
2015	833,823	6,638,625	6,482,241	5.65	12.19	1.73	– 8.27
2016	1,034,705	7,463,968	6,910,725	24.09	9.17	1.74	13.19
2017	1,190,780	8,490,054	7,326,275	15.08	10.14	1.58	3.37
2018	1,234,842	9,704,536	7,595,625	3.70	10.55	0.97	– 7.81
2019	1,288,324	10,731,739	7,837,531	4.33	7.81	0.84	– 4.31
2020	1,393,110	12,106,151	7,942,697	8.13	9.44	0.35	– 1.66
2008–2010	323,604	2,291,827	4,439,121	21.20	23.75	1.48	– 4.03
2011–2015	652,569	4,995,026	5,711,144	17.11	13.14	1.75	2.23
2016–2020	1,228,352	9,699,290	7,522,571	10.81	9.42	1.09	0.31
2008–2020	798,109	6,180,543	6,114,303	15.10	13.22	1.43	0.45

Note Numbers in periods are averages; growth over the period is the compound annual growth rate (CAGR)
Source Author's calculation using Eq. (2)

in the period of 2016–2020. However, with 0.31% for CAGR of TFP in the period of 2016–2020 compared to 0.45% for the whole period of 2008–2020, TFP growth also shows signs of slowing down.

5 Discussions

The VA growth of enterprises in the industrial and service sectors in Vietnam in the period of 2008–2020 is mainly driven by capital increase, especially in the industrial sector. This is explained first of all by the role of capital in explaining the VA of enterprise in the industrial and service sectors, which have elasticity of 0.7374 and 0.4657, respectively. However, perhaps the more important reason is that at the national level, labor resources are often limited, while capital and technology are less limited than labor resources. In addition, thanks to scientific, technical, and

Table 8 Contribution of factors to the growth of enterprise in service sector

Year/ period	Quantity			VA growth (%)	VA growth by (%)		TFP growth (%)
	VA (trillion VND)	*K* (trillion VND)	*L* (person)		*K*	*L*	
2008	183,058	3,678,799	2,061,190				
2009	263,976	4,928,248	2,378,107	44.20	15.82	8.22	20.17
2010	339,857	6,933,023	2,822,048	28.75	18.94	9.97	– 0.17
2011	379,313	8,906,797	3,304,464	11.61	13.26	9.13	– 10.78
2012	412,339	9,636,439	3,550,250	8.71	3.81	3.97	0.92
2013	459,796	11,332,306	3,635,273	11.51	8.20	1.28	2.03
2014	520,359	12,211,413	3,776,665	13.17	3.61	2.08	7.48
2015	589,645	13,337,526	3,992,700	13.32	4.29	3.06	5.96
2016	703,252	16,472,891	4,406,499	19.27	10.95	5.54	2.78
2017	889,721	19,760,159	4,796,057	26.52	9.29	4.72	12.50
2018	1,025,197	24,059,553	5,045,757	15.23	10.13	2.78	2.31
2019	1,089,665	26,791,218	5,241,564	6.29	5.29	2.07	– 1.07
2020	1,042,840	30,526,061	5,233,653	– 4.30	6.49	–0.08	– 10.71
2008–2010	262,297	5,180,023	2,420,448	36.26	17.36	9.09	9.81
2011–2015	472,290	11,084,896	3,651,870	11.65	6.51	3.84	1.30
2016–2020	950,135	23,521,976	4,944,706	12.08	8.39	2.97	0.72
2008–2020	607,617	14,505,726	3,864,940	15.60	8.98	4.31	2.31

Note Numbers in periods are averages; growth over the period is the CAGR
Source Author's calculation using Eq. (2)

technological progress, enterprises regularly innovate technology and tend to use higher capital per labor. The results from this study are quite consistent with the finding that capital has a more important role than labor in explaining VA growth in most enterprises sectors by type of ownership and region, and VA growth of Vietnamese enterprises in these types is mainly due to capital gains (Nguyen, 2021); or the growth in output of enterprises in transport modes in Vietnam is mainly due to capital increase (Nguyen, 2019a). This result implies that Vietnamese enterprises need to continue to focus on developing both capital and labor for growth, especially increasing capital in the industrial sector. It helps enterprises innovate technology, scale-up to increase labor productivity. For labor, in addition to increasing quantity, improving quality is very important because it overcomes the problem of limited social labor resources.

TFP growth in both industrial and service sectors during the research period shows that intangible factors such as technological innovation, management improvement, and improvement in labor qualifications of employees contribute increasingly on growth in these regions. TFP growth tends to slow down between periods which can be explained because when the starting point of the early period is small, the growth rate is usually high, but then the starting point size increases and it causing the growth rate to tend to decrease. Similarly, when the starting point in technology and management is low, technological innovation and management improvement often create great change, but when the level of technology and management is raised, technological innovation and management improvement are hardly creating breakthroughs. These results are considered to have certain differences from results in some previous studies, for example, TFP growth in service sectors in the period of 1976–1993 is very low, but they improved in the period of 1987–1993 (Tan & Virabhak, 1998); or the majority of service sectors in Singapore had negative TFP growth between 1975 and 1994 (Mahadevan, 2002); or growth varies across firms of different ownership types in Vietnam (Nguyen, 2021). These differences are indicative of TFP growth in different contexts.

The difference in TFP growth between the industrial and service sectors indicates the different roles of TFP growth in VA growth in these sectors over the studied periods. With a higher TFP growth, the service sector is seen as more effective in technological innovation, management improvement, and scale-up to increase labor productivity. The downward trend in TFP growth is a general rule as explained above, but the rapid decline is a sign that technological innovation and management improvement are no longer effect as strong as the first stage, especially in the service sector. It is a sign for enterprises in the industrial and service sectors to find out the causes and come up with solutions to improve. It can be seen that the influence on the growth of enterprises depends not only on the inputs belonging to their internal environment, but also on external factors. During the research period, there are two important external instability events affecting the operations of enterprises, especially the service industry. Firstly, the slight economic recession in 2008 and the recovery in 2009 made the VA growth of enterprises in the industrial and service sectors grow in 2009 by 29.91 and 44.20%, respectively. It partly helps the service sector to have a TFP growth in 2009 so up to 20.17%. Second, the outbreak of the COVID-19 pandemic in 2020 has greatly affected the VA of enterprises in the service sector with a decrease of 4.3% and a negative growth in TFP of 10.71%. In which, there are many industries with huge losses in 2020 such as air transport, accommodation services, etc.

6 Conclusions

TFP growth is considered by scholars, but there is still a lack of studies comparing TFP growth between the two largest sectors, industrial and service. This study selected the method, measured, and compared the TFP growth of firms in these two sectors in Vietnam for the period of 2008–2020. The method of parameter estimation using the Cobb–Douglas production function with two basic inputs, capital and labor, has been selected to realize the research objective. Based on the results of the stationarity test, the LS panel estimation method is applied, and the REM model is selected by the Hausman test. The panel ARDL approach is chosen as the alternative estimation method to check robustness for the final selection model. Research results have confirmed the different roles of capital and labor in explaining output value as well as different TFP growth in the two sectors of industry and services. In the industrial sector, capital plays a much more important role than labor, while in the service sector, this role belongs to labor, but the difference is not much. In other words, the gap between the roles of capital and labor in the industrial sector is much larger than the gap between the roles of labor and capital in the service sector. Output growth of Vietnamese enterprises in the industrial and service sectors in recent years has been mainly due to capital increase, especially in the industrial sector. In the period of 2008–2020, both the industrial and service sectors saw growth in TFP, but these growths tend to slow down in the period of 2016–2020 compared with the previous periods. In the two sectors, services have a higher average annual TFP growth rate than industry for the entire study period as well as the periods of 2008–2010 and 2016–2020. These results imply that in order to grow and develop, enterprises need to focus on developing both capital and labor, especially capital in the industrial sector. In addition, enterprises need to continue to focus on improving labor quality, innovating technology, improving management efficiency, and expanding scale to continuously grow TFP and promote intangible resources in a situation where resources are becoming increasingly scarce.

The contribution of this study is mainly practical. It has shown the different roles of capital and labor in the output growth of enterprises in the industrial and service sectors as well as the difference in TFP growth between these two sectors as a basis for building an appropriate development policy mechanism. In addition, this study also enriches the method of estimating, calculating TFP growth, and complements the existing literature on TFP growth in the context of Vietnam—a developing economy with updated data up to 2020. Due to the ability to collect data and for the purpose of making comparisons between the industrial and service sectors, this study only considers two basic inputs, capital and labor. In addition to the two research factors, there may be some other factors belonging to the internal and external environment that affect the outputs of enterprises. Therefore, they may also serve as an opportunity for further research on this topic.

Acknowledgements This research was funded by University of Economics and Law, Vietnam National University, Ho Chi Minh, Vietnam [grant number IV-2022].

References

Avila, A. F., & Evenson, R. E. (2010). Total factor productivity growth in agriculture: The role of technological capital. *Handbook of Agricultural Economics, 4*, 3769–3822. https://doi.org/10.1016/S1574-0072(09)04072-9

Botrić, V., Božić, L., & Broz, T. (2017). Explaining firm-level total factor productivity in post-transition: Manufacturing vs. services sector. *International Studies, 10*(3), 77–90. https://doi.org/10.14254/2071-8330.2017/10-3/6

Coelli, T. J., & Rao, D. P. (2005). Total factor productivity growth in agriculture: A malmquist index analysis of 93 countries, 1980–2000. *Agricultural Economics, 32*, 115–134. https://doi.org/10.1111/j.0169-5150.2004.00018.x

Edquist, H., & Henrekson, M. (2017). Do R&D and ICT affect total factor productivity growth differently? *Telecommunications Policy, 41*(2), 106–119. https://doi.org/10.1016/j.telpol.2016.11.010

Felipe, J. (1999). Total factor productivity growth in East Asia: A critical survey. *Development Studies, 35*(4), 1–14. https://doi.org/10.1080/00220389908422579

Felipe, J., & McCombie, J. S. (2004). To measure or not to measure TFP growth? A reply to Mahadevan. *Oxford Development Studies, 32*(2), 321–327. https://doi.org/10.1080/13600810410001700017

GSO. (2022). National ascounts and state budget. In *Gross domestic product at current prices by types of ownership and by kinds of economic activity*. Retrieved from https://www.gso.gov.vn/en/statistical-data/#tab-6d683a49362bd478160.

Haider, F., Kunst, R., & Wirl, F. (2021). Total factor productivity, its components and drivers. *Empirica, 48*(2), 283–327. https://doi.org/10.1007/s10663-020-09476-4

Kinda, T., Plane, P., & Véganzonès-Varoudakis, M. A. (2011). Firm productivity and investment climate in developing country: How does Middle East and North Africa manufacturing form? *The Developing Economies, 49*(4), 429–462. https://doi.org/10.1111/j.1746-1049.2011.00146.x

Kong, N. Y., & Tongzon, J. (2006). Estimating total factor productivity growth in Singapore at sectoral level using data envelopment analysis. *Applied Economics, 38*, 2299–2314. https://doi.org/10.1080/00036840500427544

Le, C. Q., Nguyen, T. T., & Do, N. T. (2020). State ownership, quality of sub-national governance, and total factor productivity of firms in Vietnam. *Post-Communist Economies, 33*(1), 133–146. https://doi.org/10.1080/14631377.2020.1793608

Mahadevan, R. (2002). A frontier approach to measuring total factor productivity growth in Singapore's services sector. *Journal of Economic Study, 29*(1), 48–58. https://doi.org/10.1108/01443580210414111

Mai, G. H., Tran, X. D., Bui, T. H., Mai, Q. T., & Yoshida, Y. (2018). Impact of investment climate on total factor productivity of manufacturing firms in Vietnam. *Sustainability, 12*, 1–18. https://doi.org/10.3390/su10124815

Mai, G. H., Tran, X. D., Bui, T. H., & Mai, Q. T. (2019). Total factor productivity of agricultural firms in Vietnam and its relevant determinants. *Economies, 7*(1), 1–12. https://doi.org/10.3390/economies7010004

Minister, V. P. (2018). *Decision 27/2018/QD-TTg on promulgating the system of economic sub-sectors of Vietnam (in Vietnamese)*.

Nguyen, H. Q. (2017). Business reforms and total factor productivity in Vietnamese manufacturing. *Asian Economic, 51*, 33–42. https://doi.org/10.1016/j.asieco.2017.06.003

Nguyen, O. T. (2019a). Determinants of firms' total factor productivity in manufacturing industry in Vietnam. *Asian Business and Economic Studies, 26*(S01), 04–28. Retrieved from http://jabes.ueh.edu.vn/Content/ArticleFiles/597098e5-57f0-46b4-9905-d576b5f5f96d/JABES-2018-10-E367.pdf.

Nguyen, Q. H. (2019b). Comparing total factor productivity growth rate among modes of transport in Vietnam - Measured by Cobb – Douglas production function. *Asian Economic and Business Research Journal, 30*(12), 5–19.

Nguyen, Q. H. (2021). Total factor productivity growth of Vietnamese enterprises by sector and region: Evidence from panel data analysis. *Economies, 109*. https://doi.org/10.3390/economies9030109

Nkoro, E., & Uko, A. K. (2016). Autoregressive distributed lag (ARDL) cointegration technique: application and interpretation. *Statistical and Econometric Methods, 5*(4), 63–91. Retrieved from http://www.scienpress.com/Upload/JSEM/Vol%205_4_3.pdf.

Olley, G. S., & Pakes, A. (1996). The dynamics of productivity in the telecommunications equipment industry. *Econometrica, 64*(6), 1263–1298. https://doi.org/10.2307/2171831

Pesaran, M., & Shin, Y. (1995). *An autoregressive distributed lag modelling approach to cointegration analysis*. University of Cambridge: Faculty of Economics. Retrieved from https://citeseerx.ist.psu.edu/viewdoc/download?doi=10.1.1.153.3246&rep=rep1&type=pdf

Petrin, A., & Levinsohn, J. (2012). Measuring aggregate productivity growth using plant - level data. *The Rand Journal of Economics, 43*(4), 705–725. https://doi.org/10.1111/1756-2171.12005

Tan, L. Y., & Virabhak, S. (1998). Total factor productivity growth in Singapore's service industries. *Economic Studies, 25*(5), 392–409. https://doi.org/10.1108/01443589810233694

Tran, T. D., To, T. T., Vu, C. S., Nguyen, D. A., Nguyen, H. H., Nguyen, H. T., & Nguyen, V. T. (2020). *Annual Vietnam economic review 2019 - improving labor productivity in the digital economy context*. National Economics University. (in Vietnamese).

Van Biesebroeck, J. (2007). Robustness of productivity estimates. *The Journal of Industrial Economics, 55*(3), 529–569. https://doi.org/10.1111/j.1467-6451.2007.00322.x

Vu, T. H., Peng, K. C., Ha, Q. T., & Nguyen, T. G. (2019). Evaluation of total factor productivity of foreign direct investment enterprises in Vietnam: An application of Malmquist productivity index. *International Journal of Economics, Business and Management Research, 3*(8), 59–66. Retrieved from https://ijebmr.com/uploads/pdf/archivepdf/2020/IJEBMR_426.pdf.

Wooldridge, J. M. (2009). On estimating firm-level production functions using proxy variables to control for unobservables. *Economics Letters, 104*(3), 112–114. https://doi.org/10.1016/j.econlet.2009.04.026

Wu, Y. (2011). Total factor productivity growth in China: A review. *Chinese Economic and Business Studies, 9*(2), 111–126. https://doi.org/10.1080/14765284.2011.568682

ESG Effects on Stock Prices During the Epidemic: Empirical Investigations for the Finance Holding Companies in Taiwan

Yu-Hong Lin and Yen-Ling Lin

Abstract This study investigates the impact of ESG on stock prices for finance holding companies during the epidemic. ESG has been used as a criteria indicator for investment nowadays. Many governments have also proposed standards and laws to issue ESG's important in current corporate operations. However, the suddenly coming of the COVID-19 could be a time point to examine the effects of ESG. This research examines Taiwan's finance holding companies by using the Taiwan ESG score data in 2018 to 2021, the company data from Taiwan Economic Journal and Taiwan Financial Supervisory Commission, COVID-19 data from Taiwan Ministry of Health and Welfare, and macro-economic data from Taiwan DGBAS. The results show that the ESG has a significant and positive impact on the finance holding company's revenue. It's found that Taiwan has a smaller negative impact on the finance holding company's revenue when the epidemic is well controlled. In terms of stock price, the results show that the stock price of finance holding companies inside the ESG rating list will be significantly higher than those are not on the list. This shows that when market risks occur, the degree of ESG can be regarded as an important indicator for investors' concern.

1 Introduction

The coronavirus disease-19 (COVID-19) was first reported on December 31, 2019. Since then, the epidemic quickly spread and became a global pandemic, and new variants of the virus subsequently emerged. Among people infected with COVID-19, their symptoms are concentrated in the respiratory tract. Although not everyone with COVID-19 develops severe symptoms, respiratory failure can be induced by the presence of severe symptoms, substantially increasing the mortality risk of

Y.-H. Lin
Taiwan Cooperative Bank, Taipei, Taiwan, ROC

Y.-L. Lin (✉)
Department of Economics, Tamkang University, New Taipei City, Taiwan, ROC
e-mail: yenling@mail.tku.edu.tw

A. T. Nguyen et al. (eds.), *Contemporary Economic Issues in Asian Countries: Proceeding of CEIAC 2022, Volume 1*,
https://doi.org/10.1007/978-981-19-9669-6_18

COVID-19. Based on the numbers of diagnosed cases reported by various countries between January and March 2020, the World Health Organization (WHO) announced COVID-19 as the preeminent pandemic of the century. Experience indicates that the occurrence of a global pandemic quickly causes public panic and considerably affects various markets. During the current COVID-19 pandemic, various countries adopted policies to restrict the flow of people to stop the rapid spread of the virus; consequently, the supply–demand cycle of the global market was marked affected.

In 2020, the International Monetary Fund (IMF) predicted a global economic decline by 3% because of the pandemic, which is an alarming number considering that the 2009 financial crisis only caused a global economic decline by 0.7%. This comparison reveals the considerable economic effects of the current pandemic.[1] Taking the U.S. stock market as an example, on February 24, 2020, the Dow Jones Industrial Average (DJIA) decreased by more than 1,000 points and experienced a drop of 3.5%. On the same day, the Standard and Poor's 500 Index closed with a decrease of 116.84 points and experienced a drop of 3.4%. On March 9, 2020, the DJIA decreased by 7.8%, which was the biggest decrease since the subprime mortgage crisis in 2008, and the S&P 500 index also decreased by 7%, which triggered the circuit breaker mechanism and caused stock trading in the USA to be suspended for 15 min. In March 2020, the Taiwan Capitalization Weighted Stock Index decreased by 1584.11 points and experienced a drop of 14%, highlighting the panic felt by investors because of the crisis.

In risk management, enterprises should consider possible market changes and their effects to ensure their sustainable operations and development. Therefore, in addition to "COVID-19", "environmental, social, and governance (ESG)" and "corporate social responsibility (CSR)" became popular key terms in the investment market in 2020; these concepts are crucial to the sustainable development of today's enterprises and societies. CSR can be interpreted as a company's consideration of the rights and interests of all relevant stakeholders in addition to generating additional profits for them. ESG is an indicator for measuring a company's CSR effectiveness, and it can be regarded as an indicator for assessing the soundness and stability of a company. For companies and societies, improving ESG performance and CSR effectiveness is a key component of sustainable development and merits attention.

Literature findings have revealed that a company can benefit and profit considerably from implementing ESG policies (Friede et al., 2015). During a market crisis, the performance decline is minor in companies with high CSR effectiveness. For instance, Lins et al. (2017) discover that, during the financial crisis that occurred between 2008 and 2009, the stock returns of firms with high social capital (as measured by CSR intensity) were 4 to 7% higher than those of firms with low social capital. The profitability, growth, and sales per employee of firms with high CSR engagement are also higher than those of firms with low CSR engagement. Hwang et al. (2021) examine listed companies in South Korea and discover that engagement in ESG activities increases their profitability and substantially reduces their earnings volatility. Broadstock et al. (2021) study data extracted from China's CSI 300

[1] See *2020 World Economic Outlook* published in 2020.

Index and discover that ESG engagement can help reduce stock risk during a market crisis. Therefore, ESG and CSR may be increasingly decisive influencing factors for corporate risk management.

However, not all studies agree that ESG engagement exhibits positive effects. In fact, numerous studies conducted in 2021 report the opposite. For example, Takahashi and Yamada (2021) examines the effect of the shareholding structures of the constituent stocks of the Nikkei index in Japan and the global value chain and discovers that, during the pandemic, the stock returns of Japanese companies that operate in China or the USA are lower than those of Japanese companies without ESG engagement; and the rates of return achieved by companies with high ESG scores were not significantly greater than those with low ESG scores. Demers et al. (2021) collects data from the Center for Research in Security Prices of the University of Chicago and discovers that the resistance of a company's ESG rating against the effect of the pandemic is nonsignificant.

Therefore, the literature is inconclusive regarding the roles of ESG implementation and a high CSR level in influencing a company's ability to overcome risks related to large-scale pandemics; notably, the financial sector is under analyzed by researchers. To address this research gap, this study analyzes the financial sector based on the literature findings and examines the effects of ESG and CSR engagement.

This study focuses on the role of the ESG ratings of large financial holding companies in Taiwan during the pandemic. These large financial holding companies are the objects of this study for two reasons. First, these companies emphasize ESG engagement, and comprehensive data on them are highly accessible. Second, in Taiwan, these financial holding companies are popular stock deposit targets among investors, and their ESG scores may be a key reference indicator for the investors who are watching these companies. In this study, a large financial holding company in Taiwan is defined as a company with a headcount of more than 9,000. The data used in the empirical analysis of this study are obtained from five sources, namely (1) business operation data, which include monthly revenue, stock return, total assets, and number of employees as obtained from the Taiwan Economic Journal (TEJ) database; (2) data on capital adequacy ratios as obtained from the Financial Supervisory Commission, Republic of China (FSC); (3) data related to the pandemic in Taiwan as obtained from the website of the Taiwan Centers for Disease Control (CDC) of the Ministry of Health and Welfare (MOHW); (4) data on gross domestic product (GDP) and unemployment rates as obtained from the Taiwan's Directorate General of Budget, Accounting and Statistics (DGBAS); and (5) ESG scores as obtained from the score data sheets used by Common Wealth Magazine for assessing Excellence in Corporate Social Responsibility Award nominees. The databases and the variable-related data are detailed in Sect. 4.

Because the COVID-19 broke out in early 2020, the information analyzed in this study pertains to the period between January 2018 and December 2021. This period can be divided into two subperiods, namely the pre-pandemic subperiod from January 2018 to December 2019 and the post-pandemic subperiod from January 2020 to December 2021. In this study, empirical research is conducted through a two-step analysis. In the first step, the effect of each variable on the large-scale holding companies' monthly revenue is examined through vertical and horizontal regression-based data analyses. In the second step, the predicted revenue obtained in the first stage through regression is controlled and used as an independent variable, and the difference-in-difference (DID) regression estimation method is used to estimate the pre- and post-pandemic differences in the effect of the large-scale holding companies' ESG scores on their monthly stock returns.

This study mainly explores whether ESG scores significantly influence the revenue of these large financial holding companies in Taiwan during the pandemic and whether ESG scores is a significant reference indicator for investment during a major global crisis. The expected outcome is that ESG scores affect revenue and can serve as a crucial reference indicator for investors during a crisis.

This study comprises six sections. Section 1 is the Introduction, which explains the research motivation and the background of the research problems. And shows the ESG concept and discusses the evolution and development of concepts related to sustainable management (e.g., CSR and ESG). Section 2 introduces the research data and empirical method adopted. Section 3 reports the empirical results. Section 4 offers the conclusion.

2 Data and Empirical Strategy

2.1 Data

This study examines the effect of ESG performance on large-scale financial holding companies in Taiwan during the pandemic. Large-scale financial holding companies in Taiwan with more than 9,000 employees (as determined based on the latest head-count data released by the TEJ) are selected. The related items are listed in Table 3.

The data about the holding companies mainly comprise the five components as follows: (1) operational data, (2) ESG score, (3) the supervision data of financial authorities, (4) the pandemic variable, and (5) macroeconomic variables. Accordingly, the present draws data from five sources, namely the TEJ, the CSR scores of the Common Wealth Magazine, the FSC, the CDC of the MOHW, and the macroeconomic database of DGBAS. The five sources are as follows:

- In April 1990, the TEJ established a financial and economic database that specifically provides the information required to perform the fundamental analysis of securities and financial markets; the database covers macroeconomic and financial information as well as various types of information about publicly traded companies, including their operational and financial information, security market transactions, and financial derivatives. This database is currently the largest and most comprehensive financial and economic database in Taiwan.

- The Excellence in Corporate Social Responsibility Award was established in 1994. In 2007, Common Wealth Magazine used the CSR indicator to establish the Excellence in Corporate Social Responsibility Award, selected the top 50 companies by assessing the performance of participating companies in four dimensions (i.e., corporate governance, corporate commitment, social engagement, and environmental sustainability), sorted them according to their total scores, and decided on the number of companies to be selected for each group.
- The Banking Bureau of the FSC is the authority that supervises the banking industry, trust industry, credit card and electronic ticket issuing institutions, electronic payment institutions, and financial holding companies. It provides updates on the latest banking acts, interpretation letters, financial knowledge, and financial activity information. It also announces supervisory information and financial statistical indicators pertaining to the domestic banking industry.
- On January 20, 2020, the CDC of the MOHW announced the establishment of the Central Epidemic Command Center of the National Health Command Center to collect daily information on the pandemic in Taiwan through the CDC's COVID-19 database; the information collected includes the number of diagnosed cases, number of deaths, number of persons released from quarantine, and number of persons who underwent spontaneous screening.
- The DGBAS' macroeconomic database was established in 1978 and comprises 17 data categories, which include national income, commodity price, labor, import and export trade, banking, population, economic indicators, finance, police and fire protection, securities and futures transactions, social insurance and welfare, industry and commerce, household income and expenditure, education, construction and real estate, transportation, and health. Furthermore, it is connected to the relevant statistical databases of other ministries and bureaus, which facilitates the compilation of key statistics from various fields in Taiwan and enables users to rapidly grasp key information on domestic economic and social development.

2.2 Variables

This study uses data of the period from January 2018 to December 2021. This period can be divided into two subperiods, namely the pre-pandemic subperiod from January 2018 to December 2019 and the post-pandemic subperiod from January 2020 to December 2021. For data frequency, this study uses monthly data, including the 48-month data of 13 holding companies (i.e., a sample size of 624). Variable-related data are divided into four categories, namely holding companies' background information variables, macroeconomic market variables, ESG proxy variables, and the pandemic variable, which are respectively presented as follows:

Background information variables of holding companies

1. Number of employees (staff/1,000 persons)

2. Monthly revenue (revenue/NT$1,000)
3. Total assets (asset/NT$1,000)
4. Monthly stock return (stock/%)
5. Capital adequacy ratio (car/%)

Macroeconomic variables

1. Taiwan Stock Exchange issuance-weighted stock return (taiex/%)
2. GDP growth rate (GDP/%)
3. Unemployment rate (UnempR/%)

Pandemic variables
Monthly number of diagnosed cases (COVID-19/1,000 persons)

ESG variables
The scores for the Excellence in Corporate Social Responsibility Award, which are announced annually by Common Wealth Magazine, are used as the proxy variable for ESG engagement. Since 2007, the Common Wealth Magazine has been using foreign CSR-related indicators and assessment methods to select the top 50 CSR companies in Taiwan and rate them by their performance in four dimensions, namely corporate governance, corporate commitment, social engagement, and environmental sustainability.

1. Environmental sustainability (esg_e)
2. Social engagement (esg_s)
3. Corporate governance (esg_g)
4. Corporate commitment (esg_c)
5. Total ESG score (esg_tot).

Because only the scores of the top 50 winners of the Excellence in Corporate Social Responsibility Award are announced, the scores of the holding companies selected in this study are not available for every year. For the unavailable ESG scores of the selected holding companies, the ESG score of the 51th-ranked company is predicted using the scores of the top 50 companies, and the predicted score is used as the proxy score for unavailable scores.

Table 1 lists the descriptive statistical results pertaining to the dependent and explanatory variables; it reveals that among the selected holding companies, Cathay Holdings has the most employees (58,000), whereas Sinopac Holdings has the least number of employees (9,275). The mean of the holding companies' monthly revenues is 1.43e + 07 (NT$1,000); the highest and lowest revenues are reported by Cathay Holdings and Sinopec Holdings, respectively. The mean monthly stock return is 1.3492 (%). The mean capital adequacy ratio is 14.4958; the highest and lowest ratios are reported by Yuanta Financial Holdings (16.9839) and First Financial Holding (12.9983), respectively. The mean GDP growth rate is 3.855 (%), the mean unemployment rate is 3.8110 (%), and the mean monthly number of diagnosed cases is

Table 1 Descriptive Statistics

	Mean	St.d	Min	Max
staff	18,144.77	14.7584	9,275	58,000
revenue	1.43e + 07	1.50e + 07	– 1,696,861	7.99e + 07
log(revenue)	16.0113	0.9328	14.2989	18.1966
asset	4.24e + 09	2.51e + 09	1.60e + 09	1.16e + 10
log(asset)	22.0286	0.5052	21.1963	23.1737
stock	1.3492	5.1019	– 21.1454	24.2856
car	14.4958	0.8637	12.9983	16.9839
taiex	1.2342	4.6310	– 14.0284	13.2269
GDP	3.855	2.0572	0.63	9.2
UnempR	3.8110	0.2293	3.63	4.8
COVID-19	354.7708	1,365.761	0	7,383
esg_e	8.1805	0.5073	7.49	9.41
esg_s	8.3630	0.5395	7.6933	9.44
esg_g	8.1793	0.4963	7.58	9.62
esg_c	8.3180	0.5614	7.6933	9.54
esg_tot	33.0409	1.9256	30.7734	37.65

Note The information on this table is provided by this study

354.7708; the highest monthly number of diagnosed cases (7,383) is reported in May 2020. The ESG items are scored on a 10-point scale; the mean scores for environmental sustainability, social engagement, corporate governance, and corporate commitment are 8.1805, 8.3630, 8.1793, and 8.3180, respectively.

2.3 *Empirical Method*

The empirical analysis of this study consists of two steps. In the first step, panel data analyses are performed to observe the effects of the variables on the revenues of the holding companies. In the second step, a DID analysis is performed; specifically, the holding companies are divided into two types (i.e., companies that are listed and unlisted in the ESG rankings) to examine the effects before and after the pandemic.

- Step 1: Panel data analyses

  A random effects model is used for regression analysis, and the log of the revenue of holding companies serves as the dependent variable. The independent variables are data of these holding companies, macroeconomic market variables, the proxy variable for ESG engagement, and the pandemic variable. ESG scores comprise the scores of each of the four ESG items as well as their summed score. The regression model for the summed ESG score is as follows:

$$\begin{aligned}\log(\text{revenue}_{\text{it}}) = \alpha_0 + \alpha_1\text{staff}_{\text{it}} + \alpha_2\log(\text{asset}_{\text{it}}) + \alpha_3\text{car}_{\text{it}} \\ + \alpha_4\text{taiex}_t + \alpha_5\text{gdp}_t + \alpha_6\text{unemploy}_t \\ + \alpha_7\text{esg}_{\text{tot,it}} + \alpha_8\text{covid}_t + \varepsilon_{\text{it}}\end{aligned} \tag{1}$$

where α_0 is the intercept, α_1 to α_6 are the coefficients of the control variables, α_7 is the coefficient of the total ESG score, α_8 is the coefficient of the pandemic variable, and ε_{it} is the residual. Through a regression analysis of the ESG subitems in which ESG engagement is divided into environmental sustainability (esg_e), social engagement (esg_s), corporate governance (esg_g), and corporate commitment (esg_c), the regression model is

$$\begin{aligned}\log(\text{revenue}_{\text{it}}) = \beta_0 + \beta_1\text{staff}_{\text{it}} + \beta_2\log(\text{asset}_{\text{it}}) \\ + \beta_3\text{car}_{\text{it}} + \beta_4\text{taiex}_t + \beta_5\text{gdp}_t \\ + \beta_6\text{unemploy}_t + \beta_7\text{esg}_{e,\text{it}} + \beta_8\text{esg}_{s,\text{it}} \\ + \beta_9\text{esg}_{g,\text{it}} + \beta_{10}\text{esg}_{c,\text{it}} + \beta_{11}\text{covid}_t + \varepsilon_{\text{it}}\end{aligned} \tag{2}$$

where β_0 is the intercept, β_1 to β_6 are the coefficients of the control variables, β_7 to β_{10} are the coefficients of the ESG subitems, β_{11} is the coefficient of the pandemic variable, and ε_{it} is the residual. In this stage, ESG scores are examined to determine whether they have a significant effect on the revenues of large-scale holding companies during the pandemic and whether the effects of other variables on the revenues are significant. The regression formula used in this stage allows for the prediction of fitted revenue values, which serve as a variable for analysis in the next stage.

- Step 2: DID Model

  The large-scale holding companies are divided into those that are listed and unlisted in the ESG rankings; January 2020 serves as the division point between the pre- and post-pandemic periods.

  In Table 2, $D_{\text{esg,it}} = 1$ when a holding company is listed in a given year, and such companies constitute the experimental group; $D_{\text{esg,it}} = 0$ when a holding company is unlisted in a given year, and such companies constitute the control group. Meanwhile, $D_{\text{covid,it}} = 0$ and 1 for the pre-pandemic and post-pandemic periods, respectively. The DID estimated value that is estimated in this study is the difference between listed and unlisted companies in terms of their ESG scores after the pandemic; it is obtained by applying the formula $(\beta + \gamma) - \beta = \gamma$.

Table 2 DID analysis

	$D_{\text{esg,it}} = 0$	$D_{\text{esg,it}} = 1$	Difference in ESG
$D_{\text{covid,it}} = 0$	0	α	α
$D_{\text{covid,it}} = 1$	β	$\alpha + \beta + \gamma$	$\alpha + \gamma$
Difference in COVID	β	$\beta + \gamma$	γ

In this stage, the dependent variable is the monthly stock return, and the independent variable is the predicted fitted revenue obtained through the two-line regression analysis that was performed in the first stage; furthermore, nonlinear fluctuations as well as $D_{\text{esg,it}}$, $D_{covid,it}$, and the cross term $D_{\text{esg}} \times D_{\text{covid}}$ are considered. The DID model is obtained using the following formula:

$$\begin{aligned} \text{stock} = \theta + \delta_0 \widehat{\log(\text{revenue}_{\text{it}})} + \delta_1 \widehat{\log(\text{revenue}_{\text{it}}^2)} \\ + \alpha \cdot D_{\text{esg,it}} + \beta \cdot D_{\text{covid,it}} + \gamma \cdot \left(D_{\text{esg,it}} \times D_{\text{covid,it}}\right) + \varepsilon_{\text{it}} \end{aligned} \quad (3)$$

where θ is the intercept, δ_0 is the coefficient of the estimated revenue value, δ_1 is the nonlinear effect of the capture revenue, α is the coefficient of a listed ESG score, β is the post-COVID-19 coefficient, γ is the coefficient for being listed for ESG engagement after the pandemic, and ε_{it} is the residual.

3 Empirical Analysis

3.1 Effects on Revenue

Table 3 presents the results for the effect of the companies' total ESG scores on their revenues, which reveal that this effect is positive and significant with a coefficient is 0.0278, consistent with the expectations of this study. In other words, given that all other conditions are held constant, for every 1-point increase in a financial holding company's score, its revenue increases by 2.78%. In Table 4, the effects on revenue are viewed through ESG subitem scores, and significant effects are only observed for social engagement and corporate governance scores. The coefficient of the social engagement score is 0.1695. The corporate governance score has a significant negative effect on revenue, and its coefficient is – 0.1869.

For the pandemic-related variable, the number of diagnosed cases has a negative effect on the revenue of holding companies, and its coefficients are – 0.0013 (Table 3) and – 0.0029 (Table 4); however, this effect is nonsignificant, presumably for two reasons. First, because of the stable pandemic control in Taiwan, the fluctuation in the number of diagnosed cases is small. Additionally, because holding companies mainly serve domestic enterprises and individuals, the number of diagnosed cases has a nonsignificant effect on their revenues. For the macroeconomic variables, when the monthly return of the market is higher, revenue results are significant and positive. The coefficient is 0.0111 in both Tables 3 and 4, indicating the close association of these holding companies' revenues with Taiwan's capital market; this finding is consistent with the expectations of this study. For the unemployment rate variable, when the unemployment rate is high, it has a significant and positive effect on the revenues of holding companies; this finding suggests that a larger number of unemployed persons leads to an increase in the number of persons in the market who require loans; this phenomenon is further affected by the pandemic, which causes an

Table 3 Effect of ESG on revenue—estimated with total score of ESG

	Coef	Std.E	Z value	P value
esg_tot	0.0278***	0.0102	2.71	0.007
staff	0.0471***	0.0110	4.26	0.000
log(asset)	0.0562	0.1213	0.46	0.643
stock	0.0179	0.0163	1.10	0.272
car	0.0111***	0.0020	5.35	0.000
GDP	– 0.0026	0.0066	– 0.39	0.695
UnempR	0.1120**	0.0458	2.44	0.014
COVID-19	– 0.0013	0.0084	– 0.16	0.874
Constant	12.3034***	2.4908	4.94	0.000

Note The information on this table is provided by this study. *** represents significance at the 1% level. ** represents significance at the 5% level. * represents significance at the 10% level

Table 4 Effect of ESG on revenue—estimated with separated scores of ESG

	Coef	Std.E	Z value	P value
esg_e	– 0.1869***	0.0356	– 5.24	0.000
esg_s	0.0124	0.0596	0.21	0.834
esg_g	0.1695***	0.0475	3.56	0.000
esg_c	0.0673	0.0553	1.22	0.223
staff	0.0488***	0.0094	5.20	0.000
log(asset)	0.0282	0.1240	0.23	0.820
stock	0.0144	0.0162	0.89	0.374
car	0.0111***	0.0020	5.53	0.000
taiex	0.0065	0.0069	0.95	0.344
GDP	0.1225***	0.0445	2.75	0.006
UnempR	– 0.0029	0.0081	– 0.36	0.721
Constant	13.2410***	2.5636	5.16	0.000

Note The information on this table is provided by this study. *** represents significance at the 1% level. ** represents significance at the 5% level. * represents significance at the 10% level

increase in the number of persons requiring bailout loans and explains the positive correlation between unemployment rate and the revenues of holding companies. For the GDP variable, its effect on the revenues of holding companies is nonsignificant.

Table 5 DID effect on stock—estimated from total score of ESG

	(1)	(2)
$\widehat{\log(\text{revenue})}$	0.2333 (0.2758)	59.4510** (24.1555)
$\widehat{\log(\text{revenue})}^2$	–	− 1.7683** (0.7209)
D_{esg}	− 0.9771** (0.4915)	− 1.9411*** (0.5585)
D_{covid}	− 1.0953*** (0.3104)	− 1.5550*** (0.4945)
$D_{esg} \times D_{covid}$	2.3454*** (0.6722)	3.0308*** (0.8093)
Constant	− 2.1016 (4.2612)	− 495.313*** (201.2719)

Note The information on this table is provided by this study. *** represents significance at the 1% level. ** represents significance at the 5% level. * represents significance at the 10% level

3.2 Effect on Stock Price

In Table 5, the linear model results presented in column (1) show that the coefficient of the cross term is 2.3454 (1% significance level), indicating that the post-pandemic stock return performance of the holding companies that are listed in the ESG rankings is significantly superior to that of the companies that are unlisted. The nonlinear estimation findings in column (2) of Table 5 reveal that the coefficient of the cross term is 3.0308, which is also significant and positive. The findings are consistent with the expectations of this study, and they indicate that ESG can serve as a crucial reference indicator for investors during major global crises.

In Table 5, the fitted revenue value results are nonsignificant in the linear model; however, if the squared item in the nonlinear regression is considered, significant results are observed. The coefficient is 59.4510, indicating the positive effect of revenue on stock price; however, this effect progressively decreases when revenue increases.

Table 6 presents the estimation results for the four ESG subitems, and the trend is the same trend as that observed in Table 4. For the linear model in column (1), the coefficient of the cross term is 2.3447 (1% significance level). As revealed in column (2), if the nonlinear model is considered, the coefficient of the cross term is 2.8926, which is significant. The findings pertaining to the fitted revenue values in Table 6 are the same as the results presented in Table 5, which are nonsignificant in the linear model; however, if the squared item in the nonlinear regression are considered, significant results are observed; the coefficient is 41.7285.

Table 6 DID effect on stock—estimated from separated scores of ESG

	(1)	(2)
$\widehat{\log(\text{revenue})}$	0.2131 (0.2632)	41.7285** (19.9399)
$\widehat{\log(\text{revenue})^2}$	–	– 1.2415** (0.5953)
D_{esg}	– 0.9682* (0.5158)	– 1.6830** (0.7153)
D_{covid}	– 1.0921*** (0.3085)	– 1.4058*** (0.4178)
$D_{\text{esg}} \times D_{\text{covid}}$	2.3447*** (0.6852)	2.8926*** (0.8868)
Constant	– 1.7858 (4.0599)	– 347.074** (166.0784)

Note The information on this table is provided by this study. *** represents significance at the 1% level. ** represents significance at the 5% level. * represents significance at the 10% level

4 Conclusion

With the emergence of the ESG concept in recent years, numerous investors and institutions incorporated ESG engagement in their criteria for selecting investment targets. Various countries have established relevant standards and norms; therefore, ESG engagement plays a key role in current business practices. Meanwhile, the sudden outbreak of the COVID-19 pandemic provides researchers with an opportunity to examine the role of ESG engagement in a crisis. This study conducts empirical research using data on large financial holding companies in Taiwan and discovers that ESG performance has a significant and positive effect on the revenues of these companies. Meanwhile, because of the controlled pandemic situation in Taiwan, the pandemic's effect on the revenues of the holding companies is small. Through an examination of the effect of each of ESG variable on the revenues of the holding companies, the social engagement (*S*) dimension is revealed to be the main positive influencing factor, whereas the corporate governance (*G*) dimension is revealed to have a negative effect on the revenues of holding companies. This study posits that these findings can be attributed to the particularity of the financial sector. Specifically, the loan criteria that are applied by holding companies with favorable corporate governance are stricter than those applied by companies with unfavorable corporate governance; consequently, persons who require loans in the market prioritize loan options with easier loan conditions, and companies with a higher governance (*G*) score generate less revenue than those with a lower score. The empirical results reveal that unemployment rate is a significant influencing factor for the revenues of holding companies. This study holds that an increase in unemployment rate causes an increase in the demand for loans; therefore, the revenues of the companies that provide loans increase. In terms of stock pricing, the results of the DID model-based estimations reveal that the stock prices of the financial holding companies that are listed on ESG rankings are significantly higher than those of companies that are unlisted on ESG rankings, indicating that when a crisis occurs, the ESG score is a major reference indicator for investors.

This study differs from other ESG studies in that most of these studies do not focus on the effect of ESG engagement in the financial sector in Taiwan. Moreover, this study discovers that ESG performance has a mostly positive effect on the performance of holding companies but a negative effect on corporate governance (*G*). These findings contribute to the literature on ESG engagement. This study further uses DID estimation to compare companies that are listed and unlisted on ESG rankings and reports on the effects that are neglected in the literature.

References

Broadstock, D. C., Chan, K., Cheng, L. T. W., & Wang, X. (2021). The role of ESG performance during times of financial crisis: Evidence from COVID-19 in China. *Finance Research Letters, 38*, 101716.

Demers, E., Hendrikse, J., Joos, P., & Lev, B. (2021). ESG didn't immunize stocks during the COVID-19 crisis, but investments in intangible assets did. *Journal of Business Finance Accounting, 48*(3–4), 433–462.

Friede, G., Busch, T., & Bassen, A. (2015). ESG and financial performance: Aggregated evidence from more than 2000 empirical studies. *Journal of Sustainable Finance and Investment, 5*(4), 210–233.

Hwang, J., Kim, H., & Jung, D. (2021). The effect of ESG activities on financial performance during the COVID-19 pandemic-evidence from Korea. *Sustainability, 13*(20), 11362.

Lins, L. V., Servaes, H., & Tamayo, A. (2017). Social capital, trust, and firm performance: The value of corporate social responsibility during the financial crisis. *The Journal of Finance, 74*(4), 1785–1824.

Takahashi, H., & Yamada, K. (2021). When the Japanese stock market meets COVID-19: Impact of ownership, China and US exposure, and ESG channels. *International Review of Financial Analysis, 74*, 101670.

Factors Affecting the Relationship Quality Between Coffee Farmers and Enterprises: A Case Study of the Central Highlands of Vietnam

Hoa Thi Thu Ha, Dang Bach Nguyen, Hoa Le Dang, and Nhung Thi Hong Pham

Abstract The study analyzes factors affecting the relationship quality between coffee farmers and processing and export enterprises. A sample size of 171 coffee farmers in the Central Highlands of Vietnam is used for this study. The structural equation modeling (SEM) is utilized. The results show that five factors influencing the relationship quality are collaboration, perceived price, profit and risk sharing, power asymmetry, and effective communication. Perceived price has the strongest positive influence on the relationship quality. However, the relationship quality is negatively affected by power asymmetry. At the same time, a positive relationship between coffee farmers and processing and export enterprises brings farmers' profits and positively impact on the intention to continue trading. Policymakers should focus on improving the quality of relationships through information sharing and ensuring fairness in transactions between coffee farmers and processing and export enterprises.

Keywords Coffee farmers · Processing and export enterprises · Relationship quality · Central Highlands of Vietnam

1 Introduction

Coffee is Vietnam's top export agricultural product and distributed through many different channels. Farmers can choose to deal with various buyers depending on their preferences. Coffee is a main crop to alleviate poverty for farmers in the Central Highlands, with nearly 20% of the total export value (Nguyen et al., 2021). In the early 1980s, the whole country had about 20 thousand hectares of coffee, with an annual output of 4 to 5 thousand tons of coffee beans. After more than 30 years, Vietnam's coffee growing area has reached over half a million hectares with an output of more than 1 million tons. The Central Highlands is the largest coffee producing region of Vietnam. The coffee production area of the whole region accounts for

H. T. T. Ha (✉) · D. B. Nguyen · H. Le Dang · N. T. H. Pham
Nong Lam University, Ho Chi Minh City, Vietnam
e-mail: hoaha@hcmuaf.edu.vn

A. T. Nguyen et al. (eds.), *Contemporary Economic Issues in Asian Countries: Proceeding of CEIAC 2022, Volume 1*,
https://doi.org/10.1007/978-981-19-9669-6_19

more than 90% of the coffee production area of the country (Nguyen & Bokelmann, 2019). In recent years, the government has had guidelines and policies to support investment and development of coffee production. The planning-oriented coffee planting has contributed to creating more jobs and the economic development of smallholder farmers. At present, hundreds of private (including foreign-invested) and state-owned exporters/processors are thriving in the coffee industry, making the coffee bean market highly competitive.

Relationship quality helps unite farmers and their buyers, enhancing cooperation and mitigating the disadvantages of nature (Bandara et al., 2017). Moreover, when stakeholders trust this relationship, they will actively execute transactions in the most efficient way (Baihaqi & Sohal, 2013). However, farmers still face many limitations in participating in association with processing and exporting enterprises. The production and consumption of coffee still face many difficulties because the relationship is still loose and there is no legal effect. Farmers do not have power in their relationship with enterprises, so they often suffer losses and unable to protect their own interests. Among those difficulties and challenges, the problem of building the relationship between farmers and buyers is considered as one of the key issues, having a decisive influence on coffee production and consumption.

Most of the previous studies have addressed aspects of relationship quality in different agricultural products. However, the theoretical basis and methods are not the same in these studies. Most studies unify that the quality of the relationship is perceived and assessed through the perceptions of the stakeholders (Lees & Nuthall, 2015; Schulze & Lees, 2014; Schulze et al., 2006). Furthermore, the relationship between farmers and processing and exporting enterprises in the coffee industry is also different from that of other agricultural products (Gërdoçi et al., 2017; Nandi et al., 2018). Most of the studies in Vietnam focus on analyzing difficulties in the production and consumption of agricultural products (Nga & Niem, 2017; Truc & Hanh, 2017) or understanding the preferences of farmers to choose different distribution channels (Nguyen & Bokelmann, 2019; Pham et al., 2019). In the supply chain, studies on the quality of the relationship between farmers and processing and exporting enterprises are not focused. These studies rarely mention measures to help parties improve cooperation between them. As such, this study is conducted to identify determinants of the quality of the relationship. Several policy implications are proposed to enhance mutual satisfaction, help balance power, and facilitate information sharing among parties.

2 Methodologies

2.1 Study Area

The Central Highlands consists of a series of adjacent plateaus, located west of the Annamites mountains range with an average elevation of 500–800 m. The study area covers 54,508 km2, accounting for 16.5% of the area and approximately 6.1% of the population of Vietnam. There are five provinces in the Central Highlands of Vietnam, namely Gia Lai, Kon Tum, Dak Lak, Dak Nong, and Lam Dong. With the advantage of the monsoon climate and fertile basalt soil, the region has focused on developing industrial crops such as coffee, cocoa, pepper, and mulberry with high economic efficiency for households. Cashew and rubber trees are also grown here. Coffee is the most important industrial crop in this region. Besides, the Central Highlands is the second largest rubber growing area after the Southeast. It also has a large forest area with diverse biota, rich mineral reserves, and great tourism potential. The Central Highlands can be considered as the roof of the central region with a great protective function. However, destruction of natural resources and indiscriminate exploitation of forest products can lead to the risk of depleting forest resources and changing the ecological environment. In fact, the average area of coffee cultivation land of farmers is estimated at 1–2 ha/household.

Coffee is usually picked from October to December or from November to January. Normally, the provinces of Gia Lai, Kom Tum, Dak Lak, and Dak Nong are the regions that specialize in growing Robusta coffee. Arabica coffee is mainly grown in Lam Dong province. Currently, coffee development is not sustainable, requiring the urgent restructuring of this crop in association with the application of reasonable techniques. The application of scientific and technical advances has not been synchronized. The improper investment in fertilizer compared to the recommended process has resulted in soil degradation. The waste of irrigation water leads to a serious decrease in the groundwater level. In addition, farmers use pesticides inappropriately and in the wrong dosage. Most of the coffee growing area in Vietnam has no shade trees or windbreaks. Farmers harvest coffee when the percentage of green berries is still high. The processing facilities are not built to match the scale of production, reducing the quality of Vietnam's coffee. The coffee industry is also facing the situation of "old" coffee (yield less than 1.5 tons/ha, orchards over 25 years). Although the re-cultivation of coffee in the Central Highlands has been promoted, the area of "aging" coffee still accounts for a large proportion.

Local authorities have encouraged farmers to link up with processors to improve competitiveness in the world market. Processors and exporters in the region also rely on buying agents to guarantee their supplies of dried coffee beans. Some processing and exporting companies have been trying to contact coffee farmers, but the result is unremarkable. This explains that only a small proportion of dried beans are sold directly to processors or exporters. Currently, there is fierce competition between processors and exporters in collecting coffee in the market.

2.2 Empirical Studies and Research Hypotheses

The rationale for relationship quality is based on the relationship marketing theory (Crosby et al., 1990; Dwyer et al., 1987). Therein, interconnectivity and interdependence are the essences of business relationships. A long-term business relationship helps both parties share expectations about future transactions (Schulze & Lees, 2014; Schulze et al., 2006), maintaining long-term cooperation between buyers and suppliers (Fischer, 2013). Maintaining business relationships is a major challenge in a competitive environment. The relationship between buyers and sellers is mutually binding to share benefits in future transactions (Boniface, 2011). Structural features of business relationships include continuity, complexity, symmetry, and informality. Relationships are also complex because it includes both business relationships and social relationships. This relationship concerns the interests of buyers and sellers. In the supply chain, relationship management is necessary, but it needs to be done systematically and in accordance with each specific chain.

All business relationships are based on: trust, commitment, and satisfaction. Relationship quality is related to positive levels of information sharing, quality of long-term oriented communication, and relationship satisfaction. When the transaction process meets the expectations of the parties, it thus contributes to a stable relationship (Schulze & Lees, 2014). Cooperation is based on mutual trust between the parties (Capaldo, 2014; Ebrahim-Khanjari et al., 2011). The partners will commit to the implementation of the agreements, thereby maintaining and strengthening the relationship (Chen et al., 2011; Nyaga et al., 2010). The reciprocal adaptation of the partners binds them together, and this binding manifests in the mutual commitment between the partners.

Most of the approaches referred to transaction cost economics (TCE) theory (Bhagat & Dhar, 2014; Lees & Nuthall, 2015; Nandi et al., 2018; Son, 2011) and the behavioral approach (Boniface, 2011; Le & Batt, 2012; Loc & Nghi, 2018) in studying relationship quality in the agricultural sector. Other approaches also used theories such as organizational justice theory (Mühlrath et al., 2014), commitment-trust theory (Puspitawati, 2011), random utility theory (Newman & Briggeman, 2016), life cycle approach (Rota et al., 2013), and signaling theory (Bandara et al., 2017). Most studies use a combination of qualitative and quantitative methods. Most published studies still refer to empirical investigation. There has been an increase in the use of structural equation models (SEM) in studies of relationship quality in the agricultural sector in recent years (Bandara et al., 2017; Fischer, 2013; Loc & Nghi, 2018; Nandi et al., 2018). In addition, the OLS regression method and exploratory factor analysis or component analysis are widely used (Boniface, 2011; Son, 2011; Le & Batt, 2012; Dlamini-Mazibuko et al., 2019).

Many studies based on transaction cost economics perspectives have approached different governance structures. Vertical integration is a popular form of management in the supply chain. The cooperation reflects the level of mutual understanding between partners. This is an important factor because the stakeholders can dialogue and solve problems more simply and easily with good cooperation (Lees & Nuthall, 2015). In addition, the received price affects the implementation of agreements between farmers and their partners (Jena et al., 2011). The quality of the relationship is strengthened when the parties take measures to share profits and risks in transactions (Lages et al., 2005; Sun et al., 2018). In a Business-to-Business (B2B) relationship, power partners will allow or use their influence over the actions of other individuals or partners in an imperative manner (Bandara et al., 2017; Lees &

Nuthall, 2015). Effective communication gives farmers a significant advantage by being provided with the most appropriate information. Access to information allows parties to adapt to supply and market issues more quickly. It means that the actors are informed most suitably (Kac et al., 2016). The relationship enhances interests and the desire to continue doing transactions (Jena et al., 2011). At the same time, the parties always expect to continue to receive benefits from the relationship. Therefore, the result of a quality relationship is the intention of continuity in the execution of transactions (Schulze et al., 2006). In addition, quality relationships enable farmers to reinvest in their businesses and improve their long-term economic performance (Baihaqi & Sohal, 2013). In other words, the business performance of the stakeholders is improved through the relationship. Based on the research model adapted from Hoa et al. (2022), there are seven hypotheses raised in this study:

H1: The effect of collaboration on the relationship quality is positive.

H2: The effect of perceived price on the relationship quality is positive.

H3: The effect of profit and risk sharing on the relationship quality is positive.

H4: The effect of power asymmetry on the relationship quality is negative.

H5: The effect of effective communication on the relationship quality is positive.

H6: The effect of relationship quality on farmers' profitability is positive.

H7: The effect of relationship quality on intention to maintain the relationship is positive.

2.3 Data Collection

This study uses data from the districts of Dak Lak, Lam Dong, and Gia Lai province. First, the secondary data was collected from BMTCA (Buon Ma Thuot Coffee Association), VICOFA (Vietnam Coffee Cocoa Association), WASI (Western Highlands Agro-Forestry Scientific and Technical Institute), Lam Dong DARD, Dak Lak DARD, and Gia Lai DARD. Simultaneously, the structured questionnaire was built based on typical attributes of transaction costs to collect data for the study. The minimum sample size should be 170 because the number of observations should be five times the number of variables (Hair et al., 1998). This study used a direct survey method through structured questionnaires to interview coffee farmers who sold their products to enterprises. Most of the interviewed farmers have less than 2 hectares of coffee growing area. The quota sampling method was used to select the sample size for the study based on the total number of coffee households in each district and the production area of each household. Finally, SPSS 22.0 and AMOS 20.0 software were used for statistical analysis of coffee production and regression of factors affecting the quality of the relationship between farmers and enterprises through 171 completed questionnaires.

The study used the Likert scale has five points, where 1 = fully disagreement, 2 = disagreement, 3 = neutral, 4 = agreement, and 5 = fully agreement. Exploratory factor analysis and confirmatory factor analysis are used to test the discriminant and convergence of the scales in the model after they have met the requirements for reliability. Finally, structural equation modeling was used for evaluating the hypotheses test and estimating the proposed model. In the study, the popular goodness-of-fit tests often use measures such as chi-square, CMIN/df, CFI, TLI, GFI, and RMSEA. The study consists of eight concepts: Relationship quality (RQ), Collaboration (CN), Perceived price (PP), Profit and risk sharing (RS), Power asymmetry (PA), Effective communication (EC), Relationship continuity intention (CI), and Farmers' profit (FP) (Table 4).

3 Results

3.1 Overview of Coffee Production in the Central Highlands of Vietnam

Most of the surveyed farmers grow Rubosta coffee, accounting for 90% of coffee varieties in the area. The survey results also show that interviewed coffee farmers' age ranges from 20 to 67 years, with an average age of 40.1 years. Therein, the highest proportion is the age-group from 25 to 35 years old (accounting for 39.8%) and the age-group from 35 to 45 years old (accounting for 29.2%). Concurrently, the average education level is 9.4, and most of the household heads are male (accounting for 73.7%). The percentage of interviewed farmers in high school was 35.1%, in secondary school was 30.4%, and in primary school was 10.5%. At the same time, the average yield is 2–3 tons/ha and the profit is 60–80 million VND/ha/year. The average farm size is 1.4 hectares, which hampers farmers to benefit from economies of scale and apply synchronous production technologies. Farmers with 1–2 hectares of coffee land account for 22.8% of the entire sample.

In general, the cultivated area in the Central Highlands tends to decrease due to various reasons. The small-scale production of smallholder farmers leads to difficulties such as technological improvement, application of mechanization in production and processing, linkage in production and business, and access to market information. The number of interviewed households said that difficulties encountered in the production process were lack of capital, difficulty in accessing loans (28.3%), lack of productive land (28.7%), lack of labor (22.4%), and difficulty accessing production techniques (20.6%). Lack of water for irrigation in the dry season is also one of the difficulties that people face (accounting for 6.1%). Coffee gardens are located far from water sources, so a lack of water for irrigation in the dry season is unavoidable.

Market intermediaries are actors that exist between producers and final consumers. The number of market intermediaries involved in a marketing channel varies depending on the nature of the products and the place of growers. Actors in the coffee supply chain of the Central Highlands of Vietnam comprise farmers, local traders, purchasing agents, and processors/exporters. In this way, producers can sell at high prices by developing relationships with their buyers. Coffee farmers have been informed about changes in prices and export markets. Agents buy coffee beans from farmers through verbal contracts. Some coffee farmers deposit a large percentage of their coffee because of their credit tie-up. When coffee farmers feel the price is high, they decide to sell the quantity of coffee from their consignment. In addition, some companies have recently used direct, informal, and simple production contract with coffee farmers. However, exporters and processors can buy coffee beans through intermediaries (buying agents or local traders). In addition, some exporters and processors often sign long-term production contracts directly with smallholder farmers.

3.2 The Relationship Between Coffee Farmers and Enterprises

Three different types of contracts between coffee farmers and processing and export enterprises are informal contracts, intermediary contracts, and nucleus estate contracts. In an informal contract, the exporters and processors make informal production contracts with farmers on a seasonal basis. Specifically, some companies have recently used direct, informal, and simple production contracts with coffee farmers, such as Anh Minh, Simexco Dak Lak, and Armajaro. In this case, buyers (exporters and processors) only sign a contract with farmers. Companies only provide informal arrangements that specify the share of benefits and responsibilities of the two parties. An intermediary contract has seen as a formal subcontract by exporters and processors to intermediaries (cooperative or farmer groups) where they have their arrangements with farmers. The use of cooperatives as an intermediary is typical in this contract farming model in Amazaro and Dak Man Company. In a nucleus estate contract, the exporters and processors own and manage the estate plantation. Estate is often large and close to a processing plant where the firms can guarantee close supervision of production. These companies often sign long-term production contracts with farmers (e.g., Phuoc An and Thang Loi Company).

The survey results show that the relationship between farmers and coffee processing and export enterprises is not really effective, not ensuring long-term cohesion. Households said that they are (usually and always) satisfied with the companies accounting for 15.8 and 12.9% of the total interviewed households. Table 1 also shows that 20.5 and 14.0% of respondents trust enterprises to maintain their reputation by fulfilling their commitments. In addition, 18.7 and 14.6% of farmers said that the two sides committed to fully implementing the terms of the contracts. Moreover, the data also reveals the existence of conflicts between farmers and their buyers regarding the breach of the terms of the contracts. Specifically, there are conflicts of interest,

Table 1 Statistics of farmers' assessment of the relationship with enterprises

Aspects	Never	Rarely	Sometimes	Usually	Always	Total
Satisfaction	44 (25,7)	43 (25,1)	35 (20,5)	27 (15,8)	22 (12,9)	171
Trust	33 (19,3)	48 (28,1)	31 (18,1)	35 (20,5)	24 (14,0)	171
Commitment	43 (25,1)	26 (15,2)	45 (26,3)	32 (18,7)	25 (14,6)	171

Source The authors' calculations
Note Numbers in brackets are percentages

disputes related to time for delivery, receipt, and payment, disputes over product quality, or non-compliance with commitments.

3.3 *The Reliability of Scales*

Four observed variables were removed after performing Cronbach's Alpha analysis (Gliem & Gliem, 2003). Besides, EFA analysis has a Kaiser–Meyer–Olkin (KMO) value equal to 0.854 ($0.5 < KMO < 1$) with a significance level of 0.000 (< 0.05). Therefore, the factor analysis well fits with the data. All variables in the rotation factor matrix have factor loadings higher than 0.5 (Table 2). In conclusion, the variables are strongly correlated in each construct of the proposed model. The cumulative variance of the eight factors is 71.725% (with eigenvalue $= 1.077 > 1$). Bartlett's test has Sig. $= 0.000 < 0.05$ shows that the variables are fully correlated. The results of factor analysis are acceptable (Anderson & Gerbing, 1988; Cudeck, 2000).

After the EFA step, the scales were checked again by the CFA method to ensure more certainty about the reliability and validity of the scales. The CFA results show that the appropriateness index of the theoretical model has the expected results, such as RMSEA $= 0.057$ (< 0.08); CFI $= 0.942$; TLI $= 0.933$; and GFI $= 0.827$ (Table 3). In addition, all results are acceptable with indices of conformity such as CMIN/df $= 1.556$ (< 3); chi-square $= 586.485$; df $= 377$. All factor loadings are more than 0.5; therefore, observed variables have a close relationship with their representative factors. As a result, the data are suitable to represent the research model (Steiger, 1990). Contemporary, Average Variance Extracted (AVE) is more than 0.5, Composite Reliability (CR) of the scales is more than 0.7. In addition, the squared correlation must be lower than the AVE for the two constructs. Therefore, the scales in the model achieve the value of discrimination (Fornell & Larcker, 1981).

Table 2 Results of EFA analysis

Factors	Factor loadings								
Effective communication	EC1	0.774							
	EC2	0.789							
	EC3	0.913							
	EC4	0.967							
Farmers' profit	FP1		0.728						
	FP2		0.893						
	FP3		0.796						
	FP5		0.889						
Power asymmetry	PA1			0.858					
	PA2			0.825					
	PA3			0.884					
	PP4			0.776					
Relationship continuity intention	CI1				0.674				
	CI2				0.774				
	CI3				0.837				
	CI4				0.835				
Relationship quality	RQ1					0.705			
	RQ2					0.949			
	RQ3					0.609			
	RQ4					0.938			
Perceived price	PP1						0.676		
	PP2						0.714		
	PP3						0.713		
	PP4						0.850		
Collaboration	CN1							0.854	
	CN2							0.912	
	CN3							0.725	
Profit and risk sharing	RS1								0.838
	RS2								0.939
	RS3								0.824
Eigenvalues		9.154	3.968	2.499	2.135	1.835	1.672	1.347	1.077
Cumulative variance = 71.725%		29.603	12.404	7.343	6.151	5.155	4.761	3.609	2.700

Source Survey data

Table 3 Results of convergent and discriminant validity test

Component scales	CR	AVE	MSV
Collaboration (CN)	0.912	0.776	0.490
Perceived price (PP)	0.833	0.556	0.195
Profit and risk sharing (RS)	0.903	0.758	0.183
Power asymmetry (PA)	0.911	0.721	0.299
Effective communication (EC)	0.920	0.742	0.046
Relationship quality (RQ)	0.895	0.683	0.231
Relationship continuity intention (CI)	0.881	0.650	0.490
Farmers' profit (FP)	0.901	0.695	0.187

Source Survey data

3.4 Structural Equation Modeling Analysis and Hypothesis Test

SEM was performed to determine the prefixes of relationship quality and their influence on the relationship. SEM conducts an analysis of the initially proposed research model after having the appropriate CFA results. The SEM analysis indicators confirm the theoretical model's statistically significant fit (df = 388, chi-square = 680.492, CMIN/df = 1.754 < 3). The estimated parameters of the factors affecting the relationship quality include: perceived price factor is 0.26; effective communication is 0.23; profit and risk sharing factor is 0.18; collaboration factor is 0.17; power asymmetry factor is – 0.31. At the same time, the suffixes of the relationship quality are the farmer's profit factor with the estimated parameter of 0.46 and the intention to maintain the relationship with the estimated parameter of 0.45 (Fig. 1).

The results show that 48% of the variance of relationship quality is explained by the five independent variables in the above model. In addition, the P-value of the estimated parameters in the SEM model is less than the significance level of 5%. Therefore, seven hypotheses are accepted in this study. The Bootstrap method was used with a large number of observations (including 500 random observations). It involves repeated resampling of an original dataset to verify the reliability of the estimated parameters (Schumacker & Lomax, 2004). The Bootstrap method evaluates the estimates in the model with a confidence level of 95%. The results show that the SE-Bias of the estimators is less than 0.05. Thus, these results confirm the reliability of the estimates in the model.

4 Discussions

The findings can be better explained by incorporating the contexts of practice. In Vietnam, agricultural transactions are done through the spot market. Growers are

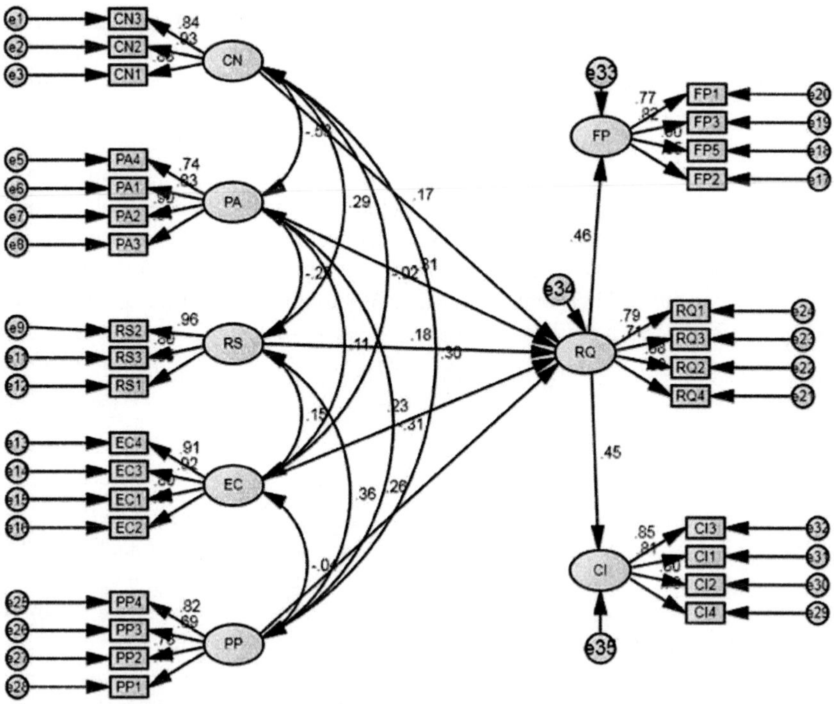

Fig. 1 Factors affecting relationship quality (*Source* Survey data)

easily attracted to buyers who offer reasonable prices. Farmers often compare the price received with the price that neighboring households sell to others. Farmers have desire to continue future transactions when they are satisfied with the price received from the processors and exporters. In addition, sustainable business relationships are enabled and enhanced through communication, which can be seen as formal and informal information sharing among chain stakeholders. Communication can be defined as the catalyst that keeps partners together. Enterprises share difficulties with farmers to help reduce uncertainty through the exchange of information on markets and production techniques.

Profit and risk sharing is an effective measure to control volatility in the market and improve exchange efficiency. Farmers will join with buyers interested in sharing profits and risks with them. Farmers are supported when crop yields are low or market prices are high. Therefore, transactions between coffee farmers and processing and export enterprises are driven by profit and risk sharing factor. Next, the cooperation is used as a basis for deciding future actions with the trading partner. Farmers cooperate directly with partners by establishing regular contacts. Thus, active cooperation reduces the probability of switching to another buyer. In fact, good relationships can survive when conflicts are resolved. Furthermore, the cooperation in business relationships primarily enhances (directly and indirectly) trust and stability. Therefore, it is clear that building trust in a business relationship is based on active cooperation.

Power asymmetry leads to unfairness in transactions, reducing the motivation for cooperation between parties. As a result, power asymmetries can reduce trust and commitment, damaging relationship quality between farmers and enterprises. Relationship quality can be negatively affected by abusing stronger market positions. Finally, the motivation to improve the relationship between farmers and coffee buyers is beneficial for the farmers. In other words, the business performance of the stakeholders is improved through the relationship. Long-term relationships are more profitable while boosting efficiency, productivity, and performance.

5 Conclusions

The study identifies determinants of the perceived relationship quality based on TCE theory. Relationship quality contributes to stabilizing coffee production in the Central Highlands. The relationship between farmers and businesses helps farmers raise their income. Processors and exporters are vital actors, enabling farmers to optimize coffee production in the Central Highlands. The study identifies five factors that influence relationship quality in the agricultural supply chain, including collaboration, profit and risk sharing, perceived price, effective communication, and power asymmetry. The perceived price factor plays the most important role, followed by effective communication. Power asymmetry is a major cause of relationship entanglement. The study also demonstrates the impact of relationship quality on profitability and intention to continue trading between coffee farmers and processing and export enterprises.

In fact, the relationship between stakeholders is not well established in the agricultural supply chain due to the lack of transparency and mutual information sharing. Policymakers should develop programs to encourage linkages and ensure legal requirements for the relationship between coffee farmers and enterprises. This study has a small sample size (171 households) and only identifies a few factors affecting the quality relationship between coffee farmers and processing and export enterprises. Many other factors were not included in this study. Future studies may consider the relationship quality with a larger sample size and use the perspectives of both sides (including farmers and buyers).

Appendix A

See Table 4.

Table 4 Interpretation of observed variables in the research model

Sign	Items	Factors
RQ1	You can trust the buyer	Relationship quality (RQ)
RQ2	You are satisfied with the transaction with the buyer	
RQ3	The relationship with the buyer meets your expectations	
RQ4	The buyer does not violate the contract/agreement/commitment with you	
CN1	The buyer can handle your inquiries	Collaboration (CN)
CN2	The buyer cooperates in solving problems with you	
CN3	There is good cooperation between the buyer and you	
PP1	Purchasing price is commensurate with coffee quality	Perceived price (PP)
PP2	Purchasing price of the buyer is reasonable	
PP3	Coffee sold to the buyer is always at good prices	
PP4	The buyer's price does not fluctuate	
RS1	The buyer is willing to share the risk with you	Profit and risk sharing (RS)
RS2	The buyer is willing to share difficulties in production and consumption	
RS3	The buyer is willing to pay more when the market price increases	
PA1	The buyer is very powerful in the relationship	Power asymmetry (PA)
PA2	The buyer controls all the information in the relationship	
PA3	The buyer has a strong influence on you	
PA4	You must follow the requirements of the buyer	
EC1	The buyer provides information about market movements for you	Effective communication (EC)
EC2	You can easily contact the buyer	
EC3	You have regular contact with the buyer	
EC4	The information provided by the buyer is timely and reliable	
CI1	You will continue to sell coffee to the buyer	Relationship continuity intention (CI)
CI2	Your relationship with the buyer is long-lasting	

(continued)

(continued)

Sign	Items	Factors
CI3	You will continue to maintain a long-term relationship with the buyer	
CI4	You will introduce the buyer to other households	
FP1	Building relationships with buyers helps improve economic efficiency	Farmers' profit (FP)
FP2	A good relationship with the buyer makes coffee easy to sell, as expected	
FP3	Selling coffee to the buyer helps farmers have a more stable income	
FP5	The relationship creates the linkage between production and consumption	

References

Anderson, J. C., & Gerbing, D. W. (1988). Structural equation modeling in practice: A review and recommended two-step approach. *Psychological Bulletin, 103*(3), 411–423. https://doi.org/10.1037/0033-2909.103.3.411

Baihaqi, I., & Sohal, A. S. (2013). The impact of information sharing in supply chains on organisational performance: An empirical study. *Production Planning & Control, 24*(8–9), 743–758. https://doi.org/10.1080/09537287.2012.666865

Bandara, S., Leckie, C., Lobo, A., & Hewege, C. (2017). Power and relationship quality in supply chains. *Asia Pacific Journal of Marketing and Logistics, 29*(3), 501–518. https://doi.org/10.1108/APJML-09-2016-0165

Bhagat, D., & Dhar, U. (2014). Relationship dynamics in the pineapple supply chain: Empirical evidence from the Garo Hills of Meghalaya. *Global Business Review, 15*(4), 747–765. https://doi.org/10.1177/0972150914543410

Boniface, B. (2011). Producer segmentation and the role of long-term relationship in Malaysia's milk supply chains. In *Australian Agricultural and Resource Economics Society AARES Conferences (55th), No. 422–2016–26898.*

Capaldo, A. (2014). Network governance: A cross-level study of social mechanisms, knowledge benefits, and strategic outcomes in joint-design alliances. *Industrial Marketing Management, 43*(4), 685–703. https://doi.org/10.1016/j.indmarman.2014.02.002

Chen, J. V., Yen, D. C., Rajkumar, T. M., & Tomochko, N. A. (2011). The antecedent factors on trust and commitment in supply chain relationships. *Computer Standards & Interfaces, 33*(3), 262–270. https://doi.org/10.1016/j.csi.2010.05.003

Crosby, L. A., Evans, K. R., & Cowles, D. (1990). Relationship quality in services selling: An interpersonal influence perspective. *Journal of Marketing, 54*(3), 68–81. https://doi.org/10.1177/002224299005400306

Cudeck, R. (2000). Exploratory factor analysis. In *Handbook of applied multivariate statistics and mathematical modeling* (pp. 265–296). Elsevier.

Dlamini-Mazibuko, B. P., Ferrer, S., & Ortmann, G. (2019). Examining the farmer-buyer relationships in vegetable marketing channels in Eswatini. *Agrekon, 58*(3), 369–386. https://doi.org/10.1080/03031853.2019.1596824

Dwyer, F. R., Schurr, P. H., & Oh, S. (1987). Developing buyer-seller relationships. *Journal of Marketing, 51*(2), 11–27. https://doi.org/10.1177/002224298705100202

Ebrahim-Khanjari, N., Hopp, W., & Iravani, S. M. R. (2011). Trust and information sharing in supply chains. *Production and Operations Management, 21*(3), 444–464. https://doi.org/10.1111/j.1937-5956.2011.01284.x

Fischer, C. (2013). Trust and communication in European agri-food chains. *Supply Chain Management: An International Journal, 18*(2), 208–218. https://doi.org/10.1108/13598541311318836

Fornell, C., & Larcker, D. F. (1981). Evaluating structural equation models with unobservable variables and measurement error. *Journal of Marketing Research, 18*(1), 39–50. http://www.jstor.org/stable/3151312

Gërdoçi, B., Skreli, E., Zhllima, E., & Imami, D. (2017). Determinants of long-term business relationships in the dairy value chain in transition countries: the case of Albania. *Studies in Agricultural Economics, 119*(3), 139–147. https://doi.org/10.22004/ag.econ.266793

Gliem, J. A., & Gliem, R. R. (2003). Calculating, interpreting, and reporting Cronbach's alpha reliability coefficient for Likert-type scales. In *Paper Presented at the 2003 Midwest Research to Practice Conference in Adult, Continuing, and Community Education.* Columbus, Ohio: Ohio State University.

Hair, J. F., Black, W. C., Babin, B. J., Anderson, R. E., & Tatham, R. L. (1998). *Multivariate data analysis.* New Jersey, USA: Prentice hall Upper Saddle River.

Hoa, H. T. T., Dang, N. B., Hoa, D. L., & Nhung, P. T. H. (2022). Factors affecting the relationship quality between coffee farmers and local traders: A case study in a highland commune of Dak Lak, Vietnam. *The Journal of Agriculture and Development, 21*(3), 1–11.

Jena, S., Guin, K. K., & Dash, S. B. (2011). Effect of relationship building and constraint-based factors on business buyers' relationship continuity intention. *Journal of Indian Business Research, 3*(1), 22–42. https://doi.org/10.1108/17554191111112451

Kac, S. M., Gorenak, I., & Potocan, V. (2016). The influence of trust on collaborative relationships in supply chains. *Ekonomika a Management, 19*(2), 120–131. https://doi.org/10.15240/tul/001/2016-2-008

Lages, C., Lages, C. R., & Lages, L. F. (2005). The RELQUAL scale: A measure of relationship quality in export market ventures. *Journal of Business Research, 58*(8), 1040–1048. https://doi.org/10.1016/j.jbusres.2004.03.001

Le, B., & Batt, P. (2012). Farmer-buyer relationships in the cut flower supply chain in Da Lat, Vietnam. In *Paper Presented at the IV International Symposium on Improving the Performance of Supply Chains in the Transitional Economies 1006* (pp. 209–216).

Lees, N., & Nuthall, P. (2015). Collaboration, cooperation and power in food supply chains. In *Paper Presented at the Contributed Paper Prepared for Presentation at the 59th AARES annual conference, Rotorua, New Zealand.*

Loc, H. T. H., & Nghi, N. Q. (2018). Factors influencing the relationship quality between shrimp farmers and processing enterprises in the Mekong Delta. *International Journal of Business and Management Invention (IJBMI), 7*(9), 45–51.

Mühlrath, D., Möller, D., & Schumacher, J. (2014). Fairness and satisfaction in business relationships: Results of a survey among Swiss organic farmers and buyers. *Building Organic Bridges, 4*, 1027–1030.

Nandi, R., Gowdru, N. V., & Bokelmann, W. (2018). Determinants of contractual choice and relationship sustainability in organic fruits and vegetable supply chains: Empirical evidence from stakeholder's survey from South India. *International Journal of Agricultural Marketing, 5*(1), 178–193.

Newman, C., & Briggeman, B. C. (2016). Farmers' perceptions of building trust. *International Food and Agribusiness Management Review, 19*(1030–2016–83136), 57–76.

Nga, D. T., & Niem, D. L. (2017). The linkage between farmers and businesses in coffee production and consumption in the Central Highlands. *Journal of Agricultural Science of Vietnam, 14*(11), 1835–2184.

Nguyen, A. H., & Bokelmann, W. (2019). Determinants of smallholders' market preferences: The case of sustainable certified coffee farmers in Vietnam. *Sustainability, 11*(10), 2897. https://doi.org/10.3390/su11102897

Nguyen, T. T. H., Bekrar, A., Le, T. M., & Abed, M. (2021). Supply chain performance measurement using SCOR model: A case study of the coffee supply chain in Vietnam. In *Paper Presented at the 2021 1st International Conference on Cyber Management and Engineering* (*CyMaEn*) (pp. 1–7). https://doi.org/10.1109/CyMaEn50288.2021.9497309

Nyaga, G. N., Whipple, J. M., & Lynch, D. F. (2010). Examining supply chain relationships: Do buyer and supplier perspectives on collaborative relationships differ? *Journal of Operations Management, 28*(2), 101–114. https://doi.org/10.1016/j.jom.2009.07.005

Pham, T. T., Theuvsen, L., & Otter, V. (2019). Determinants of smallholder farmers' marketing channel choice: Evidence from the Vietnamese rice sector. *Asian Economic Journal, 33*(3), 281–300. https://doi.org/10.1111/asej.12187

Puspitawati, E. (2011). Determinants of trust in the Indonesian potato industry: A comparison among groups of potato farmers. In *Paper Presented at the Australian Agricultural and Resource Economics Society Conference* (*55th*), *Melbourne, Australia.*

Rota, C., Reynolds, N., & Zanasi, C. (2013). Sustainable food supply chains: The role of collaboration and sustainable relationships. *International Journal of Business and Social Science, 4*(4), 45–53.

Schulze, B., Wocken, C., & Spiller, A. (2006). Relationship quality in agri-food chains: Supplier management in the German pork and dairy sector. *Journal on Chain and Network Science, 6*(1), 55–68. https://doi.org/10.3920/JCNS2006.x065

Schulze, B., & Lees, N. J. (2014). Supplier relationship quality and performance in (agribusiness) supply chains: A literature review. In *Paper presented at the 11th Wageningen International Conference on Chain and Network Management: Sustainability and Innovation in Chains and Networks, Naples, Italy.*

Schumacker, R. E., & Lomax, R. G. (2004). *A beginner's guide to structural equation modeling.* Psychology Press.

Son, N. P. (2011). Relationship quality in fish value chains: Buyer–supplier management in the pangasius industry, Vietnam. In *Environmental Change and Agricultural Sustainability in the Mekong Delta* (pp. 287–301). Springer.

Steiger, J. H. (1990). Structural model evaluation and modification: An interval estimation approach. *Multivariate Behavioral Research, 25*(2), 173–180. https://doi.org/10.1207/s15327906mbr2502_4

Sun, Y., Liu, Z., & Yang, H. (2018). How does suppliers' fairness affect the relationship quality of agricultural product supply chains? *Journal of food quality, 2018*, Article ID 9313068. https://doi.org/10.1155/2018/9313068

Truc, P. T. T., & Hanh, N. T. T. (2017). The linkage between actors in coffee value chain in the Central Highlands of Vietnam. *Journal of Science and Technology, University of Danang, 2*(111), 114–117.

How Adaptive Leadership Facilitates Employees' Career Adaptability During the COVID-19 Pandemic?

Anh Tho To

Abstract The spread of the COVID-19 epidemic has significantly influenced human resource management and organizational behavior. A lack of adaptability and failure to find solutions to change might be a leadership weakness. In the context of the pandemic in particular, and the crisis in general, the traditional leadership style has become inactive. Organizations should use an adaptive leadership style to respond to changes in the business environment. The following are research questions: What are the primary functions of an adaptive leadership style? What are the roles of adaptive leaders? Through what aspects is the career adaptability of employees demonstrated? This exploratory study is based on the author's observations and literature analysis in the context of the COVID-19 epidemic. We provide a theoretical framework of the adaptive leadership competencies required in times of crisis, offering several useful implications for human resource development practitioners.

Keywords COVID-19 · Adaptive leadership · Career adaptability

1 Introduction

Due to social isolation and required limitations, the COVID-19 pandemic has had a significant negative influence on the workplace in 2020. Businesses and employees all across the world have been severely impacted by workplace closures and necessary measures to stop the disease's spread. Due to temporary closures caused by worries about the spread of infectious illnesses, several businesses and corporations have made plans to cut hours to avoid layoffs, while others face the risk of having to stop operating permanently because of not being able to survive the pandemic. An estimation of 8.8% of total working hours has been lost internationally, which equates to the number of hours worked annually by 255 million full-time employees by 2020 (ILO, 2021).

A. T. To (✉)
University of Finance—Marketing, Ho Chi Minh City, Vietnam
e-mail: totho@ufm.edu.vn

A. T. Nguyen et al. (eds.), *Contemporary Economic Issues in Asian Countries: Proceeding of CEIAC 2022, Volume 1*,
https://doi.org/10.1007/978-981-19-9669-6_20

Employees take longer to perform tasks during the COVID-19 pandemic because they need to take the required measures, follow cleaning rules, and prevent crowding at the workplace. Given that most employees are accustomed to set procedures in organizations, these changes result in slower production and operations overall, which in turn results in a decline in productivity. Additionally, firms and employees are shifting toward more work from home due to the COVID-19 epidemic. "*Nearly 70 percent of professionals who transitioned to remote work because of the pandemic say they now work on the weekends, and 45 percent say they regularly work more hours during the week than they did before, according to a survey of 2,800 workers by Los Angeles-based staffing firm Robert Half*" (Maurer, 2020).

Many businesses think that training may help employees become more career adaptable, which will aid them in adjusting to tasks, work roles, and careers in unpredictable environments (Rasheed et al., 2020). Because career adaptability is frequently thought of as a type of human capital, Brown et al. (2012) also believe that career adaptability is acquired through time in the workplace through new knowledge, experience, and education. According to Porfeli and Savickas (2012), adaptable individuals are "*becoming concerned about the vocational future, taking control of trying to prepare for one's vocational future, displaying curiosity by exploring possible selves and future scenarios, and strengthening the confidence to pursue one's aspirations*". According to Yang et al. (2019) and Rossier et al. (2017), career adaptability could help a person determine their career path, recover from unforeseeable outcomes, eliminate decision-making difficulties, and achieve better performance. Career adaptability is associated with job and life satisfaction (Fiori et al., 2015; Guan et al., 2015).

To create career adaptability for employees, leaders need to build and maintain an adaptive leadership style. Yukl and Mahsud (2010) discovered that adaptive leaders are expected to: be able to foresee future scenarios and use a variety of suitable behaviors; and proactively construct environments and organizations that do not need monitoring closely. It is required of adaptive leaders adjust to changes in the environment and skillfully apply their knowledge to guarantee the achievement of organizational objectives. They frequently exhibit a positive outlook, encourage, and urge their staff to do better for the company (Afsar et al., 2014; Caniëls et al., 2018; Joo & Lim, 2013). Along with developing an adaptive workforce, they are also skilled at retaining highly adaptable workers that are successful at enhancing the operations of the company and contributing significantly to the organization's success (Safavi & Bouzari, 2019).

This study enriches career construction theory by developing a framework of adaptive leadership as a predictor of employees' career adaptability. By highlighting the importance of adaptive leadership in fostering employees' career adaptability throughout the COVID-19 epidemic, this study makes a significant contribution to the body of literature on leadership and career development.

2 The Impact of the COVID-19 Pandemic

During the COVID-19 epidemic, a business had to increase its disease control measures if it wanted to continue operating. While manufacturers and big businesses were investing the most in healthcare systems, other businesses had kept their employees on-site by providing food, accommodation, and medical care so they can continue to work in a COVID-free environment. To be retained on the job for a long period, nevertheless, is not something that all employees agree to and are willing to do. Because it makes people feel cut off from the community, and social interaction is a fundamental human need. Humanity is always seeking methods to adapt and maintain social contacts to recover, particularly with the growing support of information and communication technology.

The COVID-19 epidemic has also inspired a widespread trend of work-from-home solutions. Since the start of the epidemic, 91% of workgroups in the Asia Pacific have made arrangements to work remotely, and 80% of firms actively promote or request that their workers work from home (Levin, 2020). Employee autonomy and satisfaction are not, however, increased by remote working. It still has certain drawbacks such as the need for childcare, issues with internet connectivity, the need for a quiet workspace, etc.

Businesses have used digital platforms to face and overcome challenges during the epidemic. Small firms use digital technology for certain standard company operations like sales and payments. Larger-scale applications are made across the supply chain management by large businesses, from planning to controlling. Employees are under a lot of pressure to adapt and learn constantly because of this transformation. In other words, this epidemic is likely to speed up automation, the adoption of new technologies, and the transformation of the workplace. HR experts should come up with fresh ideas and efforts to put into practice the required training activities to assist employees in adjusting to the work-from-home environment and enhancing their employability skills so they are better equipped to handle new challenges (Li & Zehr, 2020).

3 Change for Adaptability

3.1 Career Adaptability

It has been defined as "*a psychosocial construct that denotes an individual's resources for coping with current and anticipated tasks, transitions, and traumas in their occupational roles*" (Savickas & Porfeli, 2012). Employees must possess career adaptability as a necessary personal skill to carry out their responsibilities efficiently and respond to changes in the always-changing workplace (Ferreira, 2019; Tladinyane & Van der Merwe, 2016). In other words, it's the psychological and resource state required to fulfill present and expected professional developments, including the attitudes, skills, and behaviors employees need to be suitable for the workplace (Savickas, 2005). It entails the capacity to change with the demands of the job, engage in ongoing learning, and manage one's professional path. Career-related changes might help employees become more adaptable (Ocampo et al., 2020). Career adaptability is required to meet the rising demand from employers for a highly flexible workforce.

Career adaptability enables workers to predict the potential for unanticipated changes and respond to them proactively to reduce risk or recover (Ginevra et al., 2018; Rudolph et al., 2017). Proactive employees are better prepared for changes in their careers (Tolentino et al., 2014). Additionally, prior research has demonstrated that career adaptability greatly enhances employee well-being and work performance (Ferreira, 2019; Tolentino et al., 2013; Yu & Zheng, 2013; Zacher et al., 2015), reduced work stress, job satisfaction, and organizational commitment (Maggiori et al., 2013; Rossier et al., 2017; Tladinyane & Van der Merwe, 2016; Zacher, 2014).

From the perspective of career-building theory, career adaptability is expressed in four main attitudes (Savickas, 2005; Savickas & Porfeli, 2012): Concern, control, curiosity, and confidence. "*Concern involves planning for the future, control implies a personal responsibility to shape the future, curiosity leads to exploring possible roles, and confidence is belief in one's ability to achieve goals and implement choices*" (Lee et al., 2021). These "4C" psychological characteristics might help people consider different options, assess their environment, create objectives, and keep trying to adapt. Guan et al. (2016) and Luke et al. (2016) stated that these components work together to motivate employees to manage and effectively adapt to the current and rapidly changing world of work.

3.2 Adaptive Leadership

Adaptive leadership is found in leaders who have a high degree of adaptability to changes and difficulties, foresee potential future developments, have a positive outlook, and are willing to accept difficulties and failure (Asaari & Hasmi, 2012). It is required to manage any adaptive changes and challenges (Randall & Coakley, 2007). The adaptive challenges require us to develop new ways of perceiving and acting, in contrast to the technical challenges which are sometimes quite obvious and may be resolved using the existing techniques (Heifetz et al., 2009).

The foundation of adaptive leadership is based on the view that leadership is a process rather than a personal quality (Heifetz et al., 2004). The best illustration of adaptive leadership is that of an educator calling on learners to discover, invent, and take collective responsibility for their situation (Heifetz & Heifetz, 1994). Adaptive leadership is helpful when no easy solution is available. It is quite important for managers who want to find innovative solutions in a world of constant change. Most adaptable leaders frequently experiment to determine the most effective means of shaping the current context.

According to Mulder (2017) and Ramalingam et al. (2020), adaptive leadership focuses on four dimensions as follows:

Ramalingam et al. (2020)	Mulder (2017)
"*Anticipation of likely future needs, trends and options*"	"Navigating business environments: *Adaptive leaders should embrace uncertainty and encourage the organization to look for new approaches*"
"*Articulation of these needs to build collective understanding and support for action*"	"Leading with empathy: *Adaptive leaders create a group attitude instead of an atmosphere of divide and conquer. An adaptive leader will understand alternative perspectives and ideas and be able to share them with others*"
"*Adaptation so that there is continuous learning and the adjustment of responses as necessary*"	"Learning through self-correction and reflection: *Adaptive leaders ensure that their organization is always aware of what's going on outside it and what customers are thinking. By identifying mistakes and problems properly and in a timely manner, they can be tracked down, filtered, and decoded —giving the organization a chance to respond*"
"*Accountability, including maximum transparency in decision making processes and openness to challenges and feedback*"	"Create win–win solutions: *An adaptive leader values platforms for cooperation and builds on them*"

3.3 *Leadership Practices for Career Adaptability in the Context of COVID-19 Epidemic*

In an ever-changing work environment, leadership serves as a guide and aids employees in adapting and working more effectively toward the general objectives of the business in a workplace that is always changing. Employee career adaptability is more likely to be improved by leadership practices that prioritize employee development, provide clear direction and expectations for the workforce, and encourage and support worker involvement at work (Bardoel et al., 2014; King & Rothstein, 2010).

Crises force leaders to take on multiple roles and tasks at once due to time pressure and limited resources. Adaptive leaders establish a clear vision for their followers and serve as facilitators, advisors, and trainers to them. According to Lacerda (2019), leaders who can envision and forecast the organization's future are effective under difficult circumstances. They provide an explanation of an organization's purpose and then empower followers to select how work should be carried out to accomplish the organization's ultimate objectives. Adaptive leaders promote more employee involvement and encourage them to develop creative solutions (Afsar et al., 2014).

Therefore, Özşahin et al. (2011) demonstrate that a leader's capacity to establish strong connections through challenging times is essential.

An organization's adaptive leadership style can lessen employee burnout and boost work satisfaction, psychological capital, organizational citizenship, and job engagement (Fiori et al., 2015; Singh et al., 2018). It could assist employees to identify their future requirements and prospects to prepare for the next assignments and job transitions. Individualized attention from adaptive leaders enables followers to feel confident in their capacity to handle their job and more readily adapt to changes (Afsar et al., 2014; Tims et al., 2011). Continuous and open communication enables followers to easily express their demands and concerns without fear (Afsar et al., 2014). Adaptive leaders make sure that followers have more resources necessary to deal with changing conditions.

Leaders should maintain some competencies in difficult times. According to Koehn (2020), it's important to acknowledge fear, give people responsibilities and goals, concentrate on learning, and pay attention to the people involved's feelings and energies. Additionally, Schwantes (2020) mentioned being adaptable, taking emotions into account, staying involved, and hearing their opinions. As shown in Fig. 1, leaders' competencies can be summarized into four parts as follows:

Responsiveness to Change Leaders need to be able to recognize changing events and take rapid action in response. According to McGuinness (2020), leaders must be able to decisively change. Leaders need to have good emotional regulation, analyze information rapidly, prioritize needs, and refrain from irrational behavior. Thereby, managers may assist staff members to become more career adaptable and successfully manage challenges in their careers (Zacher et al., 2015).

Openness to Technology Leaders must be able to adopt new technology and inspire employees to use it to carry out their responsibilities. This role might help staff successfully respond to the dynamic demands of the international market (Colville & Murphy, 2006).

Taking Care of Employees' Emotions and Well-Being Leaders must recognize and assist staff in resolving various emotional and personal issues. To do this, supervisors must regularly communicate with staff to bridge the gap between them. Especially, the effective support of technological means of communication reduces the distance barrier created by the epidemic. Besides, a steady financial condition is critical for managerial decisions. The epidemic puts significant financial pressure on each employee, reducing job productivity. In times of crisis, financial constraints have a significant impact on leaders' decision-making (Schwantes, 2020). As a result, it is necessary to share benefits between employees and leaders during times of crisis, even though these actions, such as salary cuts and reduced working hours, may have an immediate impact on employee well-being.

Effective Communication Organizations should have innovative communication strategies as part of their crisis management because, in times of crisis, scarce resources and incomplete information generate uncertainty that causes fear and

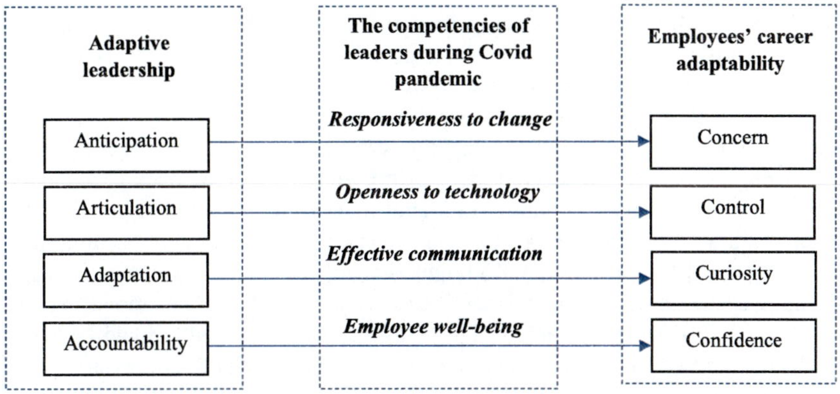

Fig. 1 A framework of the adaptive leadership competencies required in times of crisis

anxiety among employees and makes crisis management much harder (Lockwood, 2005). According to Orangefiery's (2020) survey on the communication requirements of employees, their primary concerns are: transparency about what they know and don't know; resources for coping with stress and anxiety; and the complexity of the issue. To motivate team members and come up with original solutions, there must be continuous communication (Lacerda, 2019). Effective communication places a strong emphasis on both quantity and quality of communication (Balakrishnan & Masthan, 2013). According to Mohr and Sohi (1995), five factors affect communication quality overall: completeness, reliability, accuracy, timeliness, and completeness of communication flows.

4 Discussions

This research offers several implications for managing crises and selecting the best leadership approach for an organization in trying circumstances, particularly during the COVID-19 epidemic. First, as adaptability is extremely valuable in times of crisis, firms should seek out and hire highly adaptable leaders. Second, even though personality traits are largely constant, they are not entirely static (Damian et al., 2019); therefore, leaders should maintain a spirit of self-education and self-update or take part in training programs that enhance their adaptive behavior to be better able to handle the current COVID-19 epidemic and similar situations. Third, adaptability should be fostered in an adaptive and open culture. In order for adaptive leaders to succeed, businesses must maintain a highly flexible culture and remove any elements that can undermine their drive for change (Schein, 2010).

Employers should consider employee career adaptability as a criterion to examine recruiting and retaining good workers because of the dynamic nature of work in a modern organizational context under pressure from ongoing technology application and innovation. Adaptive leadership may assist staff members in making quick adjustments to their careers and even assist pessimistic individuals in embracing change and adjusting to the workplace more successfully (Nguyen et al., 2016; Tolentino et al., 2014), thereby reducing highly adaptable employees turnover intention during a crisis (Ito & Brotheridge, 2005). To achieve this, managers should provide a positive and encouraging work atmosphere that promotes employee optimism. Such a work atmosphere should include career counseling, coaching, and mentoring (Lent & Brown, 2019; Rottinghaus et al., 2017). The interpersonal needs of employees should be frequently identified and addressed by organizational leaders (Singh et al., 2018). During this trying period, department heads should be trustworthy, empathetic, and proactive in providing their staff with the necessary assistance for their physical and emotional well-being (Ahmed et al., 2020). Through platform-based social networking technology that allows individuals to discuss and communicate their concerns, employers may also obtain more supervisory help. Companies may encourage a positive work environment that supports employees' aspirations for the future (Papagiannidis & Marikyan, 2020). Human resource development specialists should plan a variety of conversations and discussions as well as facilitate positive interactions to help people, groups, and organizations get ready for the future of work by preparing work-from-home and displaced workers for how to work with and manage organizational changes.

5 Conclusions

Human resource managers and leaders need to reconsider implementing a new leadership style that is more appropriate than conventional leadership styles in the context of the COVID-19 epidemic. According to this study, adaptive leadership is more successful in times of crisis and high uncertainty. Adaptive leadership shows adaptability to changes and difficulties, the ability to foresee potential future events, and positive attitudes. Adaptive leadership focuses on four dimensions: Anticipation, Articulation, Adaptation, and Accountability. Moreover, adaptive leaders will promote employees' involvement and encourage them to more readily adapt to changes. Employees' career adaptability is expressed in four main attitudes: planning for the future (Concern), responsibility for the future (Control), investigating potential roles (curiosity), and belief in one's capacity to accomplish goals (confidence). Our framework emphasizes that adaptive leaders will facilitate employees' career adaptability by using four main competencies: recognizing changing events and taking rapid action in response; adopting new technology and inspiring employees to use it; recognizing and assisting staff in resolving various emotional and personal issues; and building innovative communication strategies. This framework gives managerial insights that during the crisis, adaptive leadership will enhance the effectiveness of crisis management.

References

Afsar, B., Badir, F., & Saeed, B. B. (2014). Transformational leadership and innovative work behavior. *Industrial Management & Data Systems, 114*(8), 1270–1300. https://doi.org/10.1108/IMDS-05-2014-0152

Ahmed, F., Zhao, F., & Faraz, N. A. (2020). How and when does inclusive leadership curb psychological distress during a crisis? Evidence from the COVID-19 outbreak. *Frontiers in Psychology, 11*, 1898. https://doi.org/10.3389/fpsyg.2020.01898

Asaari, A. H., & Hasmi, M. (2012). *Academic leadership and work-related attitude*. Ph.D. thesis, University of Hull.

Balakrishnan, C., & Masthan, D. (2013). Impact of internal communication on employee engagement–A study at Delhi international airport. *International Journal of Scientific and Research Publications, 3*(8), 1–13.

Bardoel, E. A., Pettit, T. M., De Cieri, H., & McMillan, L. (2014). Employee resilience: An emerging challenge for HRM. *Asia Pacific Journal of Human Resources, 52*(3), 279–297.

Brown, A., Bimrose, J., Barnes, S.-A., & Hughes, D. (2012). The role of career adaptabilities for mid-career changers. *Journal of Vocational Behavior, 80*(3), 754–761.

Caniëls, M. C. J., Semeijn, J. H., & Renders, I. H. M. (2018). Mind the mindset! The interaction of proactive personality, transformational leadership and growth mindset for engagement at work. *Career Development International, 23*(1), 48–66. https://doi.org/10.1108/CDI-11-2016-0194

Colville, I. D., & Murphy, A. J. (2006). Leadership as the enabler of strategizing and organizing. *Long Range Planning, 39*(6), 663–677.

Damian, R. I., Spengler, M., Sutu, A., & Roberts, B. W. (2019). Sixteen going on sixty-six: A longitudinal study of personality stability and change across 50 years. *Journal of Personality and Social Psychology, 117*(3), 674.

Ferreira, N. (2019). Enhancing career wellbeing: The role of workplace friendship, career adaptability and organisational commitment. In *Theory, Research and Dynamics of Career Wellbeing* (pp. 197–216). Springer.

Fiori, M., Bollmann, G., & Rossier, J. (2015). Exploring the path through which career adaptability increases job satisfaction and lowers job stress: The role of affect. *Journal of Vocational Behavior, 91*, 113–121. https://doi.org/10.1016/j.jvb.2015.08.010

Ginevra, M. C., Magnano, P., Lodi, E., Annovazzi, C., Camussi, E., Patrizi, P., & Nota, L. (2018). The role of career adaptability and courage on life satisfaction in adolescence. *Journal of Adolescence, 62*, 1–8.

Guan, Y., Zhou, W., Ye, L., Jiang, P., & Zhou, Y. (2015). Perceived organizational career management and career adaptability as predictors of success and turnover intention among Chinese employees. *Journal of Vocational Behavior, 88*, 230–237. https://doi.org/10.1016/j.jvb.2015.04.002

Guan, Y., Yang, W., Zhou, X., Tian, Z., & Eves, A. (2016). Predicting Chinese human resource managers' strategic competence: Roles of identity, career variety, organizational support and career adaptability. *Journal of Vocational Behavior, 92*, 116–124.

Heifetz, R. A., & Heifetz, R. (1994). *Leadership without easy answers* (Vol. 465). Harvard University Press.

Heifetz, R. A., Kania, J. V., & Kramer, M. R. (2004). Leading boldly. *Stanford Social Innovation Review, 2*(3), 20–32.

Heifetz, R. A., Heifetz, R., Grashow, A., & Linsky, M. (2009). *The practice of adaptive leadership: Tools and tactics for changing your organization and the world*. Harvard Business Press.

ILO. (2021). Uncertain and uneven recovery expected following unprecedented labour market crisis. *International Labour Organization-ILO, 25*.

Ito, J. K., & Brotheridge, C. M. (2005). Does supporting employees' career adaptability lead to commitment, turnover, or both? *Human Resource Management, 44*(1), 5–19.

Joo, B.-K. (Brian), & Lim, T. (2013). Transformational leadership and career satisfaction: The mediating role of psychological empowerment. *Journal of Leadership & Organizational Studies, 20*(3), 316–326. https://doi.org/10.1177/1548051813484359

King, G. A., & Rothstein, M. G. (2010). *Resilience and leadership: The self-management of failure.* Edward Elgar Publishing.

Koehn, N. (2020). Real leaders are forged in crisis. *Harvard Business Review, 3,* 1–6.

Lacerda, T. C. (2019). Crisis leadership in economic recession: A three-barrier approach to offset external constraints. *Business Horizons, 62*(2), 185–197.

Lee, P. C., Xu, S. T., & Yang, W. (2021). Is career adaptability a double-edged sword? The impact of work social support and career adaptability on turnover intentions during the COVID-19 pandemic. *International Journal of Hospitality Management, 94,* 102875.

Lent, R. W., & Brown, S. D. (2019). Social cognitive career theory at 25: Empirical status of the interest, choice, and performance models. *Journal of Vocational Behavior, 115,* 103316.

Levin, G. (2020). 15 Surprising Stats on the Shift to Remote Work due to COVID-19. In *Riverbed.* Retrieved from https://www.riverbed.com/blogs/15-surprising-stats-on-remote-work-due-to-covid-19.html. Accessed on 14 Dec 2022.

Li, J., & Zehr, S. M. (2020). The wellbeing of communities, large and small: A ponderance of HRD. In *Human Resource Development International* (Vol. 23, Issue 2, pp. 103–107). Taylor & Francis.

Lockwood, N. R. (2005). Crisis management in today's business environment. *SHRM Research Quarterly, 4,* 1–9.

Luke, J., McIlveen, P., & Perera, H. N. (2016). A thematic analysis of career adaptability in retirees who return to work. *Frontiers in Psychology, 7,* 193.

Maggiori, C., Johnston, C. S., Krings, F., Massoudi, K., & Rossier, J. (2013). The role of career adaptability and work conditions on general and professional well-being. *Journal of Vocational Behavior, 83*(3), 437–449.

Maurer, R. (2020). Remote employees are working longer than before. In *SHRM.* Retrieved from https://www.shrm.org/hr-today/news/hr-news/pages/remote-employees-are-working-longer-than-before.aspx. Accessed on 16 Dec 2022.

McGuinness, J. (2020). COVID-19 leadership lessons. In *Chief Executive.* Retrieved from https://chiefexecutive.net/4-covid-19-leadership-lessons/.

Mohr, J. J., & Sohi, R. S. (1995). Communication flows in distribution channels: Impact on assessments of communication quality and satisfaction. *Journal of Retailing, 71*(4), 393–415.

Mulder, P. (2017). Adaptive Leadership. In *toolshero.* Retrieved from https://www.toolshero.com/leadership/adaptive-leadership/.

Nguyen, Q., Kuntz, J. R., Näswall, K., & Malinen, S. (2016). Employee resilience and leadership styles: The moderating role of proactive personality and optimism. *New Zealand Journal of Psychology (online), 45*(2), 13.

Ocampo, A. C. G., Reyes, M. L., Chen, Y., Restubog, S. L. D., Chih, Y.-Y., Chua-Garcia, L., & Guan, P. (2020). The role of internship participation and conscientiousness in developing career adaptability: A five-wave growth mixture model analysis. *Journal of Vocational Behavior, 120,* 103426.

Orangefiery. (2020). *Leadership communications during COVID-19: A survey of US organizations.*

Özsahin, M., Zehir, C., & Acar, A. Z. (2011). Linking leadership style to firm performance: The mediating effect of the learning orientation. *Procedia-Social and Behavioral Sciences, 24,* 1546–1559.

Papagiannidis, S., & Marikyan, D. (2020). Smart offices: A productivity and well-being perspective. *International Journal of Information Management, 51,* 102027.

Porfeli, E. J., & Savickas, M. L. (2012). Career adapt-abilities scale-USA form: Psychometric properties and relation to vocational identity. *Journal of Vocational Behavior, 80*(3), 748–753. https://doi.org/10.1016/j.jvb.2012.01.009

Ramalingam, B., Nabarro, D., Oqubuy, A., Carnall, D., & Wild, L. (2020). *Principles to guide adaptive leadership.* Harvard Business Review.

Randall, L. M., & Coakley, L. A. (2007). Applying adaptive leadership to successful change initiatives in academia. *Leadership & Organization Development Journal, 28*(4), 325–335.

Rasheed, M. I., Okumus, F., Weng, Q., Hameed, Z., & Nawaz, M. S. (2020). Career adaptability and employee turnover intentions: The role of perceived career opportunities and orientation to happiness in the hospitality industry. *Journal of Hospitality and Tourism Management, 44*, 98–107. https://doi.org/10.1016/j.jhtm.2020.05.006

Rossier, J., Ginevra, M. C., Bollmann, G., & Nota, L. (2017). The importance of career adaptability, career resilience, and employability in designing a successful life. In K. Maree (Ed.), *Psychology of career adaptability*, employability and resilience (pp. 65–82). Springer International Publishing. https://doi.org/10.1007/978-3-319-66954-0_5

Rottinghaus, P. J., Eshelman, A., Gore, J. S., Keller, K. J., Schneider, M., & Harris, K. L. (2017). Measuring change in career counseling: Validation of the career futures inventory-revised. *International Journal for Educational and Vocational Guidance, 17*(1), 61–75.

Rudolph, C. W., Lavigne, K. N., & Zacher, H. (2017). Career adaptability: A meta-analysis of relationships with measures of adaptivity, adapting responses, and adaptation results. *Journal of Vocational Behavior, 98*, 17–34.

Safavi, H. P., & Bouzari, M. (2019). The association of psychological capital, career adaptability and career competency among hotel frontline employees. *Tourism Management Perspectives, 30*, 65–74. https://doi.org/10.1016/j.tmp.2019.02.001

Savickas, M. L. (2005). The theory and practice of career construction. *Career Development and Counseling: Putting Theory and Research to Work, 1*, 42–70.

Savickas, M. L., & Porfeli, E. J. (2012). Career adapt-abilities scale: Construction, reliability, and measurement equivalence across 13 countries. *Journal of Vocational Behavior, 80*(3), 661–673.

Schein, E. H. (2010). *Organizational culture and leadership* (Vol. 2). Wiley.

Schwantes, M. (2020). 4 Signs to instantly identify a great leader during crisis. In *Inc.* Retrieved from https://www.inc.com/marcel-schwantes/great-leader-time-of-crisis.html. Accessed on 24 Mar 2022.

Singh, B., Shaffer, M. A., & Selvarajan, T. T. (2018). Antecedents of organizational and community embeddedness: The roles of support, psychological safety, and need to belong. *Journal of Organizational Behavior, 39*(3), 339–354.

Tims, M., Bakker, A. B., & Xanthopoulou, D. (2011). Do transformational leaders enhance their followers' daily work engagement? *The Leadership Quarterly, 22*(1), 121–131.

Tladinyane, R., & Van der Merwe, M. (2016). Career adaptability and employee engagement of adults employed in an insurance company: An exploratory study. *SA Journal of Human Resource Management, 14*(1), 1–9.

Tolentino, L. R., Garcia, P. R. J. M., Restubog, S. L. D., Bordia, P., & Tang, R. L. (2013). Validation of the career adapt-abilities scale and an examination of a model of career adaptation in the Philippine context. *Journal of Vocational Behavior, 83*(3), 410–418.

Tolentino, L. R., Garcia, P. R. J. M., Lu, V. N., Restubog, S. L. D., Bordia, P., & Plewa, C. (2014). Career adaptation: The relation of adaptability to goal orientation, proactive personality, and career optimism. *Journal of Vocational Behavior, 84*(1), 39–48.

Yang, X., Feng, Y., Meng, Y., & Qiu, Y. (2019). Career adaptability, work engagement, and employee well-being among Chinese employees: The role of guanxi. *Frontiers in Psychology, 10*, 1029. https://doi.org/10.3389/fpsyg.2019.01029

Yu, H., & Zheng, X. (2013). The impact of employee career adaptability: Multilevel analysis. *Acta Psychologica Sinica, 45*(6), 680–693.

Yukl, G., & Mahsud, R. (2010). Why flexible and adaptive leadership is essential. *Consulting Psychology Journal: Practice and Research, 62*(2), 81.

Zacher, H. (2014). Career adaptability predicts subjective career success above and beyond personality traits and core self-evaluations. *Journal of Vocational Behavior, 84*(1), 21–30.

Zacher, H., Ambiel, R. A., & Noronha, A. P. P. (2015). Career adaptability and career entrenchment. *Journal of Vocational Behavior, 88*, 164–173.

The Effectiveness of Intelligent Management Accounting System and Internationalization of Small and Medium Enterprises: The Mediating Role of Organizational Resilience in the Circular Economy Adoption

Huy Quang Pham and Phuc Kien Vu

Abstract Management accounting has been well acknowledged to be a paramount ingredient in determining, assessing and exploiting international entrepreneurship opportunities and has been substantiated to be positively associated with internationalization capacities (IC). Outstandingly, due to the inclusive measures for sustainability and resilience promotion in the 2030 Agenda, small and medium enterprises (SMEs) have begun to recognize the value of circular business practices in formulating organizational resilience for the target of IC enhancement. This research conceptualizes and corroborates a model concentrating on the interconnection between the effectiveness of intelligent management accounting system (EIMAS) and internationalization capacities (IC) of SMEs. Additionally, it pursues to investigate the mediation mechanism of organizational resilience in circular economy adoption (ORCEA) on the linkage between EIMAS and IC. The structural equation modeling was employed to investigate a theoretical model hypothesizing the interconnections between the abovementioned components through large-scale response data captured from convenient and snowball sample of 612 accountants within SMEs in the southern areas of Vietnam. The result analyses highlighted that EIMAS could intensify IC and ORCEA. However, the obtained findings revealed that ORCEA did not act as a mediator in the interconnection between EIMAS and IC. These observations might give rise to newer insights to practitioners to generate targeted strategies as well as future action plans to intensify IC through capturing much more advantages of EIMAS in the long run. In addition, these findings would help policy-makers to formulate prerequisite solutions to warrant the favorable environment for the digital transformation and management accounting practices implementation among SMEs.

Keywords Management accounting system · Circular economy · Internationalization · Organizational resilience · Small and medium enterprises

H. Q. Pham (✉) · P. K. Vu
School of Accounting, University of Economics Ho Chi Minh City (UEH), Ho Chi Minh City, Vietnam
e-mail: pquanghuy@ueh.edu.vn

A. T. Nguyen et al. (eds.), *Contemporary Economic Issues in Asian Countries: Proceeding of CEIAC 2022, Volume 1*,
https://doi.org/10.1007/978-981-19-9669-6_21

1 Introduction

As developing measures to overcome business frontier and operating in the international market have been pondered as the most promising ways to ensure achievement in the long term, internationalization process has been well acknowledged to generate the potential to render advantages to all types of enterprises (Kuivalainen et al., 2012). Indeed, it has enabled for the amelioration of organizational competitive advantage, sustainable performance and it has composed one of the approaches through which SMEs could take advantage to react effectively to environmental challenges to survive and develop (Andres et al., 2022; Freixanet & Renart, 2020). In this regard, numerous SMEs have aimed for business advancement through diversifying their products/services and their markets (Lin & Ho, 2019). Nevertheless, the implementation of this strategy has not been simple. Faced with ever-shifting markets, all the enterprises must ineludibly adapt to their environment to effectively respond to the emerging challenges. Additionally, the COVID-19 pandemic has magnified the complicatedness which induced the negative effects on the way of operating business cross-borders (Van Witteloostuijn, 2021). In order to minimize the adverse impacts of the COVID-19 pandemic on SMEs, the senior managers in this type of organization have been seeking for innovative approaches to carry out operations to assure their corporate survival (Khalil et al., 2022) and enhance their organizational resilience to improve as well as intensify their internationalization process. Against this backdrop, circular economy (CE) business practices have been implicitly presumed to formulate the resilience practices which focused on gaining resource efficiencies through removing the buffers, diversity and slack resources required to resist and adapt to shocks as well as disturbances (Kennedy & Linnenluecke, 2022). In addition, entering foreign markets has been linked to considerable uncertainty and significant investment (Ahsan & Musteen, 2011). To that end, managing the expenses and uncertainties of internationalization has been peculiarly relevant for SMEs (Ipsmillera et al., 2022).

Unfortunately, the capabilities related to expense management to smooth operations of SMEs could not be undertaken as well as large multinational enterprises. As such, SMEs should take the adoption of appropriate information and control systems into consideration. This was because management accounting system (MAS) enabled SMEs to allocate resources efficiently and effectively through managing information demands (Ali Qalati et al., 2020; Helgeson et al., 2022; Le & Ikram, 2022; Müller et al., 2021). Thus, in SMEs, MAS has played the paramount role in SMEs' functions. It buttressed business functions by supplying management planning and tackling with or analyzing valuable information (Müller et al., 2021). MAS could also address all demands and enable decision-making process as well as the planning process that facilitated in accomplishing economic circular targets for these organizations in long-term development (Fitria, 2021). Given that the introduction of advanced information technologies that enabled the automation of accounting processes through enhancing the speed of access to information, the security of information as well as the level of transparency, eradicating the shortcomings of the human factors in certain practices, the integration advanced information technologies into MAS has become the inevitable tendency (Yoon, 2020). In this vein, it has been prerequisite to set up a modern paradigm in MAS through adapting to digital technologies and modifying techniques (Guni, 2018) to generate an intelligent MAS for the efficiency and effectiveness of managerial decisions in terms of gaining the organizational resilience in CE adoption (ORCEA) and enhancing the internationalization capabilities (IC) of SMEs.

Startlingly, there has been a dearth of academic works that systematically investigated the influence of strengthening ORCEA for IC achievement through leveraging the advantages of intelligent MAS within SMEs in emerging countries. With this focus, the current research aimed at bridging this gap through addressing the intriguing research questions as follows.

RQ1. *What is the impact of EIMAS on IC?*

RQ2. *Does ORCEA act as a mediator of the interconnection between EIMAS on IC?*

Aside from the introduction, the other sections in this manuscript are structured in the following sequence. The literature review is demonstrated in Sect. 2. The research model and hypotheses development are outlined in Sect. 3. Afterward, the methodology is depicted in Sect. 4. This manuscript culminates with statistical analyses which are exposed in the penultimate section. The last section focuses on outlining the theoretical contributions and managerial implications as well as some recommendations for the development of future works.

2 Literature Review

2.1 Theoretical Underpinnings

Contingency theory (CT). CT was first founded by researchers from Ohio State University in 1950 (Nohria & Khurana, 2010). The fundamental idiosyncrasy of the theory was its behavioral approach that associated with the optimal fit of organizational constitution hinged on contingent circumstances (Bastian & Andreas, 2012). The structure of an organization was widely recognized to bank on numerous determinants, namely nature of organizational work, market conditions, culture, the external environment as well as technology (Hanisch & Wald, 2012). CT has been determined as a vital area of research in management accounting (Chenhall, 2006). The perspectives of a CT of management accounting emerged in the 1970s in an effort to illuminate a wide range of management accounting practices that were obvious at that time (Otley, 2016).

Dynamic capabilities theory. The term of dynamic capabilities (DCs) was first emerged in a working paper of Teece et al. (1990) and was subsequently published in 1997 (Teece et al., 1997). The DCs accentuated on the organizational capacities on such aspects as sensing and shaping chances and menaces, seizing opportunities, retaining competitiveness through gaining, integrating, safeguarding, and, when essential, reconfiguring the organizational intangible and tangible assets (Teece, 2007). DCs theory depicted path-dependent processes which allowed organizations to adapt to changing environments by establishing, combining and reconfiguring their resource and capacities portfolio in a rapid manner (Teece et al., 1997). Numerous academic notes have so far deepened the analyses on the interconnection between DCs and internationalization which thus has contributed to the strength of DCs theory (Peng & Lin, 2017).

2.2 Conceptualization of Measured Variables

Effectiveness of intelligent management accounting system. MAS referred to systematic processes of control employ to affect members within an organization to reach the organizational goals (Naranjo-Gil & Hartmann, 2007). MAS has been considered as one of the complicated administrative systems that was largely banked upon by institutions to produce helpful information about users of the internal as well as external business environment of this system (Ngo, 2021). Based on the suggestion of Chenhall and Morris (1986), the features of valuable and reliable information comprised of scope, timeliness, aggregation as well as integration. Accordingly, scope focused on the information that consisted of the organizational issues to support the managers to generate more effective solutions so that the outcomes were expected to revamp managerial performance better. Timeliness concentrated on the speed or the time span between the demands for information by the demonstration of information required by the organization to reinforce managers to cope the environmental uncertainties. Aggregation mentioned on the information that offered explicitness on the areas which were responsibility of each organizational manager consistent with their respective functions. Integration was defined as the information that encompassed facets, namely the provision of a target organization which was gauged from the percentage of the interaction between sub-department within the organization. Building on the perspectives of Ghasemi et al. (2019) and Pedroso and Gomes (2020), the effectiveness of MAS could be achieved only when each of characteristics of information, namely scope, timeliness, aggregation and integration of MAS reached the effectiveness.

The adoption big data and artificial intelligence were corroborated to bring out favorable opportunities for effective decision-making through synthesizing robust patterns from large datasets, artificial intelligence and, particularly, machine learning algorithms would enable the generation of valuable information and predictions from data (Elliott et al., 2021). With the implementation of artificial intelligence on MAS would help to minimize the human errors and frauds, gain the overall performance as well as consumers' satisfaction and generate rapid, precise, repeatable, low-expense decisions, with high quality approaching human-like intelligence (Cubric, 2020; Shrestha et al., 2019). Additionally, the supports of big data would revamp the quality and relevance of accounting information for management control, thereby perking up transparency and decision-making (Dehbi et al., 2022). Moreover, big data analytics was corroborated to make great contribution to MAS through ameliorating organizational expense containment quality and effectiveness (Fahlevi et al., 2022). In this regard, the convergence of artificial intelligence, big data and big data analytics in MAS would make MAS become much more intelligent. In other words, this convergence could help all features of MAS to perform in a more intelligent manner.

Organizational resilience. The conceptualization of organizational resilience reflected on the differential capabilities of organizations to predict, respond to, recover from and acquire from adversity (Linnenluecke, 2017). Organizational resilience was an issue of degree, such that more resilient organizations predicted adversity earlier and more completely, responded more rapidly and comprehensively, recover more entirely and grasp more intensively from the experiences (van der Vegt et al., 2015). Building on the perspectives of Ortiz-de-Mandojana and Bansal (2015), organizational resilience was not a static characteristic, but a latent, path-dependent wide range of capabilities that organizations advanced through tackling with unexpected incidents. This work included investigations of how enterprise, industries and communities could sustain resilience through CE implementation while mitigating destruction of the life-supporting basements supplied by ecosystem resilience.

Internationalization capacities. Internationalization has been determined both as a process and pertaining to the practices that enterprises carry out to reach foreign markets (Ribau et al., 2016). Numerous academic notes have so far deepened the analyses on the interconnection between DCs and internationalization (Peng & Lin, 2017). Building on the findings of Pinho and Prange (2016), IC was considered as DCs which was categorized into the aspects, namely exploitative capabilities and explorative capabilities. In this regard, exploitative capabilities reflected on the threshold capabilities and consolidation capabilities. More concretely, threshold capabilities put accent on undertaking the similar actions more effectively and associated with the capabilities of the enterprise to organize in order to function competitively in various environments (Pinho & Prange, 2016). The consolidation capabilities aimed at support the firm to generate structures and routines to concentrate on chance recognition and on formulating the skills required for local operations (Pinho & Prange, 2016). In the meanwhile, exploration capabilities comprised of the value-adding capabilities and disruption capabilities. More instrumentally, value-adding capabilities placed emphasis on the advancement of novel capabilities and disruption capabilities stressed on giving rise to the flexibility, which enabled the enterprises to rapidly grasp the competencies essential to pursue continued development (Pinho & Prange, 2016).

3 Authentication of Research Hypotheses

The differentiation strategy has been pondered as one of the most prerequisites for the accomplishment in internationalization process as this form of strategy would enable the SMEs to become burgeoningly critical and to draw concern in markets with high degree of competition (Cavusgil & Knight, 2015; Freixanet & Renart, 2020). As such, the differentiation strategy was carried out by distinction in terms of goods design, idiosyncrasies and technology to reach an enhancement in profitability (Cavusgil & Knight, 2015). To do so, SMEs should take the appropriate information and control systems into consideration to manage and govern the deficient resources. Therefore, a MAS should be set up in the way that could allow SMEs to allocate resources rationally by managing and governing information requirements in an efficient and effective manner (Ali Qalati et al., 2020; Helgeson et al., 2022; Le & Ikram, 2022; Müller et al., 2021), supporting business functions by offering management planning and tackling with or analyzing valuable information (Müller et al., 2021).

The adoption big data and artificial intelligence were corroborated to bring out favorable opportunities for effective decision-making through synthesizing robust patterns from large datasets, artificial intelligence and, particularly, machine learning algorithms would enable the generation of valuable information and predictions from data (Elliott et al., 2021). Building on the perspectives of Saggi and Jain (2018), the implementation of big data analytics techniques into information system would allow the organization to gain the predictive value of the available data. Indeed, this implementation would enable the MAS to become efficient and effective in performing expense prediction, product profitability analysis, the financial effect of manufacture change and evaluations of consumer segment profitability. With the support of advanced information technologies, the EIMAS would expand the content and processing approaches of information to supply broad and in-depth information for managers. This was because the EIMAS would focus on the external environment of the organization, ultimately analyze the market position of the organization and support managers to make effective strategic decisions. Along the similar line of reasoning, the first hypothesis in this research was posited as follows.

Hypothesis 1 (H1). *EIMAS exerts an impact on IC in a significant and positive manner*

Under the condition of advanced information technology, accounting information system would become comprehensive and open. Accordingly, it would heavily rest on the network and would be employed to analyze, design and formulate the accounting business produced by the value transfer during organization operations. It could also supply financial information, enable in financial analysis, forecast and plan as well as intensify financial control and financial decision-making (Deng et al., 2021; Wu et al., 2020). With the implementation of artificial intelligence on MAS would help to minimize the human errors and frauds, gain the overall performance as well as consumers' satisfaction and generate rapid, precise, repeatable, low-expense decisions, with high quality approaching human-like intelligence (Cubric, 2020; Shrestha et al., 2019). Additionally, the support of big data would revamp the quality and relevance of accounting information for management control, thereby perking up transparency and decision-making (Dehbi et al., 2022). Furthermore, big data analytics would facilitate the adoption of more detailed, precise and proper expense assessment systems (Fahlevi et al., 2022). In this regard, the EIMAS would buttress the organization to succeed in strategic planning, strategic cost analysis and strategic market analysis through strategic high-degree analysis and thinking which thus helped the senior leadership to raise awareness of constant concerns and develop effective strategies to resolve. Moreover, EIMAS would provide valuable information which allowed the organization to strengthen the capabilities to determine internal and external developments, deepen the capabilities to ascertain vital developments and potential threats to prepare for unexpected incidents and react to these incidents before their full influences would become visible through developing and implementing solutions. Along the similar line of reasoning, the second hypothesis in this research was conjectured as follows.

Hypothesis 2 (H2). *EIMAS exerts an impact on ORCEA in a significant and positive manner*

The differentiation strategy has been considered as the main driver for success in the internationalization process among SMEs (Cavusgil & Knight, 2015; Freixanet & Renart, 2020). This type of strategy was undertaken by dissimilarity in terms of product/service design, idiosyncrasies and technology to accomplish an amelioration in profitability (Cavusgil & Knight, 2015).

The CE has been generally pondered as an intentionally designed, restorative as well as regenerative economy (Pla-Julián & Guevara, 2019). CE practices have been implicitly presumed to establish social-ecological system resilience that has been the resilience of interacting, nested systems at numerous degrees (Kennedy & Linnenluecke, 2022). The CE strategies have focused on reconfiguring businesses and industries to turn out to be industrial systems of closed material loops with the goal of abolishing waste and pollution, facilitating multitudinous product life cycles and maximizing the employment and embedded value of products and materials as well as capabilities for resource usage (Lacy & Rutqvist, 2015). The transformation into CE would help the organization to reach the superior resilience and response capabilities. As such, the circular principles like ameliorations in the design and reusability of products could have assisted to generate the resilience. In doing so, the ORCEA would enable the organization capabilities to build, integrate and reconfigure organizational resources and competences; improve the cooperating and learning capabilities; renew, regenerate and reconfigure capital to enhance the IC. Along the similar line of reasoning, the third hypothesis in this research was posited as follows.

Hypothesis 3 (H3). *ORCEA exerts an impact on IC in a significant and positive manner* (Fig. 1)

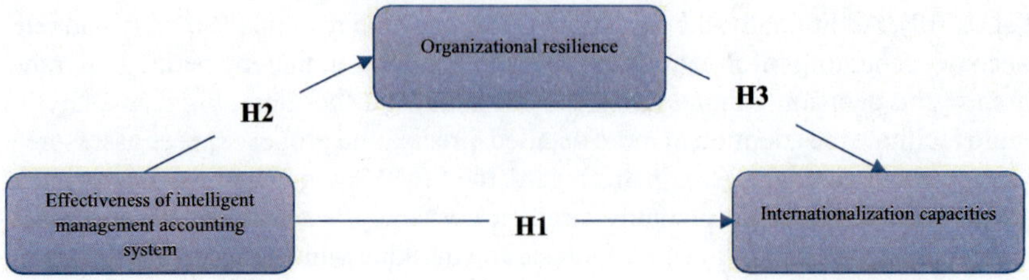

Fig. 1 Hypothesized model

4 Methodological Framework

4.1 Operationalization of Variables for Measurement

In the current study, data was procured predominantly through self-administrated questionnaire. Questionnaire survey has been considered as a commonly and extensively employed research technique for rapid collection and analysis of data from the target population (Heeringa et al., 2017). As a consequence of a wide-ranging analysis of the literature, an initial questionnaire was formulated in English. This translation process was performed to warrant that the original meanings of all the survey instruments were retained (Van de Vijver & Tanzer, 2004). As the questionnaire in this study was set up in divergent setting, both culturally and environmentally, a pre-test with 5 experts was conducted. The interview session offered a platform to entirely scrutinize and refine each question as well as the wording of particular variables. Prior to the final design of the questionnaire, a pilot test of the draft questionnaire was embarked to evaluate its suitability for the research, 30 participants with similar idiosyncrasies to the survey population were requested to involve in the small-scale pilot test. All the suggestions received during the exercise were incorporated into our final design for accuracy and reliability. After finalizing the measuring constructs, the standardized questionnaire was established in close-ended form with multiple choices on the five-point Likert scale ranging from "1 = staunchly disagree" to "5 = staunchly agree" employed in all items of the questionnaire.

Effectiveness of intelligent management accounting system. The first-order construct of EIMAS was constituted by four components, namely scope, timeliness, integration and aggregation. More instrumentally, the criteria applied to assess scope were emanated from the combination between the works of Carvalho et al. (2018), Dubey et al. (2019), Chenhall and Morris (1986), Cheng (2012), Etemadi et al. (2009), Chung et al. (2012). The criteria applied to assess timeliness were inherited from the combination between the findings of Carvalho et al. (2018), Dubey et al. (2019), Chenhall and Morris (1986), Agbejule (2005), Etemadi et al. (2009). The criteria applied to assess integration were sprung from the combination between the outcomes of Carvalho et al. (2018), Dubey et al. (2019), Chenhall and Morris (1986), Agbejule (2005), Bouwens and Abernethy (2000). The criteria applied to assess aggregation were proceeded from the combination between the contributions of Carvalho et al. (2018), Dubey et al. (2019), Chenhall and Morris (1986), Bouwens and Abernethy (2000), Agbejule (2005).

Organizational resilience in the circular economy. The first-order construct of ORCEA was formulated by three components, namely capital resilience, strategic resilience and relationship resilience. Particularly, capital resilience referred to capabilities of a business to operate in a normal manner and to recapitalize against risk during a crisis (Chen et al., 2021). Capital resilience was evaluated by 3-item scale stemmed from the findings of Välikangas (2010). Strategic resilience mentioned on the organizational capabilities in maintaining strategic consistency over time, enabling them to determine and eradicate disadvantages and to be capable of choosing the appropriate growth model (Chen et al., 2021). Strategic resilience was evaluated by 3-item scale developed from the findings of Davenport and Cronin (2000). Relationship resilience reflected on the reciprocal interconnection between business and stakeholders (Chen et al., 2021). Relationship resilience was evaluated by 3-item scale arisen from the works of Vogus and Sutcliffe (2007).

Internationalization capacities. The first-order construct of IC was constituted by three components, namely threshold capabilities, value-adding capabilities and disruption capabilities. More concretely, threshold capabilities were evaluated by 3-item scale stemmed from the findings of Atuahene-Gima (2005), Danneels (2008), Lisboa et al. (2011), Prange and Verdier (2011). Value-adding capabilities were evaluated by 3-item scale derived from the contributions of Atuahene-Gima (2005), Danneels (2008), Lisboa et al. (2011), Prange and Verdier (2011). Disruption capabilities were evaluated by 3-item scale originated from the works of Atuahene-Gima (2005), Danneels (2008), Lisboa et al. (2011), Prange and Verdier (2011).

4.2 Sampling Procedure and Data Collection

The focal point of the current study was on SMEs, which made up the vast majority of enterprises in Vietnam, as in most Asian countries, and the respondents were accountants in SMEs. One plausible for this selection lied in the fact that accountants were responsible for tackling with measuring, disclosing and assuring all the organizational information and decision-making processes. Additionally, accountants would become precondition for the auspicious result in view of the progressive permeation of digital technologies (Zybery & Rova, 2014). As such, the data were captured from the accountants in SMEs located in the southern areas in Vietnam. The survey was carried out in a voluntary manner. Accordingly, the background, objectives and methodology of the current study were noticed with target population who readily consented to partake. The respondents were also informed that all the information would be employed stringently for academic rationales and would remain anonymous. The questionnaire excluded information, namely names, race as well as religion which could be wielded to discriminate. Consequently, this research solely presented the processed data and gave rise to no specific opinions or statements. The sample of this research was formulated on the basis of convenience and snowball sampling. While Wolf et al. (2013) advocated a sample size fluctuated from 30 to 460, Urbach and Ahlemann (2010) recommended a sample size of 200–800 responses for CB-SEM technique. The data capturing process transpired during the months of March 2022 and August 2022. The final sample size left for analysis included 612 cases after screening and scrutiny of questionnaires, with a 12.57% data loss rate.

4.3 *Statistical Analysis and Calculations*

Building on the perspectives of Alzahrani et al. (2012), the software employed in this research for data analysis was SPSS with structural equation modeling (SEM-AMOS). The analysis included two stages, namely the measurement model evaluation and the structural model evaluation.

5 Results

5.1 *Statistics for Demographic Variables*

The socio-demographic profile of the sample illustrated that majority of the respondents were females, with the proportion taking up to 65.03%, while males accounted for 34.97%. Concerning to the lifespan, the group "30- below 40" amounted to 67.32% of the whole sample, which was followed by the group 40- below 50, approximately 23.20%, while the group "above 51" made up an inconsequential 9.48%, ranking last among the given groups. All of respondents in this study acquired a minimum of graduate degree qualification and obtained experience of more than 10 years working as accountants in SMEs.

5.2 *Measurement Model: Reliability and Validity*

The assessment on Cronbach's Alpha (CA) and composite reliability (CR) was performed prior to conducting the basic analysis. As such, the CA and CR should be above the suggested thresholds of 0.7 (Bagozzi & Yi, 2012). Concerning to the convergent validity, the standardized factor loading was recommended to surpass the value of 0.50 (Hair et al., 2019) and average variance extracted (AVE) exceeded the recommended value of 0.5 (Balog & Pribeanu, 2010). The outputs in

Table 1 Results summary of measurement model assessment

Construct	Items (abbreviation)	Convergent validity		Construct reliability		Discriminant validity
		Factor loadings ranges	AVE	Cronbach's Alpha	Composite reliability	
Effectiveness of intelligent management accounting system						
Scope	SC	0.750–0.819	0.593	0.851	0.853	Yes
Timeliness	TI	0.725–0.808	0.614	0.863	0.864	Yes
Integration	IN	0.731–0.779	0.566	0.838	0.939	Yes
Aggregation	AG	0.727–0.810	0.593	0.852	0.853	Yes
Organizational resilience						
Capital resilience	CR	0.674–0.899	0.673	0.848	0.859	Yes
Strategic resilience	SR	0.758–0.853	0.653	0.838	0.839	Yes
Relationship resilience	RR	0.773–0.822	0.629	0.834	0.836	Yes
Internationalization capacities						
Threshold capabilities	TC	0.732–0.891	0.649	0.844	0.847	Yes
Value-adding capabilities	VC	0.733–0.880	0.619	0.824	0.829	Yes
Disruption capabilities	DCA	0.732–0.854	0.634	0.834	0.838	Yes

Table 1 demonstrated the reliability, internal consistency and convergent validity of the measurement model.

The support for discriminant validity could be reached when the values of arithmetic square root of the AVE of a dimension were greater than the correlation coefficient between the dimension and others (Fornell & Larcker, 1981). Besides, the correlations among the constructs should be less than the recommended threshold value of 0.85. The outcomes of the comparisons in Table 2 put accent on that there was no discriminant validity problem.

5.3 *Goodness-of-Fit Index Evaluation*

The measurement model was evaluated utilizing goodness-of-fit strategies, namely chi-square/df, Root Mean Square Error of Approximation (RMSEA), Comparative Fit Index (CFI), goodness-of-fit index (GFI) and Tucker–Lewis index (TLI). The result assured the model fit within the acceptable range as recommended by Kline

Table 2 Results summary of discriminant validity using Fornell–Larcker process

	TI	SC	AG	IN	CR	TC	SR	DCA	RR	VC
TI	**1**									
SC	0.112	**1**								
AG	0.268	−0.01	**1**							
IN	0.163	0.161	−0.01	**1**						
CR	0.1	0.141	0.161	0.124	**1**					
TC	0.061	0.234	0.141	0.198	0.025	**1**				
SR	0.188	0.018	0.234	0.086	0.1	0.017	**1**			
DCA	0.08	0.221	0.018	0.101	0.069	0.093	0.102	**1**		
RR	0.221	0.018	0.221	0.046	0.024	0.099	0.092	0.123	**1**	
VC	0.123	0.173	0.018	0.133	−0.055	0.121	0.028	0.103	0.134	**1**

(2018) chi-square/df = 1.392; TLI = 0.976; CFI = 0.979; GFI = 0.942; RMSEA = 0.025. As such, the anticipated model was corroborated to fit the procured data.

5.4 Structural Model: Hypotheses Testing

Direct effect. Building on the outputs in Table 3, EIMAS had significantly positive association with IC (H1: $\beta = 0.683, p = 0.008$). As expected, EIMAS was substantiated to be related to ORCEA (H2: $\beta = 1.434, p = 0.003$) in a significantly positive manner. Thus, H1 and H2 were supported. Nonetheless, the paths linking ORCEA and IC (H3: $\beta = 1.103, p = 0.202$) were insignificant. To that end, H3 was rejected.

Indirect effect. Partial mediation was supposed when both direct and indirect effects were substantial, conversely, if indirect effect was significant, while direct effect was insignificant, thus, full mediation was considered (Cheung & Lau, 2007). Given the association between ORCEA and IC was authenticated to be insignificant, there was no mediating impact induced by ORCEA on the interlink between EIMAS and IC. Succinctly put, ORCEA did not act as a mediator of the interlink between EIMAS and IC.

Table 3 Results summary of hypotheses acceptance

Hypothesis no	Hypothesized path	Standardized	S.E	C.R	*p*-value	Status
H1	EIMAS → IC	0.683	0.476	2.661	0.008	Accepted
H2	EIMAS → ORCEA	1.434	0.423	2.998	0.003	Accepted
H3	ORCEA → IC	1.103	0.483	2.285	0.202	Rejected

Note Chi-square/df = 1.497; TLI = 0.969; CFI = 0.972; GFI = 0.933 and RMSEA = 0.029

6 Discussions

6.1 Theoretical Contribution

Although the operation on international market would supply SMEs with numerous chances, namely accessing to larger markets and modern information technologies, improvement of technical degrees, risk minimization as well as accessing to finance (Saunila, 2019), there has been a common consensus that the studies on SMEs' internationalization have been fragmented (Cavusgil & Knight, 2015; Freixanet & Renart, 2020). In order to handle the shortcomings, the current research enriched the body of literature on the internationalization of SMEs through conducting in-depth investigation on how to ameliorate and develop the IC for SMEs within emerging economies. Building on the statistical outcomes of large-scale response data captured from convenient and snowball sample of 612 accountants within SMEs in the southern areas of Vietnam, the current research pioneered research on the interconnection between EIMAS and IC. On one side, this research also bridged a research gap of prior works related to the deficiency of MAS within SMEs in developing countries. On the other side, the convergence of artificial intelligence, big data and big data analytics in MAS has made MAS become much more intelligent to improve and develop IC. The advantages of integrating artificial intelligence into MAS would mitigate frauds and increase the quality of information which thus improved the quality of decisions (Makridakis, 2017). Additionally, the support of big data would revamp the quality and relevance of accounting information for management control, thereby perking up transparency and decision-making (Dehbi et al., 2022). Remarkably, implementation would enable the MAS to become efficient and effective in performing expense prediction, product profitability analysis, the financial effect of manufacture change and evaluations of consumer segment profitability. Thus, EIMAS could facilitate decision-making processes to support the SMEs in succeeding in intensifying the IC during the internationalization process.

More importantly, in volatile economic environments, enterprises must heavily rest on the combination of MAS implementation and an increasingly CE to acquire useful information (Bai et al., 2021). As such, this study advanced the academic works of prior researchers on MAS and CE (i.e., Al-Nasrawi and Thabit (2020), Fitria (2021)) through shedding light on the interconnection between EIMAS and ORCEA. The obtained observations on ORCEA also added the body of literature on CE implementation SMEs. Admittedly, EIMAS would provide valuable information which allowed the organization to strengthen the capabilities to determine internal and external developments, deepen the capabilities to ascertain vital developments and potential threats to prepare for unexpected incidents and react to these incidents before their full influences would become visible through developing and implementing solutions. Strikingly, the originality of this study lied the fact that ORCEA did not act as a mediator on the interrelation between EIMAS and IC. In other word, the SMEs could only reach the higher IC through taking advantage of EIMAS.

6.2 Practical Implication

Resting on the managerial outlooks, the obtained findings of this study offered numerous takeaway insights to put into practice. Due to the paramount role of EIMAS in support the SME to enhance IC as well as its contribution on gaining the ORCEA, all the managers in SMEs should recognize the value in advancing the design and application of an intelligent MAS. Besides, as the investment in digital technologies has been imperative now more than ever, tangible resources, namely big data and artificial intelligence infrastructure, digital platform as well as other fundamental resources to adopt big data and artificial intelligence have been encouraged to be focused. Following on from advanced information technologies development, the accounting staff of SMEs should become conversant with operating intelligent MAS. Thus, managers should make every effort to provide their staff with appropriate training. Moreover, CE practices should be taken into consideration by all managers in SMEs as these practices would contribute to the organizational resilience for tackling with turbulences and unexpected events in advance.

The observations of this work also generated implications for policy-makers. In this respect, regulation and policies framed at the microlevel should be established an overarching orientation for SMEs to implement intelligent MAS. On one end, the governmental policies should focus on handling strategic IC which corresponded to the specific features, strategic orientations and international engagement of these enterprises. On the other end, policy-makers and governmental influencers should exert to stimulate the internationalization of SMEs through supplying much-needed reinforcement by the form of grants and loans for SMEs to invest in digital technologies and develop the capabilities of their labor force. Moreover, governments should offer an enabling business environment and minimize the adverse effect of the COVID-19 crisis on SMEs' resilience.

Information technology developers were encouraged to set up country-centered innovations which met the demands of businesses. Accordingly, developers should attempt to produce accessible, affordable as well as adaptable resolutions in terms of intelligent MAS for SMEs in developing countries.

6.3 Limitations and Agenda for Further Research

Regardless of the contributions derived from the statistical analysis, there were several inherent drawbacks in this study which could act as starting points for future works. Firstly, organizational resilience has been evolving conceptualizations with conceptual heterogeneity as well as terminological inconsistencies. In the meanwhile, the delineation of EIMAS was grounded on a semantic aggregation of constructs and demonstrative descriptions procured in literature. As such, follow-up studies could further develop these conceptualizations and substantiate the measures. Secondly, the sample of this research was drawn from SMEs located in the southern areas of

Vietnam. Although the result analyses were expected to benefit SMEs from other regions, especially those in a similar business context; however, follow-up studies were encouraged to carry out in other regions to set up whether or not the findings would hold in other environments. Thirdly, as the employment of cross-sectional research design was considered as the other shortcomings of this study, the assessment on the generalizability of these findings with longitudinal design would become a worthwhile endeavor. Lastly, it could be worth noting for future authors to keep on investigating the role of ORCEA on the interconnection between EIMAS and IC.

References

Agbejule, A. (2005). The relationship between management accounting systems and perceived environmental uncertainty on managerial performance: A research note. *Accounting and Business Research, 35*(4), 295–305. https://doi.org/10.1080/00014788.2005.9729996

Ahsan, M., & Musteen, M. (2011). Multinational enterprises' entry mode strategies and uncertainty: A review and extension. *International Journal of Management Reviews, 13*(4), 376–392. https://doi.org/10.1111/j.1468-2370.2010.00296.x

Ali Qalati, S., Li, W., Ahmed, N., Ali Mirani, M., & Khan, A. (2020). Examining the factors affecting SME performance: The mediating role of social media adoption. *Sustainability, 13*(1), 1–24. https://doi.org/10.3390/su13010075

Al-Nasrawi, S. A., & Thabit, T. (2020). The influence of the environmental factors on the adoption of the international accounting system IAS/IFRS: Case of Iraq. *Journal of Accounting, Finance and Auditing Studies, 6*(1), 66–85.

Alzahrani, A., Stahl, B. C., & Prior, M. (2012). Developing an instrument for e-public services' acceptance using confirmatory factor analysis. *Journal of Organizational and End User Computing, 24*(3), 18–44. https://doi.org/10.4018/joeuc.2012070102

Andres, B., Poler, R., & Guzman, E. (2022). The influence of collaboration on enterprises internationalization process. *Sustainability, 14*, 1–23. https://doi.org/10.3390/su14052843

Atuahene-Gima, K. (2005). Resolving the capability—Rigidity paradox in new product innovation. *Journal of Marketing, 69*(4), 61–83. https://doi.org/10.1509/jmkg.2005.69.4.61

Bagozzi, R. P., & Yi, Y. (2012). Specification, evaluation, and interpretation of structural equation models. *Journal of the Academy of Marketing Science, 40*(1), 8–34. https://doi.org/10.1007/s11747-011-0278-x

Bai, B., Guo, Z., Zhou, C., Zhang, W., & Zhang, J. (2021). Application of adaptive reliability importance sampling-based extended domain PSO on single mode failure in reliability engineering. *Information Sciences, 546*, 42–59. https://doi.org/10.1016/j.ins.2020.07.069

Balog, A., & Pribeanu, C. (2010). The role of perceived enjoyment in the students' acceptance of an augmented reality teaching platform: A structural equation modelling approach. *Studies in Informatics and Control, 19*(3), 319–330.

Bastian, H., & Andreas, W. (2012). A bibliometric view on the use of contingency theory in project management research. *Project Management Journal, 43*(3), 4–23. https://doi.org/10.1002/pmj.21267

Bouwens, J., & Abernethy, M. A. (2000). The consequences of customization on management accounting system design. *Accounting, Organizations and Society, 25*(3), 221–241. https://doi.org/10.1016/s0361-3682(99)00043-4

Carvalho, N., Chaim, O., Cazarini, E., & Gerolamo, M. (2018). Manufacturing in the fourth industrial revolution: A positive prospect in sustainable manufacturing. *Procedia Manufacturing, 21*, 671–678. https://doi.org/10.1016/j.promfg.2018.02.170

Cavusgil, S. T., & Knight, G. (2015). The born global firm: An entrepreneurial and capabilities perspective on early and rapid internationalization. *Journal of International Business Studies, 46*(1), 3–16. https://doi.org/10.1057/jibs.2014.62

Chen, R., Xie, Y., & Liu, Y. (2021). Defining, conceptualizing, and measuring organizational resilience: A multiple case study. *Sustainability, 13*(5), 2517. https://doi.org/10.3390/su13052517

Cheng, M. (2012). The joint effect of budgetary participation and broad-scope management accounting systems on management performance. *Asian Review of Accounting, 20*(3), 184–197. https://doi.org/10.1108/13217341211263256

Chenhall, R. H. (2006). Theorizing contingencies in management control systems research. *Handbooks of Management Accounting Research*, 163–205. https://doi.org/10.1016/s1751-3243(06)01006-6

Chenhall, R. H., & Morris, D. (1986). The impact of structure, environment, and interdependence on the perceived usefulness of management accounting systems. *Accounting Review, 61*(1), 16–35.

Cheung, G. W., & Lau, R. S. (2007). Testing mediation and suppression effects of latent variables. *Organizational Research Methods, 11*(2), 296–325. https://doi.org/10.1177/1094428107300343

Chung, S.-H., Su, Y.-F., & Yu-Ju, S. (2012). Broad scope management accounting system and managerial performance: The impact of role ambiguity and functional difference. *African Journal of Business Management, 6*(30), 8873–8878.

Cubric, M. (2020). Drivers, barriers and social considerations for AI adoption in business and management: A tertiary study. *Technology in Society, 62*, 1–12. https://doi.org/10.1016/j.techsoc.2020.101257

Danneels, E. (2008). Organizational antecedents of second-order competences. *Strategic Management Journal, 29*(5), 519–543. https://doi.org/10.1002/smj.684

Davenport, E., & Cronin, B. (2000). Knowledge management: Semantic drift or conceptual shift? *Journal of Education for Library and Information Science, 41*(4), 294–306. https://doi.org/10.2307/40324047

Dehbi, S., Lamrani, H. C., Belgnaoui, T., & Lafou, T. (2022). Big data analytics and management control. *Procedia Computer Science, 203*, 438–443. https://doi.org/10.1016/j.procs.2022.07.058

Deng, X., Guo, X., Wu, Y. J., & Chen, M. (2021). Perceived environmental dynamism promotes entrepreneurial team member's innovation: Explanations based on the uncertainty reduction theory. *International Journal of Environmental Research and Public Health, 18*(4), 1–12. https://doi.org/10.3390/ijerph18042033

Dubey, R., Gunasekaran, A., Childe, S. J., Bryde, D. J., Giannakis, M., Foropon, C., Roubaud, D., & Hazen, B. T. (2019). Big data analytics and artificial intelligence pathway to operational performance under the effects of entrepreneurial orientation and environmental dynamism: A study of manufacturing organisations. *International Journal of Production Economics*, 1–42. https://doi.org/10.1016/j.ijpe.2019.107599

Elliott, K., Price, R., Shaw, P., Spiliotopoulos, T., Ng, M., Coopamootoo, K., & van Moorsel, A. (2021). Towards an equitable digital society: Artificial intelligence (AI) and corporate digital responsibility (CDR). *Society, 58*(3), 179–188. https://doi.org/10.1007/s12115-021-00594-8

Etemadi, H., Dilami, Z. D., Bazaz, M. S., & Parameswaran, R. (2009). Culture, management accounting and managerial performance: Focus Iran. *Advances in Accounting, 25*(2), 216–225. https://doi.org/10.1016/j.adiac.2009.08.005

Fahlevi, H., Irsyadillah, I., Indriani, M., & Oktari, R. S. (2022). DRG-based payment system and management accounting changes in an Indonesian public hospital: Exploring potential roles of big data analytics. *Journal of Accounting & Organizational Change, 18*(2), 325–345. https://doi.org/10.1108/JAOC-10-2020-0179

Fitria, G. N. (2021). The effect of management accounting system, sustainable leadership and environmental strategy on business performance with organizational culture as moderator. *International Journal of Contemporary Accounting, 3*(1), 45–60.

Fornell, C., & Larcker, D. F. (1981). Evaluating structural equation models with unobservable variables and measurement error. *Journal of Marketing Research, 18*(1), 39–50. https://doi.org/10.2307/3151312

Freixanet, J., & Renart, G. (2020). A capabilities perspective on the joint effects of internationalization time, speed, geographic scope and managers' competencies on SME survival. *Journal of World Business, 55*(6), 1–14. https://doi.org/10.1016/j.jwb.2020.101110

Ghasemi, R., Habibi, H. R., Ghasemlo, M., & Karami, M. (2019). The effectiveness of management accounting systems: Evidence from financial organizations in Iran. *Journal of Accounting in Emerging Economies, 9*(2), 182–207. https://doi.org/10.1108/JAEE-02-2017-0013

Guni, C. N. (2018). The management accounting: The boundary between traditional and modern. *Journal of Management and Strategy, 42*, 193–199.

Hair, J. P., Black, J. P., Babin, J. P., & Anderson, R. E. (2019). *Multivariate data analysis* (8th ed.). Cengage Learning.

Hanisch, B., & Wald, A. (2012). A bibliometric view on the use of contingency theory in project management research. *Project Management Journal, 43*(3), 4–23. https://doi.org/10.1002/pmj.21267

Heeringa, S. G., West, B. T., & Berglund, P. A. (2017). *Applied survey data analysis* (2nd ed.). Chapman & Hall/CRC Press.

Helgeson, J. F., Aminpour, P., Fung, J. F., Henriquez, A. R., Zycherman, A., Butry, D., Nierenberg, C., & Zhang, Y. (2022). Natural hazards compound COVID-19 impacts on small businesses disproportionately for historically underrepresented group operators. *International Journal of Disaster Risk Reduction, 72*, 1–13. https://doi.org/10.1016/j.ijdrr.2022.102845

Ipsmillera, E., Dikova, D., & Brouthers, K. D. (2022). Digital internationalization of traditional firms: Virtual presence and entrepreneurial orientation. *Journal of International Management, 28*(4), 1–15.

Kennedy, S. & Linnenluecke, M. K. (2022). Circular economy and resilience: A research agenda. *Business Strategy and the Environment*, 1–12. https://doi.org/10.1002/bse.3004

Khalil, A., Abdelli, M. E. A., & Mogaji, E. (2022). Do digital technologies influence the relationship between the COVID-19 crisis and SMEs' resilience in developing countries? *Journal of Open Innovation: Technology, Market, and Complexity, 8*, 1–14. https://doi.org/10.3390/joitmc8020100

Kline, R. B. (2018). Response to Leslie Hayduk's review of principles and practice of structural equation modeling. *Canadian Studies in Population, 45*(3–4), 188–195. https://doi.org/10.25336/csp29418

Kuivalainen, O., Sundqvist, S., Saarenketo, S., & McNaughton, R. (2012). Internationalization patterns of small and medium-sized enterprises. *International Marketing Review, 29*(5), 448–465. https://doi.org/10.1108/02651331211260331

Lacy, P., & Rutqvist, J. (2015). *Waste to wealth: The circular economy advantage*. Palgrave Macmillan.

Le, T. T., & Ikram, M. (2022). Do sustainability innovation and firm competitiveness help improve firm performance? Evidence from the SME sector in Vietnam. *Sustainable Production and Consumption, 29*, 1–12. https://doi.org/10.1016/j.spc.2021.11.008

Lin, F. J., & Ho, C. W. (2019). The knowledge of entry mode decision for small and medium enterprises. *Journal of Innovation & Knowledge, 4*(1), 32–37. https://doi.org/10.1016/j.jik.2018.02.001

Linnenluecke, M. K. (2017). Resilience in business and management research: A review of influential publications and a research agenda. *International Journal of Management Reviews, 19*(1), 4–30. https://doi.org/10.1111/ijmr.12076

Lisboa, A., Skarmeas, D., & Lages, C. (2011). Entrepreneurial orientation, exploitative and explorative capabilities, and performance outcomes in export markets: A resource-based approach. *Industrial Marketing Management, 40*(8), 1274–1284. https://doi.org/10.1016/j.indmarman.2011.10.013

Makridakis, S. (2017). The forthcoming Artificial Intelligence (AI) revolution: Its impact on society and firms. *Futures, 90*, 46–60. https://doi.org/10.1016/j.futures.2017.03.006

Müller, J. M., Buliga, O., & Voigt, K.-I. (2021). The role of absorptive capacity and innovation strategy in the design of industry 4.0 business models—A comparison between SMEs and large enterprises. *European Management Journal, 39*(3), 333–343. https://doi.org/10.1016/j.emj.2020.01.002

Naranjo-Gil, D., & Hartmann, F. (2007). Management accounting systems, top management team heterogeneity and strategic change. *Accounting, Organizations and Society, 32*(7–8), 735–756. https://doi.org/10.1016/j.aos.2006.08.003

Ngo, Q. H. (2021). The impact of market orientation on small businesses' performance in Vietnam: The mediating effects of the management accounting system. *Entrepreneurial Business and Economics Review, 9*(3), 59–72.

Nohria, N., & Khurana, R. (2010). *Handbook of leadership theory and practice*. Harvard Business School Press.

Ortiz-de-Mandojana, N., & Bansal, P. (2015). The long-term benefits of organizational resilience through sustainable business practices. *Strategic Management Journal, 37*(8), 1615–1631. https://doi.org/10.1002/smj.2410

Otley, D. (2016). The contingency theory of management accounting and control: 1980–2014. *Management Accounting Research, 31*, 45–62. https://doi.org/10.1016/j.mar.2016.02.001

Pedroso, E., & Gomes, C. F. (2020). The effectiveness of management accounting systems in SMEs: A multidimensional measurement approach. *Journal of Applied Accounting Research, 21*(3), 497–515. https://doi.org/10.1108/jaar-05-2018-0059

Peng, Y., & Lin, K. (2017). The effect of global dynamic capabilities on internationalizing SMEs performance: Organizational culture factors as antecedents. *Baltic Journal of Management, 12*, 307–328.

Pinho, J. C., & Prange, C. (2016). The effect of social networks and dynamic internationalization capabilities on international performance. *Journal of World Business, 51*(3), 391–403. https://doi.org/10.1016/j.jwb.2015.08.001

Pla-Julián, I., & Guevara, S. (2019). Is circular economy the key to transitioning towards sustainable development? Challenges from the perspective of care ethics. *Futures, 105*, 66–77. https://doi.org/10.1016/j.futures.2018.09.001

Prange, C., & Verdier, S. (2011). Dynamic capabilities, internationalization processes and performance. *Journal of World Business, 46*(1), 126–133. https://doi.org/10.1016/j.jwb.2010.05.024

Ribau, C. P., Moreira, A. C., & Raposo, M. (2016). SME internationalization research: Mapping the state of the art. *Revue Canadienne Des Sciences De L'administration [Canadian Journal of Administrative Sciences], 35*(2), 280–303. https://doi.org/10.1002/cjas.1419

Saggi, M. K., & Jain, S. (2018). A survey towards an integration of big data analytics to big insights for value-creation. *Information Processing & Management, 54*(5), 758–790. https://doi.org/10.1016/j.ipm.2018.01.010

Saunila, M. (2019). Innovation capability in SMEs: A systematic review of the literature. *Journal of Innovation & Knowledge, 5*(4), 260–265. https://doi.org/10.1016/j.jik.2019.11.002

Shrestha, Y. R., Ben-Menahem, S. M., & von Krogh, G. (2019). Organizational decision-making structures in the age of artificial intelligence. *California Management Review*, 1-18. https://doi.org/10.1177/0008125619862257

Teece, D. J. (2007). Explicating dynamic capabilities: The nature and micro-foundations of (sustainable) enterprise performance. *Strategic Management Journal, 28*(13), 1319–1350. https://doi.org/10.1002/smj.640

Teece, D. J., Pisano, G., & Shuen, A. (1997). Dynamic capabilities and strategic management. *Strategic Management Journal, 18*(7), 509–533. https://doi.org/10.1002/(sici)1097-0266(199708)18:7%3c509::aid-smj882%3e3.0.co;2-z

Teece, D. J., Pisano, G., & Shuen, A. (1990). *Firm capabilities, resources and the concept of strategy*. Economic Analysis and Policy Working Paper EAP, 38, University of California, Oakland, CA.

Urbach, N., & Ahlemann, F. (2010). Structural equation modeling in information systems research using partial least squares. *Journal of Information Technology Theory and Application, 11*(2), 5–40.

Välikangas, L. (2010). *The resilient organization: How adaptive cultures thrive even when strategy fails*. McGraw Hill Professional.

Van de Vijver, F., & Tanzer, N. K. (2004). Bias and equivalence in cross-cultural assessment: An overview. *Revue Européenne De Psychologie Appliquée/european Review of Applied Psychology, 54*(2), 119–135. https://doi.org/10.1016/j.erap.2003.12.004

van der Vegt, G. S., Essens, P., Wahlström, M., & George, G. (2015). Managing risk and resilience. *Academy of Management Journal, 58*, 971–980.

Van Witteloostuijn, A. (2021). The post-corona crisis paradox: How labor will globalize to the benefit of emerging economies. *Management and Organization Review, 17*(1), 40–44. https://doi.org/10.1017/mor.2020.81

Vogus, T. J., & Sutcliffe, K. M. (2007, October 7–10). *Organizational resilience: Towards a theory and research agenda*. Proceedings of the 2007 IEEE International Conference on Systems, Man and Cybernetics, Montreal, QC, Canada, pp. 3418–3422.

Wolf, E. J., Harrington, K. M., Clark, S. L., & Miller, M. W. (2013). Sample size requirements for structural equation models. *Educational and Psychological Measurement, 73*(6), 913–934. https://doi.org/10.1177/0013164413495237

Wu, W., Wang, H., & Wu, Y. J. (2020). Internal and external networks, and incubatees' performance in dynamic environments: entrepreneurial learning's mediating effect. *The Journal of Technology Transfer*, 1–27. https://doi.org/10.1007/s10961-020-09790-w

Yoon, S. (2020). A study on the transformation of accounting based on new technologies: Evidence from Korea. *Sustainability, 12*(20), 1–22. https://doi.org/10.3390/su12208669

Zybery, I., & Rova, L. (2014). The role of the accountants in the framework of the modern technological developments and digital accounting systems. *European Scientific Journal, 24*, 30–48.

The Relationship Between Unemployment and Inflation in Vietnam: Does the Phillips Curve Exist?

Le Thanh Tung, Pham Thi Bich, Le Thi Phuong Anh, and Nguyen Bang Khanh

Abstract The relationship between unemployment and inflation is one of the important research topics in economics for improving the effectiveness of the macroeconomic policymaking process. However, there are few study results on this topic in Vietnam, an Asian developing country that has a high economic growth rate. This paper aims to clarify the relationship between these two macroeconomic variables in Vietnam through the Phillips curve theory with a database collected in the period of 1992–2021. The research methodologies include the Johansen-Juselius co-integration test and the Engle-Granger error correction model. The quantitative results confirm that there was no trade-off relationship between unemployment and inflation in Vietnam in both the short run and long run. Therefore, the empirical results did not support the Phillips curve theory in the Vietnamese economy. Finally, the paper suggests some policy implications for promoting sustainable economic growth in Vietnam in the future.

Keywords Unemployment · Inflation · Phillips curve · Macroeconomic policy · Policymaker

1 Introduction

Macroeconomic policies are implemented to achieve the government's objectives, for example, to reach a natural employment rate and a single-digit inflation rate (Schreiber & Wolters, 2007). Thus, it has an important reason to study and establish the frameworks of policies for stabilizing inflation and reducing unemployment (Aydin, 2017). Besides, the studies on the relationship between unemployment and inflation help to create suitable policies for growth and employment in the long term. In 1958, the economist named Phillips published a study that claimed the inverse relationship between unemployment and the corresponding rate of wage change in the UK over the 1861–1957 period. Since then, the trade-off relationship between

L. T. Tung (✉) · P. T. Bich · L. T. P. Anh · N. B. Khanh
Faculty of Economics and Public Management, Ho Chi Minh City Open University, Ho Chi Minh City, Vietnam
e-mail: tung.lt@ou.edu.vn

A. T. Nguyen et al. (eds.), *Contemporary Economic Issues in Asian Countries: Proceeding of CEIAC 2022, Volume 1*,
https://doi.org/10.1007/978-981-19-9669-6_22

the unemployment rate and the inflation rate has been known as the Phillips curve theory. It can be simply understood as rising inflation due to high economic growth creates more job opportunities, thereby reducing the unemployment rate (Phillips, 1958). On the contrary, when the unemployment rate increases, the number of labors applying for jobs exceeds the number of jobs available. Businesses are not pressured to increase wages to retain workers, and wage inflation will therefore not take place. In general, the Phillips curve has played an important role in making macroeconomic policies.

For promoting rapid and sustainable economic growth, suitable inflation and low unemployment are the principal objectives in many economies, both in developing and developed countries (DiNardo & Moore, 1999; Eser et al., 2020; Ghosh & Phillips, 1998; Tung & Thanh, 2015). The Phillips curve theory was rechecked and confirmed by some well-known economists including Samuelson and Solow (1960) and Gordon (1970). However, the hypothesis of a trade-off relationship between these two macroeconomic variables has also been harshly criticized by various economists, such as Friedman (1968) who argued that the Phillips curve was only effective in the short term. Because of some conflicts, in the 1980s, the Phillips curve theory has ignored by some economists and policymakers. However, in recent decades, some studies have been done on the unemployment–inflation nexus which applied the Philips curve theory (Ball & Mazumder, 2021; Furuoka et al., 2021; King & Watson, 1994; Schreiber & Wolters, 2007).

There are some concerns regarding the relationship between inflation and unemployment in Vietnam (Tung, 2021). Some questions are raised such as: Is there a relationship between unemployment and inflation? Is it possible to apply the Phillips curve theory to Vietnam? How can we improve the macroeconomic policymaking process in practice? This paper aims to use the Phillips curve theory in studying the relationship between unemployment and inflation in Vietnam, an Asian emerging economy (Bentzen & Tung, 2021), thereby suggesting macroeconomic policies for enhancing rapid and sustainable economic growth. The database is an annual format collected in the 1992–2021 period. This paper contributes two main directions. First, this study uses the newest data, and it contributes fresh evidence for the policymakers in this country. Second, for practical activities, the result of this study provides useful information for the Vietnamese on two macroeconomic issues: inflation, unemployment, and their interaction. Third, because Vietnam has a robust economic growth model, the policy implications can be referenced and considered by other countries which are at the same development level.

2 Literature Reviews

Because of the importance of unemployment and inflation information for the policymaking process, the Phillips curve theory has gradually become a hot topic in both academic and real practices for making macroeconomic policies for countries. The

Phillips curve was first introduced in 1958 by investigating the nexus of unemployment and wage change in the UK (Phillips, 1958). Based on this empirical finding, Samuelson and Solow (1960) suggested that expansionary monetary policy could reduce the unemployment rate by allowing inflation to rise. From the traditional Keynesian perspective, the expected inflation rate can affect the real inflation rate, and the Phillips curve does not change from time to time. Therefore, the government can use expansionary monetary-fiscal policies to stimulate consumption and support private investment for boosting the economy, thereby increasing the income of workers and reducing the unemployment rate. In other studies, Friedman (1968) and Phelps (1967) argued that the Phillips curve only worked in the short run. These monetarist economists stated that, in the long run, inflation and unemployment had no trade-off relationship.

Karanassou et al. (2005) provided empirical evidence for supporting the long-term trade-off hypothesis in the relationship between inflation and unemployment in some conditions, besides the study of Franz (2005) for the case of Germany. Some empirical models used the panel data approach to examine the Phillips curve in different countries. For example, DiNardo and Moore (1999) examined 9 OECD members and concluded regarding the Phillips curve in these countries. Besides, Turner and Seghezza (1999) applied the Phillips curve theory and investigated with a sample including 21 OECD countries. The estimated result concluded that the overall evidence provided a clear Phillips curve evidence in these OECD members. On the other hand, Akerlof et al. (2000) and Holden (2004) discussed some expanded models that mentioned long-term trade-offs between output and inflation that could exist if inflation rates were low.

However, the Phillips curve has received some conflict recommendations in some economies. Some studies showed that the Phillips curve did not exist in the short-term relationship between unemployment and inflation, but it has been found in the long-term relationship, for example, in Greece during the period 1980–2010 (Dritsaki & Dritsaki, 2013). In contrast to the case of Greece, the relationship between inflation and unemployment in Indonesia from 1987 to 2018 showed a trade-off effect in the short run but did not confirm it in the long run. Some studies indicated no evidence of a Phillips curve in the countries, such as Kuştepeli (2005) in Turkey or Wulandari et al. (2019) in Indonesia.

Therefore, with this current study, we will examine and explore whether the Phillips curve existed in Vietnam's economy in 1992–2021. Based on the estimated results, the policy implications will be provided for enhancing the efficiency of macroeconomic policies in the future.

3 Methodology

3.1 Mathematic Model

The relationship between unemployment and inflation in Vietnam is investigated by the Phillip curve theory and estimated by the error correction model (ECM) of Engle and Granger (1987). Following this econometric approach, the long-term relationship between X (independent variable) and Y (dependent variable) is described by the following equation.

$$Y_t = \varphi_0 + \varphi_1 X_t + u_t \quad (1)$$

The residual (u_t) is assumed to have the normal distribution and the white noise. Equation (1) is calculated by the ordinary least squares method (OLS) to determine long-term coefficients that represent relationships between two variables. Based on the long-term result, the u_t residual is denoted as ECM_t and used in the error correction model (Engle & Granger, 1987). Specifically, ECM_t is defined as follows: $ECM_t = Y_t - \varphi_0 - \varphi_1 X_t$. The impact of variable X on variable Y in the short-term analysis is calculated using Eq. (2) below.

$$\Delta Y_t = \alpha_0 + \alpha_{1t}\Delta X + \alpha_2 ECM_{t-1} + \varepsilon_t \quad (2)$$

The coefficient (α_2) of the ECM_{t-1} was used to measure the rate of adjustment from short-term to long-term equilibrium (Engle & Granger, 1987). The symbol "Δ" is used to refer to the first difference values of the variables used in Eq. (2). Thus, the sign and the statistical significance of the coefficient of variable X will help conclude the short-term and long-term relationships of the variables in the study model.

Based on the error correction model approach of Engle and Granger (1987), the Phillips curve model will be estimated with five formats. The dependent variable is the inflation rate, and the independent variables are the unemployment rate and economic growth rate. The lag values of the unemployment rate, inflation rate, and economic growth rate are included in the estimated models for further investigating the relationship between unemployment and inflation in Vietnam. The five equations of the Phillips curve model represent the long and short run established by the error correction model approach of Engle and Granger (1987) as follows (Table 1).

Before applying the error correction model (ECM) of Engle and Granger (1987), these time series will be checked by the augmented Dickey-Fuller (ADF) test and the Johansen-Juselius co-integration test. Finally, the estimated results of the five Phillips curve models will be carried out by the diagnostic tests to ensure the robustness of the results.

Table 1 Proposed models for the estimation process

Number of the models	Proposed model	Equations
Model 1	Long run	$\text{Inflation}_t = \alpha_1 + \beta_1\text{Unemployment}_t + \varepsilon_{1t}$
	Short run	$\Delta\text{Inflation}_t = \lambda_1 + \varphi_1\Delta\text{Unemployment}_t + \delta_1\text{ECM}_{t-1} + \epsilon_{1t}$
Model 2	Long run	$\text{Inflation}_t = \alpha_2 + \beta_2\text{Unemployment}_t + \varphi_1\text{Inflation}_{t-1} + \varepsilon_{2t}$
	Short run	$\Delta\text{Inflation}_t = \lambda_2 + \varphi_2\Delta\text{Unemployment}_t + \omega_1\Delta\text{Inflation}_{t-1} + \delta_2\text{ECM}_{t-1} + \epsilon_{2t}$
Model 3	Long run	$\text{Inflation}_t = \alpha_3 + \beta_3\text{Unemployment}_t + \beta_4\text{Unemployment}_{t-1} + \varphi_2\text{Inflation}_{t-1} + \varepsilon_{3t}$
	Short run	$\Delta\text{Inflation}_t = \lambda_3 + \varphi_3\Delta\text{Unemployment}_t + \varphi_3\Delta\text{Unemployment}_{t-1} + \omega_2\Delta\text{Inflation}_{t-1} + \delta_3\text{ECM}_{t-1} + \epsilon_{3t}$
Model 4	Long run	$\text{Inflation}_t = \alpha_4 + \beta_4\text{Unemployment}_t + \beta_5\text{Growth}_t + + \varphi3\text{Inflation}_{t-1} + \varepsilon_{4t}$
	Short run	$\Delta\text{Inflation}_t = \lambda_4 + \varphi_4\Delta\text{Unemployment}_t + \psi_1\Delta\text{Growth}_t + \omega_3\Delta\text{Inflation}_{t-1} + \delta_4\text{ECM}_{t-1} + \epsilon_{4t}$
Model 5	Long run	$\text{Inflation}_t = \alpha_5 + \beta_6\text{Unemployment}_t + \beta_7\text{Growth}_t + \varepsilon_{5t}$
	Short run	$\Delta\text{Inflation}_t = \lambda_5 + \varphi_5\Delta\text{Unemployment}_t + \psi_2\Delta\text{Growth}_t + \delta_5\text{ECM}_{t-1} + \epsilon_{5t}$

Source Authors

3.2 *Data Collection*

The paper uses an annual database sourced from the General Statistics Offices of Vietnam (GSO, 2022). The sample has 30 observations in the period of 1992–2021 (Table 2). The inflation rate is calculated by the consumer price index (CPI), and the unemployment rate is employed by the international standard of the International Labor Organization. The rate of economic growth is computed by the Gross Domestic Product (GDP) at a constant price. Some graphs of the time series variables are shown in Fig. 1.

4 Results

4.1 *Unit Root Test*

The unit root test for variables is done by the ADF method (Dickey & Fuller, 1979). In the testing process, each time series variable will be simultaneously checked with

Table 2 Statistical analysis of the variables

Variable	Max	Min	Mean	Std. Dev	Obs
Inflation	32.60	0.600	9.189	7.298	30
Unemployment	7.200	2.010	4.791	1.545	30
Growth	9.500	2.580	6.832	1.665	30
ΔInflation	13.10	−16.50	−1.060	6.383	29
ΔUnemployment	0.840	−1.170	−0.137	0.408	29
ΔGrowth	2.000	−4.110	−0.207	1.243	29

Source GSO (2022)

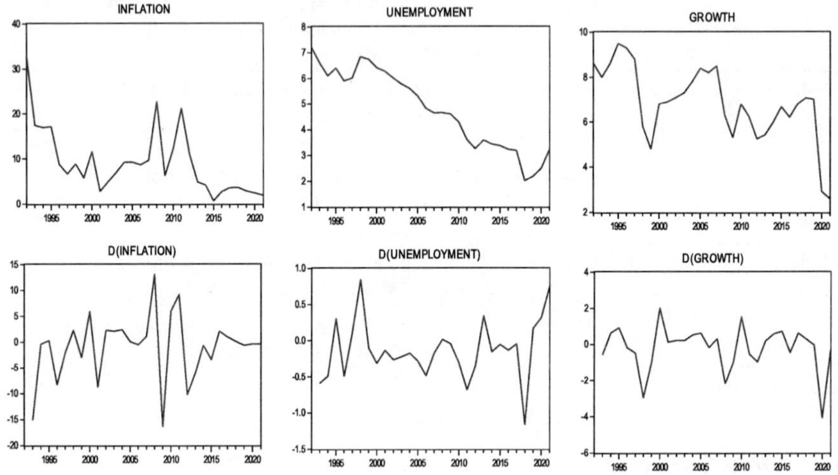

Fig. 1 Time series variables at the level and first difference values. 2022 *Source* GSO (2022)

three types, including (i) an intercept, (ii) with a trend and an intercept, and (iii) without an intercept. The testing results are shown in Table 3.

According to the results obtained after running the database on the EViews software, at the level, there is only the inflation variable that is a stationary time series at 1% and 5%. The unemployment variable is stationary at 10% (without a constant). However, the economic growth variable is not a stationary time series in all tests. Based on this result, the regression of these variables can lead to unreliable estimated outputs. In the next step, the ADF test will help to test these variables at the first difference values. The testing results confirm that these time series variables are stationary at 1%. Therefore, if regression is carried out between the first difference databases, the estimated results will be reliable and minimize the phenomenon of biased results.

Table 3 Result of the unit root test

Variable	With intercept	With trend and intercept	Without constant
Inflation	−4.256902[a]	−4.479356[a]	−2.094326[b]
Unemployment	−1.228169	−2.712873	−2.132305[c]
Growth	−1.451087	−2.400126	−1.183319
ΔInflation	−6.218959[a]	−6.143683[a]	−6.090639[a]
ΔUnemployment	−4.344433[a]	−4.268606[b]	−4.170600[a]
ΔGrowth	−4.743896[a]	−4.756801[a]	−4.729728[a]

Note [a], [b], [c] denoted significant at 1%, 5%, 10%
Source Authors

Table 4 Johansen and Juselius test for the co-integration relationship

Hypothesized full of co-integration equation(s)	Trace statistic		Max-Eigen statistic	
	Statistic	*p*-value	Statistic	*p*-value
The relationship between inflation and unemployment				
None	15.494	0.2228	14.264	0.2309
At most 1	3.8414	0.2944	3.8414	0.2944
The relationship among inflation, unemployment, and growth				
None	29.797	0.0800	21.131	0.1502
At most 1	15.494	0.2452	14.264	0.1822
At most 2	3.8414	0.9444	3.8414	0.9444

Source Authors

4.2 Co-integration Test

The paper uses the Johansen and Juselius (1990) method to check the long-term co-integration relationship between two variables including the inflation rate and the unemployment rate. Whether there is a long-term relationship between these two macro-variables in Vietnam in the period 1992–2021 or not? The testing results show that it is impossible to reject the hypothesis (H_0) of the non-existence of co-integration vectors between inflation and unemployment. The Johansen and Juselius (1990) method confirms that there is no existence of a co-integration vector among inflation, unemployment, and economic growth in Vietnam for the study period. The Johansen and Juselius (1990) tests are performed in Table 4.

4.3 Estimated Results

Based on the error correction model approach of Engle and Granger (1987), five econometric models representing five kinds of the Phillips curve will be estimated in the long-term models. The regressive results of the long-term models can be used to identify the relationships in the short term with five types of equations. The estimated results are presented in Table 5.

After conducting regression with the ECM, the research results are summarized based on five Phillips curve road regression models of Vietnam as follows. In the long-run analysis, the Phillips curve did not appear in the Vietnamese economy from 1992 to 2021. The sign of coefficients of unemployment and inflation was positive

Table 5 Long-run and short-run estimations of the Phillips curve

The long-run analysis. The dependent variable is the inflation					
Variable	Model 1	Model 2	Model 3	Model 4	Model 5
	Coefficient	Coefficient	Coefficient	Coefficient	Coefficient
Constant	−1.428	1.116	0.691	−1.714	−5.831
Unemployment	2.216[a]	0.776	−1.557	0.483	1.679[c]
Unemployment (−1)		0.382[b]	2.438		
Inflation (−1)			0.339[b]	0.364[c]	
Growth				0.646	1.020
R^2	0.2202	0.3348	0.3608	0.3605	0.2615
Adj R^2	0.1923	0.2836	0.2841	0.2837	0.2068
F-statistics	7.908[a]	6.543[a]	4.704[a]	4.697[a]	4.781[b]
Obs	30	30	30	30	30
The short-run analysis. The dependent variable is the inflation					
Constant	−1.114	−0.482	−0.565	−0.706	−1.225
ΔUnemployment	−1.029	−1.776	−1.604	−2.190	−1.117
ΔUnemployment (−1)			−0.194		
ΔInflation (−1)		0.187	0.179	0.152	
ΔGrowth				−0.351	−0.096
ECM (−1)	−0.652[a]	−0.896[a]	−0.917[a]	−0.909[a]	−0.685[b]
R2	0.4334	0.3602	0.3826	0.3872	0.4571
Adj R2	0.3898	0.2802	0.2752	0.2806	0.3919
F-statistics	9.945[a]	5.505[a]	3.563[b]	3.633[b]	7.017[a]
Obs	29	29	29	29	29
The short-run diagnostic tests					
Test	*F*-statistic	*F*-statistic	*F*-statistic	*F*-statistic	*F*-statistic
Serial Correlation	0.1135 (0.8931)	1.7634 (0.1948)	1.0782 (0.3583)	2.4779 (0.1081)	0.0605 (0.9414)
Heteroskedasticity	0.6294 (0.6791)	0.6603 (0.7332)	0.4126 (0.9436)	0.7546 (0.6969)	0.1092 (0.4122)

Notes [a], [b], [c] displays significant at 1%, 5%, 10%. Serial correlation is the Breusch-Godfrey test; Heteroskedasticity is the White test. *p*-value is in the parentheses
Source Authors

and significant in the two models. The lag value of the inflation rate variable had a positive impact on the inflation rate in the current year. The lag values of the unemployment variable had a positive effect on the inflation rate in Model 2. In general, the economic growth rate had a positive impact but was insignificant on the inflation rate in the long run.

In the short-run analysis, the estimated results showed that the Phillips curve also did not appear clearly in Vietnam in the period 1992–2021 because all coefficients of unemployment variables were not statistically significant. The lag values of the inflation rate had no discernible impact on the inflation rate of the current year. The lag values of the unemployment rate also confirmed no impact on the inflation rate of the current year as well. Economic growth was not a factor that affected the rate of inflation.

To further clarify the long-term relationship between the two macroeconomic variables for the period 1992–2021, a graph was done for showing the statistical relationship between these two variables. The graph between the inflation rate and the unemployment rate has confirmed that there was no existence of a long-term trade-off relationship between these variables in Vietnam in the period 1992–2021 (Fig. 2). Data analysis by the graphing method was consistent with quantitative analysis in the above sections.

Specifically, Fig. 2 shows a positive relationship between inflation and unemployment indicators, and we see that when inflation increased, unemployment also raised in the same period. This evidence concludes a positive relationship between these macroeconomic indicators in long run; however, it seems to contradict the inverse relationship between inflation and unemployment defined by the Phillips curve theory.

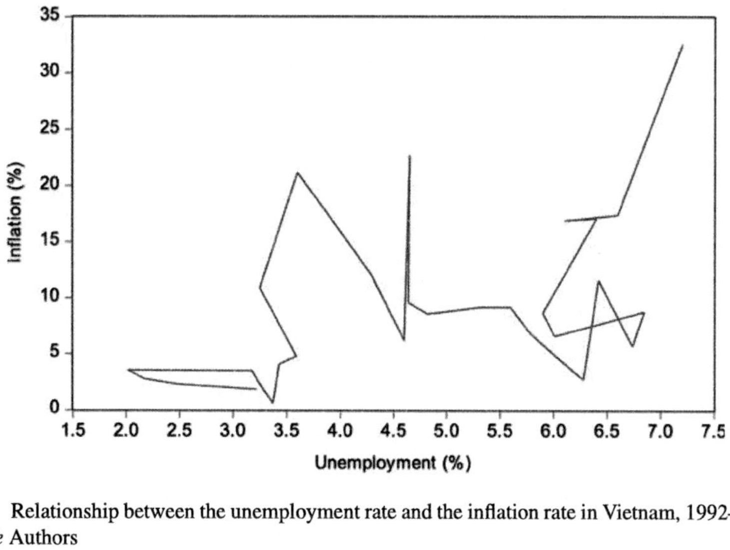

Fig. 2 Relationship between the unemployment rate and the inflation rate in Vietnam, 1992–2021. *Source* Authors

Therefore, by some quantitative methods, the empirical results have indicated that in Vietnam, the Phillips curve theory has not been confirmed. The Phillips curve is an interesting topic for researchers to make the exact policies for the countries. This paper concludes the non-existence of the Phillips curve in the case of Vietnam. Our finding is supported by some previous studies, for example, in Nigeria (Orji et al., 2015), in Indonesia (Wulandari et al., 2019), and Turkey (Kuştepeli, 2005). In addition, the Phillips curve did not exist in the short term but has long-term relationships as was the case with Greece in the period 1980–2010 (Dritsaki & Dritsaki, 2013).

5 Conclusions and Policy Implications

The study uses several quantitative methods to estimate the Phillips curve in Vietnam for the period 1992–2021. The results show that the Phillips curve theory of the relationship between unemployment and inflation is not consistent with the actual developments in Vietnam in the period from 1992 to 2021. Based on empirical results, the study reveals several important implications for policymakers in the Vietnamese economy.

Inflation and unemployment are two important factors of a country in the macroeconomic policymaking process. In the case of Vietnam, as the current study presented, there was no trade-off between unemployment and inflation both in the short and long term. Therefore, policymakers need to accurately determine the national natural unemployment rate and orient economic policies based on this rate in a reasonable way. On the other hand, they should identify signals of the economy to implement macroeconomic tools effectively where unemployment depended greatly on the state of the economy and policies on job creation. When the economy is in a recession, unemployment is quite high, and vice versa. On the other hand, inflation depends mainly on fiscal and monetary policy.

The government's job creation policies help the level of unemployment decrease, so far, and the unemployment rate tends to be close to the natural level of unemployment. If the policymakers keep the expected unemployment in the future, it requires policymakers to focus on two directions. First, attracting outside investment sources, especially foreign direct investment, when the country receives direct investment from foreign individuals or companies through the establishment of production that will help create abundant jobs for workers. Second, policymakers need to increase the number of domestic firms. Domestic businesses can create a huge number of new jobs for the economy.

Considering the expected inflation, the inverse relationship between unemployment and inflation is not shown in Vietnam when considering the period from 1992 to 2021, and this trade-off is completely unclear, even when analyzing data in the short and long term. This evidence suggests to policymakers, especially in macroeconomic policies, a message about the flexible application and independent policy offerings between the goal of curbing inflation and reducing unemployment.

References

Akerlof, G. A., Dickens, W. T., Perry, G. L., Bewley, T. F., & Blinder, A. S. (2000). Near-rational wage and price setting and the long-run Phillips curve. *Brookings Papers on Economic Activity, 2000*(1), 1–60.

Aydin, C. (2017). Inflationary threshold effects on the relationship between unemployment and economic growth: Evidence from Turkey? *International Journal of Economic Perspectives, 11*(2), 88–97.

Ball, L., & Mazumder, S. (2021). A Phillips curve for the euro area. *International Finance, 24*(1), 2–17.

Bentzen, J., & Tung, L. T. (2021). Regional income convergence in an emerging Asian economy: Empirical evidence from Vietnam. *Post-Communist Economies, 33*(1), 64–78.

Dickey, D. A., & Fuller, W. A. (1979). Distribution of the estimators for autoregressive time series with a unit root. *Journal of the American Statistical Association, 74*(366), 427–431.

DiNardo, J., & Moore, M. P. (1999). *The Phillips curve is back? Using panel data to analyze the relationship between unemployment and inflation in an open economy.* Working Paper-7328.

Dritsaki, C., & Dritsaki, M. (2013). Phillips curve inflation and unemployment: An empirical research for Greece. *International Journal of Computational Economics and Econometrics, 3*(1–2), 27–42.

Engle, R. F., & Granger, C. W. (1987). Co-integration and error correction: Representation, estimation, and testing. *Econometrica, 55*(2), 251–276.

Eser, F., Karadi, P., Lane, P. R., Moretti, L., & Osbat, C. (2020). The Phillips curve at the ECB. *The Manchester School, 88*(S1), 50–85.

Franz, W. (2005). Will the (German) NAIRU please stand up? *German Economic Review, 6*(2), 131–153.

Friedman, M. (1968). The role of monetary policy. *American Economic Review, 58*(1), 215–231.

Furuoka, F., Pui, K. L., Chomar, M. T., & Nikitina, L. (2021). Is the Phillips curve disappearing? Evidence from a new test procedure. *Applied Economics Letters, 28*(6), 493–500.

General Statistics Office (GSO). (2022). *National account.* Available at https://www.gso.gov.vn

Ghosh, A., & Phillips, S. (1998). Warning: Inflation may be harmful to your growth. *IMF Staff Papers, 45*(4), 672–710.

Gordon, R. J. (1970). Prices in 1970: The horizontal Phillips curve? *Brookings Papers on Economic Activity, 1970*(3), 449–458.

Holden, S. (2004). The costs of price stability: Downward nominal wage rigidity in Europe. *Economica, 71*(282), 183–208.

Johansen, S., & Juselius, K. (1990). Maximum likelihood estimation and inference on cointegration—With applications to the demand for money. *Oxford Bulletin of Economics and Statistics, 52*(2), 169–210.

Karanassou, M., Sala, H., & Snower, D. J. (2005). A reappraisal of the inflation—Unemployment tradeoff. *European Journal of Political Economy, 21*(1), 1–32.

King, R. G., & Watson, M. W. (1994). The post-war US Phillips curve: A revisionist econometric history. *Carnegie-Rochester Conference Series on Public Policy, 41*, 157–219.

Kuştepeli, Y. (2005). A comprehensive short-run analysis of a (possible) Turkish Phillips curve. *Applied Economics, 37*(5), 581–591.

Orji, A., Anthony-Orji, O. I., & Okafor, J. C. (2015). Inflation and unemployment nexus in Nigeria: Another test of the Phillips curve. *Asian Economic and Financial Review, 5*(5), 766–778.

Phelps, E. S. (1967). Phillips curves, expectations of inflation and optimal unemployment over time. *Economica, 34*(135), 254–281.

Phillips, A. W. (1958). The relation between unemployment and the rate of change of money wage rates in the United Kingdom, 1861–1957. *Economica, 25*(100), 283–299.

Samuelson, P. A., & Solow, R. M. (1960). Analytical aspects of anti-inflation policy. *The American Economic Review, 50*(2), 177–194.

Schreiber, S., & Wolters, J. (2007). The long-run Phillips curve revisited: Is the NAIRU framework data-consistent? *Journal of Macroeconomics, 29*(2), 355–367.

Tung, L. T. (2021). Fiscal policy, monetary policy and price volatility: evidence from an emerging economy. *Organizations and Markets in Emerging Economies, 12*(1), 57–70.

Tung, L. T., & Thanh, P. T. (2015). Threshold in the relationship between inflation and economic growth: empirical evidence in Vietnam. *Asian Social Science, 11*(10), 105–112.

Turner, D., & Seghezza, E. (1999). *Testing for a Common OECD Phillips Curve*. OECD Economics Department Working Papers.

Wulandari, D., Utomo, S. H., Narmaditya, B. S., & Kamaludin, M. (2019). Nexus between inflation and unemployment: Evidence from Indonesia. *The Journal of Asian Finance, Economics and Business, 6*(2), 269–275.

Agricultural Investment with the Growth of Agribusiness in the Red River Delta, Vietnam

Duyen Dang Thi Thuy and Thoa Le Minh

Abstract Agricultural development in general and sustainable agricultural development in particular play an important role in promoting socio-economic development in most countries. Over the past two decades, agriculture in the Red River Delta has created thousands of jobs for local workers and neighboring provinces. This study aims to examine the affecting of investment and other factors on agricultural development in the Red River Delta. The author has collected secondary data from 11 provinces and used panel data regression model to analyze the influencing factors. Using a sample data of 121 observations was performed. The experimental results prove that the independent variables explain 70% of the variation of the dependent variable, and the rest (30%) can be explained by other reasons. Research result shows that factors such as investment capital, labor, land, and fixed assets have a positive impact on agricultural development in the Red River Delta. By identifying and considering each factor, the author will expand the scope of the study further. We propose possible solutions that can ensure the attraction and rational use of investment capital in the Red River Delta. The combination of factors offers solutions to help agriculture develop more rationally and sustainably.

Keywords Agriculture development · Investment · Sustainable development

1 Introduction

Agriculture has been associated with Vietnamese life for thousands of years. Vietnam has many advantages in terms of nature such as climate, land, and soil that are maintained in agricultural development. During the economic development of Vietnam, agriculture has always been identified as the basic economic sector, the "original" economic field and the foundation for the existence and development of all economic

---

D. D. T. Thuy (✉)
Banking Academy, Hanoi, Vietnam
e-mail: duyendtt@hvnh.edu.vn

T. Le Minh
Thuyloi University, Hanoi, Vietnam

A. T. Nguyen et al. (eds.), *Contemporary Economic Issues in Asian Countries: Proceeding of CEIAC 2022, Volume 1*,
https://doi.org/10.1007/978-981-19-9669-6_23

sectors. According to statistics, more than 90% of the poor in Vietnam are living in rural and mountainous areas. This is also the reason why the fluctuations of the agricultural sector will have an important impact on the majority of the population of Vietnam. Therefore, the policy of agriculture and rural development is always concerned and focused by the state, GSO (2020).

The Red River Delta (or the Northern Delta) is the downstream area of the Red River and Thai Binh River in the north of Vietnam. It includes 11 provinces with 2 central cities, 9 provinces, and 13 provincial cities.

Geographical location of the Red River Delta: The north and west are adjacent to the Northern Midlands and Mountains. The east is adjacent to the Gulf of Tonkin. The south is adjacent to the North Central Coast. Economically, the Red River Delta is located in the center of the north, in the northern key economic region.

In terms of natural conditions and natural resources: the low terrain and mostly flat plain are very favorable for the development of all economic sectors and population concentration. The land is mainly alluvial soil of the Red River and Thai Binh River systems, which is favorable for agricultural development. There are soil types such as: Feralit soil in the area adjacent to the midland and mountainous areas of the north. Wetlands are found in Nam Dinh, Ninh Binh, Ha Nam, and Bac Ninh. Alluvial soil is in most of the provinces and occupies the largest area. Acid and saline soils are located along the Gulf of Tonkin. Gray soil is found in ancient alluvium in Vinh Phuc, Hanoi.

Climate: tropical humid monsoon with the cold winter that creates conditions for diversification of agricultural products, intensification, and increase in crops. That helps make winter crop one of the main crops of the year, Khoa and Bau (2022).

Rivers: The Red River Delta has a dense network of rivers with many large rivers belonging to the Red River and Thai Binh River systems. Rivers are accreted by alluvium, so they have provided water for irrigation, development of river transport, fisheries, and tourism for the region. Biological system: There are many national parks such as Cat Ba, Tam Dao, Ba Vi, Cuc Phuong, and Xuan Thuy that are valuable for developing eco-tourism. Beach: The Red River Delta has a 400 km coastline from Hai Phong to Ninh Binh. This is an excellent condition for aquaculture, development of sea transport, and tourism. Minerals: Mineral resources are not many, and only some valuable minerals are quarries in Hai Phong and Ninh Binh, clay and kaolin in Hai Duong, brown coal in Hung Yen, and natural gas in Thai Binh. The mineral diversity is very suitable for industrial development, Geography Textbook (2021).

Besides the favorable natural features, the region also has many difficulties. The agricultural land fund is limited, and the land in the dyke is not regularly replenished and is gradually degraded. The terrain is low and has many sunken cells, so the rainy season is prone to prolonged flooding. Toxic weather with severe cold, harmful cold, and humid tropical climate easily causes disease outbreaks and makes it difficult to maintain production equipment. The Red River Delta lacks minerals and has limited local resources, so it has to import most of its fuels from elsewhere, Geography Textbook (2021).

Along with the development of other economic regions, the Red River Delta also focuses on developing industries and services. Therefore, agriculture also tends to decrease in both size and contribution to the local economy in the past period (Fig. 1).

Both agricultural production value and net revenue tend to increase, averaging over 10% per year in the period 2010–2019. In 2020, due to the impact of the COVID-19 epidemic, most development indicators will decrease. Labor participation in agriculture tends to decrease over the years, and this is also the policy of the general

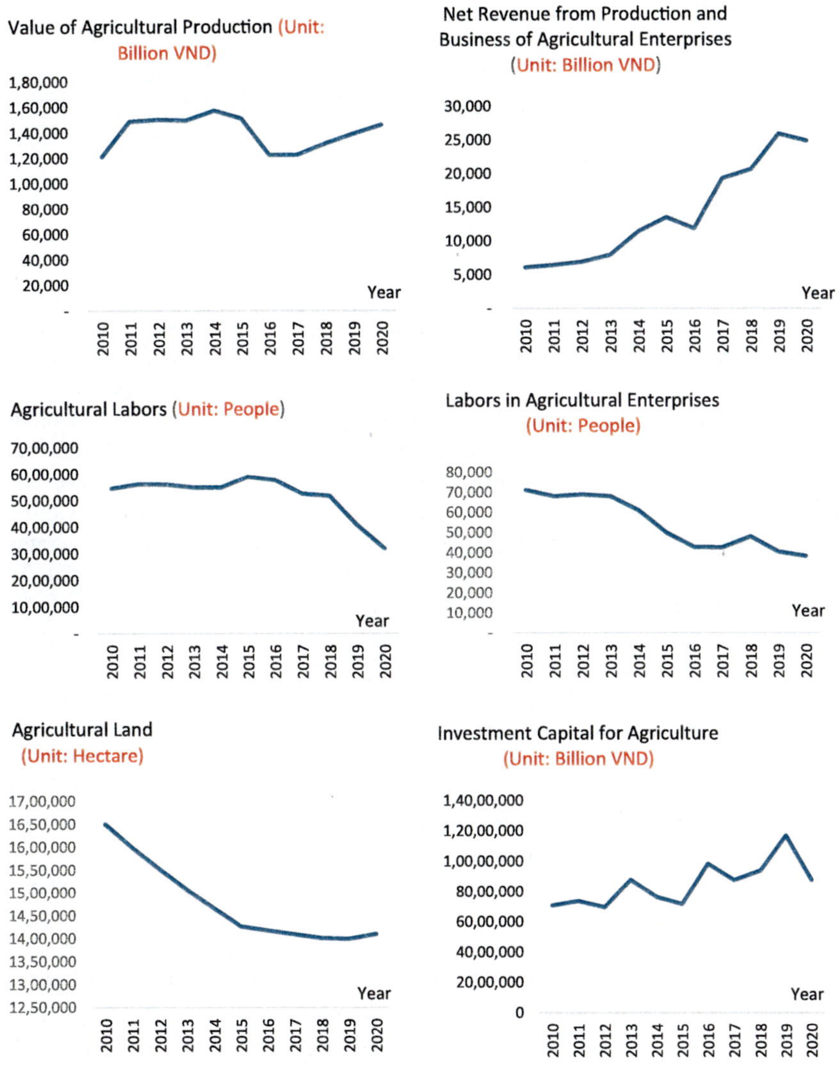

Fig. 1 Agricultural data in the Red River Delta. *Source* GSO (2020)

development plan of the provinces. Although agricultural land decreased, investment capital for agriculture tended to increase in the period 2010–2019 (an average increase of 6%/year).

2 Methodologies

2.1 Theories

The number of enterprises engaged in agricultural production has a positive impact on agricultural development in Indonesia, Soetriono et al. (2020). With the support of the government, businesses and the government set up development centers to facilitate the application of technology in agriculture and integrated crop management. The result of the study suggests that some suitable localities have many potential opportunities rice production and agribusiness development.

Labor and population are always major factors affecting economic growth in general and economic sectors in particular. Population growth can have a positive or negative impact on economic growth. Population and labor have favorable and powerful effects on economic growth both in the short and long terms. When the education system, living standards, and quality of life of the people are improved, the people's intellectual level is raised. This level has contributed greatly to economic development, Mohsen and Chua (2015). Population growth can lead to continuous economic growth, but it must improve the quality of life and education level for people, especially those of working age who participate in the labor market, Golley and Wei (2015). For developing countries, a population policy that reduces birth rates in order to reduce the population can have a negative impact on growth, Dao (2012).

Capital and fixed assets are one of the important factors contributing to the economic growth of agricultural countries, Siregar and Widjanark (2022). The formation of fixed capital (fixed assets) can boost economic growth by 3.32%, and Belloumi and Alshehry (2018) shared the view that fixed asset value has a positive impact on growth both in the short and long terms. Fixed assets also help businesses strengthen their positions, maintain business cycles, and grow steadily, Rehman and Hysa (2021). When growth is stable, it can improve the system of policies, finance, and promote investment. As such, fixed asset value can positively impact growth by promoting and increasing investment, Rinosha and Mustafa (2021).

Social capital makes an important contribution to the operation of agricultural enterprises, Zhou et al. (2022). The more an enterprise invests, the greater its competitive advantage and social responsibility are, which makes agricultural businesses more successful. Ensuring sufficient production capital in agricultural production is the top concern of agricultural enterprises in the period of disease prevention and control, Pham and Tran (2022). Especially in rural areas, capital plays a stronger role. Investment capital both helps to promote the agricultural economy and modernize rural areas, Baba et al. (2010). Initially, investment capital and labor determine the agricultural growth rate in Can Tho (Vietnam), but in 2019, investment capital and science and technology play a more important role, Thanh and Doan (2021).

Agricultural land is a long-term determinant of income in agriculture. Agricultural land, including land for growing annual crops, land for growing perennial crops, and land for aquaculture, is leased by local authorities for a long term, Pham and Tran (2022). In addition, the process of land exploitation also needs the participation of the private sector (enterprises) to promote agricultural development, Bjarstig and Sandstrom (2016). Partnerships between the state–enterprises, enterprises–enterprises, and enterprises–farmers are interested and promoted in the process of agricultural development.

2.2 *Model Building*

Based on the survey of related documents as well as previous studies, we propose a research model as shown in Fig. 2.

Model analysis of factors affecting agricultural production revenue in the Red River Delta: The study is based on data sources, which are: enterprises in agriculture (EN), fixed asset value of agricultural enterprises (AV), labors in agricultural enterprises (La), agricultural investment capital (Ia), and agricultural land (R). The data collected and calculated are eleven provinces, namely Bac Ninh, Ha Nam, Hanoi, Hai Duong, Hai Phong, Hung Yen, Nam Dinh, Ninh Binh, Thai Binh, Vinh Phuc, and Quang Ninh from 2010 to 2020 due to Provincial Statistical Offices, provided by the General Statistics Office.

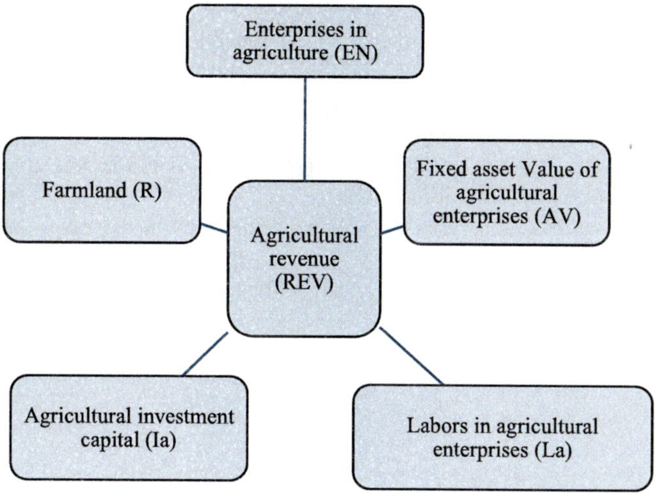

Fig. 2 Proposed research model

2.3 Research Methods

Fixed and random effects models for panel data are common in economic studies. The main advantage of the model is that it can include many variables without being constrained by the number of observations like a time series model. However, when using these models, it is important to choose whether to apply a pooled regression model (POLS), a fixed effect model (FEM), or a random effect model (REM). The research model is set up as follows:

$$\mathrm{lnREV}_{it} = \beta_1 + \beta_2\mathrm{lnEN}_{it} + \beta_3\mathrm{lnAV}_{it} + \beta_4\mathrm{lnLa}_{it} + \beta_5\mathrm{lnIa}_{it} + \beta_6\mathrm{lnR}_{it} + u_{it}$$

i: 11 Provinces 1,2…11t: Year (from 2010 to 2020).

Research hypothesis

H1: Enterprises has a positive impact on agricultural production revenue.

H2: The fixed asset value of the enterprise has a positive impact on the revenue of agricultural production.

H3: Agricultural labors have a positive impact on agricultural production revenue.

H4: Agricultural investment has a positive impact on agricultural production revenue.

H5: Agricultural land has a positive impact on agricultural production revenue.

The data used in the study were obtained from different sources, mainly from the statistical yearbooks of eleven provinces. In addition, the author also consults and takes from other scientific research projects and topics.

Description of the model variable

Enterprises in agriculture variable (EN): collected in the area (provinces), using the interval scale, unit: enterprise.

Fixed assets value of agricultural enterprises variable (AV): collected in the area (provinces), using the interval scale, unit: billion VND.

Labors variable in agricultural enterprises (La): collected in the area (provinces), using the interval scale, unit: people.

Agricultural investment capital variable (Ia): collected in the area (provinces), using the interval scale, unit: billion VND.

Agricultural land variable (R): collected in the area (provinces), using the interval scale, unit: hectare.

Descriptive statistics of variables by Stata software are as follows (Table 1).

Table 1 Descriptive statistics of variables

Variable	Obs	Mean	Std. dev	Min	Max
lnREV	121	9.25774	0.5823458	8.402894	10.73387
lnEN	121	4.661935	1.363126	1.94591	7.090077
lnAV	121	7.448824	0.7307793	6.009	9.434683
lnLa	121	8.090614	0.9076199	6.700731	10.03228
lnIa	121	6.803925	0.8025539	4.442675	8.824972
lnR	121	11.57744	0.6170192	10.7689	13.09346

Source Regression Results on Stata 14

3 Results

The model establishes the correlation coefficient matrix that represents the degree of interaction of the independent variables with each other. Based on the correlation analysis table (Table 2), it shows that all independent variables can be included in the model and ensure that the model does not have multicollinearity.

Results were performed using Stata software. The author performed a regression between the dependent variable and the variables: number of enterprises operating in agriculture (EN), fixed assets value of agricultural enterprises (AV), agricultural labors in agricultural enterprises (La), agricultural investment (Ia), and agricultural land (*R*). The result shows that the slope of the variable number of agricultural enterprises (EN) is not statistically significant due to $p > 0.05$. Therefore, using the test to remove one of the above variables from the model, it shows that the p-value of the F-test is 0.252 > 0.05, and it is necessary to remove that variable from the model. The remaining variables are fixed assets value of agricultural enterprises (AV), number of employees in agricultural enterprises (La), agricultural investment (Ia), and agricultural land (*R*). The author has performed regression in turn, according to POLS, FEM, and REM (Table 3).

Coefficient $P = 0.0000$, pooled models OLS, FEM, REM all have statistical significance. Based on POLS, FEM, and REM, it shows the fixed assets value of agricultural enterprises (AV), number of employees in agricultural enterprises (La), agricultural

Table 2 Matrix of correlation coefficients between independent variables

	lnEN	lnAV	lnLa	lnIa	lnR
lnEN	1.0000				
lnAV	0.6024	1.0000			
lnLa	0.9322	0.9322	1.0000		
lnIa	0.5984	0.8369	0.5310	1.0000	
lnR	0.2656	0.2312	0.2839	0.1207	1.0000

Source: Regression Results on Stata 14

Table 3 Regression results POLS, FEM, REM

Variable	Pooled OLS			FEM			REM		
lnREV	Coef	t	P > |t|	Regression coefficient	t	P > |t|	Regression coefficient	t	P > |t|
lnAV	1.118477	13.57	0.000	1.4471	16.20	0.000	1.310886	18.06	0.000
lnLa	0.4662082	7.11	0.000	0.1574506	1.94	0.056	0.2484789	3.32	0.001
lnla	0.172752	2.34	0.021	0.1029832	1.61	0.111	0.1318584	2.04	0.041
lnR	0.2859843	3.09	0.002	1.147953	2.03	0.044	0.1191426	0.63	0.528
cons	1.00759	1.02	0.031	18.03343	2.64	0.009	2.889085	1.32	0.018
P	$P > F = 0.0000$			$P > F = 0.0000$			$P > F = 0.0000$		

Source Regression Results on Stata 14

Table 4 Model selection test between POLS and FEM

Chi2	P
277.16	0.0000

Source Regression Results on Stata 14

Table 5 Model selection test between POLS and REM

Chi2	P
110.96	0.0000

Source Regression Results on Stata 14

investment (Ia), and agricultural land (*R*) will have an impact on agricultural revenue (REV).

From the results of the research models, the author conducts F-tests (choosing between POLS and FEM) and Hausman test (choosing between REM and FEM). Finally, the author chooses the appropriate estimation method.

The author uses Wald test to choose a model between POLS and FEM, with the assumption:

H0: POLS model is more suitable than FEM.

H1: FEM model is more suitable than POLS (Table 4).

F-test: $P = 0.0000$, $a = 0.05$.

$P = 0.000 < 0.05$, accept H0. Therefore, choosing the POLS model is appropriate.

The author uses Hausman test to choose the model between POLS and REM, with the assumption:

H0: POLS model is more suitable than REM.

H1: REM model is more suitable than POLS (Table 5).

From the results of Hausman test, $P = 0.0000 < 0.05$, rejecting H0, so we choose to use the model with fixed effect REM. Regression with the fixed-effect model REM will yield the best results.

*The regression model tests.

- Multicollinearity:

Multicollinearity is a phenomenon in which the independent variables in the model are linearly correlated with each other. The study conducted to test the hypothesis that there is no multicollinearity phenomenon by using the variance inflation factor (VIF) (Table 6).

All correlation coefficients are less than 2 indicating that multicollinearity is not serious. That shows that independent variables can be used to estimate the model.

- Heteroscedasticity test

What is heteroscedasticity or heteroskedasticity? It is the phenomenon where the residuals or errors (e) of the model after the regression do not follow a random distribution and the variance is not equal. Conduct Breusch-Pagan test with the hypothesis that the variance of the error is constant with hypothesis H0: There is no phenomenon

Table 6 Multicollinearity test

	VIF coefficient	1/VIF
lnAV	1.5	0.665518
lnLa	1.47	0.681315
lnIa	1.45	0.690125
lnR	1.35	0.742324
VIF	1.44	

Source Regression Results on Stata 14

of variance. The author conducts Breusch-Pagan test with the hypothesis that the variance of the error is constant with hypothesis H0: There is no phenomenon of variance (Table 7).

With significance level $\alpha = 5\%$, $P = 0.000 < 0.05$, hypothesis H0 be rejected: There is no phenomenon of variance. Thus, the model has the phenomenon of variance change.

– Autocorrelation test

Autocorrelation is that the errors are correlated with each other, which will make the obtained estimates stable but ineffective, so the regression coefficient tests are no longer reliable. The study conducted Wooldridge test on the autocorrelation phenomenon on panel data with hypothesis H0: no autocorrelation (Table 8).

With significance level $\alpha = 5\%$, $P = 0.0077 < 5\%$ rejects hypothesis H0. Thus, the model has the phenomenon of autocorrelation.

Regression Model Results

After testing, if the model violates assumptions such as autocorrelation and variable variance, these methods are not optimal. Therefore, a better method must be used, which is the feasible generalized least squares estimators (FGLS) method. FGLS will overcome this phenomenon to ensure that the obtained estimates are stable and efficient. The adjusted R^2 value is 70%. It means that the independent variables in the model can explain 70% of the variation of the dependent variable.

Table 7 Heteroscedasticity test

Chi2	P
110.96	0.0000

Source Regression Results on Stata 14

Table 8 Autocorrelation test

F	P
11.035	0.0077

Source Regression Results on Stata 14

Table 9 Regression results by FGLS

Chi2(3) = 158.39				P > F = 0.0000		
lnREV	Coef	Std. Err	t	P > |t|	[95% Conf.]	
lnAV	1.023846	0.0907447	11.28	0.000	0.8459898	1.201703
lnLa	0.3597459	0.0739205	4.87	0.000	0.2148644	0.5046273
lnIa	0.059434	0.0474057	1.25	0.021	0.1523475	0.0334795
lnR	0.2712204	0.1634972	1.66	0.097	0.5916689	0.0492282
Cons	0.3928886	1.767562	0.22	0.824	3.857246	3.071469

Source Regression Results on Stata 14

With significance level $\alpha = 5\%$, $P = 0.000 < 5\%$ rejects hypothesis H0. Thus, the model overcomes the phenomenon of autocorrelation and variable variance.

From the results of the research model, the authors discuss and make comments on factors such as fixed assets, labor, investment capital, and agricultural land that have affected the revenue of agricultural enterprises in the Red River Delta. Through the model test results, no multicollinearity occurs. But the model has autocorrelation and variable variance. To overcome, the author used feasible generalized least squares estimators–FGLS.

Table 9 shows that the number of agricultural enterprises has the most positive impact on the revenue of agricultural enterprises in the Red River Delta. That said the more businesses in the Red River Delta invest and operate in agriculture, the more diversified and more abundant agricultural products will be on the market. In fact, as the number of businesses increases that can bring more positive signals to consumers, they will have more options that suit their needs. This finding is consistent with Bjarstig and Sandstrom (2016), a private partnership that helps increase rural income.

Agricultural workers include local farmers and specialized agricultural workers. The higher the agricultural labor level is, the greater the contribution to agricultural revenue. In this study, an increase in the number of employees working helps to increase agricultural revenue. Because the Red River Delta is a purely agricultural area and most of the laborers work in agriculture, the contribution of labor is extremely important. More specifically, a 1% increase in labor can positively improve revenue by 0.336%. This is consistent with previous studies by Dao (2012) and Golley and Wei (2015), who confirmed that labor can have a positive and significant impact on growth.

Agricultural land is a significant contributor to agricultural revenue. The Red River Delta has a long-standing agricultural land area. Currently, agricultural land is used to grow wet rice and many vegetables and fruits. The more the potential organic farming and green agriculture are, the greater the agricultural revenue increase. Potential agricultural land is agricultural products grown on the land, which are future tourist attractions to attract tourists. Potential agricultural land is agricultural products grown on the land, which are future tourist attractions to attract tourists. It is also possible to both exploit agricultural products and let visitors experience the countryside. Table

9 shows that a 1% increase in agricultural land can positively boost sales by 0.27%. This is consistent with the studies by Pham, 2022.

Investment in agriculture is a long-term investment that requires persistence and is aimed at clean and healthy food for people. Investment in agricultural development should focus on places where it is possible to exploit and produce specific and competitive products. This makes a brand for agriculture in the Red River Delta. Although investment in agriculture is still modest in total investment in the Red River Delta, according to research, agricultural investment really has spillover effects on many other industries such as tourism and industry. Coef = 0.06, a 1% increase in investment capital can positively improve revenue by 0.06%. This is consistent with previous studies by many authors, that capital investment has a significant impact on growth and revenue. The impact of agricultural investment capital on revenue is not as large as other factors because the investment capital for agriculture in the Red River Delta is mostly domestic. Foreign investment capital in the Red River Delta is very small (mainly in large provinces).

4 Discussions and Conclusion

Research result shows that, over time, there are four factors that positively affect the revenue of agricultural enterprises. The effects in descending order are fixed assets, labor, land, and investment capital. If the value of fixed assets increases by 1%, labor increases by 1%, land increases by 1%, and investment capital increases by 1%, revenue increases by 1.02%, 0.36%, 0.27%, and 0.06%, respectively. Although the title of the study is agricultural investment with the growth of agricultural enterprises, it is possible to consider factors such as the value of fixed assets, labors, and land. Therefore, according to the author, the Red River Delta provinces need to pay attention to a number of issues to promote local agricultural development:

Firstly, the Red River Delta should focus on attracting investment capital for the asset system, high technology for agriculture in both quality and quantity. At present, the policy of the government and the provinces should promote the high-tech agriculture development. The agricultural sector needs to apply more machinery and equipment, science and technology to have clean, high-quality, and high-yield agricultural products. Agricultural enterprises should invest in automatic irrigation and fertilizer systems, in soil quality, environment, moisture analysis techniques and in technology for preserving and processing agricultural products.

Secondly, the Red River Delta needs to improve the quality of agricultural human resources. Agricultural labors play an important role in the agricultural enterprises development in the future. The urban population is increasing, while the rural population tends to decrease. In 2019, the urban population in Vietnam was 33,122,548 people compared to 26,515,900 people in the countryside, accounting for 34.4% and 30.6% of the country's population, respectively. From 2009 to 2020, the proportion of the population in urban areas will increase by 4.8%. One of the reasons for this change is the rapid and widespread urbanization process that has increased the population in urban areas, GSO (2020). Comparing annual statistics shows that the labor structure in the agricultural sector is gradually decreasing. In Vietnam, the proportion of employed workers in the agricultural, forestry, and fishery sector continuously decreased, from 53.9% in 2009 to 35.3% in 2019. Meanwhile, labors in agricultural enterprises in the Red River Delta also decreased by 15.67% in 2019 and by 5.28% in 2020 (provincial statistics, 2021).

There should be close coordination between the Ministry of Labour, Invalids and Social Affairs, the Ministry of Education and Training, and the Ministry of Agriculture and Rural Development. Vocational education and training programs for agricultural workers are in the direction of modern industry and services (agriculture must be connected with industry and services). The Red River Delta needs to improve the quality of career guidance, associated with practical experience at educational institutions and businesses in rural areas. From there, students have the right perception of the profession. This not only helps to promote classification of students according to appropriate grade-level standards and disciplines, but also avoids waste in training and building training models for high school graduates. It is possible to bring them to study at both cultural and vocational training levels.

The Red River Delta should create favorable conditions to attract qualified and dedicated teachers to agricultural training institutions. It should have policies of giving priority to students studying agriculture such as reducing tuition fees, awarding scholarships. That is to solve the difficult situation in enrollment for the agricultural sector. It is also a way to support rural children to have more learning opportunities.

The Red River Delta should strengthen training and fostering modern agricultural production knowledge for farmers. Proposed measures for the Red River Delta are: (1) promote socialization, improve the role of socio-political organizations and unions in vocational training for rural workers, (2) strengthening educational activities and transfer of scientific applications to farmers, (3) enhance productivity and quality agricultural products and income for farmers, (4) build training effective occupation models for rural workers in each industry and field, (5) pilot building online courses for rural workers, and (6) building digital resources to create favorable conditions for rural workers to participate in vocational training. Thereby, it is possible to take advantage of local agricultural labor.

The Red River Delta should promote programs, schemes, and projects on vocational training for rural workers so that they become a team of "agricultural workers" with vocational skills and scientific and technical knowledge.

Thirdly, the Red River Delta should attract diversified sources of agricultural investment capital. Agriculture is a specific industry, directly affected by risks from natural disasters, epidemics (cultivation and animal husbandry, aquaculture) that are likely to strike at any time and affect production and profits. The government has directed relevant ministries and branches to complete mechanisms and policies to support and create a favorable environment. That is to attract businesses to invest in agricultural, forestry, and fishery production in rural areas.

The government raises awareness of the importance of attracting FDI in the agricultural sector. It is necessary to have policies to encourage FDI in agriculture such as policies on infrastructure development, tax, credit, human resource development. At the same time, it is necessary to improve the quality of effective and synchronous investment promotion. The agricultural sector should give priority to foreign investors to invest in high-tech agricultural production, sustainable development, and the form of public–private partnership (PPP). However, agricultural investment capital in the Red River Delta is still mostly domestic capital, and mobilized capital from abroad is very small. The provinces in the Red River Delta need to strengthen linkages and actively seek funding from organizations, individuals, and businesses inside and outside the province for the local agricultural industry.

Finally, toward sustainable agricultural growth: The Red River Delta needs to set growth targets in line with national, regional, and local resources. The productivity of agricultural, forestry, and fishery workers must be increased in order to expand and develop markets, especially export markets. It is necessary to increase the agricultural, forestry, and fishery products export to raise people's incomes and reduce poverty sustainably. The Red River Delta should develop green and environmentally friendly agriculture. It needs to take agriculture as the basis and consider agriculture as the foundation of local economic and social stability.

References

Baba, S. H., Saini, A. S., Sharma, K. D., & Thakur, D. R. (2010). Impact of investment on agricultural growth and rural development in Himachal Pradesh: Dynamics of public and private investment. *Indian Journal of Agricultural Economics, 65*(1), 1–24.

Bac Ninh Statistical Office. (2021). *Provincial Statistical Yearbook 2010–2020.*

Belloumi, M., & Alshehry, A. (2018). The impacts of domestic and foreign direct investments on economic growth in Saudi Arabia. *Economies., 6*(1), 1–17. https://doi.org/10.3390/economies6010018

Bjarstig, T., & Sandstrom, C. (2016). Public-private partnerships in a Swedish rural context-A policy tool for the authorities to achieve sustainable rural development. *Journal of Rural Studies, 49*, 58–68.

Dao, M. Q. (2012). Population and economic growth in developing countries. *International Journal of Academic Research in Business and Social Sciences, 2*(1), 6–17. https://hrmars.com/papers_submitted/8747/population-and-economic-growth-indevelopingcountries.pdf

Fathima, K., & Mustafa, A. M. (2021). Nexus between financial development and economic growth: Evidence from Sri Lanka. *Journal of Asian Finance, Economics and Business, 8*(3), 165–170. https://doi.org/10.13106/jafeb.2021.vol8.no3.0165

Geography Textbook. (2021). Vietnamese Educational Publishing House.

Golley, J., & Wei, Z. (2015). Population dynamics and economic growth in China. *China Economic Review, 35*, 15–32. https://doi.org/10.1016/j.chieco.2015.05.005

GSO (General Statistics Office). (2020). https://www.gso.gov.vn/.

Ha Nam Statistical Office. (2021). *Provincial Statistical Yearbook 2010–2020.*

Hai Phong Statistical Office. (2021). *Provincial Statistical Yearbook 2010–2020.*

Hai Duong Statistical Office. (2021). *Provincial Statistical Yearbook 2010–2020.*

Hanoi Statistical Office. (2021). *Provincial Statistical Yearbook 2010–2020.*

Hung Yen Statistical Office. (2021). *Provincial Statistical Yearbook 2010–2020.*

Khoa, N. D., & Bau, N. D. (2022, December). Impacts of climate change on the yield of some key crops in the Red River Delta. *Industry and Trade Magazine—Scientific research results*

and technology application, Issue 27. https://tapchicongthuong.vn/bai-viet/tac-dong-cua-bien-doi-khi-hau-den-nang-suat-mot-so-nong-san-chu-luc-o-dong-bang-song-hong-86455.htm
Mohsen, A. S., & Chua, S. Y. (2015). Effects of trade openness, investment and population on the economic growth: A case study of Syria. *Hyperion Economic Journal Year, 2*(3), 14–23. http://hej.hyperion.ro/articles/2(3)_2015/HEJ%20nr2(3)_2015_A2Mohsen.pdf
Nam Dinh Statistical Office. (2021). *Provincial Statistical Yearbook 2010–2020.*
Ninh Binh Statistical Office. (2021). *Provincial Statistical Yearbook 2010–2020.*
Pham, P. N., & Tran, T. Y. (2022). Factors affecting income from public agricultural land use: An empirical study from Vietnam. *Journal of Asian Finance, Economics and Business, 9*(6), 1–9.
Quang Ninh Statistical Office. (2021). *Provincial Statistical Yearbook 2010–2020.*
Rehman, N. U., & Hysa, E. (2021). The effect of financial development and remittances on economic growth. *Cogent Economics & Finance, 9*(1), 1–15. https://doi.org/10.1080/23322039.2021.1932060
Siregar, A. P., & Widjanark, N. P. A. (2022). The impact of trade openness on economic growth: Evidence from agricultural countries. *Journal of Asian Finance. Economics and Business, 9*(3), 23–31.
Soetriono, D. S., Hani, E. S., Suwandari, A., Narmaditya, B. S. (2020). Challenges and opportunities for agribusiness development: Lesson from Indonesia. *Journal of Asian Finance, Economics and Business, 7*(9), 791–800
Thai Binh Statistical Office. (2021). *Provincial Statistical Yearbook 2010–2020.*
Thanh, N. H., & Doan, H. M. (2021, October). Factors affecting agricultural economic growth in Can Tho city. *Industry and Trade Magazine*, Issue 24. https://tapchicongthuong.vn/bai-viet/cac-nhan-to-tac-dong-den-tang-truong-king-te-nong-nghiep-thanh-pho-can-tho-86073.htm
Vinh Phuc Statistical Office. (2021). *Provincial Statistical Yearbook 2010–2020.*
Zhou, Z., Jia, H., & Yang, Q. (2022). The influence of corporate social responsibility on business performance: Evidence from agricultural enterprises in China. *Journal of Asian Finance, Economics and Business, 9*(3), 83–94.

The Regional Comprehensive Economic Partnership (RCEP): Free Trade and Investment in East Asia Under a "Single Rulebook"

Patrick Ziltener

Abstract Negotiations on the Regional Comprehensive Economic Partnership (RCEP) were launched in November 2012 between the Association of Southeast Asian Nations (ASEAN) and ASEAN's free trade agreement partners Australia, China, India, Japan, New Zealand and South Korea. After more than 30 negotiations rounds, the Agreement was signed on November 15, 2020. In November 2019, India had indicated that it is not in a position to sign the Agreement, but even without India, RCEP still is the world's largest free trade agreement, covering about 30% of global GDP. RCEP entered into force on January 1, 2022. This book chapter analyzes the economic potentiality of the agreement, focusing on the phasing out of tariffs on trade. Tariffs are expected to be reduced over 20 years, most tariffs being eliminated. The main impact will be in the cases where there has been no free trade agreement so far, which is the case between Japan and South Korea and, most importantly, between Japan and the People's Republic of China. Beyond tariffs, RCEP improves the business environment and opportunities for transnational value chains and cross-border trading for all companies operating in the RCEP area. It standardizes rules and procedures (the "single rulebook"), and it is supposed to create more transparency and increase the level of compliance to these rules and procedures. RCEP will have an impact on economic actors' behavior and ultimately increase the level of economic integration in the region.

Keywords ASEAN · East Asia · Economic integration · Free trade agreements (FTA) · Regional Comprehensive Economic Partnership (RCEP)

P. Ziltener (✉)
University of Zurich, Zurich, Switzerland
e-mail: ziltener@soziologie.uzh.ch

A. T. Nguyen et al. (eds.), *Contemporary Economic Issues in Asian Countries: Proceeding of CEIAC 2022, Volume 1*,
https://doi.org/10.1007/978-981-19-9669-6_24

1 Introduction[1]

The negotiations on the Regional Comprehensive Economic Partnership (RCEP) Agreement were launched in November 2012 between the Association of Southeast Asian Nations (ASEAN) and ASEAN's free trade partners Australia, China, India, Japan, New Zealand and South Korea. After more than 30 negotiations rounds, the Agreement was signed on November 15, 2020, through a video conference. In November 2019, India had indicated that it is not in a position to sign the Agreement.[2] But even without India, RCEP still is the world's largest free trade agreement, covering about 30% of global GDP. RCEP entered into force on January 1, 2022. RCEP has been ratified by Australia, Brunei, Cambodia, China, Japan, Laos, Malaysia, New Zealand, South Korea, Singapore, Thailand and Vietnam; in Indonesia, ratification is under way (as of October 2022). This chapter analyzes the economic potentiality of the agreement, focusing on the phasing out of tariffs on trade and the institutional changes that characterize the agreement (the "single rulebook"). What is RCEP's significance, and what will be the impact of this "mega-regional" Agreement, as it has been aptly called?

The following Sect. 2 characterizes RCEP as an agreement against its (now historical) institutional background, the *ASEAN plus*-System as it has evolved over decades. Section 3 looks at RCEP from an ASEAN perspective, both institutionally and economically. Since all bilateral relations among RCEP members were based on FTAs already, the focus is on the changes to be expected among Japan (Sect. 4), China (Sect. 5) and South Korea (Sect. 6)—after all, RCEP is the first regional agreement in Northeast Asia. Where sectoral agreements have created more or less global free trade, as in the case of information technology, RCEP's impact will be less significant (Sect. 7). Section 8 concludes this book chapter with an outlook.

2 The Character of RCEP

During the last decades, the world region East Asia has not been the place of grand designs for regional institutions (Ziltener, 2007, 2012). The historical-cultural preference is in favor of the incremental upgrading of flexible forms of cooperation toward shared objectives, rather than formally binding rules and well-defined legal procedures (cf. Peng, 2000). The instrumental use of law as core mechanism to foster integration like in the European Union is unthinkable in East Asia. The resulting predominant "Asian incrementalism" (Haas, 1989) has built on ASEAN as a multilateral organization and led to the *ASEAN plus*-System in East Asia (Ziltener, 2006),

[1] This chapter builds on a 2020 analysis by the author: *The Regional Comprehensive Economic Partnership (RCEP) as "Single Rulebook" bringing Free Trade to 30% of the World Economy and 30% of World Population–What does it mean for Swiss Companies?* (Ziltener 2020).

[2] The protocol of engagement with India which keeps open a path for eventual Indian membership in RCEP (cf. Armstrong & Drysdale, 2022: 15).

the ASEAN countries being linked through so-called *ASEAN plus one* agreements to their partners Australia, China, India, Japan, New Zealand and South Korea. *ASEAN plus 3* (ASEAN plus China, Japan and South Korea) laid the groundwork for a new multilateralism. As a result of competing institutional schemes and a period of intense diplomatic negotiations and informal talks, the first *East Asia summit* (2005) also included India, Australia and New Zealand (Ziltener, 2007). With this, the grouping of countries that would begin the process leading up to the opening of multilateral negotiations at regional level was born. However, it took more than 30 negotiations rounds (each involving up to 700 officials) until an agreement could be concluded (without India) and signed.

The Australian Department of Foreign Affairs and Trade (DFAT) characterizes RCEP as "a modern and comprehensive free trade agreement."[3] Indeed, RCEP contains 20 chapters on topics usually covered by up-to-date trade agreements, such as trade, investment, services, sanitary measures, technical standards, government procurement, intellectual property and telecommunications. Tariffs are expected to be reduced over 20 years, with a large drop upfront for some products, and most tariffs being eliminated over the next 20 years (Estrades et al., 2022). Some members have different tariff elimination schedules, depending on the partner country, and others—Australia, Brunei, Cambodia, Malaysia, Myanmar, New Zealand, Singapore and Thailand—have just one tariff schedule that is on offer for all other members (ibid.). For the digital economy, RCEP provides a chapter for multilateral rules, aiming at the liberalization of e-commerce. RCEP members committed themselves to maintain the practice of not imposing customs duties on electronic transmissions among them. There are certain provisions to prevent data localization requirements, but significant exemptions for the purpose of national security and certain public policy objectives. However, a comparison with the Comprehensive and Progressive Agreement for Trans-Pacific Partnership (CPTPP) indicates not only quite significant differences in regulatory ambitions, e.g., in the chapter on e-commerce or the one on intellectual property, but also differences in coverage: The most important issues not covered in RCEP are state-owned enterprises, labor and environmental standards, transparency and anti-corruption measures. These are topics that usually are, as in the case of the transpacific CPTPP,[4] high on the agenda of Western countries, but of lower or even no interest to many Asian countries—therefore, this is not a surprising fact. Different from most international FTA, RCEP does not include an Investor-State Dispute Mechanism (ISDS). However, the member states foresee a review of this institutional gap, five years after ratification. In this case, and also with other topics, speed in the conclusion of the agreement seems to have been more important than the level of ambition, especially regarding institutional aspects. Comparing RCEP

---

[3] https://www.dfat.gov.au/trade/agreements/not-yet-in-force/rcep, November 18, 2020.

[4] The CPTPP, which went into effect at the end of 2018, has seven members overlapping with the RCEP. The member states are: Australia, Brunei, Canada, Chile, Japan, Malaysia, Mexico, New Zealand, Peru, Singapore and Vietnam.

with the CPTPP, Schott (2021: 6) concludes that RCEP is "a big deal for intra-Asian economic relations, even though it offers a more incremental and slow-paced pathway to regional integration than its Asia- Pacific cousin."

3 The ASEAN Perspective

From the ASEAN perspective, RCEP is a modern, comprehensive, high-quality agreement "not made just for today but also…for tomorrow." In comparison with the existing *ASEAN plus one* agreements on which ASEAN has grounded its regional trade policy so far, RCEP goes *beyond*—in terms of both coverage and depth of commitments, e.g., in government procurement. Most importantly, the RCEP Agreement has "the added value of bringing together a single rulebook to help facilitate the development and expansion of regional supply chains."[5] The "single rulebook," or as Tian and Guo (2022) call it, the new "unified system of rules in the region," is widely seen as the most important element of RCEP: "Under the agreement, foreign investors will have access to a whole region when they access a single country, with broader markets, lower operating costs and easier outward investment" (ibid.). For Propper and Catarivas (2020), RCEP bridges existing FTAs among its members, it consolidates distinct agreements into a "single basket" to aid exporters, and it unifies source rules among member countries and expands preceding WTO agreements. Schott (2021: 7), being rather critical of RCEP, admits that new regional content rule is RCEP's most important achievement and predicts that it will encourage deeper integration of supply chains across the 15 markets.[6]

As all ASEAN Agreements, RCEP contains provisions for special and differential treatment for the less developed countries Cambodia, Laos, Myanmar and Vietnam, as well as technical cooperation and capacity building programs. In spite of that, the United Nations Conference on Trade and Development (UNCTAD) warned that RCEP tariff concessions could hit exports from Cambodia, Indonesia, the Philippines and Vietnam.[7] In contrast, a World Bank study suggests that lower middle-income economies such as Laos are likely to benefit far more than high-income members like Japan. In the case of Laos, for example, reducing tariffs and non-tariff barriers

[5] *Summary of the regional comprehensive economic partnership agreement*, asean.org, November 18, 2020.

[6] This is the common rule of origin for all goods trade (RoO): In order to get the preferential treatment, a product needs to either reach an RCEP regional value content (RVC) level of 40% or undergo a change in tariff heading (CTH) at the four-digit HS code level. If these conditions are met, the goods qualify for RCEP trade preferences; this is lower and therefore less restrictive than comparable provisions of the CPTPP (Schott, 2021). This common rule of origin is estimated to reduce export transaction costs between 1.4 and 5.9%, thereby boosting merchandise exports among signatories by around $90 billion on average annually (4% of 2019 intra-zone merchandise trade and 0.5% of global merchandise trade; Dib et al., 2020).

[7] Japan exports will rise 5.5% due to RCEP membership: UNCTAD, *NikkeiAsia*, December 15, 2021.

and adopting a common rule of origin would boost income by 2.1% by 2035. An estimated productivity "kick" would raise that to 2.4%. In Vietnam and Malaysia, real incomes could grow by as much as 5% (Estrades et al., 2022). Quimba et al. (2021) predict a positive impact on Philippine exports and GDP, also for Vietnam. Others see the traditional labor-intensive sectors in Southeast Asia as receiving special benefits from the RCEP:

> The supply and industry chains in some Southeast Asian countries are not complete, and they compete with one another… As the RCEP comes into force, high-end manufacturing in Singapore, Japan and South Korea together with China's complete industry chains will boost inter-regional investments, and integrate the region's manufacturing chains better.[8]

Today, half of the population in the region, or 1.1 billion people, consumes the equivalent of $10 a day or more in purchasing power terms—the rough definition of middle-class status. In the World Bank's most optimistic scenario, the impact of RCEP could lead to drawing as many as 27 million more people into the middle class by 2035 (Estrades et al., 2022). As a proportion of the population in 2035, the largest gains in the middle class would be seen in Laos (1.72%), Vietnam (1.59%) and Indonesia (1.35%, ibid.).

4 The Japanese Perspective

The Japanese Ministry of Foreign Affairs (MOFA) sees the significance of RCEP as contributing to the economic growth of Japan and the region, "by further strengthening the link between Japan and the region as a growth center of the world."[9] This "single rulebook" is also very important from a Japanese business perspective: RCEP will contribute to making supply chains already established by Japanese companies in Asia "more broad, effective and resilient."[10] For the Japanese Ministry of the Economy, Trade and Industry (METI), RCEP is a significant improvement of the conditions under which Japanese companies conduct their business in the region:

> …it will be the first time China and some ASEAN countries promise... to prohibit technology transfer requests and protect well-known trademarks... The RCEP Agreement will prohibit the participating countries from requesting technology transfers and introduce other common rules such as to protect intellectual property... Through rules like this, the agreement is expected to provide a more transparent and predictable business environment for Japanese companies that enter the countries concerned.[11]

---

[8] Quote by Zhou Shixin, Director of the Institute for Foreign Policy Studies at the Shanghai Institute for International Studies, in: RCEP starts to deliver benefits for enterprises: business leaders, *Global Times*, January 6, 2022.

[9] Press Release *The date of the entry into force of the Regional Comprehensive Economic Partnership Agreement*, November 3, 2021, www.mofa.go.jp/press/release/press4e_003051.html.

[10] Akio Mimura, Chairman of Japan Chamber of Commerce and Industry, *KYODO*, November 16, 2020.

[11] Press Conference by METI-Minister Hagiuda (Excerpt), November 5, 2021, www.meti.go.jp/english/speeches/press_conferences/2021/1105001.

Junichiro Nakatsuka, President of Mitsubishi Corp., China, sees RCEP as a help to boost complementary relationships among Japan, China and Southeast Asian countries:

> By effectively integrating the digital trend of Chinese enterprises, the advanced industry chain management of Japanese enterprises and the advantages of local enterprises in Southeast Asia, it is highly possible to achieve a win-win situation among the three parties… With the implementation of the RCEP, we hope that companies from China, Japan and South Korea will cooperate more closely. We also hope that Japanese companies can conduct comprehensive cooperation with Chinese companies and welcome more Chinese companies to enter the Japanese market.[12]

Also regarding *trade*, RCEP is highly significant for Japan: It will cover about 46% of Japan's total foreign trade. Tariffs are expected to be removed from 91.5% of goods exported from Japan to RCEP countries, as well as 98.6% of goods imported by Japan from these.[13] The exemption rate is lower than the 99.9% agreed for the CPTPP, but not much. Furthermore, RCEP includes—other than CPTPP—China and South Korea (Japan's largest and third-largest trading partner, respectively) and *all* ASEAN countries.[14] Tariffs will likely be eliminated on 86% of industrial goods exported from Japan to China, which gives Japanese manufacturers not only a significant but also long-term competitive advantage on the large Chinese market, in relation to manufacturers in the USA or in the EU, since a US-China or a EU-China FTA is, for political reasons, very unlikely to be concluded in the near future. For the exports to Korea, the proportion of tariff-free items will ultimately increase from 19% to 92%.[15] UNCTAD even sees Japan as the biggest beneficiary of tariff cuts agreed under the RCEP, estimating that Japan's exports to RCEP's 15 member countries will increase by 5.5% from 2019.[16]

The total savings potential for Japanese exporters to China is enormous: As Wardani and Cooray (2019) have calculated, this amounts to more than $8 billion annually, once tariffs have phased out completely. In the vehicles sector (HS86-89) alone, this amounts to more than $2.5 billion, in machinery (HS84-85, information technology products cut out, for reasons discussed in Sect. 7) $1.8 billion, in chemicals and pharmaceuticals (HS28-38) $724 Mio. The realization of this potential, however, needs time and patience for the Japanese manufacturers. When RCEP entered into force, China started phasing out tariffs on machinery made in Japan, in most cases over 10 years, in some cases in 5 or 20 years. That means that in the first

---

[12] RCEP starts to deliver benefits for enterprises: business leaders, *Global Times*, January 6, 2022. 13.

[13] *NikkeiAsia*, November 15, 2020.

[14] CPTPP includes only four ASEAN countries (Brunei, Malaysia, Singapore and Vietnam).

[15] Press Conference by METI-Minister Hagiuda (Excerpt), November 5, 2021, www.meti.go.jp/english/speeches/press_conferences/2021/1105001.

[16] Japan exports will rise 5.5% due to RCEP membership: UNCTAD, *NikkeiAsia*, December 15, 2021.

and second year, 2022 and 2023, most machinery tariff lines are still above 10%. Complete free trade will likely be achieved in 2031.[17]

As for Japan's exports to South Korea, 92% of goods will be waived from duties, up from the current 19%. About 80% of automotive goods, including air bags and electronic parts, will experience a gradual phasing out of tariffs. Japan itself has almost no tariffs applied on the import of industrial goods (exceptions being textiles and footwear). Japan will remove tariffs on 49–61% of agricultural and fisheries products, but managed successfully to keep import duties on five agricultural products, including rice, wheat, beef and pork, dairy products and sugar crops.

5 The Chinese Perspective

China is the economic heavyweight of RCEP. The Chinese economy is larger than the aggregate of the other 14 countries, and its total trade is almost twice that of Japan and Korea combined. The People's Republic lauds RCEP in first place as the historically first agreement among China, Japan and South Korea. It sees the three countries in different positions in the global value chain, with highly complementary economies and closely linked industrial and supply chains that are expected to further deepen under RCEP. China intends to speed up and conclude also the negotiations on a trilateral FTA among the countries mentioned. From a geopolitical perspective, RCEP is seen as a major Chinese success against US-led initiatives such as APEC and CPTPP.[18] Chinese companies, especially the internet giants, see in RCEP a way to expand e- commerce and to get better market access in Southeast Asia, since factors such as customs clearance and logistics efficiency, as well as the consumer shopping experience, will be improved and imposing taxes on electronic transmissions hindered.[19]

RCEP is expected to increase the imports of certain goods by making it cheaper for Chinese consumers, such as luxury cars and cosmetics from South Korea, or electronics, food and sake from Japan. Chinese exporters to the Japanese market are to be found mainly in the textile and footwear sector (a savings potential of more than $2.2 billion annually), in foodstuffs ($711 Mio), in leather, plastics and chemicals (Wardani & Cooray, 2019). Here, Japanese tariffs are still significant, in contrast to the main industrial sectors; therefore, the impact will be the greatest here. Other exporters to Japan, competitors with Chinese companies in Southeast Asia or in Latin America (with which Japan had already or still has bilateral FTAs in force), will feel this now increasingly competitive-level field.

[17] Analysis based on *RCEP Annex 1 Schedule of Tariff Commitments of China, Section C: For Japan.* www.dfat.gov.au, November 16, 2020.

[18] APEC's Post-2020 vision will get RCEP support as US wanes, *Global Times*, November 18, 2020.

[19] RCEP's boost to e-commerce. Benefits in the world's largest FTZ offer huge potential amid coronavirus; Asian auto sector to make major gains under RCEP, *Global Times*, November 17, 2020.

Studies suggest that China will gain the most out of RCEP implementation, in terms of increased real income: Chung et al. (2017) found in their study including India as an RCEP member that China would receive more than one-quarter of total RCEP gains. Petri and Plummer (2020) predicted that China will reap about half of the $174 billion in cumulative real income gains of the 15 RCEP countries. Kang et al. (2020) estimated potential gains in real income at $85 billion for China, $48 billion for Japan and $23 billion for South Korea, but the economies of Indonesia, Malaysia, Thailand and Vietnam are also expected to gain from RCEP. Kumagai & Hayakawa (2021) came to similar results for the three economies in Northeast Asia, but they predicted the strongest impact of RCEP as a percentage of GDP for Japan (0.66%), followed by South Korea (0.24%) and China (0.13%).

6 The Korean Perspective

Since South Korea (ROK) has "more free trade agreements than any other country,"[20] including FTAs with all RCEP members except Japan, it is argued that RCEP "offers Korean firms only incremental trade liberalization beyond the reforms already embedded in Korea's existing bilateral FTAs with the RCEP countries" (Schott, 2021: 7). Korea has concluded a bilateral FTA with China in 2015 already, covering 23.3% of Korea's total trade. RCEP incl. the trade with China will cover almost half of Korea's trade (48.0%, ibid.,3f). In some cases, Korea's older FTAs were substantially more comprehensive than RCEP (e.g., the agreements with Australia, Singapore or Vietnam). On the other side, RCEP will reduce tariffs and other barriers to trade to a significantly larger extent than the 2007 ASEAN-Korea FTA (Pardo, 2021).

The biggest economic implication of RCEP for Korea obviously is that this is its first-ever FTA with Japan.

For a long time, Korea had been reluctant to sign an FTA with Japan due to concerns about competition in high-tech sectors where Japanese companies were seen as more advanced. This concern has subsided in recent years as South Korean firms now often compete head-to-head with their Japanese counterparts, and there are also complementarities between firms in both countries (Pardo, 2021). As in the case of China, Korea expects an increase of imports from Japan as a consequence of lower tariffs, especially in high-tech and luxury goods. Also in the food sector, imports from Japan will become cheaper and increase in volume. However, regarding core products of Korean agriculture, such as beef and pork, rice and other wheat, cereals, chilies, tea, as well as many fruits and vegetables, no concessions have been made.[21]

In exports, Korea is looking forward to benefits especially for key export items including steel and cars, machine parts and medical equipment, as well as such service

[20] Korea joins RCEP, a light but significant trade agreement, *koreajoongangdaily*, February 1, 2022.

[21] Analysis based on *RCEP Annex 1 Schedule of Tariff Commitments of Korea, Section D: For Japan.* www.dfat.gov.au, November 16, 2020.

sectors as online game, animation, film and records.[22] It is projected that in RCEP's first three years, Korea's export of cultural services under RCEP will increase by 1.1% annually.[23]

7 The Case of Information Technology (IT)

Not everywhere the impact of RCEP will be significant, e.g., in the case of IT products. All RCEP members (except Myanmar, Laos and Cambodia) are participants in the WTO Information Technology Agreement (ITA). The ITA signed by 81 WTO members covers a large number of high-technology products, including computers, telecommunication equipment, semiconductors, semiconductor manufacturing and testing equipment, software, scientific instruments, as well as most of the parts and accessories of these products.[24] Tariffs on these products are abolished or phasing out, independently from RCEP. However, RCEP's "single rulebook" will have an impact in the improvement of the business environment of IT companies in East Asia, as mentioned, especially in the facilitation of the development and expansion of regional supply chains.

Similarly, the WTO Agreement on Government Procurement (GPA) provides global rules for its signatories: Japan, South Korea, Singapore, Australia and New Zealand are participants in the GPA, and China is still in the GPA accession process. In the sector of pharmaceutical products, the Agreement on Trade in Pharmaceutical Products has been concluded in 1994, creating free trade for pharmaceutical products among its signatories. But, so far, only one RCEP country has joined this Agreement under the WTO, Japan. Therefore, RCEP provisions will have greater impact here.

8 Conclusions

This book chapter analyzed the economic potentiality of RCEP, focusing on the phasing out of tariffs on trade and the institutional changes. As a starting point, it is important to realize that only in two cases of bilateral relations, free trade relations were newly established, and all other bilateral relations among RCEP members were based on FTAs already. Therefore, the main significance of RCEP lies in being the historically first FTA between Japan and South Korea and the first agreement among China, Japan and South Korea. RCEP is the first regional agreement in Northeast Asia, and it is the first plurilateral free trade agreement China has ever participated in. RCEP has developed out of the *ASEAN plus*-System, and most significant in RCEP

[22] RCEP pact to take effect for S. Korea next month, *Korea Herald*, January 27, 2022.

[23] Korea joins RCEP, a light but significant trade agreement, *koreajoongangdaily*, February 1, 2022. 24.

[24] www.wto.org/english/tratop_e/inftec_e/inftec_e.htm (October 10, 2022).

is the *regional unification* of the rules under which free trade has been conducted (the "single rulebook"). Actually it would be possible to speak of a successful multilateralization of an existing dense network of bilateral FTAs in East Asia or the Asia–Pacific region.

What impact RCEP is going to have will become visible to the degree tariffs that are dismantled, step by step over a decade, and as soon as the news rules have been effectively implemented. Economic actors— businesses, supply and industry chains—will make use of the new opportunities under a significantly increased clarity of rules. This will have an impact on the level of economic integration in an already highly integrated region, contributing to making Asia the "center of gravity for global commerce" (Sheng, 2022). What remains to be seen is whether RCEP has a significant growth effect in this "growth center of the world" (MOFA Japan) and how (un)even this impact will materialize.

References

Armstrong, S., & Drysdale, P. (2022). The economic cooperation potential of East Asia's RCEP agreement. *East Asian Economic Review, 26*(1), 3–25. https://doi.org/10.11644/KIEP.EAER.2022.26.1.403

Chung, C., Park, S. C., Park, I., Kim, M.-S., Kwak, S., & Chung, M. (2017). The impact of cumulative rules of origin on trade costs: Estimates from FTAs, economic effects and policy implications. *Korea Institute for International Economic Policy (KIEP) Policy Analysis*, 17–12. KIEP

Dib, G., Huang, F., & Poulou, A. (2020, November 15). RCEP: Common rule of origin could boost regional trade by around USD90BN annually. *Allianz Research.* www.allianz-trade.com

Estrades, C., Maliszewska, M., Osorio-Rodarte, I., & Pereira, M. S. (2022). *Estimating the economic and distributional impacts of the regional comprehensive economic partnership.* World Bank Policy Research Working Paper 9939, World Bank.

Haas, M. (1989). *The Asian way to peace. A study of regional integration.* Praeger.

Kang, J. W., Crivelli, P., & Tayag, M. C. (2020, December). *Regional comprehensive economic partnership: Overview and economic impact.* Asian Development Bank (ADB) Brief No. 164. ADB. https://doi.org/10.22617/BRF200422-2

Kumagai, S., & Hayakawa, K. (2021). *Economic impacts of the regional comprehensive economic partnership: Analysis using IDE-GSM.* Institute of Developing Economies (IDE) Policy Brief No. 147. IDE. www.ide.go.jp

Pardo, R. P. (2021, September 27). *South Korea and the RCEP.* Korea Chair Explains, 16/11/2021. https://brussels-school.be/publications/other-publications/korea-chair-explains-south-korea-and-rcep

Peng, S. (2000). The WTO legalistic approach and East Asia: From the legal culture perspective. *Asian-Pacific Law and Policy Journal, 13*, 1–34.

Petri, P. A., & Plummer, M. G. (2020). *East Asia Decouples from the United States: Trade War, COVID-19, and East Asia's New Trade Blocs.* Peterson Institute for International Economics (PIIE) Working Paper 20-9. Washington DC. www.piie.com

Propper, E., & Catarivas, D. (2020, December 1). *New trade agreement in Asia: China is in, the United States is out.* Institute for National Security Studies (INSS) Insight No. 1408. Tel Aviv: INSS, www.inss.org.il

Quimba, F. M. A., Barral, M. A. A., & Andrada, A. E. (2021). *Who benefits from RCEP? Application of trade policy tools.* Philippine Institute for Development Studies (PIDS) Discussion Paper Series No. 2021-35. PIDS.

Schott, J. J. (2021). *RCEP is not enough: South Korea also needs to join the CPTPP*. Peterson Institute for International Economics (PIIE) Policy Brief No. 21-17, Washington DC, www.piie.com

Sheng, W. (2022, January 9). Will RCEP make Asia 'center of gravity' for global commerce? *Global Times*

Tian, Z., & Guo, B. (2022, January 13). RCEP takes effect! What benefits will it bring to China? *China Europe International Business School (CEIBS) Column*. www.ceibs.edu/new-papers-columns/20730

Wardani, R. Y., & Cooray, N. S. (2019). Saving potential of regional comprehensive economic partnership (RCEP): Implication for China and Japan. *Journal of Economic Info (JEI), 6*(1), 34–42. https://doi.org/10.31580/jei.v6i1.122

Ziltener, P. (2006). Das *ASEAN plus*-system in Ostasien (The *ASEAN plus*-system in East Asia). *Bochumer Jahrbuch Zur Ostasienforschung (BJOAF), 30*(2006), 247–272.

Ziltener, P. (2007). East Asia: structure and formation of a world region. In M. Herkenrath (Ed.), *The regional and local shaping of world society* (pp. 111–149). LIT.

Ziltener, P. (2012). The East-Asian integration as a reaction to global challenges. In N. Genov (Ed.), *Global trends and regional development, Routledge Studies in Development and Society No. 28* (pp. 81–102). Routledge.

Ziltener, P. (2020). *The regional comprehensive economic partnership (RCEP) as "single rulebook" bringing Free Trade to 30% of the world economy and 30% of world population—What does it mean for Swiss Companies?* Report on behalf of Switzerland Global Enterprise, www.sg-e.ch

The Impact of Soil Conservation on Fertilizers and Pesticides Utilization and the Economic Efficiency of Rice Cultivating Households in Vietnam

Nguyen Thi Thu Thuong, Doan Thi Hong Thinh, and Nguyen Tien Long

Abstract Using data from the Vietnam Access Resources Household Survey in 2018, we employ a propensity score matching method to evaluate the impact of soil conservation measures' adoption on the use of fertilizers and pesticides and the economic efficiency of rice cultivating households in Vietnam. The empirical results show that households that adopt soil conservation measures spend between 62.30 and 86.40 USD less per hectare on chemical fertilizers and between 7.00 and 11.20 USD less per hectare on pesticides than households that did not. Additionally, soil conservation adoption increases the farming efficiency of households cultivating rice.

Keywords Vietnam · Soil conservation · Propensity score matching · Rice production · Efficiency · Fertilizers and pesticides

1 Introduction

Economic growth in many developing countries is closely related to the performance of their agricultural sector. However, this growth might be affected by the efficiency with which natural resources are managed. The availability and quality of a country's natural resources can be reduced by extreme weather events and human activities (Ashoori & Sadegh, 2016; Lal, 2013). A considerable portion of the world's agricultural land has been influenced by varying degrees of soil degradation (Bindraban et al., 2012). According to Karlen and Rice (2015), although this is not a new problem, it has now become urgent for humanity to acknowledge, mitigate, and remediate its root causes. Soil degradation can lead to poor agricultural production (Mengstie, 2009).

There are three groups of causes leading to land deterioration: climate, biogeophysical, and management (Jara-Rojas et al., 2013). These variables have strong

N. T. T. Thuong · N. T. Long (✉)
TNU-University of Economics and Business Administration, Thai Nguyen, Vietnam
e-mail: nguyentienlong@tueba.edu.vn

D. T. H. Thinh
Aix-Marseille School of Economics, CNRS & EHESS, Marseille, France

A. T. Nguyen et al. (eds.), *Contemporary Economic Issues in Asian Countries: Proceeding of CEIAC 2022, Volume 1*,
https://doi.org/10.1007/978-981-19-9669-6_25

influences on the trend toward an acceleration of soil erosion. Conventional agricultural methods where soil improvement and conservation techniques are not usually applied can degrade soil (Amsalu & De Graaff, 2007; Solís et al., 2009). In addition, in the short term, increasing productivity often goes with the intensive use of fertilizers and pesticides, leading to further deterioration in the long term (Solis & Bravo-Ureta, 2005).

In Vietnam, the agricultural sector is the third key contributor to economic growth, after the service and industry sectors. According to the General Statistical Office (GSO, 2022), agriculture accounted for roughly 12.36% of the Vietnamese gross domestic product (GDP) in 2021. In Vietnam, rice is the main consumption food, and rice farming is the essential source of income for the rural population. For this reason, large agricultural areas are devoted to rice farming. As stated by Nguyen (2017), in 2014, almost 7.8 million ha of land was used for rice cultivation. The annual value of Vietnam's rice exports reaches approximately US$3 billion, which represents 20% of the country's total agricultural exports. Although crops play a determinant role in poverty alleviation and food security, in the creation of jobs for local citizens, and lead to substantial export earnings, they have been facing many challenges. Diseases, pests, as well as climate change have contributed to saline intrusion and drought in some regions (Nguyen, 2017). Furthermore, land and soil pollution, which arise from the excessive use of fertilizers and pesticides, are emerging issues.

In addition to intensive farming tendency, the use of fertilizers and pesticides has accelerated in Vietnam. Between 1985 and 2005, the utilization rate of fertilizers such as nitrogen, phosphorous, and potassium rose by approximately 10% annually. In terms of fertilizer consumption, rice was the main consumer. This can be due to crop intensification (three rice crops per year) and soil degradation, which is the result of an absence of alluvium (Nguyen, 2017).

Inappropriate usage of fertilizers in agriculture, especially in rice production, is common in Vietnam, where fertilizers are still applied according to traditional practices with little knowledge of the quantities required for new varieties of rice. Farmers continue to spread excessive quantities of fertilizers, whereas only around 60% of the nitrogen, 40% of the phosphorous, and 50% of the potassium they contain are assimilated. The surplus is discharged into the soil and waterways, leading to unnecessary environmental pollution (Nguyen, 2017).

Very few studies have modeled the influence of soil conservation practices on the use of fertilizers and pesticides, nor their impact on economic efficiency. To the best of our knowledge, no study has used the "Average treatment effect model" to evaluate the role of soil conservation. The present study aims to estimate the outcome of adopting and non-adopting soil conservation practices in rice farming households. Soil conservation in Vietnam's agricultural sector has attracted the attention of policymakers and scientists, who need solid empirical evidence, concerning the role of soil conservation in environmental protection, social stability, and agricultural productivity in particular to make well-informed decisions. Soil conservation practices can also mitigate the excessive use of chemical fertilizers and pesticides with the potential benefits of increasing agricultural productivity and limiting environmental pollution.

The present research evaluates the influence of soil conservation on rice cultivation in Vietnam. More precisely, we examine how soil conservation affects the total value of production, net farming revenue, farming productivity, and some variable production costs. The study used propensity score matching (PSM) to control for observed differences in principal operator, geographic location of the household, and other socioeconomic factors characterizing soil conservation adopters and non-adopters.

2 Methodology

2.1 Empirical Approach

PSM is an econometric technique that can be used to deal with the selection bias problem in cross-sectional data with a non-experimental assignment and the absence of assumptions concerning the functional and distributional form of the control variables (Becker & Ichino, 2002; Dehejia & Wahba, 2002; Diagne & Demont, 2007; Rosenbaum & Rubin, 1985; Schilling et al., 2014). According to the PSM method, selection bias might be reduced by conditioning on observed characteristics. This allows us to find out the causal association between the investment in soil conservation and outcome variables of interest.

The study used the PSM method to estimate the impact of a household's investment in soil conservation on rice production per hectare, the value of rice yield per hectare, the added value of rice production per hectare, chemical fertilizer costs per hectare, pesticide costs per hectare, the total variable costs of rice production per hectare, and rice farming efficiency.

For a household i, $i = 1,....N$; $\{Y_{i0}, Y_{i1}\}$ represents the two potential outcomes: Y_{i0} as the outcome when the household does not adopt soil conservation, corresponding to the control group, and Y_{i1} as the outcome when the household adopts soil conservation, corresponding to the treated group. We apply the average treatment effect on treated (ATT) to analyze the impact of household investment in soil conservation on several economic variables related to rice production in Vietnam.

The ATT is illustrated as:

$$\text{ATT} = E[Y_{i1} - Y_{i0} | T = 1] \tag{1}$$

where T denotes the treatment status, $T = 1$ indicates treatment, and 0 otherwise.

We used probit regression to estimate the predicted probability scores of being treated. In the PSM method, based on similar characteristics, the adopter and non-adopter are matched together. This allows us to eliminate the effect of dissimilar characteristics and outliers on the estimation results. The propensity scores, described by (Rosenbaum & Rubin, 1985), are computed as follows:

$$p(X) \equiv \Pr\{T = 1|X\} = E\{T|X\} \quad (2)$$

where X is covariates that may affect the treatment status and outcome variables.

Several matching algorithms were used to ensure the robustness of the findings. The nearest neighbor matching (NN1) with replacement was used as it is the most straightforward matching estimator. However, if the closest neighbor is at a high distance, NN1 matching can lead to poor matches. Therefore, to increase matching quality (Dehejia & Wahba, 2002), radius matching with a caliper was also used. In this study, we used the calipers of 0.02 and 0.05 proposed by Attavanich (2016) and Schilling et al. (2014). As the study consisted of many comparable untreated (non-adopting household) observations in the sample, an oversampling with five matching partners (NN5) and kernel matching algorithms were also used. An optimal bandwidth was chosen, proposed by (Silverman, 1986).[1]

Covariate balancing tests were used so that PMS results can be verified (Rosenbaum & Rubin, 1985). We used a two-sample t-test for the equality of the mean values of the covariates in both groups after matching. Additionally, we applied the trimming method suggested by Crump et al. (2009) to deal with the weak common support problem.

Besides, before and after matching, the matching was examined by comparing *p*-values and the pseudo R^2 of the likelihood ratio test acquired from the probit model. The results should be insignificant, which are expressed by a lower pseudo R^2 and insignificance of matched covariates (Rosenbaum & Rubin, 1985). A visual check was done with the use of a propensity score graph to answer the question of whether the common support condition was satisfied, i.e., whether there was adequate overlap.

2.2 Data and Variables

We employed observational data obtained from the Vietnam Access Resources Household Survey (VARHS) conducted in 2018. The VARHS was carried out across 12 provinces, standing of seven regions: Red River Delta, Northeast, Northwest, North Central Coast, South Central Coast, Central Highlands, and Mekong River Delta.

The data of VARHS, where the target population of the survey is rural households, provide useful information for our study. The database includes 13,754 observations with a total of 3648 households, including 2960 rice households. Then, the household cases with missing data were eliminated; therefore, the whole sample had in the end 2010 households. Variables used in our study were obtained and constructed from this dataset. They are potential outcomes, landowner and household properties, the slope of the land, problems with the rice plot, and regional dummy variables.

---

[1] Silverman's rule of thumb:bandwidth (h) = 0.9*sample_standard_deviation*(sample size)$^{-1/5}$.

The covariates in this study are divided into approximately four sets. Firstly, the covariates include demographic variables such as age, gender, education of household head, marital status, and ethnicity. These variables were used in several studies (Ashoori & Sadegh, 2016; Gitonga et al., 2013; Jara-Rojas et al., 2013; Mishra et al., 2017; Pan, 2014).

The variables of farm characteristics constitute the second type of covariate. They are the number of household laborers, farm size, participation in the formal organization and agricultural extension services, the use of old local varieties of rice seed, and access to information on fertilizer use. These variables are supposed to increase the probability of investing in the farms' soil conservation practices (Gitonga et al., 2013; Gottlieb et al., 2015; Mishra et al., 2017; Pan, 2014; Schilling et al., 2014; Uematsu & Mishra, 2012).

The third set of control variables includes the slope of the land. If the household's rice fields are located on sloping land or in a very steep area, this can lead to greater investment in soil conservation. The fourth type of independent variable is related to the rice field's characteristics. It can be suggested that households facing issues such as gullies, dry fields, low-lying fields, sedimentation, landslides, and stony soils will invest more in soil conservation. Finally, regional dummy variables are considered, in which the Mekong Delta region is used as a reference category. These variables represent the differences in economic and social and political conditions of regions (Gottlieb et al., 2015).

The definition of outcome variables employed to compute the average treatment effects and the covariates influencing the probability of adopting soil conservation on the rice fields are provided in Table 1. The households' performance and managerial characteristics' outcome measures used to estimate the ATTs are the farming efficiency, net farm income, fertilizer costs, and pesticide costs. Three categories of explanatory variables include landowner characteristics, household characteristics, the slope of the land, problems with rice fields, and regional dummy variables.

3 Results

3.1 Descriptive Statistics

The summary statistics for the sample are shown in Table 2. From the 2010 observations retained for this study, 418 (or 20.79%) adopted soil conservation. The last column indicates the t-statistics comparing the mean values of the treated and control observations. In general, we observed that the farming efficiency and net farm income of adopting households were greater than those of non-adopting households. However, the costs of fertilizers and pesticides incurred by adopting households tended to be lower than by non-adopters. While the age of the head of the household was significantly lower for adopters than for non-adopters, the number of years of education of the former was significantly higher.

Table 1 Description of variables

Variables	Definition
Potential outcome and treatment	
The added value of rice production	Gross cash income from rice cultivation minus total rice production expenses (000 VND)
Value of rice production per hectare	Total value of rice produced per hectare (000 VND)
Rice production per hectare	Total quantity of rice produced per hectare (Kilogram)
Total production cost per hectare	Total rice production cost per hectare (000 VND)
Fertilizer cost per hectare	Cost per hectare of chemical fertilizers used during the last 12 months (000 VND)
Pesticide cost per hectare	Cost per hectare of pesticides and herbicides used during the last 12 months (000 VND)
Farming efficiency	Ratio of the gross value of production to total variable costs
Investment in soil conservation on plot	Whether the farm has invested in soil conservation since July 2016 (1 if yes/0 if no) (soil conservation practices include rock bunds, soil bunds/ grass lines, terraces, brick wall/fence)
Landowner characteristics	
hhage	Age of household head (years)
hhgen	Gender of household head (1 if male/0 if female)
hhedu	Education of household head (years)
hhmar	Marital status of household head
Ethnic	Whether the household head belongs to ethnic Kinh group (equal to 1 if yes/0 if the household head belongs to a different ethnic minority group)
Farm characteristics	
Labor	Number of household members aged 15–60 years
Credit	Whether the household has access to credit (equal to 1 if yes)
Land area	Land area used for rice cultivation (hectares)
Membership	Whether the household has joined a formal organization (1 if yes/0 if no)
Type of rice seed	Whether the household uses an old local variety of rice (1 if yes/0 if no)
Organic fertilizer	Whether the household uses organic fertilizers for rice farming (1 if yes/0 if no)
Agricultural extension service	Whether the household participates in an agricultural extension program (1 if yes/0 if no)
Information on fertilizer	Whether the household accesses information on fertilizer use from different sources such as retailers, neighbors, radio, authority, and others (1 if yes/0 if no)

(continued)

Table 1 (continued)

Variables	Definition
Slope of plot	
Flat land	Whether the plot is located on flat land (1 if yes/0 if no)
Slight slope	Whether the plot is located on slightly sloping land (1 if yes/0 if no)
Slope	Whether the plot is located on sloping land (1 if yes/0 if no)
Very slope	Whether the plot is located on very steep land (1 if yes/0 if no)
Problems with rice plot	
Gullies	Whether the household has difficulties with gullies (1 if yes/0 if no)
Dry plot	Whether the household has difficulties with dry rice plots (1 if yes/0 if no)
Low-lying plot	Whether the household has difficulties with low-lying rice plots (1 if yes/0 if no)
Sedimentation	Whether the household has difficulties with sedimentation (1 if yes/0 if no)
Landslide	Whether the household has difficulties with landslides (1 if yes/0 if no)
Stony soil	Whether the household has difficulties with stony soil on the rice plots (1 if yes/0 if no)
Plot quality	Whether the quality of the plot is better than or equal to the average land fertility in the village (1 if yes/0 if no)
Region dummy variables	Equal to 1 if the household is located in the corresponding region: – Red River Delta – Northern Midlands and Mountainous – Central Coast – Central Highlands – Southwest (Mekong Delta)

Compared to non-adopting households, the adopting households had a larger number of laborers (3.2 versus 3.6), had a lower probability of access to credit (0.86 versus 0.92), and had a higher probability of participation in official organizations such as youth unions, women's unions, farmer's unions, farmer interest groups, water user associations, business associations, cooperatives and others (0.84 versus 0.89). Adopters also used more old varieties of seeds than non-adopters (0.27 versus 0.14). Besides, they participated more frequently in agricultural extension services than non-adopters (0.77 versus 0.52), and they had access to more information on fertilizers (0.96 versus 0.91).

Table 2 Summary statistics for adopters and non-adopters of soil conservation practices

	Non-adopters		Adopters		Mean difference	Test statistics (*t*-value)
	Mean	SD	Mean	SD		
Potential outcome						
The added value of rice production	15,769	11,691	15,958	9444	−188.2	−0.304
Value of rice production	11,012	16,785	8724	14,054	2255	2.561[b]
Rice production	4975	3231	4660	3035	315.6	1.799[a]
Total production cost	14,957	8829	12,414	6743	2544	5.486[c]
Fertilizer cost per hectare	5779	5484	3971	3887	1808	6.336[c]
Pesticide cost per hectare	1353	1894	995	1345	357.7	3.629[c]
Farming efficiency	2.388	1.128	2.701	1.274	−0.313	−4.914[c]
Landowner characteristics						
hhage	50.66	12.89	48.48	11.52	2.181	3.144[c]
hhgen	0.867	0.340	0.906	0.291	−0.039	−2.196[b]
hhedu	9.519	3.535	9.667	3.574	−0.149	−0.763
hhmar	0.883	0.322	0.904	0.295	−0.022	−1.251
Ethnic	0.602	0.489	0.380	0.486	0.221	8.238[c]
Farm characteristics						
Laborers	3.226	1.369	3.600	1.509	−0.375	−4.874[c]
Credit	0.861	0.346	0.919	0.274	−0.057	-3.149[c]
Land area	0.661	1.409	0.783	2.825	−0.122	−1.236
Membership	0.837	0.369	0.888	0.316	−0.050	−2.548[b]
Types of rice seed	0.136	0.343	0.268	0.443	−0.132	-6.579[c]
Organic fertilizers	0.507	0.500	0.493	0.501	0.014	0.512
Agricultural extension service	0.522	0.499	0.770	0.421	−0.248	−9.329[c]
Information on fertilizers	0.912	0.283	0.957	0.203	−0.045	−3.039[c]
Slope of land						
Flat land	0.764	0.424	0.612	0.488	0.152	6.309[c]
Slightly sloping	0.402	0.490	0.569	0.496	−0.167	−6.195[c]

(continued)

Table 2 (continued)

	Non-adopters		Adopters		Mean difference	Test statistics (t-value)
	Mean	SD	Mean	SD		
Sloping	0.218	0.413	0.411	0.493	−0.194	−8.174[c]
Very steep	0.028	0.164	0.0096	0.097	0.181	2.155[b]
Problems with rice plots						
Gullies	0.126	0.332	0.266	0.442	−0.139	−7.083[c]
Dry plots	0.407	0.491	0.543	0.499	−0.136	−5.021[c]
Low-lying plots	0.059	0.237	0.081	0.274	−0.022	−1.609[a]
Sedimentation	0.015	0.122	0.012	0.109	0.003	0.475
Landslides	0.017	0.129	0.057	0.232	−0.041	−4.704[c]
Stony soil	0.039	0.194	0.045	0.209	−0.007	−0.602
Plot quality	0.994	0.079	0.998	0.049	−0.004	−0.959
Region dummy variables (Mekong Delta is the base category)						
Red River Delta	0.181	0.385	0.084	0.277	0.097	4.840[c]
Northern Midlands and Mountainous	0.847	0.988	1.105	0.996	−0.259	−4.752[c]
Central Coast	0.567	1.175	0.294	0.893	0.272	4.425[c]
Central Highlands	0.510	1.335	0.852	1.639	−0.342	−4.429[c]
No observations	1592		418			

[a], [b], and [c] indicate statistical significance at the 10%, 5%, and 1% levels, respectively
Source Calculation by authors based on data from VARHS 2018

Additionally, there were significant differences between adopters and non-adopters in terms of the slope of the land. Interestingly, the water and soil conservation adopters experienced more problems with their rice plots, such as the presence of gullies, dry plots, low-lying plots, and landslides than non-adopters.

3.2 Probit Model Estimation

In the first stage of the PSM, the probit model was estimated to analyze the factors affecting the decision to invest in rice plot soil conservation. The likelihood ratio statistics of − 890.9 suggests that the estimated model was highly significant at the 1% level. The value of the pseudo-R^2 which was equal to 13.30% showed that the model explained 13.3% of the variance in the decision-making concerning investment in soil conservation on the rice plot. The model performed well with the percentage of correct predictions equal to 80.75% (Table 3).

The results obtained with this model indicate that households were more likely to participate in soil conservation if they had a higher number of laborers, and they had access to credit, joined formal organizations, used old local varieties of rice

Table 3 Probit model parameter estimates

Variables	Coefficient	Standard errors
Landowner characteristics		
hhage	−0.0055[a]	0.0031
hhgen	0.1236	0.1530
hhedu	0.0152	0.0098
hhmar	−0.2107	0.1626
Ethnic	−0.1229	0.1069
Farm characteristics		
Labor	0.0558[b]	0.0247
Credit	0.2102[a]	0.1265
Land area	0.0144	0.0189
Membership	0.2672[b]	0.1045
Type of rice seed	0.3827[c]	0.0889
Organic fertilizer	−0.1442[a]	0.0769
Agricultural extension service	0.4825[c]	0.0788
Information on fertilizer	0.3178[a]	0.1646
Slope of land		
Flat land	−0.1608[a]	0.0970
Slight slope	0.1077	0.0853
Slope	0.2285[b]	0.0916
Very slope	−0.6719[b]	0.2857
Problems with rice plot		
Gullies	0.3882[c]	0.1073
Dry plot	−0.0457	0.0883
Low-lying plot	0.4281[c]	0.132
Sedimentation	0.2038	0.3119
Landslide	0.4982[b]	0.2036
Stony soil	−0.0499	0.1725
Region dummy variables		
Red River Delta	−0.1029	0.1785
Northern Midlands and Mountainous	−0.0248	0.0898
Central Coast	−0.0841	0.0601
Central Highlands	0.0613	0.0449
Number of observations		2010
Log-likelihood		−890.9
Pseudo R^2		0.1330

(continued)

Table 3 (continued)

Variables	Coefficient	Standard errors
% Correctly predicted		80.75

[a], [b], and [c] indicate statistical significance at the 10%, 5%, and 1% levels, respectively
Source Calculation by authors based on data from VARHS 2018

seeds, participated in agricultural extensions, and obtained information concerning the use of fertilizers. This outcome is consistent with that of Chesterman et al. (2019) who found that household participation in training programs had a significant effect on the use of soil and water conservation farming practices. Similarly, Asfaw and Neka (2017) and Mengistu and Assefa (2019) reported that the age of the household head, access to extension services, and access to training services influenced the adoption of soil and water conservation practices. In accordance with the present results, Jara-Rojas et al. (2013) demonstrated that farms with low-fertility soil were more likely to adopt water and soil conservation. Not surprisingly, households who faced gullies, low-lying land, and landslide problems were more likely to engage in the soil conservation program. Older household heads were less likely to participate in the soil conservation scheme. In addition, households using organic fertilizers were less likely to participate in soil conservation practices. The same pattern was held for households farming on flat land areas. Additionally, "very steep land" was the land characteristic that reduced the probability of adopting soil conservation.

3.3 Propensity Score Estimation

In the second step, propensity scores were derived for each household to obtain the predicted probability of adopting soil conservation. Then, households were matched with the use of the kernel, radius, and nearest-neighbor matching.

Considering the advantages of the trimming, concerning the common support method, we used the trimming technique to evaluate the ATT of soil conservation for each potential outcome (Table 4). Specifically, Table 4 shows differences in outcomes of soil conservation on total farm household income, the value of rice production, rice production, total production expenses, profit, farming efficiency, and some components of production costs between households adopting soil conservation and matching non-adopting households (counterfactual).

The ATT on the outcome of the added value of rice production per hectare was generally positive for all matching algorithms. However, the estimates were not statistically significant even at the 10% level. This finding is consistent with that of Schilling et al. (2014) who found that there was no statistically significant difference in profit between preserved and matching unpreserved farms. Similarly, the study also found that for all matching techniques, the differences between ATT estimates of rice value production and rice production were not statistically significant. This outcome

Table 4 Estimated average treatment effects (ATTs) of soil conservation practices with trimming on outcomes

Outcome variable	Matching algorithm	Treated	Controls	ATT	t-stat	Matched observations		
						Adopters	Non-adopters	Total
Added value of rice production per hectare (000 VND)	NN1	15,869.05	15,592.64	276.40 (957.79)	0.29	403	1.592	1.995
	NN5	15,869.05	15,097.89	771.15 (731.65)	1.05	403	1.592	1.995
	Kernel	15,884.83	15,596.30	288.52 (653.12)	0.44	406	1.592	1.998
	Radius 0.02	15,869.05	15,703.33	165.71 (646.20)	0.26	403	1.592	1.955
	Radius 0.05	15,896.00	15,490.54	405.46 (654.79)	0.62	408	1.592	2.000
Value of rice production (000 VND)	NN1	8809.59	11,379.93	−2570.33[a] (1520.198)	−1.69	403	1.592	1.995
	NN5	8809.59	9893.44	−1083.84 (989.30)	−1.10	403	1.592	1.995
	Kernel	8771.68	9718.60	−946.91 (957.32)	−0.99	406	1.592	1.998
	Radius 0.02	8809.59	9874.02	−1064.42 (949.84)	−1.12	403	1.592	1.955
	Radius 0.05	8759.37	9640.10	−880.72 (959.45)	−0.92	408	1.592	2.000
Rice production (yield per hectare) (kg.)	NN1	4674.46	5104.26	−429.79 (305.34)	−1.41	403	1.592	1.995

(continued)

Table 4 (continued)

Outcome variable	Matching algorithm	Treated	Controls	ATT	*t*-stat	Matched observations		
						Adopters	Non-adopters	Total
	NN5	4674.46	4789.46	−114.99 (206.92)	−0.56	403	1.592	1.995
	Kernel	4673.55	4791.61	−118.05 (196.88)	−0.60	406	1.592	1.998
	Radius 0.02	4674.46	4815.09	−140.63 (195.79)	−0.72	403	1.592	1.955
	Radius 0.05	4671.23	4770.10	−98.86 (197.15)	−0.50	408	1.592	2.000
Total production cost (000 VND)	NN1	12,457.39	15,004.34	−2546.92[c] (909.91)	−2.80	403	1.592	1.995
	NN5	12,457.39	14,514.61	−2057.22[c] (558.47)	−3.68	403	1.592	1.995
	Kernel	12,470.48	14,157.29	−1686.80[c] (479.14)	−3.52	406	1.592	1.998
	Radius 0.02	12,457.39	14,182.82	−1725.42[c] (474.80)	−3.63	403	1.592	1.955
	Radius 0.05	12,466.40	14,142.39	−1675.99[c] (480.56)	−3.49	408	1.592	2.000
Chemical fertilizer cost per hectare (000 VND)	NN1	3992.67	5996.13	−2003.457[c] (557.717)	−3.59	403	1.592	1.995

(continued)

Table 4 (continued)

Outcome variable	Matching algorithm	Treated	Controls	ATT	*t*-stat	Matched observations		
						Adopters	Non-adopters	Total
	NN5	3992.67	5767.82	−1775.146[c] (344.043)	−5.16	403	1.592	1.995
	Kernel	3996.92	5463.69	−1466.767[c] (287.391)	−5.10	406	1.592	1.998
	Radius 0.02	3992.67	5794.38	−1501.706[c] (284.607)	−5.28	403	1.592	1.955
	Radius 0.05	3996.94	5441.76	−1444.816[c] (288.413)	−5.01	408	1.592	2.000
Pesticide cost per hectare (000 VND)	NN1	1022.00	1282.63	−260.626[a] (144.199)	−1.81	403	1.592	1.995
	NN5	1022.00	1196.56	−174.561[a] (96.919)	−1.80	403	1.592	1.995
	Kernel	1017.27	1185.34	−168.064[a]	−1.69	406	1.592	1.998
	Radius 0.02	1022.00	1183.42	−161.420[a] (98.253)	−1.64	403	1.592	1.955
	Radius 0.05	1013.96	1192.64	−178.679[a] (99.606)	−1.79	408	1.592	2.000
Efficiency	NN1	2.695	2.477	0.217[b] (0.104)	2.09	403	1.592	1.995
	NN5	2.695	2.449	0.245[c] (0.084)	2.93	403	1.592	1.995

(continued)

Table 4 (continued)

Outcome variable	Matching algorithm	Treated	Controls	ATT	*t*-stat	Matched observations		
						Adopters	Non-adopters	Total
	Kernel	2.693	2.495	0.197^b (0.077)	2.56	406	1.592	1.998
	Radius 0.02	2.694	2.499	0.196^b (0.077)	2.55	403	1.592	1.955
	Radius 0.05	2.692	2.494	0.198^b (0.077)	2.57	408	1.592	2.000

[a], [b], and [c] indicate statistical significance at the 10%, 5%, and 1% levels, respectively, and standard errors are shown in parentheses. 1 USD = 21,387 VND

Source Calculation by authors based on data from VARHS 2018

is contrary to that of Kiboi et al. (2017) who investigated that tied ridging enhanced grain maize yields, and maize grain yields were more stable under mulching and tied ridging. Kassie et al. (2007) found that while in a low rainfall area of the Ethiopian highlands, conserved plots had a higher value of crop production per ha compared with non-conserved plots, and in the higher rainfall area, the crop production value between these plots was not statistically significantly different.

The adopters of soil conservation experienced significantly lower farming costs per hectare than matching non-adopters. The ATT on total farming costs ranged from 1675.99 to 2546.92 thousand VND per hectare (approximately 79–120 USD). This could be explained by the fact that following the adoption of soil conservation, the farmer's plots became more fertile, thus cutting the costs of fertilizers, pesticides, and other costs. Given the fact that water and soil conservation adopters had lower rice cultivation costs than counterfactual non-adopters, we estimated some components of variable costs per hectare, such as chemical fertilizers and pesticide costs. The ATT estimations for these costs were all negative and significant, showing that on average, adopting households spent between 1444.81 and 2003.45 thousand VND (68–94 USD) per hectare less on chemical fertilizers than matching non-adopting households. Similarly, our ATT estimation of pesticide costs showed that households investing in soil conservation spent between 161.42 thousand VND and 260.62 thousand VND (or 8–12 USD) per hectare less than their counterfactual non-adopters. Our results agree with Sommer et al. (2014), who found that conservation agriculture improved fertilizer use efficiency.

ATT estimates of the influence of soil conservation adoption on farming efficiency were significantly positive at 5% and 1% levels of significance, for all matching techniques. Households with soil conservation were found to be more efficient (efficiency ranging from 0.196 to 0.245) than non-adopting households. Our result confirms the existing empirical literature (Kinney et al., 2018) which showed the positive economic impact of private land conservation in Colorado, USA. Similarly, Yang et al. (2017) indicated that soil conservation practices improved farm efficiency.

3.4 Assessing the Quality of the Matching Process

The following tasks were performed to assess the quality of the estimated treatment effects. Firstly, a very significant decrease in the mean standardized bias and pseudo R^2 indicates the good quality of the results obtained, with each matching estimator (Table 5). Before matching, the standardized mean bias was 21.1%, while after matching, it ranged between 2.5% and 4.1%. By way of comparison, it is interesting to note that Rosenbaum and Rubin (1985) considers a 20% reduction in standardized bias sufficient, whereas the value reported here corresponds to a reduction of approximately 80% after matching.

Table 5 Matching quality indicators with a trimming approach

Matching algorithm	Mean bias		% |bias| Reduction	Pseudo R^2		p-value of LR	
	Before matching	After matching		Unmatched	Matched	Unmatched	Matched
NN1	21.1	4.1	80.57	0.133	0.018	0.000	0.826
NN5	21.1	2.8	86.73	0.133	0.005	0.000	1.000
Kernel	21.1	2.5	88.15	0.133	0.003	0.000	1.000
Radius 0.02	21.1	2.5	88.15	0.133	0.003	0.000	1.000
Radius 0.05	21.1	2.6	87.68	0.133	0.004	0.000	1.000

Source Calculation by authors based on data from VARHS 2018

Similarly, for all matching methods, the pseudo R^2 of the estimated probit model reduced considerably after matching. Besides, after matching, the p-value of the likelihood ratio test was rejected. This means that after matching, systematic differences in the distribution of covariates between adopters and non-adopters were eliminated.

The check of the matching process was done to identify whether the distribution of covariates in both the treatment and control groups is balanced. The covariate-balancing tests' results are reported in Table 6. After matching, the propensity score test presented a noteworthy drop in bias, and importantly, for any of the covariates, there were no differences in matched non-adopters and adopters (Table 6).

Figure 1a reveals considerable differences between the density distributions of the two groups, prior to matching. As shown in Fig. 1b, their density distributions almost overlap after matching, and a clear reduction in their deviations can be observed. This result shows that the requirement for a common support assumption was met in the present study. Figure 1b plots the pre- and post-matching kernel density functions of the treated and matching control groups.

Additionally, the PSM also relied on conditional independence. This reflects the fact that if the model takes into account all relevant characteristics, the bias in the estimated ATTs is eliminated. If the matching models fail to consider unobserved characteristics, the estimated results will be affected. Subsequently, by making use of Rosenbaum bounds with one-by-one matched pairs, we confirm that our outcomes were relatively robust to unobserved properties, in particular with respect to the results obtained for fertilizer cost per hectare (Table 7).

These results are quite robust, with the threshold gamma ranging from 1.1 to 1.58 (with a 90% confidence interval), according to different potential outcomes. We found that the results of farming efficiency outcome and pesticide cost per hectare were less robust than those of the outcomes of total production costs and chemical fertilizer costs per hectare (threshold gamma = 1.1). The statistical significance of the ATTs was probably questionable when the odds ratio of investment in soil conservation between adopting and non-adopting households was less than 1.1 for the outcomes

Table 6 Test for selection bias after matching

Variable	Matched sample		Bias		T-test p-value
	Treated (418)	Control (1592)	% bias	% bias reduction	
Landowner characteristics					
hhage	48.496	48.34	1.3	92.8	0.848
hhgen	0.9057	0.9165	−3.4	72.9	0.591
hhedu	9.6705	9.4776	5.4	−29.8	0.443
hhmar	0.9032	0.9197	−5.3	24.2	0.410
Farm characteristics					
Ethnic	0.3945	0.4176	−4.7	89.6	0.505
Labor	3.5658	3.5349	2.1	91.8	0.774
Credit	0.9156	0.9206	−1.6	91.4	0.798
Land area	0.7882	0.7847	0.2	97.1	0.983
Membership	0.8859	0.8968	−3.2	78.3	0.619
Type of rice seed	0.2531	0.2396	3.4	89.8	0.656
Organic fertilizer	0.4963	0.5031	−1.4	51.9	0.848
Agricultural extension service	0.7618	0.7571	1.0	98.1	0.876
Information on fertilizer	0.9553	0.9553	0.0	100.0	1.000
Slop of land					
Flat land	0.6328	0.6467	−3.0	90.9	0.682
Slightly sloping	0.5583	0.5677	−1.9	94.4	0.789
Sloping	0.3945	0.3898	1.0	97.6	0.891
Very steep	0.0099	0.0099	0.0	100.0	1.000
Problems with rice plot					
Gullies	0.2238	0.2462	−2.0	94.3	0.793
Dry plots	0.5261	0.5509	−5.0	81.8	0.480
Low-lying plots	0.0819	0.0950	−5.1	39.3	0.512
Sedimentation	0.0124	0.0139	−1.3	52.2	0.853
Landslides	0.0273	0.0352	−4.2	80.4	0.518
Stony soil	0.0472	0.0407	3.2	0.9	0.655
Plot quality	0.9975	0.9975	0.0	100.0	1.000

(continued)

Table 6 (continued)

Variable	Matched sample		Bias		T-test p-value
	Treated (418)	Control (1592)	% bias	% bias reduction	
Region dummy variables					
Red River Delta	0.0868	0.0705	4.9	83.1	0.388
Northern Midlands and Mountainous	1.0769	1.027	5.0	80.7	0.479
Central Coast	0.3052	0.3543	-4.7	82.0	0.458
Central Highlands	0.8735	0.9355	−4.1	81.8	0.599

Source Calculation by authors from VARHS 2018 data

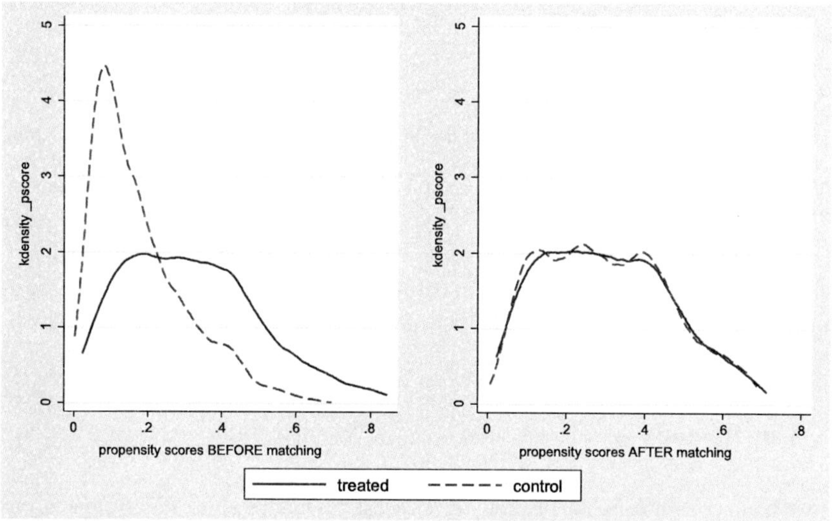

Fig. 1 Densities of p-scores before and after matching. *Source* Calculation by authors based on data from VARHS 2018

of farming efficiency and pesticide cost per hectare, 1.17 for the outcome of total production costs, and 1.58 for the outcome of chemical fertilizer cost per hectare.

Although the present study has made important contributions to empirical studies of the impact of soil conservation on economic efficiency and the use of fertilizers in rural areas of Vietnam, there are still various limitations to this study. In particular, the effect of soil conservation on the change in household fertilizer use and farming efficiency is likely to be seen in the middle or long term. Nevertheless, using a big cross-sectional dataset with checking the quality of the matching process, we

Table 7 Sensitivity analysis with Rosenbaum bounds

Gamma	Total production costs		Chemical fertilizer cost per hectare		Pesticide cost per hectare		Efficiency	
	sig+	sig−	sig+	sig−	sig+	sig−	sig+	sig−
1	0.0040	0.0040	1.E-07	1.E-07	0.0038	0.0038	0.0154	0.0154
1.1	0.0002	0.0328	1.E-09	7.E-06	0.0002	**0.0318**	**0.0914**	0.0014
1.17	3.E-04	**0.0932**	4.E-11	6.E-04	3.E-04	0.8992	0.2120	0.0002
1.2	1.E-04	0.1338	9.E-12	0.0001	1.E-04	0.1295	0.2807	8.E-04
1.3	4.E-07	0.3313	6.E-14	0.0015	4.E-07	0.3240	0.5435	4.E-06
1.4	1.E-08	0.5735	3.E-16	0.0096	1.E-08	0.5654	0.7728	1.E-07
1.5	3.E-10	0.7776	0.0000	0.0377	3.E-10	0.7714	0.9105	4.E-09
1.58	1.E-11	0.8850	0.0000	**0.0876**	1.E-11	0.8809	0.9635	2.E-10
1.6	7.E-12	0.9043	0.0000	0.1050	6.E-12	0.9007	0.9714	1.E-10
1.7	1.E-13	0.9654	0.0000	0.2230	6.E-13	0.9637	0.9924	2.E-12
1.8	2.E-15	0.9892	0.0000	0.3824	2.E-15	0.9886	0.9982	5.E-14
1.9	0.0000	0.9971	0.0000	0.5550	0.0000	0.9969	0.9996	1.E-15
2.0	0.0000	0.9993	0.0000	0.7100	0.0000	0.9992	0.9994	0.000

Gamma, log odds of differential assignment due to unobserved factors; sig+: upper bound significance level; sig−: lower bound significance level. The boxed numbers indicate the critical level of the strength of the effect, Gamma for each of the dependent variables. Source: Calculation by authors based on data from VARHS 2018

believe that our study can be used to reliably estimate the short-term and long-term improvements achieved through the implementation of soil conservation practices.

4 Conclusions

Three main conclusions can be drawn from these research results. Firstly, to analyze the factors influencing the decision to invest in soil conservation among rice farming households, the probit model was estimated with a high percentage of correct predictions (80.75%). Secondly, the empirical results using the PSM method indicate that soil conservation adoption has a positive effect in terms of a reduction in rice farming costs per hectare and overall chemical fertilizer and pesticide costs. Adopters of soil conservation experience significantly lower farming costs per hectare than matching non-adopters, with these costs ranging from 1675.99 to 2546.92 thousand VND per hectare (79–120 USD), respectively. Compared with non-adopting households, the reduction in chemical fertilizer utilization by adopting households led to monetary gains ranging from 1444.81 to 2003.45 thousand VND per hectare (68 to 94 USD), and the reduction in pesticide cost per hectare ranged between 161.42 and 260.62 thousand VND (8–12 USD) per hectare. Additionally, the empirical results indicated the positive and statistically significant influence on farming efficiency of soil conservation adoption among rice cultivating households ranging from 0.196 to 0.245. Thirdly, different tests were used to assess the quality of our statistical estimations. We found that the balancing property and common support were overall satisfactory.

This study can be used to establish significant policy recommendations. Firstly, to ensure a long-term positive effect of soil conservation on economic efficiency, together with the use of fertilizers and pesticides in rice farming households, the government should encourage them to join formal organizations where they can share their experiences in soil conservation practices with each farmer. Furthermore, there is a need for a participatory approach to farming education, through an agricultural extension system throughout the entire crop season. There is also a need for effective methods for the delivery of information concerning soil conservation and the use of fertilizers. Additionally, integrated national policies which focus both on agricultural productivity and environmental protection are extremely desirable. At the same time, the government should also introduce policies to ensure that the quantities of fertilizer applied to the soil are evaluated by using regular soil analyses. This would lead to a considerable reduction in the detrimental effects of over-fertilization.

References

Amsalu, A., & De Graaff, J. (2007). Determinants of adoption and continued use of stone terraces for soil and water conservation in an Ethiopian highland watershed. *Ecological Economics, 61*(2), 294–302.

Asfaw, D., & Neka, M. (2017). Factors affecting adoption of soil and water conservation practices: The case of Wereillu Woreda (District), South Wollo Zone, Amhara Region, Ethiopia. *International Soil and Water Conservation Research, 5*(4), 273–279.

Ashoori, D., & Sadegh, M. (2016). Adoption of conservation farming practices for sustainable rice production among small-scale paddy farmers in northern Iran. *Paddy and Water Environment, 15*(2), 237–248.

Attavanich, W. (2016). Did the Thai rice-pledging programme improve the economic performance and viability of rice farming. *Applied Economics, 48*(24), 2253–2265.

Becker, S. O., & Ichino, A. (2002). Estimation of average treatment effects based on propensity scores. *The Stata Journal, 2*(4), 358–377.

Bindraban, P. S., van der Velde, M., Ye, L., Van den Berg, M., Materechera, S., Kiba, D. I., Tamene, L., Ragnarsdóttir, K. V., Jongschaap, R., Hoogmoed, M., & Hoogmoed, W. (2012). Assessing the impact of soil degradation on food production. *Current Opinion in Environmental Sustainability, 4*(5), 478–488.

Chesterman, N. S., Entwistle, J., Chambers, M. C., Liu, H. C., Agrawal, A., & Brown, D. G. (2019). The effects of trainings in soil and water conservation on farming practices, livelihoods, and land-use intensity in the Ethiopian highlands. *Land Use Policy, 87*, 104051.

Crump, R. K., Imbens, G. W., Mitnik, O. A., & Hotz, V. J. (2009). Dealing with limited overlap in estimation of average treatment effects. *Biometrika, 96*(1), 187–199.

Dehejia, R. H., & Wahba, S. (2002). Propensity score-matching methods for nonexperimental causal studies. *The Review of Economics and Statistics, 84*(1), 151–161.

Diagne, A., & Demont, M. (2007). Taking a new look at empirical models of adoption: Average treatment effect estimation of adoption rates and their determinants. *Agricultural Economics, 37*(2–3), 201–210.

Gitonga, Z. M., Groote, H. D., Kassie, M., & Tefera, T. (2013). Impact of metal silos on households' maize storage, storage losses and food security: An application of a propensity score matching. *Food Policy, 43*, 44–55.

Gottlieb, P. D., Schilling, B. J., Sullivan, K., Esseks, J. D., Lynch, L., & Duke, J. M. (2015). Are preserved farms actively engaged in agriculture and conservation? *Land Use Policy, 45*, 103–116.

GSO (General Statistical Ofice). (2022). *Statistical Yearbook of Vietnam.* Statistical Publishing House, Hanoi, Vietnam.

Jara-rojas, R., Bravo-ureta, B. E., Engler, A., & Díaz, J. (2013). An analysis of the joint adoption of water conservation and soil conservation in Central Chile. *Land Use Policy, 32*, 292–301.

Karlen, D. L., & Rice, C. W. (2015). Soil degradation: Will humankind ever learn? *Sustainability, 7*, 12490–12501.

Kassie, M., Pender, J., Yesuf, M., Köhlin, G., Bluffstone, R. & Mulugeta, E. (2007). *Impact of soil conservation on crop production in the Northern Ethiopian Highlands.* International Food Policy Research Institute (IFPRI), No. 733.

Kiboi, M. N., Ngetich, K. F., Diels, J., Mucheru-Muna, M., Mugwe, J., & Mugendi, D. N. (2017). Minimum tillage, tied ridging and mulching for better maize yield and yield stability in the Central Highlands of Kenya. *Soil and Tillage Research, 170*, 157–166.

Kinney, D., Tanner, D., Hu, S. & Bradley, P. (2018), *Measuring impact: Evaluating the economic, social, and ecological services of the City of Ann Arbor Greenbelt program.* University of Michigan's School for Environment and Sustainability.

Lal, R. (2013). Food security in a changing climate. *Ecohydrology and Hydrobiology, 13*, 8–21.

Mengistu, F., & Assefa, E. (2019). Farmers' decision to adopt watershed management practices in Gibe basin, southwest Ethiopia. *International Soil and Water Conservation Research, 7*(4), 376–387.

Mengstie, F. A. (2009). *Assessment of adoption behavior of soil and water conservation practices in the Koga watershed, highlands of Ethiopia.* Cornell University.

Mishra, A. K., Khanal, A. R., & Mohanty, S. (2017). Gender differentials in farming efficiency and profits: The case of rice production in the Philippines. *Land Use Policy, 63*, 461–469.

Nguyen, T. H. (2017). *An overview of agricultural pollution in Vietnam: The crops sector.*

Pan, D. (2014). The impact of agricultural extension on farmer nutrient management behavior in Chinese rice production: A household-level analysis. *Sustainability, 6*(10), 6644–6665.

Rosenbaum, P. R., & Rubin, D. (1985). Constructing a control group using multivariate matched sampling methods that incorporate the propensity score. *The American Statistician, 39*(1), 33–38.

Schilling, B. J., Attavanich, W., Sullivan, K. P., & Marxen, L. J. (2014). Measuring the effect of farmland preservation on farm profitability. *Land Use Policy, 41*, 84–96.

Silverman, B. W. (1986). *Density estimation for statistics and data analysis.* Chapman & Hall.

Solís, D., & Bravo-Ureta, B. (2005). Economic and financial sustainability of private agricultural extension in El Salvador. *Journal of Sustainable Agriculture, 26*(2), 81–102.

Solís, D., Bravo-Ureta, B., & Quiroga, R. (2009). Determinants of household efficiency among small-scale hillside farmers in El Salvador and Honduras. *Agricultural Economics, 60*(1), 202–219.

Sommer, R., Thierfelder, C., Tittonell, P., Hove, L., Mureithi, J. & Mkomwa, S. (2014). Fertilizer use should not be a fourth principle to define conservation agriculture: response to the opinion paper of Vanlauwe et al. (2014)A fourth principle is required to define conservation agriculture in sub-Saharan Africa: the appropriate use of fertilizer to enhance crop productivity. *Field Crop Res, 169*, 145–148.

Uematsu, H., & Mishra, A. K. (2012). Organic farmers or conventional farmers: Where's the money? *Ecological Economics, 78*, 55–62.

Yang, Z., Mugera, A. W., Yin, N., & Wang, Y. (2017). Soil conservation practices and production efficiency of smallholder farms in Central China. *Environment, Development and Sustainability, 20*(4), 1517–1533.

Determinants of the Manipulated Financial Statements of Enterprises Listed on the Vietnam Stock Exchange

Hoang Ha Anh and Tran Minh Da Hanh

Abstract This study was conducted to analyze the factors affecting the manipulation of financial statements for listed enterprises in Vietnam by analyzing and surveying 281 financial statements of enterprises in industries classified by MSCI and S&P and binary logistic regression model. Of the 281 observations, 159 financial statements had manipulation (56.6%) and 122 financial statements had no manipulation (43.4%). The study found no difference of this ratio for different occupations and capital sizes. The indicators DSRI, SGI, SGAI, FIRMSIZE, and FIRMAGE, which represent the number of times the financial statements audited by auditing firms of the Big 4 group, differed statistically significantly between the two groups of manipulative and non-manipulated financial statements. The model's estimation results show that the statistically significant variables are: DSRI, SGAI, LVGI, FIRMAGE. In particular, the SGAI and FIRMAGE variables had the strongest impact with a marginal impact of 35.8% and 32.7%, respectively. The predictive probability of the model is 66%. Based on the results of the survey data and the results of the regression model estimate, the topic has synthesized and proposed five groups of solutions to minimize manipulation of financial statements.

Keywords Manipulation · Financial statement · Enterprises · Stock exchanges · Vietnam

1 Introduction

Today, the stock market is a channel to attract capital from investors; the information of enterprises published in their financial statements (financial statements) serves as a bridge for investors and interested parties to comprehend the financial status of enterprises. Consequently, investors will be able to make timely judgments if the enterprise's financial statements are accurate and complete in all material respects. With the growth of the economy and the intensification of rivalry between businesses,

H. H. Anh · T. M. Da Hanh (✉)
Nong Lam University, Ho Chi Minh City, Vietnam
e-mail: dahanh@hcmuaf.edu.vn

A. T. Nguyen et al. (eds.), *Contemporary Economic Issues in Asian Countries: Proceeding of CEIAC 2022, Volume 1*,
https://doi.org/10.1007/978-981-19-9669-6_26

pressure is placed on their business operations. Under the pressure of protecting the interests of shareholders, the pressure of the market, the pressure of market rivalry, and the pressure of the businesses' own interests, firms have additional incentives to manipulate financial statements (Atieh & Hussain, 2012; Kasznik, 1999; Mamo & Aliaj, 2014; Moreira & Pope, 2007).

In recent years, several incidents involving the disclosure of corporate information have negatively impacted investor trust and market stability. From 2016 through 2020, there are several issues with the financial accounts of listed companies that inflate revenue and profit. During this time period, the assets reported on the financial accounts of many listed companies, particularly those with commercial advantages, expanded significantly. After the disappearance of 1 trillion VND in inventory from the financial accounts of the Truong Thanh Wood Joint Stock Company in Vietnam in February 2016, a number of other publicly traded companies were revealed to be experiencing similar issues. The fact that businesses intentionally conceal their true financial status has resulted in significant losses for all parties involved, including insolvent businesses, declining stock prices, unstable markets, private investors losing money, and creditors unable to recover debts.

Therefore, the detection of financial statement manipulation and the identification of factors affecting financial statement manipulation become an urgent issue. Studies in Vietnam in recent years on the problem of financial statement manipulation have focused on the application of the Beneish model (Beneish, 1999) combined with the Z-score index to assess the ability to manipulate financial statements (Anh & Linh, 2016; Hoang, 2016; Tuyen, 2019; Duong, 2016). However, the majority of studies in Vietnam used the difference in profit before and after the audit (5% difference as standard) to determine whether businesses are manipulating financial statements or not. Up to now, there has been no research in Vietnam based on the opinions on the audit report to determine that the financial statements are manipulated. From that context, this study was conducted to analyze the financial statements of enterprises listed on the stock exchange, taking the case study at the Ho Chi Minh City exchange (HCM), in order to realize two objectives.

Consequently, the detection of financial statement manipulation and the identification of variables influencing financial statement manipulation have become pressing concerns. In recent years, research on the problem of financial statement manipulation in Vietnam has focused on the application of the Beneish model (1999) in conjunction with the Z-score index to evaluate the ability to manipulate financial statements (Anh & Linh, 2016; Duong, 2016; Hoang, 2016; Tuyen, 2019). The bulk of research in Vietnam, however, mostly examined the difference in profit before and after the audit (often 5%) to assess whether or not enterprises manipulate financial accounts. No study has been conducted in Vietnam based on audit report views to establish whether financial statements are altered. In this context, this study was done to assess the financial statements of publicly traded companies, using the Ho Chi Minh City exchange as a case study, in order to achieve two goals: (i) examining the disparities between two groups of financial statements with and without indicators of manipulation and (ii) examining the factors influencing the capacity to detect financial statement manipulation. The study's findings will give investors with more timely information, minimize risk when making decisions, and serve as the foundation for offering solutions to improve information transparency in financial statements.

2 Methodology

2.1 Financial Statement Manipulation

The concept of information manipulation on financial statements is the presentation of a lack of financial information by accountants and managers, resulting in an increase in net income to increase stock prices, which is accomplished by increasing revenues and reducing costs (Beneish, 1999).

Mamo and Aliaj (2014) argue that financial information manipulation is misrepresentation, misrepresentation, or lack of information and that financial performance creates a misjudgment of an organization's financial strength. It is called: revenue management, income adjustment, creative accounting practices and accounting manipulation.

Study by Spathis (2002) and Alaryan et al. (2014) has divided financial statements into two groups: manipulating and non-manipulating, corresponding to the opinions of independent auditors and the opinions of tax authorities. Financial statements that the auditor seriously doubts about the correctness of items or reports rejected by tax authorities will be classified as manipulated financial statements and vice versa.

In Vietnam, according to the regulations of the Ministry of Finance in Auditing Standards 700 and 705, there are four types of audit opinions corresponding to the level of errors and fraud of financial statements detected by independent auditors. Specifically, (i) the auditor gives a full acceptance opinion, (ii) the audit gives an audit opinion except, (iii) the auditor presents a contrary audit opinion, (iv) the auditor refused to give an opinion (Ministry of Finance, 2012a, 2012b). Thus, with the classification of audit opinions of Vietnam Auditing Standards, based on the importance of the information on the financial statements, it is possible to classify the financial statements into two groups: the group with important errors that may or may not spread and the group that prepares and presents the financial statements truthfully and reasonably in the important aspect.

In summary, the concept of manipulating financial statements in this study is that the act of reporting the financial situation of enterprises no longer reflects honestly and reasonably in key aspects because of the motives from the enterprises themselves and from shareholders. Specifically, if the financial statements have a "fully accepted opinion," it means that the financial statements are not manipulated against the audited financial statements with "excepted audit opinions," "contrary audit opinions," and "refusal to give opinions" which will be classified as manipulated financial statements.

In the field of research on the motives and fraudulent behavior of organizations, it is necessary to first mention two important studies: the fraud triangle theory (Clinard & Cressey, 1954) and the shareholder theory (Friedman, 1970). In the study, these two theories were used to explain why listed companies change their financial statements.

The fraud triangle theory emphasizes three essential traits of fraudulent individuals and organizations: motive, opportunity, and justification (Clinard & Cressey, 1954). When individuals or organizations are under financial or emotional pressure, there is an increased likelihood of fraud. Specifically, financial stress is the greatest strain. In addition, when there is pressure, if there is an opportunity, i.e., the organization is capable of executing the fraudulent scheme without being discovered, they will perform the act. After committing fraud, the perpetrator and the perpetrating organization will rationalize the wrongdoing.

In accordance with the shareholder theory, businesses are obligated to use their resources to increase profits within the framework of business law (Friedman, 1970). The premise of this approach is the identification of managers who are employed to operate on behalf of the company's shareholders. Consequently, they will plan business activities with the intention of maximizing profits for businesses.

Beneish (1999) compared the signs and information on the financial statements of businesses who manipulated their income vs those that did not manipulate their income between 1982 and 1992. The findings reveal that the variables "Receivables/Net Revenue," "Gross Profit," "Asset Quality," "Revenue Growth," and "Accrual Index" connect positively with the capacity to manipulate financial statements. In further study, Beneish et al. (2013) devised a very essential tool, the formula for generating the M-score index, to provide a baseline for identifying a company that has falsified financial statements. The M-score index has subsequently been inherited by a large number of themes that investigate financial statement manipulation.

Hansen et al. (1996) used M-score to build an overall quality feedback model to find the characteristics of enterprises that manipulated their financial statements. They found financial statement manipulation and the fisher effect in listed enterprises in Korea, Malaysia, Singapore, and Thailand. Christianto (2014) uses M-score to figure out which companies in Indonesia have manipulated their financial statements and which have not, as well as what effect manipulation has on stock profits. Empirical evidence shows that financial statements with manipulation indexes are calculated using the M-score. The higher the score, the lower the return on shares. Repousis (2016) used the M-score model to find 8486 manipulated financial statements of businesses in Greece from 2011 to 2012. Hasan et al. (2017) also used M-score as a basis for figuring out manipulation indicators for countries. The results of this study show that the countries in Asia with the highest manipulation index are China, Indonesia, Hong Kong, Singapore, Thailand, Malaysia, and Japan, in that order. There are differences in the financial statement manipulation index between countries.

Based on a survey of 76 businesses, Spathis (2002) made a model for how information can be changed on financial statements. The study made a model that can tell with 84% accuracy if financial statements can be manipulated. The regression model had 10 variables. Nine of them came from the financial statements, and the last one was a Z-score. Estimates from the model showed that the variables "Inventory/Total Revenue" and "Total Debt/Total Assets" were positively related to the ability to manipulate financial statements, while the Z-score was negatively related to the ability to do so.

Dechow et al. (2011) surveyed businesses in the USA in 2019 to find violations in financial statements in different industries. They used the F-score model to figure out how likely manipulation was. Some of the most common mistakes and omissions in financial statements were incorrectly profit statements, reporting profits that were higher than they really were, and missing expenses.

Yang et al. (2017) looked at how corporate governance and regulatory requirements affect how financial statements can be manipulated in China. The results of the research showed that corporate governance has a big effect on how financial statements are made. In particular, financial fraud is more likely to happen when regulations put more pressure on businesses with less centralized ownership, executives, presidents, directors, and shorter tenures of audit services.

In addition to variables from the financial perspective of the firm, studies in this field have expanded the research model to include non-financial corporate governance-related factors. The number of board meetings, the level of CEO independence, the CEO's capital ownership ratio, the structure of capital ownership, the internal control system, the size of the company, and the seniority of the company's operations all had an impact on the manipulation of financial statements, according to these studies (Safdar et al., 2019; Siregar & Utama, 2008; Tuyen, 2019; Xie et al., 2003).

There have also been a number of research on financial statement manipulation in Vietnam. Anh and Linh (2016) analyzed income management behavior using the M-score index to identify businesses exhibiting symptoms of income manipulation on financial statements. Combining M-score and Z-score, Tuyen (2019) developed a binary logistic regression model to assess the possibility of fraud on the financial statements of companies listed on the Ho Chi Minh City exchange. In addition, there is research by Hoang (2016) and Duong (2016).

It is clear that M-score, Z-score, and F-score models are frequently used in research on the manipulation of financial statements. Each of the aforementioned models relies on some indicator found within financial statements to determine whether an enterprise has been improperly classified as having financial statement manipulation or not. Also, prior writers have elaborated on how the DSRI, GMI, AQI, SGI, DEPI, SGAI, TATA, LVGI, and INV variables affect financial statement manipulation (Beasley et al., 1999; Beneish, 1999; Clinard & Cressey, 1954; Liu, 2012; Lu & Zhao, 2021; Repousis, 2016; Siegel, 1991).

Based on the research overview and access to data for this topic, the author uses the Beneish model (1999) to study how financial statements can be manipulated by combining the variables "Inventory," "Company Size," and "Company Age" to complete the research model. Next, this paper compares the differences between the two groups of financial statements with and without manipulation. There are eight variables in the Beneish model (1999), along with the indexes "Inventory," "Company size," "Company seniority," "Number of board meetings," and "Big 4" between these two groups. Instead of using the post-audit profit difference ratio, as was done in previous studies in Vietnam, this topic will be based on the auditor's opinion of the audited financial statements to figure out if the financial statements have been manipulated. This is also a novel point on the subject.

2.2 Data Analysis

Based on the previously described audit opinions, this study categorizes the group of financial statements with and without manipulation. To analyze the factors influencing the probability of enterprises manipulating financial statements, the logistic regression function was applied, with independent variables comprising indicators according to the Beneish (1999), as well as indicators of inventory/total revenue, operating time of the enterprise, and business size.

$$Ln\left[\frac{P(Y=1)}{P(Y=0)}\right] = \beta_0 + \beta_1 X_1 + \beta_2 X_2 + \beta_3 X_3 + \beta_4 X_4 + \beta_5 X_5 + \beta_6 X_6 + \beta_7 X_7 + \beta_8 X_8 + \beta_9 X_9 + \beta_{10} X_{10} + \beta_{11} X_{11} \quad (1)$$

where: $P(Y=1)$ is manipulated financial statements and $P(Y=0)$ is non-manipulated financial statements. The variables in the model are expressed in Table 1.

Variable X_1 indicates if revenues and receivables have been imbalanced for two consecutive years. According to Beneish (1999) and Repousis (2016), an increase in this index will enhance the possibility that financial statements have been manipulated.

$$X_1 = \text{DSRI} = \left(\frac{\text{Receivables}_t}{\text{Sales}_t}\right) / \left(\frac{\text{Receivables}_{t-1}}{\text{Sales}_{t-1}}\right) \quad (2)$$

Variable X_2 indicates a change in the gross margin from last year to this year. If the ratio is more than 1, the gross margin between the two years has decreased. According to Lev and Thiagarajan (1993) and Beneish (1999), a declining gross margin is a poor sign for the company's future. A business with weak prospects is more prone to engage in manipulative practices.

$$X_2 = \text{GMI} = \left(\frac{\text{Sales}_{t-1} - \text{Costs of goods sold}_{t-1}}{\text{Sales}_{t-1}}\right) / \left(\frac{\text{Sales}_t - \text{Costs of goods sold}_t}{\text{Sales}_t}\right) \quad (3)$$

Table 1 Variables in the regression model

Variable	Name	Explanation	Expected signs
Dependent variable			
Y	Manipulation	$Y = 1$: manipulated financial statement $Y = 0$: non-manipulated financial statement	
Explanatory variables			
X_1	DSRI	Days sales in receivables index	+
X_2	GMI	Gross margin index	+
X_3	AQI	Asset quality index	+
X_4	SGI	Sales growth index	+
X_5	DEPI	Depreciation index	+
X_6	SGAI	Sales, general, and administrative expenses index	−
X_7	LVGI	Leverage index	−
X_8	TATA	Total accruals to total assets	+
X_9	INV	Inventory over total revenue	+
X_{10}	FIRMAGE	Operating years of the company	+
X_{11}	FIRMSIZE	Company size	−

Source Summarized by authors

Variable X_3 indicates the quality of assets this year in comparison to last year. If the index is more than 1, it indicates that the corporation may delay expenses (Beneish, 1999; Siegel, 1991). The growth in asset accounts suggests a desire to capitalize the company's assets and defer expenses. Repousis (2016) discovered evidence of a link between asset quality index and income manipulation.

$$X_3 = \text{AQI} = \frac{1 - \frac{(\text{Current assets}_t + \text{PPE}_t)}{\text{Total assets}_t}}{1 - \frac{(\text{Current assets}_{t-1} + \text{PPE}_{t-1})}{\text{Total assets}_{t-1}}} \quad (4)$$

Variable X_4 displays the five t sales-to-five $t-1$ ratio. Manipulation does not indicate growth. However, analysts believe that rising organizations have more fake financial statements than the others because financial problems and capital demands put pressure on management to meet profitability targets. Furthermore, in relation to the control and proclivity to disclose delays during moments of great growth, if a corporation sees a drop in stock prices as the first warning of a collapse, it is more driven to falsify results. According to the relevant study of Lu and Zhao (2021), variable X_4 is connected with the capacity to alter financial accounts, which is compatible with Beneish's (1999) anticipation of variable X_4 and Clinard and Cressey's (1954) theory of fraud motives.

$$X_4 = \text{SGI}\frac{\text{Sales}_t}{\text{Sales}_{t-1}} \quad (5)$$

Variable X_5 represents the depreciation ratio from the prior year to the current year. A value of X_5 larger than one reflects the company's capacity to extend the usable life of assets or implement techniques to boost revenue. As a result, there is a positive relationship between X_5 and the probability of financial statement manipulation. Relevant research and development of a fraud detection model for Chinese A-share listed businesses based on M-score yielded comparable results (Lu & Zhao, 2021).

$$X_5 = \text{DEPI} = \frac{\frac{\text{Depreciation}_{t-1}}{\text{Depreciation}_{t-1} + \text{PPE}_{t-1}}}{\frac{\text{Depreciation}_t}{\text{Depreciation}_t + \text{PPE}_t}} \tag{6}$$

Variable X_6 is the ratio of this year's sales, general, and administrative costs to sales compared to the prior year. This index reflects an increase in the balance between the cost of sales and the administration of the firm compared to revenue during a two-year period (Beneish, 1999). However, the findings of this coefficient in Beneish's (1999) original model and the results of Repousis' (2016) regression analysis both show an inverse link between SGAI and the likelihood of falsifying financial accounts.

$$X_6 = \text{SGAI} = \frac{\frac{\text{SGA expense}_t}{\text{Sales}_t}}{\frac{\text{SGA expense}_{t-1}}{\text{Sales}_{t-1}}} \tag{7}$$

Variable X_7 is the ratio of this year's financial leverage to the previous year. If this ratio is more than one, the corporation is using more debt balance. According to Repousis's (2016) research, there is an inverse association between the debt balance ratio between two years and the capacity to alter financial accounts.

$$X_7 = \text{LVI} = \frac{\frac{(\text{LTD}_t + \text{Current liabilities}_t)}{\text{Total assets}_t}}{\frac{(\text{LTD}_{t-1} + \text{Current liabilities}_{t-1})}{\text{Total assets}_{t-1}}} \tag{8}$$

Variable X_8 represents accumulated income on total assets for the current fiscal year rather than cash income. Beneish (1999) suggested this variable to indicate the business's non-cash flow income. This variable is provided with the idea that an increase in accrual will be associated with a greater chance of financial manipulation. Previous findings also show a positive association between this independent variable and the dependent variable (Lu & Zhao, 2021; Repousis, 2016).

$$X_8 = \text{TATA} = \frac{\begin{array}{l}\text{Current assets}_t - \text{Cash}_t \\ -\text{Current liabilities}_t - \text{Current maturuties of LTD}_t \\ -\text{Income Tax payable}_t - \text{Depreciation and amortization}_t\end{array}}{\text{Total assets}_t} \tag{9}$$

Besides the eight variables taken from the M-score model (Beneish model), variable X_9 was synthesized from previous studies (Alaryan et al., 2014; Kaminski et al., 2004; Summers & Sweeney, 1998) and included in the model. From these studies, the "Inventory/Revenue" ratio is consistent with the possibility of manipulating financial statements. This is consistent with the studies in affirming the importance of inventory when identifying manipulation on financial statements (Beasley et al., 1999; Loebbecke et al., 1989; Persons, 1995)

$$x_9 = \text{INV} = \frac{\text{Inventory}_t}{\text{Total revenue}_t} \quad (10)$$

Variable X_{10} is the number of operating years of the business. Results from research by Safdar et al. (2019) show that corporates' age has impacts on financial statement manipulation.

$$X_{10} = \text{FIRMAGE} = \text{Year of financial statement} - \text{Year of establishment of company} \quad (11)$$

The size of the firm is variable X_{11}; research has shown that large organizations exhibit less income management behavior than small ones (Safdar et al., 2019; Siregar & Utama, 2008). In contrast, Mamo & Aliaj, (2014) discovered a positive association between the company's size and financial statement falsification.

$$X_{11} = \text{FIRMSIZE} = \ln(\text{Total asset}) \quad (12)$$

2.3 Data Collection

This paper used secondary data from openly published audited quarterly and annual consolidated financial statements, management activity reports on the exchanges. The expected sample size was selected according to the formula $n \geq 50 + k*P$ (Greene, 2012), where P is the independent variable number of the model, and k is the observed number of a variable ($k = 5$ or 10). With the number of independent variables in the regression model being 11, $k = 5$, the minimum number of samples of the topic is 105 (Đinh Phi Hổ, 2012). Samples are selected by quota sampling method. The study sample will be collected at the rate of 50% for each type of financial statement (with manipulation and without manipulation). The project collected 356 samples initially. After the screening and cleaning process, the remaining samples used for analyzing were 281.

Table 2 Categorized collected financial statements

	HNX		HOSE		OTC		Upcom		Total	
	Count	%	Count	%	Count	%	Count	%	Count	%
Non-manipulated financial statements										
Fully accepted opinion	17	50	47	60.3	7	33.3	51	34.5	122	43
Manipulated financial statements										
Excepted audit opinions	17	50	31	39.7	10	47.6	90	60.8	148	53
Contrary audit opinions	0	0	0	0	2	9.5	6	4.1	8	3
Refusal to give opinions	0	0	0	0	2	9.5	1	0.7	3	1
Total	34	100	78	100	21	100	148	100	281	100

Source Collected and calculated by authors

3 Results

3.1 Sample Characteristics

A total of 281 financial statements from four different exchanges between 2016 and 2020 are categorized as either manipulating or non-manipulating based on the auditor's assessment on the audited company's annual consolidated financial statements (Table 2), whereas, in 122 reports (43.4% of the total), the financial statements were not manipulated, which is the same as receiving an "audit opinion of complete acceptance." There were 159 false financial statements found in the dataset, representing 56.6% of the whole. Half of HNX reports were manipulated, compared to 39.7% in HOSE, 66.7% in OTC, and 65.6% in Upcom. When analyzed further by economic sector, the highest rates of altered financial statements are found in the real estate, construction, and consumer goods sectors, followed by the service sector and the remaining sectors.

3.2 Differences Between Manipulated and Non-manipulated Financial Statements

This paper utilized the t-test to investigate the financial metrics between the two groups in order to compare manipulated and unmanipulated data (Table 3). The following hypotheses were proposed:

Hypothesis H_0: There is no difference between manipulated and non-manipulated financial statements in terms of financial metrics.

Hypothesis H_1: There is a difference between manipulated and non-manipulated financial statements in terms of financial metrics.

Table 3 Comparison between manipulated and non-manipulated financial statements

	Non-manipulated		Manipulated		T-test
	Mean	S.D	Mean	S.D	Sig
DSRI	0.994	0.345	1.200	0.567	0.000
GMI	1.045	0.496	1.079	0.529	0.584
AQI	0.969	0.724	1.056	0.647	0.298
SGI	1.237	0.562	1.037	0.463	0.002
DEPI	1.008	0.572	1.051	0.599	0.547
SGAI	0.757	0.500	1.091	0.588	0.000
TATA	−0.068	0.391	−0.151	0.660	0.184
LVGI	0.968	0.207	1.002	0.224	0.180
INV	0.515	2.478	0.501	1.209	0.954
FIRMSIZE	27.655	1.769	27.249	1.523	0.044
FIRMAGE	19.787	14.553	23.560	15.691	0.038
Number of meetings of board of directors	9.860	10.026	9.586	9.586	0.853
Big 4	0.270	0.446	0.125	0.332	0.003

Source Collected and calculated by authors

This study found no statistically significant difference between manipulated and non-manipulated reports for the variables GMI, AQI, TATA, DEPI, INV, number of board of directors' meetings, and LVGI at a confidence level of 5%. In contrast, statistical differences existed between DSRI, SGI, and SGAI for the two groups. The non-manipulated group had lower DSRI (0.994 to 1.2) and SGAI (0.757 to 1.091), but greater SGI (1.237 to 1.037).

In addition to financial measures, the study investigates disparities in firm size, firm age, and audits by the "Big 4" in Vietnam. The inspection revealed that the average age of enterprises whose financial statements were falsified was 23.56 years, compared to 19.787 years for companies whose financial statements were not modified. The t-test Sig value of 0.038 indicates that there is a statistically significant difference between the operation times of these two sets of financial statements. Moreover, when comparing the average number of financial statements audited by the Big 4 auditing firms in Vietnam, the data indicate that Big 4 audited more manipulated reports than non-manipulated ones.

Consequently, based on the outcomes of verifying this difference, the H_0 hypothesis was rejected and the H_1 hypothesis was accepted. At a level of confidence of 5%, the indicators DSRI, SGI, SGAI, FIRMSIZE, and FIRMAGE, which represent the number of times an audit was conducted by a Big 4 auditing firm, differ between the two groups of financial statements.

3.3 *Factors Affecting Financial Statement Manipulation*

The results of the logistic model estimation show the degree of interpretation of the Nagelkerke model, $R^2 = 0.23$. This means that 23% of the change of the dependent variable is explained by 11 independent variables in the model. Omnibus test has a Sig significance level less than 0.05, so the independent variables have a relation to the dependent variable in the whole. Thus, the selected cell is relatively suitable.

DSRI, SGAI, LVGI, and FIRMAGE were statistically significant independent variables that correlated with the manipulation at the 10% confidence level, according to the Wald test. Also, there is no significant correlation between the variables GMI, AQI, SGI, DEPI, TATA, INV, and FIRMSIZE and manipulation (Table 4).

The marginal effects of statistically significant variables are presented in Table 5.

The DSRI variable had a positive coefficient, so it positively affected the likelihood of financial statement manipulation. If the DSRI increases by one unit and all other factors remain constant, the probability that the company's financial statements have been manipulated will increase by 21%. Revenue growth between years is a sign that the financial picture of the business is good if accompanied by good capital recovery. This is evidenced by a decline in accounts receivable/revenues between years or a DSRI of less than 1. However, if the DSRI is greater than 1 in conjunction with a low liquidity ratio, this indicates that the business is experiencing financial difficulties. The same dimensional impact of the DSRI variable on the ability to manipulate financial statements is similar to the research (Beneish, 1999; Repousis, 2016; Tuyen, 2019).

The SGAI variable correlated positively with manipulation. The probability that a company's financial statements have been manipulated increases by 35% if its SGAI increases by 1 unit, assuming that all other factors remain constant. The study's findings contradicted both initial expectations and the findings of previous research (Beneish, 1999; Repousis, 2016; Tuyen, 2019). In this indicator, sales and enterprise management expenses are compared to revenue. When this indicator is greater than 1, it indicates that the ratio of the cost of sales and management to revenue has increased between the two most recent years. Businesses that show signs of recording or distributing abnormally high costs, businesses with a significant drop in revenue between years, and businesses with weak cost control, and cost investments outside of cost that lead to revenue growth that does not match cost growth are all able to manipulate their financial statements. All of the aforementioned cases demonstrate that businesses may be pursuing the following objectives: reducing profits to reduce dividends, reducing taxes, and weak business. In contrast to previous research and expectations, the SGAI variable positively affects the ability to manipulate financial statements. This matches the actual situation in Vietnam. In accordance with Article 1, Clause 4 of Law No. 71/2014/QH13, as of 01.01.2015, the ceiling rate for sales and business management expenses related to reception, marketing support, and advertising expenses is no longer regulated. Expenses related to reception, marketing support, and sales advertising are deducted. As a result, the SGAI variable with a

Table 4 Estimated regression model

	Coefficients (S.D)
DSRI	0.907**
	(0.374)
GMI	0.270
	(0.299)
AQI	0.287
	(0.207)
SGI	−0.055
	(0.326)
DEPI	0.206
	(0.248)
SGAI	1.477***
	(0.340)
LVGI	1.351*
	(0.730)
TATA	−0.320
	(0.273)
INV	0.007
	(0.068)
FIRMAGE	0.021**
	(0.009)
FIRMSIZE	−0.084
	(0.092)
Constant	−2.321
	−2.972
Log likelihood	−164.241
Akaike information criterion	352.482

Note * $p < 0.1$; ** $p < 0.05$; *** $p < 0.01$
Source Collected and calculated by authors

Table 5 Marginal effects

	dF/dx	Standard error	z	$P > \|z\|$
DSRI	0.219795	0.089977	2.442793	0.014574
SGAI	0.357973	0.081348	4.400542	1.08E-05
LVGI	0.327329	0.176636	1.853129	0.063864
FIRMAGE	0.005140	0.002234	2.295076	0.021729

Source Collected and calculated by authors

positive influence also reflects a portion of the accounting policy of Vietnamese enterprises after 2015.

The estimated coefficient for the LVGI variable was also positive. If a company uses more financial leverage the following year than the previous year, assuming all other factors remain constant, the company's financial statements are more likely to be manipulated. In this model, the LVGI and DSRI variables both carry positive signs, indicating that businesses with financial problems, namely low liquidity ratios, using high financial leverage are more likely to manipulate financial statements than businesses without this situation. This is consistent with the theory of cheating triangles (Clinard & Cressey, 1954) from the perspective of organizational pressure and is similar to the findings of Tuyen (2019). However, some prior studies did not find a statistically significant relationship between financial leverage ratios and financial statement manipulation (Beneish, 1999; Repousis, 2016).

The FIRMAGE variable had a positive influence on the likelihood of manipulated reports. A company's financial statement is marginally more likely to be manipulated if it continues to operate for one additional year. This result differs from Orazalin's (2020) and is comparable to Sial et al. (2019) findings.

The remaining variables were not statistically significant (GMI, AQI, SGI, DEPI, TATA, INV, and FIRMSIZE). The INV variable is calculated as the inventory/revenue ratio, but the characteristics of inventory vary between business types. In the field of services and utility services, the INV variable value is either 0 or very small, indicating that there is no or very little inventory. In contrast, in the construction and real estate industry, inventory is determined as the value of real estate built but not yet delivered to customers, and revenue is determined when the entire project is delivered to customers, so the value of INV is very high for businesses in this industry. Furthermore, some companies presented data that were inconsistent between the indicators on the balance sheet and the inventory explanation on the notes to financial statements, resulting in data that did not fully reflect the inventory situation for the observations in the topic. Previous research has found a covariate relationship between inventory-to-total revenue ratios and financial statement manipulation (Alaryan et al., 2014; Spathis, 2002; Summers & Sweeney, 1998).

The model accurately predicts 51.6% or 59 out of 122 of the non-manipulated financial statements. The model correctly identified 122 of the 159 manipulated financial statements, representing a 76.7% rate of accurate predictions. The overall rate of correct predictions for the entire model was 65.8% (Table 6).

Table 6 Predictive capabilities of the model

		Manipulation		Correct ratio
		Non-Manipulated	Manipulated	
Manipulation	Non-Manipulated	63	59	51.6
	Manipulated	37	122	76.7
Overall				65.8

Source Collected and calculated by authors

Table 7 Results of financial statement manipulation forecast scenario

No	Variables	Coefficients	Value	
			Scenario 1 (Min)	Scenario 2 (Max)
1	DSRI	0.907	0.198	3.563
2	SGAI	1.477	0.000	4.451
3	LVGI	1.351	0.001	3.007
4	FIRMAGE	0.021	1	73
	Log odds		0.202	15.401
	eLog odds		1.224	4,882,778.590
	1 + eLog odds		2.224	4,882,779.590
	$P(Y/X_i)$ %		55.03	99.99

Source Collected and calculated by authors

From the model with independent variables of statistical significance, the topic proceeded to develop a manipulative forecasting scenario for the model (Table 7). Based on the maximum and minimum values, the topic developed two scenarios. When independent variables have the lowest value, the probability of a financial statement having manipulation is 55.03%, and when independent variables have the highest value, the probability of a financial statement having manipulation is 99.99%.

4 Discussions and Conclusions

Through the collection of 281 financial statements of companies listed on four exchanges HNX, HOSE, Upcom, and OTC, the study conducted an analysis of the current state of manipulation and identified factors affecting the ability of enterprises to manipulate financial statements. The DSRI, SGI, SGAI, FIRMAGE, FIRMSIZE, and Big 4 variables differ statistically significantly between the two groups of financial statements with manipulation and without manipulation.

The binary logistic model shows that the factors affecting financial statement manipulation in order of strongest and decreasing influence are SGAI, LVGI, DSRI, and FIRMAGE. These factors all have a positive impact on the probability to manipulate financial statements. At the same time, the study also developed two scenarios to predict the possibility of manipulating financial statements based on the minimum and maximum values of independent variables. When independent variables reach the minimum value, the probability of financial statements having manipulation is 55.03%. When independent variables reach their maximum value, the probability of manipulated financial statements is 99.99%. The correct prediction rate of the model is 65.8%.

From the results of the model, the topic proposes five groups of solutions to limit financial statement manipulation, enhance information transparency on the stock market from a financial perspective.

When businesses are under financial stress, they have an incentive to alter financial figures. Enterprise financial insecurity is expressed in liquidity indicators, financial leverage, and receivables turnover. As the indicator of receivables on net revenue rises year after year, auditing firms must examine the indications of liquidity, measuring the financial ability of businesses and indicating the risk of errors and fraud in financial accounts. From the standpoint of an investor, a more holistic understanding of businesses is required. Revenue and profit after tax growth over time are not a strong indicator of a company's financial health if it has problematic receivables and the load of bank loans grows year after year.

To ensure loan recovery, credit institutions should require borrowing enterprises to submit audited yearly consolidated financial statements or reviewed semi-annual financial statements with full acceptance opinions. The usefulness of financial statements in obtaining loans is also a motive for corporations to alter financial statements. When requiring financial statements to be audited with a full acceptance opinion, it means that in addition to the increase of profit after tax over each year, firms must have a transparent financial status on crucial elements such as short-term debt solvency and long-term debt solvency.

According to Article 1 Clause 4 of Law 71/2014/QH13, enterprises are no longer required to oversee sales and business management charges related to marketing and marketing advertising expenses as of January 1, 2015. As a result, businesses may inflate these expense items for tax and dividend purposes or in other words list businesses that use expenses to manipulate income. As a result of this reality, audit firms must examine and implement a tight audit procedure, altering the materiality to meet those organizations whose sales and management expenses compared to revenue have changed irregularly over the years, requiring specific accountability for such variations. At the same time, audit units must increase their assessment of issues relating to sales costs and business management, such as fixed asset depreciation, instrument allocation, inventory, cost pricing, and revenue.

Aside from auditing financial statements in accordance with legislation, exchanges and regulatory agencies must intensify their inspection of enterprises experiencing abnormal stock price swings. Unusual gains and declines in stock prices all imply significant movements in securities demand and supply. Manipulation of the value of securities has caused significant damage in Vietnam, resulting from organizations with extensive expertise trading on the stock market. To promote market transparency, it is necessary to strengthen oversight and delist businesses that have distorted financial statements (not gaining complete acceptance from the auditing company).

Current regulations compel public corporations listed on the stock exchange to publish annual consolidated financial statements. Furthermore, these companies must make public the reviewed semi-annual and quarterly financial accounts. However, auditing quarterly financial accounts is not mandatory. When the difference in profit after tax after and before the audit year is big, investors who rely on unaudited quarterly financial statements to make short-term decisions face significant risks. With this in mind, investors should conduct a comprehensive assessment of the year's financial situation, relevant issues from the perspective of corporate governance, and market signals rather than focusing solely on the enterprise's short-term financial situation in the quarterly financial statements.

Acknowledgements We would like to thank Nong Lam University for financially supporting the implementation of this research.

References

Alaryan, L. A., Abu Haija, A. A., & Alrabei, A. M. (2014). The relationship between fair value accounting and presence of manipulation in financial statements. *International Journal of Accounting and Financial Reporting, 4*(1), 222–237. https://doi.org/10.5296/ijafr.v4i1.5405

Anh, N. H., & Linh, N. H. (2016). Using the M-score model in detecting earnings management: Evidence from non-financial Vietnamese listed companies. *VNU Journal of Science: Economics and Business, 32*(2), 14–23. https://js.vnu.edu.vn/EAB/article/view/1287

Atieh, A., & Hussain, S. (2012). Do UK firms manage earnings to meet dividend thresholds? *Accounting and Business Research, 42*(1), 77–94. https://doi.org/10.1080/00014788.2012.622187

Beasley, M. S., Carcello, J., & Hermanson, D. R. (1999). *Fraudulent financial reporting: 1987–1997: An analysis* (58 p.). American Institute of Certified Public Accountants.

Beneish, M. D. (1999). The detection of earnings manipulation. *Financial Analysts Journal, 55*(5), 24–36. https://doi.org/10.2469/faj.v55.n5.2296

Beneish, M. D., Lee, C. M. C., & Nichols, D. C. (2013). Earnings manipulation and expected returns: Author response. *Financial Analysts Journal, 69*(5), 14–14. https://doi.org/10.2469/faj.v69.n5.8

Christianto, W. (2014). The effect of earnings manipulation with using m-score on stock return (empirical evidence in Indonesia listed companies on LQ45 at Indonesia stock exchange period 2009–2011). *Jurnal Ekonomi Akuntansi, 3*(1), 1–13.

Clinard, M. B., & Cressey, D. R. (1954). Other people's money: A study in the social psychology of embezzlement. *American Sociological Review, 19*(3), 362–363. https://doi.org/10.2307/2087778

Dechow, P. M., Ge, W., Larson, C. R., & Sloan, R. G. (2011). Predicting material accounting misstatements. *Contemporary Accounting Research, 28*(1), 17–82. https://doi.org/10.1111/j.1911-3846.2010.01041.x

Duong, V. M. (2016). *Using the Beneish M-score model to assess the quality of financial statements in Vietnam* [Master thesis]. University of Economics Ho Chi Minh City.

Friedman, M. (1970). A Friedman doctrine: The social responsibility of business is to increase its profits. *The New York Times Magazine.*

Greene, W. H. (2012). *Econometric analysis.* Pearson Education, Limited. https://books.google.com.vn/books?id=-WFPYgEACAAJ

Hansen, J. V., McDonald, J. B., Messier, W. F., Jr., & Bell, T. B. (1996). A generalized qualitative-response model and the analysis of management fraud. *Management Science, 42*(7), 1022–1032. https://doi.org/10.1287/mnsc.42.7.1022

Hasan, M. S., Omar, N., Barnes, P., & Handley-Schachler, M. (2017). A cross-country study on manipulations in financial statements of listed companies. *Journal of Financial Crime, 24*(4), 656–677. https://doi.org/10.1108/jfc-07-2016-0047

Hồ, D. P. (2012). *Phương pháp nghiên c´ưu định lượng và nh˜ưng nghiên c´ưu thực tiễn trong kinh tế phát triển - nông nghiệp* (447 p.). NXB Phương Đông.

Hoang, T. T. T. (2016). *Predicting the possibility of financial statement violations of companies listed on the Ho Chi Minh Stock Exchange* [Master thesis]. Da Nang University.

Kaminski, K. A., Sterling Wetzel, T., & Guan, L. (2004). Can financial ratios detect fraudulent financial reporting? *Managerial Auditing Journal, 19*(1), 15–28. https://doi.org/10.1108/02686900410509802

Kasznik, R. (1999). On the association between voluntary disclosure and earnings management. *Journal of Accounting Research, 37*(1), 57–81. https://doi.org/10.2307/2491396

Lev, B., & Thiagarajan, S. R. (1993). Fundamental Information Analysis. *Journal of Accounting Research, 31*(2), 190. https://doi.org/10.2307/2491270

Liu, J. (2012). Board monitoring, management contracting and earnings management: An evidence from ASX listed companies. *International Journal of Economics and Finance, 4*(12), 121–136. https://doi.org/10.5539/ijef.v4n12p121

Loebbecke, J. K., Eining, M. M., & Willingham, J. J. (1989). Auditors experience with material irregularities-frequency, nature, and detectability. *Auditing-A Journal of Practice & Theory, 9*(1), 1–28.

Lu, W., & Zhao, X. (2021). Research and improvement of fraud identification model of Chinese A-share listed companies based on M-score. *Journal of Financial Crime, 28*(2), 566–579. https://doi.org/10.1108/jfc-12-2019-0164

Mamo, J., & Aliaj, A. (2014). Accounting manipulation and its effects in the financial statements of Albanian entities. *Interdisiplinary Journal of Research and Development, 1*(2), 55–60.

Ministry of Finance. (2012a). *Auditing standards 700.* https://vbpl.vn/TW/Pages/vbpq-van-ban-goc.aspx?ItemID=30478

Ministry of Finance. (2012b). *Auditing standards 705.* https://vbpl.vn/TW/Pages/vbpq-van-ban-goc.aspx?ItemID=30478

Moreira, J. A., & Pope, P. F. (2007). *Earnings Management to Avoid Losses: a cost of debt explanation* (45 p.). CETE—Faculdade de Economia, Universidade do Porto.

Orazalin, N. (2020). Board gender diversity, corporate governance, and earnings management: Evidence from an emerging market. *Gender in Management: An International Journal, 35*(1), 37–60. https://doi.org/10.1108/GM-03-2018-0027

Persons, O. S. (1995). Using financial statement data to identify FFS. *Journal of Applied Business Research, 11*(3), 38–46.

Repousis, S. (2016). Using Beneish model to detect corporate financial statement fraud in Greece. *Journal of Financial Crime, 23*(4), 1063–1073. https://doi.org/10.1108/jfc-11-2014-0055

Safdar, S. M., Vinh, V. X., Al-Haddad, L., & Nguyen, T. T. (2019). Impact of female directors on the board and foreign institutional investors on earning manipulation of Chinese listed companies. *Asia-Pacific Journal of Business Administration, 11*(3), 288–300. https://doi.org/10.1108/APJBA-10-2018-0189

Siegel, J. G. (1991). *How to analyze businesses, financial statements, and the quality of earnings* (900 p.). Prentice Hall.

Siregar, S. V., & Utama, S. (2008). Type of earnings management and the effect of ownership structure, firm size, and corporate-governance practices: Evidence from Indonesia. *The International Journal of Accounting Education and Research, 43*(1), 1–27. https://doi.org/10.1016/j.intacc.2008.01.001

Spathis, C. T. (2002). Detecting false financial statements using published data: Some evidence from Greece. *Managerial Auditing Journal, 17*(4), 179–191. https://doi.org/10.1108/02686900210424321

Summers, S. L., & Sweeney, J. T. (1998). Fraudulently misstated financial statements and Insider trading: An empirical analysis. *The Accounting Review, 73*(1), 131–146.

Tuyen, P. T. M. (2019). Combine the Beneish M-Score model and the Z-Score index to identify the possibility of fraud on the financial statements of companies listed on the Ho Chi Minh Stock Exchange. *Tạp Chí Kế Toán & Kiểm Toán, 8*, 57–61.

Xie, B., Davidson, W. N., III., & DaDalt, P. J. (2003). Earnings management and corporate governance: The role of the board and the audit committee. *Journal of Corporate Finance, 9*(3), 295–316. https://doi.org/10.1016/s0929-1199(02)00006-8

Yang, D., Jiao, H., & Buckland, R. (2017). The determinants of financial fraud in Chinese firms: Does corporate governance as an institutional innovation matter? *Technological Forecasting and Social Change, 125*, 309–320. https://doi.org/10.1016/j.techfore.2017.06.035

Potential Economic Gains for Major Powers Through Mega Trade Deals: Gains for China, India and Japan

Ranti Yulia Wardani and Nawalage S. Cooray

Abstract The Regional Comprehensive Economic Partnership (RCEP) is a mega-regional trade deal initiated by ASEAN member countries. It has six Free Trade Agreements (FTAs) with India, Japan, China, South Korea, Australia and New Zealand and has been signed by 15 countries without India. India has decided to withdraw from the RCEP negotiation. The RCEP agreement faced many challenges. There are many studies about the RCEP potential economic gain, but few analyzes about the comparison of potential economic gain between Japan, India and China. Japan has signed an FTA with India, while India and Japan have no FTA with China before signing the RCEP trade deal. The objective of this chapter is to explore the potential economic gain of the FTA between Japan, China and India. The potential economic gain will be analyzed by comparing the saving potential of FTA between the three countries. Japan has a lower tariff barrier than India and China, while India has a higher tariff barrier than China and Japan. Currently, China is one of its biggest trading partners for Japan. Among the three countries, the results show that the RCEP agreement will have significant advantages for Japan's exporters because China is one of the biggest export markets for Japan. India has maintained its protectionism policy and has a high tariff barrier compared to Japan and China. India has low potential saving FTA benefit export to Japan due to Japan's low tariff barriers and India's low export value to Japan. India's export value to China is higher than India's export value to Japan. The reason is that China is India's second biggest export partner after the USA.

Keywords RCEP · ASEAN · International trade · Regionalism · China · India · Japan

---

R. Y. Wardani (✉)
International Relations Study Program, Faculty of Humanities, President University, Bekasi Regency, Indonesia
e-mail: ranti.yulia@president.ac.id

N. S. Cooray
Graduate School of International Relations, International University of Japan, Niigata, Japan

A. T. Nguyen et al. (eds.), *Contemporary Economic Issues in Asian Countries: Proceeding of CEIAC 2022, Volume 1*,
https://doi.org/10.1007/978-981-19-9669-6_27

1 Introduction

The formation of the RCEP can be described from the perspective of several international relations' theories. This chapter focuses mainly on the theories of regionalism and neoliberalism institutionalism. One question has been raised for this chapter to explore the potential gain of tariff reductions from FTAs between the major powers.

Dent (2016a) defined regionalism as the structure, process and arrangements that focuses on achieving a specific objective within a particular international region to link in terms of economic, sociocultural, political and security agreements. The RCEP has emphasized an economic partnership as a fundamental element to bind participating countries to work together and achieve their objectives. The RCEP could create an exclusive FTA economic zone in terms of economy. Regionalism is an important concept in exploring the formation of the RCEP. Many actors and non-state actors are involved in decision-making in the international arrangement. Sixteen countries with different levels of economic development, human capital and the structure of political institutions took part in the RCEP formation. These conditions have made it challenging for the negotiating parties to find a balanced compromise and consensus.

The participating countries within the RCEP formation have many different interests. However, establishing the RCEP as an international institution would create greater coherence among the participating countries through shared interests. As explained by Keohane (1989), the theory of neoliberalism institutionalism mentions the importance of mutual interests that can be achieved from cooperation. Some research has been done to explore the potential gain of FTA for the Japan–India, India–China and Japan–China (Wardani, 2020; Wardani & Cooray, 2019a, 2019b). Therefore, this chapter will enrich insight from different perspectives by bringing together the potential gain analysis among three major powers through mega trade deals.

The chapter is organized as follows. Section 2 describes the development of the RCEP. Section 3 describes the relations between China, India and Japan. Section 4 presents the potential savings gains of China, India and Japan. Section 5 provides a conclusion.

2 The Development of the RCEP

ASEAN is an international institution that promotes intergovernmental cooperation and encourages economic, political, security and sociocultural integration in East Asia. ASEAN plays the role of a catalyst within the formation of the RCEP and has attracted major power countries to participate in a regional economic partnership based on the ASEAN's FTA framework. This partnership has been based on ASEAN's norms of cooperation, namely consensus-based decision-making. ASEAN has a unique position at the center of the regional FTA. ASEAN is an international institution containing broadly weak countries, consisting of the least developed and developing countries. Nevertheless, it has attained economic cooperation involving major powers such as China, India and Japan. China and Japan have competed for FTA leadership with different agendas. In 2005, China proposed an FTA that consisted of ASEAN Plus Three (APT) members and the East Asia Free Trade Agreement (EAFTA). In 2006, Japan proposed an FTA that included APT members with India, Australia and New Zealand, the Comprehensive Economic Partnership for East Asia (CEPEA). ASEAN's leaders, under Indonesia's chairmanship in 2011, successfully bridged the two major powers' FTA agendas by proposing and agreeing on the RCEP under the ASEAN framework. Table 1 shows the emergence of the RCEP with its antecedents.

The RCEP has undergone many negotiations, including ministerial meetings, intersessional meetings and summits. These several meetings and negotiations have included different leaders, ministries and negotiators. The RCEP reached its substantial conclusion at the 3rd RCEP summit in November 2019. It took an extended period for the RCEP participating countries to reach this stage. India opted to withdraw from the RCEP at the 3rd RCEP summit, showing that different actors have found it challenging to balance and reach a consensus in the RCEP trade deal.

3 Relations Between the Three Major Power Countries

There are two perspectives on the major power countries' involvement in the RCEP: liberal and realist. The liberal perspective argues that forming regional cooperation and economic relations through international trade and FDI leads to harmony and peace (Keohane, 1983). The formation of the RCEP as an international institution reduces distrust and mistrust among participating countries. The realist perspective argues that major power countries compete with each other to reach a position of dominance. This means that in the realm of international politics, political rivalry among major powers is unavoidable. The so-called 'Asian miracle' emergence attracted major power countries such as China and Japan to compete for FTA leadership in Asia. The essential issue is how nations organize their economic relations in international politics through an outward-looking perspective in which they seek to influence, prestige and economic gains in interactions with other countries (Watson, 2008, p. 34).

China, India and Japan are the major countries involved in the RCEP negotiations. These countries have pursued their domestic interests on the negotiation table of the RCEP. China is perceived to have rivalries with Japan and India, while Japan and India have a deep and broad bilateral partnership in maritime issues, defense and economics.

Sino–Japanese relations contain distrust and mistrust that lead to complex bilateral ties, particularly tension over the Senkaku/Diaoyu Island territory. China and India also have border tensions and different perspectives over the Indian Ocean. India, Japan and the USA have a close security partnership to contain Chinese behavior and

Table 1 Antecedents and negotiations of the RCEP

Date	RCEP
2001	Based on APT group membership, the East Asia vision group proposed the EAFTA
2002	At the APT summit, EAFTA accepted as a long-term goal
2004	In November's APT summit, China proposed an EAFTA feasibility study
2005	In April, the EAFTA feasibility study began
2006	In April, Japan proposed the CEPEA
2007–2008	EAFTA and CEPEA feasibility studies were conducted
2009	EAFTA and CEPEA feasibility studies submitted to be considered at the APT level
2010	APT considered the two regional FTA proposals (EAFTA and CEPEA)
2011	In November's ASEAN summit, ASEAN proposed the RCEP
2012	• In August, the RCEP proposal was accepted and tabled during the ASEAN Economic Ministers plus FTA partner consultations • In November's ASEAN summit, the RCEP proposal was launched by the 16 national leaders
2013	• In May, 1st round of RCEP negotiations held • In August, 1st RCEP ministerial meeting held • In September, 2nd round of RCEP negotiations held
2014	• In January, 3rd round of RCEP negotiations held • In April, 4th round of RCEP negotiations held • In June, 5th round of RCEP negotiations held • In August, 2nd RCEP ministerial meeting held • In December, the 6th round of RCEP negotiations held
2015	• In February, the 7th round of RCEP negotiations held • In June, the 8th round of RCEP negotiations held • In July, 1st RCEP intersessional ministerial meeting held • In August, the 3rd RCEP ministerial meeting and 9th round of RCEP negotiations held • In October, the 10th round of RCEP negotiations held
2016	• In February, the 11th round of RCEP negotiations held • In April, the 12th round of RCEP negotiations held • In June, the 13th round of RCEP negotiations held • In August, the 4th RCEP ministerial meeting and 14th round of RCEP negotiations held • In October, the 15th round of RCEP negotiations held • In November, 2nd RCEP intersessional ministerial meeting held • In December, the 16th round of RCEP negotiations held
2017	• In February, the 17th round of RCEP negotiations held • In May, the 18th round of RCEP negotiations held • In July, the 19th round of RCEP negotiations held • In September, 5th RCEP ministerial meeting held • In October, the 20th round of RCEP negotiations held • In November, 1st RCEP summit

(continued)

Table 1 (continued)

Date	RCEP
2018	• In February, the 21st round of RCEP negotiations held • In March, 4th RCEP intersessional ministerial meeting held • In May, the 22nd round of RCEP negotiations held • In July, the 23rd round of RCEP negotiations held • In August, 6th RCEP ministerial meeting held • In October, the 6th RCEP intersessional ministerial meeting & 24th round of RCEP negotiations held • In November, 2nd RCEP summit
2019	• In February, the 25th round of RCEP negotiations held • In March, 7th RCEP ministerial meeting held • In June, the 26th round of RCEP negotiations held • In July, the 27th round of RCEP negotiations held • In August, 8th RCEP intersessional ministerial meeting held • In September, the 7th RCEP ministerial meeting and 28th round of RCEP negotiations held • In October, 9th RCEP intersessional ministerial meeting held • In November, 3rd RCEP summit
2020	• In February, the RCEP special meeting of the trade negotiation committee • In November, signing the RCEP agreement

Sources Dent (2016b, pp. 201–202), Ministry of Foreign Affairs of Japan (2019), ASEAN secretariat (2020)

to maintain the balance of power within the Asia–Pacific and Indo–Pacific regions. China's claims to the South China Sea territory under the 'nine-dash line' have raised tensions with Southeast Asian countries such as Brunei, Vietnam, Malaysia, the Philippines and Indonesia. These disputes could have had negative consequences on the development of the RCEP. Nevertheless, the RCEP has not involved negotiations related to territorial or security issues, with the leaders of participating countries preferring to settle their disputes outside the RCEP negotiation table.

Some of the participating countries have issues over territory and security. However, China and Japan have increased economic interdependence over the past 20 years. Economic interdependence is essential for bilateral relationship development. China is Japan's largest export destination, accounting for 22.05% of its exports in 2020 (Table 2). India has a trade deficit with China, while at the same time, China is the second-largest market for India's exports, with a share of 6.9% of India's total export (Table 2).

India decided to withdraw from the RCEP until its remaining issues were solved. Japan initiated the formation of a regional partnership that involved ASEAN + 3 together with India, Australia and New Zealand. India is an important economic and strategic partner for Japan; therefore, Japan has encouraged India to rejoin the RCEP to contain China's conduct in developing the RCEP. The domestic interests of three major countries, namely Japan, India and China, have influenced leaders' commitment to the RCEP agreement (Wardani, 2021).

Table 2 Top three trading partners of China, India and Japan in 2020

Country	Trade partner	Exports (US$ Thousand)	Export partner share (%)
China	USA	452,492,876	17.48
China	Hong Kong, China	271,708,494	10.49
China	Japan	142,596,624	5.51
India	USA	49,320,596	17.9
India	China	19,008,266	6.9
India	United Arab Emirates	17,953,335	6.52
Japan	China	141,399,199	22.05
Japan	USA	118,793,185	18.53
Japan	Korea, Rep	44,688,451	6.97

Source World Integrated Trade Solution (WITS, 2022)

4 Saving Potential of the RCEP: China, India and Japan

Due to intergovernmental dialogs and negotiations, the RCEP was initiated as a comprehensive economic partnership by states. Scholars from the neoliberal institutionalism perspective stated that there would be no cooperation without mutual interests (Keohane, 1989). Economic development is one of the fundamental mutual interests for the RCEP to form collaboration.

The RCEP was proposed as a comprehensive regional economic partnership with 16 participating countries at the beginning of the negotiations. The dismantling of tariffs is one of the main issues negotiated in international trade agreements such as the RCEP. All participating countries are seeking their gains from the trade negotiations. In forming the RCEP, it is important to highlight the actors' interests and the potential gains from economic cooperation.

The RCEP is based on the ASEAN + 1 FTA framework. The ASEAN + 1 FTAs created a 'noodle-bowl' situation regarding varying regulations and tariffs. The RCEP attempts to overcome the 'noodle-bowl' situation and create one set of principles, rules, regulations and commitments. The major countries have different trade arrangements; while China and Japan have no FTA, both countries are involved in an FTA negotiation with South Korea, the China–Japan–South Korea FTA. China, Japan and South Korea have held several stages of FTA negotiations since 2012, encouraged by the notion of an 'East Asian identity' (Ahn, 2020). China and India have no FTA yet, while India and Japan signed an FTA on February 16, 2011.

Table 2 shows the top three trading partners China, India and Japan. Japan is China's third-largest export partner, with a share of 5.51% and a total export value of 142.6 US$ billion. China accounts for 6.9% of India's exports, while China is Japan's largest export market with an export value of 141 US$ billion, accounting for 22.05% of its total exports.

The following section describes the saving potential of China, India and Japan from involvement in the RCEP. The saving potential methodology will be used to calculate the potential gain from the tariff reduction of FTA (Ziltener, 2016). The calculation of saving potential used the most favored nations (MFN) data from the World Trade Organization (2018) and excluded the six-digit level Harmonized System (HS) code clusters 71, 93, 97 and 99. The saving potential of an FTA could be considered the potential benefit of the FTA. The saving potential was calculated based on each commodity cluster. The potential savings from each commodity cluster will make it easier for policymakers or exporters to understand current trade conditions. The following formula will be used to calculate the potential gain (Wardani, 2020, p. 6):

$$D\% = \frac{Dt}{Tv}$$

$D\%$: duties' percentage.
Dt: total duties.
Tv: total trade value

$$\mathrm{SP}(n) = \mathrm{Tv} \times (1 + \mathrm{Gr})^{(n-1)} \times (n) \times (\mathrm{Utr})$$

SP: saving potential in year (n).
Tv: the total of export trade value.
Utr: utilization rate percentage of FTA (scenario-based).
Dty: dismantling tariff years (scenario-based).
Gr: export growth.

4.1 The Saving Potential of the China–Japan FTA

China is one of the Japan's largest trading partners. China and Japan have maintained economic interdependence over the last 20 years, although they have not yet agreed to an FTA. It is necessary to measure the saving potential of an FTA between Japan and China if both countries are included in the RCEP. The results of the saving potential analysis help to understand which country will gain the most.

The measurement shows that Japan has a higher saving potential than China. The total saving potential of Japan's export to China is 9 US$ billion, with 2016 as the base year for the measurement. The saving potential of China's exports to Japan is 3.7 US$ billion. This implies that overall, Japan has lower tariff barriers than China.

The transportation commodity has the highest saving potential for Japan's exports to China, with an overall trade value of 14.3 US$ billion and duties of 2.5 US$ billion. Even though the trade value of machinery and electronics is higher, at 67.7 US$ billion, the duties are lower at 2.3 US$ billion. This shows that China has high tariffs on transportation commodities. The highest saving potential for Japan's transportation commodities based on the six-digit HS code is for vehicle parts (HS code 870,323), with a trade value of 4.9 US$ billion and a duty value of 1.2 US$ billion. The vehicle parts' commodity with the HS code 870,324 holds the second-highest saving potential with a trade value of 2.5 US$ billion and a duty value of 636 US$ million. The vehicle parts' commodity with HS code 870,840 holds the third-highest saving potential with a trade value of 4.1 US$ billion and a duty value of 370 US$ million.

The highest total duties for China's export to Japan are for the textile commodity. China's total textiles' export trade value is 21.6 US$ billion, with a total duty value of 1.8 US$ billion. The total trade value for machinery and electrical goods is 71.2 US$ billion, with a duty value of 2 US$ million. The duty value of the textile commodity is higher than the machinery/electrical commodity, which shows that Japan has high tariffs for textiles compared to other commodities. The highest saving potential for China's textile commodities is in jerseys, pullovers and cardigans of human-made fibers (HS code 611,030), with a trade value of 1.9 US$ billion and a saving potential of 206 US$ million. The second-highest potential saving in the textile commodity is jerseys, pullovers and cardigans made of cotton (HS code 611,020), with a trade value of 721 US$ million and a duty value of 75 US$ million. The third-highest potential saving is for t-shirts, singlets and other cotton vests (HS code 610,910) with a trade value of 614 US$ million and a 10% duty with a duty value of 60 US$ million.

4.2 The Saving Potential of the China–India FTA

Relations between China and India play an essential role in the international system. The countries have a dynamic relationship that needs to be analyzed in the RCEP development. The domestic interests of China and India have shaped their behavior in the international playing field. China has no FTA with India. The RCEP is a multilateral trade deal, but the negotiations between India and China have been at the heart of the RCEP trade deal because China and India had difficulty finding a balance between them. The regionalism theory emphasizes international cooperation in terms of trade. Therefore, this study explores the saving potential of an FTA between India and China.

The total saving potentials of India and China have been measured with 2017 as the base year for the measurement. India's exports to China have a value of 15.9 US$ billion with a saving potential of 537 US$ million. China's exports to India have a value of 59.9 US$ billion, with a saving potential of 3.1 US$ billion. India's trade deficit with China has become one of its primary considerations in the RCEP negotiations, as India is wary of its markets being flooded by China imports. An FTA risks increasing India's trade deficit with China.

The mineral product commodity cluster is the most valuable of India's exports to China, with a total trade value of 7.4 US$ billion and total duties of 74 US$ million. On the other hand, while the total trade value of the textiles' cluster is 1.5 US$ billion, the total duty is 94 US$ million. This shows that the duties for the textile cluster are higher than for the mineral products cluster. Based on the six-digit level HS code, salt is the highest trade value commodity in India's exports to China, with a total trade value of 6.8 US$ billion and a duty value of 50.7 US$ million. The second-highest value trade commodity is copper and cathode components for the electronics industry, an important sector in China. The total trade value of copper and cathode components is 2.1 US$ billion, with entire duties of 42.8 US$ million.

For China, the machinery or electrical products' commodity cluster holds the largest trade value in its exports to India. The total value of the machinery/electrical commodity cluster is 36.6 US$ billion, with total duties of 1 US$ billion. The chemicals and allied industries' commodity cluster has the second-highest trade value, at 7.5 US$ billion, with total duties of 572 US$ million. The metals' commodity cluster has the third-highest trade value, at 4.4 US$ billion, with total duties of 389.9 US$ million. The highest trade value based on the six-digit level HS code is telephone sets (HS code 851,770), with a total trade value of 9.4 US$ million. The second-highest trade value is electrical apparatuses (HS code 854,140), with a total trade value of 3.9 US$ million. The third-highest trade value is for telephones for cellular networks (HS 851,712), with a trade value of 3.3 US$ million. The tariffs for these three commodity codes are zero. India could consider lowering its tariffs for other products, such as raw materials, that are needed for India's industries. For example, the chemical and allied industries' cluster could be considered for negotiation. Kalirajan (2022) study shows that India has potential gain by having trade with China if the constraints that influence India's 'behind the border' are dismantled.

4.3 The Saving Potential of the India–Japan FTA

The total value of India's exports to Japan is 4.9 US$ billion, and the total saving potential is 108 US$ million. Japan's total exports to India have a value of 10.2 US$ billion. Japan's exports to India are more than double those of India to Japan, and the total saving potential for Japan's exports to India is 801 US$ million. This shows that in general, India's import tariffs are higher than those of Japan.

The mineral products' commodity cluster has the highest trade value of India's exports to Japan at 1.5 US$ billion, with total duties of 16 US$ million. The chemicals' and allied industries' cluster is the second largest, with a total value of 870 US$ million and total duties of 15 US$ million. The third-largest export commodity cluster is animal and vegetable products, with a trade value of 709 US$ million and total duties of 13.9 US$ million. Based on the six-digit level HS code, the petroleum oils' and oils' commodity (HS code 271,012) has the highest trade value at 1.2 US$ billion with total duties of 15.6 US$ million. The second highest is frozen shrimps and prawns (HS code 030,617), with a trade value of 340 US$ million and duties of 3.4 US$ million. The third-highest commodity is iron ores and concentrates (HS code 260,111), with a trade value of 176.4 US$ million. The tariff line for this product is zero.

The highest trade value for Japan's export to India is in the machinery or electrical commodity cluster, with a total trade value of 3.8 US$ billion and a total duty value of 251 US$ million. The metals' commodity cluster is the second-highest value at 1.9 US$ billion with total duties of 177 US$ million. The chemicals' and allied industries' commodity cluster have the third-highest trade value at 1.1 US$ billion with total duties of 87 US$ million. The highest saving potential in Japan's exports to India is in the machinery or electrical cluster, with the total duties equaling 251 US$ million. The metal commodity cluster has the second-highest saving potential, with total duties of 177 US$ million. The transportation commodity cluster has the third-highest saving potential, with a total of 103 US$ million. Based on the six-digit level HS code, the vehicle parts' commodity (HS code 870,840) has the highest trade value for Japan at 250 US$ million with total duties of 25 US$ million. The iron or non-alloy steel commodity (HS code 720,839) is the second highest, with a trade value of 227 US$ million and a total duty value of 22.7 US$ million. The vehicle parts' and accessories' commodity (HS code 870,899) has the third-highest trade value, at 223.8 US$ million, with total duties of 22.4 US$ million.

4.4 The Ten-Year Saving Potential Projection for China, India and Japan

The RCEP has tried to gain a commitment to reduce 90–95% of existing tariffs. This study created three scenarios based on different utilization rates: a low utilization rate at 33%, a medium utilization rate at 66% and a maximum utilization rate at 100%; the scenarios for the projection of saving potential start in the year of the FTA enforcement. The data are accessed through (UN Comtrade, 2019).

Figure 1 shows the maximum saving potential at a 100% utilization rate for China, India and Japan. In year 5, China's exports to Japan are 2.3 US$ billion, and the saving potential in exports to India is 1 US$ billion. Japan's saving potential in its exports to China is 4.98 US$ billion and the saving potential of exports to India is 251 US$ million. India's saving potential in its exports to China is 205 US$ million and the potential savings for exports to Japan is 41 US$ million. In year 10, China's saving potential in its exports to Japan is 6.1 US$ billion and the saving potential for exports to India is 3.1 US$ billion. Japan's saving potential in its exports to China is 12.96 US$ billion and the saving potential of exports to India is 665 US$ million. India's saving potential in its exports to China is 699 US$ million and the saving potential of exports to Japan is 141 US$ million.

Figure 2 shows ten years of projections of each country's total saving potential of exports to the two other countries at a 100% utilization rate. In year 5, China's total saving potential is 3.4 US$ billion, Japan's total saving potential is 5.2 US$ billion and India's total saving potential is 247 US$ million. In year 10, China's total saving potential is 9.2 US$ billion, Japan's total saving potential is 13.6 US$ billion and India's total saving potential is 839 US$ million.

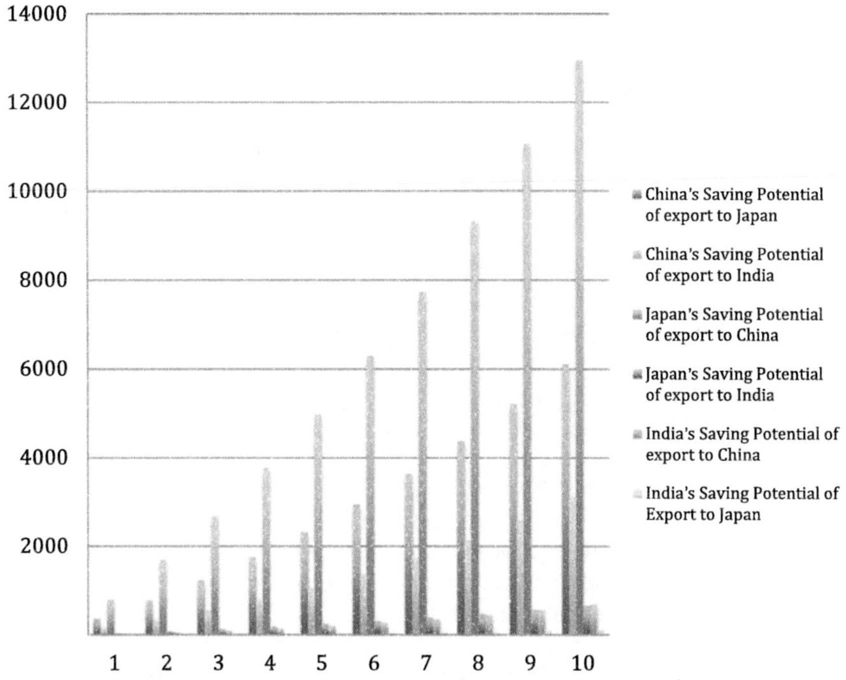

Fig. 1 Maximum saving potential, ten years' projection, at the 100% utilization rate (US$ million). Data source: UN Comtrade 2018, authors' calculations

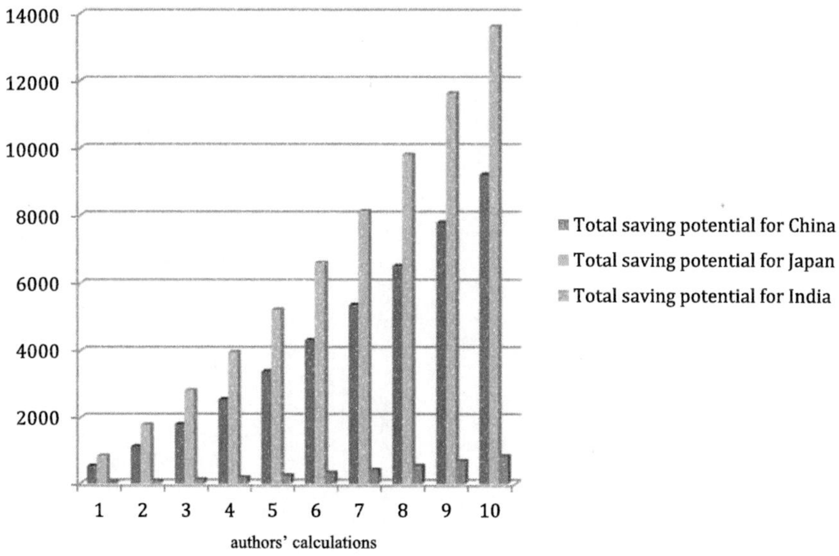

Fig. 2 Total saving potential, ten years' projection, at the 100% utilization rate (US$ million). Data source: UN Comtrade 2018, authors' calculations

Figure 3 shows a ten-year projection of the saving potential at a 66% utilization rate for China, India and Japan. In year 5, China's saving potential is 1.5 US$ billion in its exports to Japan and 701 US$ million in its exports to India. Japan's saving potential is 3.3 US$ billion in its exports to China and 166 US$ million in its exports to India. India's saving potential is 136 US$ million in its exports to China and 27 US$ million in its exports to Japan. In year 10, China's saving potential is 4 US$ billion in its exports to Japan and 2 US$ billion in its exports to India. Japan's saving potential is 8.6 US$ billion in its exports to China and 439 US$ million in its exports to India. India's saving potential is 461 US$ million in its exports to China and 93 US$ million in its exports to Japan.

Figure 4 shows the ten years' projection of each country's total saving potential in exports to the two other countries at a 66% utilization rate. In year 5, China's total saving potential is 2.2 US$ billion, Japan's total saving potential is 3.5 US$ billion and India's total saving potential is 163 US$ million. In year 10, China's total saving potential is 6.1 US$ billion, Japan's total saving potential is 9 US$ billion and India's total saving potential is 554 US$ million.

Figure 5 shows the ten years' projection of saving potential at a 33% utilization rate for China, India and Japan. In year 5, China's saving potential is 768 US$ million in its exports to Japan and 350 US$ million in its exports to India. Japan's saving potential is 1.6 US$ billion in its exports to China and 83 US$ million in its exports to India. India's saving potential is 68 US$ million in its exports to China and 14 US$ million in its exports to Japan. In year 10, China's saving potential is 2 US$ billion in its exports to Japan and 1 US$ billion in its exports to India. Japan's saving

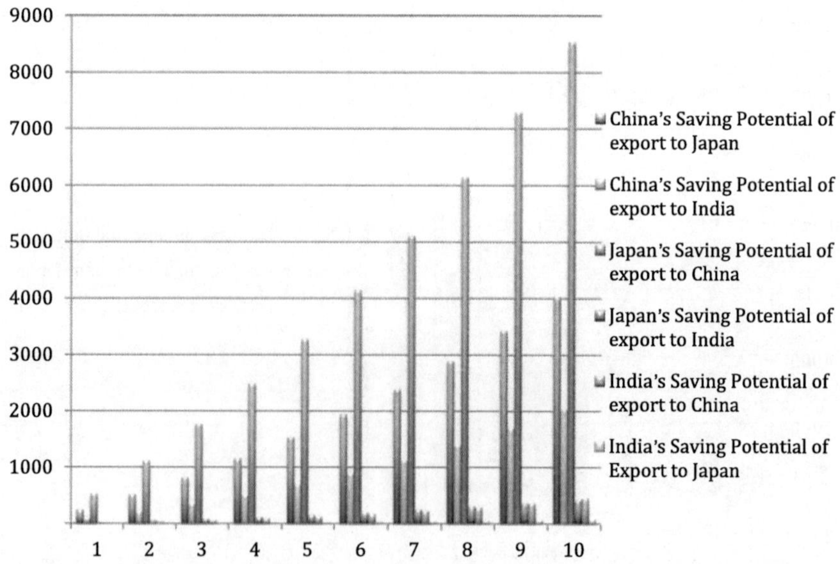

Fig. 3 Maximum saving potential, ten years' projection, at the 66% utilization rate (US$ million). Data source: UN Comtrade 2018, authors' calculations

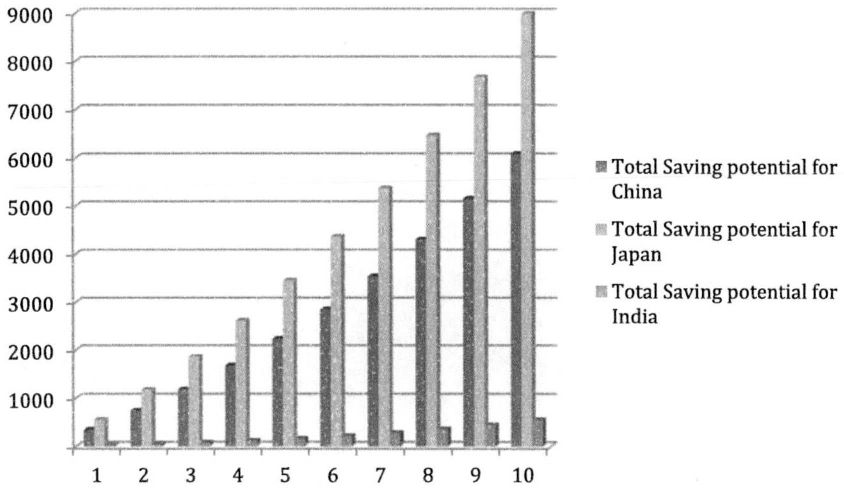

Fig. 4 Total saving potential, ten years' projection, at the 66% utilization rate (US$ million). Data source: UN Comtrade 2018, authors' calculations

potential is 4.3 US$ billion in its exports to China and 220 US$ million in its exports to India. India's saving potential is 231 US$ million in its exports to China and 46 US$ million in its exports to Japan.

Figure 6 shows the ten years' projection of each country's total saving potential in exports to the two other countries at a 33% utilization rate. In year 5, China's total saving potential is 1.1 US$ billion, Japan's total saving potential is 1.7 US$ billion

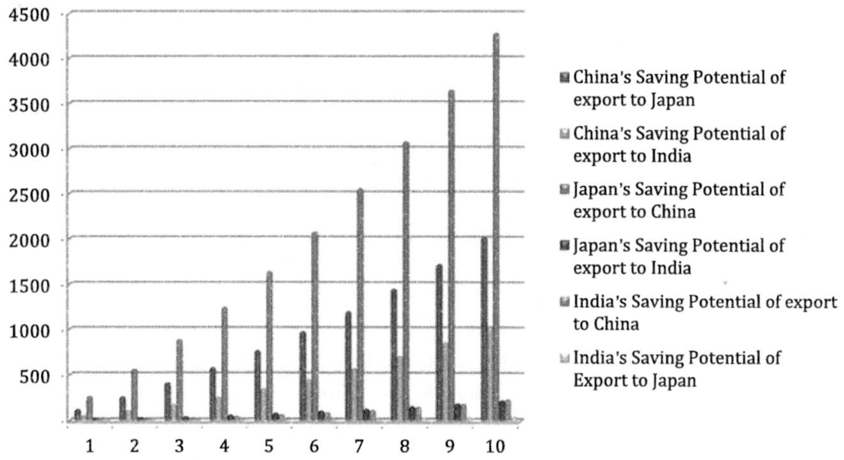

Fig. 5 Saving potential, ten years' projection, at the 33% utilization rate (US$ million). Data source: UN Comtrade 2018, authors' calculations

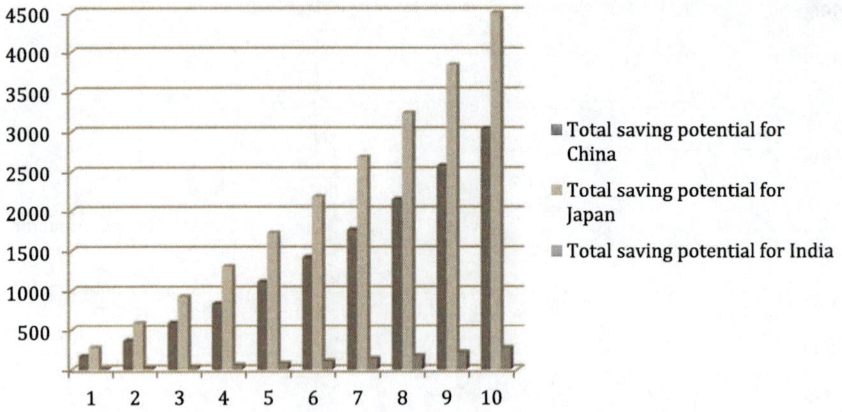

Fig. 6 Total saving potential, ten years' projection, at the 33% utilization rate (US$ million). Data source: UN Comtrade 2018, authors' calculations

and India's total saving potential is 81 US$ million. In year 10, China's total saving potential is 3 US$ billion, Japan's total saving potential is 4.5 US$ billion and India's total saving potential is 277 US$ million.

5 Conclusion

The RCEP negotiations began at the ASEAN summit in November 2012. The RCEP reached a 'substantial conclusion' with 20 chapters in November 2019 and was signed in November 2020. India decided to stay out of the RCEP formation and needs updates. The RCEP negotiations challenged finding balance and consensus among the participating countries. The existence of the three major power countries, China, India and Japan, in the RCEP formation is one of its main challenges. Many different countries' interests have been discussed on the RCEP negotiation table. Cooperation is possible among many different actors if there are common interests. The RCEP will be formed as an international institution with shared norms, rules, principles and decision-making procedures. The formation of the RCEP as an international institution is important to increase trust among the participating countries and achieve its objectives.

The RCEP would contribute to gains from trade for the participating countries. While there are many measurements to calculate the benefits of an FTA, this study focused on estimating the saving potential for the three major countries within the RCEP. China, India and Japan are the biggest trading partners among them. China is the biggest trading partner for Japan, accounting for US$ 141 billion (22.05% Japan's export partner share) in 2020. Japan is the third-biggest trading partner for China, accounting for US$ 142 billion (5.51% of China's export partner share). China is its second-biggest trading partner for India, accounting for US$ 19 billion (6.9% of India's export partner share).

The potential gain of FTA for Japan's exports to China is US$ 9 billion and the potential gain of FTA China's exports to Japan is 3.7 billion (2016 base year of calculation). The potential gain of FTA for India's exports to China is US$ 537 million and the potential gain of FTA for China's exports to India is US$ 3.1 billion (2017 base year of calculation). The potential gain of FTA for India's exports to Japan is US$ 108 million and the potential gain of FTA for Japan's exports to India is US$ 801 million (2017 base year of calculation). Japan's export value is double India's export value. It means that India has higher import tariffs than Japan.

This showed that Japan has the most significant saving potential of the major countries. Japan already has an FTA with India, and the existence of China within the RCEP will have significant benefits for Japan's exporters. As a dominant exporter to other RCEP countries, China will also benefit from the tariff reductions under the deal. India could have the potential gain from the trade deal with China if India would consider the importance of dismantling tariff barriers for production materials needed to produce some products such as the chemical and allied industries' cluster and machinery or electrical commodity cluster. It will enhance the competitiveness level of India's production.

References

Ahn, C. Y. (2020). *Chengdu Trilateral Summit still to deliver*. Retrieved February 4, 2020, from https://www.eastasiaforum.org/2020/02/11/chengdu-trilateral-summit-still-to-deliver/

ASEAN Secretariat. (2020). *ASEAN hits historic milestone with signing of RCEP*. Retrieved September 1, 2022, from https://asean.org/asean-hits-historic-milestone-with-signing-of-rcep/

Dent, C. M. (2016a). East Asian regionalism. *East Asian Regionalism*. https://doi.org/10.4324/9781315717258

Dent, C. M. (2016b). Trade diplomacy and East Asian regionalism. In *East Asian Regionalism* (pp. 174–218). Routledge.

Kalirajan, K. (2022). Will free trade agreement between India and China reduce India's trade deficit? *Journal of Social and Economic Development*. https://doi.org/10.1007/s40847-022-00203-2

Keohane, R. O. (1983). Cooperation and international regimes. *After Hegemony* (pp. 49–64). Princenton University Press.

Keohane, R. O. (1989). Neoliberal institutionalism: A perspective on world politics. In *International institutions and state power* (pp. 1–20). Westview Press.

Ministry of Foreign Affairs of Japan. (2019). *Regional comprehensive economic partnership (RCEP)*. Retrieved February 21, 2020, from https://www.mofa.go.jp/policy/economy/page2e_000001.html

UN Comtrade. (2019). *UN Comtrade*. Retrieved December 20, 2019, from https://comtrade.un.org/data

Wardani, R. Y. (2020). India's readiness for the regional comprehensive economic partnership through the India–Japan partnership. *Management and Economics Research Journal*, *6*(2), 1–15. https://doi.org/10.18639/MERJ.2020.9900017

Wardani, R. Y. (2021). The interests of China, India, and Japan in the negotiations of the Regional Comprehensive Economic Partnership: Theoretical perspectives. *International Social Science Journal*. https://doi.org/10.1111/issj.12272

Wardani, R. Y., & Cooray, N. S. (2019a). Saving potential of regional comprehensive economic partnership (RCEP): Implication for China and Japan. *Journal of Economic Info*, *6*(1), 34–42. https://doi.org/10.31580/jei.v6i1.122

Wardani, R. Y., & Cooray, N. S. (2019b). The savings potential of Sino-Indian free trade agreement within regional comprehensive economic partnership initiatives. *Journal of Reviews on Global Economics, 8*, 739–754. https://doi.org/10.6000/1929-7092.2019.08.64

Watson, M. (2008). Theoretical traditions in global political economy. In J. Ravenhill (Ed.), *Global political economy* (2nd ed., pp. 27–66). Oxford University Press.

WITS. (2022). *World integrated trade solution.* Retrieved August 24, 2022, from https://wits.worldbank.org

World Trade Organization. (2018). *Trade and tariff data.* Retrieved October 20, 2018, from https://www.wto.org/english/res_e/statis_e/statis_e.htm

Ziltener, P. (2016). *The potential of free trade agreements with ASEAN countries and BRAZIL for Swiss exporters.* Zurich. Retrieved from https://www.s-ge.com/sites/default/files/cserver/static/downloads/potential-fta-asean-countries-brazil-swiss-exporters-s-ge-2016.pdf

Economic Zone Development Model of Some Countries in the World and Implications for Vietnam

Khanh Cuong Le and Thi Thuong Vu

Abstract This study has studied the economic zone development models of some countries in the world and pointed out the limitations and weaknesses in the development of economic zones in Vietnam in recent years. In addition, the study has given policy implications for Vietnam to innovate the economic zone development model, such as renovating the economic zone development model toward synchronous investment attraction, exploiting resources from public–private partnership, greening production, developing tourism-class services, modernizing the administration, and privatizing governance. The research, experience acquisition, and theoretical application to continue building, developing and renovating the economic zone development model, contributing to the implementation of the guidelines and policies of the Party and State of Vietnam on economic renewal could help to create motivation for localities to grow and develop their economies quickly and sustainably in the coming time.

Keywords Development model of economic zones · Special economic zones · Vietnam

1 Introduction

Vietnam is still in the early stages of economic zone construction and development. Many countries and territories such as China, Korea, the United Arab Emirates, Malaysia, and Singapore have successfully implemented and applied many economic zone models such as special economic zones, free economic zones, and free trade zones. Economic zones are built with the same purpose of attracting resources (modern technology, capital, high-quality human resources, and development ideas, etc.) from the outside in order to create growth poles, activate economic zones, and develop economies. Economic zone is also a place to test new institutions, mechanisms, and policies that are expected to create breakthroughs. The development of

K. C. Le (✉) · T. T. Vu
VNU University of Economics and Business, Hanoi, Vietnam
e-mail: cuonglk@vnu.edu.vn

A. T. Nguyen et al. (eds.), *Contemporary Economic Issues in Asian Countries: Proceeding of CEIAC 2022, Volume 1*,
https://doi.org/10.1007/978-981-19-9669-6_28

economic zones has been effective and is having positive effects, expanding international economic exchanges, promoting economic growth, and creating changes in social life and economic development in different economic zones in Vietnam. However, the construction of the economic zone models across the country in general has not achieved the expected results. The common difficulties and limitations are as follows: (1) the management model is still inadequate; the development model has not achieved the desired effect; (2) low investment quantity and quality; the number of strategic investors, investment capital, technology quality and human resource quality are not high; (3) there are still many problems and inadequacies in terms of site clearance and infrastructure; (4) lack of value chain linkages, industry linkages, and regional linkages; (5) the mechanisms and policies still have problems and need be improved further, etc. Therefore, studying the experiences of some countries in the world on building and renewing the economic zone development models will partly provide policy-makers with suggestions to flexibly apply to each locality to build an economic zone model suitable for each locality, to promote the strength and potential of each locality, to help the locality develop sustainably, meeting the goals of socioeconomic development.

2 Literature Reviews

2.1 Economic Zones and the Role of Economic Zones in the Socioeconomic Development of a Locality

2.1.1 Definition of Economic Zone

The term "Economic zone" appeared in the late 70s of the twentieth century along with the views of economists around the world on the effect of economic concentration and new geo-economics. Until now, there have been many different definitions and ways of calling economic zones, but the international common name is "Economic zones". In a broad sense, an economic zone is a defined area where special economic policies are applied. More specifically, an economic zone is an area with a defined geographical boundary, a separate economic space, and special policies applied to attract investment capital, technology, management knowledge, and other resources for economic development with a structure of industries with certain advantages, high openness, and a focus on external economy (Anh, 2016).

From a development perspective, the economic zone presents the goal of creating a breakthrough, creating a driving force to promote the development of the region through attracting financial, human, scientific and technological resources, and investment, etc., by applying different measures on special mechanisms and policies. From a management perspective, economic zones are a form of organization in the direction of concentration and specialization, showing the basic characteristics of the organization of industrial production by territory (Porter, 2008).

From the above interpretations, "economic zone" can be understood as an area with a separate economic space, defined geographical boundaries, certain advantages in terms of geographical location, favorable investment and business environment for investors, is allowed to be built and developed by the State, operated by a favorable and open legal framework according to international practices.

2.1.2 The Role of Economic Zones in Socioeconomic Development in a Locality

(1) Attracting investment, especially foreign direct investment; (2) Developing infrastructure; (3) Supporting a broader economic reform strategy; (4) Playing the role of a "laboratory" for new policies and approaches, creating a driving force to promote the industrialization, modernization, and transformation of the local economic structure; (5) Relieving a part of the pressure of population growth and job demand.

It is shown that economic zones (EZs) have become a very important factor for the development of the economies, are large openings to attract external resources and create outstanding growth points with strong pervasive power. The EZs were established with the main objective of stimulating local production resources and attracting production resources from outside. It is the nucleus and driving force of socioeconomic development at the provincial or regional scale in the national development strategy. The objectives of the development of these EZs are to attract investment, create jobs, increase foreign currency earnings, develop exports, upgrade domestic technology, transfer technology, and develop modern technologies, provide supports to underdeveloped regions, and even help to kick start the entire economy.

The development of EZs is not only for the economic reasons of that particular area, but has much greater significance. It is the common interest of the whole economy in terms of job creation, attraction of development resources, expansion of exports, transfer of technology and management knowledge, and integrated regional development and industrialization–modernization, etc.

In the process of globalization, EZs act as a stepping stone between the protectionist economy and the free market economy. The incentives and autonomy of EZs develop and expand in the direction of increasing liberalization, reducing State management, increasing competition, and developing market economic relations further at the national, regional, and international levels.

At a regional scale, EZs create spillover effects, so the planned development of EZs helps in the uniform development of the economy by region, because they can promote comparative initial advantages (static advantage) and be further enhanced by policy advantages (dynamic advantage) as the leverage for the selected area to become an area with outstanding resilience.

2.1.3 Some Initial Selection Criteria to Make a Proposal to Establish an Economic Zone

- Geostrategic location: border-gate area, seaport, airport, or near large consumer market, etc., in order to minimize investment and operating costs for businesses. The location in the coastal area has great potential for the development of tourism, services, processing industry, manufacturing industry, and fishery logistics, etc.
- Having convenient infrastructure, such as road system, national power grid system, water supply and drainage system, information and telecommunications system, and shipping system.
- Having conditions for the development of resources such as abundant and diverse human resources; advantages in financial mobilization, access to science and technology, training and development of human resources.
- Having conditions or potential to develop social infrastructure that is attractive to investors, such as a system of hotels, supermarkets, and different services including education, health care, amusement, and entertainment services reaching international standards, as well as financial services.
- The investment and operation of economic activities such as investment, production, and business are entitled to a number of preferential mechanisms and policies through the regulations of the governments, and regulations and commitments of local authorities.

The above conditions must be considered in terms of the present and the maintenance of that possibility in the future.

2.2 *Economic Zone Development Model in a Locality*

When considering the model in general, researchers have shown that the model is a specific object, system, or an original alternative concept; a structure or a diagram that describes an image that is simplified according to the characteristics or behavior of an object, concept, or system. One of the purposes of the model is to serve the research and experiment to find out the operating rules of the original (original) object or system (Hien, 2006).

Currently, there is no unified understanding of the economic zone development model; at the same time, there has not been a complete study on the design of economic zone development model. According to the specialized approach to development economics, studying the methods of using resources to meet development goals, the *economic zone development model* is understood as the economic entity model, showing the systematic relationship between the factors affecting the development of the economic zone, the elements constituting the economic zone and the economic zone development objectives.

The selection of an appropriate development model first comes from the economic zone development goal. The objectives can be listed, such as to promote the potential and strengths of the locality; meeting the trend is also the requirement of "green growth" and "greening" in the production for the economy; to participate actively and comprehensively in order to exploit the benefits of globalization, international economic integration and the industrial revolution 4.0, etc. Depending on the level of organization, development scale and conditions influencing factors (such as natural conditions, socioeconomic institutions, economy, society, environment, human resources, and science and technology, etc.) to apply economic zone development models appropriately or implement the innovation of the development model in order to attain a more efficient, progressive, and sustainable model. Each economic zone development model can correspond to a separate development stage, following the trend of development from reciprocal to counterbalanced, from passive to active in order to bring into play all static and dynamic competitive advantages of the locality and the country during the integration and development. Thus, the economic development model differs from the pure economic model in the sense that there is a definite goal, which determines the elements of the development model.

Provincial locality is a concept related to territorial management, which is a fundamental issue for countries. The organization of administrative boundaries for territorial management is associated with the organization of the apparatus, human resources, legal framework, and management mechanism, etc. The construction and development of the economic zone model not only depends on the proposal of the locality at the provincial level, but also considers the inter-sectoral, inter-regional (provincial localities together), and even the national levels (Table 1).

Table 1 Approaches to economic zone models

Economic zone models			
By forms	By fields	By space	By management forms
• Special economic zones • Tax-suspension zones • Manufacturing zones • Free (open) economic zones • Coastal economic zones • Border-gate economic zones	• Commercial economic zones • Industrial economic zones • General economic zones	• Linear model • Propeller model • Spreading model	• State management • Enterprise management • Public–private partnership management

(*Source* Author's compilation through literature review)

3 Experiences with Economic Zone Development Model of Some Countries in the World and Limitations and Weeknesses of Economic Zone Development in Vietnam

3.1 Experiences in Developing Economic Zones of Some Countries in the World

In the current literature, there are at least 20 different terms that describe the economic zones, such as free trade zones, free ports, export processing zones, and special economic zones. These zones exist in different forms with different names by country and by periods. In addition to these models, some countries have also built models of special administrative zones such as Hong Kong and Macau in China, Jeju in Korea, and Trento in Italy. In fact, the economy of free economic zones, special economic zones, and special administrative zones, etc., in some countries has a relatively fast and sustainable growth rate thanks to special regulations and mechanisms (Table 2).

3.2 Some Factors Contributing to the Success of Economic Zones in the World

Recent international studies and development experiences in the world over the past 30 years show that the success of EZs in some countries is often associated with the following factors:

Firstly, the geo-economic, political-strategic, characterized, and special locations and favorable natural conditions (coastal areas, border areas, poor but potential areas, in addition to the developed economy; capable of attracting human resources and connecting internationally). Most of the successful zones are in strategic locations, typically the Shenzhen Economic Zone (China) is located adjacent to Hong Kong—a developed economy, which has created favorable conditions for Shenzhen to strongly absorb investment capital from Hong Kong; behind Shenzhen is a large and potential market with more than a billion people; being located away from the cities that are the centers of the planned economy made it easier for Shenzhen to adopt and transition to a market economy.

Secondly, the determination to innovate, the reform mindset, the courage to go bold, the courage to experiment, and the strong political commitment of the country's top leaders are of importance. The experience in building and developing Shenzhen EZ, China, is one of the typical examples, demonstrating the determination to innovate and the reform ideas; demonstrating the correctness and success of the EZ policy: The central government gave the policy; the local government had to implement it, then report the results.

Table 2 Some experiences in the management, institutions, and policies for foreign economic zone development in some countries in the world

Economic zone	Management model	Institutions and policies
China (Shenzhen)	– The government of the special zone is in line with that of a city in Guangdong province – The administrative apparatus of the special zone has been streamlined, leaving only 4 agencies (economic development, commercial development, transportation, and agriculture) – Centralized state management focuses on planning, separating from the business function. The government of the special zone only regulates macro-issues	– Exemption of import and export taxes for goods imported and exported – Corporate income tax, personal income tax, and value added tax, etc., are all lower than domestic taxes (about 15%) – High-tech industries are exempted from corporate income tax (CIT) for 2 years and entitled to a 50% reduction of CIT for the next 8 years. Newly established enterprises are entitled to a 50% reduction in land rent. High-tech enterprises are exempted from property tax for 5 years – Subsidies for research, development, and training for the creation of highly qualified workers – Individuals and organizations may lease land for a term of no more than 70 years. Preferential land rental price, equal to only 30–50% of domestic land rent

(continued)

Table 2 (continued)

Economic zone	Management model	Institutions and policies
China (Hong Kong)	– The government is responsible for defense and diplomacy – Characterized in high autonomy with a degree of liberalization that is clearly superior to China's inland and to the institution of the United Kingdom before. The government merely intervenes in market activities and civil society	– Preferential policies: low and effective tax rates, no duplication of tax collection – Providing foreign investors with the freedom to choose investment categories, move capital and profits; import and export; ownership and business; hire and fire employees, etc – The CIT rate is 17–18%; capital use tax is not collected; progressive tax is not applied; import and export taxes are exempted except for some special products. All goods transferred through the custom border must go through customs procedures, but are not entitle to taxes – Land lease rights are allowed to be transferred at preferential prices to the investors
Cayman Islands (United Kingdom)	– Overseas territories of the United Kingdom. Has complete autonomy, except for two fields, namely defense and diplomacy (run by the United Kingdom) – Having its own administrative apparatus, with legislative and executive independence, headed by a governor appointed by the Queen	– Financial center – Tax heaven. No direct taxation (no profit tax, income tax, or corporate tax) – Having its own currency circulating in the archipelago

(continued)

Table 2 (continued)

Economic zone	Management model	Institutions and policies
Jeju (Korea)	– A free international city under the direct control of the Central Government – The Government established a 30-member City Support Committee chaired by the Prime Minister – The operation pf the city is managed by its Local Council, headed by the Governor, with full powers except defense and diplomacy (decentralization of 1,336 jurisdictions)	– Domestic investment projects are exempted from CIT for 3 years (foreign investors are exempted for 5 years), entitling to a 50% tax reduction for 2 years – Projects are entitled to lease land for 50–100 years (can be extended), exempted from real estate tax for 10–15 years, entitled to a 50–100% reduction of land rent, and exempted from import and export taxes – In addition, businesses also enjoy many incentives and support when investing in culture, information technology, knowledge industry, and high technology such as a subsidy of 10% of the cost of building factories, a subsidy of 50% of land rent, and supports in labor training
United Arab Emirates	– The UAE's free zones are built and owned by the government – At the ministerial level, the Ministry of Finance and the Ministry of Industry are responsible for the formulation and implementation of industrial development plans – Government corporations have legal ownership of free zones. For the zones attached to ports or airports, the investor of zone infrastructure is usually the Port Authority	– UAE has the most competitive incentives in the world, with the CIT, personal income, and customs tax rates of 0%; no import and export quota; and no controls on foreign exchange; – 100% of capital and profits are repatriated without paying any taxes or fees; – No restriction on hiring foreign workers; – Reasonable land rent with lease extension for a long time; 30% reduction in costs for technology companies – Allowing foreign companies to have full ownership of real estate since 2002. Investors are required to rent land in free zones

(*Source* Author's compilation through literature review)

Third, new institutions and mechanism must be boldly piloted, not trying to be perfect, being implemented and improved at the same time. It is important to allow local authorities to be active in their implementation, to continue to replicate what can be done, and to find solution to improve weaknesses, and not strive to obtain perfection.

Fourth, there must be an outstanding, strong and stable institution, and be given a high degree of autonomy. Most of the EZs in the world are developed and managed by their own laws, for example: Korea has a special law on the establishment and development of free economic zones and a special law applies to the free international city of Jeju; the Hong Kong special zone applies the Basic Law; India, Philippines, Singapore, Japan, Malaysia, Laos, and Poland, etc., all have their own laws for economic zones, free economic zones, and free trade zones.

Fifth, a global competition policy mechanism is applied. One of the factors that determines the success of the free zones of the United Arab Emirates (UAE) is the most competitive incentives in the world, namely 0% income tax; 0% corporate income tax; 0% customs duty; no quota; no controls on foreign exchange; 100% foreign ownership; and 100% of capital repatriated without paying any taxes or fees.

Sixth, low input costs and flexible regulations on employment (salary, payroll, recruitment, and dismissal, etc.*).* In Dubai, there are no restrictions on foreign workers, and properly handled disputes can be made anywhere without the presence of a government agency representative. In the strategy of developing the economic zone model, in order to catch up with thousands of businesses looking for new business locations due to the cost of production in China in particular and Asia in general, the South African Government decided to set the low labor cost level and introduce new mechanisms and policies to develop industries that employ large numbers of low-skilled workers.

Seventh, there is initial support from the Government for infrastructure and human resource development; Close cooperation between the government and the private sector, in the form of "public leadership, private governance". The governments, including Korea, United Arab Emirates, and Malaysia, etc., spend separate budgets to support human resource training for EZs; currently, the operation and ownership of economic zones around the world has gradually shifted from public to private; this is the result of the realization that the zones can be very profitable when an operator understands the elements of regional governance, taking advantage of economies of scale.

Eighth, the administrative apparatus is streamlined and efficient; administrative procedures are simple, transparent, and public. Strategic solutions to possible challenges are provided timely: Most successful EZs have a government-authorized and site-based one-door unit to quickly proceed the procedures, and to support and accompany enterprises. One of the factors that contribute to Dubai's attractive investment environment is the establishment of e-government with strategic projects such as "e-citizen", "e-worker", "e-library", and "Electronics for All". Therefore, the administrative procedures here are solved quickly, for example, it takes less than an hour to gain a visa.

Ninth, identify the type and development goals of the EZ clearly and target specific sectors and strategic partners. Poland identifies strategic investors for the EZs, including the US, Germany, Japan, and Italy, thereby planning the EZs oriented to attract the world's famous automobile manufacturing groups.

Besides the successes, there are also some factors behind the failures of EZs in the world, as follows: (1) The policies and privileges in the EZs are limited; (2) The government is not proactive in creating favorable conditions for investors through the construction of essential infrastructure for EZs; (3) Complex and complicated administrative procedures; very high infrastructure costs (such as transport, electricity, and water), and restrictive labor regulations; (3) The administrative organization model is not streamlined and superior; (4) The government is not proactive in creating favorable conditions for investors by building essential infrastructure for EZs and often waits (in vain) to find financial companies who would invest in electricity, water, and telecommunications systems inside the EZ; (5) Companies also face complex and bureaucratic regulations and procedures, very high costs for infrastructure (such as transportation, electricity, and water) and labor restrictions. For example, India (EZs were established widely) (up to 2010, 577 EZs were established; currently, only 124 EZs remain in operation; economic and administrative institutions do not have any clear superiority); South Africa (lack of active government involvement, lack of a comprehensive policy framework leading to weaknesses in governance, planning, and management), etc.

3.3 Some Initial Successes and the Limitations and Weaknesses in Economic Zone Development in Vietnam in Recent Years

3.3.1 Forming a Legal Basis for the Establishment and Development of Economic Zones in Vietnam

Vietnam is a country that has diverse topography and ecology although the natural area is not large; having deltas, mountains and seas. At the same time, Vietnam also shares borders with a number of other countries with a long border. The mountainous region has many border gates, while the sea and coastal areas have many large ports, which are convenient for domestic and international economic and trade exchanges. In the context of globalization, these regions have been and are being paid great attention by the Communist Party and State as well as border and coastal provinces for the formation of economic zones to exploit the available potentials and create motivation for the local and national socioeconomic development. In recent years, there have been many documents issued by the Party, National Assembly, Government, ministries, and provinces related to the planning, construction, operation, and management of economic zones in the border and coastal areas. Some relevant documents can be listed our such as:

For border-gate economic zones: Resolution 470/NQ-UBTVQH13 on the supervision results of policies and laws on the construction and development of economic zones and border-gate economic zones; Decision 72/QD-TTg dated November 26, 2013, of the Prime Minister on financial mechanisms and policies for border-gate economic zones; Decision No. 52/2008/QD-TTg dated April 25, 2008, approving the Master plan on development of border-gate economic zones in Vietnam up to 2020; and Decision No. 1531/QD-TTg of the Prime Minister approving the Project "Reviewing and adjusting the planning for the development of border-gate economic zones in Vietnam up to 2020 with a vision to 2030", etc.

For marine economic zones: A policy framework for the development of Vietnam's marine economic zones has been developed. Specifically, at the 4th Conference of the 8th Central Committee of the Party (1997) mentioned the piloting of special economic zones and coastal free trade zones in localities. At the same time, it was further clarified in the 4th Conference of the Central Executive Committee, term X. Then, there were a number of other relevant documents on the development of the marine economic zone such as the guidance in the Notice of the Politburo No. 79-TB/TW; Notice No. 155-TB/TW of the Central Committee; Resolution No. 39-NQ/TW of the Politburo dated August 16, 2004; and Decision No. 148/2004/QD-TTg of the Prime Minister, etc.; all of which agreed to develop marine economy in general and marine economic zone in particular, etc. These documents can help to summarize the stages of implementing, building, and developing economic zones models in Vietnam (as shown in Table 3).

Table 3 Development process of economic zone models in Vietnam

Period	Progress
1991–1994	In the first stage of the Doi Moi (1991–1994), in order to attract foreign investment, promote export activities, and create a premise for the integration of Vietnam's economy with the world, the models of export processing zones (EPZs) were formed with the establishment of Tan Thuan EPZ in 1991
1994–1997	Forming industrial zones (IZs) and converting some EPZs into IZs to promote investment attraction, diversify and develop industries, especially light industries, toward export orientation
1997–2003	Forming high-tech zones, piloting and establishing border-gate economic zones with the establishment of Mong Cai border-gate economic zone in 1996 and Hoa Lac high-tech zone in 1998
From 2003 to present	Piloting the implementation of an open economic zone and developing coastal economic zones to create driving force zones for economic development, creating conditions for the development of heavy industries in coastal areas. At the same time, in this period, in order to promote the application of high technology in a number of important fields such as information technology, agriculture, new models such as concentrated information technology zones, and high-tech agricultural zones, etc., have been established

(*Source* Ministry of Planning and Investment, 2018)

Up to now, nationwide, there have been 16 coastal economic zones, 28 border-gate economic zones have been established. The current operation of economic zones shows the following main contents:

– *Regarding the planning and establishment of economic zones: (1) Border-gate economic zones:* according to the Vietnam's economic zone development planning up to 2020 approved by the Prime Minister in Decision No. 52/2008/QD-TTg dated April 25, 2008, the whole country had 28 SEZs with a total area of more than 660 thousand hectares in 21 out of 25 in-land border provinces; (2) *Coastal economic zones*: since the establishment of the first coastal economic zone, Chu Lai Open Economic Zone, in 2003, by the end of 2018, 16 economic zones had been established, including 2 EZs in the River Delta region, namely Van Don (Quang Ninh province) and Dinh Vu–Cat Hai (Hai Phong city); 11 EZs in the Central Coast, namely Nghi Son (Thanh Hoa province), Southeast Nghe An (Nghe An province), Vung Ang (Ha Tinh province), Hon La (Quang Binh province), Chan May-Lang Co (province). Thua Thien Hue), Chu Lai (Quang Nam province), Dung Quat (Quang Ngai province), Nhon Hoi (Binh Dinh province), Van Phong (Khanh Hoa province), South Phu Yen (Phu Yen province), and Southeast Quang Tri (Quang Tri province); and 3 economic zones in the South, namely Phu Quoc island economic zone and Nam An Thoi island cluster (Kien Giang province), Dinh An (Tra Vinh province), and Nam Can island (Ca Mau province).

3.3.2 About the Economic Zone Development Models in Vietnam

Vietnam's economic zone models are aimed at developing different industries and fields, but the competitive advantages of the models are built on the principle of promoting the advantages of scale (economics of scale) into concentrated production zones (clusters).

Regarding the target-based development model, economic zones in the country are determined to exploit to the fullest the potential and strengths of the locality, and at the same time, create a growth pole as a driving force for local development at the provincial and regional levels. The advantage of this development model is that it is possible to bring into play the internal potentials of each region for socioeconomic development, creating dynamic general economic areas with widespread influence to other regions. This model will be highly effective when geo-economic advantages are realized by important infrastructure projects and works. In fact, economic zones with concentrated investment capital for infrastructure development such as Dung Quat, Chu Lai, and Nghi Son have attracted large-scale investment projects and have a better development rate.

The industry-specific economic zone development model of the EZs in our country is currently based on multi-sectoral development, with a focus on industry and services. The development of the economic zones is carried out mainly through promoting the advantages of geo-economics (having a seaport, a road transport system connected to convenient key economic zones, having favorable conditions to attract and implement dynamic economic projects). This model has been concretized in Decree No. 29/2008/ND-CP of the Government, regulations on the operation of each economic zone. At present, the economic zones of Dung Quat, Nghi Son, Van Phong, Vung Ang, Dinh Vu–Cat Hai, Southeast Nghe An, Hon La, Chan May–Lang Co, Nhon Hoi, Chu Lai, Dinh An, and South Phu Yen are following this model. Some other economic zones are oriented to develop based on tourism and services such as Van Don economic zone, Phu Quoc Island, and Nam An Thoi island cluster.

Regarding the management model of economic zones: The management model of EZs is unified from the central to local levels, in the direction of increasing decentralization and authorization among ministries, central branches, and provincial People's Committees for the Management Boards of economic zones, simplifying administrative procedures in the state management of economic zones in the direction of "one-door, on-site". Specifically: (i) The government performs the unified state management of economic zones nationwide, directs the formulation and implementation of development master plans and plans, and promulgates policies and legal documents on EZs; (ii) The Prime Minister directs Ministries, provincial People's Committees and EZ Management Boards to implement laws and policies on EZs; approves the master plan on economic zone development; makes decisions on the establishment of an economic zone; and approves the general planning on construction of economic zones; (iii) Ministries, branches, and provincial-level People's Committees, within the ambit of their functions, tasks and powers, are responsible for performing the function of state management of the sectors, fields, and territorial administration for EZs; guide or authorize the management boards of the EZs to organize the performance of the function of providing public administrative services and other supporting services related to investment and production and business activities for investors in the EZs; (iv) The organizational structure of the Economic Zone Management Board is to be established by the Prime Minister's decision; directly under the direction and management of the Provincial People's Committee in terms of organization, staffing, program, work plan, and operating budget; is subject to the direction, inspection, and professional guidance of the concerned ministries and branches. The EZ Management Board has legal status, has an account and a seal bearing the national emblem (Table 4).

3.3.3 Initial Successes, Limitations and Weaknesses in the Formation and Development of Economic Zones in Vietnam

a. *Initial successes*: A number of border-gate economic zones have been established across the country so far. First of all, there are border-gate economic zones such as Mong Cai (Quang Ninh), Moc Bai (Tay Ninh), Lao Bao (Quang Tri), Lao Cai (Lao Cai), Cau Treo (Ha Tinh), and Dong Dang (Lang Son). In addition, the important coastal economic zones can be listed out such as Van Don (Quang Ninh), Dinh Vu–Cat Hai (Hai Phong), Nghi Son (Thanh Hoa), Vung Ang (Ha Tinh), Chan May-Lang Co (Thua Thien Hue), Chu Lai (Quang Nam), Dung Quat (Quang Ngai), and Van Phong (Khanh Hoa).

Table 4 Three-level management model for economic zones in Vietnam

Management agency	Competences
Central level	The government promulgates mechanisms and policies applicable to EZs. The Prime Minister makes decision on the establishment, expansion, and promulgation of regulations on operation of economic zones. Ministries, according to their functions and tasks, advise the government and the Prime Minister to exercise the abovementioned powers
Local level	Provincial People's Committees are responsible for the state management of economic zones in the area, performing a number of direct tasks such as: implementing the economic zone development planning approved by the Prime Minister and approving the detailed planning of EZs
Direct management agency	They are the economic zone management board of coastal economic zones and border-gate economic zones; and management board of the high-tech economic zones. These agencies are decentralized and authorized by the Provincial People's Committee, specialized departments and branches to carry out the state management functions and tasks for EZs

(*Source* Ministry of Planning and Investment, 2010)

According to information at the Vietnam Industrial Park Forum 2022 with the theme "Opening new investment capital sources", the country currently has 26 border-gate economic zones and 18 coastal economic zones established; at the same time, there are 403 industrial zones. The system of industrial zones, border-gate and coastal economic zones located in 61 out of 63 provinces and cities across the country has attracted 10,000 projects with domestic investment capital and over 11,000 valid projects with foreign direct investment capital with a total registered capital of 340 billion USD, of which FDI capital is about 230 billion USD. Many EZs have been built and put into operation, bringing positive socioeconomic impacts not only in the localities where the EZs are located but also spreading to the surrounding areas and the whole country. The EZs revive the socioeconomic life in the region, contribute to the transformation of the local economic structure, increase the budget revenue; create jobs to increase income, improve the living standards of residents in the EZ and surrounding areas; and, at the same time, make an important contribution to ensuring national security. In addition, the economic zones which have been built and put into operation have led to the development of economic-technical infrastructure of the localities such as traffic, electricity, information and communication, and associated services for the residents, etc. Border-gate or coastal economic zones also create a new look for the areas that used to be difficult, such as remote areas, border areas, or coastal areas.

b. *Limitations and weaknesses*: (1) The EZs lack strategic planning, so they are spread out, divided, fragmented and slow in determining the priority order, so many zones operate at low efficiency; (2) Infrastructure is still insufficient, not synchronous, and does not meet development requirements; (3) The EZs have not attracted strategic investors in science, technology, finance, and management level; (4) Human resources have not responded in time to the requirements of investors; (5) The investment environment has not been really attractive; the administrative procedures are complex.

Causes: In addition to the positive results, the development of economic zones has many limitations due to: (1) The fundamental contradiction between the development of production relations has not been resolved (institutions, mechanisms, and policies, etc., are limited) with the development of productive forces (resources, human resources, and the potential strength in mineral resources, etc., are abundant); (2) Innovation is not synchronized, only focusing on economic innovation, not focusing on administrative reform and organizational arrangement; (3) Institutions are not strong enough (the establishment and operation of economic zones and border-gate economic zones are only adjusted by Decree No. 29/2008/ND-CP, there has been any law regulating them); (4) The mechanisms and policies to attract all resources to develop infrastructure and human resources, especially external resources, are not competitive enough at regional and international levels; (5) The organizational model of the apparatus is still overlapping, with may difficulties in the coordination, being significant barriers (EZs located entirely in an administrative unit such as Van Don, Mong Cai, and Phu Quoc. have two economic management apparatus, People's Committee and Management Board of EZs); (6) The State's preferential and support policies for EZs are still scattered, lacking in concentration and a focus; (7) Some zones were established when they did not really fully meet the necessary factors for development, such as investment and business environment, living and learning conditions, educational levels, expertise, and foreign language qualifications, etc.; (8) Infrastructure and human resources are both inadequate and weak, especially transport infrastructure and high-quality human resources (Table 5).

4 Some Recommendations to Innovate the Economic Zone Development Models in Vietnam

In order to successfully build an EZ, it is crucial to attract development investment resources. To do this, it is of importance to have attractive mechanisms and policies to attract investors, especially large, potential, and experienced investors. Below are the policy recommendations for Vietnam to renew the economic zone development models in the coming time:

(1) Build and perfect the national software system for monitoring economic zones; thereby providing state management agencies with adequate, timely, and necessary information to carry out the state management process, and, at the same time, convenient information for the exploitation and use which promptly serves the direction, administration, and state management of the government and competent state agencies as well as meets the information needs of agencies, organizations, and individuals regarding economic zones.

Table 5 Factors affecting success or failure in the process of developing the economic zone models

Factors	Success	Failure
Institutions	Institutions are strong enough, promulgating specific preferential policies, including investment in the development of technical and social infrastructure; tax incentives, competitive fees; special investment incentives for foreign investors in the industries and trades prioritized for development	The policies are not competitive, do not contain outstanding incentives, or mainly consist of tax, rigid, and complicated regulations, with inconsistent implementation
Leadership and orientation	The leadership is synchronous and decisive. There is a consistent policy path. The development strategy and goals are clear and aimed at the priority industries and occupations with comparative advantages	Lack of long-term supportive political commitment. The state intervenes too deeply, aiming at political interests rather than sustainable competitive advantages
Organizational apparatus	The administrative management apparatus is streamlined, effective and efficient, being decentralized, and strongly autonomous; the management staff has professional qualifications. The apparatus directs timely, solves difficulties, decides major and interdisciplinary issues	The management model and organization apparatus are bulky. Management board lacks capacities, resources, or authority. The decentralization is not appropriate, and the management is overlapping
Natural conditions	Strategic geo-economic location (near important traffic routes, easy connection to the region and the world), adjacent to economic development areas, capable of attracting investment, especially foreign investment and high-quality human resources	Located in remote areas, no domestic-foreign connection, and it is expensive to invest in the construction of economic zones and infrastructure. Natural conditions are not favorable
Developing and implementing plans	The planning is modern, synchronous, and unified between the master plan and the sectoral planning, regional planning, economic zone planning, and land use planning, etc	Lack of planning or inconsistent planning

(continued)

Table 5 (continued)

Factors	Success	Failure
Infrastructure	The basic infrastructure meets and supports initial investment from the government to build important infrastructure; there are support policies during infrastructure construction	Infrastructure is lacking or unsynchronized. Lack of government commitment in establishing essential infrastructure. Infrastructure does not meet requirements investors' demand
Administrative reform	Strong administrative procedure reform; simplification of administrative procedures and removal of barriers to business investment. The investment environment is favorable and open, accelerating the progress of putting enterprises into production and business operation, reducing investment costs and increasing business profits	Administrative procedures are complicated and lack transparency. The administrative procedures are highly time- and cost-consuming. The duration before starting a business or investment is extended
Human resources	Human resources meet development requirements and future development prospects. A labor force with sufficient labor capacity (physical capabilities, professional qualifications, and spirit) is the foundation for the sustainable development of enterprises. The quantity and quality assurance, the skills of the workers in general, as well as the labor with high gray matter content in particular, are the premise to build the success of an EZ	Lack of labor resources to meet the development needs of enterprises, in which high-quality and skilled labor are required. Lack of capacity to provide human resources or provide vocational training to meet the current and future needs of EZ activities
Information technology	Having the environment and ability to apply advanced science and technology is an important factor affecting the competitiveness of enterprises, creating new production possibilities, speeding up the development of some industries, promoting the acquisition and application of advanced technology with high gray matter contents, promoting creativity and generating high added value	Limited technological conditions. Low science and technology application capacity. Insufficient human resources and equipment to keep up with the scientific and technological revolution of the age

(continued)

Table 5 (continued)

Factors	Success	Failure
Attracting investment	The political and social situations of the country are stable. A transparent business investment environment and preferential policies for international competition, specifically: There is an application of special incentive socioeconomic policies that are outstanding, globally competitive, stable and satisfying the requirements of investors according to international standards; allowed to experiment with new institutions and policies. Complete, amend, and supplement policies to create an increasingly open and favorable environment for investors	The business investment environment is not favorable. High production and business costs. Investment and business activities are not paid much attention and supported. The difficulties of the businesses are not paid enough attention to be solved properly
Economic development	Objectives and motivations for economic development are of great significance to the development of the EZs. As the economy develops, the amount of goods circulating on the market increases. The wider the consumption radius with the cores being commercial centers with economic potential, the faster they will develop, thereby forming poles and points in trade between countries. At the same time, the level of economic development also affects the structure of import and export goods, bilateral and multilateral exchange turnovers, and the scale and radius of goods spread	The economy lacks prospects or the economic model is misleading or inadequate. The competitiveness of the economy is low. Lack of supporting production base system. The economic development goals and orientations are not suitable with the competitive advantages of the economy or are too dependent on one industry (cheap labor) for a long time
Socio-cultural factors	The cultural and social environment is an important factor for the selection and development of an economic zone. Construction of economic zones is associated with the development of urban areas, service and industrial centers, along with the construction of housing areas for workers, a component part of the urban housing system and of the industrial zone	Lack of community cultural identity. Social instability or unattractiveness for long-term settlement of investors and workers. Lack of cultural conditions to meet the requirements of the community in the EZs

(continued)

Table 5 (continued)

Factors	Success	Failure
Environment	The system of collecting and draining wastewater and rainwater must be calculated to ensure enough for the demand for water collection and drainage. Wastewater treatment and waste treatment stations must be built and operated to ensure hygiene and environmental standards. The natural environment and landscape are preserved. Solutions to respond to climate change are actively implemented	The natural environment is damaged. Capacity to manage and handle the challenges of climate change is low. Lack of necessary waste and wastewater treatment system
International cooperation	Actively and proactively promoting international cooperation to promote regional and world connectivity, the ability to attract projects and investment works that are motivational and important for industry and sector development of the EZ	There are no input and output links with the world economy. Lack of training activities and external exchanges for social and economic development, science and technology, culture and foreign affairs
National defense and security	The relationship with the neighboring countries sharing the borders is stable, helping to ensure long-term and sustainable investment and business activities	Neighborhood instability. Border and internal security are not well controlled

(*Source* Author's compilation through literature review)

(2) Promulgate specific mechanisms and policies to effectively mobilize and use resources, encourage and attract investment to ensure proper implementation of socioeconomic development goals, tasks, and directions of the EZs.
(3) Improve management capacity for the economic zone steering committees. Consolidate, strengthen, and improve the capacity and stabilize the organizational structure and personnel model of the EZ management agency and local authorities to operate according to the appropriate model.
(4) Develop the plans and balance capital to support investment in the technical infrastructure system of the economic zones in accordance with the provisions of the law on investment and the state budget.
(5) Solve problems brought about by the process of promoting the new economic zone development models.
(6) In addition, research on the privatization model can be piloted to exploit the strengths of the dynamic governance role of the private sector.
(7) Develop science and technology and provide high-quality human resources training, promote innovation and creativity, take advantage of scientific achievements, new technologies and advanced technologies, attract leading experts and scientists and high-quality human resources to develop economic zones.
(8) Develop multi-purpose, synchronous infrastructure and transport network connecting major economic centers of the country and economic zones with the development of sea areas and seaports based on economic and natural ecosystems, and ensure North–South and East–West strategic connections among different regions in the country and with the world.

5 Conclusion

Since 1986, Vietnam began the period of Doi Moi (Reform), opening up the economy and starting to implement new economic models. The deployment of EZ models across the country has set an objective requirement for profound innovations, especially in the current period of international economic integration. From the study of economic zone development models of some countries in the world such as China, the United Kingdom, Korea, and the United Arab Emirates, etc., the research has also shown the limitations and weaknesses in the development of economic zones in Vietnam in recent years: (1) EZs lacked strategic planning, so they were spread out, divided, fragmented and slow in determining the priority order; hence, many zones had low efficiency operation; (2) Infrastructure was still lacking, not synchronous, not meeting development requirements; (3) Not attracting strategic investors in science, technology, finance, and management level; (4) Human resources did not responded in time to the requirements of investors; (5) The investment environment was not really attractive; administrative procedures were complicated.

The study has provided policy implications for Vietnam to innovate the economic zone development models, such as: (1) Developing and completing the Socio-Economic Development Master Plan, EZ Sustainable Development Strategy in the period up to 2025, with a vision to 2030, the project on training and development of human resources of the EZ in the period to 2030; (2) Promulgate specific mechanisms and policies to effectively mobilize and use resources, encourage and attract investment to ensure the proper implementation of socioeconomic development goals, tasks, and directions of the EZ; (3) Improve management capacity for economic zone steering committees; (4) Formulate a plan and balance capital to support investment in the technical infrastructure system of the economic zone in accordance with the law on investment and the state budget; (5) Solve problems brought about by the process of promoting the new economic zone development model; (6) In addition, research on the privatization model can be piloted to exploit the strengths of the dynamic governance role of the private sector, etc. The study and acquisition of experiences, application of theory to continue building, developing and renovating the economic zone development models, contributing to the implementation of the guidelines and policies of the Party and State of Vietnam on economic renewal, can help to open up the positive and promising prospects, creating new impetus for localities to grow and develop their economies quickly and sustainably in the coming time.

References

Anh, V. T. T. (2016). Chu Lai Open Economic Zone: Laboratory for Institutional and Policy Innovation in a centrally planned economy.

Hien, N. C. (2006). *System modeling and simulation*, Hanoi Science and Technology Publishing House.

Ministry of Planning and Investment. (2010a). *Summary of construction and development of coastal economic zones and border gate economic zones in Vietnam.*

Ministry of Planning and Investment. (2010b). *Summary report of 20 years of construction and development of industrial zones and export processing zones in Vietnam.*

Ministry of Planning and Investment. (2016, 2018). *Foreign investment website: Developing economic zones: International experiences and implications for Vietnam.*

Porter, M. E. (2008). *Competitive advantage of nations*. Publishing House. Young, Ho Chi Minh City.

Unraveling the Intelligent Dynamic Accounting Information System and Circular Economy Capabilities as the Enablers on Route to Reaching Sustainability-Oriented Innovation

Huy Quang Pham and Phuc Kien Vu

Abstract Although circular economy (CE) has been well-acknowledged as the predominant part of advanced approach to reach sustainability paradigm, CE implementation has been still in the nascent stage of development. Research has neglected to connect CE and sustainability-oriented innovation (SOI). While CE studies focused in enterprise and supply chain, there has been limited works delving into how public sector organizations (PSOs) could prepare for CE adoption to achieve the SOI. This research aims at exploring the mechanism of how intelligent dynamics accounting information system capabilities (IDAISC) promotes circular economy capabilities (CEC) which, in turn, facilitates SOI achievement in PSOs. Methodologically, this research was quantitative with a correlational scope and cross-sectional design. The significance and relevance of a series of research hypotheses in the proposed model were investigated through leveraging the partial least square structural equation modeling (PLS-SEM) approach on the statistical data captured from the convenient and snowball sample of 512 respondents. Analytical outcomes discerned that the fixable accounting information system (FAIS) and business process capabilities (BPC) of IDAISC were the enablers of CEC enhancement. This research further adduced that pursuit of CEC would result in reaching the SOI. Beyond the provision of scientific understandings through broadening the prevailing frontiers of this research string, the culmination of this study would serve as a stable cornerstone to generate research pointers for follow-up works. In the practical points of view, such valuable understandings could allow to formulate targeted strategies as well as rules and regulations for promoting the SOI in PSOs.

Keywords Intelligent dynamic accounting information system · Circular economy capabilities · Public sector · Sustainability-oriented innovation

H. Q. Pham (✉) · P. K. Vu
School of Accounting, University of Economics Ho Chi Minh City (UEH), Ho Chi Minh City,, Vietnam
e-mail: pquanghuy@ueh.edu.vn

A. T. Nguyen et al. (eds.), *Contemporary Economic Issues in Asian Countries: Proceeding of CEIAC 2022, Volume 1*,
https://doi.org/10.1007/978-981-19-9669-6_29

1 Introduction

Sustainability-oriented innovation (SOI) has typically been recommended as a main strategic approach for all organizations to generate huge contribution to sustainable development (Fichter & Clausen, 2016). Nevertheless, the advancement of SOI induced a comparatively complicated endeavor, as the corresponding results must meet the demands on namely environmental facet, social facet, and economic facet in simultaneous manner (Lubberink et al., 2017). Against this backdrop, the circular economy (CE) has been acknowledged as a promising solution to harmonize economic development and environmental defense (Yu et al., 2022). As such, it could supply the underpinning to conduct SOI completely through a constant process to gain the efficient and effective utilization of resources (Lüdeke-Freund et al., 2018). To that end, the in-depth comprehension on the interlink between CE and SOI has become valuable in reinforcing the transformation of organizations to sustainable practices (Brown et al., 2019).

The public sector organization (PSO) has been supposed to demonstrate significant sustainability-related effects on environmental, social, and economic matters which has needed managing (Nogueiro & Ramos, 2014). Given that PSO has been pondered as the provider of services directly and indirectly giving rise to substantial material as well as energy input/output flows (Shrake et al., 2011), PSO has become the primary actor in driving the CE implementation (Khan et al., 2020). Although the sustainability strategies and CE practices have been undertaken in PSO (Kristensen et al., 2021), this implementation has appeared to be in the early stages of development (Ghisellini et al., 2016).

With respect to this, the intelligent dynamic accounting information system capabilities (IDAISC) would become a prerequisite to assist PSO to reach the goal of CE implementation for SOI achievement. Given that data were the paramount driver of the Fourth Industrial Revolution (Industrie 4.0) (Dubey et al., 2019), the technological initiatives of Industrie 4.0 could be taken advantage to establish, manage, and govern all organizational operations in an intelligent manner. Indeed, big data analytics (BDA) has been proved to revamp the expense performance as well as operational performance of an entity (Dubey et al., 2019) while the integration between BDA and artificial intelligence (AI) instigated a positive impact on the inclusive health of the entity (Yasmin et al., 2020). With the support of BDA-AI, the dynamic accounting information capabilities (DAIC) would become a paramount resource of the PSO as it could provide real-time information and value feedback for intelligent decision-making, transformation of prevailing strategies, construction, and innovations to leverage valuable chances or prevent threats for value creation.

Surprisingly, although the issues in terms of the interconnection between SOI and CE as well as the development of accounting information system (AIS) in the era of Industrie 4.0 have been growing and drawn burgeoning attention in academician and practitioner communities, a holistic picture on the interrelation between IDAISC, CE capabilities (CEC) and SOI has been a deficiency. Thus, a crucial gap remained under-researched: an in-depth analysis of the how to intensify CEC to achieve SOI through IDAISC. Additionally, this theoretical gap inspired the intriguing research questions as follows.

RQ1. *What is the effect of each component of IDAISC on CEC?*

RQ2. *What is the effect of CEC on SOI?*

This current manuscript is divided into the following disparate sections. Section 2 briefly provides the scientific underpinning resulting in the research framework and hypotheses. The conceptual model and associated hypotheses are elaborated in Sect. 3. The methodological process is delineated in Sect. 4, followed by Sect. 5, in which the research analyses are illuminated. The theoretical contribution and managerial implications along with drawbacks and further scope of study are summarized in Sect. 6.

2 Theoretical Inputs

2.1 Adoption Model

Dynamic Capabilities Theory. Dynamic capabilities theory (DCT) was introduced as both an enlargement to and a response against the incapacity of the resource-based view to elucidate the advancement and readvancement of resources and capacities to handle quickly changing environments (Bleady et al., 2018). DCT offered in-depth insights of the paramount role of organization capabilities and their shifts in formulating organizational behaviors and overall performance (Teece et al., 1997; Wang et al., 2014). The underlying idea of DCT lied in the fact that all organizations should integrate, cultivate, and restructure internal and external resources in response to environmental shifts to advance the core competencies (Teece et al., 1997). Additionally, all organizations should take importance of innovation for long-term surviving accomplishment (Teece et al., 1997).

2.2 Conceptual Framework

Intelligent dynamics accounting information system capabilities. AIS has been well-acknowledged as a computer-based software application to input and process financial data (Al-Hattami et al., 2021) as well as manage and control issues in terms of the organizational financial and economic operations (Grande et al., 2011) to process into information and distribute to the end users (Hall, 2010). As proposed by Andiana (2022), AIS reflected on a set of data and processing procedures which procured, categorized, analyzed, stored credentials to give rise to and communicate requested information to internal and external stakeholders for decision-making. Based on the viewpoints of Al-Matari et al. (2022), DAIC referred to an exceptional type of resource formulated from several components related to AIS established to condition for chances and threats determination; these capabilities transferred prevailing strategies, configurations as well as innovations to make use of opportunities or prevent threats. In the meanwhile, Qiu (2016) suggested an intelligent dynamic accounting information platform relied on the internet of things comprised data acquisition, expense accounting as well as accounting document creation and report output. Nonetheless, the era of Industrie 4.0 has noticed significant advancement in computational facets and an increase in big data accessibility that has further enhanced concerns in AI (Baryannis et al., 2018). The implementation of BDA-AI could drive intelligent production (Zhou et al., 2019) and formulate of intelligent AIS. The AI was formulated by several building blocks, namely structured and unstructured data, main processes, knowledge base as well as information (Paschen et al., 2019) while the advantages of BDA lied in the capacities to supply valuable insights for high-quality decision-making and enable the organization to modify the strategies in appropriate with the business environment (Bag et al., 2020). As such, the integration of BDA-AI into DAIC would open up valuable chance to generate an IDAISC. Taking these analyses into consideration, the IDAISC in this study reflected on the integration between the power of BDA-AI and the components of DAIC including fixable accounting information system (FAIS), accounting information system-related human competency (AISHC), and business process capabilities (BPC). In this vein, IDAISC would become a unique resource for organization which was fortified and intensified by computing strength, algorithm as well as data embedded in the real and virtual world, harmonious symbiosis, and resonance with similar frequency.

Circular economy capabilities. The concentration of CE was placed on the flow of products and resources over time (Bocken et al., 2016). In this regard, the CE fostered the recirculation of resources in the ecosystem (Ormazabal et al., 2018) through resource flows narrowing, resource loops slowing resource loops, or resource loops closure (Rodríguez-Espíndola et al., 2022). CEC was supposed to be a common terminology performing the principles, namely reduction, reuse as well as recycle for organization (Andersen, 2006). These capabilities covered with interrelated CE practices to reach the overall goals. Based on the perspective of Zeng et al. (2017), CEC referred to all economic operations comprised manufacture, distribution, utilization, and waste recycling.

Sustainability-oriented innovation. Sustainability has been a general and normative definition, with an issue- and process-directed application that has evolved toward a desire to warrant that both existing and future generations could meet their own demands without affecting planetary life support systems (Nagatsu et al., 2020). In the meanwhile, innovation mentioned on the sustainability required the explicit contemplation of sustainability's defining idiosyncrasies (Wilkerson & Trellevik, 2021). In this research, SOI referred to the product/service, processes, and operations that an organization must employ in order to reach the environmental goals together with financial returns (Adams et al., 2015; Dey et al., 2020; Klewitz & Hansen, 2014).

3 Hypotheses' Development and Model Specification

The application of BDA-powered AI was corroborated to strengthen the CEC (Gupta et al., 2018). Admittedly, AI and machine learning adoption would offer numerous advantages, namely lowering expenses, gaining quality as well as accelerating responsiveness (Lee & Shin, 2020; Munoko et al., 2020). Almost all of organizations would have coped with the challenges in big data management which could be addressed through the implementation of BDA approaches, namely descriptive analytics, predictive analytics as well as prescriptive analytics (Sivarajah et al., 2017). AI has demonstrated its capabilities to contribute significantly to minimize the percentage of losses related to producing false and imprecise accounting information which supported the organizational leaders to make effective accounting and financial decisions in light of the quality and efficiency of the outcomes of AIS (Askary et al., 2018). In the same vein, the adoption of BDA in AIS would ameliorate the quality of information, reinforce integrated reporting more efficiently and effectively, increase risk management, enhance the efficiency and accuracy of predictive analytics, forecast future consequences as well as intensify automate non-routine financial tasks. As the FAIS could become facilitated for the ambidexterity and reconfiguration of organizational processes (Rai & Tang, 2010), with the support of BDA-AI, it would become much more intelligent enough to intensify the innovation capabilities of the organization and rapidly adapt to new operational mechanisms or new paradigm. Additionally, BPC has been pondered as the other important component in DAIC as business process comprised internal and external information, business market information as well as analysis (Lönnqvist & Pirttimäki, 2006). The adoption of BDA-AI in DAIC would make the BPC become much more intelligent to help the organization achieve the better analytical capabilities. Last but not least, the competency of human resource has been deemed as the main factor of contributed enormously to DAIC development (Yoshikuni et al., 2022). In this regard, the integration of BDA-AI into DAIC would enable the organizational staff to coordinate real-time responses to unexpected incidents or indirectly coordinate these responses through a general reference point for new operation. All of these achievements in IDAISC would provide the organization with perfect and well-timed information for decision-making to enhance the CEC. Logically, the above evidence resulted in the hypothetical statements as follows.

Hypothesis 1 **(H1)**. *FAIS instigates an impact on CEC in a significant and positive manner.*

Hypothesis 2 **(H2)**. *AISHC instigates an impact on CEC in a significant and positive manner.*

Hypothesis 3 **(H3)**. *BPC instigates an effect on CEC in a significant and positive manner.*

As sustainability has been considered as the main goal of CE model (Lazarevic & Valve, 2017), the CE implementation would foster the adjustment of processes and operations targeting at reaching the sustainable growth (Brown et al., 2019). Building

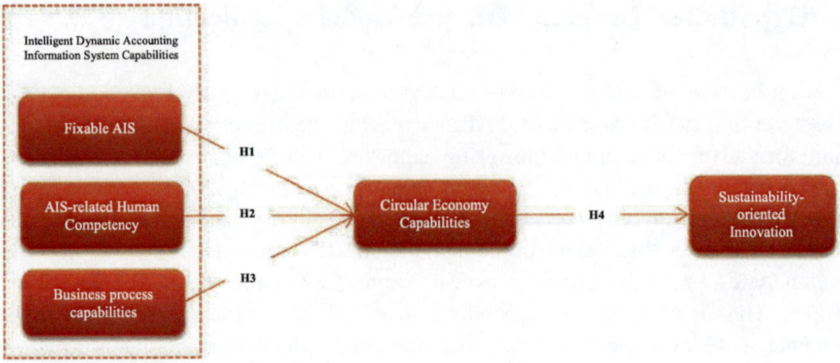

Fig. 1 Hypothesized model

on the perspectives of Klewitz and Hansen (2014), SOI could be reinforced by the utilization of closed-loop manufacture at the process level due to the benefit of transforming from linear models to more innovative configurations buttressed by CE (Adams et al., 2015; Brown et al., 2019). On the other hand, CE was authenticated to be capable of triggering imaginaries leading to actions that could enable sustainable innovation processes (Narayan & Tidström, 2019). Additionally, the implementation of CE principles could enable to produce innovative solutions balancing sustainability and organizational growth (Rattalino, 2017). Logically, the above evidence resulted in the hypothetical statement as follows.

***Hypothesis 4* (H4).** *CEC instigates an effect on SOI in a significant and positive manner* (Fig. 1).

4 Materials and Methods

4.1 Operationalization of the Measured Constructs

A questionnaire survey was made use in this study due to the scarcity of readily available secondary database on digital transformation within PSOs, especially in developing nations like Vietnam. The survey questionnaire items were composed through reviewing of the related literature and taking advantage of relevant ideas obtained from the precedent findings. The questionnaire was first formulated in English and redesigned in mother tongue language by means of a back translation procedure to prevent the discrepancies of translation (Cai et al., 2010). Since the questionnaire in the current research was set up in various contexts—both culturally and environmentally, a pretest with 5 experts was undertaken to mitigate unanticipated complexity. Rested on their feedbacks, the constructs were adjusted to abolish ambiguous measuring items. In order to bolster the substantiation of inputs, 30 respondents with similar idiosyncrasies to the target population were requested to participate in the small-scale pilot test. Based on their responses, the questionnaires and several items were revised where suitable.

Intelligent dynamics accounting information system capabilities. The measurement scale for IDAISC in this research comprised three ingredients, namely FAIS, BPC as well as AISHC. More concretely, the criteria applied to evaluate FAIS were stemmed from the integration of the findings of Prasad and Green (2015), Uwizeyemungu and Raymond (2012), and the works of Carvalho et al. (2018) and Dubey et al. (2020). The criteria applied to evaluate BPC were emanated from the integration of the findings of Prasad and Green (2015), Torres et al. (2018), and the works of Carvalho et al. (2018) and Dubey et al. (2020). The criteria applied to evaluate AISHC were derived from the findings of Prasad and Green (2015), Terry and Douglas (2000), and the works of Carvalho et al. (2018) and Dubey et al. (2020).

Circular economy capabilities. The criteria applied to evaluate CEC were taken as reference from the contributions of Zhu et al. (2005), Unob et al. (2007), Lee et al. (2007), French and LaForge (2006).

Sustainability-oriented innovation. The criteria applied to evaluate SOI were designed from those propounded by Dey et al. (2020), Tellis et al. (2009), Zeng et al. (2017), Kannan et al. (2014).

The five-point Likert scale to evaluate the degree of the respondents' approval with the statements (1 = "vigorously disagree", 5 = "vigorously agree") was employed for all the measurement.

4.2 Sampling Procedure and Data Collection

The unit of analysis was PSOs, and the participants were accountants. The statistical data in this research were procured through a paper-and-pencil questionnaire. As such, the questionnaires were circulated in person to respondents by the researchers. The sample of this research was established on the basis of convenience and snowball sampling. Building on the recommendation of Abdul-Rashid et al. (2017), the volume of responses should be at least 10 times larger than the maximum volume of arrowheads directing at latent variables, while Yuan et al. (2011) argued that the proper sample size for SEM analysis should range from 300 to 400. As suggested by Hair et al. (2022), the responses suffered from more than 15% missing items would be eliminated from the dataset. Therefore, all the responses suffered from 15% of missing data would be eradicated. The data gathering process took place during the months of April 2022 and September 2022. The final sample size left for analysis included 512 cases after screening and scrutiny of questionnaires, with a 14.67% data loss rate.

4.3 *Model Assessment and Modification*

The analysis of statistical data was undertaken with the buttress of partial squares structural equation modeling (PLS-SEM) and gauged by software deployment Smart-PLS 3.3.9. Initially, the measurement model was established to assess the reliability and validity of the constructs and measurement items. Subsequently, the structural model was evaluated to investigate the research hypotheses.

5 Results

5.1 *Descriptive Statistics of Informants' Demographics*

Predominantly, female accounted for the greatest proportions of accountants invited to partake in the current study, at around 68.16%, while males made up 31.84%. Regarding to the lifespan, the groups "31–40" amounted to 75.20% of the whole sample, which was followed by the groups "41–50", approximately 17.97%, whereas the group "above 50" made up 6.83%, ranking last among the given groups. All of participants in the current research obtained a minimum of graduate degree qualification and had experience of more than 10 years working as accountants in PSOs.

5.2 *Establishing Validity and Reliability*

The content validity mentioned on the items of questionnaire conveyed the similar meanings as integrated in certain concepts Ur Rehman et al. (2019) which was evaluated by means of cross-loading. By doing so, the value of an assessed construct had to be higher than others in the same rows and columns (Hair et al., 2010). Based on the values of all measured constructs as illustrated in Table 1, it could conclude that the measurement scale of this study obtained the perfect content validity.

The reliability was evaluated through Cronbach's alpha and composite reliability. As such, these criteria should surpass the suggested threshold value of 0.7 (Hair et al., 2016). Simultaneously, the value of Rho_A for all constructs was recommended to be greater than the cutoff score of 0.7 (Henseler et al., 2016). Convergent validity was determined through average variance extracted (AVE) and outer loadings. Accordingly, the outer loadings of all constructs were found to be above the critical value of 0.70 (Sarstedt et al., 2016), while the AVE was higher than the threshold values of 0.50 (Hair et al., 2022). Building on the statistical corroboration of these criteria in Table 2, the constructs were substantiated to reflect the good internal consistency reliability and convergent validity.

Table 1 Results' summary for the content validity

	AISHC	BPC	CEC	FAIS	SOI
AISHC1	0.876	0.100	0.085	0.073	0.094
AISHC2	0.837	0.081	0.081	0.096	0.135
AISHC3	0.879	0.156	0.111	0.122	0.075
BPC1	0.069	0.862	0.186	0.169	0.076
BPC2	0.136	0.828	0.183	0.227	0.061
BPC3	0.120	0.809	0.144	0.208	0.070
BPC4	0.128	0.722	0.091	0.145	0.067
CEC1	0.073	0.145	0.706	0.141	0.169
CEC2	0.073	0.180	0.709	0.154	0.181
CEC3	0.024	0.121	0.748	0.113	0.255
CEC4	0.091	0.116	0.747	0.078	0.218
CEC5	0.118	0.143	0.745	0.139	0.237
CEC6	0.092	0.165	0.744	0.221	0.260
CEC7	0.093	0.076	0.743	0.168	0.218
CEC8	0.087	0.193	0.744	0.169	0.217
CEC9	0.061	0.181	0.761	0.172	0.240
CEC10	0.083	0.160	0.735	0.146	0.202
CEC11	0.081	0.105	0.731	0.152	0.236
CEC12	0.080	0.129	0.711	0.153	0.149
FAIS1	0.140	0.196	0.157	0.844	0.131
FAIS2	0.087	0.176	0.166	0.856	0.114
FAIS3	0.080	0.233	0.210	0.901	0.129
SOI1	0.094	0.070	0.251	0.162	0.872
SOI2	0.103	0.086	0.266	0.115	0.904
SOI3	0.104	0.068	0.268	0.105	0.876

Discriminant validity reflected the circumstance, in which a latent variable was differentiated from the others pertaining to statistics (Ur Rehman et al., 2019). For assessing the discriminant validity, two criteria are used: (i) Fornell–Larcker criterion and (ii) heterotrait–monotrait (HTMT) ratio. Regarding to the Fornell and Larcker criterion, the square root of AVE of extracted in each latent variable should be larger than the rest of the other correlations relating that latent variable (Fornell & Larcker, 1981). Given that the correlation matrix illustrated in Table 3 satisfied the Fornell–Larcker criterion, all the constructs in the proposed model demonstrated discriminant validity for the empirical data.

Concerning to the heterotrait–monotrait (HTMT) ratio criterion, the HTMT has been well-regarded as a substitute and more impeccable index for discriminant validity due to the demands on reckoning bootstrapping confidence intervals with

Table 2 Results' summary for the measurement model

Constructs and operationalization		Convergent validity		Construct reliability			Discriminant validity
		Outer loadings	AVE	Cronbach's alpha	Composite reliability	Rho_A	
Fixable accounting information system	FAIS	0.844–0.901	0.753	0.837	0.901	0.859	*Yes*
Accounting information system-related human competency	AISHC	0.837–0.879	0.747	0.833	0.899	0.856	*Yes*
Business process capabilities	BPC	0.722–0.862	0.651	0.825	0.881	0.858	*Yes*
Circular economy capabilities	CEC	0.709–0.761	0.541	0.923	0.934	0.926	*Yes*
Sustainability-oriented innovation	SOI	0.872–0.904	0.781	0.860	0.915	0.861	*Yes*

Table 3 Results' summary for discriminant validity using Fornell–Larker scale

	AISHC	BPC	CEC	FAIS	SOI
AISHC	0.865				
BPC	0.135	0.807			
CEC	0.109	0.196	0.735		
FAIS	0.115	0.235	0.208	0.867	
SOI	0.114	0.084	0.296	0.143	0.884

Table 4 Results' summary for discriminant validity using heterotrait–monotrait ratio

	AISHC	BPC	CEC	FAIS	SOI
AISHC					
BPC	0.163				
CEC	0.121	0.212			
FAIS	0.138	0.275	0.229		
SOI	0.138	0.101	0.328	0.170	

5,000 resamples (Hair et al., 2017). In doing so, the HTMT score was proposed drastically below the cutoff value of 0.85 for shaping up the discrimination between two latent variables (Henseler et al., 2016). Given that the correlation matrix illustrated in Table 4 satisfied the HTMT criterion, all the constructs in the proposed model demonstrated discriminant validity for the empirical data.

5.3 *Fitting and Analyzing the Hypothesized Model*

The model fit was evaluated through standardized root mean square residual (SRMR). The value of SRMR in this study was 0.046, which authenticated an appropriate fit for PLS path models because it was lower than the suggested cutoff, the degree of 0.08 (Hair et al., 2017). Building on the output in Table 5, VIF values of each construct were below 3.3 (Hair et al., 2018). The R^2 value for CEC and SOI was 0.071 and 0.088, respectively, which was equal to approximately 0.1 (Falk & Miller, 1992). The f^2 values for CEC and SOI were greater than the recommended threshold value of zero (Dubey et al., 2020). This study also fortified gratifying predictive relevance as the Q^2 values for CEC and SOI were 0.036 and 0.066, respectively, which were well above zero (Hair et al., 2020). Bootstrapping with the commonly 5,000 sub-samples (Kaya et al., 2020) was employed in this study. Based on the output reported in Table 5, FAIS had substantially positive relationship with CEC (**H1**: $\beta = 0.165$; *t*-value $= 3.696$; *p*-value $= 0.000$). Besides, the BPC was substantiated to be associated with CEC (**H3**: $\beta = 0.148$, *t*-value $= 3.384$; *p*-value $= 0.001$) in a markedly positive manner. The investigation on the relationship between CEC and SOI (**H4**) denoted

Table 5 Results of hypotheses' testing

Relevant hypothesis	Relevant path	Path coefficient	SE	95% Confidence interval	*t*-value	*p*-value	VIF	Result
H1	FAIS → CEC	0.165	0.003	[0.082–0.256]	3.696	0.000	1.066	Undergirded
H2	AISHC → CEC	0.070	0.005	[−0.006–0.159]	1.579	0.114	1.026	Rejected
H3	BPC → CEC	0.148	0.007	[0.068–0.242]	3.384	0.001	1.072	Undergirded
H4	CEC → SOI	0.296	0.004	[0.215–0.384]	6.876	0.000	1.000	Undergirded
R^2	$R^2_{CEC} = 0.071$; $R^2_{SOI} = 0.088$							
f^2	$f^2_{FAIS \Rightarrow CEC} = 0.028$; $f^2_{AISHC \Rightarrow CEC} = 0.005$; $f^2_{BPC \Rightarrow CEC} = 0.022$; $f^2_{CEC \Rightarrow SOI} = 0.096$							
Q^2	$Q^2_{CEC} = 0.036$; $Q^2_{SOI} = 0.066$							

that the path coefficient (β) was 0.296 with t-value was 6.876 and p-value was 0.000. Contrary to the expectation of the researchers, the paths linking AISHC and CEC (**H2**: $\beta = 0.070$; *t*-value $= 1.579$; *p*-value $= 0.114$) were insignificant. Thus, H1, H3, and H4 were supported, while H2 was rejected.

6 Discussions and Conclusions

6.1 Theoretical Contribution

When the organizational operations have targeted at generating solutions to addressing the social and environmental challenges, these could boost effective sustainability innovation (Van Holt et al., 2020). The demands to orient business operations toward sustainability utilizing innovation have developed from eco-innovations—pondering predominantly environmental concerns—to SOI (Klewitz & Hansen, 2014). SOI involved product/service, process, and organizational renovation to create social and environmental value in line with financial advantages (Adams et al., 2015; Dey et al., 2020; Klewitz & Hansen, 2014). With this recognition, the current research aims at exploring the mechanism of how IDAISC promoted CEC which, in turn, facilitates SOI achievement in PSOs. The acquired findings of this research enlarged the body of literature on SOI through in-depth investigation on how to achieve SOI within PSOs in developing countries. The statistical analyses in this research contributed to the reinforcement for assumption on the interconnection between CE and SOI reported in prior studies (i.e., Rodríguez-Espíndola et al. (2022)) through adding novel and important proposals on the interconnection between CEC and SOI. As the implementation of CE was appropriate with the formulation of innovation at the product/service, process, and organizational degree (Klewitz & Hansen, 2014), the intensification in CEC was authenticated to be a promising solution for the achievement in SOI. In doing so, this study widened the frontier of knowledge on CE literature how to strengthen the CEC in PSO. Accordingly, the current study provided an understanding of the evolving logics between IDAISC and CEC, which, thus, broadened the research string on the integration advanced information technologies into AIS. More instrumentally, the intelligent FAIS would supply an intelligent platform for evaluating the proper data and shaping a network communication system for effective communication with the others. Additionally, intelligent BPC provided intelligent analytical capabilities manage enormous organization-related information. However, the intelligent AISHC was reported to induce insignificant impact on CEC although it acted as the mechanism for obtaining enhanced organizational productivity. One plausible reason for this result lied in the fact that advanced information technologies' application and adjusted the operational process for the CE implementation appeared to be the most priorities of PSO to handle for CE implementation prior to managing human resource. Taken together, with the support of BDA-AI, IDAISC would generate to a unique organizational resource which enabled the organization intelligent decision-making, transformation of prevailing strategies, construction, and innovations to improve and increase CEC.

6.2 *Managerial and Policy Implications*

Pertaining to practical implications, the current research generated several takeaway points. Firstly, all the leaders in PSOs should recognize that IDAISC implementation has been paramount for CE implementation and SOI. More instrumentally, the result analyses of the current study highlighted that BPC and FAIS were two top essential drivers to IDAISC implementation. These empirics could enable leaders in PSO to integrate IDAISC into their business routines to ameliorate and advance CEC through devising how to efficiently and effectively take advantage of two forms of IDAISC. Simultaneously, this could assist the leaders to decipher how to manage and govern the FAIS to accomplish the excellence in CEC enhancement through placing their focuses on tangible resources, namely BDA-AI infrastructure, digital platform, advanced information technologies, and other fundamental resources to deploy the BDA-AI projects. Even though AISHC was reported with the lowness in the presence of CEC enhancement, it was also indispensable that leaders in PSOs should concentrate on improving and developing workforce proficiencies through appropriate training programs to keep them abreast of the cutting-edge BDA-AI programming techniques. Additionally, seminars, expert visits as well as campaigns on the importance of intelligent AIS implementation and CE practices could be launched and evolved for organizational responsiveness to intelligent AIS implementation and CE practices. The current research also casted light on the fact that amelioration and development acquired from the enhancement of CEC could allow organizations to reinforce product/service, process as well as organizational innovation to succeed in SOI. Furthermore, CEC was considered as the precondition to assure the application of digital platforms to buttress SOI. As such, these findings underlined to all leaders in PSOs on the magnitude of a transformation from the linear to the CE paradigm to reinforce the transition of PSOs toward sustainability.

These observations could enable policymakers and governmental influencers to comprehend the impact of IDAISC with the goal of gaining the CEC for SOI achievement. In this regard, governmental supports should emphasize on the investment in digital platforms and skills' development as well as providing financial sponsor to promote the implementation of sustainable activities.

Information technology developers and providers were encouraged to formulate country-centered innovations which were appropriate with the requirements of businesses. Against this backdrop, developers and providers should exert themselves to generate accessible, affordable as well as adaptable resolutions in relation to intelligent AIS for PSOs in developing countries.

6.3 Limitations and Avenues for Future Research

There were several inevitable limitations encountered in the current study since no academic works could flawlessly handle all the questions in terms of the phenomenon investigated (Lutfi et al., 2022). The first drawbacks were the relatively small size of the sample as the pandemic has been still an impact and challenged to capture primary data. In addition, the accountants were selected as the target population, so their feedbacks could not reflect the actual situation of the PSOs. Consequently, the increase in sample size and incorporation of opinions of other organizational departments should be taken into consider to improve accuracy. Secondly, this research was performed in Vietnam where digital transformation has still been at a nascent stage. Besides, with this cross-sectional design approach, there would be no opportunities of causality being inferred. As the scenario might be different in other regions, follow-up studies were encouraged to explore these issues in emergent economies with the application of longitudinal design. Fourthly, although there were only two components of IDAISC which were corroborated to effectively enhance CEC, namely BPC and FAIS, the interconnection between AISHC and CEC was proved to be insignificant. Additional investigation on this interconnection should be carried out by potential researchers to lay down and enlighten its benefits and its role in CEC enhancement. Finally, the notion of IDAISC introduced in the current manuscript was predominantly rested on a semantic aggregation of authenticating illuminations gathered from prior literature. To that end, it would conspicuously not be a drawback in the follow-up works when the numerous meticulous investigations would be undertaken to obtain a more vigorous clarification for this concept.

References

Abdul-Rashid, S. H., Sakundarini, N., Ariffin, R., & Ramayah, T. (2017). Drivers for the adoption of sustainable manufacturing practices: A Malaysia perspective. *International Journal of Precision Engineering and Manufacturing, 18*(11), 1619–1631. https://doi.org/10.1007/s12541-017-0191-4

Adams, R., Jeanrenaud, S., Bessant, J., Denyer, D., & Overy, P. (2015). Sustainability-oriented Innovation: A Systematic Review. *International Journal of Management Reviews, 18*(2), 180–205. https://doi.org/10.1111/ijmr.12068

Al-Hattami, H. M., Hashed, A. A., Alnuzaili, K. M., & Alsoufi, M. A. Z. (2021). Effect of risk of using computerized AIS on external auditor's work quality in Yemen. *International Journal of Advanced and Applied Sciences, 8*(1), 75–81.

Al-Matari, A. S., Amiruddin, R., Aziz, K. A., & Al-Sharafi, M. A. (2022). The impact of dynamic accounting information system on organizational resilience: The mediating role of business processes capabilities. *Sustainability, 14*, 1–22. https://doi.org/10.3390/su14094967

Andersen, M. S. (2006). An introductory note on the environmental economics of the circular economy. *Sustainability Science, 2*(1), 133–140. https://doi.org/10.1007/s11625-006-0013-6

Andiana, B. D. L. (2022). Recording of cash expenditures and receipts in school operational assistance funds (BOS) based on accounting information systems at MA Plus Al-Ittihad. *Budapest International Research and Critics Institute-Journal (BIRCI-Journal), 5*(1), 3241–3247.

Askary, S., Abu-Ghazaleh, N., & Tahat, Y. A. (2018). Artificial intelligence and reliability of accounting information. *Challenges and Opportunities in the Digital Era*, 315–324. https://doi.org/10.1007/978-3-030-02131-3_28

Bag, S., Pretorius, J. H. C., Gupta, S., & Dwivedi, Y. K. (2020). Role of institutional pressures and resources in the adoption of big data analytics powered artificial intelligence, sustainable manufacturing practices and circular economy capabilities. *Technological Forecasting and Social Change*, 1–14. https://doi.org/10.1016/j.techfore.2020.120420

Baryannis, G., Validi, S., Dani, S., & Antoniou, G. (2018). Supply chain risk management and artificial intelligence: state of the art and future research directions. *International Journal of Production Research*, 1–24. https://doi.org/10.1080/00207543.2018.1530476

Bleady, A., Ali, A. H., & Ibrahim, S. B. (2018). Dynamic capabilities theory: Pinning down a shifting concept. *Academy of Accounting and Financial Studies Journal, 22*(2), 1–16.

Bocken, N. M. P., de Pauw, I., Bakker, C., & van der Grinten, B. (2016). Product design and business model strategies for a circular economy. *Industrial and Production Engineering, 33*(5), 308–320. https://doi.org/10.1080/21681015.2016.1172124

Brown, P., Bocken, N., & Balkenende, R. (2019). Why Do Companies pursue collaborative circular oriented innovation? *Sustainability, 11*(3), 1–23. https://doi.org/10.3390/su11030635

Cai, S., Jun, M., & Yang, Z. (2010). *Implementing supply chain information integration in China: the role of institutional forces and trust.*

Carvalho, N., Chaim, O., Cazarini, E., & Gerolamo, M. (2018). Manufacturing in the fourth industrial revolution: A positive prospect in Sustainable Manufacturing. *Procedia Manufacturing, 21*, 671–678. https://doi.org/10.1016/j.promfg.2018.02.170

Dey, P. K., Malesios, C., De, D., Chowdhury, S., & Abdelaziz, F. B. (2020). The impact of lean management practices and sustainably-oriented innovation on sustainability performance of small and medium-sized enterprises: Empirical evidence from the UK. *British Journal of Management, 31*(1), 141–161. https://doi.org/10.1111/1467-8551.12388

Dubey, R., Gunasekaran, A., Childe, S. J., Bryde, D. J., Giannakis, M., Foropon, C., Roubaud, D., & Hazen, B. T. (2020). Big data analytics and artificial intelligence pathway to operational performance under the effects of entrepreneurial orientation and environmental dynamism: A study of manufacturing organisations. *International Journal of Production Economics, 226*, 1–42. https://doi.org/10.1016/j.ijpe.2019.107599

Dubey, R., Gunasekaran, A., Childe, S. J., Fosso Wamba, S., Roubaud, D., & Foropon, C. (2019). Empirical investigation of data analytics capability and organizational flexibility as complements to supply chain resilience. *International Journal of Production Research*, 1–19. https://doi.org/10.1080/00207543.2019.1582820

Falk, R. F., & Miller, N. B. (1992). *A primer for soft modeling*. University of Akron Press.

Fichter, K., & Clausen, J. (2016). Diffusion Dynamics of sustainable innovation—Insights on diffusion patterns based on the analysis of 100 sustainable product and service innovation. *Innovation Management, 4*(2), 30–67. https://doi.org/10.24840/2183-0606_004.002_0004

Fornell, C., & Larcker, D. F. (1981). Evaluating structural equation models with unobservable variables and measurement error. *Journal of Marketing Research, 18*(1), 39–50. https://doi.org/10.2307/3151312

French, M. L., & LaForge, R. L. (2006). Closed-loop supply chains in process industries: An empirical study of producer re-use issues. *Journal of Operations Management, 24*(3), 271–286. https://doi.org/10.1016/j.jom.2004.07.012

Ghisellini, P., Cialani, C., & Ulgiati, S. (2016). A review on circular economy: The expected transition to a balanced interplay of environmental and economic systems. *Journal of Cleaner Production, 114*, 11–32. https://doi.org/10.1016/j.jclepro.2015.09.007

Grande, E. U., Estébanez, R. P., & Colomina, C. M. (2011). The impact of accounting information systems (AIS) on performance measures: Empirical evidence in spanish SMEs. *The International Journal of Digital Accounting Research, 11*(1), 25–43.

Gupta, S., Chen, H., Hazen, B. T., Kaur, S., & Santibañez Gonzalez, E. D. R. (2018). Circular economy and big data analytics: A stakeholder perspective. *Technological Forecasting and Social Change*, 1–9. https://doi.org/10.1016/j.techfore.2018.06.030

Hair, J. F., Anderson, R. E., Tatham, R. L., & Black, W. C. (2010). *Multivariate data analysis* (7th ed.). Pearson.

Hair, J. F., Howard, M. C., & Nitzl, C. (2020). Assessing measurement model quality in PLS-SEM using confirmatory composite analysis. *Business Research, 109*, 101–110. https://doi.org/10.1016/j.jbusres.2019.11.069

Hair, J. F., Hult, G. T. M., Ringle, C. M, & Sarstedt, M. (2016). *A primer on partial least squares structural equation modeling (PLS-SEM).* Sage

Hair, J. F., Hult, G. T. M., Ringle, C. M., & Sarstedt, M. (2017). *A primer on partial least squares structural equation modeling (PLS-SEM)* (2nd ed.). Sage.

Hair, J. F., Hult, G. T., Ringle, C. M., & Sarstedt, M. (2022). *A primer on partial least squares structural equation modeling (PLS-SEM).* Sage.

Hair, J. F., Sarstedt, M., Ringle, C. M., & Gudergan, S. P. (2018). *Advanced issues in partial least squares structural equation modeling (PLS-SEM).* Sage.

Hall, J. A. (2010). *Accounting information systems.* South-Western Cengage Learning.

Henseler, J., Hubona, G., & Ray, P. A. (2016). Using PLS path modeling in new technology research: Updated guidelines. *Industrial Management & Data Systems, 116*(1), 2–20. https://doi.org/10.1108/imds-09-2015-0382

Kannan, D., de Jabbour, A. B. L., & Jabbour, C. J. C. (2014). Selecting green suppliers based on GSCM practices: Using fuzzy TOPSIS applied to a Brazilian electronics company. *European Operational Research, 233*(2), 432–447. https://doi.org/10.1016/j.ejor.2013.07.023

Kaya, B., Abubakar, A. M., Behravesh, E., Yildiz, H., & Mert, I. S. (2020). Antecedents of innovative performance: Findings from PLS-SEM and fuzzy sets (fsQCA). *Journal of Business Research, 114*, 278–289. https://doi.org/10.1016/j.jbusres.2020.04.016

Khan, O., Daddi, T., & Iraldo, F. (2020). The role of dynamic capabilities in circular economy implementation and performance of companies. *Corporate Social Responsibility and Environmental Management*, 1–16. https://doi.org/10.1002/csr.2020

Klewitz, J., & Hansen, E. G. (2014). Sustainability-oriented innovation of SMEs: A systematic review. *Cleaner Production, 65*, 57–75. https://doi.org/10.1016/j.jclepro.2013.07.017

Kristensen, H. S., Mosgaard, M. A., & Remmen, A. (2021). Integrating circular principles in environmental management systems. *Cleaner Production, 286*, 125485. https://doi.org/10.1016/j.jclepro.2020.125485

Lazarevic, D., & Valve, H. (2017). Narrating expectations for the circular economy: Towards a common and contested European transition. *Energy Research & Social Science, 31*, 60–69. https://doi.org/10.1016/j.erss.2017.05.006

Lee, I., & Shin, Y. J. (2020). Machine learning for enterprises: Applications, algorithm selection, and challenges. *Business Horizons, 63*(2), 157–170. https://doi.org/10.1016/j.bushor.2019.10.005

Lee, J., Song, H. T., & Yoo, J.-M. (2007). Present status of the recycling of waste electrical and electronic equipment in Korea. *Resources, Conservation and Recycling, 50*(4), 380–397. https://doi.org/10.1016/j.resconrec.2007.01.010

Lönnqvist, A., & Pirttimäki, V. (2006). The measurement of business intelligence. *Information Systems Management, 23*(1), 32–40. https://doi.org/10.1201/1078.10580530/45769.23.1.20061201/91770.4

Lubberink, R., Blok, V., van Ophem, J., & Omta, O. (2017). Lessons for responsible innovation in the business context: A systematic literature review of responsible, social and sustainable innovation practices. *Sustainability, 9*(5), 1–31. https://doi.org/10.3390/su9050721

Lüdeke-Freund, F., Bohnsack, R., Breuer, H., & Massa, L. (2018). Research on sustainable business model patterns: Status quo, methodological issues, and a research agenda. *Sustainable Business Models*, 25–60. https://doi.org/10.1007/978-3-319-93275-0_2

Lutfi, A., Al-Okaily, M., Alsyouf, A., & Alrawad, M. (2022). Evaluating the D&M IS success model in the context of accounting information system and sustainable decision making. *Sustainability, 14*, 1–17. https://doi.org/10.3390/su14138120

Munoko, I., Brown-Liburd, H. L., & Vasarhelyi, M. (2020). The ethical implications of using artificial intelligence in auditing. *Business Ethics*, 1–26. https://doi.org/10.1007/s10551-019-04407-1

Nagatsu, M., Davis, T., DesRoches, C. T., Koskinen, I., MacLeod, M., Stojanovic, M., & Thorén, H. (2020). Philosophy of science for sustainability science. *Sustainability Science*, 1–11. https://doi.org/10.1007/s11625-020-00832-8

Narayan, R., & Tidström, A. (2019). Circular economy inspired imaginaries for sustainable innovations. *Seamless Learning, 393–413*,. https://doi.org/10.1007/978-3-319-97385-2_21

Nogueiro, L., & Ramos, T. B. (2014). The integration of environmental practices and tools in the Portuguese local public administration. *Cleaner Production, 76*, 20–31. https://doi.org/10.1016/j.jclepro.2014.03.096

Ormazabal, M., Prieto-Sandoval, V., Puga-Leal, R., & Jaca, C. (2018). Circular economy in Spanish SMEs: Challenges and opportunities. *Cleaner Production, 185*, 157–167. https://doi.org/10.1016/j.jclepro.2018.03.031

Paschen, J., Kietzmann, J., & Kietzmann, T. C. (2019). Artificial intelligence (AI) and its implications for market knowledge in B2B marketing. *Journal of Business & Industrial Marketing, 34*(7), 1410–1419. https://doi.org/10.1108/jbim-10-2018-0295

Prasad, A., & Green, P. (2015). Organizational competencies and dynamic accounting information system capability: Impact on AIS processes and firm performance. *Journal of Information Systems, 29*(3), 123–149.

Qiu, F. (2016). Overall framework design of an intelligent dynamic accounting information platform based on the internet of things. *International Journal of Online Engineering, 12*(5), 14–16.

Rai, A., & Tang, X. (2010). Leveraging IT capabilities and competitive process capabilities for the management of interorganizational relationship portfolios. *Information Systems Research, 21*(3), 516–542. https://doi.org/10.1287/isre.1100.0299

Rattalino, F. (2017). Circular advantage anyone? Sustainability-driven innovation and circularity at Patagonia, Inc. *Thunderbird International Business Review, 60*(5), 747–755. https://doi.org/10.1002/tie.21917

Rodríguez-Espíndola, O., Cuevas-, A., Chowdhury, S., Díaz-Acevedo, N., Albores, P., Despoudi, S., Malesios, C., & Dey, P. (2022). The role of circular economy principles and sustainable-oriented innovation to enhance social, economic and environmental performance: Evidence from Mexican SMEs. *International Journal of Production Economics, 248*, 1–18. https://doi.org/10.1016/j.ijpe.2022.108495

Sarstedt, M., Diamantopoulos, A., Salzberger, T., & Baumgartner, P. (2016). Selecting single items to measure doubly concrete constructs: A cautionary tale. *Journal of Business Research, 69*(8), 3159–3167. https://doi.org/10.1016/j.jbusres.2015.12.004

Shrake, S. O., Landis A. E., & Bilec, M. M. (2011). *Greening the service industries: A case study of a United States engineering consulting firm.* Proceedings of the 2011 IEEE International Symposium on Sustainable Systems and Technology, Chicago, IL, USA, pp. 1–6. https://doi.org/10.1109/issst.2011.5936881

Sivarajah, U., Kamal, M. M., Irani, Z., & Weerakkody, V. (2017). Critical analysis of Big Data challenges and analytical methods. *Journal of Business Research, 70*, 263–286. https://doi.org/10.1016/j.jbusres.2016.08.001

Teece, D. J., Pisano, G., & Shuen, A. (1997). Dynamic capabilities and strategic management. *Strategic Management Journal, 18*(7), 509–533. https://doi.org/10.1002/(sici)1097-0266(199708)18:7%3c509::aid-smj882%3e3.0.co;2-z

Tellis, G. J., Prabhu, J. C., & Chandy, R. K. (2009). Radical innovation across nations: The preeminence of corporate culture. *Journal of Marketing, 73*(1), 3–23. https://doi.org/10.1509/jmkg.73.1.003

Terry, A. B., & Douglas, E. T. (2000). Measuring the flexibility of information technology infrastructure: Exploratory analysis of a construct. *Journal of Management Information Systems, 17*(1), 167–208. https://doi.org/10.1080/07421222.2000.11045632

Torres, R., Sidorova, A., & Jones, M. C. (2018). Enabling firm performance through business intelligence and analytics: A dynamic capabilities perspective. *Information & Management, 55*, 1–65. https://doi.org/10.1016/j.im.2018.03.010

Unob, F., Wongsiri, B., Phaeon, N., Puanngam, M., & Shiowatana, J. (2007). Reuse of waste silica as adsorbent for metal removal by iron oxide modification. *Journal of Hazardous Materials, 142*(1–2), 455–462. https://doi.org/10.1016/j.jhazmat.2006.08.049

Ur Rehman, S., Bhatti, A., & Chaudhry, N. I. (2019). Mediating effect of innovative culture and organizational learning between leadership styles at third-order and organizational performance in Malaysian SMEs. *Journal of Global Entrepreneurship Research, 9*(1), 1–24. https://doi.org/10.1186/s40497-019-0159-1

Uwizeyemungu, S., & Raymond, L. (2012). Impact of an ERP system's capabilities upon the realisation of its business value: A resource-based perspective. *Information Technology and Management, 13*(2), 69–90. https://doi.org/10.1007/s10799-012-0118-9

Van Holt, T., Statler, M., Atz, U., Whelan, T., Loggerenberg, M., & Cebulla, J. (2020). The cultural consensus of sustainability-driven innovation: Strategies for success. *Business Strategy and the Environment*, 1–11. https://doi.org/10.1002/bse.2584

Wang, C. L., Senaratne, C., & Rafiq, M. (2014). Success traps, dynamic capabilities and firm performance. *British Journal of Management, 26*(1), 26–44. https://doi.org/10.1111/1467-8551.12066

Wilkerson, B., & Trellevik, L.-K.L. (2021). Sustainability-oriented innovation: Improving problem definition through combined design thinking and systems mapping approaches. *Thinking Skills and Creativity, 42*, 1–12. https://doi.org/10.1016/j.tsc.2021.100932

Yasmin, M., Tatoglu, E., Kilic, H. S., Zaim, S., & Delen, D. (2020). Big data analytics capabilities and firm performance: An integrated MCDM approach. *Journal of Business Research, 114*, 1–15. https://doi.org/10.1016/j.jbusres.2020.03.028

Yoshikuni, A. C., Galvão, F. R., & Albertin, A. L. (2022). Knowledge strategy planning and information system strategies enable dynamic capabilities innovation capabilities impacting firm performance. *VINE Journal of Information and Knowledge Management Systems, 52*(4), 508–530. https://doi.org/10.1108/VJIKMS-07-2020-0128

Yu, Y., Junjan, V., Yazan, D. M., & Iacob, M.-E. (2022). A systematic literature review on Circular Economy implementation in the construction industry: A policy-making perspective. *Resources, Conservation & Recycling, 183*, 1–17. https://doi.org/10.1016/j.resconrec.2022.106359

Yuan, K.-H., Wu, R., & Bentler, P. M. (2011). Ridge structural equation modelling with correlation matrices for ordinal and continuous data. *British Journal of Mathematical and Statistical Psychology, 64*(1), 107–133. https://doi.org/10.1348/000711010x497442

Zeng, H., Chen, X., Xiao, X., & Zhou, Z. (2017). Institutional pressures, sustainable supply chain management, and circular economy capability: Empirical evidence from Chinese eco-industrial park firms. *Cleaner Production, 155*, 54–65. https://doi.org/10.1016/j.jclepro.2016.10.093

Zhou, G., Zhang, C., Li, Z., Ding, K., & Wang, C. (2019). Knowledge-driven digital twin manufacturing cell towards intelligent manufacturing. *International Journal of Production Research*, 1–18. https://doi.org/10.1080/00207543.2019.1607978

Zhu, Q., Sarkis, J., & Geng, Y. (2005). Green supply chain management in China: Pressures, practices and performance. *International Journal of Operations & Production Management, 25*(5), 449–468. https://doi.org/10.1108/01443570510593148

Foreign Ownership and Corporate Governance on Firm Risk in ASEAN Countries: In the Context of the US-China Trade War

Do Thi Hai Yen and David Han-Min Wang

Abstract Many multinationals recently tend to divert their investments to emerging economies in the Association of Southeast Asian Nations (ASEAN) during the US-China trade war. In order to ensure effective investment, overseas investors pay more attention not only to efficient corporate governance but also firm risk. The present research adopts fuzzy-set qualitative comparative analysis (fsQCA) to test whether foreign ownership and corporate governance could affect firm risk of listed firms in four ASEAN countries over the US-China trade war. In addition, we compare the effects of foreign ownership and corporate governance between developed and developing ASEAN countries. Our findings have practical implications for investors in making overseas investments and managers in performing strategic management in ASEAN countries.

Keywords Corporate governance · Foreign ownership · Fuzzy-set qualitative comparative analysis · US-China trade war

1 Introduction

A notable point in globalization nowadays is the reorganization of economic activities in countries and regions. The established global supply chains raise interdependence among firms all over the world. The ongoing US-China trade war started in 2018 has caused a serious impact on the business operations of multinational companies in the world. The trade war offers the opportunity of boosting the Southeast Asian supply chains which generate large earnings for ASEAN. Investment diversion of multinationals along with foreign investment flows will be beneficial for ASEAN countries, which are the essential partners of the US and China on trading (U.S. Mission to ASEAN, 2019).

D. T. H. Yen (✉)
University of Transport and Communications, Hanoi, Vietnam
e-mail: yendth_ph@utc.edu.vn

D. H.-M. Wang
Feng Chia University, Taichung, Taiwan

A. T. Nguyen et al. (eds.), *Contemporary Economic Issues in Asian Countries: Proceeding of CEIAC 2022, Volume 1*,
https://doi.org/10.1007/978-981-19-9669-6_30

The economic growth of ASEAN is ranked 5th in the world's largest economy and the 3rd largest market in Asia (ASEAN Secretariat, 2018) with GDP reaching US$2.8 trillion in 2017. The ASEAN Investment Report (2018) states that multinational enterprises (MNEs) plan to raise the level of their investments in ASEAN in the next five years. Due to geographical location, the ASEAN region is heavily affected by capital movements under the US-China trade war. Although the inflow of foreign investment is associated with greater advantages for ASEAN countries, potential risks should be considered during the trade war. The important role of foreign capital in promoting performance has been confirmed by numerous empirical studies (Burker et al., 2013; Globerman et al., 1994). However, unpredictable risks are also contained in foreign investment (Chen et al., 2013; Fiala & Havranek, 2017).

ASEAN countries have striking characteristics, including both developed and developing countries with different regulations and business environments to other countries. The US-China trade war turns ASEAN markets gradually to be the destinations for overseas investors. Corporate governance plays a crucial role in investment decision-making. Johnson et al. (2000) and Gibson (2003) have argued that poor governance and transparency should be avoided in emerging markets. Managers in firms with better corporate governance would reduce short-sighted decisions by boosting R&D expenditures and capital investments. The increase in capital expenditures has a conclusive impact on business performance and firm value (Stein, 1988, 1989). Khanchel (2007) reveals that corporate governance ratings help investors make effective investment decisions. Additionally, mandatory disclosure policies and governance mechanisms safeguard overseas investors in developing countries (Mangena & Tauringana, 2007).

As mentioned above, foreign investment and corporate governance influence the management effectiveness. The roles of overseas investment and corporate governance in ASEAN countries have attracted much attention after the US-China trade war. The diversion of investment due to the trade war has affected the ownership structure and corporate governance of firms in the host countries. Beatson and Chen (2018) argue that foreign investors are considered as a significant governance mechanism in enhancing firms' efficiency. Furthermore, firms with good corporate governance, derived from foreign investors, tend to be passionate about high-risk projects (John et al., 2008). Thus, foreign ownership and corporate governance could affect business risk, especially in the unstable business environment of developing countries in ASEAN.

Considering the risks suffered from foreign investors, previous studies primarily discussed the issues for individual countries over the sample periods. Previous global financial crises have differential impacts on firms in ASEAN countries at different time scales. Therefore, it is interesting to further explore the effects of foreign ownership and corporate governance on term firm risk during the trade war. Hence, this study attempts to investigate the impacts of foreign ownership and corporate governance on firm risk and to further examine whether there exist different effects between developed and developing ASEAN countries amid the US-China trade war. Our empirical findings would help managers to understand the roles of foreign ownership and corporate governance during the trade war and to provide the guidelines in building their investment policies in ASEAN countries in the future.

The paper is structured as follows. Section two reviews the literature and hypothesis development. Section three presents the data and research methods, followed by empirical results. The next section offers a discussion and conclusion. Finally, the last section is limitations and future directions.

2 Hypothesis Development

The contest on whether foreign investors create stability for domestic securities has not been concluded. Opening up the local market to overseas shareholders is a fundamental matter in economic research. It could have both positive and negative reactions on economic gain and firms who accepted international participants. The US-China trade war has created a large change in the flow of foreign capital that leads to concerns for firms with foreign capital. Risks are significant interest for investors, managers, shareholders since it shows the possibility of survival or bankruptcy. According to agency theory, risk-taking is hugely impacted by capital structure.

Foreign capital is increasingly becoming a significant resource to economic growth in many developing countries. Given the existence of market imperfection in developing countries, the impact of foreign capital on firm risk is still debatable. Wright et al. (1996) argue that shareholders with significant stakes could shape the property of firm risk-taking, which impact firm's competitiveness and its existence. Naufa et al. (2019) ascertain that foreign investment attenuates return volatility, asset, and equity risk in six ASEAN markets. However, Chen et al. (2013) indicate that return fluctuations increased with the development of overseas investment in China due to high liquidity. Foreign ownership can cause business risks, which have been observed during financial crises (Fiala & Havranek, 2017). Nguyen et al. (2019) posit that stock risk will be higher when foreign investment capital increases.

The US-China trade war has induced a large change in the flow of foreign capital that leads to grave concerns about stock return volatility with foreign ownership. Based on above arguments, the following hypothesis is suggested:

Hypothesis 1: *Foreign ownership could affect firm risk in the ASEAN countries during the US-China trade war.*

Some researchers argued that corporate governance relates to increasing risks. Whitley (2000) elicits that the differences in governance practices have substantial influence on firm's risks, based on the ability to innovate. John et al. (2008) imply that the acceptance of higher-risk projects will increase in foreign-invested firms with better corporate governance. Investors are inevitably interested in firms with good governance, therefore, the increasing frequency of stock trading makes the price fluctuate, and the risk is also higher. Huang and Wang (2015) report a negative relation between board size and firm risk-taking in China. Hatane et al. (2019) find that board independence has a negative impact on total risks in Indonesia. Bernile et al. (2018) also evidence that greater board diversity leads to lower volatility and better performance of firms. Especially, Andries et al. (2020) posit an inverted U-shaped relationship between governance quality and investment risk.

In the context of the US-China trade war, the change of investment capital flows can create instability in corporate governance, leading to corporate risks. Accordingly, this research proposes the following hypothesis:

Hypothesis 2: *Corporate governance could affect firm risk in the ASEAN countries during the US-China trade war.*

3 Methodology

3.1 Fuzzy-Set Qualitative Comparative Analysis (fsQCA) Method

In this paper, we analyze data by applying fuzzy-set qualitative comparative analysis (fsQCA) which has more advantages than regression analysis in the extraction of outcomes. The commonly used analysis techniques such as analysis of variance (ANOVA) or structural equation model (SEM) treat variables independently and separately. The fsQCA analysis makes conditions combine with each other and incorporates ingredients conditions to results (Woodside & Zhang, 2013). Consequently, fsQCA provides logically possible causal combinations on variables to overcome the limitations of quantitative methods (Simón-Moya & Revuelto-Taboada, 2016). Furthermore, Fiss (2011) and Woodside (2012) demonstrate that fsQCA is well suited for both small and large samples. Therefore, this method is appropriate for our study with limited sample sizes of ASEAN countries.

The fsQCA software provides embracing of complex, parsimonious, and intermediate solutions (Fiss, 2011). Due to its high intelligibility, the intermediate solution is the preferred solution for interpreting the results (Ragin, 2008). Hence, this research chooses the intermediate solution for analyzing outcomes. We provide solutions with the parameters of consistency and raw coverage. Consistency of solution terms or whole solutions measures the extent of agreement among cases sharing a specific conjunction of conditions in describing target outcomes. Raw coverage measures experience according to importance, it evaluates the level to which an account associate's cause or effect to the case of a result or "how much of the outcome is covered (or explained) by each solution term or whole solutions" (Ragin, 2017). A certain causal combination may have small insurance with certain paths leading to a specific outcome. In other words, consistency is the same as the level of agreement between cases in describing the target outcome. Coverage shows how important empirical relevance is and that a combination might have a small raw coverage (Mas-Verdú et al., 2015; Ragin, 2008).

3.2 Sample and Data Collection

This research identifies the sample from the Datastream Thomson Reuters database over the period 2015–2018. Due to the data availability of corporate governance in some ASEAN countries, our final sample includes non-financial listed firms from Indonesia (29 firms), Philippines (18 firms), Thailand (49 firms), and Singapore (35 firms).

3.3 Variables Measurement

Risk (RISK) The standard deviation of daily returns as a measurement of total risk—a commonly used indicator in finance research (Herremans et al., 1993; McGuire et al., 1988; Rego et al., 2009).

Foreign ownership (FO) This study obtains foreign ownership data from Datastream, which is the percentage of strategic shareholdings held by foreign investors. It captures only strategic holdings that are required to report in public.

Corporate Governance Index (CGI) Abdullah (2004), Dahya et al. (2008) state that the board of directors is a key governance mechanism in business management. This study constructs a Corporate Governance Index, including board size, board diversity, split chairman/CEO role, independent board members, and non-executive members, to provide a comprehensive description of firm-level corporate governance. We first set (0, 1) scores for each component based on the level of (negative or positive) influence on corporate governance (Singareddy et al., 2018). Then, we calculate an overall corporate governance score for each firm after combining scores of five governance components to form the Corporate Governance Index. The Corporate Governance Index includes five components in Table 1 description of corporate governance construction variables.

In addition, we use three control factors—additional firm-specific variables for firm risk, including firm size, firm age, and leverage in our analysis.

The size of firm is computed by the logarithm of the total assets (Alonso et al., 2005). Leng (2004) argues larger firms are predicted to have greater profits due to the economies of scale and produce better firm performance. Firm's behavior can be impacted by firm size, and large firms avoid risks by diversifying risks with different investments or product lines.

Firm age is calculated by the logarithm of listing years. According to Gomes and Schmid (2010), managers of young firms prioritize growth strategies, leading to higher idiosyncratic risk. Firms with advanced age would have more experience in dealing with the financial crisis and related issues, capturing possible risks for adjustments to reduce risk to a minimum level.

Leverage is computed by the ratio of the total debts to its total assets. Leverage could prevent managers from engaging in low-profit investments (Jensen & Meckling, 1976). Leverage is one of corporate governance mechanisms that could develop

Table 1 Description of corporate governance construction variables

Variable	Description	Related research
Board size	The total number of board members	Khanchel (2007), Mak and Kusnadi (2005), Brown and Caylor (2009)
Board diversity	The percentage of female members on the board	Omar and Davidson (2001), Smith et al. (2006)
Split chairman/CEO role	The separation roles between chairman and CEO	Goyal and Park (2002), Khanchel (2007), Brown and Caylor (2009)
Independent board members	The percentage of independent board members on the board	Morck and Nakamura (1999), Elloumi and Gueyié (2001), Brown and Caylor (2009)
Non-executive members	The percentage of non-executive board members on the board	O'Sullivan and Wong (1999), Bova et al. (2015)

Source Compiled by the authors

performance. In addition, the finance theories have linked the relationship of leverage with a firm's ability on paying debt, the smaller the rate, the less risk firms face.

4 Results

4.1 Descriptive Statistics

The descriptive statistics for the model variables are presented in Table 2—descriptive statistics in ASEAN countries 2015–2018. After winsorizing outliers, the sample consists of 524 observations from 2015 to 2018. Table 2 discloses that the mean of all variables is positive. Among them, Singapore has the highest mean of foreign ownership ratio (0.207). The high level of foreign investment in Indonesia can be explained by the inward flows of foreign direct investment that achieved an impressive 460% increase from 2016 to 22 million US$ in 2018 (ASEAN Stats Data Portal, 2020).

During the period of US-China trade war, investors not only shifted investment locations but also restructured investment portfolios. Figure 1—foreign ownership rate by industry amid the trade war—reports the average percentage of foreign ownership for firms in ten industries (excluding the financial sector). Technology industry has experienced tremendous growth amid the trade war (the average of foreign ownership over the period 2015–2017 is 6.11%, 13.6% in 2018). Declines in foreign ownership rate are found in the fields of consumer discretionary, energy, health care, real estate, telecommunications, and utilities.

Table 2 Descriptive statistics in ASEAN countries 2015–2018

Variable	Mean	Std. dev	Min	Median	Max
Indonesia					
Firm risk (RISK)	2.364	0.543	1.310	2.330	3.640
Tobin's Q	2.674	3.104	0.593	1.248	11.894
Return on assets (ROA)	0.114	0.102	−0.006	0.076	0.396
Return on equity (ROE)	0.203	0.257	−0.165	0.151	1.215
Foreign ownership (FO)	0.191	0.294	0.000	0.000	0.790
Corporate Governance Index (CGI)	2.181	0.741	1.000	2.000	3.000
Firm size (SIZE)	6.384	0.334	5.602	6.355	6.962
Firm age (AGE)	1.259	0.206	0.778	1.362	1.447
Leverage (LEV)	0.201	0.148	0.000	0.213	0.481
Philippines					
Firm risk (RISK)	1.692	0.336	1.136	1.719	2.451
Tobin's Q	1.276	0.480	0.619	1.252	2.748
Return on assets (ROA)	0.062	0.020	0.026	0.061	0.095
Return on equity (ROE)	0.144	0.057	0.047	0.130	0.296
Foreign ownership (FO)	0.048	0.122	0.000	0.000	0.470
Corporate Governance Index (CGI)	2.611	0.832	1.000	3.000	4.000
Firm size (SIZE)	6.872	0.336	6.237	6.948	7.521
Firm age (AGE)	1.325	0.172	0.954	1.380	1.623
Leverage (LEV)	0.370	0.132	0.126	0.351	0.661
Thailand					
Firm risk (RISK)	1.596	0.891	0.000	1.525	3.890
Tobin's Q	1.952	1.411	0.639	1.439	5.872
Return on assets (ROA)	0.081	0.048	−0.012	0.078	0.180
Return on equity (ROE)	0.167	0.117	−0.053	0.164	0.404
Foreign ownership (FO)	0.011	0.040	0.000	0.000	0.230
Corporate Governance Index (CGI)	3.224	0.716	2.000	3.000	5.000
Firm size (SIZE)	6.642	0.468	5.768	6.739	7.329
Firm age (AGE)	1.315	0.122	1.041	1.342	1.477
Leverage (LEV)	0.315	0.159	0.015	0.315	0.577
Singapore					
Firm risk (RISK)	1.411	0.464	0.758	1.302	2.755
Tobin's Q	1.129	0.612	0.557	0.881	2.768
Return on assets (ROA)	0.051	0.032	0.002	0.045	0.127
Return on equity (ROE)	0.097	0.079	−0.031	0.080	0.323
Foreign ownership (FO)	0.207	0.289	0.000	0.060	0.840

(continued)

Table 2 (continued)

Variable	Mean	Std. dev	Min	Median	Max
Corporate Governance Index (CGI)	3.364	0.841	1.000	4.000	5.000
Firm size (SIZE)	6.809	0.583	5.564	6.836	7.929
Firm age (AGE)	1.346	0.195	1.041	1.362	1.653
Leverage (LEV)	0.243	0.143	0.008	0.239	0.515

Source Compiled by the authors

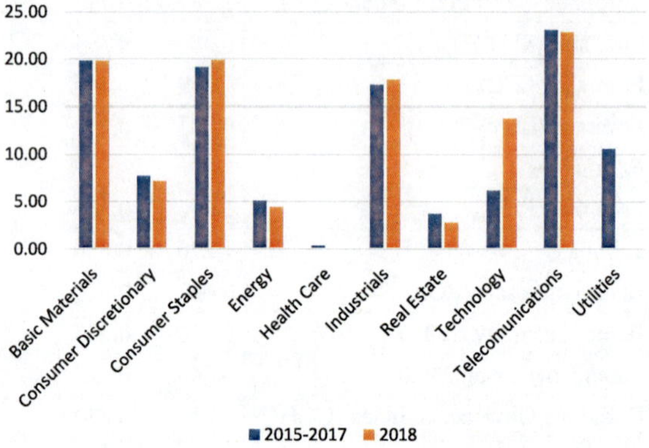

Fig. 1 Foreign ownership rate by industry amid the trade war. *Source* Compiled by the authors

Figure 2 average of Corporate Governance Index 2015–2018 by country—shows the rank of the Corporate Governance Index by country. It reveals that the developed country (Singapore) has better corporate governance than the developing ones. The quality of corporate governance depends on the level of economic and political development, as well as the governance requirements of country.

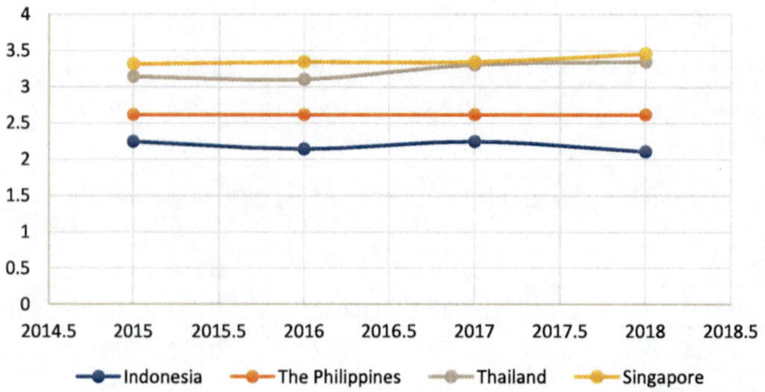

Fig. 2 Average of corporate governance index 2015–2018 by country. *Source* Compiled by the authors

4.2 Foreign Ownership and Corporate Governance on Firm Risk Amid the US-China Trade War

Before the US-China trade war

For the effect on firm risk before the US-China trade war, fsQCA yields solution terms with consistency (SC) greater than 0.9 (minimum consistency = 0.906, maximum = 0.964) (Table 3—FsQCA results—foreign ownership and corporate governance on firm risk). It demonstrates that all solutions give strong and necessary support to firm risk. Foreign ownership has significantly positive impacts on the risk of listed firms in Indonesia. The positive effects of foreign ownership are also confirmed in the remaining three countries—Philippines, Thailand, and Singapore. Our findings are generally consistent with Nguyen et al. (2019).

Table 3 reports different relationships for the Corporate Governance Index among countries. The variations of corporate governance cause more risks in Indonesia before the trade war. However, mixed effects on firm risks (cg_c or ~cg_c) are found in other countries. The rising of foreign ownership with low or high quality of corporate governance leads to higher firm risk in the Philippines, Thailand, and Singapore. The mixed effects on firm risk are also shown for the control variables of firm size and listing age.

The findings for developed country like Singapore are similar to developing ASEAN countries. Foreign ownership in Singapore also has a positive impact on firm risk before the trade war. The fsQCA results for firm risk during the period before the trade war support Hypotheses 1 and 2.

After the US-China trade war

Compared to the results for the period before the trade war, our findings for the period after the trade war have become more complicated. Although most solutions after the trade war are necessary, some cases are sufficient in Indonesia and Philippines (the lowest consistency in Indonesia is 0.784). The differences between Singapore and developing countries (Indonesia, Philippines, and Thailand) are revealed more clearly. Both foreign ownership and corporate governance in Singapore have positive effects on firm risk, while they produce mixed effects in other ASEAN countries. The findings after the trade war also support Hypotheses 1 and 2 that foreign ownership and corporate governance could affect firm risk.

The causal combinations in three developing countries of Indonesia, Philippines, and Thailand indicate that foreign ownership could produce mixed (positive or negative) effects on firm risk after the trade war. The perfect subset is especially found in the Philippines after the trade war ($C = 0.202$, SC = 1.000). Compared to the solutions before the trade war, increased foreign capitals in developing ASEAN countries during the war could play multiple roles by reducing or increasing firm risk. However, the role of foreign capitals in the developed ASEAN country—Singapore—remains the same before or after the trade war. The results imply that the US-China trade war has no effect on the status of foreign ownership in the developed ASEAN country. Table 3 also indicates that the roles of corporate governance in the sample countries

Table 3 FsQCA results—foreign ownership and corporate governance on firm risk

Country	2015–2017			2018		
	Causal recipe	Raw coverage	Consistency	Causal recipe	Raw coverage	Consistency
Indonesia	fo_c*cg_c*~size_c*~age_c*~lev_c fo_c*cg_c*size_c*~age_c*lev_c	0.134 0.136	0.920 0.906	~cg_c*~size_c*~age_c*~lev_c fo_c*~size_c*~age_c*~lev_c ~fo_c*cg_c*~size_c*~age_c*lev_c	0.366 0.225 0.235	0.784 0.828 0.843
	Solution coverage: 0.188 Solution consistency: 0.886			Solution coverage: 0.557 Solution consistency: 0.812		
Philippines	fo_c*~cg_c*~size_c*age_c*lev_c fo_c*cg_c*size_c*age_c*lev_c	0.145 0.157	0.963 0.953	~fo_c*cg_c*~size_c*~lev_c ~fo_c*cg_c*size_c*~age_c fo_c*~size_c*age_c*lev_c	0.434 0.383 0.202	0.951 0.872 1.000
	Solution coverage: 0.195 Solution consistency: 0.936			Solution coverage: 0.695 Solution consistency: 0.925		
Thailand	fo_c*cg_c*~size_c*age_c*~lev_c fo_c*~cg_c*size_c*age_c*~lev_c	0.103 0.087	0.909 0.907	~fo_c*~cg_c*size_c*~age_c*~lev_c fo_c*~cg_c*~size_c*age_c*lev_c	0.245 0.081	0.905 0.942
	Solution coverage: 0.119 Solution consistency: 0.908			Solution coverage: 0.278 Solution consistency: 0.915		
Singapore	fo_c*~cg_c*size_c*lev_c fo_c*size_c*age_c*lev_c fo_c*cg_c*~size_c*~age_c*~lev_c	0.172 0.181 0.133	0.944 0.930 0.964	fo_c*cg_c*~size_c*~age_c*~lev_c fo_c*cg_c*size_c*age_c*lev_c	0.196 0.144	0.952 0.909
	Solution coverage: 0.260 Solution consistency: 0.939			Solution coverage: 0.264 Solution consistency: 0.933		

Source Extracted from fsQCA software

have changed amid the trade war. Under the investment relocations from the trade war, ASEAN countries have encountered fluctuations as well as business risks due to unstable corporate governance.

This study demonstrates that fsQCA can better comprehend the drives affecting firm risk in the ASEAN region during the US-China trade war. In addition, fsQCA allows us to analyze the effects of combination factors on firm risk. fsQCA, compared to regression methods, not only provides a wide variety of combinations but also offers a set of effective management solutions. Moreover, fsQCA results could evidence inefficient combinations which managers should avoid engaging.

5 Discussions and Conclusions

This research attempts to study the combined impacts of overseas ownership and corporate governance on firm risk of ASEAN countries amid the US-China trade war. To conduct the analysis, fsQCA method is applied to investigate the impact of foreign ownership and governance on firm risk with the data from four ASEAN countries.

The role of overseas investment in business operations is indisputable. In the sample of four ASEAN countries, the increases in foreign investment led to more threats before the trade war. The trade war started by the US government had an impact on ASEAN member states due to tight relationships of ASEAN countries and both US and China. After the trade war, risks in developing countries tend to decrease due to the benefits from foreign investors in developing countries. Sharing financial risks and the supportive contribution in business management from foreign ownership is a key factor in reducing risks and enhancing business performance in these regions. By contrast, Singapore is a country with a majority of foreign investment firms that may lack business motivation by guaranteed foreign capital. In addition, local firms in Singapore may struggle to find opportunities to develop among firms with large capital investments in their countries. As our empirical results, corporate governance directly affects firm risk during the US-China trade war in ASEAN countries. This study indicates the relative advantages of corporate governance in mitigating risks. However, these effects differ depending on the development level of each ASEAN country.

Foreign investment from strategic investors is not merely capital but is accompanied by advanced technologies and management skills, allowing the creation of new products and opening new markets for receiving firms. Good corporate governance is central to sustaining the growth of business, especially in ASEAN countries. To close the gap and catch up with developed countries, firms in ASEAN developing countries need to quickly access advanced technologies, strategically manage foreign capitals, and develop a stable corporate governance system for the long run. Developing countries can draw on their experience in managing foreign investment capital and building corporate governance according to developed countries in the same region to limit risks.

This research contributes to the literature by providing possible combinations of foreign ownership and corporate governance lead to business risk. Our findings would fill the gap in understanding the effects of the location shift of foreign capitals during the trade war. It would help investors and managers with better understanding on investing and corporate governance situation in ASEAN markets. The implications of the results from the qualitative model provide useful guidelines to researchers and practitioners.

Our study has some limitations. Firstly, the sample listed firms include only four countries in the ASEAN (Indonesia, Philippines, Thailand, and Singapore) due to the availability of data. As a result, our findings may not be valid for all ASEAN countries. The future research could expand the samples by including more ASEAN countries if data are available. Secondly, the data are taken from 2015 to 2018, including only one year after the US-China trade war. Potential studies could examine datasets with extended periods after the trade war. Thirdly, this study uses a sample of listed firms. It is somewhat uncertain whether the conclusions we have obtained hold for private firms. Future studies on private firms will give us a more comprehensive picture of these relationships in ASEAN. Finally, the Corporate Governance Index used in this context captures only the board structure variables of firms. Other corporate governance mechanisms could be considered in the future research.

References

Abdullah, S. N. (2004). Board composition, CEO duality and performance among Malaysian listed companies. *Corporate Governance, 4*(4), 47–61. https://doi.org/10.1108/14720700410558871

Alonso, P. D. A., Iturriaga, F. J. L., & Sanz, J. A. R. (2005). Financial decisions and growth opportunities: A Spanish firm's panel data analysis. *Applied Financial Economics, 15*(6), 391–407. https://doi.org/10.1080/09603100500039201

Andries, A. M., Balutel, D., Ihnatov, I., & Ursu, S. G. (2020). The nexus between corporate governance, risk taking, and growth. *PLoS ONE, 15*(2), e0228371. https://doi.org/10.1371/journal.pone.0228371

ASEAN Investment Report. (2018). https://asean.org/storage/2018/11/ASEAN-Investment-Report-2018

ASEAN Secretariat. (2018). Asean Key Figures. *The Statistics Division (ASEANstats) of the ASEAN Secretariat*, 1–51. https://www.aseanstats.org/wp-content/uploads/2018/12/ASEAN-Key-Figures-2018.pdf

ASEAN Stats Data Portal (2020). https://data.aseanstats.org/fdi-by-hosts-and-sources.

Beatson, S. A., & Chen, J. (2018). Foreign investment, corporate governance and performance in the Chinese listed a share companies. *Journal of Chinese Economic and Business Studies, 16*(1), 59–93. https://doi.org/10.1080/14765284.2017.1346931

Bernile, G., Bhagwat, V., & Yonker, S. (2018). Board diversity, firm risk and corporate policies. *Journal of Financial Economics, 127*, 588–612. https://doi.org/10.1016/j.jfineco.2017.12.009

Bova, F., Kolev, K., Thomas, J. K., & Zhang, F. (2015). Non-executive employee ownership and corporate risk. *The Accounting Review, 90*(1), 115–145. https://doi.org/10.2308/accr-50860

Brown, L. D., & Caylor, M. L. (2009). Corporate governance and firm operating performance. *Review of Quantitative Finance and Accounting, 32*, 129–144. https://doi.org/10.1007/s11156-007-0082-3

Burker, M., Franco, C., & Minerva, G. A. (2013). Foreign ownership, firm performance, and the geography of civic capital. *Regional Science and Urban Economics, 43*(2013), 964–984. https://doi.org/10.1016/j.regsciurbeco.2013.09.009

Chen, Z., Du, J., Li, D., & Ouyang, R. (2013). Does foreign institutional ownership increase return volatility? Evidence from China. *Journal of Bank and Finance, 37*(2), 660–669. https://doi.org/10.1016/j.jbankfin.2012.10.006

Dahya, J., Dimitrov, O., & McConnell, J. J. (2008). Dominant shareholders, corporate boards, and corporate value: A cross-country analysis. *Journal of Financial Economics, 87*(1), 73–100. https://doi.org/10.1016/j.jfineco.2006.10.005

Elloumi, F., & Gueyié, J. (2001). Financial distress and corporate governance: An empirical analysis. *Corporate Governance, 1*(1), 15–23. https://doi.org/10.1108/14720700110389548

Fiala, T., & Havranek, T. (2017). The sources of contagion risk in a banking sector with foreign ownership. *Economic Modelling, 60*, 108–121. https://doi.org/10.1016/j.econmod.2016.08.025

Fiss, P. C. (2011). Building better causal theories: A fuzzy set approach to typologies in organization research. *Academy of Management Journal, 54*(2), 393–420. https://doi.org/10.5465/amj.2011.60263120

Gibson, M. S. (2003). Is corporate governance ineffective in emerging markets? *Journal of Financial and Quantitative Analysis, 38*, 231–250. https://doi.org/10.2307/4126771

Globerman, S., Ries, J. C., & Vertinsky, I. (1994). The economic performance of foreign affiliates in Canada. *The Canadian Journal of Economics, 27*(1), 143–156. https://doi.org/10.2307/135807

Gomes, J., & Schmid, L. (2010). Levered returns. *Journal of Finance, 65*, 467–494. https://doi.org/10.1111/j.1540-6261.2009.01541.x

Goyal, V., & Park, C. W. (2002). Board leadership structure and CEO turnover. *Journal of Corporate Finance, 8*(1), 49–66. https://doi.org/10.1016/S0929-1199(01)00028-1

Hatane, S. E., Supangat, S., Tarigan, J., & Jie, F. (2019). Does internal corporate governance mechanism control firm risk? Evidence from Indonesia's three high-risk sectors. *Corporate Governance, 19*(6), 1362–1376. https://doi.org/10.1108/CG-02-2019-0071

Herremans, I. M., Akathaporn, P., & McInnes, M. (1993). An investigation of corporate social responsibility reputation and economic performance. *Accounting, Organizations and Society, 18*, 587–604. https://doi.org/10.1016/0361-3682(93)90044-7

Huang, Y. S., & Wang, C.-J. (2015). Corporate Governance and risk-taking of Chinese firms: The role of board size. *International Review of Economics and Finance, 37*, 96–113. https://doi.org/10.1016/j.iref.2014.11.016

Jensen, M. C., & Meckling, W. H. (1976). Theory of the firm: Managerial behavior, agency costs, and capital structure. *Journal of Financial Economics, 3*(4), 305–360. https://doi.org/10.1016/0304-405X(76)90026-X

John, K., Litov, L., & Yeung, B. (2008). Corporate governance and risk-taking. *The Journal of Finance, 63*(4), 1679–1728. https://doi.org/10.1111/j.1540-6261.2008.01372.x

Johnson, S., Boone, P., Breach, A., & Friedman, E. (2000). Corporate governance in the Asian Financial Crisis. *Journal of Financial Economics, 58*, 141–186. https://doi.org/10.1016/S0304-405X(00)00069-6

Khanchel, I. (2007). Corporate governance: Measurement and determinant analysis. *Managerial Auditing Journal, 22*(8), 740–760. https://doi.org/10.1108/02686900710819625

Leng, A. C. A. (2004). The impact of corporate governance practices on firms' financial performance: Evidence from Malaysian Companies. *ASEAN Economic Bulletin, 21*(3), 308–318. *Project MUSE* muse.jhu.edu/article/387800

Mangena, M., & Tauringana, V. (2007). Disclosure, corporate governance and foreign share ownership on the Zimbabwe stock exchange. *Journal of International Financial Management and Accounting, 18*(2), 53–85. https://doi.org/10.1111/j.1467-646X.2007.01008.x

Mak, Y., & Kusnadi, Y. (2005). Size really matters: Further evidence on the negative relationship between board size and firm value. *Pacific-Basin Finance Journal, 13*, 301–318. https://doi.org/10.1016/j.pacfin.2004.09.002

Mas-Verdú, F., Ribeiro-Soriano, D., & Roig-Tierno, N. (2015). Firm survival: The role of incubators and business characteristics. *Journal of Business Research, 68*(4), 793–796. https://doi.org/10.1016/j.jbusres.2014.11.030

McGuire, J. B., Sundgren, A., & Schneeweis, T. (1988). Corporate social responsibility and firm financial performance. *Academy of Management Journal, 31*, 854–872. https://doi.org/10.5465/256342

Morck, R., & Nakamura, M. (1999). Banks and corporate control in Japan. *The Journal of Finance, 54*(1), 319–339. https://doi.org/10.1111/0022-1082.00106

Naufa, A. M., Lantara, W. N., & Lau, W.-Y. (2019). The impact of foreign ownership on return volatility, volume, and stock risks: Evidence from ASEAN countries. *Economic Analysis and Policy, 64*, 221–235. https://doi.org/10.1016/j.eap.2019.09.002

Nguyen, A. P., Nguyen, H. A., Ho, T. H. M., & Ngo, P. T. (2019). Risk and returns of different foreign ownership portfolios: Evidence from Vietnam stock market. *Cogent Economics and Finance, 7*, 1–12. https://doi.org/10.1080/23322039.2019.1589412

Omar, A., & Davidson, M. J. (2001). Women in management: A comparative cross-cultural overview. *Cross Cultural Management: An International Journal, 8*(3/4), 35–67. https://doi.org/10.1108/13527600110797272

O'Sullivan, N., & Wong, P. (1999). Board composition, ownership structure and hostile takeovers: Some UK evidence. *Accounting and Business Research, 29*, 139–155. https://doi.org/10.1080/00014788.1999.9729575

Ragin, C. C. (2008). *Redesigning social inquiry: Fuzzy sets and beyond*. Wiley Online Library.

Ragin, C. C. (2017). *User's guide to fuzzy-set/qualitative comparative analysis*. Department of Sociology, University of California.

Rego, L. L., Billett, M. T., & Morgan, N. A. (2009). Consumer-based brand equity and firm risk. *Journal of Marketing, 73*, 47–60. https://doi.org/10.1509/jmkg.73.6.47

Simón-Moya, V., & Revuelto-Taboada, L. (2016). Revising the predictive capability of business plan equity for new firm survival using qualitative comparative analysis. *Journal of Business Research, 69*(4), 1351–1356. https://doi.org/10.1016/j.jbusres.2015.10.106

Singareddy, R. R. R, Chandrasekaran, S., Annamalai, B., & Ranjan, P. (2018). Corporate governance data of 6 Asian economies (2010–2017). *Data in Brief, 20*, 53–56. https://doi.org/10.1016/j.dib.2018.07.048

Smith, N., Smith, V., & Verner, M. (2006). Do women in top management affect firm performance? A panel study of 2500 Danish firms. *International Journal of Productivity and Performance Management, 55*(7), 569–593. https://doi.org/10.1108/17410400610702160

Stein, J. (1988). Takeover threats and managerial myopia. *Journal of Political Economy, 96*, 61–80. https://doi.org/10.1086/261524

Stein, J. (1989). Efficient capital markets, inefficient firms: A model of myopic corporate behavior. *Quarterly Journal of Economics, 104*, 655–669. https://doi.org/10.2307/2937861

U.S. Mission to ASEAN (2019). The United States and ASEAN: An enduring partnership. https://asean.usmission.gov/the-united-states-and-asean-an-enduring-partnership/

Whitley, R. (2000). The institutional structuring of innovation strategies: Busines-systems, firm types and patterns of technological change in different market economies. *Organization Studies, 21*, 855–886. https://doi.org/10.1177/0170840600215002

Woodside, A. G. (2012). Proposing a new logic for data analysis in marketing and consumer behavior: Case study research of large-N survey data for estimating algorithms that accurately profile X (extremely high-use) consumers. *Journal of Global Scholars of Marketing Science, 22*(4), 277–289. https://doi.org/10.1080/21639159.2012.717369

Woodside, A. G., & Zhang, M. (2013). Cultural diversity and marketing transactions: Are market integration, large community size, and world religions necessary for fairness in ephemeral exchanges? *Psychology and Marketing, 30*(3), 263–276. https://doi.org/10.1002/mar.20603

Wright, P., Ferris, S. P., Sarin, A., & Awasthi, V. (1996). Impact of corporate insider, blockholder, and institutional equity ownership on firm risk taking. *Academy of Management Journal, 39*(2), 441–458. https://doi.org/10.5465/256787

Enhancing Social Housing Governance in Vietnam Through Dynamic Performance Management—Toward Sustainable Social Housing

Minh Quang Le and Kieu Van Nguyen

Abstract The overarching objective of social housing is to meet housing needs, particularly those of the vulnerable households—low- and middle-income families. However, there is an evidence to show that social housing is not adequately supported to achieve sustainable goals despite its significance for addressing the housing crisis. This paper firstly discusses the understanding of concept sustainable social housing, then illustrates how the "Dynamic Performance Management" framework is able to support policy networks to pursue sustainable outcomes in social housing governance. Main drivers of social housing have been recognized in the increasing population size, as well as the in the urbanization of big cities, the growing number of nuclear families, affordability, and land management and urban planning. The management of social housing—as a global concern—requests that local communities and regional government are ready to tackle such a wicked problem effectively. This concern is the motivation of this research. Therefore, the work would offer a contribution to improve the social housing governance and to support policymakers in adopting a more integrated approach to the national housing sector.

Keywords Social housing · Sustainable social housing · Dynamic performance management

1 Introduction

The concept of sustainable development (SD) can be defined as the achievement of a better quality of life through the efficient use of resources, which realizes continued social progress while maintaining stable economic growth and caring for the environment (Essa & Fortune, 2008). The United Nations' report defines SD as "development that meets the needs of the present without compromising the ability of future generations to meet their own needs" (United Nations General Assembly, 1987). The two vital common concepts contained in the above two SD definitions are: the concept of

M. Q. Le (✉) · K. Van Nguyen
Department of Leadership Studies and Public Policy, Academy of Politics Region I, 15 Khuat Duy Tien, Thanh Xuan Dist., Hanoi, Vietnam
e-mail: lequangminh.hcma1@gmail.com

A. T. Nguyen et al. (eds.), *Contemporary Economic Issues in Asian Countries: Proceeding of CEIAC 2022, Volume 1*,
https://doi.org/10.1007/978-981-19-9669-6_31

needs, which seeks to ensure that the essential needs of the poor are adequately met; and the need for addressing every limitation arising from the use of technology and activities of social elements affecting the environment's ability to meet the present and future needs.

Based on the aforementioned two concepts, social housing should adequately meet the housing needs of the vulnerable households on a continuous basis and, at the same time, consider the environmental limitations while meeting such needs both in the present and future in relation to the development techniques and social components. However, it requires economic means to provide social service or actualize not-for-profit motive in making it available to beneficiaries and putting environmental protection into consideration while taking social housing development decisions. Sustainability issues are bound to arise where appropriate measures are not adequately and adequately linked together in social housing delivery. Therefore, sustainable social housing (SSH) is described as housing that is environmentally friendly, built from recycled materials or other natural resources, and energy-efficient by using alternatives such as solar power (Shelter England, 2016). Oyebanji et al. (2013) view SSH as a form of:

> Housing that is made available by governments and/or non-profit organizations through various assisted housing programmes, built with environmental friendly and sustainable materials, have a long-term economic, environmental and social benefits without an increased life-cycle cost, and allowing not only the present but also the future generations to meet their housing needs on the overall social value basis.

Thus, given the nature of providers, the state and not-for-profit organizations with social or non-profit motives, economic sustainability in social housing can be achieved through various reliable schemes like affordable rents, purchase through mortgage loans at low-interest rates, and other forms of subsidies (Cooper & Jones, 2008). This can facilitate SSH delivery on a continuous basis and be financially sustainable over the long term for both the providers and beneficiaries. It can help to avoid waste of natural resources by ensuring that sufficient land of the right type is available in the right places and at the right time to support growth and innovation; and by identifying and coordinating development requirements, including the provision of infrastructure (Department of Communities & Local Government, 2016).

Achieving economic sustainability in social housing has some advantages not only for increasing the supply but for promoting the country's economic growth (Higgins, 2013). For example, the more the supply of SSH, the more the development of the nation's assets, provision of job opportunities, the use of recyclable materials, and modern technology for adequate provision of social housing to meet housing needs. From an environmental perspective, achieving sustainability in social housing requires taking cognizance of the fact that housing and the environmental impact on each other. This happens in a number of significant ways in terms of carbon emissions, land take, water usage, sewerage, and flooding (Shelter England, 2007). Achieving sustainability in social housing also requires the understanding that human beings and the environment are two inseparable components that must support each other in the SD process (Pattinaja & Putuhena, 2010). Environmental sustainability in social housing can help to improve biodiversity, use natural resources prudently, minimize waste and pollution, and mitigate and adapt to climate change, including moving to a low-carbon economy (Department of Communities & Local Government, 2016).

In addition, the concept of social sustainability in social housing seeks to recognize the diverse nature of the occupiers in terms of cultural backgrounds, the pattern of lives, size of households, and housing needs (Oyebanji, 2014). Achieving SSH is all-embracing in the social context as it gives room for social interaction, security and convenience, access for spiritual development, education, public health facilities, and natural resources (Pattinaja & Putuhena, 2010). SSH has the capacity for supporting strong, vibrant, and healthy communities by providing the supply of housing required to meet the needs of the present and future generations; and by creating a high-quality built environment, with accessible local services that reflect the community's needs and support its health, social, and cultural wellbeing (Department of Communities & Local Government, 2016). Furthermore, social sustainability combines the design of social housing with a focus on how the people living and using it relate to each other and function as a community, including the provision of the right infrastructure to support a healthy social and cultural life, opportunities for people to get involved, and scope for the place and the community to evolve (Dixon & Woodcraft, 2016).

2 Framing Social Housing Governance from a Sustainable Outcome-Based Perspective

In order to analyze social housing governance from a sustainable outcome-based perspective, a multidisciplinary approach is required (Guerrero et al., 2013; Marshall & Farahbakhsh, 2013; Morrisey & Browne, 2004). In social housing governance, environmental, social, and economic issues are highly intertwined. Framing and understanding the long-term effect of policies on wider-system performance are crucial for a sustainable approach to social housing management.

Davies (2008) conceptualized the relationship between governance and social housing as social relations, which also includes the "roles and responsibilities of actors and institutions." The perspective of governance "allows for consideration not only of technical matters or scientific analyses but also of the social, cultural, political and economic contexts and networks" (Davies, 2008, p. 15). Sustainability is clearly defined by Diesendorf (2000) as the endpoint of sustainable development, which corresponds to "types of economic and social development that protect and enhance the natural environment and social equity." It was the World Commission on Environment and Development which published its report to put the concept of "sustainable development" into the public debate. The concept was defined as "development that meets the needs of the present without compromising the ability of future generations to meet their own needs" (World Commission on Environment & Development, 1987). However, there is no unanimously approval on the concept of sustainability. Within social housing governance studies, sustainability has been described as a socio-political construct, which is different from the definition provided in the World Commission on Environment and Development Report (World Commission on Environment & Development, 1987). Indeed, it has become common in the language of governance, but it has led to alternative approaches toward its core meaning: from an eco-centric to an anthropocentric view. Such an alternative has led to contradiction to what should be considered the leading solution in terms of sustainability (Kaika & Swyngedouw, 2000).

This flaw was clearly recognized by Swyngedouw and Heynen (2003). The authors stated that "there is no such thing as an unsustainable city in general. Rather, there are a series of urban and environmental processes that negatively affect some social groups while benefiting others." Therefore, a sustainable program can be compromised by conflicting beliefs, leading to consensual conclusions that potentially move away from original sustainable policy intentions. Today, societies and communities ask that social housing management should be not only safe but must also be sustainable. According to this political request, sustainable social housing management must be:

- economically affordable;
- socially acceptable;
- environmentally effective.

As Fig. 1 shows, policymakers must consider each dimension equally; otherwise, the whole system will become unbalanced (McDougall et al., 2008).

Traditionally, the social housing management system was managed mainly according to service concern, which took over the other aspects of the service. The management of social housing excludes all development activities related to

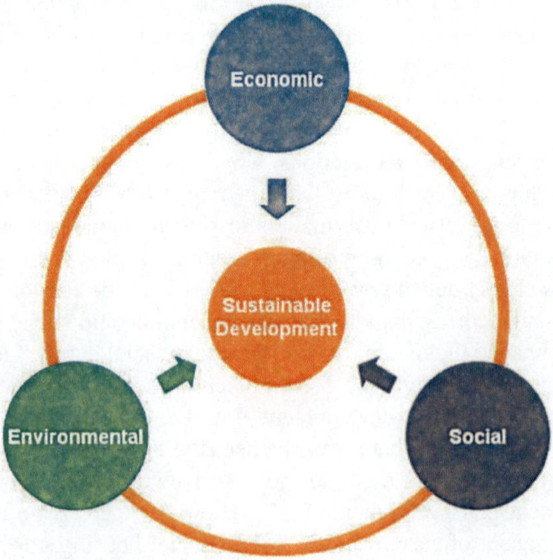

Fig. 1 Three dimensions of sustainable development. *Source* Authors' drawing

the production of housing services from new social housing (Priemus et al., 1999). In the last ten years, environmental and social matters have acquired increasing attention and become more prominent in the political debate. Social aspects of social housing management have an influence on the decision-making process; however, few researchers have investigated how to measure it. Environmental issues in social housing management can be characterized into two streams: environmental protection and appropriate land-use plan.

To connect the abovementioned pillars, it is necessary that the social housing management system adopts the perspective of sustainability. To this end, it is necessary to understand those main issues which affect social housing management. This is a prerequisite to embody the concept of sustainability within the integrated social housing management system.

3 A Dynamic Performance Management Approach to Enhance Sustainability in the Social Housing Management System

At the core of dynamic performance management, there is the instrumental views of performance, which provide a framework to assess performance sustainability.

Dynamic performance management is an approach that matches traditional performance management methods and techniques with system dynamics modeling (Bianchi, 2002, 2012, 2016). Based on a learning-oriented approach, dynamic performance management may support decision-makers to frame better the policymakers involved in a policy field (i.e., the relevant system) and to design sustainable policies concerning outcomes.

Dynamic performance management moves from synthesis to analysis and is a conceptual framework based on three layers: end-result, performance drivers, and strategic resources (Fig. 2).

The first step in applying dynamic performance management is the identification of end-results (both outcomes and outputs). If the first step in applying the *instrumental view* is the identification of end- results, the second phase requires recognizing performance drivers. Performance drivers compare the actual performance in terms of efficiency, productivity, and effectiveness in the use of the set of strategic resources against a benchmark: for instance: *skills/desired skills* affect service delivery failure rate; *actual service time/expected users waiting time* affects the users' satisfaction; the number of *administrative tasks/administrative tasks threshold* affects *the costs per paperwork.*

As a third step, dynamic performance management supports decision-makers to outline the policies to adopt in order to affect the strategic resources (i.e., the stocks of tangible and intangible factors to build-up and deploy together with others) that will influence performance drivers, and through them, the end-results will feedback on the strategic resources making policy sustainable (Bianchi, 2016).

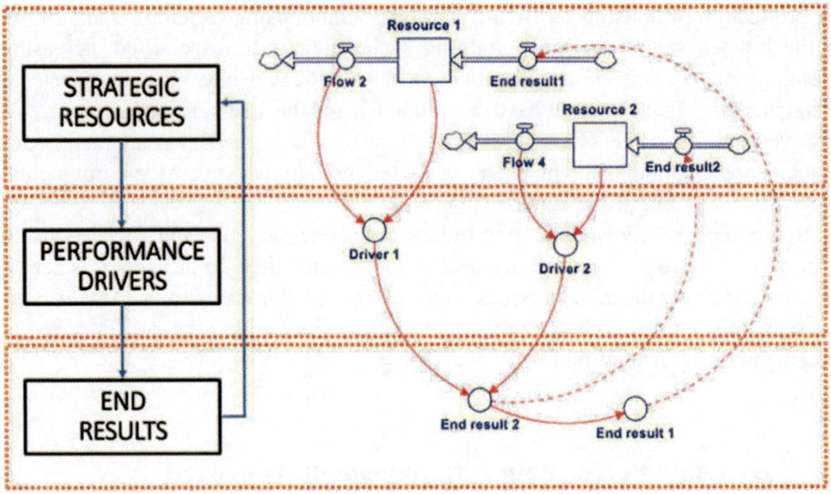

Fig. 2 Dynamic view of performance management. *Source* Authors' drawing

In complex governance settings, such as social housing management, immigration, or health care, strategic resources are rarely managed or controlled by the same institution (e.g., municipality, hospital, ministers); they are rather managed by the different organizations and players. Dynamic performance management may support each organization in a network in identifying how it may contribute to the achievement of wider outcomes in the relevant system (Bianchi, 2016; Bianchi et al., 2017; Cosenz & Bivona, 2018). As Bianchi (2016) stated: "the instrumental view implies that alternative means for improving performance be made explicit. In this regard, it is necessary to identify both end-results and their respective drivers. To affect such drivers, each responsibility area must build-up, preserve, and deploy a proper endowment of strategic resources that are systemically linked to each other." End-results provide endogenous sources of accumulation and/or depletion of resources which are strategic for the performance, e.g., cash flow accumulates into the bank account; the rate of the problem solved at customer services depletes the backlog of problems to be solved.

End-results are flows which capture both output and outcomes, and they can be modeled as in-and-out flows of strategic resources. Strategic resources can be classified in physical resources referring to the ones which can be purchased on the market (inventory, employee, capacity) and resources generated by management (internal) routines (reputation, organizational climate, skills, solvency) that can be obtained only through efficiency or effectiveness of operations.

4 Applied Dynamic Performance Management and Suggested Model for Achieving Sustainable Social Housing in Vietnam

The overall concept of the model comes from the common failures of policies to build social housing in the past. This policy was originally intended to address the housing shortage caused by many urban dwellers unable to afford commercial housing. The root cause of this problem is that high population growth rate in a small part of city and the city expansion planning has not met this population growth.

The local authorities have initially addressed this problem by planning the construction of social housing in the suburbs. The outstanding features of these planned areas are long distance from city center, low land prices but lack of basic infrastructure (public transport, schools, community areas, parks, hospitals, etc.) and public services. When building social housing massively in these areas, local authorities encountered a problem that social housing was not filled by people who lack housing, but they continued to choose the solution of renting houses in urban area. For examples, in Hanoi, there is a paradox that many social housing projects have been currently open for sale multiple times, but social houses still cannot be sold, such as the Bamboo Garden social housing project (Quoc Oai, Hanoi), of CEO Group, after 19 sale launches, has not yet sold out despite the price below 10 million VND/m^2. And the social housing project, AZ Thang Long (Hoai Duc, Hanoi), invested by Thang Long Confectionery Co., Ltd. must also open for sale for the 11th time. Or the social housing project in plot 5. B2 (Dong Hoi, Dong Anh, Hanoi), developed by Thang Long Co., Ltd., total of 504 apartments cannot be sold over 7 sales launches (Trang, 2020).

Therefore, the initial lack of social housing problem has not only been solved, but also raised another problem. The absence of people in these new social housing area leads to the deterioration of buildings and the deterioration of the value of the buildings themselves. This failure is the waste of public investment that will undermine the credibility of local governments. It is important to understand that public investments related to social housing in the future depend significantly on the reputation of local authorities. For example, in Vietnam, when local governments receive funding from the state to build social housing but these projects are ineffective (social housing cannot be sold), then the possibility of receiving funds books to continue building social housing will be lower in the coming years. This budget will be distributed to other localities that have more efficient social housing.

From the above fact, we argue that the policy on building social housing cannot be considered as a single problem, only solved by one agency/ministry. This issue requires the cooperation and settlement of responsible agencies to formulate policies.

Social housing must be located in an ecosystem full of essential infrastructure and public services mentioned above, in which public transport, community living areas, schools, hospitals play a key role. The fulfillment of these infrastructure areas must be done in parallel with social housing. The argument for a policy implying a systematic investment in social housing and related essential infrastructure may be subject to criticism for its impracticality, due to, for instance, inadequate capital. I argue that the problem of lack of capital cannot be solved purely by public investment efforts. Social housing itself should not only be considered as an area for low-income people, but it must have its own economy.

The mechanism that makes up the economy of social housing areas is essential infrastructure not only serving the needs of social housing buyers, but also for the other stakeholders in the area. Furthermore, we cannot look at the public infrastructure and services that accompany social housing in a closed ecosystem but in an open system. On the one hand, public infrastructure and services enhance the quality of life, which in the medium term increases the value of housing complexes in general, including social and commercial housing; infrastructure and public services in social housing areas will attract businesses to invest.

Through taxation, initial public investment can be counterbalanced by the two mechanisms mentioned above, and this amount can continue to be reinvested to continue building social housing according to the above ecosystem idea.

The following section will discuss how DPM can build a sustainable social housing policy for Vietnam. In this section, computer simulation modeling will not be used as a tool to develop policy. As remarked by Wolstenholme (1999) "computer simulation modeling adds significant value to qualitative mapping by enabling deeper and more rigorous analysis." However, "the problem with computer simulation is that data are often not available and hence models are still speculative. Furthermore, they are often idealized representations of the world constrained by the restrictive nature of the feedback paradigm and the language of stock-flow diagrams [...] there is always a tendency to produce models which are too detailed and complex to sufficiently validate them against the mental models of their creator."

In order to feed performance management with simulation models that may portray suggested policies through DPM charts and reports, a more pervasive effort than qualitative modeling is needed. This does not only require detailed data, but also an active collaboration of decision-makers with basic system dynamics modeling skills.

4.1 A Causal Perspective of Social Housing in Vietnam

Figure 3 portrays main feedback loops describing the development and potential decline process of social housing in Vietnam. The liquidity comes from three sources, the revenue from selling social houses, taxation of commercial houses sale, and public funding from central government. The key point is instead of using liquidity to build more social houses, the liquidity will be invested in infrastructure and public services. The reinforcing loop R1 describes how to raise the liquidity through enhancing city attractiveness and quality of life (Fig. 4).

The higher the city attractiveness and quality of life are, the higher price of commercial houses will be, and by taxation of commercial houses sales, the liquidity increases. According to the study of two Vietnamese authors Hang and Tran, the

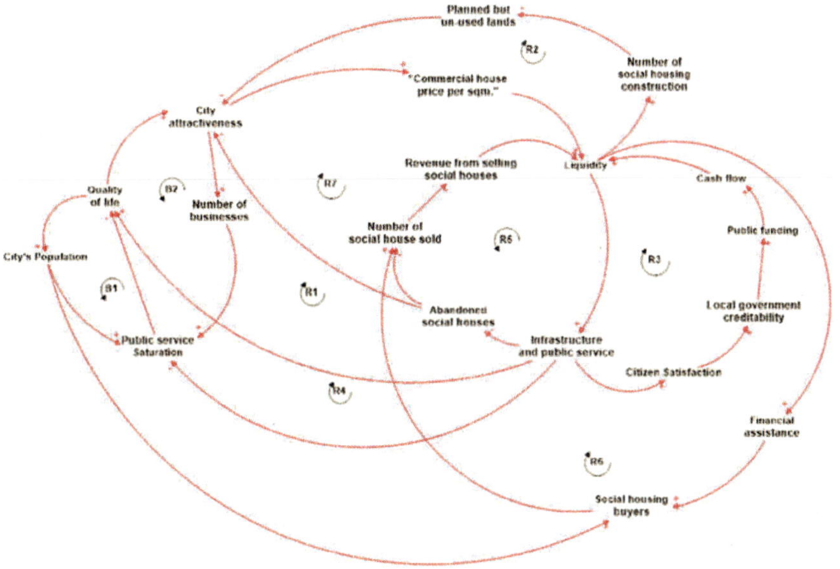

Fig. 3 Main feedback loops describing the development of social housing in Vietnam. *Source* Authors' drawing

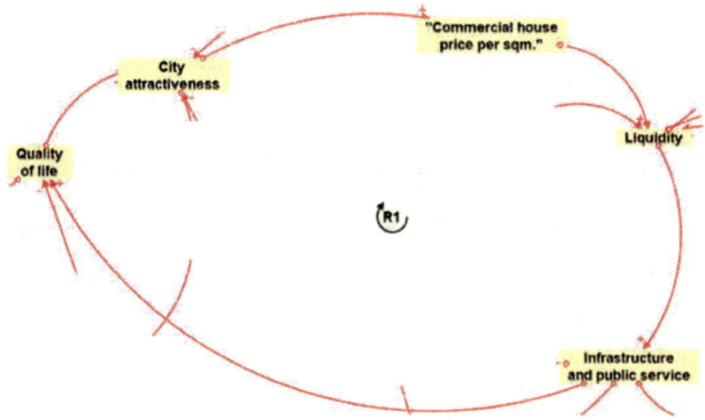

Fig. 4 Reinforcing loop R1. *Source* Authors' drawing

completion of infrastructure and public services is the second most important factor (after the location of the real estate), having an impact on the value of the real estate. This factor can increase the value of real estate by up to 200% (Hang & Tran, 2020). The liquidity generated by taxation of commercial houses is reinvested in developing infrastructure and services which foster the quality of life and attract residents. Also, an increased city attractiveness encourages a higher number of businesses in the area. Investing in infrastructure and public services also decreases the number of abandon

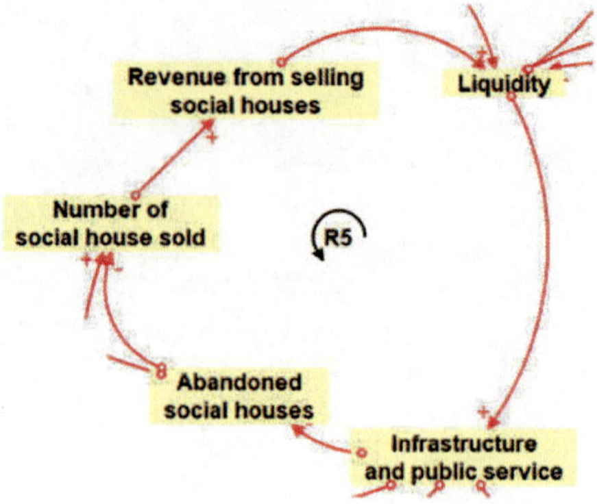

Fig. 5 Reinforcing loop R5.
Source Authors' drawing

social houses that will stimulate the social housing buyers (reinforcing loop R5). The higher the social houses sold, the higher the revenue generated from sales that will add up to the liquidity (Fig. 5).

Investing in infrastructure and public services not only improves the quality of life for social housing buyers but also increases citizen satisfaction in the area. For example, the complete development of infrastructure and public services in Linh Dam new urban area (Hanoi) has brought great satisfaction to people living in this area. Linh Dam new urban area has become a model urban area in Hanoi (Hieu, 2015). The higher the citizen satisfaction is, the higher the local government's credibility which increases the possibility of receiving public fund of the central government. The reinforcing loop R3 presents how the local government can mobilize the public fund from increasing its creditability through investing in infrastructure and public services (Fig. 6).

However, such growth might be tackled by public service saturation. In fact, a too fast growth in population and businesses would gradually reduce the capability of infrastructure and public service capacity to satisfy resident needs. For example, in the Trung Hoa–Nhan Chinh urban area (Hanoi), an overload situation has created

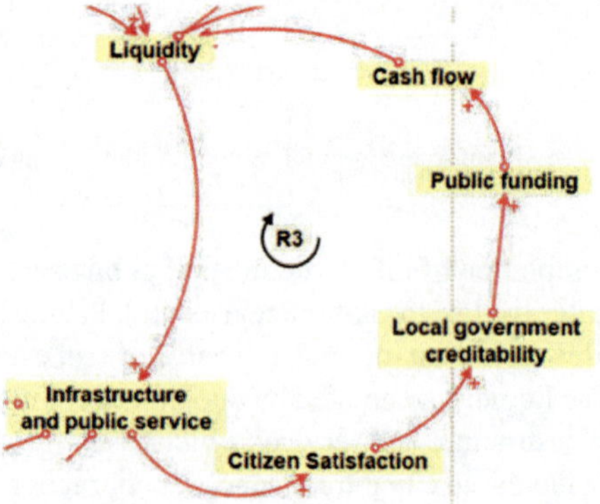

Fig. 6 Reinforcing loop R3.
Source Authors' drawing

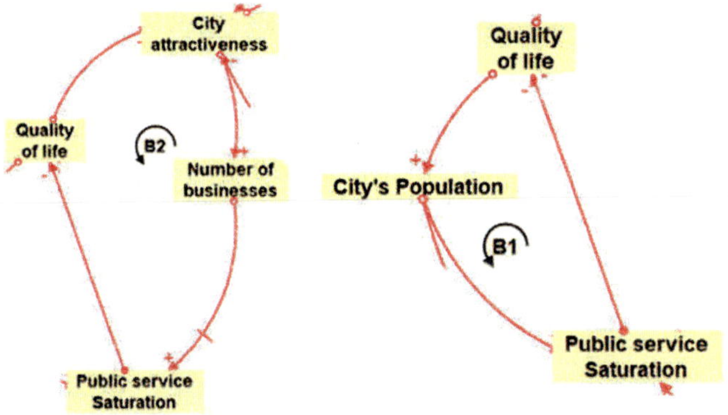

Fig. 7 Balancing loops B1 and B2. *Source* Authors' drawing

parking lots on roadbed, on sidewalks, obstructing traffic and causing urban disorder and unsafety, besides, space for walking is very limited and unsafe (Hieu, 2015). This condition would reduce perceived quality of life, which would in turn lead to a decline of residents (balancing loop B1) and to a drop in city attractiveness (balancing loop B2). Such potential limits to growth have been faced by city governance (Fig. 7).

In this regard, the reinforcing loop R7 shows how to prevent public service saturation through investing infrastructure in abandoned social housing area. That investment will help to relocate part of city population to suburbs and deal with public service saturation in city center. This investment also increases sale of social houses in that area, contributing the cash flow to liquidity. By keeping pace with rapid population growth through such investments, it has been possible to improve the local area attractiveness, so to generate further commercial houses sales and cash flows (reinforcing loop R7) (Fig. 8).

In summary, the liquidity from public funding should not be spent in only building more social houses, which will create more abandoned houses in suburb area; instead, it should be invested infrastructure and public services to stimulate public value (quality of life and city attractiveness). The focus of this view is on local area performance, i.e., the aptitude of stakeholders in a region (e.g., a city neighborhood) to collaborate for developing common goods that may generate public value, which may provide better conditions for local organizations to pursue sustainable development (Bianchi et al., 2019). So that it will solve the problem of abandoned social houses, and also the commercial house price will increase, and through taxation of commercial house sale and the social house sales, the liquidity will be increased.

The causal perspective here illustrated provides a sound basis for designing a dynamic performance management (DPM) chart to illustrate the benefits that policy design and implementation might have gained in attaining sustainable outcomes in the case at hand.

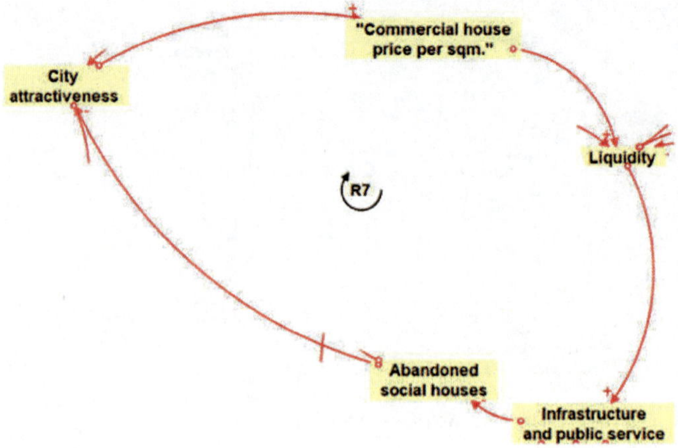

Fig. 8 Reinforcing loop R7. *Source* Authors' drawing

Reframing social housing development in Vietnam through dynamic performance management

Figure 9 shows the different levels of related outcomes associated with the described policy. Such outcomes are changes in the endowment of strategic resources in the area that cannot be purchased on the market. Therefore, it is expected that such resources are built up as ultimate consequences of network policy implementation. Among them, an important role is played by shared resources (e.g., quality of life and city attractiveness). The competence of network governance to attain such resources provides a synthesis of policy sustainability. This entails balancing financial with competitive and social goals in the short and long run. In this perspective and in the investigated context, financial outcomes are framed at the same time as an end and as a means for developing quality infrastructure and service capacity and fostering social inclusiveness.

The final outcomes are the: change in city attractiveness and change in liquidity. These performance measurements determine the capability of the city to be a target destination for a population which consists of residents and businesses but also solving the problem of social housing in city. The city attractiveness is one of the final outcomes since it synthetizes the impact of network policies on the conditions that make the area appealing and therefore may retain or increase people, who may contribute to further development. The change in liquidity is considered a financial outcome as it helps to mobilize funds to invest in infrastructure and public services and develop social houses project in the future.

To attain the first final outcome, the important role is played by two first-level intermediate outcomes, which is the change in quality of life (referred to residents) and change in abandoned social houses. The first outcome qualifies the core factors attracting residents in the area, based on the current living conditions provided by public services and the quality of infrastructures. A declining stock of residents, due

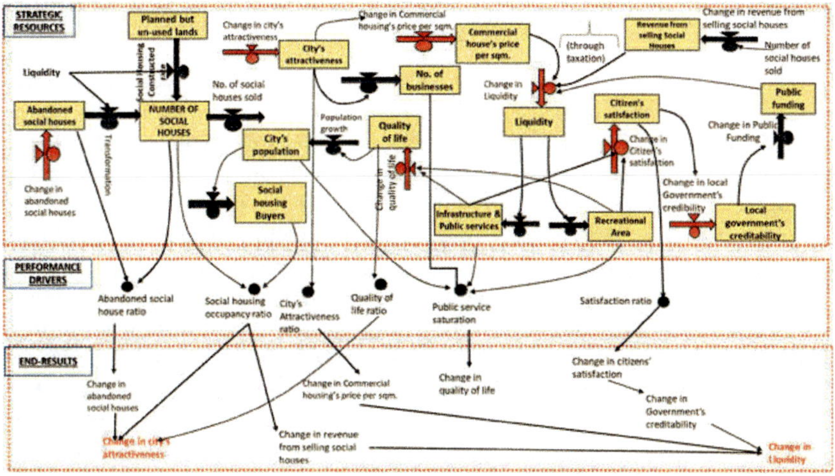

Fig. 9 Dynamic performance management view of sustainable social housing development in Vietnam. *Source* Authors' drawing

to a low quality of life, would make it hard to attract businesses in the city. Therefore, to generate conditions for improving quality of life, proper investments in service capacity are needed. The second outcome shows that the lower abandoned social houses will create the higher city attractiveness.

To achieve the second final outcome, the first-level intermediate outcomes should be considered which are change in citizens' satisfaction, change in government's creditability, change in revenue from selling social houses, and change in commercial housing price per square meter. Change in citizens' satisfaction will lead to change in government's creditability which helps local government to mobilize capital from public fund. An increase in change in commercial housing price will also stimulate the liquidity (through taxation).

In the design and implementation of a sustainable policy for social housing development, performance management should first focus on outcome measures, such as those previously described. To attain them, a number of performance drivers (i.e., intermediate short-term results) can be identified.

Two main drivers affect the change in city attractiveness: quality of life ratio and social housing occupancy ratio. Such performance measures imply that, although a precondition to attract people in the area is residents' quality of life, the construction factor is relevant. Also, an increase in the social housing occupancy ratio increases a place attractiveness since it reduced the abandoned houses in city.

Public service saturation is the main performance driver which inversely affects change in quality of life. Such driver is a result of available capacity, in terms of infrastructures, public services, recreational spaces compared with the current request for such services. An increase in public service saturation (due to over exploitation of the area's carrying capacity by business and overpopulation) reduces the quality of life.

Abandoned social houses ratio is the other main performance driver which affects the change in abandoned social houses. This driver shows how many abandoned houses compared with number of social houses in the city neighborhood. The lower this ratio (due to the development of infrastructure and public services) will lead to decrease in change in abandoned social houses that will increase city attractiveness.

Three main performance drivers affect the change in liquidity: city attractiveness ratio, citizens' satisfaction ratio, and social housing occupancy ratio. City attractiveness ratio indirectly affects cash flows; this is affected by the level of city attractiveness. When such performance driver increases, one may expect that the asset price per square meter will rise. The increase citizens' satisfaction ratio will increase local governments' creditability which helps local governments to receive more public fund from central government. The social housing occupancy ratio relates to number social house over total social houses. The higher such ratio is, the higher the selling rate of social houses, and this will add up to the liquidity.

4.2 Stakeholders Network

Around social housing development in Vietnam, a net of stakeholders has played an active role. As Fig. 10 shows, different actors can be positioned on concentric circles around NHA (Vietnam National Housing Association). Each circle represents different levels of relevance and involvement in the design and implementation of the development process.

Those stakeholders positioned in the inner circles have a higher level of proximity to NHA, relevance, power, and commitment in the network, with respect to those in the outer circles.

At the core of the network are NHA counterparts: the provincial (local) governments. The NHA would act as an overarching coordinating body, across ministries and local governments, to oversee housing sector interventions. The NHA need not be a new institution; it could be anchored under a lead ministry, determined by GoV, or by establishing a new executive body at a higher level of government, or by expanding the mandate and capacity of the existing Central Steering Committee (CSC).

Such an agency could have the following specific set of functions:

- Coordinate and convene public agencies
- Carry out policy analysis and development
- Channel subsidies and support to local governments
- Monitor and evaluate programs
- Develop a common targeting framework
- Oversee housing and real estate market information systems.

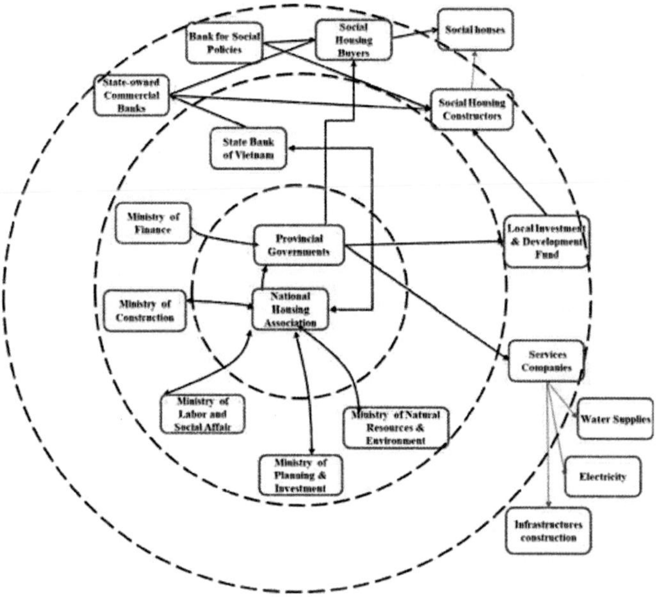

Fig. 10 Stakeholder network. *Source* Authors' drawing

The NHA would also engage and support local governments toward preparation and execution of the Flagship Initiatives. Local governments play an important role in carrying out local needs assessments, identifying target households, facilitating land provision, administering subsidies, engaging private sector, and reporting back results for the monitoring of programs. By creating common delivery systems, clear performance indicators and platforms for training and peer-to-peer sharing, central government can support and incentivize local governments to become more effective agents in social housing provision.

The NHA also works directly with other ministries to foster policy collaboration, such as:

- State Bank of Vietnam: for banking sector regulations for social housing development
- Ministry of Construction: for social housing policies and development plan
- Ministry of Planning and Investment: for local socio-economic planning and investment
- Ministry of Natural Resources and Environment: for local land management and land-use planning for social housing
- Ministry of Labor and Social Affairs: for determining social welfare policies for social housing buyers.

The provincial governments receive budget allocation from Ministry of Finance for developing social house.

The third circle determines the stakeholders that involve directly with local social housing development. Their roles are:

- State-owned commercial banks and Bank for Social Policies will provide social housing finance (e.g., low-interest loans) for both contractors and social housing buyers.
- Local Investment and Development Fund is responsible for targeting and selection, capital subsidy allocation for social housing constructors.

The Services Companies manages in providing public utilities drinkable water, fire services, natural gas, electric power, cable television, telephone, and internet.

5 Conclusions

Solving the problem of sustainable social housing just by increasing the number of social houses cannot solve the root problem. The reason is that focusing solely on the increase in the quantity of houses, without accompanying infrastructure and public services, will lead to abandoned housing area, which creates a huge waste. Instead, it is necessary to see social housing having its own economy. By investing in development of infrastructure for cities, it will contribute to solving the problem of sustainable social housing. Infrastructure development will enhance the attractiveness of urban areas, which will raise the prices of commercial real estate in urban areas; through the taxation of these real estate sales, the state budget will increase significantly. In addition, the development of infrastructure in abandoned housing areas will reduce the population density in the city center, while at the same time addressing the social housing areas that are still abandoned. Lastly, the development of infrastructure will increase the satisfaction of the people and lead to an increase in the local governments' creditability and increase the opportunity to access the central budget for social housing development.

References

Bianchi, C. (2002). Introducing SD modelling into planning and control systems to manage SMEs' growth: A learning-oriented perspective. *System Dynamics Review, 18*(3), 315–338. https://doi.org/10.1002/sdr.258

Bianchi, C. (2012). Enhancing performance management and sustainable organizational growth through system-dynamics modelling. In S. Grösser, R. Zeier (Eds.), *Systemic Management for Intelligent Organizations*. Springer, Berlin, Heidelberg. https://doi.org/10.1007/978-3-642-29244-6_8

Bianchi, C. (2016). *Dynamic Performance Management*. Zurich: Springer International Publishing. https://doi.org/10.1007/978-3-319-31845-5

Bianchi, C., Bovaird, T., & Loeffler, E. (2017). Applying a dynamic performance management framework to wicked issues: How coproduction helps to transform young people's services in surrey county council, UK. *International Journal of Public Administration, 40*(10), 833–846. https://doi.org/10.1080/01900692.2017.1280822

Bianchi, C., Vignieri, V., Bereciartua, P., & Cohen, A. (2019, September 30). Enhancing urban brownfield regeneration to pursue sustainable community outcomes through dynamic performance governance. *International Journal of Public Administration*, 2–15. https://doi.org/10.1080/01900692.2019.1669180

Cooper, J., & Jones, K. (2008, October). *Sustainability and social housing maintenance, phase 1—Results of a questionnaire survey, research under the EPSRC (Programme: Sustainable Urban Environment) Programme: Sustainable urban environment.* Retrieved from Innovation in Design, Construction and Operation of Buildings for People (IDCOP): http://www.eco-networks.org/idcop/outcomes/IDCOP%20WP%202.1%20Questionnaire-Analysis.pdf

Cosenz, F., & Bivona, E. (2018). Designing outcome-based performance management systems to assess policies impacting on caesarean section rate: An analysis of the sicilian maternity pathway. In E. Borgonovi, E. Anessi-Pessina, & C. Bianchi (Eds.), *Outcome-Based Performance Management in the Public Sector* (pp. 63–77). Zurich: Springer International Publishing.

Davies, A. R. (2008). *The geographies of garbage governance: Interventions, interactions, and outcomes.* Routledge.

Department of Communities and Local Government. (2016, February 25). *House building: December Quarter 2015, England.* Retrieved from Department for Communities and Local Government Website: https://assets.publishing.service.gov.uk/government/uploads/system/uploads/attachment_data/file/502930/House_Building_Release_Dec_Qtr_2015.pdf

Diesendorf, M. (2000). Sustainability and sustainable development. In D. Dunphy, J. Benveniste, A. Griffiths, & P. Sutton (Eds.), *Sustainability: The corporate challenge of the 21st century* (pp. 19–37). Allen & Unwin.

Dixon, T., & Woodcraft , S. (2016, January 15). *Creating strong communities—measuring social sustainability in new housing development.* Retrieved from Designing Buildings Wiki Website: https://www.designingbuildings.co.uk/wiki/Creating_strong_communities_%E2%80%93_measuring_social_sustainability_in_new_housing_development

Essa, R., & Fortune, C. (2008). Pre-construction evaluation practices of sustainable housing projects in the UK. *Engineering, Construction and Architectural Management, 15*(6), 514–526. https://doi.org/10.1108/09699980810916960

Guerrero, L. A., Maas, G., & Hogland, W. (2013). Solid waste management challenges for cities in developing countries. *Waste Management, 33*(1), 220–232.

Hang, Q. N., & Tran, T. T. (2020, August 27). *Criteria to access the value of Vietnamese house price (Tiêu chí để đánh giá giá nhà ở Việt Nam cao hay thấp).* Retrieved from Vietnam Industry and Trade Magazine: http://tapchicongthuong.vn/bai-viet/tieu-chi-de-danh-gia-gia-nha-o-viet-nam-cao-hay-thap-74376.htm

Hieu, T. T. (2015, December 29). *New urban development experience & practice (Vì sao nhà ở xã hội vẫn "dậm chân tại chỗ"?).* Retrieved from Architecture magazine of Vietnam Association of architects : https://www.tapchikientruc.com.vn/chuyen-muc/kinh-nghiem-thuc-tien-phat-trien-do-thi-moi.html

Higgins, K. (2013, May 16). *Economic growth and sustainability—are they mutually exclusive?* Retrieved from Elsevier Website: https://www.elsevier.com/connect/economic-growth-and-sustainability-are-they-mutually-exclusive

Kaika, M., & Swyngedouw, E. (2000). Fetishizing the modern city: The phantasmagoria of urban technological networks. *International Journal of Urban and Regional Research, 24*(1), 120–138.

Marshall, R. E., & Farahbakhsh, K. (2013). Systems approaches to integrated solid waste management in developing countries. *Waste Management, 33*(4), 988–1003.

McDougall, F. R., White, P. R., Franke, M., & Hindle, P. (2008). *Integrated solid waste management: A life cycle inventory.* John Wiley & Sons.

Morrisey, A. J., & Browne, J. (2004). Waste management models and their application to sustainable waste management. *Waste Management, 24*(3), 297–308.

Oyebanji, A. O., Akintoye, A., & Liyanage, C. L. (2013). *Barriers to sustainable social housing provision*. Retrieved from http://www.conference.net.au/cibwbc13/papers/cibwbc2013_submission_444.pdf

Oyebanji, A. O. (2014, July). *Development of a framework for Sustainable Social Housing Provision (SSHP) in England*. Retrieved from CLOK—Central Lancashire Online Knowledge administration Website: http://clok.uclan.ac.uk/11321/2/Oyebanji%20Final%20e-Thesis%20(Master%20Copy).pdf

Pattinaja, A. M., & Putuhena, F. J. (2010). Study on the requirements for sustainable settlement development for low income community in Indonesia. *Journal of Environmental Science and Engineering, 4*(5), 78–84.

Priemus, H., Dieleman, F., & Clapham, D. (1999). Current developments in social housing management. *Journal of Housing and the Built Environment, 4*(3), 211–223. https://doi.org/10.1007/BF02496678

Shelter England. (2007, July). *Policy: Briefing delivering environmentally sustainable housing growth. A discussion about how to meet housing needs while protecting and sustaining the natural environment*. Retrieved from Shelter, the National Campaign for Homeless People Limited Website: https://england.shelter.org.uk/__data/assets/pdf_file/0004/39469/Delivering_Environmentally_Sustainable_Housing_Growth.pdf

Shelter England. (2016). *Why we need more social housing*. Retrieved from Shelter: http://england.shelter.org.uk/campaigns_/why_we_campaign/Improving_social_housing/Why_we_need_more_social_housing

Swyngedouw, E., & Heynen, N. C. (2003). Urban political ecology, justice and the politics of scale. *Antipode, 35*(5), 898–918.

Trang, N. (2020, September 17). *Why are social housing projects still "stalled"? (Vì sao nhà ở xã hội vẫn "dậm chân tại chỗ"?)*. Retrieved from thoibaonganhang.vn: https://thoibaonganhang.vn/vi-sao-nha-o-xa-hoi-van-dam-chan-tai-cho-106522.html

United Nations General Assembly. (1987). *Development and International Economic Co-Operation: Environment Report of the World Commission on Environment and Development forty-second session, item 83 (e) of the provisional agenda note by the Secretary-General GENERAL A/43/4274*. United Nations.

Wolstenholme, E. F. (1999). Qualitative vs quantitative modelling: The evolving balance. *Journal of the Operational Research Society, 50*, 422–428. https://doi.org/10.1057/palgrave.jors.2600700

World Commission on Environment and Development. (1987). *Our Common Future*. Oxford University Press.

Factors Affecting Non-performing Loans of Vietnamese Banks in the Context of the Covid-19 Pandemic

Lai Cao Mai Phuong

Abstract The purpose of this study is to determine the factors affecting non-performing loans (NPLs) of banks in the context of the COVID-19 pandemic. The feasible generalized least square (FGLS) regression method is used to quantify the impact of groups of macro-factors, specific characteristics of banks, and the impact of the COVID-19 pandemic on the NPLs of 25 banks listed in Vietnam in the period 2016–2021 according to three research models. The results show that higher unemployment rate, higher bank size, and higher profitability of banks lead to lower NPL ratio, but higher credit growth rate and higher credit provision ratio lead to higher NPL ratio. The study found that the COVID-19 pandemic in Vietnam has a significant negative impact on the NPL ratio in 2020, a significant positive impact on the NPL ratio in 2021, but the coefficient is negative but not statistically significant when combining the effects of both 2020 and 2021, has shown the complex and unexpected impact of the pandemic on the NPL ratio of the banking system. This result is an early warning and suggests some suggestions for regulators and banks to control the NPL ratio in the near future.

Keywords Credit growth · Return on equity · Provision for credit risk · Unemployment · Vietnam

1 Introduction

Non-performing loans (NPLs) are the biggest risk of commercial banks when targeting credit growth (Tanasković & Jandrić, 2015). This risk is even more concerned by regulators and banks when the context of the economy is adversely affected by financial crises or epidemics. There are many factors that can affect NPLs (such as macro-factors and specific characteristics of the bank), so in order to effectively manage and handle these debts, it is very important to study the factors affecting NPLs (Phuong, 2022a).

L. C. M. Phuong (✉)
Industrial University of Ho Chi Minh City, Ho Chi Minh City, Vietnam
e-mail: laicaomaiphuong@iuh.edu.vn

A. T. Nguyen et al. (eds.), *Contemporary Economic Issues in Asian Countries: Proceeding of CEIAC 2022, Volume 1*,
https://doi.org/10.1007/978-981-19-9669-6_32

The previous financial and non-financial crises have resulted in an increase in underperforming loans in many countries around the world (Bayar, 2019; Goodell, 2020; Turan & Koskija, 2014). It was found that the COVID-19 pandemic was the cause of strong fluctuations in financial markets and also in banking stocks in many countries (Contessi & De-Pace, 2021; Phuong, 2021, 2022b). The application of unprecedented measures to prevent the spread of the COVID-19 pandemic in countries has negatively affected the business activities of many businesses, leading to the increased risk of bad debts at banks (Ari et al., 2021; Demary, 2021). However, the number of studies on the impact of COVID-19 on NPLs to date is still quite limited.

This fact shows the need to study the factors affecting non-performing loans of banks in the context of the COVID-19 pandemic. Therefore, this topic will be studied for banks listed in Vietnam in the period 2016–2021.

2 Literature Review

There are many different expressions when defining NPLs, but organizations such as the International Monetary Fund and the Basel Committee on Banking all agree that loans should be classified as NPLs when payment of interest and principal is past due by more than three months (90 days) or more (Basle Committee, 2016; IMF, 2019).

The groups of factors were studied by the authors about NPLs which are as follows: analysis of macro-factors, analysis of micro-factors, and a combination of analysis of macro- and micro-factors affecting NPLs (Žunić et al., 2021). Among these factors, macro-factors (external factors) have stronger influence than micro-factors (internal factors or specific characteristics of banks) (Klein, 2013; Sanjeev, 2007).

2.1 Macro-Factors Affecting Non-performing Loans of Banks

Espinoza and Prasad (2010) studied 80 banks in the Gulf Cooperative Council (GCC) region in the period 1995–2008, showing macro-factors (removing oil from GDP and interest rates) and specific factors (bank size, credit growth, management efficiency) all contributed to the increase in NPLs in these countries. The financial accelerator theory is a widely used theoretical framework to link NPLs to a country's macroeconomic environment (Bernanke & Gertler, 1989; Kiyotaki & Moore, 1997). Lawrence (1995) proposed the theory of the life cycle consumption pattern as a way to determine the probability of default. The default probability model implies that NPLs are associated with uncertainty about future earnings. Low-income borrowers have higher default rates as the risk of unemployment increases. In addition, banks charge higher interest rates to riskier customers, so the probability of default depends on current income and the unemployment rate.

Studies show both economic growth (RGDP) and unemployment rate as two important factors affecting NPLs. Low RGDP and high unemployment negatively affect the wealth and ability of borrowers to repay, so under these macro-conditions NPLs are forecast to increase higher (Bernanke & Gertler, 1989; Bernanke et al., 1999; Kiyotaki & Moore, 1997). An economy with a high RGDP and a falling unemployment rate can lead to an excessive credit boom that reduces credit quality and causes higher NPLs (Kirti, 2018; Schularick & Taylor, 2012).

Studies in developed and emerging countries have shown that NPLs are affected by macro-factors. Babouček and Jančar (2005) studied the Czech economy and credit channel activity for 11 years from 1995 to 2004 and found that unemployment and inflation are positively correlated with banks' NPLs, and vice versa, RGDP is negatively correlated with NPLs. Similar to Babouček and Jančar (2005), Klein (2013) studies 16 countries in Central, Eastern, and South-Eastern Europe (CESEE) for the period 1998–2011 also showing RGDP, unemployment, and inflation significant impact on the asset quality of banks. In addition, Klein (2013) also shows that a falling exchange rate increases NPLs in CESEE countries. Ghosh (2015) investigated state-by-state macroeconomic factors in the United States during the period 1984–2013 that affected the NPLs of commercial banks and thrift institutions. The results show that the state's RGDP reduces NPLs, while inflation, state unemployment, and US public debt significantly increase NPLs (Ghosh, 2015).

Research in emerging countries, Radivojevic and Jovovic (2017) and Bayar (2019) both show that RGDP has an important role in bank asset quality. A decrease in RGDP will increase the banking NPL ratio in emerging countries (Bayar, 2019; Radivojevic & Jovovic, 2017). Using the generalized method of moments to study the banking sector, in addition to RGDP, Bayar (2019) also shows that macro-factors such as inflation and economic freedom have a negative impact on NPLs, and unemployment has a positive impact on NPLs. Analysis of inflation: Ari et al. (2021) argue that inflation can affect the NPL ratio in two opposite directions. Rising inflation can reduce NPLs as rising inflation reduces the real value of debt and makes it easier to pay off debt in local currency. Conversely, rising inflation can also lead to higher nominal and real interest rates, increasing the cost of debt services. In this case, high inflation may also be accompanied by macroeconomic instability that exacerbates the NPL ratio (Ari et al., 2021).

2.2 *Internal Factors or Specific Characteristics Affecting Non-performing Loans of Banks*

The relationship between bank size and NPL ratio is often explained by the moral hazard hypothesis and the "too big to fail" view. Studying commercial banks and thrifts in the period 1984–2013 to determine the factors affecting NPL at the state level in the United States, Ghosh (2015) shows the size of the banking industry as well and capitalization of banks affects NPLs. Low-cap banks have more problems with NPLs (Berger & DeYoung, 1997) because managers of these banks have a moral hazard incentive to engage in risky lending practices (Keeton & Morris, 1987). Therefore, the moral hazard hypothesis implies an inverse relationship between bank capitalization and NPLs. However, there is also a view that large-cap banks will be more relaxed in credit scoring and monitoring customers because they believe in the "too big to fail" concept of Rajan (1994). In this case, the relationship between bank capitalization and NPLs is positive. Thus, the relationship between bank capitalization or bank size and NPLs has not been consistent so far.

Some studies have emphasized that the increase in NPLs of banks is due to poor credit quality control (Keeton, 1999; Asfaw et al., 2016). Survey of Federal Reserve Senior Lenders for the 1967–1983 and 1990–1998 periods, and statewide data for the 1982–1996 period, Keeton (1999) points to as credit growth faster and excessive relaxation of credit standards will result in higher loan losses, especially if the economy is in recession. Asfaw et al. (2016) studied the factors affecting NPLs of the Development Bank of Ethiopia, Central Region by surveying 43 borrowers and 55 employees. The authors found that poor credit ratings and poor credit monitoring were responsible for the increase in NPLs at this bank. Asfaw et al. (2016) suggested that banks should strengthen the borrower's screening criteria and credit assessment before and after disbursement. Disbursement of capital for the right purposes should be focused through strengthening credit supervision.

In addition, large-scale banks often increase excessive leverage and loosen credit quality, leading to increased NPLs (Stern & Feldman, 2004). The loosening of credit quality causes the risk of loan portfolio to increase rapidly, and banks will have to make higher provisions for credit risks (Keeton & Morris, 1987). This can be explained by the moral hazard hypothesis of bank managers.

The hypothesis of "bad management" and the concept of "liberal credit policy" are often applied to explain the impact of management quality on banks' NPLs. Management quality is often measured by the profitability or performance of a bank (Bayar, 2019; Berger & DeYoung, 1997; Klein, 2013; Rajan, 1994). The "bad management" hypothesis of Berger and DeYoung (1997) suggests that banks with high profitability are less likely to engage in high-risk activities so their NPLs will be low. In other words, profitability is negatively related to NPLs. However, Rajan (1994) argues that banks with higher profitability can also increase bad debt because from the point of view of "liberal credit policy". This policy implies that bank managers may try to make adjustments to increase current earnings to preserve their short-term reputation but will lead to increased NPLs in future.

A study of 43 banks in the Czech Republic, Podpiera and Weill (2008) showed that poor management was the cause of reduced cost efficiency leading to high NPL ratio in the period 1994–2005. Based on these results, Podpiera and Weill (2008) suggest that for countries in transition such as the Czech Republic, banking supervisors should focus on improving the cost-effectiveness of banks in order to reduce the NPL ratio. In agreement with Podpiera and Weill (2008), studies show that higher bank management quality (return on equity in the previous period) can improve lower NPL ratios in CESEE countries (Klein, 2013), increased bank profits reduced NPLs at US commercial banks and thrifts over the period 1984–2013 (Ghosh, 2015), higher return on assets, and return on equity have the effect of reducing bank NPLs in emerging countries (Bayar, 2019).

In addition, low equity and high loan-to-total assets ratio tend to exacerbate bad debt (Klein, 2013), increased liquidity risk, poor credit quality and less efficient costs significantly increase NPLs (Ghosh, 2015), credit growth, increasing cost-to-income ratio, and financial crisis will increase NPLs (Bayar, 2019).

2.3 Impact of Crises on Non-performing Loans of Banks

Using quarterly data from 2003 to 2013 from the Albanian bank, Turan and Koskija (2014) demonstrated that NPLs started to increase shortly after the outbreak of the financial crisis in 2008, but strong growth in NPLs occurred a year later when RGDP in most of the CESEE countries fell. Using annual data for 14 emerging economies from 2000 to 2013, Bayar (2019) also reaffirms that the financial crisis will cause NPLs to increase rapidly in these countries.

Goodell (2020) believes that the COVID-19 pandemic can affect banks because it is one of the causes of the decline in loan portfolio quality, which in turn leads to an increase in NPLs. Demary (2021) warns that NPLs are likely to increase in the euro area due to the impact of the COVID-19 pandemic, especially in countries where the bankruptcy system is inefficient. Therefore, Demary (2021) emphasizes that the extent to which the negative impact of the COVID-19 pandemic will depend heavily on the effectiveness of countries' restructuring measures.

3 Methodologies

Based on theory and the previous studies, in order to determine the factors affecting NPLs of Vietnamese banks in the context of the COVID-19 pandemic, the article proposes the following research model:

$$\mathrm{NPL}_{\mathrm{i,t}} = \alpha_0 + \sum_{l=1}^{L} \beta_l \mathrm{Macro}_{\mathrm{i,t}} + \sum_{m=1}^{M} \beta_m \mathrm{Banks\ Specific}_{\mathrm{i,t}} + \sum_{k=20y}^{21y,20.21} \beta_k D^{\mathrm{COVID}} + \varepsilon_{i,t} \quad (1)$$

The model includes 3 groups of variables: macro-factors, specific characteristics of banks, variables affected by the COVID-19 pandemic.

The dependent variable NPL is calculated by the annual total outstanding loans for each bank.

Independent variables:

The group of macro-variables consists of three variables, namely annual GDP growth (RGDP), annual inflation rate (INF), and annual unemployment rate (UNEMP).

The group of variables with specific characteristics of banks includes the following variables:

- Bank size (size) is the natural logarithm of total assets
- Credit growth (CREDIT) is the growth of outstanding loans of year t compared to year (t-1)
- Financial leverage (FLEV) is calculated as total debt to total assets
- Profitability (ROE) is calculated as profit after tax on equity
- Provision for credit risk (PCR) is provision for credit risk on total assets.

The group of dummy variables measuring the impact of the COVID-19 pandemic includes:

The impact of the COVID-19 pandemic on Vietnam in 2020 and 2021 is very different. Vietnam is known as one of the best countries to control the spread of COVID-19 in the world in 2020 (Phuong, 2022b; World Bank, 2020). However, with the emergence of many new strains of the corona virus spreading at a fast rate while the source of a vaccine against COVID-19 is limited, Vietnam's economy in 2021 has suffered a much more negative impact than in 2020. In 2020, the distance due to COVID-19 in Vietnam is about 1 month, and there are no deaths, but until 2021, Southern provinces had to be separated for the last 6 months of 2021 with the number of deaths from COVID increasing rapidly in Ho Chi Minh City, Binh Duong, Dong Nai, and Long An.[1]

To remove difficulties for businesses and borrowers due to the impact of COVID-19, the State Bank has issued Circular No. 01/2020/TT-NHNN, Circular 14/2021/TT-NHNN, Decree 31/2022/ ND-CP to restructure debt repayment terms, exempt and reduce interest charges, and support interest rates so that businesses can recover their production and business activities soon. Therefore, this paper uses three dummy variables to quantify the impact of the COVID-19 pandemic as follows:

$$D^{\text{COVID20}y} = \begin{cases} = 1 \text{ if year} = 2020 \\ = 0 \text{ otherwise} \end{cases} \tag{2}$$

$$D^{\text{COVID21}y} = \begin{cases} = 1 \text{ if year} = 2021 \\ = 0 \text{ otherwise} \end{cases} \tag{3}$$

$$D^{\text{COVID20.21}} = \begin{cases} = 1 \text{ if year} = 2020, 2021 \\ = 0 \text{ otherwise} \end{cases} \tag{4}$$

All research data was collected during the period from 2016 to 2021. The macro-data group is collected from the General Statistics Office of Vietnam, the data group

[1] Vietnam Business Operations and the Coronavirus: Updates, link https://www.vietnam-briefing.com/news/vietnam-business-operations-and-the-coronavirus-updates.html/.

on the characteristics of banks, and the dependent variable is collected in the annual audit reports of 25 listed banks in Vietnam.

Estimation method: The article uses the feasible generalized least square (FGLS) estimation method for the regression equations. This method has the advantage that it can overcome the phenomenon of variable variance and autocorrelation compared with conventional estimation methods on panel data such as pool original least square, fixed effects model, and random effects model.

4 Results

With 25 listed commercial banks in the period 2016–2022, the research data includes 150 observations. Table 1 shows the results of descriptive statistics of the variables included in the research models (Table 2).

The average non-performing loan ratio of listed banks is 1.8%, which is a relatively safe level. However, the statistical results in Table 1 show that the risk of NPLs is high for KLB bank when the NPL ratio is very high at 7.62% in 2020. Therefore, to ensure safety, banks need to come up with synchronous measures to reduce the NPL ratio below 3%.

Due to the impact of COVID-19, the lowest GDP growth is in 2020, the lowest inflation is in 2021, and the highest unemployment rate is 3.22% in 2021. The bright spot in 2021 is VIB's ROE of 26.39%, the highest of all banks in the research period.

The correlation coefficient of the variable COVID with the variables covid20 and covid21 is 0.633 so these variables will be regressed separately for each regression model. The remaining pairs have correlation coefficients in the range ±0.57, so the

Table 1 Descriptive statistics of variables in the model and VIF

Stats	*N*	Mean	Min	p50	Max	VIF	1/VIF
NPL	150	0.0180	0.0047	0.0166	0.0762		
GDP	150	−0.1109	−0.5855	−0.0394	0.0966	2.17	0.4602
INF	150	0.5113	−0.4303	0.0803	3.2222	1.31	0.7655
UNEMP	150	0.0234	0.0199	0.0225	0.0322	1.58	0.6340
SIZE	150	14.2566	13.2799	14.2178	15.2460	1.78	0.5618
CREDIT	150	0.2220	−0.9374	0.1658	7.5545	1.08	0.9235
FLEV	150	0.9173	0.8155	0.9244	0.9570	1.25	0.8018
ROE	150	0.1154	0.0003	0.1100	0.2639	1.66	0.6017
PCR	150	0.0088	0.0000	0.0059	0.0505	1.08	0.9227
COVID20y	150	0.1667	0.0000	0.0000	1.0000	2.32	0.4314
COVID21y	150	0.1667	0.0000	0.0000	1.0000	2.32	0.4314
COVID20.21	150	0.3333	0.0000	0.0000	1.0000	3.76	0.2663
Mean VIF						1.8	

Table 2 Results of correlation matrix between pairs of variables

	GDP	INF	UNEMP	SIZE	CREDIT	FLEV	ROE	PCR	COVID20y	COVID21y
INF	0.068	1								
UNEMP	−0.170	−0.151	1							
SIZE	−0.110	−0.124	0.000	1						
CREDIT	0.024	0.172	0.002	0.168	1					
FLEV	0.077	0.090	−0.175	0.310	0.130	1				
ROE	−0.144	−0.212	−0.078	0.563	0.086	0.037	1			
PCR	0.037	−0.016	0.066	0.160	0.017	−0.121	0.138	1		
COVID20y	−0.485	−0.181	−0.198	0.088	−0.059	−0.014	0.066	0.013	1	
COVID21y	−0.409	−0.255	0.541	0.148	−0.039	−0.083	0.193	−0.023	−0.200	1
COVID20.21	−0.706	−0.345	0.271	0.186	−0.077	−0.077	0.205	−0.008	0.633	0.633

variables are linearly independent (Farrar & Glauber, 1967) and are suitable for use in regression models based on panel data.

Breusch & Pagan test results from Table 3 show that models have heteroskedasticity but not autocorrelation according to Wooldridge test. Therefore, the heteroskedasticity phenomenon is overcome on the models, and the FGLS regression results after overcoming are shown in Table 3.

Unemployment rate: The results from Table 3 show that unemployment rate has a significant negative impact on NPLs in all three research models. It rejects Lawrence's (1995) view of the life cycle consumption model due to the characteristics of unemployment and loan structure in Vietnam during the research period. The unemployment rate in Vietnam in the period 2016–2021 remains low at less than 3.22%. In addition, personal loans account for a low percentage of total outstanding loans as well as in NPLs. This result is similar to banks in 14 developed countries for the period 1870–2008 (Schularick & Taylor, 2012) and bond-derived NPLs in 38 countries with the longest being in the range of 1980–2016 (Kirti, 2018).

Bank size: At 1% significance level, bank size (SIZE) has a negative impact on NPLs. This result rejects the "too big to fail" notion that large banks have high NPLs (Ghosh, 2015). In Vietnam, large-scale banks are those with better management experience and better loan quality, thereby limiting NPLs. The negative relationship

Table 3 FGLS regression results and tests

Variable	FGLS-COVID20y	FGLS-COVID21y	FGLS-COVID20.21
GDP	−0.0035	−0.0004	−0.0024
INF	−0.0005	−0.0002	−0.0003
UNEMP	−0.3471**	−0.3480***	−0.2402*
SIZE	−0.0041***	−0.0045***	−0.0042***
CREDIT	0.0010*	0.0010*	0.0010*
FLEV	−0.0265	−0.0244	−0.0300
ROE	−0.0394***	−0.0394***	−0.0373***
PCR	0.2550**	0.2654***	0.2524**
COVID20y	−0.0027*		
COVID21y		0.0025*	
COVID20.21			−0.0010
_cons	0.1105***	0.1149***	0.1126***
N	150	150	150
Breusch & Pagan test			
Chibar2(01)	32.64	33.46	33.26
Prob > chibar2	0.0000	0.0000	0.0000
Wooldridge test			
$F(1, 24)$	0.069	0.061	0.097
Prob > F	0.7955	0.8068	0.7584

between bank size and NPLs in Vietnam is similar to that of banks in the CCC region in the period 1995–2008 (Espinoza & Prasad, 2010).

Credit growth: With 90% confidence, credit growth (CREDIT) has a positive impact on NPLs. This result implies that when the economy grew in the face of competitive pressure, some banks may have experienced rapid credit growth along with loosened credit standards, weak credit monitoring will increase NPLs.

The positive relationship between credit growth and NPLs in this study has added empirical evidence in Vietnam besides the studies of Keeton (1999), Asfaw et al. (2016) and Bayar (2019).

Profitability: Profitability (ROE) has a significant negative impact on NPLs. The negative relationship between profitability and NPLs has refuted the "liberal credit policy" view as in Rajan's (1994) study and supported the "bad management" view that Berger and DeYoung (1997) proposed. This shows that banks with high profitability have less incentive to relax credit standards, so the NPLs of these banks are lower. This result is similar to Klein (2013), Bayar (2019), Ghosh (2015) but does not support the results of Rajan (1994).

PCR: Provision for credit risk has a positive impact on NPLs at 5% significance level. Bank credit quality is positively correlated with NPLs supporting the "moral hazard" hypothesis first discussed by Keeton and Morris (1987). Banks with low credit quality often have a moral hazard incentive by accepting a higher level of risk when lending, leading to high NPL ratios, while banks with high credit quality often avoid high-risk activities, so the NPL ratio is low. This result is similar to the study of Keeton (1999), Asfaw et al. (2016), and Bayar (2019).

Both the COVID20y and COVID21y variables are statistically significant at 10%, demonstrating that the NPLs of banks are affected by the COVID-19 pandemic in both 2020 and 2021. However, the regression coefficients of these two variables have opposite signs. This shows the mixed effects of COVID-19 developments in Vietnam in these two years and can be explained by many reasons.

In 2020, Vietnam is one of the countries that has successfully controlled the COVID-19 pandemic and has no deaths. Socio-economic activities were interrupted in April 2020 for social distancing, and activities have returned to almost normal for the rest of 2020. In addition, the Vietnamese government has timely provided support for those infected with COVID-19 and those who are in difficulty due to COVID-19, so the NPLs of banks in 2020 have even decreased.

In 2021, the COVID-19 pandemic in Vietnam spread rapidly with high speed, the number of deaths increased rapidly in Ho Chi Minh City and southern provinces. Therefore, there have been 4 consecutive rounds of social distancing in Ho Chi Minh City starting from May 31, 2021 to the end of 2021. The large-scale separation in the second half of 2021 in the southern provinces (the largest economic center of Vietnam) due to COVID-19 has severely affected almost all businesses and people in the southern economic region. As a result, the impact of the COVID-19 pandemic in 2021 is an important reason for the rapid increase in NPLs of Vietnamese banks. This result is similar to the impact of previous crises (Bayar, 2019; Turan & Koskija, 2014) and the impact of COVID-19 (Demary, 2021; Godell, 2020) on banks' NPLs.

When quantifying the aggregate impact in both 2020 and 2021, the COVID variable has a negative regression coefficient, showing the trend of the impact of this pandemic on banks' NPLs. This regression coefficient is not statistically significant, so there is not enough evidence to conclude, but its sign is also a sign for banks to be cautious.

5 Conclusions and Recommendations

5.1 Conclusions

The COVID-19 pandemic spreading on a global scale has affected the production and business activities of many businesses, thereby leading to the risk of non-performing loans (NPLs) for banks. In order to manage NPLs well, it is very necessary to identify the factors affecting the bank's NPLs. Therefore, this article studies 25 banks listed in Vietnam in the period 2016–2021 to determine the factors affecting the NPLs of these banks in the context of the COVID-19 pandemic. FGLS regression results from three research models show that both macro-factors, specific characteristics of banks, and the COVID-19 pandemic are important factors affecting the NPL ratio of commercial banks. Unemployment rate, bank size, and profitability are factors that negatively affect NPL ratio. Factors including credit growth and credit provision have a positive impact on NPL ratio. The study found that the COVID-19 pandemic had a significant negative impact on the NPL ratio in 2020, a significant positive impact on this ratio in 2021, but when quantifying the combined impact of both years, the coefficient regression is negative but not statistically significant. This result is a warning sign for regulators, and banks need to pay special attention to the impact of the COVID-19 pandemic on banks' NPLs in order to have a solution to this problem.

5.2 Recommendations

Based on the research results, this article makes some recommendations for regulators and banks to manage NPLs of banks in the context of COVID-19.

Based on the actual situation of commercial banks, the state bank of vietnam (SBV) needs to carefully consider and realize the merger of banks with weak business results and high NPL ratio into the top good banks in Vietnam where the ownership rate of State shareholders is dominant. This way brings many benefits to the whole banking system. Firstly, the proactive merger of weak banks will ensure the safety of the entire banking system, the interests of customers and related parties. Second, the merged bank can apply its skills and experience to restructure the merged bank to operate more efficiently. Besides, the merger increases the scale and improves the competitiveness of the banking system, thereby helping to reduce the NPL ratio.

Banks with NPL ratio above 3% (the level is considered safe according to the Vietnam currency advisory council): It is necessary to review to give synchronous measures to handle NPLs, to avoid the situation of old NPLs not being handled, new NPLs have increased.

Commercial banks need to increase the level of provision for risks commensurate with the increase in NPL ratio to have a source to handle NPLs when Circular 14/2021/TT-NHNN on rescheduling, giving exemption from or reduction of loan interests and charges, and maintaining classified loan groups in order to help their clients affected by the COVID-19 pandemic expires.

In addition, to avoid adverse effects from NPLs risk, banks should increase the "buffer zone" for existing NPLs, actively deduct additional capital for new loans when the NPL ratio exceeds an acceptable level. However, banks should not deduct more than necessary because it will affect the efficiency of capital use of banks.

Commercial banks should also pay attention not to relax credit standards compared to current regulations in order to create a competitive advantage for credit growth. Because the pursuit of "liberal credit policy" will lead to deterioration in credit quality, potentially increasing the risk of PCR, failed to bring about effective credit growth and sustainable profits for banks, especially after the crisis.

Commercial banks need to reduce costs by effectively applying effective information technology in services to reduce lending interest rates to support customers to quickly recover production and business after the pandemic. Because the faster and more effectively the customer's production and business activities recover, the less risky the bank's profit will be, helping to limit NPLs.

Effectively and practically deploying credit support policies from the state budget to remove difficulties for customers due to the COVID-19 pandemic. Specifically, at the end of May 2022, the Government issued Decree 31/2022/ND-CP on the state budget spending VND 40,000 billion, which will support 2% of interest rates for businesses to develop the economy, restore production, and support economic growth. Due to the impact of the COVID-19 epidemic in the two years 2022–2023, but by the end of July 2022, businesses have not yet been able to access this support package. Therefore, the State Bank of Vietnam needs to give specific and detailed instructions and clearly divide the responsibilities of coordinating agencies, industry associations, so that the right businesses can quickly access this low-interest capital source. In addition, in order to avoid profiting from support policies during the implementation process, the SBV needs to inspect to ensure that the support policies are objective, transparent, and to the right audience. On the side of commercial banks: It is necessary to quickly determine the list of eligible customers (for example: Issuing internal documents to have clear criteria when determining the group of subjects eligible for interest rate support under Decree No. 31/2022/ND-CP, the evaluation criteria of debt repayment ability, ability to recover and use loans for the right purposes after the pandemic, ...) to support interest rates in accordance with regulations and create convenient for customers.

In addition, the SBV can refer to the European Union's bad debt directive (Directive (EU) 2021/2167[2]) effective from December 28, 2021 to handle NPLs of the banking system under the impact of the pandemic during the COVID-19 epidemic, thereby proposing to the National Assembly on a legal framework for credit institutions to handle bad debts on the balance sheet and reduce the risk of NPLs accumulation in future.

The National Assembly of Vietnam needs to continue to pay attention to and support the NPLs settlement process of the banking system through decisions on objectives, policies, basic task of annual, and long-term economic development of the country.

References

Ari, A., Chen, S., & Ratnovski, L. (2021). The dynamics of non-performing loans during banking crises: A new database with post-COVID-19 implications. *Banking & Finance, 133*, 106140. https://doi.org/10.1016/j.jbankfin.2021.106140

Asfaw, A. S., Bogale, H. N., & Teame, T. T. (2016). Factors affecting non-performing loans: case study on development bank of Ethiopia central region. *International Journal of Scientific and Research Publications, 6*(5), 656–670.

Babouček, I., & Jančar, M. (2005). *Effects of macroeconomic shocks to the quality of the aggregate loan portfolio* (Vol. 22, pp. 1–62). Czech National Bank.

Bayar, Y. (2019). Macroeconomic, institutional and bank-specific determinants of non-performing loans in emerging market economies: A dynamic panel regression analysis. *Central Banking Theory and Practice, 8*(3), 95–110. https://doi.org/10.2478/jcbtp-2019-0026

Bemanke, B., & Gertler, M. (1989). Agency costs, net worth, and business fluctuations. *American Economic Review, 79*(1), 14–31.

Berger, A. N., & DeYoung, R. (1997). Problem loans and cost efficiency in commercial banks. *Journal of Banking & Finance, 21*(6), 849–870. https://doi.org/10.1016/S0378-4266(97)00003-4

Bernanke, B. S., Gertler, M., & Gilchrist, S. (1999). The financial accelerator in a quantitative business cycle framework. *Handbook of Macroeconomics, 1*, 1341–1393.

Basle Committee. (2016). *Prudential treatment of problem assets–definitions of non-performing exposures and forbearance.* Bank for International Settlements, Basel, Switzerland.

Contessi, S., & De Pace, P. (2021). The international spread of COVID-19 stock market collapses. *Finance Research Letters, 42*, 101894. https://doi.org/10.1016/j.frl.2020.101894

Demary, M. (2021). *Will COVID-19 cause insolvencies, zombification or debt deleveraging?* (No. 3/2021). IW-Kurzbericht.

Espinoza, M. R. A., & Prasad, A. (2010). Nonperforming loans in the GCC banking system and their macroeconomic effects. *International Monetary Fund, 10*(5089/9781455208890), 001.

Ghosh, A. (2015). Banking-industry specific and regional economic determinants of non-performing loans: Evidence from US states. *Financial Stability, 20*, 93–104. https://doi.org/10.1016/j.jfs.2015.08.004

Goodell, J. W. (2020). COVID-19 and finance: Agendas for future research. *Finance Research Letters, 35*, 101512. https://doi.org/10.1016/j.frl.2020.101512

International Monetary Fund (IMF). (2019). *Financial soundness indicators compilation guide.* Prepublication Draft, https://www.imf.org/Files/Data/fsicg2019-prepublicationfinal-042519.

Keeton, W. R. (1999). Does faster loan growth lead to higher loan losses? *Economic Review-Federal Reserve Bank of Kansas City, 84*(2), 57.

[2] https://eur-lex.europa.eu/eli/dir/2021/2167.

Keeton, W. R., & Morris, C. S. (1987). Why do banks' loan losses differ. *Economic Review, 72*(5), 3–21.
Kirti, D. (2018). Lending standards and output growth. *International Monetary Fund, 10*(5089/9781484339671), 001.
Kiyotaki, N., & Moore, J. (1997). Credit cycles. *Political Economy, 105*(2), 211–248.
Klein, N. (2013). Non-performing loans in CESEE: Determinants and impact on macroeconomic performance. *International Monetary Fund, 10*(5089/9781484318522), 001.
Lawrence, E. C. (1995). Consumer default and the life cycle model. *Money, Credit and Banking, 27*(4), 939–954.
Phuong, L. C. M. (2021). How COVID-19 impacts Vietnam's banking stocks: An event study method. *Banks and Bank Systems, 16*(1), 92–102. https://doi.org/10.21511/bbs.16(1).2021.09
Phuong, L. C. M. (2022b). Industry-level stock returns response to COVID-19 news. *Finance: Theory and Practice, 26*(1), 103–114. https://doi.org/10.26794/2587-5671-2022-26-1-103-114
Phuong, L. C. M. (2022a). Determination of factors affecting non-performing loans of banks in Vietnam. In *International conference proceedings: local economic and social development in the post Covid-19 era* (pp. 125–136). Publishing house of the National Economics University.
Podpiera, J., & Weill, L. (2008). Bad luck or bad management? Emerging banking market experience. *Financial Stability, 4*(2), 135–148. https://doi.org/10.1016/j.jfs.2008.01.005
Radivojevic, N., & Jovovic, J. (2017). Examining of determinants of non-performing loans. *Prague Economic Papers, 26*(3), 300–316.
Rajan, R. G. (1994). Why bank credit policies fluctuate: A theory and some evidence. *The Quarterly Journal of Economics, 109*(2), 399–441. https://doi.org/10.2307/2118468
Sanjeev, G. M. (2007). Bankers' perceptions on causes of bad loans in banks. *Management Research (09725814), 7*(1), 40–46.
Schularick, M., & Taylor, A. M. (2012). Credit booms gone bust: Monetary policy, leverage cycles, and financial crises, 1870–2008. *American Economic Review, 102*(2), 1029–1061. https://doi.org/10.1257/aer.102.2.1029
Stern, G. H., & Feldman, R. J. (2004). *Too big to fail: The hazards of bank bailouts.* Brookings Institution Press.
Tanasković, S., & Jandrić, M. (2015). Macroeconomic and institutional determinants of non-performing loans. *Central Banking Theory and Practice, 4*(1), 47–62. https://doi.org/10.1515/jcbtp-2015-0004
Turan, G., & Koskija, A. (2014). Nonperforming loans in Albania. *Academic Journal of Interdisciplinary Studies, 3*(3), 491–491. https://doi.org/10.5901/ajis.2014.v3n3p491
World Bank. (2020). *Taking Stock, July 2020: What will be the new normal for Vietnam? The economic impact of COVID-19* (pp. 1–33). World Bank. https://doi.org/10.1596/34268
Žunić, A., Kozarić, K., & Dželihodžić, E. Ž. (2021). Non-performing loan determinants and impact of covid-19: Case of Bosnia and Herzegovina. *Central Banking Theory and Practice, 10*(3), 5–22. https://doi.org/10.2478/jcbtp-2021-0021

Intellectual Capital and Banking Financial Performances: A Systematic Review

Sushmita Prasad, Pankaj Anand, Jitendra Kumar, Subhadeep Mukherjee, and Harishankar Vidyarathi

Abstract The current study reviews the contribution of intellectual capital on banking financial performance. In today's economic context, intangible assets along with the various tangible assets play an important role in the performance of any sector. Intellectual capital refers to that intangible asset that contributes to the bottom line (organizational processes, expertise of employees, and the sum of knowledge) of any company. Intellectual capital is identified to be one of the vital forms of assets as the conventional methods are lacking behind in measuring the intangible dimensions of performance of the company. In the current study, we will focus on reviewing past literatures on intellectual capital and banking financial performance. This study will help banking sectors in identifying the factors of intellectual capital and also the performance parameters of banking sectors and their financial performance. It will also identify how intellectual capital can increase the performance of a companies in the banking sector.

Keywords Intangible assets · Intellectual capital · Banking · Financial performance · VUCA analysis · Systematic review

1 Introduction

With the development of an information and knowledge intensive economy, the interest concerning intellectual capital is drawn (Petty & Guthrie, 2000). The word 'intellectual capital (IC)' was originally named by renowned Economist 'John

S. Prasad (✉) · P. Anand
Sikkim Manipal University, Sikkim, India
e-mail: sushmitaprasad16@yahoo.com

J. Kumar · S. Mukherjee
Sikkim Manipal Institute of Technology, Sikkim, India

H. Vidyarathi
Institute of Public Enterprise, Hyderabad, India

A. T. Nguyen et al. (eds.), *Contemporary Economic Issues in Asian Countries: Proceeding of CEIAC 2022, Volume 1*,
https://doi.org/10.1007/978-981-19-9669-6_33

Kenneth'. IC is well explained as knowledge correlated intangible assets of any organization (Kehelwalatenna & Premaratne, 2014). Software, goodwill, patents, trademarks, infrastructure, and databases are some intangible assets which are mentioned in the asset side of the firm/company's balance sheet which is recognized by the accounting standards (Ting & Lean, 2009).

Intellectual Capital (IC) is intangible assets that generate worth for an organization (Hunter et al., 2005). Stewart (1997) described IC as 'packaged useful knowledge'. Kim et al. (2011) stated that intellectual capital is a type of non-monetary asset that generates economic benefits for the world economy. Edvinson (1997) and Sullivan (1999) in their study mentioned that intellectual capital is a combination of experience, ability, customer relationships, knowledge, human competencies, and skills that helps a particular company or sector gain competitive advantage.

Intellectual capital when studied in detail is a broader concept, which in general is categorized as human capital, structural capital, and relational capital (Bansal & Singh, 2020).

Human capital (HC) is the explicit and implicit knowledge and facts of personnel. It also includes abilities, know-hows, and expertise of personnel, which they practice to carry out companies' activities (Komnenic & Pokrajcic, 2012). HC is the core component of IC, but it leaves as employees leave the organization (Chen et al., 2005). Human capital leads to generation of novelty as newer products and services.

Structural capital (SC) contributes to the knowledge which stays with the companies even when the employees leave the companies, and it includes databases, cultures, procedures, and systems, making it more like organizational capital (Jashapara, 2004; Petty & Guthrie, 2000). Bontis (1998) states that structural capital is that contrivance in companies that helps in optimizing intellectuals, hence increasing performance.

Relational capital or in other words capital employed explains the relationship of a company with its customers, partners, and investors. A company's profit depends on customer relationship, satisfaction of customers, customer loyalty, and firm's performance (Kaplan and Norton, 1996). Customer commitment and loyalty can be estimated by the dimension of employees' commitment and loyalty (Horibe, 1999) (Fig. 1).

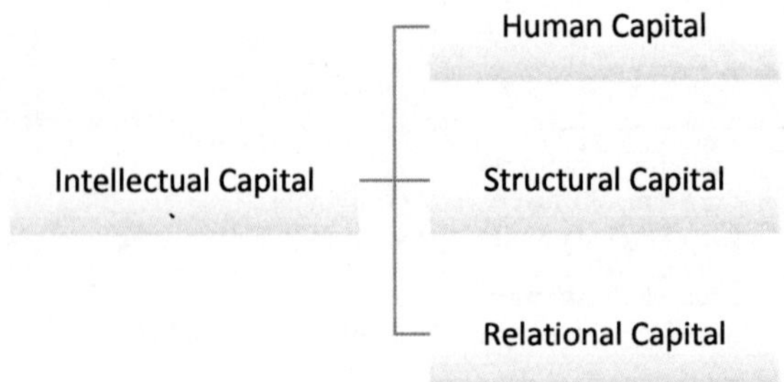

Fig. 1 Composition of organizational performance. Adapted from Chahal and Bakshi (2017)

Researchers around the world have a view that if intellectual capital is used efficiently, and effectively, it leads to improvement of financial performance of any companies (Ramandeep & Narwal, 2016). Additionally, firms which are determined by knowledge can improve on the production leading to profit maximization and enhancing market value (Pulic, 1998; Stewart, 1997). Many researchers like Mavridis (2004), El-bannany (2008), and Mondal and Ghosh (2012) have an opinion that the banking sector is a very knowledge driven industry, and it further depends on the factors of intellectual capital rather than physical or financial capital. Many researchers are keen on exploring the consequences of intellectual capital in relation to the banking segment (Singh et al., 2016).

After the adoption of the rule of liberalization, privatization, and globalization (LPG) in 1991, the banking sector has turned out to be one of the firmest growing industries in India. Banks contribute to the economic development of any country especially the developing countries by interim like a financial arbitrator between different other sectors (Goh, 2005). Indian banking sectors are one of the strong industries as it was very fewer affected by the universal economic crisis in the year 2008. Right after the Global Economic Crisis, in the year 2010 and 2011, the productivity and thus the profitability of the banking sector in India increased, which was among the utmost performance indicators around the world (Mohan, 2012).

Type of banks	Number
Public sector banks	12
Private sector banks	21
Small finance banks	12
Regional rural banks	43
Payments banks	4
Foreign banks	45
Total	137

Current distribution of banks in India: *Source* RBI (2022)

However, this brings a huge competition among the banks to be more efficient, profitable, and productive to survive which brings in the role of intellectual capital. The main motivation behind the study is to discover how intellectual capital has impacted banking performance, keeping in point the part played by the banks in the survival of the economy.

With the transformation and steps of India from developing country to developed country, as banks play a vital role as it is one of the industries that survived the global crisis and should have a durable and strong base. In addition to this, the Indian government had taken many steps for the same adding to the revival of the Indian Economy after the crisis (Wequar et al., 2018).

The current study focuses on a systematic review of the previous research done on intellectual capital (IC), pointing to identify the different types of capitals involved in developing IC and in view of the explanation of IC and its factors. It is a challenging effort to see the ways of measuring components of IC. With the past studies, it is found that there are three dimensions as follows; organizational, national, and regional. Now, the dimension brings in a different path to explore the connection of IC and banking performance (Bansal & Singh, 2020).

This current study adds to the research and helps in understanding in what way to estimate IC performance of the Indian banking sector. Furthermore, it also recognizes the increased role of IC in a bank's economic performance. The current study also contributes to the continuing research on the factors of banking effectiveness (Garcia & Guerreiro, 2016; Menicucci & Paolucci, 2016; Petria et al., 2015). The study not only helps banks mend their profitability and will also support the policy creators in achieving the goals of financial stability.

2 Literature Review

The vital area of interest for a researcher before 1990 was measurement of intellectual capital. Roslender et al. (2006) in their study showcased that the financials of the firms should have declaration on IC along with other assets of the company. It was also suggested that whenever a company is taken into consideration, all the complete value and present performance should be considered. Hence, along with computation of intellectual capital, information should be utilized efficiently.

Studies in view of the impact of IC on bank performance are talked about below.

2.1 Overview of the Intellectual Capital

IC is one of the core outcomes of human information. It plays a vital part in the approach accomplishment to achieve a competitive advantage among its competitors and to enhance presentation. Steward (1997) in his study stated that IC can be anything from intellectual information, to knowledge of material to intellectual property rights and experience, knowledge, and skills of employees that can ultimately be used to enhance firm performance leading to more profit creation.

In general, studies related to intellectual capital research are uncommon in developing markets or developing countries (Abeysekera, 2007; Khan & Ali, 2010). One such instance is the case of Malawi, and there is no empirical study related to intellectual capital and bank performances. When going through the literature, it was found that one study had been conducted with a view to just see how intellectual capital can be seen in the annual financial reports of any company, but the study could not analyze intellectual capital (Lipunga, 2014). Lipunga (2014) studied the valuation of IC on the listed companies of the Malawi Stock Exchange during the years 2008–2012 focusing on the value of IC and not emphasizing on the efficiency part. However, as per the market-to-book (MTB) value method, the outcomes of the research advocated IC to be an important portion of the selected companies in their study.

Public (2000) developed a method for measuring the intellectual capital, he stated that value creation of any company is done by both intangible and tangible assets, and in intangible assets, intellectual capital plays an important role. He also mentioned that value creation is done through IC and its components, viz., human capital, structural capital, and relational capital, and addition to it is the capital employed. He discussed value creation of a company, and value creation comprises both tangible and intangible assets. He named the method as value added intellectual capital (VAIC), the method uses human capital efficiency, capital employed efficiency, and structural capital efficiency. Like other methods, VAIC has its limitations too, but when compared to other methods, VAIC is an ideal method as it has reliability, simplicity, and indelibility (Maditinos et al., 2011). VAIC is one the approved methods for measuring intellectual capital as it is based on the accounting data that can be observed and verified as other models are customized, and they have difficulties in generalizing as per the data (Firer & Williams, 2003).

Sherif and Elsayed (2016) explained that IC is an imperative aspect in backing up the organization's performance. Intellectual capital should be managed in a very efficient and effective way in order to compete and gain better firm performances. Khalique et al. (2015) in his study stated that in IC is increasingly achieving more significance among the knowledge based economy, as a precarious strategic asset. Various scholars (Chowdhury et al., 2019; Komnenic & Pokrajcic, 2012; Kweh et al., 2019; Osinski et al., 2017; Ousama & Fatima, 2015; Pulic 1998, 2004) also bring up that intellectual capital comprises human capital, structural capital, and relational capital which is more than just knowledge.

As per Cheng et al. (2010), IC is an important basis of presentation and value formation, and IC helps in gaining a competitive advantage. Competitive advantage does not rely upon, as conventionally supposed, such things as economies of scale, technology, or natural resources, as they are progressively easier to copy (Kamukama, 2013). These are certainly among the set of assets proposed by Stewart (1997) as 'invisible assets', which are IC resources in reality.

Competitive advantage will contribute in making the company more valuable than its competitors in response to the resource-based theory anticipated by Penrose (1959) which is a company that has competitive advantage. A firm is supposed to have a competitive advantage if it can generate advanced economic and monetary value resulting from their commercial resources than the other firms/companies in its sector (Porter, 1985). According to Pulic and Kolakovic (2003), each company has its exclusive values, skills, knowledge, and other intangible resources which helps in increasing its market value. Thus, the researchers have considered competitive advantage as an intermediary between IC and company's financial performance (Kamukama & Sulait, 2017). Due to the compatibility among the individual's values and a firm's values, the employees trust each other and thus try to achieve the end goals by sharing knowledge and working as a team. And, this synergic effect (employee interpersonal relationship) makes the company/firm exclusive and permits it to build the competitive spot in the current market (Bontis, 1998).

2.2 Overview of the Indian Banking Industry

Since the emergency in 1991, because of the 'balance of payment', the Indian banking system and its competency have gained ample amounts of significance (Ahluwalia, 2002). The underperformance of the Indian Banking System in India is due to interference of the Government and their excessive control over the banks. Before 1991, majority of the banking sector was controlled by the state-owned banks, which brought out the government's outline of industrialization and socio-economic development (Bowers et al., 2003).

Competitiveness among the banking sector has increased during the recent years, due to an increased number of private and foreign banks. Due to the Indian economy's increased openness to private and foreign banks, there have been several changes in product innovation, service quality, and the usage of knowledge and skills in the Indian banking sector has increased several folds (Ataullah & Le, 2006). Economic liberalization, structural reforms, deregulation, and technological acceptance have suggestively brought in a major transformation to the banking sector in India. The banking sector has perceived a pattern change, and it transmuted into a more constant and well-organized financial system in recent times (Mohan, 2007).

2.3 Overview of Intellectual Capital and Banking Financial Performance

There has been a lot of curiosity since years among the researchers in relation to the impact of IC on bank's performances. It is also found in many studies that intellectual capital has a very vital aspect in the financial advancement of the Indian Banking System. Smriti and Das (2018) and Khalique et al. (2018) in their studies found that intellectual capital promotes the firm's performance in the service economy.

While studying the IC and firm performance dynamics, Mention and Bontis (2013) suggested that human capital was indirect as well as direct contributor to the business performance in their study where they had taken 200 banks from Luxembourg and Belgium. It was also found that relational and structural capitals were positively correlated to company's performance. Results by Mohiuddin et al. (2006) indicated a significant correlation among the 17 selected commercial banks of Bangladesh during the year 2002–2004. Kamal et al. (2012) in their study confirmed that the various measures of IC are related with bank's financial performance. The results showed that the IC components, namely value added capital employed and value added human capital, have a very important impact on the bank's performance. Also, Kamal et al. (2012) established that IC and its components are related to the productivity of the firm (Table 1).

Table 1 Literature review

Sl. No.	Title	Author & Year	Inference	Relevance with present work
1	The impact of intellectual capital on firms' market value and financial performance	Maditinos et al. (2011)	The researchers have briefly discussed intellectual capital, market value and financial performance	Effects of structural capital, human capital, and relational capital on financial performance are studied in detail
2	Back to the future of intellectual capital research: A systematic literature review	Pedro et al. (2018)	The researchers showcased the importance of using different measures to calculate intellectual capital	IC- its metrics and indicators are studied in detail
3	Exploring the effect of intellectual capital on financial performance: A study of Indian banks	Weqar et al. (2020)	The research inspects the effect of IC on the financial performance of Indian banks	Methods and techniques to calculate intellectual capital
4	Intellectual capital and firm performance: Evidence from certified firms from the EFQM excellence model	Yousaf (2021)	The study investigates impacts of IC on a firm's performance	Usage of IC on measuring performance of industries
5	The influence of intellectual capital on Indian firms' financial performance	Weqar et al. (2022)	The study explores the impact of IC and the components on the financial performance of Indian banking sector firms	Intellectual capital effects and its various dimensions

3 Methodology

The current study talks about the introduction of intellectual capital and the importance of a bank's performance. Relevant studies from the year 2009 to 2021 were taken into consideration for this study. The papers showed the shift of interest to intellectual capital and its part. It also focuses on the concept of banking performance—pre-liberalization, privatization, and globalization and post-liberalization, privatization, and globalization, and how intellectual capital has enhanced the performance of the banking sector (Fig. 2).

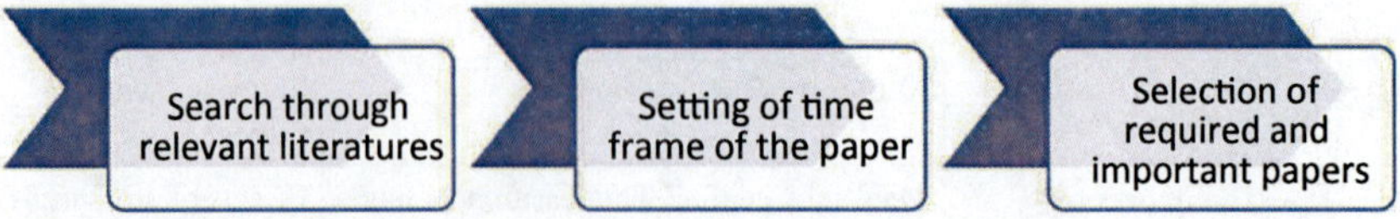

Fig. 2 Data selection procedure. *Source* Own elaboration

- Papers from various journals using the combination of keywords such as 'intellectual capital', 'banking performance', and 'systematic review' were searched from 1992 to 2022.
- Time frame of the studies was set from 2009 to 2022.
- Important relevant papers were selected as per the requirements of the research, and the intellectual capital and banking performance dynamics were studied in detail.

After reviewing literature, Table 2 is prepared to elaborate the evolution of intellectual capital.

As presented in Table 2, the evolution of intellectual capital follows the trend of different research areas. During the first phase, a theoretical model was created, and the next phase is value addition to the theoretical body in the first phase and third phase division of intellectual capital concept, which are various knowledge and dimensions (Pedro et al., 2018).

Also, we studied the impact of intellectual capital with respect to VUCA concept. With a lot of change in the dynamics of the world economy and living standards of the employees, we saw a gradual rise in the concept of VUCA. VUCA comprises four different words together as follows: volatility, uncertainty, complexity, and ambiguity.

VUCA	Description in relationship to the study
Volatility	Any company must have a vision, and that vision leads to achievement of goals. A company must have a set of goals to be achieved and think about the future of the company
Uncertainty	In any business, uncertainty should be expected but also handled without any flaws and disruption of the process
Complexity	Any process in a company is complex but understanding the process can ease out the process of any company, working with co-ordination between the employees
Ambiguity	Ambiguity is something to do with the interpersonal and way of doing work with the co-ordination. Everyday can be a new day for learning and experience

With reference to intellectual capital and banking performance, some points are discussed below:

Impact of IC on economic performance of banking sector: Any company is required to manage its resources to bring out best for them, also keeping sustainability in consideration in the VUCA era. Here, VUCA helps in optimization of financial performance.

Table 2 Phases of intellectual capital

Phase	Time period	Area of research	References
1st phase: Evolution of a hypothetical framework	During the year 1980 to 1999	This phase concentrated on awareness of intellectual capital and its importance in creating and managing a competitive advantage	Kaplan and Norton (1992, 1996), Stewart & Losee (1994), Czerniawska (1999), Guthrie and Petty (2000)
2nd phase: Evolution supported by proofs	During the period 2000–2003	This is a phase in which various approaches to measure, communication, and management of intellectual capital are validated with conceptualization and empirical proof such as accounting and different reports	Mouritsen et al. (2000), Baum et al. (2000), Andriessen & Tiessen (2000), Sullivan (2000)
3rd phase: Evolution of applications of Intellectual Capital	During the period 2004 till present	Studies during this stage were focused on a change of approaches in order to differentiate among the various drivers of wealth creation, while maintaining a balance of all relevant measures, resulting in creation of an incremental view of the national innovation capacity	Bontis (2004), Andriessen (2004), Seetharaman et al. (2004) Pasher & Shachar (2005), Barathi Kamath (2007), Schiuma & Lerro (2008), Edvinsson & Lin (2009), Salonius & Lönnqvist (2012), Roos & O'Connor (2015), Kamath (2015), Weqar et al. (2020), Weqar & Haque (2022), Alvino et al. (2020),Mohapatra et al. (2019), Nadeem et al. (2017), Rochmadhona et al. (2018), Singh and Rao (2016), Tran and Vo (2018)and Tjahjadi (2020)

Source Guthrie et al. (2012), Dumay & Garanina (2012, 2013), Labra & Sánchez (2013), Roos & O'Connor (2015), Pedro et al. (2018)

Result of ownership on the connection between IC and banking financial performance: Due to the level of management pyramid, it becomes easier for the decision makers, and also independent decisions can be taken as per the need of the banks. Impact of good decision can lead to better economic performance of the banks.

4 Conclusion

The current study shows that the literatures have aspects in relation to intellectual capital and banking performance. It is also recognized on improving the organizational processes and performance of the banking sector. It was also observed that IC was more into trend only after LPG and world economic crisis where role and importance of banking industries have been considered.

Therefore, it is also found that intangible resources are important in the growing of any company from any sector.

References

Alvino, F., Di Vaio, A., Hassan, R., & Palladino, R. (2020). Intellectual capital and sustainable development: A systematic literature review. *Journal of Intellectual Capital, 22*(1), 76–94. https://doi.org/10.1108/jic-11-2019-0259

Bansal, D., & Singh, S. (2020). Impact of intellectual capital on financial performance of the Indian software sector. *Asia-Pacific Management Accounting Journal, 15*(1), 67–95. https://doi.org/10.24191/apmaj.v15i1-04

Barathi Kamath, G. (2007). The intellectual capital performance of the Indian banking sector. *Journal of Intellectual Capital, 8*(1), 96–123. https://doi.org/10.1108/14691930710715088

Bontis, N. (1998). Intellectual capital: An exploratory study that develops measures and models. *Management Decision, 36*(2), 63–76. https://doi.org/10.1108/00251749810204142

Czerniawska, F. (1999). Intellectual capital. *Management Consultancy in the 21st Century*, 212–221. https://doi.org/10.1007/978-1-349-14873-8_21

Goh, P. C. (2005). Intellectual capital performance of commercial banks in Malaysia. *Journal of Intellectual Capital, 6*(3), 385–396. https://doi.org/10.1108/14691930510611120

Guthrie, J., & Petty, R. (2000). Intellectual capital: Australian annual reporting practices. *Journal of Intellectual Capital, 1*(3), 241–251. https://doi.org/10.1108/14691930010350800

Kamath, G. B. (2015). Impact of intellectual capital on financial performance and market valuation of firms in India. *International Letters of Social and Humanistic Sciences, 48*, 107–122. https://doi.org/10.18052/www.scipress.com/ilshs.48.107

Kamukama, N. (2013). Intellectual capital: Company's invisible source of competitive advantage. *Competitiveness Review: An International Business Journal, 23*(3), 260–283. https://doi.org/10.1108/10595421311319834

Kehelwalatenna, S., & Premaratne, G. (2014). Intellectual capital performance and its long-run behavior: The US banking industry case. *New Zealand Economic Papers, 48*(3), 313–333. https://doi.org/10.1080/00779954.2013.867796

Komnenic, B., & Pokrajčić, D. (2012). Intellectual capital and corporate performance of MNCs in Serbia. *Journal of Intellectual Capital, 13*(1), 106–119. https://doi.org/10.1108/14691931211196231

Maditinos, D., Chatzoudes, D., Tsairidis, C., & Theriou, G. (2011). The impact of intellectual capital on firms' market value and financial performance. *Journal of Intellectual Capital, 12*(1), 132–151. https://doi.org/10.1108/14691931111097944

Mohapatra, S., Jena, S. K., Mitra, A., & Tiwari, A. K. (2019). Intellectual capital and firm performance: Evidence from Indian banking sector. *Applied Economics, 51*(57), 6054–6067. https://doi.org/10.1080/00036846.2019.1645283

Mondal, A., & Ghosh, S. K. (2012). Intellectual capital and financial performance of Indian banks. *Journal of Intellectual Capital, 13*(4), 515–530. https://doi.org/10.1108/14691931211276115

Nadeem, M., Gan, C., & Nguyen, C. (2017). The importance of intellectual capital for firm performance: Evidence from Australia. *Australian Accounting Review, 28*(3), 334–344. https://doi.org/10.1111/auar.12184

Norton, A. (1996). Social ties and intellectual knots: The weaknesses of the communitarian case against markets. *The Australian Quarterly, 68*(1), 53. https://doi.org/10.2307/20634717

Pedro, E., Leitão, J., & Alves, H. (2018). Back to the future of intellectual capital research: A systematic literature review. *Management Decision, 56*(11), 2502–2583. https://doi.org/10.1108/md-08-2017-0807

Petty, R., & Guthrie, J. (2000). Intellectual capital literature review. *Journal of Intellectual Capital, 1*(2), 155–176. https://doi.org/10.1108/14691930010348731

Ramandeep, & Narwal, K. P. (2016). Investigating the impact of intellectual capital on financial performance in Indian manufacturing sector. *JIMS8M: The Journal of Indian Management & Strategy, 21*(4), 28. https://doi.org/10.5958/0973-9343.2016.00030.2

Rochmadhona, B. N., Suganda, T. R., & Cahyadi, S. (2018). The competitive advantage between intellectual capital and financial performance of banking sector in ASEAN. *Jurnal Keuangan dan Perbankan, 22*(2). https://doi.org/10.26905/jkdp.v22i2.2060

Seetharaman, A., Lock Teng Low, K., & Saravanan, A. (2004). Comparative justification on intellectual capital. *Journal of Intellectual Capital, 5*(4), 522–539. https://doi.org/10.1108/14691930410566997

Singh, B., & Rao, M. K. (2016). Intellectual capital measure. *PsycTESTS Dataset*. https://doi.org/10.1037/t56555-000

Soewarno, N., & Tjahjadi, B. (2020). Measures that matter: An empirical investigation of intellectual capital and financial performance of banking firms in Indonesia. *Journal of Intellectual Capital, 21*(6), 1085–1106. https://doi.org/10.1108/jic-09-2019-0225

Ting, I. W. K., & Lean, H. H. (2009). Intellectual capital performance of financial institutions in Malaysia. *Journal of Intellectual Capital, 10*(4), 588–599. https://doi.org/10.1108/14691930910996661

Tran, D. B., & Vo, D. H. (2018). Should bankers be concerned with intellectual capital? A study of the Thai banking sector. *Journal of Intellectual Capital, 19*(5), 897–914. https://doi.org/10.1108/jic-12-2017-0185

Weqar, F., Khan, A. M., & Haque, S. M. (2020). Exploring the effect of intellectual capital on financial performance: A study of Indian banks. *Measuring Business Excellence, 24*(4), 511–529. https://doi.org/10.1108/mbe-12-2019-0118

The Effects of Middle Managers' Leadership Practices on Strategy Implementation: Evidence from Vietnamese Information Technology Enterprises

Van Thi Nguyen and Duong Thi Thuy Luu

Abstract A middle manager is a company official who is in charge of a specific function. Therefore, at various strategic levels, a middle manager's leadership practice has a significant impact on the strategy implementation process. This study focuses on the influence of middle managers' leadership practices on the outcomes of corporate and functional strategy implementation in relation to Vietnamese information technology (IT) enterprises. The study employs quantitative research techniques to evaluate the research model by surveying 220 middle managers, and the data is analyzed using PLS-SEM. The results show the positive influence of five leadership practice areas highlighted by the Leadership Practices Inventory (LPI) on functional and enterprise-level strategy implementation. The research results assist IT businesses in developing solutions to strengthen middle managers' leadership practices in order to improve the outcomes of their strategy implementation.

Keywords Leadership practices · Middle manager · Strategy implementation

1 Introduction

The science of strategic management has developed since the 1960s and is increasingly contributing to the long-term development of enterprises. In particular, the focus of the field of strategic management has shifted from strategy formulation to strategy implementation (Obeidat et al., 2017); because according to some researchers, strategy implementation has become and is becoming the most important management challenge and the key to the survival of any organization (Jaoua, 2018). Many studies have demonstrated that despite having well-planned strategies, organizations fail in the strategy implementation phase.

V. T. Nguyen (✉) · D. T. T. Luu
Thuongmai University, Hanoi, Vietnam
e-mail: vannguyen@tmu.edu.vn

D. T. T. Luu
e-mail: duongqtcl@tmu.edu.vn

A. T. Nguyen et al. (eds.), *Contemporary Economic Issues in Asian Countries: Proceeding of CEIAC 2022, Volume 1*,
https://doi.org/10.1007/978-981-19-9669-6_34

A study conducted by Fortune Magazine in 2009 found that 90% of strategies fail, and the most important cause of this is attributed to the poor strategy implementation (Waterman et al., 1980). Čater and Pučko (2010) concluded that while 80% of organizations have the appropriate plans, just 14% of them have been successfully executed. Bell et al. (2010) also argued that the task of strategy implementation is often the most complex and time-consuming part of strategic management. Therefore, in order to enhance the effectiveness of strategy implementation, assessing the results of strategy implementation in enterprises as well as identifying and analyzing factors affecting strategy implementation is a particularly important task, especially factors that come from human resources.

As a result of our review, many studies on this topic have been conducted around the world, with the following trends emerging: the role of middle managers in strategy implementation (Jaoua, 2018; Raes et al., 2011; Wooldridge & Floyd, 1990; Wooldridge et al., 2008); and the impact of leadership practices on strategy implementation or firm performance (Avery & Bergsteiner, 2011; Cowden et al., 2011; Forner et al., 2012; Goldman, 2012; Lee, 2018; Nawab & Asad, 2020; Thoonen et al., 2011). The majority of studies on leadership practice, however, have focused on senior leaders with little or no consideration for middle managers in strategy implementation.

On the other hand, studies on middle managers primarily focus on their involvement in effective strategy implementation and the overall performance of the organization/enterprise without establishing the impact of middle managers on strategy implementation at the corporate and functional levels. This research will look at the role of middle managers in strategy implementation in the context of the IT industry, which is a crucial economic sector and a significant contributor to Vietnam's GDP development. Firms' strategies, as influenced by their leadership practices, impact strategy execution at the corporate and functional levels, and the relationship between these two levels is investigated.

2 Theoretical Framework

2.1 *Middle Manager*

There have been many previous studies on middle managers, but there has yet to be a universally accepted formal concept for the term "middle manager." Through the overview from the previous studies, there are two approaches to middle managers: (1) the approach to management positions and levels in the enterprise and (2) the middle manager's functional task-based approach. According to the first approach, middle managers are individuals who operate in positions below senior managers and above junior managers (Dutton & Ashford, 1993; Wooldridge et al., 2008). However, the distinctive feature of middle managers is not their position in the companies' organizational chart.

Instead, the capacity to engage in the activities of senior managers, paired with knowledge of operational activities, is the characteristic that makes them stand out and unique (Wooldridge et al., 2008). This combination enables them to function as intermediaries between the organization's strategy and day-to-day operations (Nonaka, 1994). This position allows them to exert some influence on strategic and operational decisions (Schilit, 1987) and thus on organizational performance (Currie & Procter, 2005; Floyd & Wooldridge, 1994; Nonaka, 1994).

The important role of middle management in strategy implementation has been confirmed for many years and in many research publications. According to Raes et al. (2011), a middle manager is central person to implement strategies effectively. Floyd and Wooldridge (1992) defined this role as a manager's intervention that aligns the organization's actions with the strategic goals of senior management. Middle managers implement strategy by converting the company's strategy into action plans and personal objectives (Currie & Procter, 2005). Minarro-Viseras et al. (2005) found that middle managers have goal orientation, organizational skills, flexibility, patience, perseverance, the ability to release the energy of team members, and the ability to let team members engage in decision-making. These elements are key factors in the success of strategy implementation.

2.2 Leadership Practices

According to Northouse (2018), leadership practices involve the processes by which an individual influences a group of other individuals to achieve a common goal. Similarly, Crevani (2015) also stated that leadership practice is "as any practice that offers direction to a group or organization." Some other authors clarify the connotation of this concept, such as Forner et al. (2012): "Leadership practices are understood as daily actions, activities, and habits that leaders use to actively pursue leadership priorities," or as Sprigghr (2020): Practice leadership refers to the activities and strategies that leaders will care about and undertake to always help their team better themselves and accomplish progress. As can be seen from the preceding definitions: Leadership practice is a process that occurs in a space where managers and other people in the organization engage to enhance the performance of those individuals and organizations.

Our review of leadership practices reveals that there are several approaches to this issue, including trait approach, skill approach, behavioral approach, situational approach, and leadership style approach (Northouse, 2018). Among them, the most popular ones are the leadership style approach (Cowden et al., 2011; Levesque-Côté et al., 2021; Nawab & Asad, 2020; Thoonen et al., 2011) and the behavioral approach (Forner et al., 2012; Waters & Marzano, 2006). However, Cowden et al. (2011) also argued that leadership practice is not always defined according to a specific leadership behavior or style because each leadership style has certain advantages and disadvantages (Cowden et al., 2011), and leadership practices also need to be flexible in response to different contexts (Forner et al., 2012). In this study, we use the Leadership Practices Inventory (LPI) approach of Posner and Kouzes (1993). Accordingly, LPI emphasizes the value and vision of the manager, the relationship between the manager and others, and the specific actions of the manager to inspire, motivate, and develop employees.

2.3 Corporate and Functional-Level Strategy Implementation

A business's strategy exists at several levels, ranging from corporate-level strategy, business-level strategy to functional-level strategy (Beard & Dess, 1981; Hill et al., 2014). In particular, corporate-level strategy includes senior management making strategic decisions regarding the scale of operations, values, and approaches to help the business maximize profits in the long term (Hill et al., 2014). Business-level strategy concerns how a firm competes successfully in a particular industry (Beard & Dess, 1981; Hill et al., 2014). Functional-level strategy is concerned with the use of resources and capabilities within the enterprise to improve operational efficiency and the competitive position of the enterprise (Hill et al., 2014).

Also according to Hill et al. (2014), these three strategic levels are inextricably related to each other, in which the corporate-level strategy plays a role in directing the company's business model over time, and at the same time, determining the type of business strategy and functional-level strategy that managers will choose to maximize the long-term profit. Many previous studies have indicated that the successful implementation of the enterprise's strategy is related to many functional activities such as operational planning, resource allocation, human resource management, finance, public administration, technology, and media (David & David, 2020; Okumus, 2003; Obeidat et al., 2017), and performance at the functional level has a positive influence on the overall performance of the organization (Rapert et al., 1996). In this study, we aim to test the relationship between the implementation of functional-level strategy and corporate-level strategy through leadership practices of middle managers, because functional-level strategy implementation will be associated with middle managers who are in charge of specific functions in the business, thereby contributing to the achievement of corporate-level strategy implementation goals.

3 Research Hypothesis

3.1 *The Relationship Between Leadership Practices and Corporate-Level Strategy Implementation*

The previous research on managers has revealed that middle managers play an expanding role when they not only play an essential part in strategy implementation but also become increasingly involved in the corporate strategy formulation process (Jaoua, 2018). This role expansion causes middle managers to increase their commitment to strategy implementation (David & David, 2020). Empirical research by Jaoua (2018) has proven the strategic role of middle managers in the successful implementation of the strategy and the performance of the business. Several other studies have also shown a relationship between leadership practices and work environment increase and employee retention (Cowden et al., 2011; Nawab & Asad, 2020), resilience enhancement (Avery & Bergsteiner, 2011), and organizational performance improvement (Jaoua, 2018). This allows us to put forward arguments for the relationship between leadership practices and corporate-level strategy implementation. Therefore, the hypothesis H1 is proposed as follows:

H1 Leadership practices of middle managers have a positive effect on the corporate-level strategy implementation of enterprises.

3.2 *The Relationship Between Leadership Practice and Functional-Level Strategy Implementation*

Successful strategy implementation requires a combination of factors, including strategic leadership and internal communication (Okumus, 2003; Obeidat et al., 2017). Dziekoński (2017) asserted that important competencies of project managers including decision-making, communication, motivating team members, problem-solving, conflict resolution, negotiation, etc., will have an important influence on project management. Lee (2018) proved that leaders' leadership practices will positively affect the performance of each department by promoting departmental collaboration, defining roles and developing staff skills and knowledge. Another experimental study conducted by Rapert et al. (1996) demonstrated that the consensus and commitment of individuals in the functional department can lead to improved functional performance, thereby improving organizational performance. Meanwhile, leadership practices according to the Leadership Practices Inventory approach support the environment and organizational culture toward positivity and cooperation in business (Goewey, 2012). Based on the above, the hypothesis H2 is proposed as follows:

H2 Leadership practices of middle managers have a positive effect on the functional strategy implementation of Vietnamese IT enterprises.

3.3 The Relationship Between Functional-Level Strategy Implementation and Corporate-Level Strategy Implementation

Specific functional units and areas are constituent parts of an enterprise. Successful strategy implementation of an enterprise is highly related to functional activities (David & David, 2020; Okumus, 2003; Obeidat et al., 2017), and functional-level performance has a positive impact on the overall performance of the organization (Rapert et al., 1996). Research by Slater and Olson (2001) studied marketing functions and demonstrated the positive influence of marketing on implementing business strategy. Goedhuysa and Veugelers (2012) demonstrated the influence of innovation strategy on the expansion and development of enterprises through the impact on product and process innovation. In another aspect, Grundy (1998) demonstrated that human resources strategy plays an important role not only in corporate strategy formulation but also in corporate strategy implementation. The effectiveness of a business might be significantly impaired without a good human resources strategy. Therefore, it is expected that enhanced functional strategy implementation will positively impact the corporate strategy implementation, and the hypothesis H3 is proposed as follows:

H3 The functional-level strategy implementation has a positive effect on corporate-level strategy implementation.

4 Research Methods

4.1 Measures

The authors have adopted and adapted research scales for variables in the research model of leadership practices of middle managers on the theoretical basis of the Leadership Practices Inventory questionnaire (Posner and Kouzes 1993). This survey tool is designed to provide leaders with critical feedback on their leadership abilities to serve their organizations. Posner and Kouzes (1993) research identified five factors that are important to most experiences. The best leadership includes (1) challenging the process, (2) inspiring a shared vision, (3) enabling others to act, (4) modeling the way, and (5) encouraging the heart.

Functional-level strategy implementation (FIS) is adapted from research by Lee (2018). Corporate-level strategy implementation scale (CSI) with five questions is used from the studies of Lu et al. (2015) and Alexander (1985). After in-depth interviews, the scales were adjusted to suit the research context.

4.2 *Sample Size and Data Collection*

The subjects of the investigation are middle managers who are in charge of a specific function in enterprises, including heads and deputy heads of functional departments working in Vietnamese IT enterprises. They are suitable subjects with enough information to evaluate not only the leadership practices themselves but also the results of the strategy implementation at the functional and corporate levels. The authors used the convenience sampling method. The research was selected to survey IT enterprises in 4 regions including Hanoi, Ho Chi Minh City, Da Nang, and Hai Phong. These are the provinces that account for over 60% of the total number of IT enterprises in the country and have the highest ranking results in the IT industry index (Vietnam Ministry of Information & Communications, 2020). Surveyed enterprises were diverse in size, type, and field of operation. The survey subjects were also diverse in age, gender, seniority, management seniority, and fields of responsibility.

Regarding sample sizes, according to Hair et al. (2010), there is a general rule that shows that the ratio should never go below 5:1, meaning that five observations are made for each independent variable in the variate. We proceeded to build a scale of 38 observed variables. On the basis of inheriting the concept of choosing the sample size from Hair et al. (2010), the minimum number of samples that need to be investigated is 190. This study used partial least squares-SEM (PLS-SEM) structural model analysis to test the research model. This analysis does not require highly normally distributed data and can be performed with small sample sizes, with sample sizes ranging from 100 to 200 usually being a good start to perform path modeling (Hoyle, 1995). Starting from the above theoretical basis and the actual conditions, we aim for the survey scale from 200 to 250.

4.3 *Questionnaire and Survey Design*

After adjusting the questionnaire according to the pre-validation results, the survey was designed to include two parts as follows: Part 1: General information. This section was designed with 2 groups of information: information about businesses and information about respondents. Part 2: Evaluation of leadership practices and strategy implementation outcomes. Each item will be rated on a 7-point Likert scale ranging from 1—"strongly disagree" to 7—"strongly agree" and 4—no opinion or neutral.

The author employed both the direct approach and the indirect strategy to contact the survey participants. With the indirect approach, the author made phone calls and sent emails to invite them to participate in the survey. After obtaining consent, the author proceeds to send the survey through the Google Doc link sent via email or other electronic means of exchange such as Zalo and Facebook. The author makes phone calls and email reminders to increase the response rate of administrators.

4.4 Data Analysis Methods

After cleaning, the data will be processed and analyzed. The authors use the structural modeling technique PLS-SEM to test the measurement model and test the proposed hypotheses. The process of data analysis on the basis of using Smart PLS software is carried out in three steps as follows: (1) data encryption and descriptive statistics; (2) evaluation of the measurement model; and (3) evaluation of the structural model.

5 Results

5.1 Survey Sample Statistics

In order to analyze and evaluate the influence of middle managers' capacity on the strategy implementation of Vietnamese enterprises in the IT industry, the author conducted a survey with the subjects of middle-level managers working in the IT industry and collected 220 valid votes after screening the votes and processing them.

According to the IT White Book (Vietnam Ministry of Information & Communications, 2020), the IT industry is divided into 4 industry groups: hardware businesses, software businesses, digital content businesses, and IT service businesses. Surveyed subjects are distributed among IT industry groups, with the largest concentration in IT service enterprises (71.1%), followed by software enterprises (42.2%), hardware enterprises (31.3%), and the group of digital content businesses (19.3%).

In 220 satisfactory questionnaires, male middle managers made up 56.4% and female ones accounted for 43.6%. Middle managers belong to many different age groups, but the most concentrated is the age group from 36 to 45 years old (accounting for 43.6%), then the age group from 25 to 35 years old (accounting for 40.9%). The remaining groups accounted for a relatively low proportion, the group under 25 years old accounted for 12.3%, and the group from 46 to 55 years old accounted for only 3.2%. 55.45% of middle managers had more than 10 years of experience, followed by those with seniority from 5 years to less than 10 years accounting for 25.45%. The percentage of young managers with less than 1 year of seniority is very low, accounting for only 4.09%. Middle managers surveyed are in charge of numerous distinct departments/units that perform various company operations such as production, engineering, marketing-communications, human resources, finance, business sales, supply, administration, etc.

5.2 Evaluation of the Measurement Model

Since the latent variables in the research model are reflective variables, the quality of the observed variables is assessed through the strategy implementation coefficient of the observed variables. This is an index showing the degree of association between the observed variable and the latent variable.

The results of the outer loading factor test showed that some scales did not meet the conditions for the quality of observed variables such as the scales DO1, DO6, DO7, DO17, DO28, DO29, DO15, and DO25, due to the small outer loading coefficient being lower than 0.7. The author removed these observed variables and retested them for the second time. The results of data analysis are shown in Table 1, showing that all observed variables ensure the reliability of Cronbach's alpha > 0.7, CR values > 0.7, AVE > 0.5.

To ensure the discriminant value between the variables in the research model, according to Henseler et al. (2015), the discriminant value was established between the structures when the HTMT index was below 0.9. The HTMT index of each structure, shown in Table 2, is less than 0.9, thus meeting the strict criteria of Henseler et al. (2015). Therefore, with the above research results, the requirements for the reliability and validity of the variables in the measurement model all satisfy the requirements given by the previous studies (Hair et al., 2016; Henseler et al., 2015).

5.3 *Assessment of the Structural Model*

To evaluate and test the structural model, the author performed bootstrapping estimation with a magnification factor of 1000 samples with the significance level of the test at 5%. Before evaluating the model and testing the hypotheses, according to Hair et al. (2016), the model needs to ensure that there is no multicollinearity problem, so the statistical value of variance inflation factor (VIF) must be smaller than 5. In this study, the VIF for both latent variables is 1.481, which is less than 5. Therefore, the model does not have the phenomenon of multicollinearity between latent variables and the test of the correlation between the latent variables to ensure the reliability of the study.

Based on the hypothesis testing criteria synthesized by Hair et al. (2016), the results of PLS-SEM analysis are shown in Table 3, showing the existence of hypotheses H1, H2, H3. This demonstrates that middle managers' leadership practices have a significant impact on both strategy implementation at the corporate and functional levels with β coefficients of 0.399 and 0.570, respectively, with a P-value of less than 0.05. Deploying a functional-level strategy also has a great contribution to the results of implementing a corporate-level strategy with a β coefficient of 0.481 (Fig. 1).

6 Discussions

The purpose of the study is to test the impact of middle managers' leadership practices with functional and corporate-level strategy implementation and also examine the

Table 1 Assessment of measurement model

	Indicators	Outer loadings	Cronbach's alpha	Composite reliability (CR)	Average variance extracted (AVE)
Leadership practices of middle managers (DO)			0.974	0.976	0.646
Model the way (DO_LG)	DO11	0.810	0.870	0.911	0.720
	DO16	0.783			
	DO21	0.810			
	DO26	0.773			
Inspire a shared vision (DO_TN)	DO2	0.813	0.902	0.932	0.773
	DO12	0.860			
	DO22	0.849			
	DO27	0.850			
Challenge the process (DO_TT)	DO3	0.728	0.897	0.924	0.708
	DO8	0.858			
	DO13	0.792			
	DO18	0.835			
	DO23	0.817			
Enable others to act (DO_NV)	DO4	0.780	0.898	0.925	0.710
	DO9	0.757			
	DO14	0.810			
	DO19	0.803			
	DO24	0.813			
Encourage the heart (DO_NH)	DO5	0.821	0.855	0.902	0.697
	DO10	0.825			
	DO20	0.749			
	DO30	0.731			
Corporate strategy implementation(CSI)	CSI1	0.845	0.861	0.900	0.644
	CSI2	0.784			
	CSI3	0.748			
	CSI4	0.822			
	CSI5	0.810			
Functional strategy implementation (FSI)	FSI1	0.807	0.795	0.880	0.710
	FSI2	0.909			
	FSI3	0.808			

Source Processing results by Smart PLS

Table 2 Discriminant validity

HTMT	Leadership practices of middle managers (DO)	Functional strategy implementation (FSI)
Leadership practices of middle managers (DO)		
Functional strategy implementation (FSI)	0.641	
Corporate strategy implementation (CSI)	0.734	0.849

Source Processing results by Smart PLS

Table 3 Assessment of the structural model

Hypotheses	β	T-value	P-value	f^2	Decision
H1: DO → CSI	0.399	7.148	0.000	0.274	Significant
H2: DO → FSI	0.570	11.672	0.000	0.481	Significant
H3: FSI → CSI	0.481	9.456	0.000	0.398	Significant

Source Processing results by Smart PLS

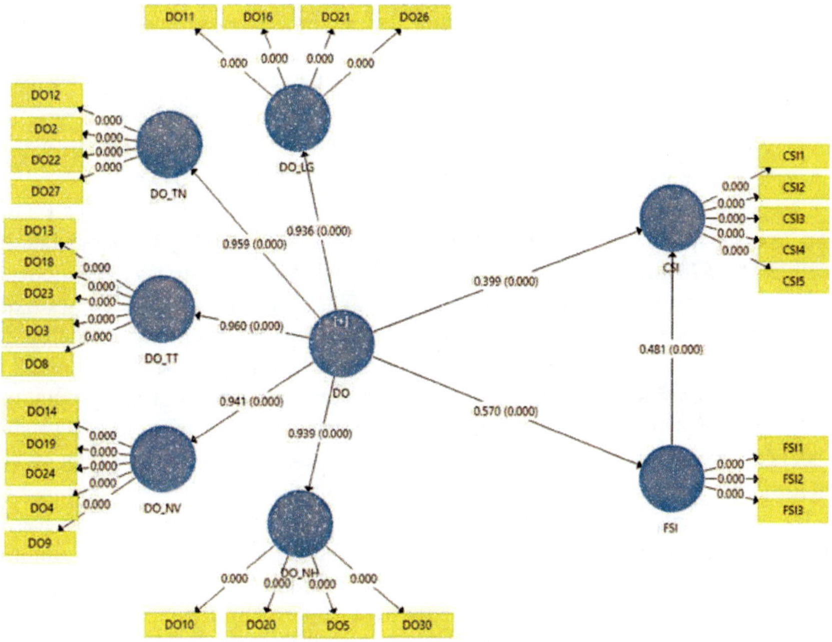

Fig. 1 Structural model. *Source* Processing results by Smart PLS

relationship between functional-level and corporate-level strategy implementation. The results of this study are important to demonstrate the positive and direct impact of middle managers' leadership practices and strategy implementation at different levels in Vietnamese IT enterprises. This is the first study to evaluate the aspects of middle managers' leadership practices in IT enterprises in a developing country where the IT industry is becoming a spearhead industry, affecting all industries and other sectors in society, especially as Vietnam is actively implementing digital transformation, taking advantage of the fourth industrial revolution to promote growth. This research has contributed to improving the understanding of the importance of middle managers' leadership practices in the strategy implementation of enterprises.

Overall, the research findings supported the H1 and H2 hypotheses, reaffirming the beneficial association between leadership practice and strategy implementation of enterprises. This finding is in line with the findings of other earlier research conducted throughout the globe, which found a link between leadership practice and strategy implementation (Jaoua, 2018) and business performance (Avery & Bergsteiner, 2011; Cowden et al., 2011; Nawab & Asad, 2020). However, the difference is that in this study, we clarify the relationship between middle managers' leadership practices and strategy implementation at each specific level. The focus here is on corporate-level strategy and functional-level strategy. In which, the leadership practices of middle managers affect the implementation of the corporate-level strategy to a smaller extent (with β of 0.399 and f^2 of 0.274) than the impact on the implementation of the functional-level strategy (with β of 0.399 and f^2 of 0.274). β is 0.570, and f^2 is 0.481. This finding further reinforces the centrality of middle managers in strategy implementation, as their position allows them to have proximity between the organization's strategy and day-to-day operations (Nonaka, 1994).

The novel contribution of this study is to show the positive and strong influence of middle managers' leadership practices on functional-level strategy implementation (with β of 0.570 and f^2 of 0.481). To our best knowledge, there are very few studies looking into this relationship around the world. The work of the author Lee (2018) is a rare example of studying the influence of leadership behavior on organizational performance through leadership functions, but this study has not focused on middle managers. Therefore, it can be affirmed that for the first time, our study demonstrates the role of middle managers in implementing functional-level strategy through leadership practices—an important finding for management decision-making of IT enterprises in Vietnam in strategy implementation at all levels.

Research results provide an important basis for middle managers to perfect leadership practices to enhance the effectiveness of the organization's strategy implementation. The authors give some recommendations that focus on considering two approaches: managers' self-consciousness about improving their own leadership practices and from a second perspective, corporate, and senior managers' policies to increase the leadership practices of middle managers. Specifically: Firstly, for middle managers themselves, it is necessary to increase awareness of leadership practice and its impact on the units' and enterprises' strategy implementation results. At the same time, they need to self-train and foster leadership methods that are suitable to the characteristics of the unit and the business. Some suggestions to increase the leadership practices of middle managers are as follows: increasing the sharing of strategic vision, the ability to transmit enthusiasm, accepting challenges and becoming ideal role models for subordinates. These suggestions will help them to be the managers who have a positive impact on the results of strategy implementation at both functional and corporate levels. Second, senior managers in the enterprise also need to have reasonable policies to improve the leadership practices of middle managers, such as strengthening policies on training, fostering, and developing leadership practices for middle managers, developing appropriate policies for evaluation, remuneration, and appointment of middle managers, as well as developing policies to attract high-quality IT human resources.

The study has achieved the research objective, but there are still certain limitations related to the small sample size and the study only focuses on the IT industry, so it is too early to generalize the results of this study for other industries, and further studies may be needed to establish the generality of the study's findings. Second, the research only focuses on two levels of strategy implementation, which are corporate-level and functional-level, but has not studied strategy implementation at the SBU level. Studying all three levels of strategy in an enterprise can yield more comprehensive results when considering strategy implementation. These can be suggestions for future research directions.

References

Alexander, L. D. (1985). Successfully implementing strategic decisions. *Long Range Planning, 18*(3), 91–97.

Avery, G. C., & Bergsteiner, H. (2011). Sustainable leadership practices for enhancing business resilience and performance. *Strategy & Leadership, 39*(3), 5–15.

Beard, D. W., & Dess, G. G. (1981). Corporate-level strategy, business-level strategy, and firm performance. *Academy of Management Journal, 24*(4), 663–688.

Bell, P., Dean, G., & Gottschalk, P. (2010). Information management in law enforcement: The case of police intelligence strategy implementation. *International Journal of Information Management, 30*(4), 343–349.

Čater, T., & Pučko, D. (2010). Factors of effective strategy implementation: Empirical evidence from Slovenian business practice. *Journal of East European Management Studies, 15*(3), 207–236.

Cowden, T., Cummings, G., & Profetto-Mcgrath, J. (2011). Leadership practices and staff nurses' intent to stay: A systematic review. *Journal of Nursing Management, 19*(4), 461–477.

Crevani, L. (2015). Is there leadership in a fluid world? Exploring the ongoing production of direction in organizing. *Leadership, 14*(1), 1–27.

Currie, G., & Procter, S. J. (2005). The antecedents of middle managers' strategic contribution: The case of a professional bureaucracy. *Journal of Management Studies, 42*(7), 1325–1356.

David, F. R., & David, F. R. (2020). *Strategic management: A competitive advantage approach, concepts and cases*. Pearson (17th ed.). Pearson Education, Inc.

Dutton, J. E., & Ashford, S. J. (1993). Selling issues to top management. *Academy of Management Review, 18*, 397–428.

Dziekoński, K. (2017). Project managers' competencies model for construction industry in Poland. *Procedia Engineering, 182*, 174–181.

Floyd, S. W., & Wooldridge, B. (1992). Middle management involvement in strategy and its association with strategic type: A research note. *Strategic Management Journal, 13*(S1), 153–167.
Floyd, S. W., & Wooldridge, B. (1994). Recognizing or dynamos? Recognizing middle management's strategic role. *Academy of Management Executive, 8*(4), 47–57.
Forner, M., Bierlein-Palmer, L., & Reeves, P. (2012). Leadership practices of effective rural superintendents: Connections to Waters and Marzano's Leadership correlates. *Journal of Research in Rural Education, 27(8),* 1–8.
Goedhuysa, M., & Veugelers, R. (2012). Innovation strategies, process and product innovations and growth: Firm-level evidence from Brazil. *Structural Change and Economic Dynamics, 23*(4), 516–529.
Goewey, D. F. (2012). Examining the Kouzes and Posner leadership practices of elementary principals in Central New York. *Education Doctoral*, Paper 9.
Goldman, E. F. (2012). Leadership practices that encourage strategic thinking. *Journal of Strategy and Management, 5*(1), 25–40.
Grundy, T. (1998). How are corporate strategy and human resources strategy linked? *Journal of General Management, 23*(3), 49–72.
Hair, J. F., Black, W. C., Babin, B. J., & Anderson, R. E. (2010). *Multivariate data analysis* (7th ed.). Pearson.
Hair, J. F., Hult, G. T. M., Ringle, C. M., & Sarstedt, M. (2016). *A primer on Partial Least Squares Structural Equation Modeling (PLS-SEM)* (2nd ed.). SAGE.
Henseler, J., Ringle, C. M., & Sarstedt, M. (2015). A new criterion for assessing discriminant validity in variance-based structural equation modeling. *Journal of the Academy of Marketing Science, 43*(1), 115–135.
Hill, C., Jones, G., & Schilling, M. (2014). *Strategic management: Theory: An integrated approach* (11th ed.). Cengage Learning.
Hoyle, R. H. (1995). Structural equation modeling: Concepts, issues, and applications. In *The structural equation modeling approach: Basic concepts and fundamental issues* (pp. 1–15). Sage Publications, Inc.
Jaoua, F. (2018). Impact of strategic roles of middle managers on the relationship between successful strategy implementation and organisational performance. *International Journal of Business Performance Management, 19*(4), 476499.
Lee, H.-W. (2018). Linking leadership practices to performance of the US federal agencies. *International Journal of Manpower, 39*(3), 434–454.
Levesque-Côté, J., Fernet, C., Morin, A. J. S., & Austin, S. (2021). On the motivational nature of authentic leadership practices: A latent profile analysis based on self-determination theory. *Leadership and Organization Development Journal, 42*(2), 178–194.
Lu, K., Zhu, J., & Bao, H. (2015). High-performance human resource management and firm performance. *Industrial Management & Data Systems, 115*(2), 353–382.
Minarro-Viseras, E., Baines, T., & Sweeney, M. (2005). Key success factors when implementing strategic manufacturing initiatives. *International Journal of Operations & Production Management, 25*(2), 151–179.
Nawab, A., & Asad, M. M. (2020). Leadership practices of school principal through a distributed leadership lens: A case study of a secondary school in urban Pakistan. *International Journal of Public Leadership, 16*(4), 411–422.
Nonaka, I. (1994). A dynamic theory of organizational knowledge creation. *Organization Science, 5*, 14–37.
Northouse, G. P. (2018). *Leadership: Theory and practice* (8th ed.). Sage Publications.
Obeidat, B. Y., Al-Hadidi, A., Tarhini, A., & Masa'deh, R. (2017). Factors affecting strategy implementation: A case study of pharmaceutical companies in the middle east. *Review of International Business and Strategy, 27*(3), 386–408.
Okumus, F. (2003). A framework to implement strategies in organizations. *Management Decision, 41*(9), 871–882.

Posner, B. Z., & Kouzes, J. M. (1993). Psychometric properties of the leadership practices inventory-updated. *Educational and Psychological Measurement, 53*(1), 191–199.

Raes, A. M. L., Heijltjes, M. G., Glunk, U., & Roe, R. A. (2011). The interface of the top management team and middle managers: A process model. *Academy of Management Review, 36*(1), 102–126.

Rapert, M. I., Lynch, D., & Suter, T. (1996). Enhancing functional and organizational performance via strategic consensus and commitment. *Journal of Strategic Marketing, 4*(4), 193–205.

Schilit, W. K. (1987). An Examination of the influence of middle-level managers in formulating and implementing strategic decision. *Journal of Management Studies, 24*(3), 271–293.

Slater, S. F., & Olson, E. M. (2001). Marketing's contribution to the implementation of business strategy: An empirical analysis. *Strategic Management Journal, 22*(11), 1055–1067.

Sprigghr. (2020). *Twelve good leadership practices*. https://sprigghr.com/blog/leaders/12-good-leadership-practices/

Thoonen, E. E. J., Sleegers, P. J. C., Oort, F. J., Peetsma, T. T. D., & Geijsel, F. P. (2011). How to improve teaching practices: The role of teacher motivation, organizational factors, and leadership practices. *Educational Administration Quarterly, 47*(3), 496–536.

Vietnam Ministry of Information & Communications. (2020). *White Book of Viet Nam information and communication technology 2020*.

Waterman, R. H., Peters, T. J., & Phillips, J. R. (1980). Structure is not organization. *Business Horizons, 23*(3), 14–26.

Waters, T., & Marzano, R. (2006). The effect of superintendent leadership on student achievement. *Mid-Continent Research for Education and Learning, 7*, 454–487.

Wooldridge, B., & Floyd, S. W. (1990). The strategy process, middle management involvement, and organizational performance. *Strategic Management Journal, 11*(3), 231–241.

Wooldridge, B., Schmid, T., & Floyd, S. W. (2008). *The middle management perspective on strategy process: Contributions, synthesis, and future research. Journal of Management* (Vol. 34).

Factors Influencing Bank Users' Behavioral Intention Toward Usage Likelihood of Fintech Services in Vietnam

Van Duong Ha

Abstract Employing the second generation of the unified theory of acceptance and use of technology (UTAUT2) model to determine the factors influencing bank users' behavioral intention toward usage likelihood of financial technology (Fintech) services in Vietnam. The methods of Cronbach's alpha test, exploratory factor analysis (EFA), confirmatory factor analysis (CFA), and structural equation modeling (SEM) analysis were used in this study. The findings indicate that there are five factors that have a positive influence on behavioral intention such as effort expectancy, social influence, hedonic motivation, user's knowledge, and experimental development. There are three factors that have a positive influence on behavior of using Fintech services, such as facilitating conditions, behavior intention, and government policies. Besides, habits, aspirational innovation, and policy synchronization are the factors that have a negative influence on behavior of using Fintech services. This study may prove diagnostically useful to the behavioral intention and behavior of using Fintech services in Vietnam.

Keywords Banks · Behavioral intention · Fintech services · UTAUT · Vietnam

1 Introduction

The term Fintech originates from the combination of two fields, such as "financial" and "technology". Combining these two words is called Fintech and is interpreted as financial technology. In Fintech research and application, there are many concepts under many different aspects and contexts, in which Fintech is a term used to describe the use of technology in financial services. Fintech companies are often startups using new business models and new products that compete with banking, insurance, or payment services. Fintech stands for financial technology, which is a broad term used primarily to refer to companies that are using technology-based systems in some way to provide direct or indirect financial services and trying to make the

V. D. Ha (✉)
Saigon Institute of Economics and Technology, Ho Chi Minh City, Vietnam
e-mail: dhv05@yahoo.com

A. T. Nguyen et al. (eds.), *Contemporary Economic Issues in Asian Countries: Proceeding of CEIAC 2022, Volume 1*,
https://doi.org/10.1007/978-981-19-9669-6_35

financial system more efficient. Originally, the term referred to technology applied to support the operation of consumer and trade financial institutions. Today, the interpretation of Fintech has expanded to include any technological innovation in the financial sector, including innovation in financial education and training, retail banking, investment, or operational innovation. Fintech's movement and expression have also become synonymous with the emerging financial services industry in the twenty-first century. In this context, Fintech encompasses a wide range of services and products, such as wireless payments, cash, lending, trading, consulting, fundraising, and digital currency, and is expected to expand further in the coming years (Karakas & Stamegna, 2017).

Fintech products, services, and business models are terms applied in financial technology activities that are the combination of finance and technology to form innovative solutions in financial services, including technology applications that are increasingly deepening in the design and delivery of financial services. Fintech is financial technology, a term used to describe the use of technology in financial services. Fintech is the main driver for financial inclusion, focusing on providing all financial services at a low cost with a larger scale and broader reach. Fintech has brought a new paradigm for design strategies. Design and implementation of financial inclusion through the application of mobile networks and the Internet for investment services and mobile banking are technologies that make financial services more accessible to the public. Therefore, it can be understood in a uniform and universal way about Fintech as a business activity that applies technology to provide services in order to make financial and banking services safer, more efficient, and convenient for users (Ha Van Duong, 2022a).

In Vietnam, changes in bank users' behavioral intention toward usage likelihood of Fintech services promise to promote the Fintech services to a new height. Major trends in Vietnam's Fintech market include the trend of strong developments in terms of quantity, diversity in products and services, and attracting investment capital, as well as the Fintech market witnessing the growth in both the quantity and quality of Fintech startups in this field. Many bank users in Vietnam spend more time on the commerce and financial transaction platforms, and they are willing to place orders of greater quantity and value. At the same time, diversifying payment methods and online banking transactions brings convenience and safety for bank users. E-money payments and online banking transactions are becoming more popular and are likely to dominate over the previous payments, financial, and banking transactions.

2 Methodology

2.1 The UTAUT2 Model

The UTAUT model was an integrated framework of eight-related technology acceptance models. Those models include the theory of reasoned action (TRA), the theory of planned behavior (TPB), the technology acceptance model (TAM), the combining technology acceptance model and the theory of planned behavior (C-TAM-TPB), the motivation model (MM), the model of personal computer utilization (MPCU), innovation diffusion theory (IDT), and social cognitive theory (SCT). The UTAUT model showed that the use of technology is defined by behavioral intention. In addition to the four most important constructs of performance expectancy, effort expectancy, social influence, and facilitation conditions, the other constructs, which include behavioral intention and use behavior, were also included in this model. Besides, the effect of predictors was moderated by age, gender, experience, and voluntariness of use (Venkatesh et al., 2003).

The UTAUT model was a comprehensive integrated model, and it was developed to better understand consumers' acceptance of new technology or systems. In order to enhance the prediction ratio of technology acceptance, the researchers consider three types, which include (i) examining consumers' new technology acceptance in a variety of contexts, such as culture and population, (ii) considering different concepts to extend the theoretical relationships of the UTAUT model, (iii) considering synthesize new predictions of the variables into the UTAUT model. They also examined consumer behavior associated with the previous studies and established a new predictive framework, namely the UTAUT2 model. Some variables are added to the UTAUT model. The hedonic motivation construct was considered as an important predictor and was integrated into the UTAUT2 model. The product and service quality, cost, and price will influence behavior intention, so the price value construct was also regarded in the UTAUT2 model. The habit is regarded as prior behavior. It can be defined as the degree to which people believe behavior to be automatic. They, thus, incorporated the habit construct into the UTAUT2 model. At the same time, the UTAUT2 model removed voluntariness of use from the demographic variables in the original UTAUT model (Venkatesh et al., 2012) as shown in Fig. 1.

2.2 *The Proposed Model*

In order to gain insight into the factors influencing bank users' behavioral intention toward usage likelihood of Fintech services in Vietnam, based on the theoretical basis of the UTAUT2 model, and synthesizing the theoretical basis related to the factors influencing bank users' behavioral intention toward usage likelihood of Fintech services in Vietnam, this study supplemented the UTAUT2 framework by combining the constructs that related to the usage likelihood of Fintech services, which are aspirational innovation and government policies. After adding those factors, the proposed research model is shown in Fig. 2.

Performance expectancy (PE): It is described as the extent to which utilizing technology brings advantages to consumers (Venkatesh et al., 2012). In the context of using Fintech, performance expectancy expresses the benefits that bank users get when using Fintech. The findings of Mulyana et al. (2020) suggest that performance expectancy has significant positive influences on behavioral intention in using

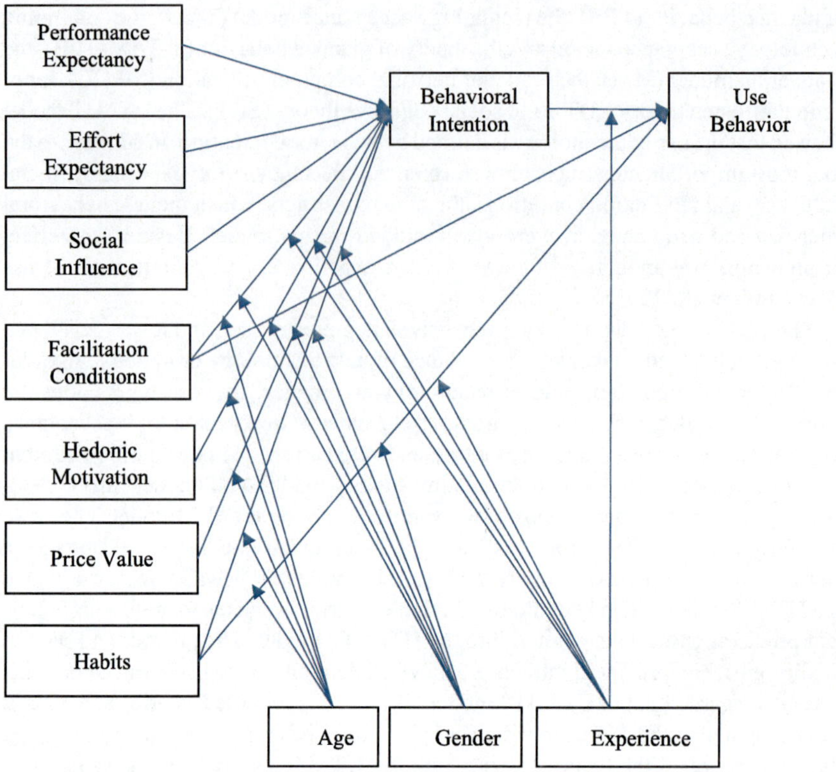

Fig. 1 UTAUT2 model. *Source* Venkatesh et al. (2012)

Fintech. Fintech and financial innovation play an integral role in the formation and transformation of banking. Technological advancement leads to the birth of new business models, modern banking products, and services. Fintech better solves customer needs and makes the process of providing banking products and services more efficient (Ha Van Duong, 2022a). Fintech is driving fundamental changes in the banking sector, providing users with widespread use of information technology and rapid innovation in banking products and services, and driving efficiencies in transactions of bank users (Ha Van Duong, 2022b). Therefore, we formed a hypothesis:

Hypothesis 1 (H1): Performance expectancy has a positive impact on the bank users' behavioral intention to use Fintech services in Vietnam.

Effort expectancy (EE) is the degree of convenience associated with the consumers' use of technology (Venkatesh et al., 2012). For an activity like using Fintech, effort expectancy is one of the important factors for the bank users' behavioral intention to apply technology. The study results of Khatun and Tamanna (2020) show that effort expectancy is an important factor affecting the users' behavioral intention to use Fintech services, and effort expectancy has a positive influence on

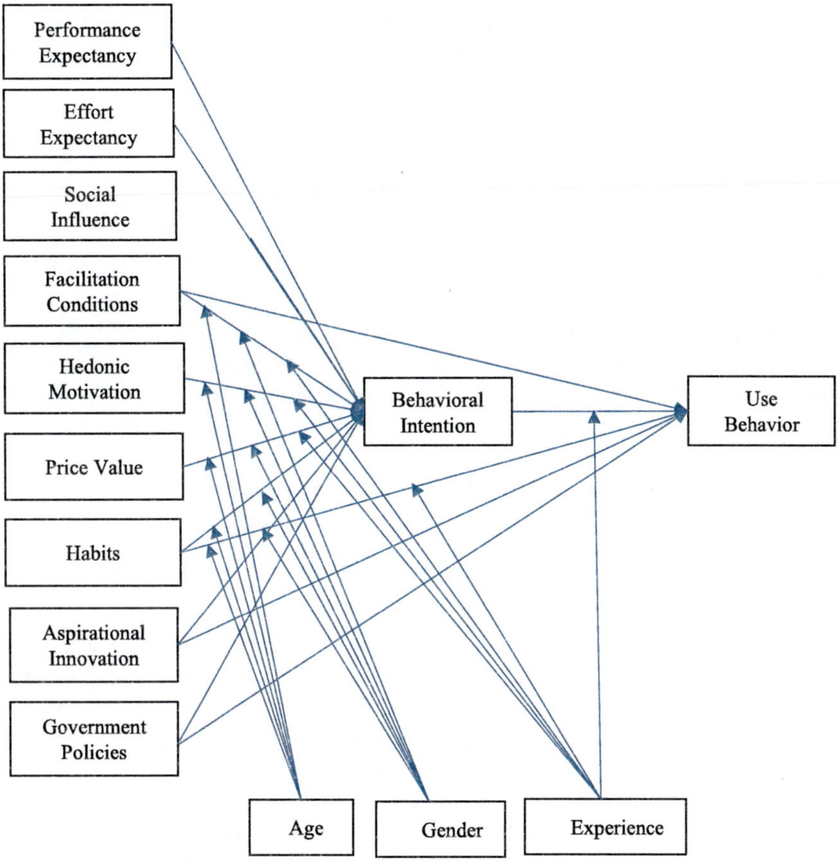

Fig. 2 Proposed model

the behavioral intention of the users to adopt Fintech. The findings of Mulyana et al. (2020) also suggest that performance expectancy has significant positive influences on behavioral intention in using Fintech. When Fintech services are used for financial transactions, it is helpful in speeding up and simplifying complex processes. This will also ensure the correct transfer of information as the transaction will only be approved if all the conditions are met. Furthermore, the information is visible to all parties involved in the transaction, so there are fewer errors at the time of the transaction (Ha Van Duong, 2022b). Fintech services create a new approach and expand the provision of financial and banking services to users. It benefits users in terms of accessibility, speed, quality, price, and increased choice in transactions (Ha Van Duong, 2022c). Hence, the hypothesis for this structure is stated as follows:

Hypothesis 2 (H2): Effort expectancy has a positive impact on the bank users' behavioral intention to use Fintech services in Vietnam.

Social influence (SI): It is defined as the amount to which a consumer believes that another important character believes the consumer needs to utilize the new technology. It expresses consumer's adoption of using technology because they are influenced by important people around them, like family and friends, who encourage them to use technology (Venkatesh et al., 2012). The findings of Mulyana et al. (2020) revealed that social influence has a significant positive effect on behavioral intention in using Fintech. The study results of Khatun and Tamanna (2020) found that social influence has a positive influence on the behavioral intention of the users to adopt Fintech. The results of quantitative analysis by Al Nawayseh (2020) indicated that social influence has a positive effect on users' behavioral intention regarding Fintech applications. The empirical findings of Submitter et al. (2021) illustrate that social influence is the most influential determinant that affects bank users' behavioral intention to use Fintech services. For social influence, marketing effect, and reputation, Fintech schemes are seen by many users as an innovative and interesting financial service. The users may be attracted to these programs not only by the novelty of Fintech, but also by the social influence, marketing effects, and reputation (Ha Van Duong, 2022d). Therefore, the following hypothesis is proposed:

Hypothesis 3 (H3): Effort expectancy has a positive impact on the bank users' behavioral intention to use Fintech services in Vietnam.

Facilitating conditions (FC) show that the users are aware of the resources and support available to perform their behavior. These are defined as the degree to which users think that organizational and technical infrastructure exists to support the use of the system (Venkatesh et al., 2012). The study results of Khatun and Tamanna (2020) found that facilitating condition has a positive influence on the behavioral intention of the users to adopt Fintech. The results of Rahardjo et al. (2020) show that facilitation conditions are factors that influence the intention and use of e-money as a Fintech service. The research findings show that facilitating condition has a positive influence on the behavioral intention of using Fintech services via digital wallets (Sukaris et al., 2021). The results showed that facilitating conditions have a positive and significant effect on behavior intention of using mobile payment as a Fintech service (Latifah et al., 2021). For technical concerns, Fintech services are accepted when users believe that the technical and organizational infrastructure will support the adaptation and convenient use of Fintech services (Ha Van Duong, 2022d). At the same time, there is a positive and significant relationship between the facilitating conditions with the behavioral intention to adopt Fintech (Al Rubaiai & Pria, 2022). Therefore, we make the hypothesis H4a and H4b as following:

Hypothesis 4a (H4a): Facilitating conditions have a positive effect on the bank users' behavioral intention to use Fintech services in Vietnam.

Hypothesis 4b (H4b): Facilitating conditions have a positive effect on the bank users' behavioral to use Fintech services in Vietnam.

Hedonic motivation (HM): It has been considered to play a crucial part in terms of users' accepting and using technology. This has been described as the users' pleasure or fun derived from using a technology (Venkatesh et al., 2012). For the setting of using Fintech services, the research findings show that hedonic motivation has a positive influence on the behavioral intention of using digital wallets as a Fintech service (Sukaris et al., 2021). The results showed that hedonic motivation has a positive and significant effect on behavior intention of using Fintech services via mobile payment (Latifah et al., 2021). Besides, there is a positive and significant relationship between the hedonic motivation with the behavioral intention to adopt Fintech (Al Rubaiai & Pria, 2022). Thus, the following hypothesis has been developed:

Hypothesis 5 (H5): Hedonic motivation has a positive impact on the bank users' behavioral intention to use Fintech services in Vietnam.

Price value (PV): This is defined as users' cognitive trade-off in terms of the benefits from using technology and the costs of using it. Cost and price are structures, and these structures may have a significant impact on the use of technology (Venkatesh et al., 2012). The research empirically by Marlina et al. (2020) found that price value was a factor that significantly influenced users to use Fintech lending. Fintech applications have a number of advantages for users, such as providing more innovative, faster, and more cost-effective services than traditional banking services. Therefore, the application of software for online transactions will save costs and provide users with more personalized services through Fintech services (Ha Van Duong, 2022b). Fintech services can help users reduce transaction fees compared to other payment methods. For this reason, Fintech services can be an attractive alternative for some individuals or organizations, especially in cross-border payments that often involve paying high fees. Additionally, many Fintech services do not require an intermediary to facilitate payments, which can affect processing costs (Ha Van Duong, 2022d). So, if the costs and fees of Fintech services are low and possess good quality compared to the traditional banking services, users will opt to adopt Fintech services. This is hypothesized as follows:

Hypothesis 6 (H6): Price value has a positive effect on the bank users' behavioral intention to use Fintech services in Vietnam.

Habits (HAs) are defined as the volume to which individuals tend to perform behaviors automatically as a result of learning. The empirical findings on the role of habit in technology show that habits influence users' behavioral intentions toward technology use (Venkatesh et al., 2012). The behavioral intention to use Fintech was influenced by the user's habits (Marlina et al., 2020). The research results of Rahardjo et al. (2020) show that habits are factor that influence the intention and use of e-money as a Fintech service. The results showed that habits have a positive and significant effect on behavior intention of using Fintech services via mobile payment (Latifah et al., 2021). Fintech services are used to provide banking and financial services based on transactions with various options and habits for transferring the ownership of currency from one person to another (Ha Van Duong, 2022d). At the same time, there is a positive and significant relationship between the habits with the behavioral intention to adopt Fintech (Al Rubaiai & Pria, 2022). Hypotheses are as follows:

Hypothesis 7a (H7a): Habits have a positive effect on the bank users' behavioral intention to use Fintech services in Vietnam.

Hypothesis 7b (H7b): Habits have a positive effect on the bank users' behavioral to use Fintech services in Vietnam.

Aspirational Innovation (AI): It is the degree to which a person is receptive to new ideas, and this is a characteristic of an individual's personality. The research by Hu et al. (2019) shows that user innovation is individual acceptance of new technology or new services. The willingness to accept the presence of new technology is a key factor for technology adoption. In the context of using Fintech, the empirical research by Kim et al. (2010) argues that most people do not have enough specialized knowledge about many types of mobile payment services, like one type of Fintech service. Their personal innovation plays an important positive role in their intention to use this service, which is considered as one of the increasing requirements for the use of Fintech services. Innovations in the quality and speed of financial and banking services through Fintech services will contribute to an increase in the needs of banking users, especially young and millennials. This is one of the powerful catalysts that make positive changes to the behavioral intention and behavior of using Fintech services of banking users (Ha Van Duong, 2022b). Thus, the following hypothesis is developed:

Hypothesis 8a (H8b): Aspirational innovation has a positive effect on the bank users' behavioral intention to use Fintech services in Vietnam.

Hypothesis 8b (H8b): Aspirational innovation has a positive effect on the bank users' behavioral to use Fintech services in Vietnam.

Government policies (GPs) are a set of decisions by governments to express the goals and actions for political, social, and economic management. It influences, changes, or frames a problem or issue that has been recognized in the political realm by policy makers and the wider public (Hassel, 2015). According to Yang (2017), the outstanding growth of Fintech services can be contributed to its connectivity to global financial markets, easy networking with other Fintech hubs, and government policy on building a business-friendly regulatory environment. Since then, Fintech services create a deep level of participation as well as increase the main attractions for many large international investors and attract users in this market. Fintech services started playing an important role in reaching users through online financial mobile applications. Convenient and affordable services in Fintech started with the help of investors and the encouragement of government policies to exploit new untapped opportunities in the financial and banking sectors (Kukreja et al., 2021). The research results by Noreen et al. (2022) show that government policies are adopted as a catalyst to expand the scope of financial services and access to innovation in banking and financial transactions through Fintech services. Fintech creates a new approach to providing banking products and services. The changes pose problems for policymakers and regulators alike as they seek to strike a balance between supporting innovation, protecting the banking system in general, and individual users in particular. At the same time, the development of Fintech in banking activities has led to the synchronous formation of many policies including Fintech development policies and directions, competition policy, risk management, consumer protection, etc. Policies and orientations for priority areas and development potential of Fintech services will create favorable conditions in all aspects to attract Fintech users, positively impact behavioral intentions, and behaviors of using Fintech services by bank users and contribute to the development of Fintech services in the banks (Ha Van Duong, 2022b). Thus, the following hypothesis has been developed:

Hypothesis 9a (H9a): Government policies have a positive effect on the bank users' behavioral intention to use Fintech services in Vietnam.

Hypothesis 9b (H9b): Government policies have a positive effect on the bank users' behavioral to use Fintech services in Vietnam.

Behavioral intention (BI) is the user's level of trust and confidence to use the technology in future. Therefore, the behavioral intention determines the technology use behavior (Venkatesh et al., 2012). The users' behavioral intention of using Fintech services has a significant correlation with users' behavior (Ha Van Duong, 2022a). This shows that the users' behavior of using Fintech services in Vietnam is affected by the behavioral intention of using Fintech services. We make the hypothesis as following:

Hypothesis 10 (H10): Behavioral intention has a positive effect on the bank users' behavioral to use Fintech services in Vietnam.

According to Venkatesh et al. (2012), the demographic factors in this model are gender, age, and experience. The impact of facilitating conditions on behavioral intentions was moderated by age, gender, and experience. Therefore, this impact will be stronger for older users in the early stages of experience with Fintech services. The impact of hedonic motivation on behavioral intentions was moderated by age, gender, and experience. Hence, the effect is stronger on young users in the early stages of their experience with Fintech services. The impact of price value on behavioral intentions was moderated by age, gender, and experience. Thus, the impact will be stronger for older users. The impact of habits on behavioral intentions and the impact of habits on using Fintech services were moderated by age, gender, and experience. Therefore, the impact will be stronger for older users who have a high level of experience with Fintech services. The experience moderated the impact of behavioral intention on using Fintech services, so the impact will be stronger for users with less experience.

2.3 Research Design

This study uses research methods including qualitative research and quantitative research. Qualitative research is used to explore the components of bank users' behavioral intention toward usage likelihood of Fintech services in Vietnam. This study was conducted through direct discussions with experts to adjust and supplement the observed variables used to measure them in the research model. Quantitative research uses quantitative data collected from primary sources to examine certain populations or samples. Through the application of the UTAUT2 model, this study was conducted to determine the factors influencing bank users' behavioral intention toward usage likelihood of Fintech services in Vietnam. The independent and dependent variables are shown in Table 1.

2.4 Data Collection

The purpose of this study was to determine the factors influencing bank users' behavioral intention toward usage likelihood of Fintech services in Vietnam. Therefore, active users of banks were selected as the survey subjects in this study. The survey subjects were randomly selected users who had used the mobile banking, Internet banking, and other Fintech services of the banks in the southern regions of Vietnam. This region of Vietnam is the most important center of economic activity in the country. This is a leading zone for advanced investments in knowledge-based and high-tech industries and services, as well as the most important financial and trading hub of the whole country. Regarding the sample size, according to Hair et al. (2015), if the number of the population is unknown, then sampling is done using size 5–10 multiplied by the question item, since the number of item questions is 58, so the number of samples in this study ($10 \times 58 = 580$) respondents. Therefore, the number of 652 samples collected from customers ensures a sufficient sample size for this study.

Data was collected by distributing questionnaires between May 2022 and September 2022, and a pilot survey was first conducted in September 2022 to validate and accuracy of the questionnaire. Feedback from respondents in the pilot survey helped to improve the final survey questionnaire, especially in replacing unclear words in measurement statements with simple, clear words, and more specific words. The questionnaire was then distributed to 720 respondents, and after data cleaning, 652 samples were completed for data analysis. The questionnaire was designed based on a theoretical literature review to measure all research items, which were measured on a 5-point Likert's scale from strongly disagree (1) to strongly agree (5), and it was derived from the constructs as shown in Table 1. The items of the questionnaire have been carefully reviewed in and considered that each item will evolve according to the operating context of the research variables. The questionnaire was divided into two main parts. The first part includes the respondent's demographics, and the second part was used to collect data on the proposed model.

The research scale was evaluated for its reliability and validity before being used for structural modeling. At the same time, this study uses quantitative research with a SEM approach and conducts the analysis method through Cronbach's alpha coefficient. EFA and CFA were also analyzed using SPSS 25 software. The SEM was established, and CFA diagram and hypothesis verification analysis were conducted using SPSS AMOS 24 software.

Table 1 Independent, moderating, and dependent variables in the research

No	Code	Item
Performance expectancy		
1	PE1	I use Fintech services, because I easily access financial and banking services anywhere
2	PE2	I use Fintech services, because I can understand the interaction of financial and banking services easily and clearly
3	PE3	I believe that the use of Fintech services brings about the expected effect in financial and banking transactions
4	PE4	By using Fintech services, it will allow me to do financial and banking transactions faster
5	PE5	The use of Fintech services is in accordance with all aspects of my work and my work style
6	PE6	By using Fintech services, it will increase my comfort in the financial and banking transactions
7	PE7	By using Fintech services, it will bring me more convenience, because I can access financial and banking transactions in 24 h
Effort expectancy		
8	EE1	The use of Fintech services can help me speed up financial and banking transactions
9	EE2	The use of Fintech services can significantly improve the quality of payment transaction results because it does not take much time
10	EE3	The use of Fintech services can help me make the financial and banking transactions more secure
11	EE4	The use of Fintech services makes it easy for me to find out financial and banking transaction information
12	EE5	The use of Fintech services helps me always have full payment information
13	EE6	Fintech services was practical, and it helps me increase efficiency in the financial and banking transactions
Social influence		
14	SI1	Many people advised me that I should use Fintech services in the financial and banking transactions
15	SI2	People who influence me think that I should use Fintech services in the financial and banking transactions
16	SI3	People who are familiar with me think that I should use Fintech services in the financial and banking transactions
17	SI4	I use Fintech services in the financial and banking transactions because my coworkers use it
18	SI5	My family supports me to use Fintech services in the financial and banking transactions

(continued)

Table 1 (continued)

No	Code	Item
19	SI6	My neighborhood helped me in using Fintech services in the financial and banking transactions
Facilitating conditions		
20	FC1	I have control over the use of Fintech services in the financial and banking transactions
21	FC2	The use of Fintech services in the financial and banking transactions increases my productivity because I am capable of doing various financial and banking transactions anywhere
22	FC3	The existence of evidence on each payment transaction that is completed with the time of the payment is a clear proof of Fintech services payment
23	FC4	I am certain that my financial and banking transactions are guaranteed the transaction conditions when I use Fintech services in the financial and banking transactions
24	FC5	I am certain that the financial and banking transactions are safe when I use Fintech services in the financial and banking transactions
25	FC6	I have smartphone and have received support from the payment service suppliers to use Fintech services in the financial and banking transactions
Hedonic motivation		
26	HM1	I feel comfortable when I use Fintech services in the financial and banking transactions
27	HM2	I feel excited when I use Fintech services in the financial and banking transactions
28	HM3	I feel blessed when I use Fintech services in the financial and banking transactions
29	HM4	I feel happy when I use Fintech services in the financial and banking transactions
30	HM5	I feel entertained when I use Fintech services in the financial and banking transactions
Price value		
31	PV1	I can save time in doing transactions when I use Fintech services in the financial and banking transactions
32	PV2	I can save a lot of transaction costs when I use Fintech services in the financial and banking transactions
33	PV3	By using Fintech services, I agree to pay the costs incurred for internet registration
34	PV4	I do not have to pay fees to check my financial and banking transactions when I use Fintech services in the financial and banking transactions

(continued)

Table 1 (continued)

No	Code	Item
35	PV5	By using the Fintech services, I do not have to pay any extra costs in my financial and banking transactions
Habits		
36	HA1	I have enough knowledge to use Fintech services in the financial and banking transactions
37	HA2	I control my financial and banking transactions well when using Fintech services in financial and banking transactions
38	HA3	I have received the directly guidance from the financial service providers and banks related to the use of Fintech services
39	HA4	There are instructions on using Fintech services in financial and banking transactions on the Websites of financial service providers and banks
40	HA5	I can use Fintech services in financial and banking transactions when no one is simulating it
Aspirational innovation		
41	AI1	I expect more and more new banking and financial services based on technology to bring the best options to users
42	AI2	I know about a new technology-based banking and financial service. I am going to try it out
43	AI3	Among my family and friends, I am often the first to use new technology-based banking and financial services
44	AI4	I believe that financial service providers and banks always have plans to innovate and develop new financial and banking services to meet the users' needs of Fintech services
45	AI5	I believe that the Fintech services develop in parallel with the application of modern and new technology to meet the users' needs
46	AI6	I believe that there are more and more successful experiments of new Fintech services
Government policies		
47	GP1	I believe that the government policies contribute to support and improve the use of Fintech services
48	GP2	I believe that the government has introduced favorable regulations and policies for users of Fintech services
49	GP3	I believe that the government is actively setting up directions to attract and facilitate users of Fintech services

(continued)

Table 1 (continued)

No	Code	Item
50	GP4	I believe that government policies related to Fintech services are aimed at encouraging people to use Fintech services
51	GP5	I believe that government policies are set in sync to promote the use and development of Fintech services
52	GP6	I believe that government policies are well established to promote the use of information technology, the Internet, e-commerce, online banking, and Fintech services
Behavioral intention		
53	BI1	I intend to continue to use Fintech services in the financial and banking transactions in future
54	BI2	I will use Fintech services in the financial and banking transactions if I need to do financial and banking transactions
55	BI3	I will recommend Fintech services in the financial and banking transactions to other people
Use behavior		
56	UB1	There are certain people and community for helping me when there are difficulties in using Fintech services in the financial and banking transactions
57	UB2	I am able to carry out financial and banking transactions with Fintech services without help from other people
58	UB3	I am able to use Fintech services in the financial and banking transactions even though I have never used it beforehand

Source Venkatesh et al. (2012), Ha Van Duong (2022a, 2022b, 2022c) and the authors' suggestions

3 Research Results

3.1 Demographic Statistics

The demographic description of respondents showed that the respondents' age range from 25 years old and up, and a large number of them are a young generation with the age range of 25–40 years. In terms of gender, the percentage of women participating in the survey was higher than that of men with 58.53% of the respondents were female. About education, a majority of respondents held a college degree, comprising about 75.44% of respondents. The respondents also found that 23.82% are at a vocational college graduation and the rest at postgraduate level. The respondents' experience of using Fintech services turned out to be the biggest answer at over 1 year to 2 years, comprising about 46.47% of respondents. Regarding income, the largest income group included people with monthly income from 500 USD to 1000 USD, accounting for the largest proportion with 386 people at 56.77%; the group of over

Table 2 Demographic profile

Item	Optional	Frequency	Percentage
Age	25–30 year old	258	37.94
	31–40 year old	156	22.94
	41–50 year old	128	18.82
	51–60 year old	86	12.65
	61 years old and up	52	7.65
Gender	Male	282	41.47
	Female	398	58.53
Experience	Start–6 Months	52	7.65
	Over 6 months–1 year	116	17.06
	Over 1 year–2 year	316	46.47
	Over 2 year	196	28.82
Education level	Vocational college graduation	162	23.82
	College graduation	513	75.44
	Postgraduate level	5	0.74
Income level	500 USD–1000 USD/month	386	56.77
	Over 1000 USD–1500 USD/month	196	28.82
	Over 1500 USD–2000 USD/month	82	12.06
	Over 2000 USD/month	16	2.35

Source Primary data, processed in 2022

1000 USD to 1500 USD per month at 28.82%; followed the group of over 1500 USD to 2000 USD per month at 12.06%. The smallest was the income group of over 2000 USD per month at 2.35% (see Table 2).

3.2 *Cronbach's Alpha Reliability Analysis*

In EFA, Cronbach's alpha was used to assess the internal consistency of the questionnaire. It is also a convenient test used to assess the reliability of scales. Therefore, Cronbach's alpha will be high if the correlation between each item in the respective questionnaire is high (Osborne, 2008).

When a variable in the measurement has a corrected item total correlation $\geq$ 0.3, Cronbach's alpha coefficient is reliable and statistically valid. If it is greater than 0.60, then that variable meets the requirements. The value of Cronbach's Alpha ranges from 0 to 1, in which the value from 0.7 to 0.9 shows that the scale has good reliability. The value of Cronbach's alpha from 0.6 to 0.7 describes an acceptable level of confidence, and 0.8 or more is very good (Hulin et al., 2001). All scales of this study were eligible to perform EFA. Since, variables have alpha greater than 0.6 and total correlation greater than 0.3 (see Table 3).

3.3 Exploratory Factor Analysis

The purpose of the EFA method is to check the suitability of the latent factors, respond to the correlation between the data and divide the observed variables into groups that have a unifying factor as well as perform corrective factors if necessary. There are several standards applied in the Kaiser–Meyer–Olkin (KMO) measure of sampling adequacy, in which KMO < 0.50 is considered unacceptable, KMO ≥ 0.50 is regarded bad, KMO ≥ 0.60 is considered mediocre, KMO ≥ 0.70 is considered acceptable, KMO ≥ 0.80 is regarded good, KMO ≥ 0.90 is considered very good (Kaiser, 1974). Besides, the significant value of Bartlett's test of sphericity is less than 0.05 for acceptable factor analysis (Field, 2000).

For independent variables, the total of 52 items proved reliability through Cronbach's alpha test. In the EFA technique, this analysis was carried out using SPSS 25 with principal component method and Varimax rotation. According to Samuels (2016), factor loading is also very important in EFA, and it is recommended to remove factor loading less than 0.3. Retained factors must have at least three items with loads greater than 0.4 for variables to be considered significant. The KMO test achieved a value of 0.807, which showed that the sampling was adequate. The Bartlett's test achieved a value of 12,185.532 ($p = 0.000$), ensured moderate inter-correlations between items, and suggested its suitability for factor analysis. The total of 52 observed variables were initially grouped into 12 groups. The total value of variance extracted is 60.131%, which is greater than 50%, and EFA is considered satisfactory. Thus, twelve factors explain 60.131% of the variability of the data. The eigenvalues of the factors are all high (>1). The 12th factor has the lowest eigenvalues of $1.114 > 1$. Therefore, twelve factors are defined as shown in Table 4.

After performing the rotation (absolute value below: 0.3) for the independent variables, twelve factors were found together with the factor loading factors all that are greater than 0.5. This result also indicated there were three new factors (with the observed variables: HA1 and HA2; GP5 and GP6; AI5 and AI6) as shown in Table 5. The characteristics of two observed variables in the factor HA associated with user's knowledge (HA1 and HA2 as can be seen in Table 1) in using the e-commerce platforms for online shopping and e-payment transactions. Therefore, this new factor is named user's knowledge (UK). The hypothesis for UK is similar to the hypothesis for HA (UK has a positive effect on users' behavioral intention of using the Fintech services, and UK has a positive effect on users' behavior of using the Fintech services in Vietnam). The characteristics of two observed variables in the factor GP are associated with policy synchronization (GP5 and GP6 as can be seen in Table 1). Therefore, this new factor is named policy synchronization (PS). The hypothesis for PS is similar to the hypothesis for GP (PS has a positive effect on users' behavioral intention of using the Fintech services, and PS has a

Table 3 Item reliability test

Item	Scale mean if item deleted	Scale variance if item deleted	Corrected item-total correlation	Cronbach's alpha if item deleted
Performance expectancy: Cronbach's alpha = 0.873				
PE1	20.69	16.361	0.718	0.846
PE2	20.69	16.422	0.683	0.851
PE3	20.38	19.182	0.598	0.864
PE4	20.47	18.345	0.576	0.865
PE5	20.54	17.290	0.626	0.859
PE6	20.63	16.031	0.777	0.837
PE7	20.59	17.628	0.611	0.861
Effort expectancy: Cronbach's alpha = 0.829				
EE1	18.06	10.713	0.691	0.780
EE2	17.83	11.922	0.541	0.813
EE3	17.92	11.574	0.629	0.795
EE4	18.00	11.829	0.585	0.804
EE5	18.06	11.584	0.597	0.801
EE6	17.93	11.745	0.549	0.811
Social influence: Cronbach's alpha = 0.838				
SI1	18.17	15.542	0.671	0.800
SI2	18.11	15.641	0.608	0.815
SI3	17.76	17.496	0.601	0.815
SI4	18.00	17.488	0.582	0.818
SI5	17.96	17.232	0.574	0.820
SI6	18.26	15.832	0.665	0.801
Facilitating conditions: Cronbach's alpha = 0.831				
FC1	12.89	7.170	0.679	0.788
FC2	12.88	7.258	0.583	0.809
FC3	12.74	7.401	0.608	0.803
FC4	12.82	7.620	0.589	0.807
FC5	12.81	7.502	0.607	0.803
FC6	12.83	7.656	0.554	0.814
Hedonic motivation: Cronbach's alpha = 0.816				
HM1	14.54	3.612	0.543	0.798
HM2	14.53	3.564	0.523	0.805
HM3	14.41	3.351	0.668	0.761
HM4	14.46	3.490	0.575	0.789
HM5	14.46	3.266	0.729	0.743

(continued)

Table 3 (continued)

Item	Scale mean if item deleted	Scale variance if item deleted	Corrected item-total correlation	Cronbach's alpha if item deleted
Price value: Cronbach's alpha = 0.779				
PV1	14.18	3.252	0.557	0.737
PV2	13.92	2.870	0.633	0.710
PV3	13.93	3.297	0.668	0.703
PV4	14.93	3.747	0.415	0.779
PV5	13.91	3.588	0.515	0.750
Habit: Cronbach's alpha = 0.678				
HA1	14.47	3.528	0.283	0.695
HA2	14.71	3.501	0.379	0.649
HA3	14.69	3.254	0.503	0.597
HA4	14.45	3.028	0.507	0.591
HA5	14.50	3.218	0.508	0.594
Aspirational innovation: Cronbach's alpha = 0.663				
AI1	17.70	7.472	0.410	0.623
AI2	17.72	7.568	0.428	0.623
AI3	17.75	7.724	0.367	0.636
AI4	17.69	7.420	0.462	0.613
AI5	17.78	5.346	0.463	0.604
AI6	17.69	5.162	0.447	0.621
Government policies: Cronbach's alpha = 0.670				
GP1	20.65	7.484	0.416	0.635
GP2	20.67	7.481	0.439	0.631
GP3	20.69	7.595	0.389	0.641
GP4	20.63	7.403	0.465	0.625
GP5	21.70	5.097	0.489	0.605
GP6	21.62	4.973	0.464	0.627
Behavioral intention: Cronbach's alpha = 0.759				
BI1	7.31	1.325	0.593	0.675
BI2	7.54	1.530	0.579	0.692
BI3	7.31	1.370	0.601	0.663
Use behavior: Cronbach's alpha = 0.701				
UB1	7.36	1.268	0.486	0.649
UB2	7.26	1.177	0.531	0.593
UB3	7.21	1.216	0.537	0.586

Source The authors' calculation from SPSS 25

Table 4 Exploratory factor analysis for independent variables

Component	Initial eigenvalues			Extraction sums of squared loadings			Rotation sums of squared loadings	
	Total	% of variance	Cumulative %	Total	% of variance	Cumulative %	Total	Cumulative %
1	6.492	12.484	12.484	6.492	12.484	12.484	4.092	7.869
2	3.565	6.855	19.339	3.565	6.855	19.339	3.447	14.497
3	3.274	6.296	25.635	3.274	6.296	25.635	3.337	20.914
4	3.001	5.772	31.406	3.001	5.772	31.406	3.317	27.294
5	2.637	5.070	36.477	2.637	5.070	36.477	2.992	33.048
6	2.473	4.756	41.233	2.473	4.756	41.233	2.753	38.341
7	2.398	4.611	45.844	2.398	4.611	45.844	2.308	42.780
8	2.184	4.199	50.043	2.184	4.199	50.043	2.306	47.216
9	1.645	3.163	53.206	1.645	3.163	53.206	2.155	51.360
10	1.312	2.523	55.729	1.312	2.523	55.729	1.563	54.366
11	1.174	2.259	57.988	1.174	2.259	57.988	1.504	57.259
12	1.114	2.143	60.131	1.114	2.143	60.131	1.494	60.131
13	0.914	1.758	61.888					
Extraction Method: Principal Component Analysis								

Source The authors' calculation from SPSS 25

positive effect on users' behavior of using the Fintech services in Vietnam). At the same time, the characteristics of two observed variables in the factor AI associated with experimental development (AI5 and AI6 as can be seen in Table 1). Therefore, this new factor is named experimental development (ED). The hypothesis for ED is similar to the hypothesis for AI (AI has a positive effect on users' behavioral intention of using the Fintech services, and AI has a positive effect on users' behavior of using Fintech services in Vietnam).

For dependent variables, the total of 6 items proved reliability through Cronbach's alpha test. In the EFA technique, this analysis was carried out using SPSS 25 with principal component method and Varimax rotation. The KMO test achieved a value of 0.679, which showed that the sampling was adequate. The Bartlett's test achieved a value of 890.763 ($p = 0.000$), ensured moderate intercorrelations between items, and suggested its suitability for factor analysis. The total of 6 observed variables were initially grouped into 2 groups. The total value of variance extracted is 65.230%, which is greater than 50%, and EFA is considered satisfactory. Thus, twelve factors explain 65.230% of the variability of the data. The eigenvalues of the factors are all high (>1). The 2sd factor has the lowest eigenvalues of 1.568 > 1. Therefore, two factors are defined as shown in Table 6.

After performing the rotation (absolute value below: 0.3) for the dependent variables, two factors were found together with the factor loading factors all that are greater than 0.5, as shown in Table 7.

Table 5 Rotated component matrix for independent variables

Variables	Component											
	1	2	3	4	5	6	7	8	9	10	11	12
PE6	0.834											
PE1	0.780											
PE7	0.728											
PE2	0.722											
PE3	0.718											
PE5	0.711											
PE4	0.630											
SI6		0.774										
SI1		0.756										
SI3		0.716										
SI4		0.684										
SI2		0.681										
SI5		0.645										
EE1			0.800									
EE3			0.764									
EE5			0.713									
EE4			0.712									
EE6			0.693									
EE2			0.638									
FC1				0.790								
FC3				0.744								
FC5				0.734								
FC4				0.711								
FC2				0.703								
FC6				0.689								
HM5					0.854							
HM3					0.807							
HM4					0.727							
HM1					0.701							
HM2					0.681							
PV3						0.820						
PV2						0.781						
PV1						0.731						
PV5						0.687						
PV4						0.586						

(continued)

Table 5 (continued)

Variables	Component											
	1	2	3	4	5	6	7	8	9	10	11	12
GP2							0.792					
GP4							0.783					
GP3							0.745					
GP1							0.581					
AI2								0.783				
AI4								0.740				
AI3								0.729				
AI1								0.631				
HA5									0.833			
HA4									0.800			
HA3									0.760			
GP5										0.788		
GP6										0.781		
HA1											0.818	
HA2											0.768	
AI6												0.776
AI5												0.741

Extraction Method: Principal Component Analysis
Rotation Method: Varimax with Kaiser Normalization

Source The authors' calculation from SPSS 25

Table 6 Exploratory factor analysis for dependent variables

Component	Initial eigenvalues			Extraction sums of squared loadings			Rotation sums of squared loadings	
	Total	% of variance	Cumulative %	Total	% of variance	Cumulative %	Total	Cumulative %
1	2.346	39.094	39.094	2.346	39.094	39.094	2.032	33.873
2	1.568	26.136	65.230	1.568	26.136	65.230	1.881	65.230
3	0.655	10.909	76.139					

Extraction Method: Principal Component Analysis

Source The authors' calculation from SPSS 25

3.4 Confirmatory Factor Analysis

To evaluate the construct validity, CFA was performed using SPSS Amos 24. This analysis is an important tool for SEM techniques that allows to check the consistency between observed data and hypothesized conceptual models, which define hypothetical relationships between latent factors (Brown, 2015).

Table 7 Rotated component matrix for dependent variables

Variable	Component	
	1	2
BI3	0.830	
BI1	0.815	
BI2	0.809	
UB3		0.810
UB2		0.788
UB1		0.764
Extraction Method: Principal Component Analysis		
Rotation Method: Varimax with Kaiser Normalization		

Source The authors' calculation from SPSS 25

In the CFA technique, this analysis was carried out using SPSS 25 with principal axis factoring method and promax rotation (absolute value below: 0.3). The KMO test achieved a value of 0.800, and the Bartlett's test achieved a value of 13,675.659, which showed that CFA is suitable for real data. At the same time, Sig = 0.000 < 0.05 shows that the observed variables are correlated with each other in terms of the population. The results of CFA showed that the number of observations is 652, as can be seen in Fig. 3. After linking e4 and e5, e10 and e12, e28 and e30, e38 and e39, e42 and e43 to correct for covariance show that this model has a Chi-square = 2531.511, with 1502 degrees of freedom (df); Chi-square/df = 1.685 < 3 with p-value = 0.000 and other indicators such as CFI = 0.917; TLI = 0.909; GFI = 0.884; RMSEA = 0.032 < 0.06; PCLOSE = 1.000 > 0.05. Thus, the CFA results are satisfactory, supporting the subsequent structural model analysis. The model fits the market data and it can be concluded that the factors are convergent because the standardized and unstandardized coefficients are greater than 0.5, and the total variance values are greater than 0.5 in this research.

3.5 *Structural Equation Modeling*

The SEM analysis was conducted with SPSS AMOS 24 to test the relationships proposed in the hypothesized model, as can be seen in Fig. 4. The indicators include Chi-square = 2554,435; df = 1507; p = 0.000; Chi-square/df = 1.695; CFI = 0.916; TLI = 0.908; GFI = 0.883; RMSEA = 0.033; PCLOSE = 1.000. Thus, the structural model is deemed acceptable.

The analysis was conducted to examine the proposed causal relationships between constructs. There were 9 out of 20 hypotheses that were not supported in this study. They are the performance expectancy, facilitating conditions, price value, habits, aspirational innovation, government policies, policy synchronization, which do not

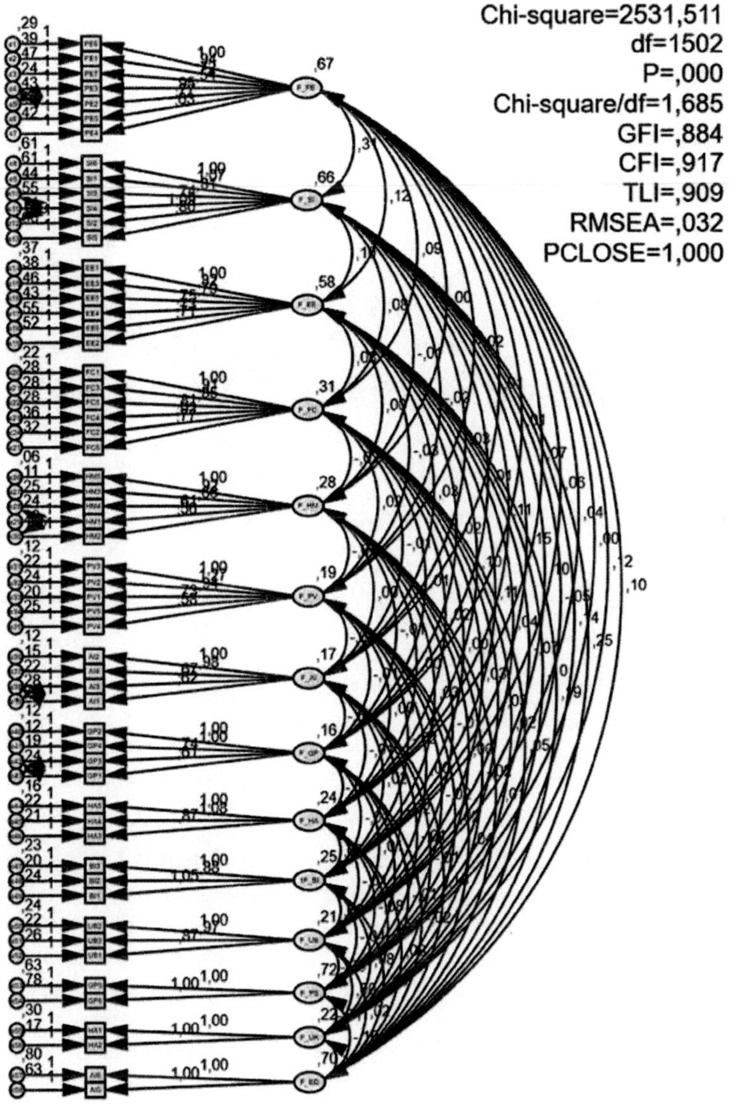

Fig. 3 Confirmatory factor analysis. *Source* The authors' calculation from AMOS 24.0

positively influence the behavior intention of using Fintech services; the user's knowledge, experimental development do not positively influence the behavior of using Fintech services, as shown in Table 8.

The factors that were hypothesized in the conceptual model to positively influence behavioral intention of using Fintech services such as effort expectancy ($\beta = 0.159$; p-value = 0.003), social influence ($\beta = 0.261$; p-value < 0.001), hedonic

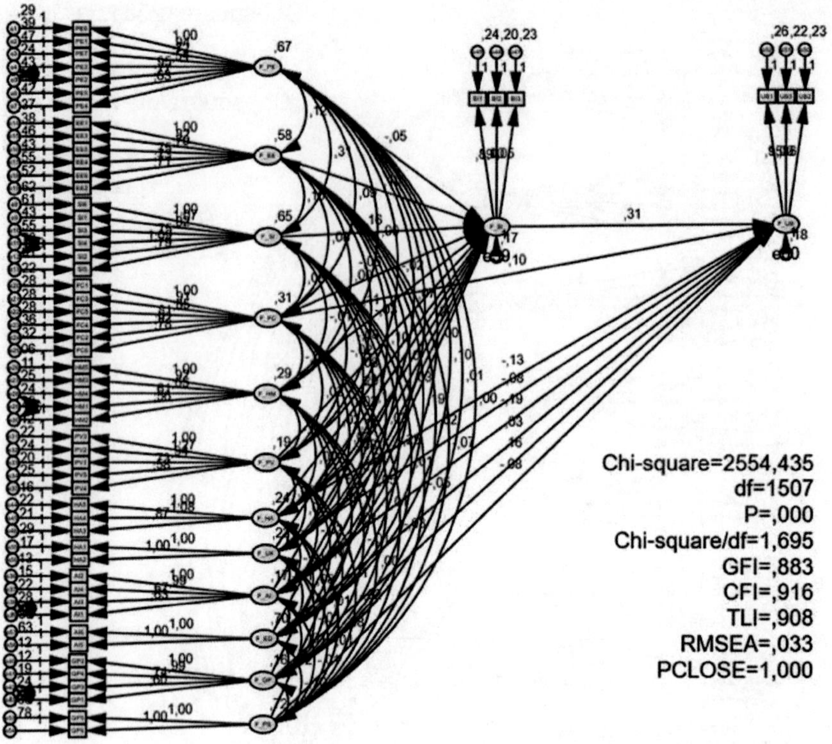

Fig. 4 Structural equation modeling. *Source* The authors' calculation from AMOS 24.0

motivation ($\beta = 0.121$; p-value $= 0.006$), user's knowledge ($\beta = 0.189$; p-value $= 0.004$), experimental development ($\beta = 0.190$; p-value $= 0.011$). The factors that were hypothesized in the conceptual model to positively influence behavior of using Fintech services, such as facilitating conditions ($\beta = 0.118$; p-value $=$ 0.021), behavior intention ($\beta = 0.337$; p-value < 0.001), government policies ($\beta =$ 0.138; p-value $= 0.032$). The factors were hypothesized in the conceptual model to negatively influence behavior of using Fintech services such as habits ($\beta = -0.118$; p-value $= 0.019$), aspirational innovation ($\beta = -0.172$; p-value $= 0.011$), and policy synchronization ($\beta = -0.140$; p-value $= 0.046$), as shown in Table 8.

The bootstrap method was employed to estimate summary statistics such as the mean or standard deviation to evaluate the robustness of the theoretical model. This study is a resampling method, used to estimate statistics, and used the number of replicate samples $N = 2600$. Compare the C.R value with 1.96 (since 1.96 is the value of the normal distribution at 0.9750). The results show that C.R < 1.96, p-value > 5%, so the estimated model can be reliable as can be seen in Table 9.

Table 8 Regression weights

		Unstandardized coefficients					Standardized coefficients	Results
			Estimate	S.E	C.R	P	Estimate	
F_BI	<–	F_PE	−0.052	0.032	−1.639	0.101	−0. 086	Not supported
F_BI	<–	F_EE	0.104	0.035	2.954	0.003	0.159	Supported
F_BI	<–	F_SI	0.161	0.038	4.232	***	0.261	Supported
F_BI	<–	F_FC	−0.059	0.043	−1.391	0.164	−0.066	Not supported
F_BI	<–	F_HM	0.112	0.041	2.736	0.006	0.121	Supported
F_BI	<–	F_PV	−0.081	0.056	−1.464	0.143	−0.072	Not supported
F_BI	<–	F_HA	0.080	0.055	1.460	0.144	0.078	Not supported
F_BI	<–	F_UK	0.201	0.069	2.909	0.004	0.189	Supported
F_BI	<–	F_AI	−0.027	0.073	−0.369	0.712	−0.022	Not supported
F_BI	<–	F_ED	0.114	0.045	2.547	0.011	0.190	Supported
F_BI	<–	F_GP	−0.079	0.074	−1.072	0.284	−0.063	Not supported
F_BI	<–	F_PS	0.005	0.037	0.145	0.885	0.009	Not supported
F_UB	<–	F_FC	0.098	0.042	2.303	0.021	0.118	Supported
F_UB	<–	F_HA	−0.132	0.056	−2.340	0.019	−0.140	Supported
F_UB	<–	F_BI	0.311	0.062	5.039	***	0.337	Supported
F_UB	<–	F_UK	−0.079	0.067	−1.188	0.235	−0.080	Not supported
F_UB	<–	F_AI	−0.191	0.075	−2.552	0.011	−0.172	Supported
F_UB	<–	F_ED	0.031	0.043	0.726	0.468	0.057	Not supported
F_UB	<–	F_GP	0.161	0.075	2.146	0.032	0.138	Supported
F_UB	<–	F_PS	−0.076	0.038	−1.996	0.046	−0.140	Supported

Source The authors' calculation from AMOS 24.0

3.6 Testing for Differences in Demographic Characteristics

For the variables of facilitating conditions, hedonic motivation, price value, and habits, the sig value of Levene's test is greater than 0.05, the variance between the two genders is not different, and the sig value of T-Test is greater than 0.05. Thus, there is no statistically significant difference in these variables of the different genders' respondents toward behavioral intention of using Fintech services, as shown in Table 10.

For the variables of facilitating conditions, hedonic motivation, price value, and habits, the sig value in the age test is greater than 0.05, the variance between the choices of these variables is not different, and the sig value in the ANOVA test is greater than 0.05. Hence, there is no statistically significant difference in these variables of the different ages' respondents toward behavioral intention of using Fintech services, as shown in Table 11.

Table 9 Bootstrap method on SEM

Parameter			SE	SE–SE	Mean	Bias	SE-Bias	C.R = Bias/SE-Bias
F_BI	<–	F_PE	0.058	0.001	−0.087	−0.001	0.001	−1.0
F_BI	<–	F_EE	0.062	0.001	0.157	−0.001	0.001	−1.0
F_BI	<–	F_SI	0.069	0.001	0.263	0.001	0.001	1.0
F_BI	<–	F_FC	0.051	0.001	−0.065	0.001	0.001	1.0
F_BI	<–	F_HM	0.045	0.001	0.122	0.001	0.001	1.0
F_BI	<–	F_PV	0.051	0.001	−0.071	0.000	0.001	0
F_BI	<–	F_HA	0.053	0.001	0.079	0.001	0.001	1.0
F_BI	<–	F_UK	0.074	0.001	0.189	0.000	0.001	0
F_BI	<–	F_AI	0.065	0.001	−0.023	−0.001	0.001	−1.0
F_BI	<–	F_ED	0.078	0.001	0.194	0.003	0.002	1.5
F_BI	<–	F_GP	0.065	0.001	−0.064	−0.001	0.001	−1.0
F_BI	<–	F_PS	0.071	0.001	0.012	0.001	0.001	1.0
F_UB	<–	F_FC	0.053	0.001	0.118	−0.001	0.001	1.0
F_UB	<–	F_HA	0.065	0.001	−0.141	−0.001	0.001	−1.0
F_UB	<–	F_BI	0.070	0.001	0.335	−0.002	0.001	−1.5
F_UB	<–	F_UK	0.081	0.001	−0.083	−0.003	0.002	−1.5
F_UB	<–	F_AI	0.077	0.001	−0.178	−0.003	0.002	−1.5
F_UB	<–	F_ED	0.096	0.001	0.062	0.002	0.002	1.0
F_UB	<–	F_GP	0.073	0.001	0.141	0.003	0.001	1.5
F_UB	<–	F_PS	0.081	0.001	−0.142	−0.002	0.002	−1.0

Source The authors' calculation from AMOS 24.0

Table 10 Independent samples test for genders

		Levene's test for equality of variances		*T*-test for equality of means		
		F	Sig	*t*	df	Sig. (2-tailed)
FC	Equal variances assumed	0.003	0.954	−1.257	650	0.209
	Equal variances not assumed			−1.261	576.310	0.208
HM	Equal variances assumed	2.656	0.104	1.472	650	0.142
	Equal variances not assumed			1.447	534.624	0.149
PV	Equal variances assumed	1.844	0.175	−0.753	650	0.451
	Equal variances not assumed			−0.757	579.353	0.449
HA	Equal variances assumed	0.496	0.481	−1.157	650	0.248
	Equal variances not assumed			−1.168	588.099	0.243

Source The authors' calculation from SPSS 25

Table 11 Test of homogeneity of variances and ANOVA

Test of homogeneity of variances			ANOVA			
	Levene statistic	Sig	Sum of squares	Mean square	*F*	Sig
FC	2.059	0.104	1.571	0.524	1.827	0.141
HM	0.532	0.661	0.086	0.029	0.138	0.937
PV	1.030	0.379	0.232	0.077	0.389	0.761
HA	3.581	0.014	1.202	0.401	2.107	0.098

Source The authors' calculation from SPSS 25

Table 12 Test of homogeneity of variances and ANOVA for experience

Test of homogeneity of variances			ANOVA			
	Levene statistic	Sig	Sum of squares	Mean square	*F*	Sig
FC	0.601	0.615	1.029	0.343	1.194	0.311
HM	2.230	0.083	0.094	0.031	0.151	0.929
PV	1.584	0.192	0.254	0.085	0.426	0.734
HA	2.037	0.107	0.533	0.178	0.928	0.427
BI	0.620	0.602	1.739	0.580	1.852	0.137

Source The authors' calculation from SPSS 25

For the variables of facilitating conditions, hedonic motivation, price value, and habits, the sig value in the experience test is greater than 0.05, the variance between the choices of these variables is not different, and the sig value in the ANOVA test is greater than 0.05. Therefore, there is no statistically significant difference in these variable of the different experiences' respondents toward behavioral intention of using Fintech services. At the same time, for the behavioral intention, the sig value in the experience test is greater than 0.05, the variance between the choices of these variables is not different, and the sig value in the ANOVA test is greater than 0.05. Thus, there is no statistically significant difference in these variable of the different experiences' respondents toward behavior of using Fintech services, as shown in Table 12.

4 Discussions

4.1 Factors Affecting Behavioral Intention

Social influence turned out to be the most positive influential factor on behavioral intention of using Fintech services ($\beta = 0.261$; p-value < 0.001). The result conforms to the expected signs and hypotheses, and it was consistent with the findings of Venkatesh et al. (2012), Mulyana et al. (2020), Khatun and Tamanna (2020), Al

Nawayseh (2020), Submitter et al. (2021), and Ha Van Duong (2022d). This result shows that bank users tend to get advice from relatives as well as absorb reviews from influencers and previous users before choosing the Fintech services.

Experimental development was the second influential factor positively affecting behavioral intention of using Fintech services ($\beta = 0.190$; p-value $= 0.011$), and this was consistent with the findings of Ha Van Duong (2022b). As the application of emerging technologies, the most prominent features of Fintech services are novelty and originality. Therefore, it is necessary to promote the experimental development scheme; from research and practical experience, Fintech services need to target new products and services to meet the needs of using Fintech services for bank users.

User's knowledge was found to be the third influential factor which has a positive impact on the behavioral intention of using Fintech services ($\beta = 0.189$; p-value $=$ 0.004). The result was consistent with the study of Ha Van Duong (2022d). This can be explained by the fact that the user's knowledge can be considered as the degree of knowledge that users have about various Fintech services, and it would help users make informed their decisions about using Fintech services. This is a predictive factor of bank users' behavioral intention to use Fintech services. Bank users have better knowledge of Fintech services, who will have a higher intention to participate in the Fintech services market.

Effort expectancy was the fourth influential factor positively affecting bank users' behavioral intention of using Fintech services ($\beta = 0.159$; p-value $= 0.003$). This result agreed with the expected signs and hypotheses, and it supports the idea of Venkatesh et al. (2012), Khatun and Tamanna (2020), Mulyana et al. (2020), Ha Van Duong (). Banking users are demanding and efficient because they want Fintech services to benefit them in terms of accessibility, speed, quality, price, and increased choice in transactions. This shows that the ease of using the features in Fintech services affects the behavior intention of using Fintech services, as well as the easier the apps are to be used, the higher bank users' acceptance of using Fintech services.

Hedonic motivation is the last factor that has a positive influence on the behavior intention of using Fintech services ($\beta = 0.121$; p-value $= 0.006$). The study results are in line with the previous research conducted by Venkatesh et al. (2012), Sukaris et al. (2021), Latifah et al. (2021), Al Rubaiai and Pria (2022), and it agreed with the expected signs and hypotheses in this study. The results showed this predictor has an impact on bank users' behavior intention of using Fintech services over their banks while getting specific services.

4.2 Factors Affecting Use Behavioral

Behavior intention turned out to be the most positive influential factor on behavior of using Fintech services ($\beta = 0.337$; p-value < 0.001). This result is in line with research conducted by the previous researchers (Venkatesh et al., 2012; Ha Van Duong, 2022a), and it agreed with the expected signs and hypotheses in this study. This means that good behavioral intentions can influence the behavior of using

Fintech services. The more the bank users' behavioral intention of using Fintech services, the greater the desire to use Fintech services of Vietnamese bank users.

Government policies were the second influential factor positively affecting behavior of using Fintech services ($\beta = 0.138$; p-value $= 0.032$). This has been examined and confirmed by many studies, such as Yang (2017), Kukreja et al., (2021), Noreen et al. (2022), Ha Van Duong (2022b). This result is consistent with the expected signs and hypotheses in this study about government policies that have a positive impact on the behavior of using Fintech services of banking users in Vietnam. This showed that the government policies could become a catalyst for the implementation of Fintech services in Vietnam.

Facilitating conditions was the last influential factor positively affecting behavior of using Fintech services ($\beta = 0.118$; p-value $= 0.021$). This result is also consistent with the expected signs and hypotheses about government policies that have a positive impact on the behavior of using Fintech services of banking users in Vietnam. At the same time, the outcome is similar to the conclusion of Venkatesh et al. (2012), Rahardjo et al. (2020), and Ha Van Duong (2022d). This showed that bank users have safely interacted with Fintech services and received support from payment service providers to use Fintech services as well as control the use of Fintech services.

Habit has a negative effect on bank uses' behavior of using Fintech services ($\beta = -0.118$; p-value $= 0.019$). This result is not consistent with the previous research by Venkatesh et al. (2012), Rahardjo et al. (2020), Ha Van Duong (2022d), and it disagreed with the expected signs and hypotheses in this study. Fintech service is one of the options for modern and convenient financial transaction methods. However, in Vietnam, the habit of using cash from Vietnamese consumers is also a big barrier to the development of Fintech services. Although the government has made many efforts to promote non-cash payment, the rate of cash use in Vietnam is still high compared to other countries in the region. Therefore, habits have a negative impact on the behavior of using Fintech services by Vietnamese bank users.

Aspirational innovation, as one of the critical factors in research within new technology or new services, has been examined and confirmed by many studies by Hu et al. (2019), Kim et al. (2010), Ha Van Duong (2022b). However, the aspirational innovation's direct influence on using behavior was not supported in this study ($\beta = -0.172$; p-value $= 0.011$). Aspirational innovation can make the Fintech services more accessible to underserved segments of the population and improve lives. But aspirational innovation sometimes cannot lead to improving bank users' behavior in using Fintech services. Because Fintech in Vietnam is in the early stages of development, it is still facing many challenges and difficulties, and the Fintech ecosystem is not yet complete, affecting the innovation process of Fintech services.

Policy synchronization has a negative effect on bank users' behavior when using Fintech services ($\beta = -0.140$; p-value $= 0.046$). The outcome is not similar to the conclusion of Ha Van Duong (2022b), and it is not consistent with the expected signs and hypotheses in this study about government policies that have a positive impact on the behavior of using Fintech services of banking users in Vietnam. Fintech services in Vietnam have made remarkable progress in terms of quantity, variety in products and services, and attracting investment capital. However, the legal framework governing the Fintech sector has not been synchronized, which is considered one of the barriers affecting the behavior of using Fintech services.

5 Conclusions and Recommendation

Besides application and inheriting the UTAUT2 model, this study has supplemented the UTAUT2 model by adding other related constructs. The theoretical contribution of this study is the addition and discovery of new factors such as user's knowledge, experimental development, aspirational innovation, and policy synchronization to the UTAUT2 model. At the same time, the present study critically examined the factors influencing bank users' behavioral intention toward usage likelihood of Fintech services in Vietnam. The findings indicate that there are five factors that have a positive influence on behavioral intention, such as effort expectancy, social influence, hedonic motivation, user's knowledge, and experimental development. There are three factors that have a positive influence on behavior of using Fintech services, such as facilitating conditions, behavior intention, and government policies. Besides, habits, aspirational innovation, and policy synchronization are the factors that have a negative influence on behavior of using Fintech services. This study has some important managerial implications, and it helps managers to develop Fintech services and offers valuable insights for the banks, and the approach taken in this study may prove diagnostically useful to the behavioral intention and behavior of using Fintech services for Vietnamese bank users. At the same time, based on the study results, this study provides a relevant recommendation to increase the behavioral intention and behavior of using Fintech services for Vietnamese bank users as follows.

Firstly, social influence has a significant positive effect on the behavioral intention to use Fintech services of bank users in Vietnam. Therefore, banks should make Fintech application platforms popular and fashionable, such as paying attention to building a bank's image, using modern means to advertise, promote, and take care of customers in a professional manner. This will help Fintech application platforms spread to many bank users, and maintaining a large user base as well as attracting new bank users is very important.

Secondly, the policy makers soon have the experimental development schemes as well as issued a controlled testing mechanism for Fintech services to promote innovation and create a testing environment to assess risks and costs, the benefits of Fintech solutions, support the construction, and development of Fintech solutions in accordance with regulatory frameworks and market needs. Therefore, the experimental development schemes support credit institutions that can test Fintech solutions, innovate and develop Fintech services, and contribute to promoting the bank users' behavioral intentions of using Fintech services in Vietnam.

Thirdly, improving users' knowledge of Fintech services is considered one of the national financial education programs. In addition to training in-depth knowledge of Fintech services at training institutions, it is necessary to combine with knowledge dissemination programs about Fintech services to people. These programs contribute to improving the users' knowledge of Fintech services, positively affecting the bank users' behavioral intention to use Fintech services in Vietnam.

Fourthly, the Fintech service providers should research and develop Fintech service platforms with high compatibility with many devices and diversify the utility of online financial transactions to suit many users, as well as the ability to expand and develop Fintech service platforms in future. These can help bank users speed up and significantly improve the quality of online financial transactions, as it takes less time to conduct more financial transactions online. Hence, the improvement in the use of Fintech service platforms will help bank users to achieve higher efficiency without much effort in financial transactions and increase the behavioral intention of using Fintech services by Vietnamese bank users.

Fifthly, the Fintech service providers should understand the extent of the hedonic motivation toward the bank users' behavioral intention of using Fintech services in Vietnam. They should regularly upgrade the Fintech service application platforms to ensure that the transaction is done quickly, accurately in financial transactions or immediately handle problems arising if any, as well as meet all the needs of bank users and be ready to support them when needed. These help bank users feel excited, comfortable, blessed, entertained, happy when they use Fintech services, help bank users beyond merely providing Fintech services, and make them feel as efficient and comfortable as possible. Consequently, hedonic motivation plays an important role in making bank users increase their behavioral intention of using Fintech services.

Sixthly, behavioral intention serves as one of the important factors influencing bank users' behavior when using Fintech services in Vietnam. At the same time, the Fintech service providers aim to motivate bank users to use Fintech services. Thus, they should improve the usefulness of the Fintech services, respond to the bank users' needs, and bring more convenience, because users can access Fintech services in 24 h. These results contribute to improving bank users' behavior of using Fintech services in Vietnam.

Seventhly, the government needs to have the right balance between promoting innovation and managing, monitoring Fintech services. The pace of innovations in the Fintech services, including some truly ground breaking ones, has picked up, heightening the urgency of addressing the safe financial transaction needs for bank users. The government needs to have strict rules and regulations in place to ensure data security and privacy protection. The policy makers should establish some guiding principles regarding the design of convenient Fintech services with greater emphasis on expandability and adaptability. Besides, the government has continued to make many efforts to promote non-cash payment to improve the rate of non-cash use in Vietnam and promote the completion of the Fintech ecosystem. These types of policy responses could help improve bank users' behavior of using Fintech services in Vietnam.

Eighthly, the Fintech service provider's degree of Fintech service's readiness or provider's technical support for the use of Fintech services should ensure safety and security for bank users. They should ensure the conditions for using Fintech services in bank users' financial transactions. Various Fintech service needs from bank users could be conveniently met with the amelioration of using conditions and the coverage of all possible scenarios. Strengthening partnerships between commercial banks, Fintech companies, and telecom operators will help improve the technical infrastructure needed for widespread use. The favorable conditions are put on top to create real behavior intention and positively impact on bank users' behavior of using Fintech services in Vietnam.

References

AL Nawayseh, M. K. (2020). Fintech in COVID-19 and beyond: what factors are affecting customers' choice of fintech applications? *Journal of Open Innovation: Technology, Market, and Complexity, 6*(4), 153

AL Rubaiai, I. R., & Pria, S. (2022). Customer usage behaviour of fintech products in Sultanate of Oman. *International Journal of Research in Entrepreneurship & Business Studies, 3*(3), 11–24.

Brown, T. A. (2015). *Confirmatory factor analysis for applied research.* Guilford publications.

Field, A. (2000). *Discovering statistics using SPSS for Windows: Advanced techniques for beginners.* Sage Publication.

Hà Văn Dương. (2022a). *Giáo trình Fintech và dịch vụ tài chính số* (pp. 6–51). Nhà xuất bản Kinh tế Thành phố Hồ Chí Minh, Việt Nam.

Hà Văn Dương. (2022b). *Giáo trình Fintech trong hoạt động ngân hang* (pp. 11–49). Nhà xuất bản Kinh tế Thành phố Hồ Chí Minh, Việt Nam.

Hà Văn Dương. (2022c). *Giáo trình Fintech trong hoạt động tài chính vi mô* (pp. 146–172). Nhà xuất bản Kinh tế Thành phố Hồ Chí Minh, Việt Nam.

Hà Văn Dương. (2022d). *Giáo trình Tiền tệ kỹ thuật số và công nghệ Blookchain.* Nhà xuất bản Kinh tế Thành phố Hồ Chí Minh, Việt Nam, pp. 8–26.

Hair, J. F., Celsi, M., Money, A., Samouel, P., & Page, M. (2015). *The essentials of business research methods* (3rd ed.) Routledge.

Hassel, A. (2015). Public policy. In J. Wright (Ed.), *International encyclopedia of the social & behavioral sciences* (2nd ed., pp. 569–575). Published by Elsevier.

Hu, Z., Ding, S., Li, S., Chen, L., & Yang, S. (2019). Adoption intention of Fintech services for bank users: An empirical examination with an extended technology acceptance model. *Symmetry, 11*(3), 340.

Hulin, C., Netemeyer, R., & Cudeck, R. (2001). Can a reliability coefficient be too high? *Journal of Consumer Psychology, 10*(1), 55–58.

Kaiser, H. F. (1974). An index of factorial simplicity. *Psychometrika, 39*(1), 31–36.

Karakas, C., & Stamegna, C. (2017). *Financial technology (FinTech): Prospects and challenges for the EU* [online]. Briefing March 2017. Available at: http://www.europarl.europa.eu/RegData/etudes/BRIE/2017/599348/EPRS_BRI(2017)599348_EN.pdf. Accessed September 7, 2022.

Khatun, N., & Tamanna, M. (2020). Factors affecting the adoption of Fintech: A study based on the financial institutions in Bangladesh. *Copernican Journal of Finance & Accounting, 9*(4), 51–75.

Kim, C., Mirusmonov, M., & Lee, I. (2010). An empirical examination of factors influencing the intention to use mobile payment. *Computers in Human Behavior, 26*(3), 310–322.

Kukreja, G., Bahl, D., & Gupta, R. (2021). The impact of Fintech on financial services in India: Past, present, and future trends. In *Innovative strategies for implementing Fintech in banking* (pp. 191–200). IGI Global.

Latifah, N., Widayani, A., & Rachmawati, I. (2021). Adoption of mobile payment approach extended the UTAUT 2. *Journal of Management and Business, 4*(2), 156–175.

Marlina, S., Abdillah, F., Kurniadi, A. R., & Gunadi, W. (2020). Acceptance of Fintech lending in Indonesia: The borrower's perspective. *Psychology and Education, 57*(9), 1071–1087.

Mulyana, A., Disman, D., Wibowo, L., & Hurriyati, R. (2020). Application of customer behavior in using fintech as business media based on the unified theory of acceptance and use of technology model. In *3rd Global conference on business, management, and entrepreneurship (GCBME 2018)* (pp. 69–75). Atlantis Press.

Noreen, M., Mia, M. S., Ghazali, Z., & Ahmed, F. (2022). Role of government policies to Fintech adoption and financial inclusion: A study in Pakistan. *Universal Journal of Accounting and Finance, 10*, 37–46.

Osborne, J. W. (Ed.). (2008). *Best practices in quantitative methods* (pp. 12–59). Sage Publications, Inc.

Rahardjo, B., Akbar, B. M. B., & Novitaningtyas, I. (2020). The analysis of intention and use of financial technology. *Journal of Accounting and Strategic Finance, 3*(1), 88–102.

Samuels, P. (2016). *Advice on exploratory factor analysis.* Working Paper. Centre for Academic Success, Birmingham City University. Available at: https://doi.org/10.13140/RG.2.1.5013.9766. Accessed September 7, 2022.

Submitter, G. A. T. R., Peong, K. K., Peong, K. P., & Tan, K. Y. (2021). Behavioural intention of commercial banks' customers towards financial technology services. *Journals and Peong, Kwee Kim and Peong, Kwee Peng and Tan, Kui Yean, Behavioural Intention of Commercial Banks' Customers Towards Financial Technology Services (March 31, 2021). Reference to this paper should be made as follows: Peong, KK*, (pp. 10–27).

Sukaris, S., Renedi, W., Rizqi, M. A., and Pristyadi, B. (2021, February). Usage behavior on digital wallet: perspective of the theory of unification of acceptance and use of technology models. *Journal of Physics: Conference Series, 1764* (1) 012071. IOP Publishing.

Venkatesh, V., Morris, M., Davis, G., & Davis, F. (2003). User acceptance of information technology toward a unified view. *MIS Quarterly, 27*(3), 425–478.

Venkatesh, V., Thong, J. Y. L., & Xu, X. (2012). Consumer acceptance and use of information technology: Extending the unified theory of acceptance and use of technology. *MIS Quarterly, 36*(1), 157–178.

Yang, H. (2017). The UK's Fintech industry support policies and its implications. *KIEP Research Paper, World Economy Brief*, 17-05.

Green Brand Equity and Its Antecedents: A Non-parametric Approach

Truong Thi Hue and Nguyen Thi Thanh Huyen

Abstract In developing countries today, there are urgent calls for green transformation, including the development of green brands, so as to address the environmental concerns in recent years. Research on green branding for sustainable development in these countries is, however, still in its infancy. Furthermore, the parametric approach has been adopted for most existing studies. This study aims to fill the gap by using the Delphi method to investigate into green food industry in Vietnam which is a developing country. Included in the survey sample are professionals with at least five years of experience shopping for green foods and vast knowledge about the products. Via purposeful sampling, the invited experts participated in a two-round Delphi survey which revealed 18 key factors affecting green branding. The factors are further classified into five clusters, namely "green brand image", "green satisfaction", "green trust", "brand perceived quality", and "green perceived value". In the participants' perception, these are the key antecedents of green brand equity. The findings give a scientific basis for policy-makers, and managers in Vietnam in their effort to enhance the green brand equity of enterprises and the nation as a whole.

1 Introduction

Severe pollution resulted from the production and consumption of industrial products has attracted increasing social concerns nowadays (Chen & Chang, 2013). Rising consumers' scrutinization and stricter environmental regulations together have put companies under ever stronger pressure to act in eco-friendly ways, so as to reduce the negative ecological impact in the production and distribution of products (Kordshuli & Buzejani, 2011). Besides, the growing popularity of eco-friendly products has created stiff competition in the green market (Delafrooz & Goli, 2015). Green marketing thus becomes inevitable for companies to survive and thrive today's rigorous competition (Chen, 2010). Marketing and branding are inherently interdependent because ultimately, marketing efforts will be reflected in the

T. T. Hue · N. T. T. Huyen (✉)
VNU School of Interdisciplinary Studies, Vietnam National University, Hanoi, Vietnam
e-mail: huyennguyen510@vnu.edu.vn

A. T. Nguyen et al. (eds.), *Contemporary Economic Issues in Asian Countries: Proceeding of CEIAC 2022, Volume 1*,
https://doi.org/10.1007/978-981-19-9669-6_36

brand (Keller, 1993). As a result, eco-friendly businesses are focusing on building and managing brands to effectively differentiate their products or services. In this context, further clarification of the concept and more empirical studies on green branding are meaningful both academically and practically.

Same as other developing countries, Vietnam has faced many environmental problems in its development process (Pham & Bui, 2022). According to the Ministry of Industry and Trade of Vietnam (2021), green transformation has thus become a trend attracting growing attention from both the government and the people. Toxic chemical residues on vegetables and fruits because of rampant use of pesticides have caused intense anxiety among consumers (Vu, 2015). Meanwhile, the development of green brands in general and in the food industry in particular in Vietnam has not received due attention and investment from many businesses. It is thus urgent enterprises become more aware about the development of GBE.

Despite the importance and urgency of GBE both in theory and in practice, understandings in this area remain sparse. To date, the majority of researches have focused on exploring the relationships among factors in brand equity structure through empirical evidences. For example, Gupta et al. (2019) study the types of values affecting GBE; Ho et al. (2019) investigate into the role of many factors for GBE such as image, quality, value, behavioral intention, green advertising, awareness, and intention through words of mouth, loyalty, and social responsibility. festival. Li et al. (2019) explore the impact of brand image, trust, satisfaction, and loyalty with GBE. Delafrooz and Goli (2015) finds a positive association between quality, perceived value, and brand image with GBE. In the context of Vietnam, Ha (2020) studies the role of brand image, trust, and satisfaction with GBE, with the survey subjects being university students who had experience in consuming green products in general. However, in most of the previous studies on GBE, quantitative methods, especially structural equation modeling (SEM), are often applied. Non-parametric studies are needed to gain insights that cannot be assembled using quantitative methods. Furthermore, there is hardly any non-parametric research conducted in a specific industry in Vietnam so far.

This study aims to fill this research gap through using the Delphi method to identify the antecedents of GBE and find out which factors receive the highest consensus rate among the invited experts. The paper is structured as follows: the theoretical background (Sect. 2), research methodology (Sect. 3), survey results (Sect. 4), and discussions (Sect. 5). Limitations and conclusions of the study are presented in Sect. 6.

2 Theoretical Background

2.1 *Brand Equity and Green Brand Equity*

Different definitions of brand equity have evolved through previous studies. For example, Chen (2010) and Yasin et al (2007) propose that brand equity be understood as the additional value derived from the brand name. Yoo et al. (2000), meanwhile, suggest that brand equity is measured by the difference between the total value of a brand and its tangible components. In general, there are three main approaches to brand equity: from a financial perspective, from a consumer point of view, and from an integration of the two (Ha, 2020). One of the most widely used consumer-based brand equity concepts is the one introduced by Aaker (1991). According to Aaker (1991), brand equity is determined by five factors: perceived quality, brand awareness, brand loyalty, brand associations, and other proprietary brand assets. As such, brand equity can bring value to consumers by enhancing their trust and satisfaction. Keller (1993) develops the concept of consumer-based brand equity which reflects the differential influence brand knowledge has on a consumer's notion and reactions toward that brand's marketing activities.

Green brand is a way to call brands associated with consumer benefits in environmental terms, such as energy efficiency, green energy use, organic origin, and environment-friendliness (Doszhanov & Ahmad, 2015; Valentine & Savage, 2010). Green brand image is determined by consumers' perceptions of the environmental benefits and values related to the brand (Doszhanov & Ahmad, 2015; Lin & Zhou, 2022). Today, the term "green brand" should be used to refer to all its constituent elements (Górska-Warsewicz et al., 2021). It is the perception of a brand, its symbol, its identity system, its image in the minds of consumers, its value system, its personality, and its relationships.

Suryawan (2019) suggests seeing GBE as an organization's asset integrated into its brand reflecting the environmental concerns attached to its names, logos, and symbols that can either boost or undermine the environmental value found in the products. Górska-Warsewicz et al. (2021) are in agreement with the majority of studies on the concept of "consumer-based green brand equity" as being a set of consumer "perceptions, influences and behaviors regarding a brand's environmental responsibilities and concerns", which enhances its usability and value. The consumer's feelings toward a brand thus make the focus in this approach. In this study, GBE would be inspected from such a perspective, being defined as "a set of consumer perceptions, behaviors, and influences related to green concerns, and environmental liabilities associated with a brand, its name, and symbol on the other".

2.2 Antecedents of Green Brand Equity

Previous studies have suggested a number of antecedents of green brand equity such as green brand image, green satisfaction, and brand perceived quality. Brand image is defined as the "perception about a brand as reflected by the brand associations held in consumer's memory" (Keller, 1993, p. 3). Chen (2010, p. 309) similarly posits that brand image is "a set of perceptions of a brand in a consumer's mind that is linked to environmental commitment and environmental concerns". Wang (2010), meanwhile, emphasizes the general public impression of a company or its brand. Javed et al. (2020) conduct a survey into consumers in fashion in China and find that green image bears influence on GBE. Khandelwal et al. (2019) achieve similar findings when investigating into purchases made by customers in big shopping centers in India. Empirical studies in Vietnam (Ha, 2020), Thailand (Pechyiam & Jaroenwanit, 2014), and Iran (Delafrooz & Goli, 2015) also yield consistent evidences of the relationship between brand image and GBE.

Brand trust can be seen through confidence that customers have in the trustworthiness, competence, honesty, responsibility, and flexibility of the brand (Doney & Cannon, 1997), or the level of promise delivery and commitments attached to the brand (Sirdeshmukh et al., 2002). Chen and Chang (2013) warns that for green brands, consumers are more likely to be skeptical of its environmental commitments, especially in light of the recent rise of greenwashing in some countries. If the brand does not keep its promises, or even cheats, not only will customer trust in the brand decline, but even the green market as a whole will be adversely affected (Chen et al., 2014). Conversely, high green trust helps boost green brand value (Akturan, 2018). Several studies have explored this positive relationship in different contexts around the world (Chen, 2010; Doszhanov & Ahmad, 2015; Ha, 2020; Lin & Zhou, 2022).

Brand satisfaction can be seen as the level of enjoyment that consumers gain when experiencing a brand, when their needs and expectations are met (Martenson, 2007), or the level of customer satisfaction when consuming a product (Paulssen & Birk, 2007). Some studies have looked at satisfaction as a key component to developing brand trust and credibility as it is formed from customers' actual experiences with brands (Lankton et al., 2010).

Javed et al. (2020) show that brand satisfaction is related to GBE in the case of fashion consumers in China. Sozer (2020) also finds that in Turkey, this factor is connected to GBE in the case of the cosmetic industry. Several other empirical studies have shown similar results in different research contexts (Chen, 2010; Ha, 2020; Martenson, 2007; Paulssen & Birk, 2007).

Brand perceived quality is defined as a consumer's assessment of the superiority or overall excellence of a product, based on their own perception (1988b; Zeithaml, 1988a). Thus, brand perceived quality is not objective (as declared or advertised by the manufacturer), but it is associated with customer experience and perception. Previous studies on GBE tend to focus on the quality of green brand perception and the quality of green brand experience (Górska-Warsewicz et al., 2021). Delafrooz and Goli (2015) finds evidences in the effect that brand perceived quality has on

GBE in the case of electronic products in Iran. Ho et al. (2019) also show similar findings in the case of paper tissue products in Taiwan. Several other studies have also uncovered this relationship in different contexts and industries (Amegbe & Hanu, 2016; Chen et al., 2014; Ishaq, 2021).

Previous studies tend to base investigation of the abovementioned GBE antecedents using quantitative methods such as SEM, regression, ANOVA and focus on consumers in general. This approach has yielded certain theoretical and practical contributions, providing valuable empirical evidences for our understanding of GBE and its driving factors. However, further qualitative study would give deeper insights into the drivers of GBE. Besides, few studies so far have combined different factors for GBE in their study models. Therefore, non-parametric approaches are needed to predict and detect antecedents of GBE which have not been found before.

3 Methodology

The Delphi method was developed in the 1960s by the RAND Corporation to explore ideas and seek consensus among a group of experts (Gordon, 2003). Today, this method is widely used in a variety of fields, such as marketing (Bonnemaizon et al., 2007), education (Perveen et al., 2017), and tourism (Huang et al., 2014). This method usually begins with an interview to get experts' initial opinions on the issues raised. Based on the results of the first round of interviews and combined with studying the literature, the researchers then design a questionnaire for the next round. In the subsequent round, specialists are required to complete questionnaires using an assessment scale. They are also asked to explain their responses and to suggest adjustments and make supplements to the questionnaires if necessary. This process may further be iterated for many rounds in which feedbacks from the previous round shall become input for adjustment in the next until a consensus of a certain level is reached among the experts (Irvine, 2005). The Delphi method allows identification and ascertaining of factors of importance which receives consensus through rounds of surveys among participants. Unlike other qualitative methods such as interview or focus group discussion, the Delphi pushes participants into rethinking and refining their opinions through rigorous reflection process (Barrett & Heal, 2020).

This study uses the modified Delphi method proposed by Murry and Hammons (1995) in which a structured questionnaire shall be used instead of the first interview to help experts maintain focus during the survey rounds (Min, 2016). The initial structured questionnaire will then be followed by open-ended questions to save time (Irvine, 2005). In this study, the questions were designed based on what were used in previous studies. The selected questions were translated into Vietnamese.

Purposeful sampling method was used, and the surveys were conducted offline during August–September of 2022 in Hanoi Vietnam. In the first round of Delphi, 36 people who satisfied the above criteria and belonged to the authors' personal contact were invited. Of these people, 30 agreed to participate into the survey which accounts for 76%. Endacott et al. (1999) recommended that in the Delphi technique, the number of experts should be between 20 and 50. Thus, the number of participants retained for the study was appropriate. Besides, the high acceptance rate of 76% was decisive (McKenna, 1994), which was thanks to the close personal contact the authors possessed with these experts.

The questionnaire used in the first round consists of three parts. The first part deals with participants' personal characteristics. Particularly, participants had to tick 'yes' to the two following criteria before allowed to move on with the rest of the questions: (1) Have at least five years of experience in regularly consuming green food; (2) Understand and love green food.

The second part consists of 17 questions on a five-point Likert scale (see Table 2). The items selected for this part were adapted from on previous researches on GBE. Specifically, included in there were five items on green brand image (based on Chen, 2010), three on green trust (based on Ng et al., 2014), four on green satisfaction (based on Chen, 2010), and five on brand quality (based on Ng et al., 2014).

The third part of the questionnaire consisted of two open-ended questions, one of which asked for respondent's opinion whether the contents of the 17 questions in part 2 needed to be revised, the other left open space for respondent to list their suggested items and to key in other comments. Respondents to the open-ended questions were also asked to explain any suggestions, whether they are willing to be involved in modifying the existing items or developing new items.

4 Results

4.1 Delphi Round 1

Table 1 presents the personal characteristics of the survey participants. Specifically, all 30 respondents in the first round are women because in Vietnam, most women are responsible for housework and food purchases in their families. About 56.7% of the participants were between the ages of 31 and 50.

Table 1 Personal characteristics of the participants

Characteristics of participants	Round 1		Round 2	
	Frequency	%	Frequency	%
Gender				
Male	0	0.0%	0	0.0%
Female	30	100.0%	21	100.0%
Age				
18–30	10	33.3%	6	28.6%
31–50	17	56.7%	12	57.1%
51 and up	3	10.0%	3	14.3%

Table 2 Results of Delphi rounds

Factors	Items		Round 1			Round 2		
			Average	CVR	Result	Average	CVR	Result
Green brand image (GBI)	GBI1	This brand is associated with protecting the environment	4.17	0.53	Accepted	4.10	0.52	Accepted
	GBI2	This brand is regarded as the best benchmark of environmental commitments	4.00	0.47	Accepted	4.05	0.43	Accepted
	GBI3	This brand is professional about environmental reputation	4.00	0.40	Accepted	4.05	0.43	Accepted
	GBI4	This brand is well established about environmental concern	3.87	0.33	Accepted	3.90	0.43	Accepted
	GBI5	This brand is trustworthy about environmental promises	3.57	0.20	Rejected			
Green trust (GTR)	GTR1	This brand's environmental commitments are generally reliable	4.20	0.53	Accepted	4.24	0.52	Accepted
	GTR2	This brand's environmental argument is generally trustworthy	3.87	0.47	Accepted	3.95	0.52	Accepted
	GTR3	This brand keeps its promises and commitments for environmental protection	3.77	0.33	Accepted	3.81	0.43	Accepted

(continued)

Table 2 (continued)

Factors	Items		Round 1			Round 2		
			Average	CVR	Result	Average	CVR	Result
Green satisfaction (GSA)	GSA1	I am happy about the decision to choose this brand because of its environmental commitments	4.07	0.53	Accepted	4.05	0.52	Accepted
	GSA2	I believe that it is a right thing to purchase this brand because of its environmental performance	3.90	0.40	Accepted	3.90	0.43	Accepted
	GSA3	Overall, I am glad to buy this brand because it is environmentally friendly	4.20	0.53	Accepted	4.19	0.52	Accepted
	GSA4	Overall, I am satisfied with this brand because of its environmental concern	3.73	0.40	Accepted	3.76	0.43	Accepted
Brand perceived quality (BPQ)	BPQ1	The brand is of high quality	4.00	0.53	Accepted	4.10	0.62	Accepted
	BPQ2	The likely quality of this brand is extremely high	3.73	0.33	Accepted	3.90	0.52	Accepted
	BPQ3	The likelihood that this brand would be functional is very high	3.77	0.33	Accepted	3.86	0.43	Accepted

(continued)

CVR or the mean, and content validity ratio was used to analyze the collected data. Lawshe (1975) suggests to adopt CVR threshold of 0.33 for a pool of 30 experts and to introduce three options for each item including *essential*, *useful but non-essential*, and *unnecessary*. As this questionnaire was designed using the five-point Likert scale, Lawshe's scale had to be integrated into the assessment for consistency as per

Table 2 (continued)

Factors	Items		Round 1			Round 2		
			Average	CVR	Result	Average	CVR	Result
	BPQ4	The likelihood that this brand is reliable is very high	3.70	0.27	Rejected			
	BPQ5	This brand must be of very good quality	3.83	0.33	Accepted	3.90	0.43	Accepted
Green perceived value (GPV)	GPV1	The green functions of this brand meet my requirements				4.05	0.52	Accepted
	GPV2	This brand gives me a sense of pride as a humane consumer				3.86	0.43	Accepted
	GPV3	This brand gives me a sense of pride as a smart consumer				3.95	0.43	Accepted

suggested by Nguyen et al. (2022). Concretely, *slight important* and *unimportant* were equaled to *not necessary*, *moderately important* to *useful but non-essential* while *extremely important* and *very important* to *essential.* Thus, items with mean lower than 3.5 or with CRV below 0.33 were eliminated.

The main results of Round 1 are presented in Table 2. Of the original 17 items, two were eliminated and not used for the next round. These were GBI5 (This brand is trustworthy about its environmental promises) and BPQ4 (The likelihood that this brand is reliable is very high) due to CRV point of lower than 0.33. Three items (GPV1, GPV2 and GPV3) were recommended by the experts for inclusion in the next round of the study (see Table 2).

4.2 Delphi Round 2

The questionnaire in this round includes 18 items, of which 15 items were originally from Round 1 and three were newly added as per suggested by experts in Round 1. The additional items were GPV1 (The green functions of this brand meet my requirements), GPV2 (This brand gives me a sense of pride as a humane consumer), and GPV3 (This brand gives me a sense of pride as a smart consumer). In this round, only 21 out of the 30 people who participated in the previous round agreed to continue. The acceptance rate for Round 2 was thus 70%.

Table 1 lists the personal characteristics of the 21 experts who agreed to be further surveyed in Round 2. 57.1% of Round 2 participants were between the ages of 31 and 50. Since the number of panelists was 21, the CVR threshold this time was calculated at 0.42 (as suggested by Lawshe (1975). Items with mean lower than 3.5 or CVR below 0.42 were thus subjected to disqualification.

The survey results of Round 2 are presented in Table 2. Accordingly, consensus were reached for all 18 items of Round 2 (including 3 newly added items), as they all possessed mean of above 3.5 and CRV above 0.42.

5 Discussions and Policy Implications

After two rounds of survey, 18 items were agreed by experts as antecedents to GBE. These factors can be grouped into five clusters: green brand image, green trust, green satisfaction, brand perceived quality, and green perceived value.

5.1 Green Brand Image

Green brand image can be seen as a driving factor of GBE, as agreed by experts in the two Delphi research rounds. Within this cluster, the item of "This brand is associated with protecting the environment" achieved a CVR in Round 2 of 0.52, which is a relatively high level. The remaining items in the cluster were all with a CVR of 0.43 in Round 2.

This finding is consistent with previous research papers on GBE. Górska-Warsewicz et al. (2021) assert that the green brand image is one of the common antecedents as found in the GBE literature. The study conducted by Javed et al. (2020) on the textile industry in China shows that brand image affects trust and brand loyalty. Delafrooz and Goli (2015) finds the same association in their research into low-power electronics in Iran.

In Vietnam, the role of the green brand image in GBE has been a focus for investigation in recent years in the face of numerous environmental and social problems (Pham & Bui, 2022). As a result of the great efforts by the government and state agencies in raising the awareness of people and businesses about environmental protection and sustainable development, the Vietnamese public has become quite acquainted with the green brand image (Ministry of Industry & Trade of Vietnam, 2021).

5.2 *Green Trust and Green Satisfaction*

Green trust and green satisfaction were also found antecedents of GBE by experts in this study. The two green items under the cluster of green trust were (1) "This brand's environmental commitments are generally reliable" and (2) "This brand's environmental argument is generally trustworthy". These both achieved a relatively high CVR in Round 2, at 0.52. Among the four accepted items under green satisfaction, two items also achieved high consensus: (1) "I am happy about the decision to choose this brand because of its environmental commitments" and (2) "Overall, I am glad to buy this brand because it is environmentally friendly" with the final CVR of 0.52.

These findings are also consistent with the results of some previous studies. Similar results on the role of green trust have also been found either in the investigation into specific industries or over all sectors in the Chinese context (Javed et al., 2020; Li et al., 2019), in Vietnam (Ha, 2020), in India (Khandelwal et al., 2019) and some other countries (Amegbe & Hanu, 2016; Doszhanov & Ahmad, 2015; Gupta et al., 2019; Lee & Chen, 2019; Lin & Zhou, 2022; Ng et al., 2014; Tsai et al., 2020). Green satisfaction has also been found to be a significant factor affecting GBE in industry-specific studies in some countries around the world (Bekk et al., 2016; Javed et al., 2020; Kang & Hur, 2012; Khandelwal et al., 2019; Lee & Chen, 2019; Li et al., 2019).

The findings of this study are explicable when looking into the context of Vietnam. Responding to the increasing awareness of the government and people about the environment, businesses are also making efforts to create green trust and green satisfaction in their products. The process of international economic integration along with the strong wave of investment from developed countries such as Japan, Korea, the US, and Europe into Vietnam has also caused changes in people's awareness and trust toward consumer products. Businesses operating in the country thus are under strong pressure for green transformation. For example, Vinamilk pioneers in setting commitments toward environment and energy saving. The company has made public its achievements in sustainable development yearly. Samsung Vina, meanwhile, claims to participate in activities to create a healthy ecosystem and preserve diversity. Such efforts by enterprises to go green have built initial green trust and satisfaction in their brands.

5.3 *Brand Perceived Quality*

Brand perceived quality was also found to have a positive effect on GBE. Among the four accepted items of brand perceived quality, two outstanding one included (1) "The brand is of high quality" with a CVR of 0.62—which was highest among all items, and (2) "The likely quality of this brand is extremely high" with a CVR of 0.52. The remaining items of this factor all reached the final CVR of 0.43 or above.

This result suggests that GBE approach based on customer perception cannot neglect the element of brand perceived quality (beside the three common factors found in the literature such as green brand image, green trust, green satisfaction).

The same finding can be seen in a number of studies in different contexts around the world. For example, Ishaq (2021) proves that brand perceived quality is one of the determinants of GBE in telecommunication and home appliance industries in Pakistan. Similar suggestions are found in Ho et al. (2019) in Taiwan, Amegbe and Hanu (2016) in Ghana, and Delafrooz and Goli (2015) in Iran.

The experts further explained why green perceived quality was important. One said that "even though it is a green product, we still care first about product quality, including product quality, including the quality of green functions" while another asserted that "organic food still needs to be fresh to ensure customer loyalty". The above results are understandable in the context of Vietnam where green brand consumers are attaching strong emphasí on quality as their living standards increase.

5.4 Green Perceived Value

The two Delphi rounds yielded a new antecedent of GBE: the green perceived value. Consensus were reached for all three new survey questions added in the 2nd Delphi Round as per suggestion by the experts in Round 1. Concretely, the CVR of GPV1 on "The green functions of this brand meets my requirements" was relatively high, at 0.52. The three newly added items fit into the cluster of "green perceived value" which is one element of customer-based brand equity (Aaker, 1991).

This finding may be supported by several previous studies. For example, Gupta et al. (2019) show that perceived values, including green utilitarian value, green hedonic value, green social value, and green altruistic value, affect the GBE of green tourist products in India. A number of other studies have shown a similar association between green perceived value and GBE in different contexts (Delafrooz & Goli, 2015; Ng et al., 2014; Pechyiam & Jaroenwanit, 2014).

An interesting finding of this study is that the non-parametric method helped uncover an emotional and psychological value factor that may not be explained by quantitative methods using structured questions. Living standards are being improved in Vietnam which means customers' requirements of the value offered by green products may also be heightened. Perhaps because of that, the value that green products bring is not only from the usage, but also from the feelings (such as joy and pride). An expert who participated in the Delphi Round 1 proposed to add items about green perceived value attached an explanation that "consumption of green products gives me a sense of satisfaction that I am a humane consumer with a sense of responsibility". Another expert also said that "green food meets the first need to ensure the health, safety and peace of mind for my family, which also gives me a pride being a smart trendy consumer". Another expert revealed that "I buy green foods first because they are safe, without harmful chemicals that affect health. They also provide a joyful feeling that we are savvy consumers who care about future generations". Thus, whether a product brings real GBE or not depends on the experiences, feelings and evaluations of customers about the green values that it bears, especially the emotion-related values that it brings to life.

5.5 *Factors on Which Consensus was Not Reached*

Two items were eliminated after the 1st Delphi round. The item "This brand is trustworthy about environmental promises" was disqualified because its CVR was 0.2, even though the mean was 3.57. The item "The likelihood that this brand is reliable is very high" was also excluded because the CVR was only 0.27 although the mean was 3.7. These results are different from the findings of some previous studies such as Chen (2010), Ha (2020), and Delafrooz and Goli (2015).

There are a number of reasons for this difference. First of all, most of the previous studies used the SEM method which may limit flexibility and do not allow experts to give further explanation for application in a new research context. Additionally, differences in research context (in Vietnam) and research category (food) may result in different findings. Green transformation in general and GBE development in particular is still a relatively new trend in recent years in Vietnam. Brand commitments will become trustworthy or reliable for customers once they more life experience, better understanding of the brand and its promises. Thus, customers need more time to experience with the products of a brand before giving their endorsement. Besides, consumers are often more skeptical against the levels of brand promise delivery and trustworthiness. Chen and Chang (2013) notes that due to the rise of greenwashing in recent years, consumers have a tendency to be cautious of the brand's environmental promises. Chen et al. (2014) also warn that the dishonesty in green branding risks reducing consumer confidence in the green market as a whole.

5.6 *Policy Implications*

Based on the research results, some implications can be made for enhancing GBE in the current context. Firstly, businesses still need to invest in building a green brand image to communicate and raise public awareness of their brand. Many businesses have done the same things as shown in studies in different countries around the world. Secondly, brand managers should focus on building green trust and green satisfaction for their company's brand next to the green brand image. The mean and CVR recorded for these items show that enterprises should simultaneously pay attention to environmental argument, commitments, and performance. It suggests that consumers choose a green brand because it is effectively attached to environmentally friendly activities and offers a feeling of trust and satisfaction (not just feelings about the mage, commitments, and views stated by the businesses).

Thirdly, the brand perceived quality, and green perceived value cannot be neglected in the construction and development of GBE. Survey results and expert opinions (expressed through open-ended questions) have shown that green branding is also closely associated with improving quality and customer experience. If the product only focuses on green brand image, trust, and satisfaction while ignoring perceived quality and value, GBE may still be limited. In particular, the findings from this work's non-parametric approach emphasize the importance of the emotional value of green brands which has seldomly mentioned in previous studies. Finally, governments, ministries, and policy-makers in general can also consider the introduction of set of items uncovered in this study in their effort to build a national green brand development strategy. Same implications apply to organizations, and enterprises of all economic sectors. In other words, the brand development strategy at the macro-level may also take into full considerations of the factors suggested in this study, especially the brand perceived quality and green trust.

6 Conclusions

The purpose of this study is to discover the determinants of GBE in the food sector in Vietnam. Research results show that in order to build and develop a strong GBE in the current context, the following factors should be focused: green brand image, green trust, green satisfaction, brand perceived quality, and green perceived value. Among the above, brand perceived quality and green trust are the factors that should get the most attention.

This study has made significant scientific and practical contributions. Through rigorous application of the Delphi method, the study has put forward a comprehensive set of indicators, grouped into five clusters of antecedents for GBE. Delphi method has not been implemented much in the literature in general and could help to overcome some of the limitations found in the previous quantitative studies. The study reinforces findings in existing studies in which GBE is determined by green brand image, green trust, green satisfaction, and brand perceived quality in previous studies using conventional approach. It supplements and highlights a new factor which is green perceived value. The study thus assists government, ministries, enterprises, policy-makers, brand, and marketing managers in their understanding the significance of GBE, providing the evidences needed for setting strategies and long-term visions for national and business brand development in the context when environmental issues are becoming increasingly urgent in developing countries like Vietnam.

This study has certain limitations. First of all, the Delphi technique with the application of mean and CVR indexes has helped to provide a complete set of items for investigation into GBE antecedents. However, it should be combined with other parametric approaches such as Analytical Hierarchy Process (AHP) to determine the relative importance of each item, or be combined with SEM, or regression method for convergent validity, discriminant validity, and model fit check. At the same time, techniques such as T-Test, Anova, and multigroup analysis (MGA) should be combined with the mentioned quantitative methods to examine differences between sample groups. Second limitation of this study lies in its scope focusing only into the food industry which hinders applicability in other industries. Thirdly, some other elements of brand equity according to Aaker's customer-based approach such as brand awareness, brand associations, and brand loyalty have not been considered in this study. Future studies, thus, can combine with other methods, take into considerations of other categories, and add other factors of brand equity in the research model to bring about more meaningful contributions both in theory and in practice.

References

Aaker, D. A. (1991). *Managing brand equity: Capitalizing on the value of a brand name.* Maxwell Macmillan International.

Akturan, U. (2018). How does greenwashing affect green branding equity and purchase intention? An empirical research. *Marketing Intelligence and Planning, 36*(7), 809–824. https://doi.org/10.1108/MIP-12-2017-0339

Amegbe, H., & Hanu, C. (2016). Exploring the relationship between green orientation, customer based brand equity (CBBE) and the competitive performance of SMEs in Ghana. *Journal of Marketing Development & Competitiveness, 10*(1), 80–93.

Barrett, D., & Heale, R. (2020). What are Delphi studies? *Evidence-Based Nursing, 23*(3), 68–69. https://doi.org/10.1136/ebnurs-2020-103303

Bekk, M., Spörrle, M., Hedjasie, R., & Kerschreiter, R. (2016). Greening the competitive advantage: Antecedents and consequences of green brand equity. *Quality and Quantity, 50*(4), 1727–1746. https://doi.org/10.1007/s11135-015-0232-y

Bonnemaizon, A., Cova, B., & Louyot, M.-C. (2007). Relationship Marketing in 2015. *European Management Journal, 25*(1), 50–59. https://doi.org/10.1016/j.emj.2006.12.002

Chen, Y. S. (2010). The drivers of green brand equity: Green brand image, green satisfaction, and green trust. *Journal of Business Ethics, 93*(2), 307–319. https://doi.org/10.1007/s10551-009-0223-9

Chen, Y. S., & Chang, C. H. (2013). Greenwash and green trust: The mediation effects of green consumer confusion and green perceived risk. *Journal of Business Ethics, 114*(3), 489–500. https://doi.org/10.1007/s10551-012-1360-0

Chen, Y. S., Lin, C. L., & Chang, C. H. (2014). The influence of greenwash on green word-of-mouth (green WOM): The mediation effects of green perceived quality and green satisfaction. *Quality and Quantity, 48*(5), 2411–2425. https://doi.org/10.1007/s11135-013-9898-1

Delafrooz, N., & Goli, A. (2015). The factors affecting the green brand equity of electronic products: Green marketing. *Cogent Business and Management, 2*(1), 1079351. https://doi.org/10.1080/23311975.2015.1079351

Doney, P. M., & Cannon, J. P. (1997). An examination of the nature of trust in buyer-seller relationships. *Journal of Marketing, 61*(2), 35–51. https://doi.org/10.2307/1251829

Doszhanov, A., & Ahmad, Z. A. (2015). Customers' intention to use green products: The impact of green brand dimensions and green perceived value. *SHS Web of Conferences, 18*, 01008. https://doi.org/10.1051/shsconf/20151801008

Endacott, R., Clifford, C. M., & Tripp, J. H. (1999). Can the needs of the critically ill child be identified using scenarios? Experiences of a modified Delphi study. *Journal of Advanced Nursing, 30*(3), 665–676. https://doi.org/10.1046/j.1365-2648.1999.01116.x

Gordon, T. J. (2003). The Delphi method. In *Futures research methodology V2.0*; Glenn, J. C., Gordon, T. J. (Eds.). American Council for the UNU, 2003

Górska-Warsewicz, H., Dębski, M., Fabuš, M., & Kováč, M. (2021). Green brand equity - empirical experience from a systematic literature review. *Sustainability (Switzerland), 13*(20), 11130. MDPI. https://doi.org/10.3390/su132011130

Gupta, A., Dash, S., & Mishra, A. (2019). Self/other oriented green experiential values: Measurement and impact on hotel-consumer relationship. *International Journal of Hospitality Management, 83*, 159–168. https://doi.org/10.1016/j.ijhm.2019.05.010

Ha, M. T. (2020). Investigating green brand equity and its driving forces. *Management Science Letters, 10*(10). https://doi.org/10.5267/j.msl.2020.2.025

Ho, T. N., Wu, W. Y., Nguyen, P. T., & Chen, H. C. (2019). The moderating effects for the relationships between green customer value, green brand equity and behavioral intention. *Academy of Strategic Management Journal, 18*(4), 1–21.

Huang, Y. C., Yang, M., & Wang, Y. C. (2014). Effects of green brand on green purchase intention. *Marketing Intelligence and Planning, 32*(3), 250–268. https://doi.org/10.1108/MIP-10-2012-0105

Irvine, F. (2005). Exploring district nursing competencies in health promotion: The use of the Delphi technique. *Journal of Clinical Nursing, 14*(8), 965–975. https://doi.org/10.1111/j.1365-2702.2005.01193.x

Ishaq, M. I. (2021). Multidimensional green brand equity: A cross-cultural scale development and validation study. *International Journal of Market Research, 63*(5), 560–575. https://doi.org/10.1177/1470785320932040

Javed, T., Yang, J., & Gilal, W. G. (2020). The sustainability claims' impact on the consumer's green perception and behavioral intention: A case study of H&M. *Advances in Management & Applied Economics, 10*(2), 1–22.

Kang, S., & Hur, W. M. (2012). Investigating the antecedents of green brand equity: A sustainable development perspective. *Corporate Social Responsibility and Environmental Management, 19*(5), 306–316. https://doi.org/10.1002/csr.281

Keller, K. L. (1993). Conceptualizing, measuring, and managing customer-based brand equity. *Journal of Marketing, 57*(1), 1–22. https://doi.org/10.2307/1252054

Khandelwal, U., Kulshreshtha, K., & Tripathi, V. (2019). Importance of consumer-based green brand equity: Empirical evidence. *Paradigm, 23*(1), 83–97. https://doi.org/10.1177/0971890719844428

Lankton, N. K., Wilson, E. V., & Mao, E. (2010). Antecedents and determinants of information technology habit. *Information and Management, 47*(5–6), 300–307. https://doi.org/10.1016/j.im.2010.06.004

Lawshe, C. H. (1975). A quantitative approach to content validity. *Personnel Psychology, 28*(4), 563–575. https://doi.org/10.1111/j.1744-6570.1975.tb01393.x

Lee, Y. H., & Chen, S. L. (2019). Effect of green attributes transparency on wta for green cosmetics: Mediating effects of CSR and green brand concepts. *Sustainability (switzerland), 11*(19), 5258. https://doi.org/10.3390/su11195258

Li, G., Li, J., & Sun, X. (2019). Measuring green brand equity in relationship interactions and its impact on brand loyalty. *Revista de Cercetare Si Interventie Sociala, 66*, 78–297. https://doi.org/10.33788/rcis.66.16

Lin, J., & Zhou, Z. (2022). The positioning of green brands in enhancing their image: The mediating roles of green brand innovativeness and green perceived value. *International Journal of Emerging Markets, 17*(6), 1404–1424. https://doi.org/10.1108/IJOEM-06-2020-0657

Martenson, R. (2007). Corporate brand image, satisfaction and store loyalty: A study of the store as a brand, store brands and manufacturer brands. *International Journal of Retail and Distribution Management, 35*(7). https://doi.org/10.1108/09590550710755921

McKenna, H. P. (1994). The Delphi technique: A worthwhile research approach for nursing? *Journal of Advanced Nursing, 19*(6), 1221–1225. https://doi.org/10.1111/j.1365-2648.1994.tb01207.x

Min, J. C. H. (2016). Guiding the guides: Developing indicators of tour guides' service quality. *Total Quality Management & Business Excellence, 27*(9–10), 1043–1062. https://doi.org/10.1080/14783363.2015.1060123

Ministry of Industry and Trade of Vietnam. (2021). *The trend of green consumption is becoming more and more important* (Xu hướng tiêu dùng xanh ngày càng được coi trọng). https://moit.gov.vn/phat-trien-ben-vung/xu-huong-tieu-dung-xanh-ngay-cang-duoc-quan-tam-nhieu-hon.html

Murry, J. W., & Hammons, J. O. (1995). Delphi: A versatile methodology for conducting qualitative research. *The Review of Higher Education, 18*(4). https://doi.org/10.1353/rhe.1995.0008

Ng, P. F., Butt, M. M., Khong, K. W., & Ong, F. S. (2014). Antecedents of green brand equity: An integrated approach. *Journal of Business Ethics, 121*(2), 203–215. https://doi.org/10.1007/s10551-013-1689-z

Nguyen, T. A. van, Tucek, D., & Pham, N. T. (2022). Indicators for TQM 4.0 model: Delphi method and analytic hierarchy process (AHP) analysis. *Total Quality Management and Business Excellence*. https://doi.org/10.1080/14783363.2022.2039062

Paulssen, M., & Birk, M. M. (2007). Satisfaction and repurchase behavior in a business-to-business setting: Investigating the moderating effect of manufacturer, company and demographic characteristics. *Industrial Marketing Management, 36*(7 SPEC. ISS.), 983–997. https://doi.org/10.1016/j.indmarman.2007.05.011

Pechyiam, C., & Jaroenwanit, P. (2014). The Macrotheme review a multidisciplinary journal of global macro trends the factors affecting green brand equity of electronic products in Thailand. *The Macrotheme Review, 3*(9), 1–12.

Perveen, S., Kamruzzaman, Md., & Yigitcanlar, T. (2017). Developing policy scenarios for sustainable urban growth management: A Delphi approach. *Sustainability, 9*(10), 1787. https://doi.org/10.3390/su9101787

Pham, V. T., & Bui, T. A. (2022). Economic growth and environmental pollution—An empirical study in ASEAN countries. *Journal of Asian Business and Economic Studies, 33*(2), 5–23.

Ranai Kordshuli, H., & Yari Buzejani, A. (2011). Effects of green marketing mix on green purchasing decisions of consumers (Case study: Consumers of Pegah dairy company in the city of Shiraz). *The Quarterly Journal of Modern Marketing Research, 4*, 165–180.

Samsung. (n.d). Sustainability *(Sự bền vững)*. https://www.samsung.com/vn/sustainability/overview/

Sirdeshmukh, D., Singh, J., & Sabol, B. (2002). Consumer trust, value, and loyalty in relational exchanges. *Journal of Marketing, 66*(1), 15–37. https://doi.org/10.1509/jmkg.66.1.15.18449

Sozer, E. G. (2020). 1. The effects of green marketing practices on green equity and customer tolerance. *Pressacademia, 7*(3), 102–111. https://doi.org/10.17261/pressacademia.2020.1283

Suryawan, I. N. (2019). Effect of utilitarian benefit on green brand equity with green brand image. *Jurnal Manajemen, 23*(2), 306–320. https://doi.org/10.24912/jm.v23i2.480

Tsai, P. H., Lin, G. Y., Zheng, Y. L., Chen, Y. C., Chen, P. Z., & Su, Z. C. (2020). Exploring the effect of Starbucks' green marketing on consumers' purchase decisions from consumers' perspective. *Journal of Retailing and Consumer Services, 56*, 102162. https://doi.org/10.1016/j.jretconser.2020.102162

Valentine, S. V., & Savage, V. R. (2010). A strategic environmental management framework: evaluating the profitability of being green. *Sustainability Matters*, 1–32. World scientific. https://doi.org/10.1142/9789814322911_0001

Vinamilk. (n.d.). *Sustainable development (Phát triển bền vững)*. https://www.vinamilk.com.vn/vi/phat-trien-ben-vung

Vu, H. (2015). Concerns about the status of pesticide residues on green vegetables (Nỗi lo lắng từ thực trạng tồn dư thuốc bảo vệ thực vật trên rau xanh). https://dangcongsan.vn/ban-doc/y-kien-ban-doc/noi-lo-lang-tu-thuc-trang-ton-du-thuoc-bao-ve-thuc-vat-tren-rau-xanh-359692.html

Wang, C. Y. (2010). Service quality, perceived value, corporate image, and customer loyalty in the context of varying levels of switching costs. *Psychology and Marketing, 27*(3), 252–262. https://doi.org/10.1002/mar.20330

Yasin, N. M., Noor, M. N., & Mohamad, O. (2007). Does image of country-of-origin matter to brand equity. *Journal of Product and Brand Management, 16*(1), 38–48.

Zeithaml, V. A. (1988a). Consumer perceptions of price, quality and value: A means-end model and synthesis of evidence. *Journal of Marketing, 52*(July), 2–22.

Zeithaml, V. A. (1988b). Consumer perceptions of price, quality, and value: A means-end model and synthesis of evidence. *Journal of Marketing, 52*(3), 2–22. https://doi.org/10.1177/002224298805200302

An Assessment on Factors Affecting Stay Services Quality of Muong Thanh Hotels in Ha Long

Hoa Thi Minh Tran and Nhung Thi Hong Pham

Abstract The paper systematized the theoretical basis, combined with expert consultation, to select the expected research model by combining research methods from qualitative to quantitative. The research model consists of nine scales and 59 independent observed variables, as well as four dependent variables, for assessing factors influencing the quality of accommodation services at Muong Thanh hotels in Ha Long. Based on this model, the study conducted a survey of 350 tourists with SPSS 23.0 software and a sample size of 350 votes. Cronbach's alpha and exploratory factor analysis were used to assess the scales' reliability. Since then, the official research model has been constructed with eight scales, 44 independent observed variables, and four dependent observed variables. Regression analysis yielded regression equations with standardized beta coefficients of varying magnitudes. This demonstrates that the factors have varying degrees of influence on the quality of accommodation services provided by the Muong Thanh hotel system in Ha Long. However, the study only surveyed domestic tourists and did not solicit feedback from international visitors.

Keywords Service quality · Accommodation service quality · Models · Ha Long · Vietnam

1 Introduction

The hotel is one of the specific and important business areas of the tourism industry, which has grown rapidly and profitably. To meet the increasing demand of tourists, large hotels, as a service complex, "*a miniature city*", not only provide accommodation and food services, but also increasingly diversify additional services (Manh & Huong, 2013). In which case, accommodation is the fundamental and primary service of the hotel industry. As a result, "*service quality is becoming increasingly a top*

H. T. M. Tran
Phenikaa University, Hanoi, Vietnam

N. T. H. Pham (✉)
Thai Nguyen University of Sciences, Thai Nguyen, Vietnam
e-mail: nhungpth@tnus.edu.vn

A. T. Nguyen et al. (eds.), *Contemporary Economic Issues in Asian Countries: Proceeding of CEIAC 2022, Volume 1*,
https://doi.org/10.1007/978-981-19-9669-6_37

concern" (Mackay & Crompton, 1988), as a tool to attract guests, satisfy needs, and bring visitors satisfaction. With the strong development of the accommodation business market, improving the quality of services in general and accommodation services in particular will mean improving competitiveness and promoting business efficiency of each hotel. As a result, the issue of service quality and accommodation service quality has become an important and necessary research topic in the hotel and tourism industries, attracting the attention of many scholars worldwide. According to some studies, "*the determinant of competitiveness is service quality*" (Brown & Ragsdale, 2002; Wong & Kwan, 2001). At the same time, Panda (2003) stated that "*the success of a service provider depends on a high-quality service relationship with customers*", thereby increasing customer loyalty and satisfaction goods (Kheng et al., 2010); lower the cost of acquiring new customers (Kotler, 2010); and increase customer repeat purchase behavior (Kheng et al., 2010). Because it is a difficult factor to quantify, studies primarily focus on assessing service quality using a variety of models and from the standpoint of customer perception. Service quality is not a random and natural result, but rather "*the result of the simultaneous impact of a series of factors*", but studies on factors affecting service quality in customers. There are not many hotels (Chen & Chen, 2014). The article aims to define a research framework with standards and criteria for assessing the influence of environmental factors and constitutive factors on the quality of accommodation services of the hotel Muong Thanh hotel system in Ha Long destination, based on the fact that there is a gap in this research.

Destination Ha Long is a well-known tourist destination not only in Quang Ninh, but also as a tourist image of Vietnam for international visitors. The Ha Long accommodation system has grown rapidly in recent years, with the appearance of many large-scale hotels with high-class brands such as Wyndham, Royal lotus, Novotel, and Vinpearl … Simultaneously, recognizing that the accommodation industry plays an important role in enhancing the destination's image, local governments, and businesses have many solutions to improve the quality of accommodation services in order to create attractiveness and bring satisfaction to visitors. Muong Thanh Group has grown into the largest private hotel chain in Indochina, with 60 international hotels of 3-4-5-star standards (accounting for 10% of the total number of rooms in the country), located in 40 provinces of Vietnam and the capital of Laos, including the Ha Long tourism center (Muong Thanh Hotel Group Office, 2018).

In the Ha Long destination, the Group has opened six large-scale hotels under the luxury, grand, and holiday brands. Muong Thanh's hotel system is all in a prime location, with modern and unique architecture, as well as quality, classy, and diverse products and services to meet the needs of visitors. Throughout the development process, the Group and each hotel determine that customer satisfaction with service quality is the most important goal that determines the business's success. As a result, each hotel is constantly striving to improve the overall quality of services and the quality of accommodation services in order to establish itself as an ideal destination for tourists visiting the World Natural Heritage. However, the Group's "rapid" development and the difficulty in filling positions with qualified candidates have significantly impacted the caliber of lodging services. Additionally, the COVID-19 pandemic's long-term effects on a business can result in facility degradation, which poses a few challenges for hotels as they work to recover. Overall service quality and lodging service quality have both been significantly impacted by facilities, staffing shortages, and changes in customer behavior. Therefore, the study and general evaluation of the elements impacting service quality conducted in the current time provides the scientific basis for the Group's hotels in Ha Long to deliver high-quality services and delight guests.

2 Methodologies

To conduct the research, the authors collected and processed two types of data: secondary data (articles, documents, reports, statistics, business data, etc.) from the locality and primary data (of 5 hotels). Primary data is the result of in-depth interviews with experts and questionnaire surveys of guests staying at the Muong Thanh hotel system in Ha Long. Scientists, lecturers; managers of the Department of Tourism of Quang Ninh province; tourism associations; Hotel Association; leaders of Muong Thanh Group; CEO of Muong Thanh hotels in Ha Long; executive director, department manager of several 5-star hotels participated in in-depth interviews. Closed questions about 09 standards, 63 evaluation criteria, and 04 criteria to evaluate the influence of factors on the quality of accommodation services were designed into the interview contents; open-ended questions to collect opinions on identifying survey subjects by questionnaire, solutions to improve the quality of accommodation services in the "new normal" state were designed into the interview contents. The questionnaire results will be coded and built into a matrix table: the horizontal row will be the evaluation criteria and criteria; the vertical column will be the expert opinion, and the cell will be the intersection of the horizontal row and the column. The vertical line represents expert agreement. Mark (x) if the expert agrees with that criterion or evaluation criterion; otherwise, the box will be left blank. Following that, the topic will synthesize expert opinions based on each criterion and evaluation criteria. Standards and evaluation criteria with a 50% retention rate will be retained and incorporated into scales and variables in the research framework (Huong, 2018). The findings of in-depth interviews with experts serve as the foundation for proposing a research framework that includes 09 standards (scales), 59 evaluation criteria (independent observed variables), and 04 criteria (dependent observed variables) to assess factors influencing the quality of accommodation services provided by the Muong Thanh hotel system in Ha Long. This proposed research framework will serve as the foundation for designing the questionnaire, developing the content of the questions, and conducting the survey to collect primary data from the guests who will be staying at this location.

Because of the nature of the topic, the survey object is the guest staying (beneficiary, using the service) at Muong Thanh hotels in Ha Long from January 2020 to April 2022 evaluation of the COVID-19 pandemic's impact. This subject selection also received the approval of 13 experts (representing 65% of the total number of

Table 1 Questionnaire collection results by hotel

Numerical order	Hotel	Total votes	Ratio (%)
1	Muong Thanh Luxury Quang Ninh	108	31.56
2	Muong Thanh Luxury Ha Long Centre	90	26.33
3	Muong Thanh Grand Hạ Long	72	21.05
4	Muong Thanh Grand Bai Chay	36	10.53
5	Muong Thanh Holiday Suoi Mo	36	10.53
Total		342	100

Source Summary of the author's survey results, 2021

opinions). The study also employs a quantitative sampling formula based on Bollen's standard 5:1 (1998), which states that at least 5 observations are required for each measurement variable and that the number of observations should not be less than 100. The formula is $N = n * 5$ (where N is the sample size and n is the factor or criterion scale) (Huong, 2018). With 63 independent observed variables used in factor analysis and 10 scales in this study, the minimum sample size required is as follows: $63 * 5 = 315$ observations. The questionnaire is divided into two sections, one of which contains the respondents' personal information (such as age, gender, income, and educational level); the main section includes questions about guests' perspectives on factors influencing the quality of accommodation services provided by the Muong Thanh hotel system in Ha Long. These questions are graded on a Likert scale of 1 to 5, with 1 being the lowest level and 5 being the highest. Since then, the topic has conducted a survey with 350 guests who use lodging services. After conducting inspection and classification, Muong Thanh Luxury Ha Long Center Hotel, Muong Thanh Luxury Quang Ninh, and Muong Thanh Grand Ha Long received 342 valid votes. The total number of valid votes is distributed to each hotel in accordance with the size and operating time of each unit, as shown in Table 1.

To process survey results and conduct analysis and evaluation, the topic employs IBM SPSS Statistics 23.0 software in conjunction with Excel tools. The paper ballots are checked, and invalid votes are removed before entering data (answers with insufficient information, omitted mandatory questions, or incorrect answers). Following that, all survey results (both online and on paper) are entered into the software and checked for data reasonableness and blank data. To determine the influence of factors on the accommodation service quality of Muong Thanh hotels in Ha Long, the data table will be encrypted, and quantitative analysis steps such as descriptive statistics, scale reliability testing, exploratory factor analysis (EFA) correlation, and multivariable regression will be performed.

3 Theoretical Foundation and Research Design

3.1 Theoretical Foundation

3.1.1 The Concept of Lodging Service Quality

Service of lodging

With the variety of activities, services are growing rapidly and playing an important role in the national economy. The concept of service is now examined from the authors' various points of view. The concept of service is considered in terms of the "*values*" defined by consumers. A service "*is any action and outcome that one party can provide to another that is essentially intangible and does not result in ownership of something*" (Kotler, 2000). As a result, a service is a unique product that is difficult to quantify and detect with the naked eye.

Each business industry will have its own product system with distinct features. The hotel industry is a service industry with products that are primarily intangible and intangible in order to always meet the maximum needs of customers. Hotel products are defined as all services and goods that enterprises provide to the market in order to meet the needs of customers for accommodation, meals, and other additional needs for profit (Manh & Huong, 2013). The essence of the hotel industry is the provision of lodging services. This is the fundamental service of the hotel industry that distinguishes it from other businesses, additional tourism services (Khuong, 2019).

Accommodation service quality

Because this factor determines the industry's business performance, the strong transformation of the world economy to services has created a driving force for research on service quality. For a long time, researchers have attempted to define and measure service characteristics because they are intangible and service quality is a difficult factor to assess. The Oxford Dictionary defines quality as "*a level of perfection, a comparative or absolute characteristic, a specific sign, facts, and basic parameters*" (Thang, 2013). Service quality is defined as "*the degree of difference between consumers' expectations of a service and their perception of service outcomes*" (Leonard et al., 1988). Many scholars believe that service quality is determined by what customers "*perceive*" based on their individual needs, rather than what the service provider claims.

3.1.2 Factors Influencing the Quality of Lodging Services

The quality of hotel services in general, and specifically accommodation services, is quite complex and constantly fluctuating, with characteristics that depend on many factors. This complexity results from the synthesis of three groups of factors:

economic, technical, social, which can be further subdivided into external factors and internal factors. Groups of external factors are objective factors that are beyond the hotel's control, and each factor affects service quality in different ways and at different levels. This group of factors includes the politics and legal policies of the State, the needs of the market, customers, epidemics, technology, competitors, natural factors, economic factors, and other factors. culture—society … directly or indirectly affect the quality of accommodation services in the direction of promoting or limiting investment, improving quality.

The division of internal factors is only conventional because of the complex interrelationships in the overall unified impact on the quality of accommodation services. This group of factors directly affects the quality of accommodation services such as technical facilities; human resources, price; qualifications and methods of organization and management of the hotel business; corporate cultural environment; capital capacity and financial potential of the hotel; service process and coordination between departments… As a result, environmental factors affecting service quality are diverse and complex, so only a few basic factors suitable for the chosen research model, as well as research conditions, are chosen within the framework of the study.

3.2 *Models for Research*

Service quality measurement is difficult due to its unique characteristics: intangibility, heterogeneity, and inseparability (Bolton & Drew, 1991). However, given the importance of service quality in business activities, this issue has piqued the interest of many academics in recent years. Through these findings, the authors developed a research model to evaluate service quality and influencing factors that can be applied to lodging services, like Gronroos' model, Parasuraman's SERVQUAL model, and Cronin and Taylor's model (SERVPERF). These studies all agree that customers determine quality, and the factors influencing service quality are also considered from the customer's perspective. However, because each model has advantages and disadvantages, flexible coordination and the selection of appropriate evaluation factors from each model is required. As a result, the initial research model was developed based on the combined results of the elements constituting the service quality assessment model, as well as an analysis of factors influencing the quality of accommodation services.

The original proposed research model is based on the inheritance of three research models: (1) Gronroos's research model on corporate image, (2) Cronin and Taylor's SERVPERF research model on 05 factors components of service quality, and (3) LODGSERV model with 5 factors to evaluate service quality in hotel experience. SERVPERF is an important model to build the topic's research model because it allows for the objective and accurate assessment of factors affecting service quality through customer perception. The research model is expected to be determined after synthesizing expert opinions on evaluation criteria and criteria, with 9 scales, 59 independent observed variables, and 04 dependent variables (Fig. 1).

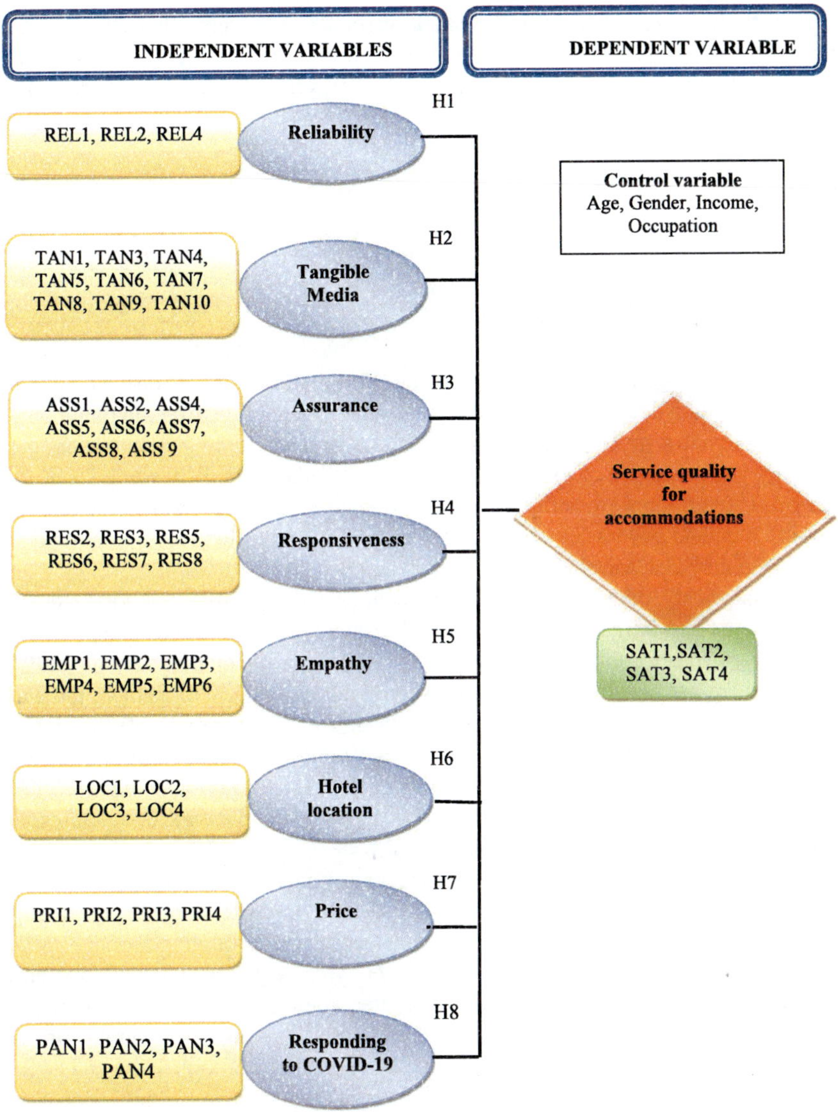

Fig. 1 Research model adjusted after qualitative research. *Source* Author's compilation from research results, 2022

The multivariable regression equation then looks like this: Y (SAT) = 1 (reliability) + 2 (tangible media) + 3 (assurance) + 4 (reaction capacity) + 5 (empathy) + 6 (hotel location) + 7 (price) + 8 (brand) + 9 (COVID-19 response) + e. In which: 9 assessment factors (scales) are recognized as being relevant to the dependent variable in the model; e is the error, and 1, 2, 3, 4, 5, 6, 7, 8, 9 are the overall regression coefficients Y with the relevant independent variables or the importance and weight coefficients (Chen & Chen, 2014). To create a formal research model, this model will be evaluated for validity using SPSS 23.0 software's EFA (exploratory factor analysis).

4 Results

4.1 Study Sample Description

Regarding the personal characteristics of the survey sample, it was discovered that there was no statistically significant difference in the gender distribution of the visitors, with females making up 53.9% and males 46.1% (Table 2). Customers are primarily young; 65.9% of them are between the ages of 26 and 45, while 2.6% are over 60. These figures are entirely compatible with the fact that clients between the ages of 26 and 45 have been employed for some time, have an income, are in good health, and may also travel or satisfy demands.

Simultaneously, the majority of customers have undergraduate and postgraduate degrees, with universities accounting for 51.3% and graduate students accounting for 23.5%; lower secondary education has the lowest rate, accounting for only 1.9%. Because there are usually people with a degree, which equates to a stable job and income, the need for rest and travel will be greater. There is no discernible difference in customer income levels, with the level of 5–10 million VND accounting for the highest proportion (32.5%), followed by the level of 15–20 million VND at 21.5%. Furthermore, due to the context of the Covid-19 pandemic, the topic only addresses domestic guests (only one case is an exception).

Simultaneously, the majority of customers have undergraduate and postgraduate degrees, with universities accounting for 51.3% and graduate students accounting for 23.5%; lower secondary education has the lowest rate, accounting for only 1.9%. Because there are usually people with a degree, which equates to a stable job and income, the need for rest and travel will be greater. There is no discernible difference in customer income levels, with the level of 5–10 million VND accounting for the highest proportion (32.5%), followed by the level of 15–20 million VND at 21.5%. Furthermore, due to the context of the Covid-19 pandemic, the topic only addresses domestic guests (only one case is an exception). Simultaneously, because these are all high-end hotels with ideal conditions for developing MICE products, they have attracted a large number of guests who stay when attending conferences and seminars at the hotel. Furthermore, the Quang Ninh Steering Committee for COVID-19 Prevention and Control permitted three Muong Thanh 4-star hotels to accept quarantined guests (accounting for 5.2% of the survey samples). However, accessing and surveying this customer is difficult due to regulations designed to ensure epidemic safety, as well as the fact that most customers refuse to cooperate.

Table 2 Descriptive information of the study sample

Criteria	Classification	Frequency	Rate (%)
Sex	Male	157	46.1
	Female	185	53.9
Age	18–25	42	12.3
	26–45	225	65.9
	45–60	65	19.2
	Over 60	10	2.6
Academic level	Below high school	6	1.9
	High school	32	9.4
	Intermediate college	47	13.9
	University	175	51.3
	After university	82	23.5
Job	Public servants	69	19.5
	Office staff	105	30.8
	Businessmen	51	15.2
	Labor in General	26	7.8
	Student	91	26.7
	Other professions	0	0
Income	<5 million VND	34	10.0
	5–10 million VND	112	32.8
	10–15 million VND	61	19.8
	15–20 million VND	67	21.5
	>20 million VND	68	15.9
Living area	Northern	240	70.45
	Central	61	20.13
	South	40	9.19
	Canada	01	0.03
Purpose	Travel, entertainment	241	70.5
	Conferences	67	19.8
	Visiting friends and relatives	6	1.7
	Job	6	1.7
	Isolation and control of the COVID-19 epidemic	18	5.2
	Another purpose	4	1.1

Source Summary of author's survey, 2022

4.2 Testing for Reliability and Exploratory Factor Analysis (EFA)

To keep the satisfactory variables, the scales are tested for reliability using the Cronbach's alpha tool. The Cronbach's alpha coefficients of the 10 scales and 63 variables are all greater than 0.8, according to the test results. Many researchers agree that a scale with a coefficient of 0.8 or greater to 1 is a good scale (Huong, 2018). As a result, the scale's reliability in measuring the influence of factors on the quality of accommodation service provided by the Muong Thanh hotel system in Ha Long is high. Furthermore, the observed variables must have a total correlation coefficient (Corrected item-total correlation) of 0.3 in order to proceed with the exploratory factor analysis (EFA). Analysis of test results for each scale found that 06 independent observed variables of 05 scales were excluded because they did not meet the standard of total variable correlation. With the scales with observed variables to ensure reliability, Cronbach's alpha has been reanalyzed for the second time, and the results have obtained 09 independent variable scales with 53 observed variables, 01 dependent variable scale, and 04 variables continue to perform the exploratory factor analysis step.

Because the topic's research model identified independent and dependent variables, Hair et al. (2015) believe that EFA analysis must be the same independent or the same type of dependency without pooling. Both types of variables have something in common (Khuong, 2019). As a result, the topic performs exploratory factor analysis on each variable type in turn.

Factor analysis with independent variable

KMO and Bartlett's test for 53 observed variables showing the factors affecting the quality of stay of Muong Thanh hotel system in Ha Long obtained KMO of 0.982 > 0.5 with a significance level of 0 (Sig. = 0.000) (Table 3). With this result, the application of exploratory factor analysis in this scale is completely appropriate.

Next, to perform factor analysis, the topic uses the characteristic value (Eigenvalue) to determine the number of factors. With this criterion, only factors with Eigenvalue ≥ 1 are kept in the analytical model. There are only 8 factors to ensure with 53 observed variables of the independent variable with corresponding Eigenvalue (Appendix 28), but with the standard value must be 1, the remaining 44 factors will not be used in this research models. That means that 53 observed variables are

Table 3 Results of KMO and Bartlett's test with independent variables

Kaiser–Meyer–Olkin measure of sampling adequacy		0.982
Bartlett's test of sphericity	Approx. Chi-Square	21,309.942
	df	778
	Sig.	0.000

Source Results of author processing survey data using SPSS, 2022

Table 4 Results of the analysis of the characteristic values (Eigenvalues) of the independent variables

Variable	Initial characteristic value			Total square of rotation factor load		
	Total	% Of the variability	% Incremental	Total	% Of the variability	% Incremental
1	17.688	33.359	71.109	17.688	33.359	71.109
2	2.750	5.186	73.278	2.750	5.186	73.278
3	2.340	4.411	75.241	2.340	4.411	75.241
4	2.019	3.807	76.785	2.019	3.807	76.785
5	1.811	3.379	78.127	1.811	3.379	78.127
6	1.642	3.063	79.338	1.642	3.063	79.338
7	13.59	2.535	80.452	1.359	2.535	80.452
8	1.276	2.381	81.552	1.276	2.381	81.552

Source Results of author processing survey data using SPSS, 2022

reduced to 8 factors with a cumulative total variance of 82.505%. At the same time, the total squared value of the rotation factor load is also 82.505%. This value means that using 8 factors representing 53 observed variables can explain 82.505% of all scales. For research in the socioeconomic field, the coefficient of squared load of the rotation factor only needs to reach 50% to meet the requirements (Tho, 2011). Therefore, the topic can completely use 8 factors to express the information provided from 53 observed variables (Table 4).

In order to perform the next regression analysis, the topic employs a factor analysis technique (Factor analysis) to reduce and group into a more meaningful factor. The first criterion with a factor loading is to ensure 0.5. With this standard, all 53 variables meet the requirements of factor analysis. However, in order to ensure the quality of the next step of analysis, the subject conducts analysis and standard comparison to further eliminate bad variables. The results of factor analysis (Appendix 28) show that the observed variable is uploaded in both factors, so the difference in the factor loading coefficient of the observed variable between the two groups of factors must be examined. If the load factor difference is greater than 0.3, the observed variable must be removed (Tho, 2011). There are 9 observed variables that upload both factors but have a difference of 0.3, so they will be excluded from the research model under this standard. As a result of the loading of two factors and the factor loading coefficient of 0.3, four observed variables of the brand scale were excluded from the research model.

If the load factor difference is greater than 0.3, the observed variable must be removed (Tho, 2011). There are 9 observed variables that upload both factors but have a difference of 0.3, so they will be excluded from the research model under this standard. As a result of the loading of two factors and the factor loading coefficient of 0.3, four observed variables of the brand scale were excluded from the research model.

Table 5 KMO and Bartlett's test results with dependent variable

Kaiser–Meyer–Olkin measure of sampling adequacy		0.846
Bartlett's test of sphericity	Approx. Chi-Square	1118.056
	df	6
	Sig.	0.000

Source Results of author processing survey data using SPSS, 2022

Factor analysis involving a dependent variable

It is also similar with four dependent variables to test the KMO and Bartlett's values to determine the appropriateness of the factor analysis step. The result of the analysis is 0.846, which is guaranteed to be greater than 0.5 and less than 0.1; the Sig. value is also less than 0.05 (Table 5).

With the Eigenvalue reflecting the explanatory factor for variation, the standard guarantee is 3.399 > 1. Therefore, in this study, the characteristic value has the effect of summarizing information from an original variable. At the same time, this factor explains 84.982% of all dependent variables (the sum of squares of the rotation factor load is 84.982%). Therefore, these 4 dependent observable variables ensure the standard when testing, so the model should be kept. The study obtained Factor loading factor loading with a value greater than 0.9 using factor rotation; ensure the standard is greater than 0.5 and the sample size is less than 350 (Khuong, 2019). Thus, the 4 proposed observed variables all meet the standards, so they will be included in the official research model.

Correction of research model after EFA analysis

Through Cronbach's alpha test and EFA analysis, removing bad variables, the research model is left with 8 scales but with 44 independent observed variables and 4 output dependent observed variables (Table 6). The research hypothesis is still the correlation between the independent variables and the dependent variables. Thus, the formal research model includes the following scales and variables:

4.3 Regression Analysis is Used to Assess the Extent to Which Various Variables Have an Impact on the Caliber of Accommodations Services in the Investigated Hotel System

To implement the multivariable regression model, the Pearson coefficient must be used to consider the interrelationships between the observed variables (*r*). The first factor to consider before considering the Pearson coefficient is the Sig. value, which shows the statistical significance of the relationship between the observed variables. With both Sig. values of 0.000, it means that not only does the independent variable

Table 6 Research framework on factors affecting the quality of accommodation services of Muong Thanh hotel system in Ha Long

Scale	Independent observed variable
1. Reliability (REL)	REL1: Staff always inform guests clearly and in detail about the hotel's accommodation services REL2: Staff always inform customers of service time REL4: The hotel's accommodation service is done right from the very beginning
2. Tangible media (TAN)	TAN1: The hotel has beautiful architecture, with its own characteristics TAN3: The hotel has a spacious and airy public area TAN4: The hotel's Internet, Wi-Fi is fast, good connection TAN5: Spacious and clean room TAN6: The equipment in the hotel room is luxurious, modern, beautifully decorated TAN7: Furniture and items in the room ensure quality standards TAN8: The hotel offers good choices of food TAN9: The hotel has a full range of convenient services such as swimming pool, spa, gym, bar ... TAN10: Staff has neat, polite and beautiful uniforms
3. Assurance (ASS)	ASS1: The hotel's security system is good ASS2: Guest information and launch key are well secured ASS4: Customers feel secure when using the hotel's services ASS5: Staff with in-depth knowledge of the hotel ASS6: Staff is friendly and polite to guests ASS7: Employee's ability to handle situations is good ASS8: Hotels provide the best solutions when guests complain ASS9: The staff advises well places to visit, shop, and entertain guests while staying
4. Responsiveness (RES)	RES2: Customers are served the right services, quickly RES3: Various additional services, to meet the maximum requirements of guests during their stay RES5: The staff doesn't say no to the service the hotel provides RES6: Hotel staff is always available when customers have problems RES7: The staff's foreign language ability is good RES8: The hotel strives to provide personalized services to guests
5. Empathy (EMP)	EMP1: Staff always has a friendly smile accompanied by a warm greeting when welcoming guests EMP2: Staff often asks and takes care of guests during their stay EMP3: Staff can judge the needs of customers EMP4: Staff always listens actively when communicating with customers EMP5: Employees adjust their behavior and attitude to suit the situation and mood of the guest EMP6: Operating hours and providing services of the hotel in accordance with the needs of guests
6. Hotel location (LOC)	LOC1: Hotel in easy to find location LOC1: The hotel is conveniently located for public transport LOC3: The location of the hotel is convenient to move to attractions, entertainment areas LOC4: The hotel is in a beautiful location with a beautiful view

(continued)

Table 6 (continued)

Scale	Independent observed variable
7. Price (PRI)	PRI1: The hotel's room rate is reasonable compared to the market PRI2: The price of additional services of the hotel is suitable for guests PRI3: The level of payment to customers is commensurate with the quality of service provided PRI4: Apply flexible pricing policy to attract customers
9. Responding to COVID-19 (PAN)	PAN1: Take safety measures when serving guests according to the regulations of the health industry PAN2: The hotel has applied the handling process when there is a case that an employee or guest is F0, F1 and is suspected of having the disease PAN3: Promote the application of digital transformation when providing services to customers PAN4: The hotel has provided accommodation services quickly and flexibly according to the requirements of guests
10. Quality of hotel accommodation service (SAT)	SAT1: Guests are satisfied with their stay service experiences SAT2: Guests have good reviews about the hotel on online channels SAT3: Guests actively recommend hotels to others SAT4: Guests intend to return to use the hotel's accommodation services

Source Compiled and proposed by the author, 2022

have a linear correlation with the dependent variable, but there is also a correlation between the independent variables. Therefore, the correlation between the independent variables and between the independent and dependent variables should be on average (Table 7).

The variables after being tested to ensure the correlation will be put into the multivariate regression model MVR (Multi-variate regression).

Table 7 Results of multivariable regression analysis

Model		Unstandardized coefficients		Standardized coefficients	*t*	Sig.	Collinearity statistics	
		B	Std. Error	Beta			Tolerance	VIF
1	(Constant)	0.088	0.097		0.906	0.00		
	REL	0.002	0.454	0.035	0.345	0.00	0.231	1.798
	TAN	0.010	0.053	0.205	0.184	0.00	0.171	1.836
	ASS	0.244	0.063	0.032	3.863	0.00	0.112	1.968
	RES	0.334	0.062	0.331	5.389	0.00	0.109	1.147
	EMP	0.004	0.060	0.004	0.064	0.00	0.116	1.584
	LOC	0.045	0.044	0.056	1.026	0.00	0.202	1.957
	PRI	0.063	0.056	0.093	1.123	0.00	0.129	1.722
	PAN	0.242	0.050	0.244	4.855	0.00	0.164	1.083

[a]Dependent variable SAT
Source Survey data processing results using SPSS, 2022

Check the fit of the model

The F-test Sig. value is equal to 0.000 < 0.05, that is, rejecting the hypothesis that all regression coefficients are zero (Huong, 2018). Therefore, there is a relationship between the independent variable and the dependent variable, so the regression model is suitable for the collected survey data system. In addition, the commonly used linear regression model fit measures are the coefficients of determination R^2 (R square) and R^2 adjusted (adjusted R^2). The analysis results show that $R^2 = 0.837$; adjusted $R^2 = 0.778$ is closer to 1, so the independent variables explain more for the dependent variable. The adjusted R-squared value of 0.778 shows that the independent variables included in the regression analysis affect 77.8% of the variation of the dependent variable; the remaining 22.2% are due to out-of-model variables and errors, random number.

About the independence of errors: The results of this table also give Durbin–Watson (DW) values to test the correlation of adjacent errors, i.e., the phenomenon of first-order series autocorrelation. The value of DW = 2.151 should be in the range of 1.5–2.5, so the results do not violate the assumption of first-order series autocorrelation. The assumptions of the regression model are tested; the results are not violated, and there is no multicollinearity because the variance exaggeration factor VIF is less than 2.

Hypothesis testing

Adjusted $R^2 = 0.778$ reflects the influence of the independent variables on the dependent variable, i.e., 09 factors affect 77.8% to the quality of accommodation services; the remaining 22.2% are due to external variables, model and random error (Table 8).

According to the findings of multivariable regression analysis, all nine factors included in the analytical model have a positive impact on the quality of accommodation services provided by Muong Thanh hotels in Ha Long, making the hypothesis proposed in the proposed research model appropriate. That is, the independent variables were positively correlated with the independent variable, indicating that the factors affecting the quality of accommodation services were affecting it in a positive direction but to varying degrees. Because the study considers the observed variables from the perspective of the response to the hotel pandemic, the positive effect persists with the total disease variable COVID-19.

Determining the influence of factors on the quality of accommodation services in the Muong Thanh hotel system

With the research content on the influence of factors on the quality of accommodation services, the topic chooses a standardized regression equation to know which variable X has a strong or weak influence on the Y variable based on the regression coefficient, standardization. Then, the larger the absolute value of the normalized regression coefficient, the greater the importance of the variable for Y, the stronger the effect of that variable. The regression model to evaluate the influence of factors on the quality of accommodation services of the Muong Thanh hotel system in Ha Long destination is as follows:

Table 8 Research hypothesis test results

TT	Hypothesis	Beta	Sig.	Inspection results
1	Reliability is correlated with accommodation service quality	0.035	0.000	Accept
2	Tangibles have a correlation with the quality of accommodation services	0.205	0.000	Accept
3	Assurance is correlated with accommodation service quality	0.032	0.000	Accept
4	Responsiveness is correlated with accommodation service quality	0.331	0.000	Accept
5	Empathy is correlated with accommodation service quality	0.004	0.000	Accept
6	The location of the hotel has a correlation with the quality of accommodation services	0.056	0.000	Accept
7	Price is correlated with the quality of accommodation service	0.093	0.000	Accept
8	Responding to COVID-19 is correlated with accommodation service quality	0.244	0.000	Accept

Source Survey data processing results using SPSS, 2022

$$\begin{aligned}\text{SAT} = {} & 0.035{*}\text{REL} + 0.205{*}\text{TAN} + 0.032{*}\text{ASS} + 0.331{*}\text{RES} + 0.004{*}\text{EMP} \\ & + 0.056{*}\text{LOC} + 0.093{*}\text{PRI} + 0.244{*}\text{PAN}\end{aligned}$$

As a result, the impact of each factor on the quality of accommodation services of this snap guest system varies: *the Satisfactory factor* has the highest impact coefficient in the analytical model of 0.331, indicating that it is factors that have the greatest influence on the quality of accommodation services at the Muong Thanh hotels studied. *The COVID-19 response factor* has the second highest beta at 0.244. This factor has a significant impact on the quality of accommodation services from the beginning of 2020 to the present, with the requirements of safety assurance, suitability with customer psychology during the pandemic of research hotels. *Empathy* (0.04) has the lowest standardized regression coefficient in the analytical model, indicating that it has the least influence on the quality of accommodation services for the system Muong Thanh hotels in Ha Long. Although all of these hotels are in the luxury segment (4 stars, 5 stars), the customer segment is primarily group and tour guests, and the number is large, making it difficult to approach and care for guests' "*personalization*". Furthermore, with a beta coefficient of 0.032, the *guaranteed factor* has a low impact factor. As a result, this factor has little influence on the quality of accommodation services provided by the hotels under consideration.

5 Conclusion

Based on a combination of qualitative and quantitative research methods, the article has solved a number of issues: Overview, systematization of the theoretical basis, combined with consultation with selected experts. The research model includes 09 scales and 59 independent observed variables, 04 dependent variables to assess the influence of factors on the quality of accommodation services of Muong Thanh hotels in Ha Long. On that basis, the research results have quantitatively evaluated the influence of factors on the quality of accommodation services of these hotels through primary data from guest reviews. The results of guests' evaluations of factors influencing the quality of accommodation services were gathered via a survey with a sample size of 350 votes and analyzed using SPSS 23.0 software. To ensure accuracy, the data processing process is carried out step by step, beginning with descriptive statistics analysis of the research sample and evaluating the mean value of the scales. The calculation results show that the scales' average value is high. Following that, the results of Cronbach's alpha coefficient testing, exploratory factor analysis (EFA removed 9 independent observed variables that were not suitable such as: REL4, RES1, ASS3, RES 4, BRA, BRA2, BRA3, BRA4, and PAN5). The official research model consists of eight scales, 44 independent observed variables, and four dependent observed variables. Strong correlation following the correlation test the regression analysis yielded a regression equation with a standardized beta coefficient: $SAT = 0.035*REL + 0.205*TAN + 0.032*ASS + 0.331*RES + 0.004*EMP + 0.056*LOC + 0.093*PRI + 0.244*PAN$.

This finding indicates that the factors have varying degrees of influence on the quality of accommodation services provided by the Muong Thanh hotel system in Ha Long. In which case, responsiveness has the most influence, and empathy has the least influence on the quality of these hotels' accommodation services. However, due to time, capacity, and objective conditions of the epidemic, the new study focuses on quantitative analysis of tourist evaluations without contacting tourists, from an employee perspective, within the framework of the thesis. In particular, the study focused solely on domestic tourists, with no consideration given to international visitors.

References

Bolton, R. N., & Drew, J. H. (1991). A multistage model of customers' assessments of service quality and value. *Journal of Consumer Research, 17*, 375–384.

Brown, J. R., & Ragsdale, T. (2002). The competitive market efficiency of hotel brands: An application of data envelopment analysis. *Journal of Hospitality and Tourism Research, 26*, 332–360.

Chen, W. J., & Chen, M. L. (2014). Factors affecting the hotel's service quality: Relationship marketing and corporate image. *Journal of Hospitality Marketing & Management, 23*, 77–96.

Huong, T. Q. N. (2018). *Improving competitiveness of tourist destination Ha Long, Quang Ninh.* Doctoral thesis, University of Commerce (in Vietnamess).

Kheng, L. L., Mahamad, O., & Ramayah, T. (2010). The Impact of service quality on customer loyalty: A study of banks in Penang, Malaysia. *International Journal of Marketing Studies, 2*, 57–66.

Khuong, X. D. (2019). *Service quality of specialized supermarkets in Hanoi city*. Doctoral thesis, National Economics University, Hanoi (in Vietnamess).

Kotler, K. (2000). *Marketing management* (10th edn).

Kotler, K. (2010). *Principles of marketing: A South Asian perspective*. Pearson Education India.

Leonard, L. B., Parasuraman, A., & Zeithaml, V. A. (1988). SERVQUAL: A multiple—item scale for measuring consumer perceptions of service quality. *Journal of Retailing, 6*, 12–37.

Mackay, K. J., & Crompton, J. L. (1988). A conceptual model of consumer evaluation of recreation service quality. *Leisure Studies, 7*, 40–49.

Manh, V. N., & Huong, T. L. H. (2013). *Hotel business administration, national economics*. University Publishing House.

Office of Muong Thanh Hotel Group. (2018). *About us: Muong Thanh hotel group*. http://muongthanh.com/ve-chung-toi.html, December 18, 2021 (in Vietnamess).

Thang, Q. N., (2013), *Hotel business administration (From theory to practice)*. Hanoi Financial Publishing House (in Vietnamess).

Tho, D. N. (2011). *Scientific research methods in business*. Labor Publishing House (in Vietnamess).

Trong, H., & Ngoc, M. C. N. (2008). *Analysis of research data SPSS*. Statistical Publishing House, Volume 1 (in Vietnamess).

Wong, K. K. F., & Kwan, C. (2001). An analysis of the competitive strategies of hotels and travel agents in Hong Kong and Singapore. *International Journal of Contemporary Hospitality Management, 13*, 293–303.

Festival Tourism and Socioeconomic Development Case of Culture Traditional Festival of Dong Hoi City—Vietnam

Do Thi Thao and Nguyen Hoang Dong

Abstract This paper analyzes the influence of tourism festivals on socioeconomic development. At the same time, the assessment is like a tool for the socioeconomic development of the people of Dong Hoi city. This study is guided by social development theory; using a regression model, the data is surveyed through 253 questionnaires of people living in Dong Hoi city, tourists, and some opinions of experts. The results show that festival tourism has a positive impact on community cohesion, social benefits, and economic benefits. Moreover, festivals tourism brings negative impacts on the socioeconomic development of this region. The study also offers some solutions to minimize the negative impacts of tourism festivals on tourism local economy, culture, and society. Application of the findings to other festivals will require a larger sample size for generalization to be made.

Keywords Festival tourism · Socioeconomic development · Economic benefits · Social benefits · Vietnam

1 Introduction

Festival tourism will study tourists participating in recreational and festive activities during festivals by Kruger and Viljoen (2021). This type of activity is usually held only once or several times a year and lasts for a limited time, providing attendees with unusual networking opportunities (O'Sullivan & Jackson, 2002a, 2002b). Festival tourism activities are built based on local cultural factors, unique experience atmosphere, and entertainment space to attract tourists by Lin (2020). Nagy (2013) in a study on the importance of festival tourism in the economics of Hungary shows analysis of the trends of Hungarian tourism, focusing on festivals, and monitor changes in tourism strategy, as well as the direct and indirect effects of tourism development on the country's economic growth. At the same time, the research has shown the

D. T. Thao (✉) · N. H. Dong
School of Hospitality and Tourism, Hue University, Thua Thien Hue, Vietnam
e-mail: dtthao@hueuni.edu.vn; dothao19091987@gmail.com

A. T. Nguyen et al. (eds.), *Contemporary Economic Issues in Asian Countries: Proceeding of CEIAC 2022, Volume 1*,
https://doi.org/10.1007/978-981-19-9669-6_38

consequences for the field and identified recommendations for further developments because they believe that festivals can be the solution for sustainable economic and social development, associations of some rural areas, bringing income to a very large part of the rural population. Currently, festival tourism is a growing tourism model in many different countries around the world. The meaning of the festival is assessed in terms o its unique value, the value of national identity that it brings (Getz et al., 2004). A study by Picard and Robinson (2006) indicates that carnival performances provide an appreciation for the historical and landscape design of a place.

In addition, there is a study by Kozlovska (2021) on "Holiday performances in the sociocultural practices of the present festival tourism." Research suggests that festivals are the main form of presentation of cultural and recreational programs of modern tourism in festival tourism. The primary purpose of the decision was to create a festive spectacle, taking into account the historically defined worldview; variety of organizational and artistic media; change; uniqueness; openness and dialogue; continuity and adaptability; situational and improvisation; interactivity; humor.

Sustainable outcomes from small-scale events are primarily contributed through the use of existing facilities, infrastructure, and appropriate representation of local communities in terms of participation and distribution, benefit. Festival tourists show a better sense of spending on local products, which is encouraged by better knowledge of the local culture inculcated through participation in cultural festivals localization (Nurse, 2001).

A study of the impact of small-scale sporting events found that such events stimulate community pride and are recognized for their authenticity and proximity to local culture; in contrast, major sporting events are highly commercialized and globalized (Malchrowicz-Mośko & Poczta, 2018).

According to the research paper about festival tourism and economic development: a case of Kwahu traditional areas of Ghana, Doe et al. (2020) have shown the impact of festival tourism on the socioeconomic of the people in the capital area. Traditional areas of Kwahu (Ghana) and how festival tourism in the region benefits businesses and communities economically.

The findings in the study of Raj (2003) show us that festivals have contributed to the development of cultural tourism. Festivals draw cultural tourists to local community events to promote rich exchanges between tourists and residents. In the Edinburgh Festival and Leeds West Festival case studies, the Indian Carnival that the carnival has become a major tourist attraction for local, regional, and international visitors.

With theory and research related to festival tourism and socioeconomic development at home and abroad, it is more and more confirmed the close relationship and interaction between these two topics. The author has developed hypotheses and conducted empirical investigations to answer hypotheses about the impact of festival tourism and socioeconomic development in Dong Hoi—Vietnam.

2 Method

2.1 Study Area

Dong Hoi is a city directly under Quang Binh province in the North Central region of Vietnam. This city is located between National Highway 1A and Ho Chi Minh Road and next to the Hanoi–Ho Chi Minh City railway line with Dong Hoi station as one of the main stations, with the Nhat Le River flowing through. A city with long beaches, and white sand, this is the ideal location for festival tourism and marine tourism. Dong Hoi has many typical traditional cultural festivals associated with Quang Binh tourism development in general as well as Dong Hoi tourism in particular, such as the traditional boat racing festival on Nhat Le River, Bao Ninh fishing festival, and the festival. Harvest festival, lantern festival, village festival… These festivals are based on the principle of preserving traditional values, gradually eliminating backward elements, and exploiting the festival's potential to develop tourism and protect, and maintain the true value of the festival, do not create according to feelings and interfere with the ceremony, which is considered the core and soul of the festival.

In recent years, Dong Hoi has well-organized a cultural and tourism festival, which is a new type of festival, combining festivals from ancient to modern, creating attraction, creating a brand, and awakening tourism potential sea calendar, so that each festival season is an ideal time for domestic and foreign tourists to come to Dong Hoi, stay and develop. According to research by Mair and Whitford (2013), festival tourism was called "an emerging giant" more than 10 years ago and highlights the global attention being paid to the venue of the festivals in the national economy. Festival tourism is considered a new type of tourism, promising and affecting the economy and society.

In addition, there is a study by Li and Lau (2022). This study aims to investigate two festivals held annually in Hong Kong to explore the meaningful nature of the festival. The article assesses the extent to which sociocultural benefits bring to the community. In addition, the article also discovers festivals that clearly show the characteristics of the locality and destination.

Research by Dychkovskyy and Ivanov (2020), Festivals have become a very important part of the tourism industry. Cultural heritage, sociology, anthropology, and historical events influence the festival in socioeconomic atmospheres around the world. The potential of the traditional cultural festivals of Dong Hoi is very large; it is one of the factors stimulating the socioeconomic development of the locality, bringing a large source of revenue to the city budget from tourism activities, bringing revenue for service activities associated with tourism activities such as catering services, hotel services, passenger transport, and souvenir services. If festival tourism is well exploited, awakening festival activities will be a great source of income, and economic sectors will participate in the service business and service activities. At the same time, the festival also creates jobs for residents and the community through festival services. Traditional cultural festivals not only carry the deep national cultural identity of each village but also preserve many traditional contents with moral and esthetic values, customs and habits, the good and the bad, beauty, its own, and its identity, thereby expressing the economic and cultural potential of the locality. There are a number of studies that point to the success and growth of modern festival tourism as a manifestation of the Leisure Age, along with the effects of increasing globalization on the word tourism industry. It also affirms the link between tourism and other industries. At the same time, festival tourism is considered indispensable in local life and economic development by Bernick et al. (2013).

At the same time, when festival tourism is created, there is a potential to reposition the city as a tourist destination (Felsenstein & Fleischer, 2003). These festivals leave an impression and are effective not only in preserving the culture and identity of the countryside, and developing sustainable tourism but also bringing in income for local people, contributing to the introduction of cultural values, cultural values of the local people. According to research by Whitford and Dunn (2014), it is again confirmed that local festivals and events are used to attract tourists who, in turn, contribute to the development of the community. Currently, there are very few domestic and foreign studies on the relationship between festival tourism and socioeconomic development. Therefore, the objectives of this study will be to assess the impact of festival tourism on the sociocultural and economic well-being of the Dong Hoi people and to identify the negative impacts of festival tourism (if any) for the infrastructure and landscape society of Dong Hoi—Vietnam. The author hopes that the findings in this study will provide the necessary impetus for shaping policies and strategies for the development of festival tourism in Quang Binh and other localities.

2.2 Data Collection

Data is encrypted and processed for analysis. With quantitative data, the author uses statistical analysis software SPSS 26.0 at 95% confidence to perform the following techniques in turn: descriptive statistical analysis, testing the reliability of the scale, exploratory factor analysis, one-way ANOVA clarifies the difference in evaluation between different groups of subjects according to independent variables and regression analysis to determine the degree of effects of the independent variables. Opinions collected through in-depth interviews are evaluated using a content analysis method—that is, interviews are coded and grouped into topics or keywords; then, appropriate citations are used to clarify the analytical contents of the study.

The study used the method of document synthesis, data collection method, sampling survey method, data processing and analysis method, in which, combining between qualitative research and quantitative research, taking quantitative research as the main. First, through an overview of studies and field surveys in Dong Hoi City, cultural festivals have had many impacts on the region's socioeconomic development. On the basis of synthesis and inheritance from related studies to build 4 hypotheses see Fig. 1

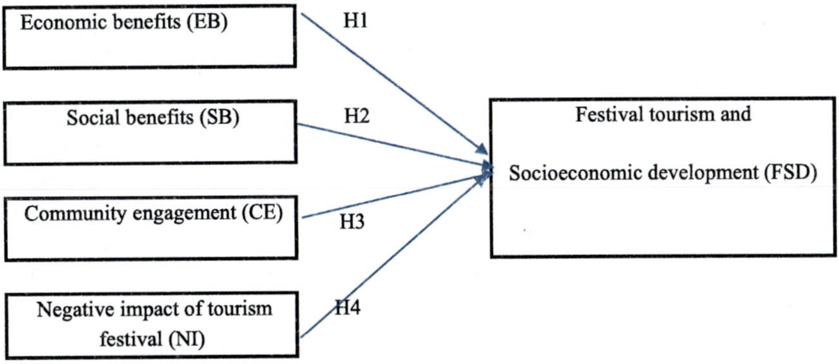

Fig. 1 Research model

H1: Economic benefits from festival tourism increase socioeconomic development.

H2: Social benefits from festival tourism increase socioeconomic development.

H3: Community cohesion from festival tourism increases socioeconomic development.

H4: Negative impacts from festival tourism reduce socioeconomic development.

Festival tourism in the world in general and in Vietnam, in particular, is likened to a still unspoiled gold mine. In Vietnam, each year there are more than 8000 festivals, but they have not been effectively exploited and attracted. Research by Courtney (1995) see the importance of tourism in stimulating economic growth, and social stability, and preserving and promoting national cultural identity in Hoi An. Tourism can create mutually beneficial conditions for tourists and local communities. Tourists benefit from their vacation, but they need travel, food, and accommodation, which will generate business profits for the local community. The presence of tourists in underdeveloped areas is seen as a driver of investment in the service and infrastructure sectors: electricity and water, healthcare facilities, security, fire prevention, etc. can therefore also be significantly improved. The income from tourism services can also be used to reinvest, supporting service costs for the strong tourist community. According to a study by Gursoya et al. (2004), the development of festival tourism provides opportunities for the innovative and creative creation of indigenous products and services that translate into financial returns for existing businesses and startups, which generates a significant source of income for local business development (Derrett, 2005) as well as revenue generation for both local and state governments, especially through prolonging the duration of tourism (O'Sullivan & Jackson, 2002a, 2002b).

Another study by Porfido (2022) examined the contribution of festival tourism to sustainable local economic development, using three types of festivals. Here, three festivals that fit their type are analyzed based on a model of sustainable local economic development. The results show that although festivals may have the potential to provide opportunities for sustainable local economic development, these opportunities are not yet fully developed. At the same time, the author has proposed a number of development policies for the type of festival tourism.

Another study by Chebotibin et al., (2018a, 2018b) also found from their study conducted in Kenya that there is a symbiotic relationship between cultural tourism and economic development. At the same time, festival tourism brings various economic benefits such as creating jobs for people and improving the living conditions of local people Agbabiaka et al. (2017). Currently, festival tourism in many countries around the world has made a new step and brought significant economic benefits.

3 Results of Research

The level of participants' opinions on socioeconomic development as measured by 22 items is presented in Table 1. To collect data for this study, the researchers approved and revised the data. One scale was developed by Gursoya et al. (2004).

Based on the results of Table 2, we can see the description of the investigated object by gender; it can be seen that the ratio of men and women is almost equal. Of the 253 surveyed subjects, 121 male subjects (accounting for 47.82%) and 132 female subjects (52.18%). By age, through the survey results in the table above, the age group accounted for the largest proportion when there were 69 respondents (27.27%). In addition, groups are followed by groups of 35 to 54 years old. This shows that the interests and interest in festival tourism are interested and noticed by all ages.

For education, the university rate accounts for the highest rate of 44.27%. This shows that the intellectual level here is quite high. In addition, the main occupation status is workers and employees. In addition, the married and single status has a higher rate of being married than being single.

Before entering the steps of data analysis, the study conducted a step to test the reliability of the scale through Cronbach's alpha coefficient. Cronbach's alpha must be performed first to remove extraneous variables (Garbage Items) before exploratory factor analysis EFA. The research topic uses a scale consisting of 4 independent variables including: economic benefits (EB); social benefits (SB); community engagement (CE); negative impact of tourism festivals (NI); and dependent variable: Festival tourism and socioeconomic development (FSD) which are the results of Cronbach's alpha test are summarized in the table below.

The results of Table 3 with summary results of factors: factor loading, Cronbach's alpha, Eigenvalues, and variance are explained for all index variables. With the above results, we see that both independent and dependent variables have Cronbach's alpha values greater than 0.6, indicating that there is reasonableness and consistency in the items of the structure.

From the results of multivariate regression analysis, the Durbin–Watson value is 1.908. According to the regression condition, the value of Durbin–Watson is between

Table 1 Coding scale of influencing factors and measurement variables

Design the questionnaire	
I. Economic benefits (EB) Alexander Chidakel et al. (2020), Chebotibin et al., (2018a, 2018b), Figini and Patuelli (2022)	
1-EB1	Increase the number of jobs for people
2-EB	Increase the quantity and quality of infrastructure
3-EB	Enhancing the living standard of local people increase local sales
4- EB	Encourage local development of new services
5-EB	Increase local sales
II. Social benefits (SB) Okyere-Manu and Antwi (2016), Kugbonu et al. (2018), Ciascai et al (2022)	
6-SB	Increase the connection between residents and visitors provide more entertainment opportunities
7-SB	Promote and create opportunities for cohesion between organizations and businesses
8-SN	Offer family-based recreational activities
9-SB	Enhance community image with outsiders
10-SB	Education and engagement help people understand more about culture
III. Community engagement (CE) Gursoya et al. (2004), Humphrey (2001), Kugbonu et al. (2018)	
11-CE	Community building
12-CE	Enhance cohesion in the community
13-CE	Increase local influence in the community
14-CE	Help local develop and preserve culture
IV. Negative impact of tourism festival (NI) Boley et al. (2014), Boo et al. (2005)	
15-NI	Increasing crime rate
16-NI	Increasing traffic congestion
17-NI	Increasing social evils (addiction, gamble, etc.)
18-NI	Noise pollution
V. Festival tourism and Socioeconomic development (FSD) Mair and Whitford (2013), Nunkoo and Gursoy (2012)	
19-FSD	I believe the number of tourists coming to the festival is increasing
20-FSD	I believe that festival tourism activities positively affect quality and employment
21-FSD	I support the celebration of this festival in my community
22-FSD	I think this festival is the best strategy for socioeconomic development for Dong Hoi people

Table 2 Demographic characteristics of survey subject

Demographic characteristics/classification	No.	%
Sex total:	*253*	
Male	121	47.82
Female	132	52.18
Age		
20 and under	23	9.09
20–30	50	19.76
31–55	94	37.16
>55	86	33.99
Education level		
Uneducated	22	8.70
High school and under	26	10.28
Vocational school	35	13.83
Vocational colleges	38	15.02
Bachelor's degree	112	44.27
Master's degree and above	20	7.90
Employment status		
Workers and employees	78	30.83
Officer	27	10.67
Business	15	5.92
Worker	38	15.02
Students	27	10.67
Retire	51	20.16
Housewife	17	6.72
Marital status		
Single	117	46.25
Married	136	53.75

1.6 and 2.6; the variables will not have the phenomenon of autocorrelation with each other. Thus, the model does not violate the assumption of autocorrelation. Check for multicollinearity: With large acceptance (Tolerance) and variance exaggeration (VIF) of very small variables are all below 2, so the regression model does not violate the phenomenon of multicollinearity. The regression model violates multicollinearity if and only if the variables have a variance exaggeration factor (VIF) greater than or equal to 10. Based on the analysis of multivariate regression results, we see the Sig value of all independent variables is less than 0.05. Therefore, all the above factors are significant in the model and affect socioeconomic development.

Thus, from the results of the regression analysis of Table 4, we have the following equation:

Table 3 EFA and Cronbach's alpha results

Category	Independent variable								Dependent variable	
	EB		SB		CE		NI		FSD	
	Items	Factor loading	Items	Factor loading	Items	Factor loading	Items	Factor loading	Items	Factor loading
	1-EB	0.840	6-SB	0.730	11-CE	0.638	15-NI	0.610	19-FSD	0.750
	2-EB	0.789	7-SB	0.682	12-CE	0.669	16-NI	0.787	20-FSD	0.659
	3-EB	0.774	8-SB	0.871	13-CE	0.740	17-NI	0.672	21-FSD	0.674
	4-EB	0.751	9-SB	0.654	14-CE	0.697	18-NI	0.742	22-FSD	0.841
	5-EB	0.673	10-SB	0.742						
Eigen Value		6.142		3.171		2.139		2.410		7.201
Total variance explained (%)		52.121		6.182		11.291		5.192		76.192
Cum. variance explained (%)		51.121		62.120		51.212		46.192		76.192
Cronbach's alpha		0.811		0.743		0.816		0.701		0.893

Table 4 Results of regression analysis

Relationship between variables	B	β	t-Value	P-Value	Tolerance	VIF
EB → FSD	0.951	0.931***	6.624	0.000	0.956	1.046
SB → FSD	0.769	0.664***	3.707	0.000	0.929	1.077
CE → FSD	0.871	0.746***	4.732	0.040	0.938	1.066
NI → FSD	−0.5120	−0.399***	3.727	0.020	0.923	1.083
R	$\boldsymbol{R^2}$	**Adjusted $\boldsymbol{R^2}$**	**Std. Error**	**Durbin–Watson**		
0.789[a]	0.723	0.606	0.330	1.908		

Notes "***" indicates-$p < 0.001$

$$\text{ESD} = 0.931 \times \text{EB} + 0.664 \times \text{SB} + 0.749 \times \text{CE} - 0.399 \times \text{NI}$$

Through the normalized regression coefficients of the model, we can know the importance of the variables participating in the regression equation. Specifically, the factor EB ($\beta = 0.931$) has the greatest impact on socioeconomic development. Next, in descending order from important to least important, including factors CE ($\beta = 0.749$), SB ($\beta = 0.664$), and NI ($\beta = -0.399$) have a negative impact on socioeconomic development in Dong Hoi city.

In general, all four factors above have certain effects on the socioeconomic development of Dong Hoi city. Through the research results, we can see that any change of one of the four factors above can make a change to the socioeconomic development situation of Dong Hoi city.

The results show that there is a significant difference between the groups of interviewees by education level, age, and employment status related to their perception of the influence of festival tourism on development, economy—society of Dong Hoi city. This is clearly shown in Table 5. Through the collected results, we can see that people with education levels feel and evaluate differently about the level of influence. Through a number of tests, we find that highly educated people perceive social and economic impacts more positively and assess negative impacts more clearly and rigorously than do, people with low education. In addition, in terms of age, the impact assessment is also different, which shows that people in the age group 31–55 years old evaluate the impact higher than other groups, this is also the most vulnerable group by local tourism festivals (p.value = 0.006 (SB); p.value = 0.003 (CE); p.value = 0.002 (NI). Therefore, this is a potential group of people, contributing to the socioeconomic development of the locality; this is also the main employee of the family and society. This resource group seems to have quite positive comments about the impact of festival tourism on socioeconomic development and has a liberal view on the negative impacts of festival tourism on development. socioeconomic development.

Table 5 ANOVA test results

Independents	Factors affecting Socioeconomic development			
	EB	SB	CE	NI
Age	–	0.006*	0.003**	0.002***
Education	0.002	0.001***	0.007*	0.002
Employment status	–	–	0.004***	–

Notes "***" indicates-$p < 0.001$

In addition, occupational status affects people's assessment of community cohesion; the group of workers has low evaluation of community cohesion compared to other groups of subjects (p.value = 0.004).

4 Discussions and Conclusions

The objective of the study is to examine the impact of festival tourism on the economy and sociocultural well-being of the residents of Dong Hoi city. With the results of the questionnaire survey, through data collection and processing, the research has shown that economic benefits, social benefits, and community cohesion as well as its positive impacts bring about affecting the socioeconomic development of Dong Hoi city. This has also been confirmed in some previous studies. According to research by McGehee (2004) has shown that tourism activities create opportunities for businesses and increase economic benefits, job opportunities research by Diedrich et al (2009). Therefore, this finding is a confirmation of previous research. With the collected results, this is also a strategy for leaders and local authorities to realize the economic benefits to have appropriate directions.

The festival always plays the role of a link connecting the community, creating a cultural space that is both solemn, sacred, jubilant, and exciting, where the public comes to the history of ancestors and returns to the roots of the people, people, remembering the merits of the ancestors, wishing for good things. One of the important contributions is that community cohesion in festival tourism is also a factor bringing about the socioeconomic development of the locality. Each festival that was born and existed is associated with a certain land. So, the festival in any region has the nuances of that region. The locality of the festival is what proves that the festival is closely linked with the people's life; it meets the spiritual and cultural needs of the people, not only in the content of the festival but also in the festival, festival style too. For this study, the author found the impact of community cohesion on the socioeconomic development of Dong Hoi city. These results are similar to those of *Margaret* Deery et al (2012) who studied the impact of tourism on Ghanaian communities. Similar findings were reported in the study by Gursory et al (2004). Community cohesion is a concept with content quite similar to the concepts of each individual's perception of the value of places and places and social relationships there, such as:

sense of community, sense of attachment, place-feelings, and attachment to where you live neighborhood attachment. With cultural festivals, the annual cultural festival in Dong Hoi city promotes unity and pride among the people. A study by Andereck et al. (2005) showed that pride is one of the immediate feelings experienced as a result of people choosing their destination not only to spend money but also to socialize and explore the culture. The previous finding was corroborated in this study. There are also studies by Haley et al. (2005) and Woosnam et al. (2009) that recognized that tourism enhances social life.

In addition to the positive effects, the study also found many remaining problems such as overuse of public infrastructure that can wear down facilities. At the same time, it is the increase in prices of goods and services that reduce the quality of life of local people, overcrowding, and improper tourist behavior. According to the study, the impacts of festival tourism also bring opposite impacts on the development, socioeconomic of the locality. For example, Kugbonu et al. (2018) lamented the degradation and pollution of the environment due to the impacts of tourism. In addition, research by Fredline et al. (2006) indicates the impact of tourism on traffic congestion and antisocial behaviors Diedrich and Garcia-Buades (2009).

Dong Hoi's traditional cultural festival is a valuable asset, a unique human resource, a "living museum" in the heart of the community, and a potential to be awakened for tourism development. Currently, Dong Hoi is inheriting a number of unique and special traditional cultural festivals but lacks a professional promotion strategy. Recently, Dong Hoi has had creative activities in organizing festivals for tourism. However, in the short term and in the long term, it is necessary to have a reasonable strategy in investing in festivals. Selectively invest in a number of unique and impressive festivals, thereby building it into the main product and offering an appropriate promotion strategy. Maintain, nurture, and promote the identity of coastal festivals such as the Boat Racing Festival on Nhat Le River, combining folk games in the festival to attract tourists.

Tourism associated with festivals is an effective way to exploit the inherent potential advantages of the locality. However, how and to what extent need to be considered and properly oriented. Attention should be paid to the subjects and people organizing the festival. The traditional cultural festival itself takes place in the correct order and ritual that has been preserved and maintained for generations without the intervention or impact of the event organization companies, but it still exists in history and satisfy the needs of the local community as well as the discovery of visitors. These festivals, mainly focusing on the ritual part, account for a large proportion and often take place in communes and wards. Some festivals such as the fishing festival, the harvest festival, the village festival, the Nghinh Ong festival, etc., need to improve the cultural values inherent in each ritual, find and choose an appropriate organization method, design. The lively model of the festival's activities creates an attractive factor for visitors, creating a spillover effect beyond the community.

Awakening the potential of traditional festivals is to fully exploit the specific cultural strengths of the festival to develop tourism, gradually eliminating outdated elements, and building new and appropriate cultural criteria. Categorize festivals, strengthen management, research, organize seminars make a list of festival registrations, select typical festivals, and create scientific ranking records, so that the exploitation of the festival's potential is the more scientific, the more meaningful. Appreciating the specificity of each type of festival, exploiting the potential of traditional cultural festivals is also one of the measures to preserve cultural identity and develop sustainable tourism, creating favorable conditions for economic development. Dong Hoi city's tourism economy, affirming a city with a lot of potential for festival tourism, has been awakened along with other tourism products to attract tourists to the city. Arousing the potential of festival tourism in order to seize opportunities for socio-economic development, hunger eradication and poverty alleviation, incresing global integration. This present many opportunities but also challenges with negative effects it brings. This puts the people of Dong Hoi to constantly preserve, develop and bring cultural values to tourists.

References

Agbabiaka, H. I., et al. (2017). Factors influencing patronage of traditional festivals in Sub-Sahara Africa. *International Journal of Tourism Cities, 3*, 363–381.

Andereck, K. L., Valentine, K. M., Knopf, R. C., & Vogt, C. A. (2005). Residents perception of community tourism impacts. *Annals of Tourism Research, 32*, 1056–1076.

Bernick, L. N. et al. (2013). Festival tourism and the entertainment age: Interdisciplinary thought on an international travel phenomenon. *International Journal of Culture Tourism and Hospitality Research, 7*, 13–23.

Boley, B. B., et al. (2014). Empowerment and resident attitudes toward tourism: Strengthening the theoretical foundation through a Weberian lens. *Annals of Tourism Research, 49*, 33–50.

Boo, S., et al. (2005). Impact analysis of a tourism festival on tourists destination images. *Event Management, 9*, 223–237.

Chebotibin, D., et al. (2018a). Influence of cultural festivals and custom artifacts on tourism development in Baringo County, Kenya. *Journal of Hospitality and Tourism Management, 1*(1), 1–20.

Chebotibin, D., et al. (2018b). Festival tourism and socioeconomic development: Case of Kwahu traditional areas of Ghana. *International Hospitality Review, 36*, 174–192.

Chidakel, A., et al. (2020). The comparative financial and economic performance of protected areas in the Greater Kruger National Park, South Africa: Functional diversity and resilience in the socio-economics of a landscape-scale reserve network. *Journal of Sustainable Tourism, 28*, 1100–1119.

Ciascai, O. R., et al. (2022). Cycling tourism: A literature review to assess implications, multiple impacts, vulnerabilities, and future perspectives. *Journal Sustainability, 14*, 2–18.

Courtney, J. (1995). Book reviews: learning to live with tourism by Stephen Craig-smith and Christine French. *Journal of Vacation Marketing, 1*, 401–403.

Deery, M., et al. (2012). Rethinking social impacts of tourism research: A new research agenda. *Tourism Management, 33*(1), 64–73.

Derrett, R. (2005). Why do regional community cultural festivals survive?. In *The impacts of events: triple bottom line evaluation and event legacies: Third International Event Management Conference.*

Diedrich, A., & Garcia-Buades, E. (2009). Local perceptions of tourism as indicators of destination decline. *Tourism Management, 30*(4), 512–521.

Diedrich, A., et al. (2009). Local perceptions of tourism as indicators of destination decline. *Tourism Management, 30*(4), 512–521.

Doe, F., et al. (2020). The future of youth tourism in Ghana: Motives, satisfaction and behavioural intentions. *Journal of Tourism Futures, 10*, 158–171.

Dychkovskyy, S., & Ivanov, S. (2020). Festival tourism as part of international tourism and a factor in the development of cultural tourism. *Vilnius University, 89*, 327–345.

Felsenstein, D., & Fleischer, A. (2003). Local festivals and tourism promotion: The role of public assistance and visitor expenditure. *Journal of Travel research, 6*, 253–269.

Figini, P., & Patuelli, R. (2022). Estimating the economic impact of tourism in the European Union: Review and computation. *Journal of Travel Research, 61*(6), 1409–1423.

Fredline, L., Deery, M., & Jago, L. (2006). *Development of a scale to assess the social impacts of tourism within communities*. Queensland, Australia: CRC for Sustainable Tourism Pty Limited.

Gursoya, D., et al. (2004). Perceived impacts of festivals and special events by organizers: An extension and validation. *Tourism Management, 47*, 171–181.

Haley, A. J., Snaith, T., & Miller, G. (2005). The social impacts of tourism a case study of Bath, UK. *Annals of tourism research, 32*(3), 647–668.

Humphrey, C. (2001). *The politics of carnival: festive Misrule in Medieval England*. Manchester University Press.

Kozlovska, M. V. (2021). Holy sights in social and cultural practices off the present: festival tourism. *Bulletin of the National Academy of Core Personnel of Culture and Science*, (4), 56–60.

Kruger, M., & Viljoen, A. (2021). Terroir wine festival visitors: uncorking the origin of behavioural intentions. *Current Issues in Tourism, 24*, 616–636.

Kugbonu, M. A., et al. (2018). Event waste management: A study of a religious event in Ho. *Journal of Tourism, Hospitality and Sports, 33*, 40–46.

Li, Y., & Lau, C. (2022). Analyzing the effects of an urban food festival: A place theory approach. Annals of Tourism Research, 74, 43–55.

Lin, C. L. (2020). Establishing environment sustentation strategies for urban and rural/town tourism based on a hybrid MCDM approach. *Current Issues in Tourism, 23*(19), 2360–2395.

Mair, J., & Whitford, M. (2013). An exploration of events research: event topics, themes and emerging trends. *International Journal of Event and Festival Management, 4*, 6–30.

Malchrowicz-Mośko, E., & Poczta, J. (2018). A small-scale event and a big impact—Is this relationship possible in the world of sport? The meaning of heritage sporting events for sustainable development of tourism—Experiences from Poland. *Sustainability, 10*(11), 4289.

Milman, A. P. (1988). Social impacts of tourism on Central Florida. *Annals of Tourism Research, 15*, 191–204.

Nagy, H. (2013). The importance of festival tourism in the economic development of Hungary". *Visegrad Journal on Bioeconomy and Sustainable Development, 2*, 57–59.

Nancy, G. M. (2004). Factors predicting rural residents' support of tourism. *Journal of Travel Research, 43*(2), 131–140.

Nunkoo, R., & Gursoy, D. (2012). Residents' support for tourism. An identity perspective. *Annals of Tourism Research, 39*, 243–268.

Nurse, K. (2001). Festival tourism in the Caribbean: An economic impact assessment. *Report prepared for Inter-American Development Bank.*

Okyere-Manu, B., & Antwi, J. (2016). Cultural and social festivity as a silent contributor to HIV infections: A Moral Challenge Of The easter festivity to the Kwahuman Leadership in Ghana. *Alternation Journal, 23*(2), 236–249.

O'Sullivan, D., & Jackson, M. J. (2002b). Festival tourism: a contributor to sustainable local economic development. *Journal for Sustainable Tourism, 10*, 325–342.

O'Sullivan, D., & Jackson, M. J. (2002a). Festival tourism: A contributor to sustainable local economic development? *Journal of Sustainable Tourism, 10*, 325–342.

Picard, D., & Robinson, M. (2006). *Festivals, tourism and social change: Remaking worlds*. Channel view publications.

Porfido, E. (2022). Rural festival and event tourism in Albania. In *Handbook of Niche Tourism* (pp. 124–139). Edward Elgar Publishing.

Raj, R. (2003). The impact of festivals on cultural tourism. In *The 2nd DeHaan Tourism Management Conference. Developing Cultural Tourism, Nottingham*, UK.

Whitford, M., & Dunn, A. (2014). Papua new Guinea's indigenous cultural festivals: Cultural tragedy or triumph?". *Ingentaconnect, 18*, 265–283

Woosnam, K. M., Norman, W. C., & Ying, T. (2009). Exploring the theoretical framework of emotional solidarity between residents and tourists. *Journal of Travel Research, 48*(2), 245–258.

Factors Affecting the Quality of Lodging Services of 3 to 5 Star Hotels in Hanoi

Nguyen Minh Tuan

Abstract This study determines the standard of hotel accommodations in Hanoi. The survey approach was used to obtain information on customer evaluations based on SERVQUAL theory for lodging visits in Hanoi. The findings indicate that customers are placing a greater emphasis on service quality and pricing. Enterprises in the hotel sector should focus on enhancing service quality in order to preserve and increase customer loyalty and satisfaction based on the aforementioned characteristics.

Keywords Service quality · Hotel · Hanoi

1 Introduction

Business hotel and accommodation services contribute significantly to the growth of tourism, the economy, and the creation of employment in society. In Hanoi, the number and quality of lodging places have expanded fast.

Service quality is one of the most important components of a successful business. Service quality is defined as the gap between the customer's expectations for service performance before to accessing the service and the actual service performance experienced by the customer (Asubonteng et al., 1996).

According to De Ruyter and Bloemer (1998), service quality is the primary determinant of customer loyalty and retention. Customers' perceptions of their loyalty will be inaccurate if the firm fails to give high-quality customer service. Consequently, service quality is comprised of several facets relating to client requirements, and no single service is deemed adequate for every customer. To generate customer loyalty, it is crucial to increase service quality and strengthen marketing efforts, and it is essential to reduce the gap between consumer expectations and perceptions.

To improve service quality, hotels must analyse what occurs before to, during, and after a guest's interaction with the hotel. Customer happiness is heavily dependent

N. M. Tuan (✉)
Hanoi University of Natural Resources and Environment, Hanoi, Vietnam
e-mail: nmtuan.kt@hunre.edu.vn

A. T. Nguyen et al. (eds.), *Contemporary Economic Issues in Asian Countries: Proceeding of CEIAC 2022, Volume 1*,
https://doi.org/10.1007/978-981-19-9669-6_39

on service quality. Therefore, this article gives a more in-depth approach to service quality by studying the influence of tangible characteristics; dependability; guarantee; empathy and lodging rates in order to enhance the quality of hotel enterprises' future tourist accommodation services.

2 Methodologies

2.1 *Tangibility*

Tangibility is the physical look of the buildings, equipment, and staff utilised to offer services to clients (Parasuraman et al., 1988). A thing's tangibility encompasses its cleanliness, space, atmosphere, and position. In addition, it refers to the physical appeal of the salesman, including face attractiveness and dress style.

Bostanji (2013) performed and disseminated a self-administered survey to 500 clients of a five-star hotel in Riyadh. Using multiple regressions, the association between tangibles and customer loyalty was shown to be positive and statistically significant.

Additionally, another experimental investigation was conducted. Multiple regression analysis on 395 self-administered surveys disseminated throughout the Thai hotel system found that tangibility had a favourable effect on customer loyalty (Polyorat & Sophonsiri, 2010).

In addition, postgraduate students at the University of Utara Malaysia received a total of 341 questions. All valid replies were assessed using multiple linear regression analyses, and the results revealed a favourable correlation between customer connections and loyalty (Mokhtar et al., 2011).

2.2 *Reliability*

Reliability is a company's capacity to perform the promised service flawlessly (Malik et al., 2011). That signifies the firm will maintain its word and execute the objective correctly. In the context of services, the first organisation that can deliver a suitable service is deemed reliable. Important aspect in assessing service quality is dependability. Yuen and Chan (2010) delivered 447 questionnaires to Hunter Douglas clients in Hong Kong. Multiple linear regression reveals that shop dependability has a favourable influence on consumer loyalty and has no effect on staff.

Using correlation analysis and multiple regression, Yousaf et al. (2013) also found results comparable to those of prior studies. Primary method (interviews and surveys with bus passengers) and secondary method data were obtained (documents from previous researchers).

2.3 Enthusiasm

Using multiple regression analysis, Al-Rousan and Mohamed (2010) revealed that strong staff responsiveness can lead to client loyalty. 322 guests staying at Marriott hotels in Jordan were provided with self-administered questionnaires.

Using Pearson correlation and linear multiple regression, Auka et al.'s (2013) study likewise revealed a favourable association between customer responsiveness and customer loyalty. Existing clients of commercial banks in Kenya received a total of 384 self-administered surveys.

According to Amirosadat et al. (2013), responsiveness positively influences consumer loyalty. 250 questions were returned and evaluated using structural equation modelling after 260 questionnaires were given to Internet users and purchased online.

2.4 Guarantee

Assurance is described as the kindness and expertise of personnel who inspire clients' trust and confidence. In British banking, reassurance refers to personnel who are kind, pleasant, competent, and eager to offer financial advice.

Badara et al. (2013) administered 209 questionnaires to Nigerian students attending the University of Utara Malaysia, and the results demonstrated that assurance had a beneficial impact on loyalty. Standard regression weighted analysis of customer satisfaction in Islamic banking in Nigeria.

2.5 Empathy

Empathy refers to businesses that provide unique attention and personalised care to their clients. It is mirrored in the service provider who offers the consumer with access, communication, and comprehension. It refers to the company's level of knowledge, care, and focus on its consumers (Parasuraman et al., 1988).

Using multiple regression analysis, Obeidat et al. (2012) conducted a study demonstrating that empathy is positively connected with customer loyalty. A total of 422 survey questions were collected and sent to clients of Umnniah Company's retail location.

By delivering 422 survey questions to Riyadh bank customers, Albarq (2013) demonstrated that customer empathy and loyalty were strongly connected. Multiple regression analyses were employed to examine the validity of replies.

2.6 Service Price

According to Martins and Monroe (1994), the perception of an improper pricing will have an impact on consumer satisfaction and desire to repurchase. Similarly, Xia et al. (2004) concluded in their study that customers' evaluations of product value and satisfaction are influenced by their sense of a suitable price. In addition, research demonstrates that negative emotion-generating perceptions vary in intensity and occur in various categories, and that customer value judgments and negative emotions are mediating variables that influence various behavioural actions, such as purchase intention and negative speech.

Research hypothesis is as below (see Fig. 1).

H1: There is a positive relationship between tangibles and accommodation service quality.

H2: There is a positive relationship between reliability and accommodation service quality.

H3: There is a positive relationship between enthusiasm and quality of accommodation services.

H4: There is a positive relationship between assurance and quality of accommodation services.

H5: There is a positive relationship between empathy and accommodation service quality.

H6: There is a positive relationship between price and quality of accommodation services.

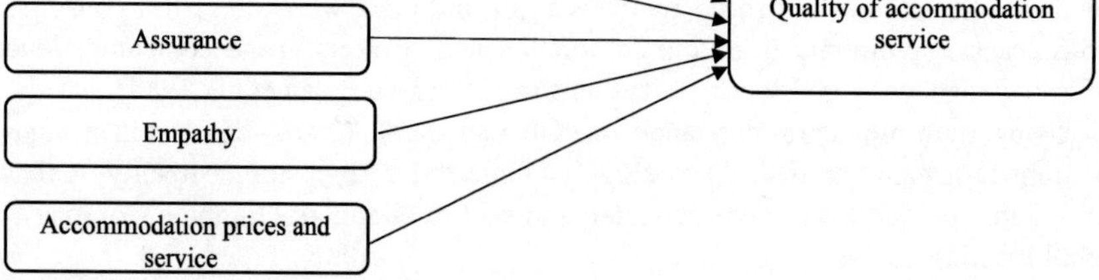

Fig. 1 Proposed research model

2.7 Methods

Based on the foregoing theoretical foundation, the author conducted a survey of Hanoi hotel guests who had utilised the services of three- to five-star hotels. The survey questionnaire is divided into four sections. The first step is to determine the requirements of hotel guests in Hanoi. The second section examines customer expectations for hotel services in Hanoi. The final section focuses on the sentiments and experiences of hotel guests in Hanoi. The final section describes the customer's personal details, including gender, employment, and age.

Because the overall target is too diversified and vast, i.e. individuals who have recently stayed at 3- to 5-star hotels in Hanoi, convenience sampling is employed for this study (non-probability). To conduct the study, the sample is selected based on the number of 3- to 5-star hotels in the districts and cities of Hanoi, such as Hoan Kiem: 71; Ba Dinh: 43; Nam Tu Liem: 114; and Tay Ho: 81 (see Table 1).

Table 1 Characteristics of sample

Categories	Quantity	Percentage (%)
1. Areas		
Hoan Kiem	71	23.0
Ba Dinh	43	13.9
Tay Ho	81	26.2
Nam Tu Liem	114	36.9
2. Gender		
Male	181	58.6
Female	128	41.4
3. Age		
Under 30	118	38.2
From 30 to 39	108	35.0
From 40 to 49	63	20.4
Above 50	20	6.5
4. Occupation		
Regular employees	81	26.2
Managers	69	22.3
Researchers	53	17.2
Others	106	34.3
5. Nationality		
Vietnam	102	33.0
Asia	112	36.2
USA	42	13.6
Europe	53	17.2

Source Compiled by the author, 2022

The research subject focuses on Hanoi hotel guests staying at hotels with 3 to 5 stars. Data were acquired using a combination of convenient sampling and norm sample through direct interviews and the distribution of questionnaires at the front desk with 309 hotel clients. March to June 2022 constitutes the survey period.

Interviews and direct surveys with hotel guests in Hanoi were employed in the data collecting procedure. 309 is the number of collected data that fulfils the parameters for reliability and objectivity for data processing. Statistical programme SPSS 22.0 is used to process the data.

The study used qualitative methodologies to conduct a preliminary analysis of the factors influencing the quality of hotel accommodations in Hanoi. In addition, the study employs quantitative analysis using the Cronbach's alpha test and a series of regression models to determine the influence of tangible aspects; dependability; guarantee; empathy and pricing on the quality of lodging service.

The characteristics of the sample are listed in Table 1, where the proportion of male and female visitors polled is nearly equal. The percentage of male clients is 58.6%, while the percentage of female customers is 41.4%; the percentage of 30 years old using hotels with 3 to 5 stars in Hanoi is 38.2%, followed by 30–39 years old who account for 35%. The occupational group utilising the hotel might relate to the group of employees (26.2%) and managers (22.3%), respectively, followed by other groups (34.3%). To objectively evaluate the service quality, pricing, and payment terms of certain hotels, the interviewees represent a wide variety of nationalities. 33% are of Vietnamese nationality, followed by 36.2% of Asian nationality and 17.7% of European nationality.

3 Results

3.1 *Customer Evaluations on Hanoi Hotels*

*** Tangibility**

The average score of the indexes in this group is 2.66, showing that customers are not really satisfied with tangible factors (see Table 2).

The hotel's instructions were clear and comprehensive, together with the look and clothing of the courteous hotel employees, received the highest average rating of 2.69 out of a possible 7. With 2.64 points, the view and space of the magnificent and appealing hotel were scored the lowest.

Conversely, the number of consumers who absolutely agreed with them was just 10.36%, while the number of customers who completely disagreed was 6.98%. Despite the fact that the overall assessment of the quality of housing services in the group of tangibles is at a moderate level, it is clear that these are not very encouraging signs for the quality of accommodation services in Hanoi.

Table 2 Average scores of observed variables in the aspect tangibles

Observed variables	Totally disagree	Disagree	Agree	Totally agree	Mean
HH01	26	101	161	21	2.68
HH02	44	65	173	27	2.69
HH03	20	127	140	22	2.64
HH04	16	97	170	26	2.67
HH05	19	130	114	46	2.60
HH06	11	67	180	51	2.69
HH07	15	90	173	31	2.67
Total	151	677	1111	224	2.66
Percentage (%)	6.98	31.30	51.36	10.36	

Source Compiled by the author, 2022

*** Reliability**

In the evaluations of lodging service quality qualities in terms of dependability, only, 42.98 per cent of visitors agree that the quality of service at three, and only, 9.97 per cent agree totally. This is a low proportion, suggesting that the service quality is average.

The overall average score of the observed variables in terms of reliability is only 2.54, and there are two observed variables with average scores below 2.5, namely, the hotel always provides timely services such as: committed and the service at the hotel is exactly as advertised, the provided information. In addition, the variable with the highest rating, 2.64 points, is that the hotel is constantly interested in resolving client issues (see Table 3).

On the other hand, the proportion of customers who assess the quality of the dependability aspect at the absolute level (strongly agree) is quite low, while the percentages of customers who rate the quality at the levels of entirely disagree and disagree are 14.43% and 32.62%, respectively.

Table 3 Average scores of observed variables in the aspect reliability

Observed variables	Totally disagree	Disagree	Agree	Totally agree	Mean
TC01	65	88	133	23	2.39
TC02	73	84	134	18	2.56
TC03	42	87	150	30	2.64
TC04	23	117	134	35	2.48
TC05	20	128	113	48	2.54
Total	223	504	664	154	2.54
Percentage (%)	14.43	32.62	42.98	9.97	

Source Compiled by the author, 2022

Table 4 Average scores of observed variables in the aspect reliability

Observed variables	Totally disagree	Disagree	Agree	Totally agree	Mean
NT01	40	105	137	27	2.49
NT02	29	90	157	33	2.65
NT03	22	125	129	33	2.54
NT04	33	89	152	35	2.49
NT05	26	127	106	50	2.65
Total	150	536	681	178	2.56
Percentage (%)	9.71	34.69	44.08	11.52	

Source Compiled by the author, 2022

*** Enthusiasm**

Only, 44.08% of visitors agree with the service quality at level 3 and the overall average score of the observable variables in the method's evaluation of lodging service quality features in terms of excitement. There are two observed characteristics with the highest average score, namely, the hotel staff is always eager to assist customers when necessary, and the front desk personnel ensures the highest quality services. Check-in and check-out service that swiftly reaches 2.65 points for convenience. At 2.49 points, the lowest-rated observable variable is that hotel personnel always responds swiftly to all customer enquiries and is always available when customers need them (see Table 4).

Conversely, the number of customers who assessed the quality in terms of excitement at the absolute level (absolutely agree) was 11.52%, while the number of customers who rated the quality as totally disagreeing and disagreeing was 9.71% and 34.69%, respectively.

*** Guarantee**

In terms of guarantee, the hotel ensures security work; safety … for visitors during their stay and usage of the hotel's services.

In terms of assurance, only 47.99% of visitors concur with the service quality at level 3 and the overall mean score of the method's observed factors. There are two observable factors with the highest average score in the assurance aspect, namely that the hotel assures security and safety for visitors and that customer information and room keys are effectively secured, with corresponding average scores of 2.84 and 2.71. In contrast, the characteristic with the lowest rating, 2.67 points, is that hotel personnel is knowledgeable about tourist destinations and tourism.

On the other side, 12.85% of customers assessed the quality as "absolute assurance" (strongly agree), but the percentage of customers who rated the quality differently was entirely different. The percentages of those who agree and disapprove are 7.40% and 31.76%, respectively (see Table 5).

We can see that the assurance levels of hotels for hotels are relatively good.

Table 5 Average scores of observed variables in the aspect guarantee

Observed variables	Totally disagree	Disagree	Agree	Totally agree	Mean
ĐB01	26	101	161	21	2.68
ĐB02	38	84	128	59	2.70
ĐB03	27	88	143	51	2.84
ĐB04	16	97	170	26	2.67
ĐB05	19	130	114	46	2.71
ĐB06	19	96	150	44	2.69
ĐB07	15	91	172	31	2.67
Total	160	687	1038	278	2.71
Percentage (%)	7.40	31.76	47.99	12.85	

Source Compiled by the author, 2022

* Empathy

Only 41.29 per cent of customers agree with the empathy aspect of service quality at level 3, and the average point of the observed variables in the reliability dimension is only 2.72. There are two observed variables with the highest average score, namely, the staff takes care of customers and cares about guests, and the hotel staff tries to understand the needs of guests, with scores of 2.79 and 2.77, respectively. In contrast, the criteria that received the lowest rating, 2.64 points, were that personnel understand consumer demands (see Table 6).

The percentage of customers who assessed the quality of the empathy element as "absolute" (totally agree) was 18.19%, while the percentage of customers who rated the quality as totally disagree and disagree are 5.89% and 34.63%, respectively.

Thus, it can be shown that clients who score this element are often pleased with the hotel's empathy during their stay and service consumption.

Table 6 Average scores of observed variables in the aspect empathy

Observed variables	Totally disagree	Disagree	Agree	Totally agree	Mean
ĐC01	28	93	120	68	2.75
ĐC02	17	89	145	58	2.79
ĐC03	13	121	134	41	2.66
ĐC04	14	107	124	64	2.77
ĐC05	19	125	115	50	2.64
Total	91	535	638	281	2.72
Percentage (%)	5.89	34.63	41.29	18.19	

Source Compiled by the author, 2022

Table 7 Average score of observed variables in the aspect prices at the hotel

Observed variables	Totally disagree	Disagree	Agree	Totally agree	Mean
GC01	18	81	137	73	2.86
GC02	17	85	152	55	2.82
GC03	13	90	155	51	2.72
GC04	14	101	127	67	2.68
GC05	14	87	158	50	2.67
Totally	76	444	729	296	2.75
Percentage (%)	4.92	28.74	47.18	19.16	

Source Compiled by the author, 2022

*** Service price**

Only 47.18 per cent of customers agree with the price aspect of service quality at level 3, and the average point of the observed variables in the reliability dimension is only 2.75 points, with the highest average score belonging to the price of the hotel room and the price of the hotel's additional services. In addition, the observed variable that is overestimated by 2.67 points is the satisfaction of unique demands at a fair cost.

The percentage of customers who assessed the quality of the empathy component at the absolute level (strongly agree) was 19.16%, and the percentage of customers who rated the quality at the absolute level of empathy was 19.16%. The percentages of those who agree and disagree are 4.92% and 28.74%, respectively (see Table 7).

Consequently, it appears that the customers who evaluate this feature are somewhat pleased with the hotel's pricing throughout their stay and usage of its services.

3.2 *The Impact of Factors on Customer Satisfaction for Hotels in Hanoi*

3.2.1 Testing the Scale by Cronbach Alpha Reliability Coefficient

*** Independent variables**

Testing the reliability coefficients of the observed variables is crucial to the precision and applicability of the study findings. On the one hand, it helps to eliminate insufficiently trustworthy observable variables, and on the other, it assists in adjusting and developing official scales.

The Cronbach alpha reliability coefficient is used to reject unsuitable variables, as these variables might lead to the formation of pseudofactors during EFA analysis. To be utilised, the Cronbach alpha coefficient must be at least 0.6, and observed variables having correlation coefficients between the total variable and the resultant table that are less than 0.3 will be eliminated. When assessing the reliability of independent variables with 38 indices, encompassing seven aspects, and eliminating unreliable variables (see Table 8).

From the test results, it can be seen that the Cronbach alpha coefficients of all variables are greater than 0.6.

*** Dependent variable**

The KMO coefficient and Bartlett's test of the independent margins with reliability (Cronbach's Alpha) of 0.736 are satisfactory, corresponding to 5 observed variables belonging to the means of accommodation service quality (see Table 9).

3.2.2 Results of EFA

To determine the number of factors in the exploratory factor analysis scale is required for both independent and dependent variables.

*** Independent variables**

After exploratory factor analysis, the following results were obtained (see Table 10).

KMO coefficient greater than 0.5 and Bartlett's test for Sig. value. < 0.05 indicates that the factor analysis is appropriate.

The extracted variance table shows that the total variance extracted is 63.68% (greater than 50%), so the observed variables are formed from 6 factors.

According to (Appendix G), the author performs analysis according to principal components with varimax rotation. As a result, 34 observed variables were initially grouped into 6 groups. Total value of variance extracted $= 63.68\% > 50\%$, satisfactory then it can be said that these 6 factors explain 63.68% of the variation of the data. The Eigenvalues of the factors are all high (>1); the 7th factor has the lowest Eigenvalues of $1.520 > 1$.

*** Dependent variable**

Kaiser–Meyer–Olkin measure of sampling adequacy (KMO coefficient) $= 0.815 > 0.5$, so factor analysis is appropriate; Barlett's test result is 382,196 with Sig. significance level. (Bartlett's Test) $= 0.000$ (sig. < 0.05) shows that the observed variables are correlated with each other in the population (see Table 11).

The author performs analysis according to principal components with varimax rotation. The results show that 5 observed variables were initially grouped into 1 group. Total value of variance extracted $= 61.68\% > 50\%$, satisfactory; then, it can be said that these 5 factors explain 61.68% of the variation of the data. The Eigenvalues of the factors are all high (>1); the 5th factor has the lowest Eigenvalues of $4.318 > 1$ (see Table 12).

From the analysis of the correlation between the independent variables and the dependent variable of service quality (CL), it is shown that Sig. is less than 0.05, indicating that all variables are correlated. The correlation coefficient between the independent variables and the service quality variable is positive.

Table 8 Results of evaluating the reliability of the scale of independent variables

Observed variables	Scale average if variable type	Scale variance if variable type	Total variable correlation	Cronbach Alpha coefficient if variable type
1. Tangibility (HH) Cronbach Alpha = 0.824, N = 7				
HH01	25.3215	13.832	0.684	0.863
HH02	25.4859	13.235	0.647	0.796
HH03	25.3658	12.308	0.632	0.797
HH04	25.3698	11.906	0.635	0.698
HH05	25.3214	13.069	0.653	0.792
HH06	25.3654	12.634	0.689	0.791
HH07	25.6987	12.369	0.664	0.815
2. Reliability (TC) Cronbach Alpha = 0.835, N = 5				
TC01	25.3579	10.305	0.663	0.865
TC02	25.6657	11.603	0.653	0.817
TC03	25.4859	12.236	0.634	0.806
TC04	25.3656	11.624	0.606	0.812
TC05	25.3657	10.326	0.693	0.845
3. Enthusiasm (NT) Cronbach Alpha = 0.819, N = 5				
NT01	24.4563	11.645	0.668	0.708
NT02	25.6342	12.634	0.653	0.740
NT03	25.3649	12.763	0.653	0.694
NT04	24.2469	12.796	0.668	0.726
NT05	24.5346	11.036	0.654	0.694
4. Enthusiasm (NE) Cronbach Alpha = 0.827, N = 7				
ĐB01	23.5684	10.369	0.634	0.736
ĐB02	24.3654	14.153	0.645	0.765
ĐB03	24.3265	13.953	0.654	0.756
ĐB04	24.6325	12.965	0.695	0.696
ĐB05	24.1546	14.263	0.657	0.742
ĐB06	23.2365	13.369	0.698	0.709
ĐB07	24.1543	13.065	0.657	0.756
5. Empathy Cronbach Alpha = 0.836, N = 5				
ĐC01	10.3056	12.763	0.657	0.756
ĐC02	10.1062	12.796	0.656	0.809
ĐC03	10.4530	11.036	0.654	0.726
ĐC04	11.2635	14.154	0.646	0.789
ĐC05	12.3694	13.963	0.656	0.784

(continued)

Table 8 (continued)

Observed variables	Scale average if variable type	Scale variance if variable type	Total variable correlation	Cronbach Alpha coefficient if variable type
6. Price (GC) Cronbach Alpha = 0.801, N = 5				
GC01	25.3246	12.965	0.665	0.673
GC02	24.2364	14.263	0.665	0.663
GC03	24.6387	13.369	0.647	0.763
GC04	24.1556	13.065	0.696	0.746
GC05	23.3244	12.796	0.678	0.634

Source Compiled by the author, 2022

Table 9 KMO coefficients and Bartlett's test of dependent boundaries

Observed variables	Quality of accommodation service	Total degree of correlation between variables (Total correlation)	Reliability (Cronbach's Alpha)
CL01	(a) Hotel room service is very good	0.69	0.736
CL02	(b) The quality of the hotel's additional services is very good	0.73	
CL03	(c) The quality of the hotel's services is generally very good	0.62	
CL04	(d) I am satisfied and very satisfied with this hotel	0.77	
CL05	(e) The hotel is ranked high in service quality	0.78	

Source Compiled by the author, 2022

Table 10 KMO coefficients and Bartlett's test of independent boundaries

KMO and Bartlett's test		
Kaiser–Meyer–Olkin measure of sampling adequacy		0.897
Bartlett's test of sphericity	Approx. Chi-Square	5321.611
	df	821
	Sig.	0.000

Source Compiled by the author, 2022

Table 11 KOM coefficient and Bartlett's test of service quality

KMO and Bartlett's test		
Kaiser–Meyer–Olkin measure of sampling adequacy		0.815
Bartlett's test of sphericity	Approx. Chi-Square	382.196
	df	21
	Sig.	0.000

Source Compiled by author, 2020

Table 12 Total explained variance for service quality

Total variance explained						
Component	Initial eigenvalues			Extraction sums of squared loadings		
	Total	% of Variance	Cumulative %	Total	% of Variance	Cumulative %
1	4.318	61.679	71.679	4.318	61.679	61.679
2	0.614	8.771	83.450			
3	0.418	5.975	89.971			
4	0.365	5.213	95.184			
5	0.337	4.816	100.000			
Extraction Method: Principal Component Analysis						

Source Compiled by the author, 2022

4 Discussions and Solutions

Firstly, upgrade the hotel's physical and technological facilities, as they play a crucial role in the creation and consumption of products/services, as well as in determining the extent to which the hotel's ability to meet consumer demands is used. In addition to focusing on the policy of training and developing human resources in order to satisfy consumer expectations and increase customer satisfaction, enterprises in the hotel industry must also focus on enhancing their technological infrastructure.

Secondly, human resources are a vital asset in the process of enhancing service quality; they are directly involved in quality enhancement in order to increase customer satisfaction. This human resource, comprising managers and staff, has a significant impact on the service quality. The hotel industry's products and services have both tangible and immaterial components. The evaluation of service quality is based on customer perception. Non-physical aspects comprise mostly service strategies and communication abilities of hotel workers (receptionist, room service, restaurant …). Therefore, in order to improve the quality of service in the hotel, it is necessary to improve the quality of human resources. Businesses must have strategies and policies that are suited to their size, culture, and circumstances, thereby creating new positions and forces to develop in line with the overall economic trend.

Thirdly, human resource management and training policies encompass policies on recruiting, employment, training, and human resource development. The recruiting policy must guarantee that the appropriate individuals with the proper skills and credentials are recruited and put in the right roles at the right time to achieve the organisation's goals. To do this, tourist firms must develop recruiting procedures that are tailored to their specific needs. Improving service quality necessitates recruiting the appropriate people and allocating the proper roles in accordance with the capability and quality of personnel.

Fourthly, enhancing the occupational and working skills of employees is always regarded as a top priority. The professional and job abilities of the personnel will enhance the quality of the service. Employees' abilities to execute service process tasks are frequently seen as a key indicator of human capital. With a huge work force that is continuously renewed each year, the Hanoi housing market has an abundant human resource that ranks amongst the best in the nation. Due to the disparity in training, the fraction of technical employees meeting business demands is insufficient. In certain hotels, the number of unskilled personnel is extremely considerable. The professional and job abilities of the personnel will enhance the quality of the service.

Fifthly, increase professional competence. Professional competence is a component of intelligence. It consists of intellectual capacity, professional knowledge, the ability to conduct work in accordance with professional standards, cognitive ability, and creative thinking in order to adapt to the job of workers. Referring to the professional element is the mental factor, the cultural and educational level, the employee's knowledge and abilities, which express in the employee's capacity to apply the material and spiritual conditions to the operation practise to attain high efficiency, as well as the capacity to orient the values of one's own operations to meet the aims of the individual employees and the enterprise's commercial goals.

Finally, in order to enhance the pricing strategy, firms must provide a flexible pricing structure and a variety of payment options for clients, such as cash, credit card, Moto Payment by vouchers/barter, Foc/Hu, or city ledger (the following transfer method).

References

Albarq, A. N. (2013). Applying a SERVQUAL model to measure the impact of service quality on customer loyalty among local Saudi banks in Riyadh. *American Journal of Industrial and Business Management, 3*, 700–707.

Al-Rousan, M. R., & Mohamed, B. (2010). Customer loyalty and the impacts of service quality: The case of five star hotels in Jordan. *International Journal of Business, Economics, Finance & Management Sciences, 2*(3), 886–892.

Amirosadat, S. N., Esfahani, M. T., & Moshkforosh, A. (2013). An empirical study of customers' perceptions of service quality and loyalty with using structural equation modeling. *International Journal of Academic Research in Economics and Management Sciences, 2*(4), 15–26.

Asubonteng, P., MacCleary, K. J., & Sawn, J. E. (1996). SERQUAL revisited: A critical review of service quality. *Journal of Service Marketing, 10*(6), 62–81.

Auka, D. O., Bosire, J. N., & Matern, V. (2013). Perceived service quality and customer loyalty in retail banking in Kenya. *British Journal of Marketing Studies, 1*(3), 32–61.

Badara, M. S., Mat, N. K. N., Mujtaba, A. M., Al-Refai, A. N., Badara, A. M., & Abubakar, F. M. (2013). Direct effect of service quality dimensions on customer satisfaction and customer loyalty in Nigerian Islamic bank. *Management, 3*(1), 6–11.

Bostanji, G. M. A. (2013). The impact of service quality on customers loyalty: A study on five stars hotel's customers in Riyadh, KSA. *European Journal of Business and Management, 5*(31), 230–240.

De Ruyter, K., & Bloemer, J. (1998). Customer loyalty in extended service settings: The interaction between satisfaction, value attainment and positive mood. *International Journal of Service Industry Management, 10*(3), 320–336.

Malik, M., Naeem, B., & Arif, Z. (2011). Impact of perceived service quality on banking customers' loyalty. *Interdisciplinary Journal of Contemporary Research in Business, 3*(8), 637–645.

Martins, M., & Monroe, K. B. (1994). Perceived price fairness: A new look at an old construct. *Advances in Consumer Research, 5*(1), 7–16.

Mokhtar, M., Sanuri, S., & Maiyaki, A. (2011). The relationship between service quality and satisfaction on customer loyalty in Malaysian mobile communication industry. *School of Doctoral Studies (European Union) Journal*, (3), 32–38.

Obeidat, B. Y., Sweis, R. J., Zyod, D. S., Masa'deh, R., & Alshurideh, M. (2012). The effect of perceived service quality on customer loyalty in internet service providers in Jordan. *Management Research, 4*(4), 224–242.

Parasuraman, A., Zeithaml, V. A., & Berry, L. (1988). SERVQUAL: a multiple-item scale for measuring consumer perception of service quality. *Retailing 64*(1), 35–48.

Polyorat, K., & Sophonsiri, S. (2010). The influence of service quality dimensions on customer satisfaction and customer loyalty in the chain restaurant context: A Thai case. *Global Business & Technology, 6*(2), 64–76.

Xia, L., Monroe, K. B., & Cox, J. L. (2004). The price is unfair! A conceptual framework of price fairness perceptions. *Marketing, 68*(4), 1–15.

Yousaf, U., Yousaf, J., Altaf, M., & Mehmood Bagram, M. M. (2013). Studying customer loyalty at Daewoo Express Bus Service, Pakistan. *Scientific Journal of Logistics, 9*(3), 145–152.

Yuen, E. F., & Chan, S. S. (2010). The effect of retail service quality and product quality on customer loyalty. *Database Marketing & Customer Strategy Management, 17*(3), 222–240.